P9-CEG-489

Principles of Wireless Networks

Prentice Hall Communications Engineering and Emerging Technologies Series

Theodore S. Rappaport, *Series Editor*

DOSTERT *Powerline Communications*

GARG *Wireless Network Evolution: 2G to 3G*

GARG *IS-95 CDMA and cdma2000: Cellular/PCS Systems Implementation*

GARG & WILKES *Principles and Applications of GSM*

HAĆ *Multimedia Applications Support for Wireless ATM Networks*

KIM *Handbook of CDMA System Design, Engineering, and Optimization*

LIBERTI & RAPPAPORT *Smart Antennas for Wireless Communications: IS-95 and Third Generation CDMA Applications*

PAHLAVAN & KRISHNAMURTHY *Principles of Wireless Networks: A Unified Approach*

RAPPAPORT *Wireless Communications: Principles and Practice, Second Edition*

RAZAVI *RF Microelectronics*

STARR, CIOFFI & SILVERMAN *Understanding Digital Subscriber Line Technology*

PRINCIPLES OF WIRELESS NETWORKS

A Unified Approach

Kaveh Pahlavan
Worcester Polytechnic Institute
University of Oulu

Prashant Krishnamurthy
University of Pittsburgh

PH PTR
Prentice Hall PTR
Upper Saddle River, New Jersey 07458
www.phptr.com

A CIP catalog record for this book can be obtained from the Library of Congress.

To those whose love kept us writing, those whose work inspired our learning, and those with whom we learned.

Acquisitions editor: *Bernard Goodwin*
Marketing manager: *Dan DePasquale*
Manufacturing buyer: *Alexis Heydt-Long*
Editorial/production supervision: *Jessica Balch (Pine Tree Composition)*
Production coordinator: *Anne R. Garcia*
Cover design director: *Jerry Votta*
Cover design: *Anthony Gemmellaro*
Art director: *Gail Cocker-Bogusz*
Art studio: *LP Graphics*

Prentice-Hall, Inc.
Upper Saddle River, New Jersey 07458

Prentice Hall books are widely used by corporations and government agencies for training, marketing, and resale.

The publisher offers a discount on this book when ordered in bulk quantities. For more information, contact: Corporate Sales Department, Phone: 800-382-3419; Fax: 201-236-7141; E-mail: corpsales@prenhall.com; or write: Prentice Hall PTR, Corp. Sales Dept., One Lake Street, Upper Saddle River, NJ 07458

Printed in the United States of America
10 9 8 7 6 5 4 3 2 1

ISBN 0-13-093003-2

Pearson Education Ltd., *London*
Pearson Education Australia Pty, Limited, *Sydney*
Pearson Education Singapore, Pte. Ltd.
Pearson Education North Asia Ltd., *Hong Kong*
Pearson Education Canada, Ltd., *Toronto*
Pearson Educación de Mexico, S.A. de C.V.
Pearson Education—Japan, *Tokyo*
Pearson Education Malaysia, Pte. Ltd.

CONTENTS

Preface xi

CHAPTER 1 Overview of Wireless Networks 1

1.1 Introduction 2
1.2 Different Generations of Wireless Networks 12
1.3 Structure of the Book 21
Appendix 1A Backbone Networks for Wireless Access 26
Appendix 1B Summary of Important Standards Organizations 33
Questions 34

PART ONE PRINCIPLES OF AIR-INTERFACE DESIGN 37

CHAPTER 2 Characteristics of the Wireless Medium 39

2.1 Introduction 40
2.2 Radio Propagation Mechanisms 44
2.3 Path-Loss Modeling and Signal Coverage 46
2.4 Effects of Multipath and Doppler 58
2.5 Channel Measurement and Modeling Techniques 68
2.6 Simulation of the Radio Channel 71
Appendix 2A What is dB? 76
Appendix 2B Wired Media 77
Appendix 2C Path Loss Models 79
Appendix 2D Wideband Channel Models 79
Questions 80
Problems 81

CHAPTER 3 Physical Layer Alternatives for Wireless Networks 85

3.1 Introduction 86
3.2 Applied Wireless Transmission Techniques 91
3.3 Short Distance Baseband Transmission 92
3.4 UWB Pulse Transmission 94
3.5 Carrier Modulated Transmission 96
3.6 Traditional Digital Cellular Transmission 96
3.7 Broadband Modems for Higher Speeds 108
3.8 Spread Spectrum Transmissions 111
3.9 High-Speed Modems for Spread Spectrum Technology 118
3.10 Diversity and Smart Receiving Techniques 120
3.11 Comparison of Modulation Schemes 133
3.12 Coding Techniques for Wireless Communications 137
3.13 A Brief Overview of Software Radio 142
Appendix 3A Performance of Communication Systems 143
Appendix 3B Coding and Correlation 150
Questions 155
Problems 156

CHAPTER 4 Wireless Medium Access Alternatives 159

4.1 Introduction 160
4.2 Fixed-Assignment Access for Voice-Oriented Networks 161
4.3 Random Access for Data-Oriented Networks 179
4.4 Integration of Voice and Data Traffic 201
Questions 214
Problems 217

PART TWO PRINCIPLES OF WIRELESS NETWORK OPERATION 221

CHAPTER 5 Network Planning 223

5.1 Introduction 224
5.2 Wireless Network Topologies 225
5.3 Cellular Topology 229
5.4 Cell Fundamentals 234
5.5 Signal-to-Interference Ratio Calculation 237
5.6 Capacity Expansion Techniques 240
5.7 Network Planning for CDMA Systems 260
Questions 263
Problems 263

CHAPTER 6 Wireless Network Operation 265

6.1 Introduction 266
6.2 Mobility Management 266
6.3 Radio Resources and Power Management 284
6.4 Security in Wireless Networks 297
Appendix 6A The Diffie-Hellman (DH) Key Exchange Protocol 311
Appendix 6B Nonrepudiation and Digital Signatures 312
Questions 313
Problems 313

PART THREE WIRELESS WANS 317

CHAPTER 7 GSM and TDMA Technology 319

7.1 Introduction 320
7.2 What Is GSM? 321
7.3 Mechanisms to Support a Mobile Environment 327
7.4 Communications in the Infrastructure 332
Questions 346
Problems 346

CHAPTER 8 CDMA Technology, IS-95, and IMT-2000 349

8.1 Introduction 350
8.2 Reference Architecture for North American Systems 351
8.3 What Is CDMA? 355
8.4 IMT-2000 371
Questions 376
Problems 376

CHAPTER 9 Mobile Data Networks 379

9.1 Introduction 380
9.2 The Data-Oriented CDPD Network 383
9.3 GPRS and Higher Data Rates 394
9.4 Short Messaging Service in GSM 405
9.5 Mobile Application Protocols 407
Questions 410
Problems 411

PART FOUR LOCAL BROADBAND AND AD HOC NETWORKS 413

CHAPTER 10 Introduction to Wireless LANs 415

10.1 Introduction 416
10.2 Historical Overview of the LAN Industry 416
10.3 Evolution of the WLAN Industry 420
10.4 New Interest from Military and Service Providers 426
10.5 A New Explosion of Market and Technology 430
10.6 Wireless Home Networking 431
Questions 444
Problems 445

CHAPTER 11 IEEE 802.11 WLANs 447

11.1 Introduction 448
11.2 What Is IEEE 802.11? 448
11.3 The PHY Layer 452
11.4 MAC Sublayer 460
11.5 MAC Management Sublayer 466
Questions 470
Problems 471

CHAPTER 12 Wireless ATM and HIPERLAN 473

12.1 Introduction 474
12.2 What Is Wireless ATM? 475
12.3 What Is HIPERLAN? 481
12.4 HIPERLAN-2 485
Questions 496
Problems 497

CHAPTER 13 Ad Hoc Networking and WPAN 499

13.1 Introduction 500
13.2 What Is IEEE 802.15 WPAN? 500
13.3 What Is HomeRF? 501
13.4 What Is Bluetooth? 503
13.5 Interference between Bluetooth and 802.11 520
Questions 530
Problems 531

CHAPTER 14 Wireless Geolocation Systems 533

14.1 Introduction 534
14.2 What Is Wireless Geolocation? 534
14.3 Wireless Geolocation System Architecture 536
14.4 Technologies for Wireless Geolocation 538
14.5 Geolocation Standards for E-911 Services 546
14.6 Performance Measures for Geolocation Systems 547
Questions 550
Problems 551

Acronyms and Abbreviations 553

References 561

Index 573

About the Authors 583

PREFACE

Wireless networking has emerged as its own discipline over the past decade. From cellular voice telephony to wireless access to the Internet and wireless home networking, wireless networks have profoundly impacted our lifestyle. After a decade of exponential growth, today's wireless industry is one of the largest industries in the world. At the time of this writing, close to one billion people subscribe to cellular services, close to 200 billion GSM short messages are exchanged yearly, and the penetration of the cellular telephone in Finland exceeded 75%, the highest in the world. In response to this growth, a number of universities and other educational institutions have started wireless research and teaching programs and a number of engineers and scientists are re-educating themselves in this field. There are a number of recent textbooks in the general area of networking that also address some aspects of wireless networks. The treatment in these books is not adequate because design and analysis of wireless networks are very different from wired networks. In wireless networks the complexity resides in the design of air-interface and support of mobility, neither of which play a dominant role in wired networks. Therefore, we have always needed a comprehensive textbook on wireless networks that provides a deeper understanding of the issues specific to the wireless networks.

In 1995 when wireless networking was an emerging discipline, the principal author, along with Allen Levesque, wrote the first comprehensive textbook in *Wireless Information Networks* that addressed cellular and PCS systems as well as mobile data and wireless LANs. Wireless-related books published prior to that book were focused on analog cellular systems. *Wireless Information Networks* covered 2G digital cellular systems, had significant emphasis on physical layer issues, and was written for students with background in electrical engineering, especially communications and signal processing. With the growth of the wireless industry in the latter part of the past decade, several books have emerged that explain the latest developments of specific standards or groups of standards like GSM, IS-95, W-CDMA, wireless LANs and Bluetooth. However, there is no textbook that integrates all the aspects of current wireless networks together. In this book, like the previous book, we address the need for a comprehensive treatment that provides a unified foundation of principles of all voice- and data-oriented wireless networks. The novelty of this book is that it covers 3G and wireless broadband ad hoc networking as well as 2G legacy systems, places emphasis on higher-layer communications issues, and is written for software and systems engineers as well as modern telecommunications engineers with electrical engineering or computer science backgrounds.

Traditionally, voice-oriented wireless networks have been the focus of books on wireless systems. However, with the exponential growth of the Internet, wireless data-oriented networks are also becoming very popular. The third generation (3G) wide area cellular systems are designed to support several hundreds of kbps with comprehensive coverage and up to 2 Mbps for local selected zones. Even before the emergence of 3G services, mobile data networks such as the general packet radio service (GPRS) over TDMA systems and high-speed packet data over CDMA systems are becoming increasingly popular. At the same time, after the introduction of Bluetooth technology in 1998, local broadband and ad hoc wireless networks have attracted tremendous attention. This sector of the wireless networking industry includes the traditional wireless local area networks (WLANs) and the emerging wireless personal area networks (WPANs). Wireless broadband and ad hoc networking is expected to create a revolution in the future of Internet access, home networking, and wireless consumer products. While there is a plurality of standards and a differentiation between voice and data networks, the essential aspects of wireless systems remain the same. We see that wireless networks share a common foundation in the design of the physical layer, medium access, network planning and deployment, and network operation. *Principles of Wireless Networks: A Unified Approach* emphasizes this similarity hidden in the diversity of wireless networks.

The structure and sequence of material for this book was first formed in a lecture series by the principal author at Digital Equipment Corporation in 1996. The principal author also taught shorter versions of the course focused on broadband and ad hoc networking in several conferences and industrial forums. The core of the book is based on presentation material and reference papers prepared by the principal author for a course called "Wireless Mobile Data Networks" first taught in spring 1999 as a fourteen-week, 3 hour/week graduate course at Worcester Polytechnic Institute, Massachusetts. In summer 1999, he also taught another ten-week, 3 hour/week, version of the course titled "Advances in Wireless Networks" at the University of Oulu, Finland. The co-author of the book has taught material from this book in spring 2000 and summer 2001 at the University of Pittsburgh in a course entitled Mobile Data Networks. These courses were taught for students with electrical engineering, computer science, and networking backgrounds, both from the academia and the industry.

Providing an overall organization for understanding both legacy voice-oriented and emerging data-oriented wireless networks for a diverse audience comprising managers, engineers, scientists, and students who need to understand this industry is very challenging. If we provide an in-depth treatment of specific topics related to the air-interface, such as channel modeling or modem design, we lose the overall systems-engineering perception. If we avoid the details of air-interface, there will be no wireless content, as this forms the core of the difference between wireless and wired networks. Our approach has been to try striking a balance between intuitive understanding of the wireless medium and detailed aspects of the system. We divide the topics in three categories: (1) overview, comparative evaluation, and logical classification of important standards, (2) explanation of principles of design and analysis of wireless networks, and (3) detailed description of important wireless systems. Overview of the popular standards is treated in Chapter one. Principles of wireless network design and analysis is divided into principles of air-interface design (Chapters 2–4) and principles of network deployment and operation (Chapters 5–6). System descriptions are divided into two parts covering legacy wide area wireless networks (Chapters 7–9) and emerging broadband local and ad hoc wireless networks (Chapters 10–14). The partitioned structure of the book allows flexibility in teaching the material and makes it easier for the text to be used as a reference book as well. Therefore, depending on selection of the material, depth of the coverage, and background of the students, this book can be used for senior undergraduate or first- or second-year graduate courses in computer science (CS), telecommunications, electrical and computer engineering (ECE), or electrical engineering (EE) departments as one course or a sequence of two courses.

In the last offering of the course at WPI to ECE and CS students, the first two weeks were devoted to the introduction of wireless networks. The first week was a lecture entitled "Overview of Wireless Networks" from Chapter 1 that provided the overall structure of the standards and trends in wireless networks. The second week was a lecture on "Overview of Networking Aspects" from Chapter 5 that clarified the technical issues that are related to wireless networks. The next part of the course (about six weeks) involved the detailed technical aspects of wireless networks. This part began with a lecture on "Characteristics of Wireless Medium" from Chapter 2. The next lectures in this sequence were from Chapter 5, "Principles of Network Planning" that spanned two weeks. This was followed by "PHY Layer Alternatives" from Chapter 3 that together with the next lecture on "Medium Access Alternatives" took three weeks. At this stage students were ready to understand the details of standards so that a two-week lecture on "GSM—An Example of TDMA Technology" could follow. This was a detailed treatment of an overall structure of GSM from Chapter 7. This lecture was followed by a one-week lecture on "CDMA Technology: IS-95 and IMT-2000" from Chapter 8. Wireless broadband and ad hoc networks formed the last four weeks of the course with the first lecture on "Wireless LANs and IEEE 802.11" from Chapters 10 and 11 followed by a one-week lecture from Chapters 12 and 13 on "Voice-Oriented HIPERLAN-2 and Bluetooth." We spent approximately two weeks on administration of the exams and presentations by students. The students had weekly homework that consisted of a set of questions and some problems relevant to the lecture. They were also required to do a mandatory project on handoff and an optional term paper.

The Finnish version taught at the University of Oulu to EE and CS students roughly covered the same material in ten three-hour lectures. In a period of several months after the lectures, two examinations were arranged, and the students performed a project mandated for completion of the course. The emphasis on lectures in Finland was on the last part of the book as students had prior exposure to GSM and W-CDMA systems. The principal author has also used the last five chapters of this book with parts of Stalling's book *Local & Metropolitan Area Networks* in a 30-hour lecture course entitled "Wired and Wireless LANs" at the University of Oulu for students from university and the industry in the summer of 2001. The flexibility in using the book can be seen from the course "Mobile Data Networks" taught by the co-author of the book in spring 2000 and summer 2001 at the University of Pittsburgh. The course spanned a quick review of the first part of the book and some parts of Chapter 7 followed by Chapters 9–14 that primarily considered wireless data networks.

Most of the writing and preparations of the principal author took place in several months of his teachings and scholarship at the University of Oulu, Finland, as the first non-Finn Nokia Fellow in 1999 and the first Fulbright-Nokia scholar in 2000. He would like to express his deep appreciation to Nokia Foundation, Fulbright Foundation, University of Oulu, and Worcester Polytechnic Institute for allowing these opportunities to occur. In particular he appreciates the help of Prof. Pentti Leppanen, director of the Telecommunication Laboratory at the University of Oulu, for his continual encouragement and creative administrative support and Dr. Yrjo Neuvo, Executive VP and CTO of Nokia Mobile Phone, for his support through the Nokia Foundation. Also, he appreciates Prof. John Orr, Head of the ECE department, and Provost John Carney, Vice President of Academic Affairs of WPI for their support and understanding, in particular during the Fulbright-Nokia scholarship period. The co-author's involvement in this project began during his stay at CWINS in WPI and extended into his current position at the University of Pittsburgh. At the University of Pittsburgh, he has been involved in the Wireless Information Systems track in the Master of Science in Telecommunications with Profs. David Tipper and Joseph Kabara, who have helped him immensely in his teaching and research work. The director of the Telecommunications Program at the University of Pittsburgh, Prof. Richard Thompson, and the chair of the Department of Information Science and Telecommunications, Prof. Martin Weiss, have fostered an environment that supports innovation and freedom, which has been a significant reason behind the co-author's ability to contribute to this book. He would also like to thank his colleagues, Profs. Sujata Banerjee and Taieb Znati for their support during the last two years in his academic efforts.

The authors would like to thank Dr. Bernard Sklar and Dr. Jacques Beneat, for careful review of the manuscript and their useful comments, and Phongsak Prasithsangaree, Xinrong Li, and Wiklom Teerapabkajorndet for careful proofreading of the manuscript.

In addition, the authors would like to express their appreciation to Dr. Allen Leveque, Xinrong Li, Dr. Ali Zahedi, Jeff Feigin, Dr. Aram Falsafi, Dr. Robert Tingley, Hamid Hatami, Yan Xu, Dr. Jim Matthews, and other recent affiliates of the CWINS Laboratory and Dr. Jaakko Talvitie, Jari Vallstrom, Dr. Matti Latva-Aho, Dr. Roman Pichna, Mika Ylianttila, and Juha-Pekka Makela of the CWC, University of Oulu who have directly or indirectly helped them to learn more in this field and shape their thoughts for preparation of this book. Finally, we would like to thank DARPA, DoD, and NSF in the United States, TEKES, Nokia, and Sonera in the Finland, and others who funded our research projects in the past few years. A substantial part of the new material in this book was the fruit of these research efforts.

Kaveh Pahlavan
Prashant Krishnamurthy

CHAPTER 1

OVERVIEW OF WIRELESS NETWORKS

1.1 Introduction

1.1.1 Information Network Infrastructure
1.1.2 Overview of Existing Network Infrastructure
1.1.3 Four Market Sectors for Wireless Applications
1.1.4 Evolution of Voice-Oriented Wireless Networks
1.1.5 Evolution of Data-Oriented Wireless Networks

1.2 Different Generations of Wireless Networks

1.2.1 1G Wireless Systems
1.2.2 2G Wireless Systems
1.2.3 3G and Beyond

1.3 Structure of the Book

1.3.1 Part One: Principles of Air-Interface Design
1.3.2 Part Two: Principle of Wireless Network Operation
1.3.3 Part Three: Wireless Wide Area Networks
1.3.4 Part Four: Local Broadband and Ad Hoc Networks

Appendix 1A Backbone Networks for Wireless Access

1A.1 Evolution of PSTN and Cellular Telephony
1A.2 Emergence of Internet
1A.3 The Cable TV Infrastructure

Appendix 1B Summary of Important Standards Organizations

Questions

1.1 INTRODUCTION

Technological innovations of engineers during the 20th century have brought a deep change in our lifestyle. Today, when we fly over a modern city at night we see the earth full of footprints made by engineers. The glowing lights remind us of the impact made by electrical engineers; the planes we fly in and the moving cars below remind us of the contributions of mechanical engineers; and high-rise buildings and complex roads remind us of what civil engineers have accomplished. From the eyes of an engineer, the glow of light, the movement of cars, and the complexity of civil infrastructure relates to the challenges in implementation, size of the market, and the impact of the technology on human life. There is, however, one industry whose infrastructure is not seen from the plane because it is mostly buried under the ground, but it is the most complex, it owns the largest market size, and it has enabled us to change our lifestyle by entering the information technology age. This industry is the *telecommunications networking industry.*

To have an intuitive understanding of the size of the telecommunications industry, it is good to know that the size of the budget of AT&T in the early 1980s, before its divestiture, was nearly the budget of the fifth largest economy of the world. AT&T was the largest telecommunications company in the world, and its core revenue at that time was generated from plain old telephone service (POTS) that was first introduced in 1867.

During the past two decades, the cellular telephone industry augmented the income of the prosperous voice-oriented POTS with the subscriber fees of about 1 billion cellular telephone users worldwide [EMC01]. Today the income of the wireless industry has already surpassed the income of the wired telephone industry, and this income is by far dominated by the revenue of cellular phones. In the mid 1990s, the data-oriented Internet brought the computer communications industry from the of-

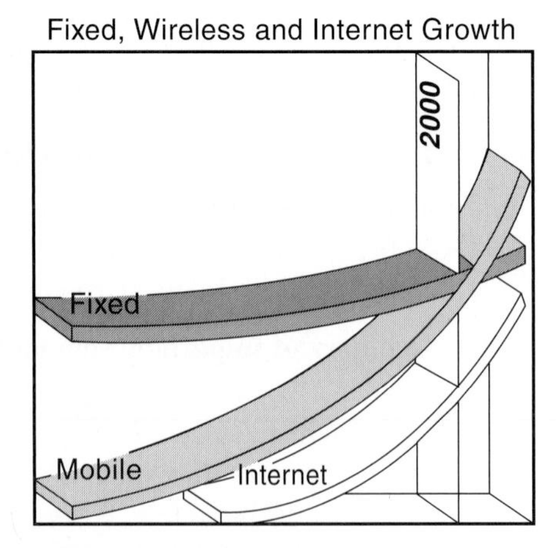

Figure 1.1 Worldwide growth of the fixed, wireless, and Internet communication industries in the past decade.

fice to the home, which soon generated an income comparable to that of the voice-oriented POTS and wireless industry. Figure 1.1 illustrates the growth of the fixed (POTS), wireless, and Internet industries in recent years. At the time of this writing, the *information exchange industry* that includes the fixed and wireless telephone as well as Internet access industries has an annual revenue of a few trillion dollars and is by far the largest industry in the world. The wireless networking industry makes up a third of the revenue of the information industry, and its share of the overall market is growing. Today this income is dominated by revenue from cellular telephone applications. The future of this industry relies on broadband wireless Internet access that is expected to develop a large market for emerging multimedia applications.

The purpose of this book is to provide the reader with a text for understanding the principles of wireless networks, which include the cellular telephone and wireless broadband access technologies. Wireless networking is a multidisciplinary technology. To understand this industry, we need to learn aspects of a number of disciplines to develop an intuitive feeling of how these disciplines interact with one another. To achieve this goal, we provide an overview of the important wireless standards and products, describe and classify their underlying technologies in a logical manner, give detailed examples of successful standards and products, and provide a vision of evolving technologies. In this first chapter, we provide an overview of the wireless industry and its path of evolution. The following five chapters describe principles of technologies that are used in wireless networks. The next three chapters discuss the details of wireless wide area networks (WANs), and the last four chapters describe short-range broadband and ad hoc wireless networks.

In this chapter we first provide an overview of the evolution of the wireless information network industry. We describe the meaning of a wireless network, and we give a summary of the important standards. Then we discuss the technical aspects and general structure of a wireless network. Finally we give an outline of the chapters of the book and how they relate to one another.

1.1.1 Information Network Infrastructure

An information network infrastructure interconnects telecommunication devices to provide them with means for exchanging information. Telecommunication devices are terminals allowing users to run applications that communicate with other terminals through the information network infrastructure. The basic elements of an information network infrastructure are a number of switches or routers that are connected via point-to-point links. Switches include fixed and variable rate voice-oriented circuit switches and routers that are low speed and high speed data-oriented packet switches. The point-to-point links include a variety of fiber, coaxial cable, twisted pair wires, and wireless connections.

To support transmission of voice, data, and video, several wired information network infrastructures have evolved throughout the past century. Wireless networks allow a mobile telecommunications terminal to access these wired information network infrastructures. At first glance it may appear that a wireless network is only an antenna site connected to one of the switches in the wired information infrastructure which enables a mobile terminal to be connected to the backbone network. In reality, in addition to the antenna site, a wireless network may also need to add its own mobility-aware switches and base station control devices to be able to support mobility when a mobile terminal changes its connection point to the network. Therefore, a wireless network has a fixed infrastructure with mobility-aware switches and point-to-point connections, similar to other wired infrastructures, as well as antenna sites and mobile terminals.

Example 1.1: PSTN and Cellular

Figure 1.2 shows the overall picture for the wired and wireless telephone services. The public switched telephone network (PSTN), designed to provide wired telephone services, is augmented by a wireless fixed infrastructure to support mobility of the mobile terminal that communicates with several base stations mounted over antenna posts. The PSTN infrastructure consists of switches, point-to-point connections, and computers used for operation and maintenance of the network. The fixed infrastructure of the cellular telephone service has its own mobility-aware switches, point-to-point connections, and other hardware and software elements that are needed for the mobile network operation and maintenance. A wireless telecommunications device, such as a cordless telephone, can connect to the PSTN infrastructure by replacing the wire attachment with radio transceivers. But, for the mobile terminal to change its point of contact (antennas) the PSTN switches must be able to support mobility. Switches in the PSTN infrastructure were not originally designed to support mobility. To solve this problem, the cellular telephone service providers add their own fixed infrastructure with mobility-aware switches. The fixed infrastructure of the cellular telephone service provider is an interface between the base stations and the PSTN infrastructure that implements the requirements to support mobility.

In the same way a telephone service provider needs to add its own infrastructure to allow a mobile telephone to connect to the PSTN, a wireless data network provider needs its own infrastructure to support wireless Internet access.

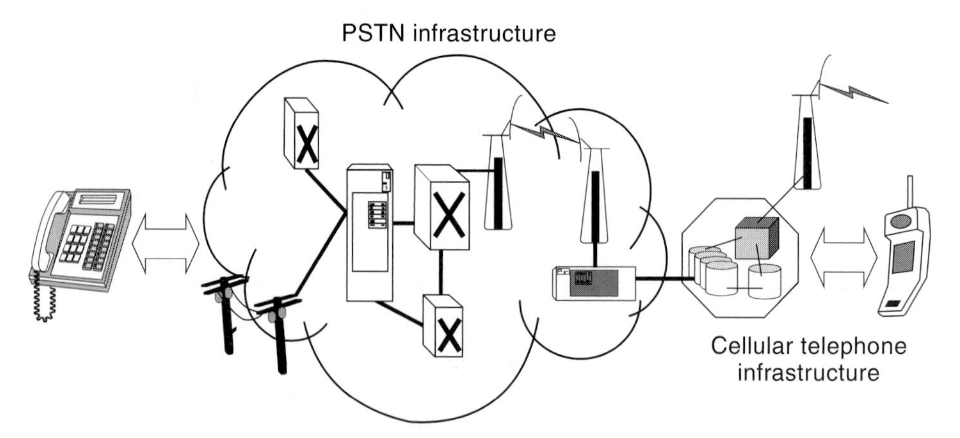

Figure 1.2 PSTN and its extension to cellular telephone services.

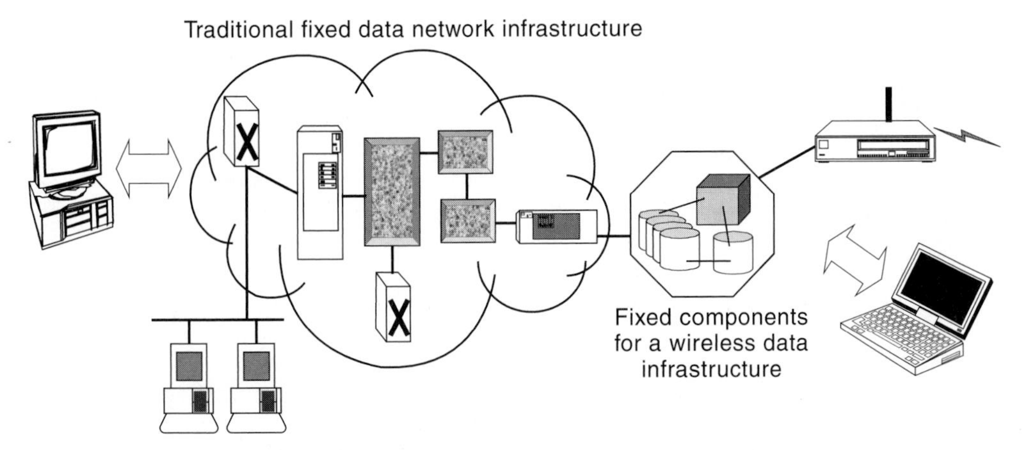

Figure 1.3 Internet and its extension to wireless data services.

Example 1.2: Wireless Internet

Figure 1.3 shows the traditional wired data infrastructure and the additional wireless data infrastructure that allow wireless connection to the Internet. The traditional data network consists of routers, point-to-point connections, and computers for operation and maintenance. The elements of a wireless network include mobile terminals, access points, mobility-aware routers, and point-to-point connections. This new infrastructure has to support all the functionalities needed to support mobility.

The difference between Examples 1 and 2 is that the wireless network in Figure 1.2 is a connection-based, voice-oriented network, and the wireless network in Figure 1.3 is a connectionless data-oriented network. A voice-oriented network needs a dialing process, and after the dialing, a certain quality of service (QoS) is guaranteed to the user during the communication session. In data-oriented networks there is no dialing, and terminals are always connected to the network, but a definite QoS is not guaranteed.

1.1.2 Overview of Existing Network Infrastructure

Because the existence of the wireless networks heavily depends on the wired infrastructure that they connect to, in this section we provide an overview of the important types of wired infrastructure. More details on the evolution of these wired backbone infrastructures are provided in Appendix 1A. The most commonly used wired infrastructures for wireless networks are the PSTN, the Internet, and hybrid fiber coax (HFC), originally designed for voice, data, and cable TV distribution applications, respectively. Figure 1.4 provides an overall picture of these three networks and how they relate to other wired and wireless networks.

The main sources of information transmitted through telecommunications devices are voice, data, and video. Voice and video are analog in nature, and data is digital. The dominant voice application is telephony, which is a bidirectional, symmetric, real-time conversation. To support telephony, telephone service

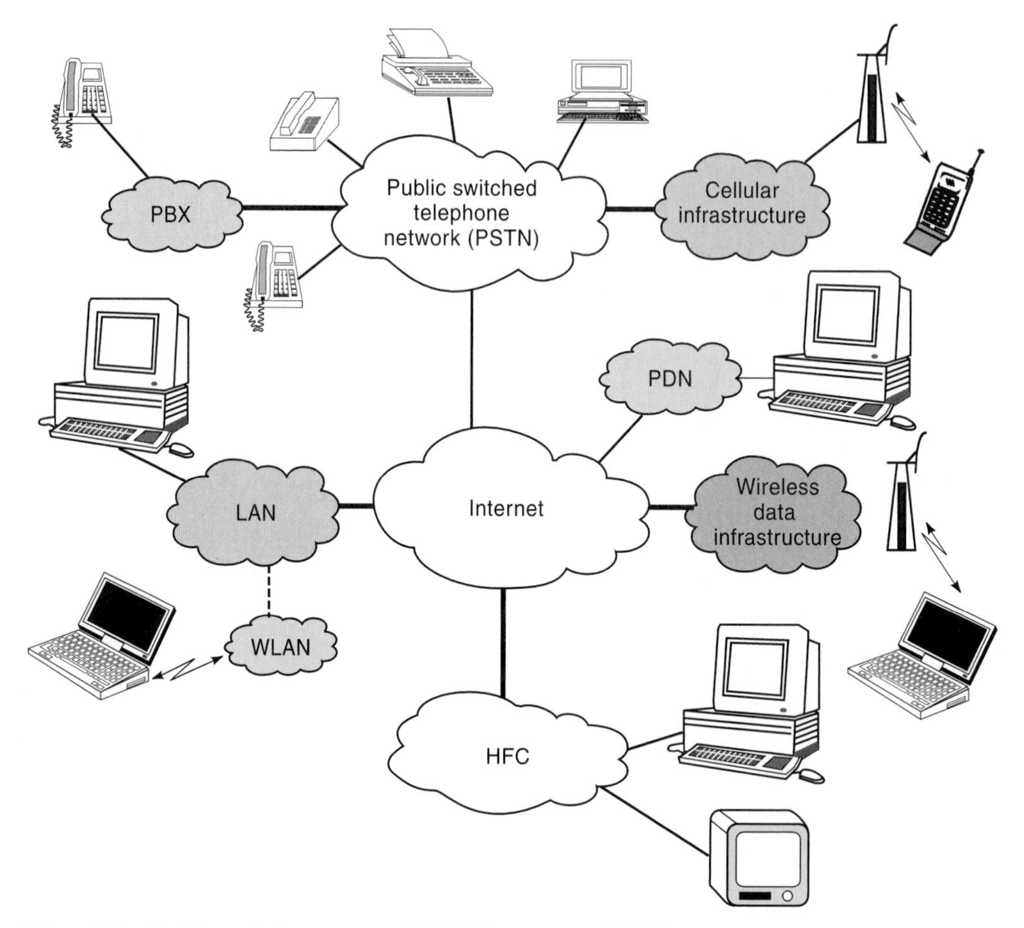

Figure 1.4 Backbone infrastructures: PSTN, Internet, and HFC.

providers have developed a network infrastructure that establishes a connection for a telephone call during the dialing process and disconnects it after completion of the conversation. As we saw in Example 1, this network is referred to as the PSTN. As shown in the top of Figure 1.4, the cellular telephone infrastructure provides a wireless access to the PSTN. Another network attached to the PSTN is the private branch exchange (PBX), which is a local telephone switch privately owned by companies. This private switch allows privacy and flexibility in providing additional services in an office environment. The PSTN physical connection to homes is a twisted-pair analog telephone wiring that is also used for broadband digital services. The core of the PSTN is a huge digital transmission system that allocates 64 kbps channels for each direction of a telephone conversation. Other network providers often lease the PSTN transmission facilities to interconnect their nodes.

The infrastructure developed for video applications is cable television, shown in the lower part of Figure 1.4. This network broadcasts wideband video signals to residential buildings. A cable goes from an end office to residential areas, and all users are provided service that is tapped from the same cable. The set-top boxes leased out by cable companies provide selectivity of channels depending on the charged rates. The end offices, where a group of distribution cables arrive, are connected to one another with fiber. For this reason, the cable TV network is also called hybrid fiber coax (HFC). More recently cable distribution has also been used for broadband home access to the Internet.

The data network infrastructure was developed for bursty applications and evolved into the Internet that supports Web access, email, file transfer, and telnet applications, as well as multimedia (voice, video, and data) sessions with a wide variety of session characteristics. The middle part of Figure 1.4 shows the Internet and its relation to other data networks. From the user's, point of view, the data-oriented networks are always connected, but they only use the transmission resources when a burst of information needs it. Sessions of popular data communications applications, such as Web browsing or file transfer protocol (FTP) are often asymmetric, and a short burst of upstream requests results in a downstream transmission of a large amount of data. Symmetric sessions such as Internet Protocol (IP) telephony over data networks are also becoming popular, providing an alternative to traditional telephony. The Internet access to home is a logical access that is physically implemented on other media such as cable TV wiring or telephone wiring. The distribution of the Internet in office areas is usually through local area networks (LANs). Wireless LANs (WLANs) in the offices are usually connected to the Internet through the LANs. These days all other private data networks, such as those used in the banks or airline reservation industries, are also connected to the Internet. As we saw in Example 2, the Internet is also the backbone for wireless data services.

1.1.3 Four Market Sectors for Wireless Applications

The market for wireless networks has evolved in four different segments that can be logically divided into two classes: *voice-oriented* market and *data-oriented* market. The voice-oriented market has evolved around wireless connections to the PSTN. These services further evolved into local and wide area markets. The local voice-oriented market is based on low-power, low-mobility devices with higher quality of voice, including cordless telephone, personal communication services (PCS), wireless PBX, and wireless telepoint. The wide area voice-oriented market evolved around cellular mobile telephone services that are using terminals with a higher power consumption, comprehensive coverage, and lower quality of voice. Figure 1.5(a) compares several features of these two sectors of the voice-oriented market. The wireless data-oriented market evolved around the Internet and computer communication network infrastructure. The data-oriented services are divided into broadband local and ad hoc and wide area mobile data markets. The wide area wireless data market provides for Internet access for mobile users. Local broadband and ad hoc networks include WLANs and wireless personal area networks (WPANs) which provide for high-speed Internet access, as well as evolving ad hoc wireless consumer products. Figure 1.5(b) illustrates several differences among the local- and wide-area wireless data networks.

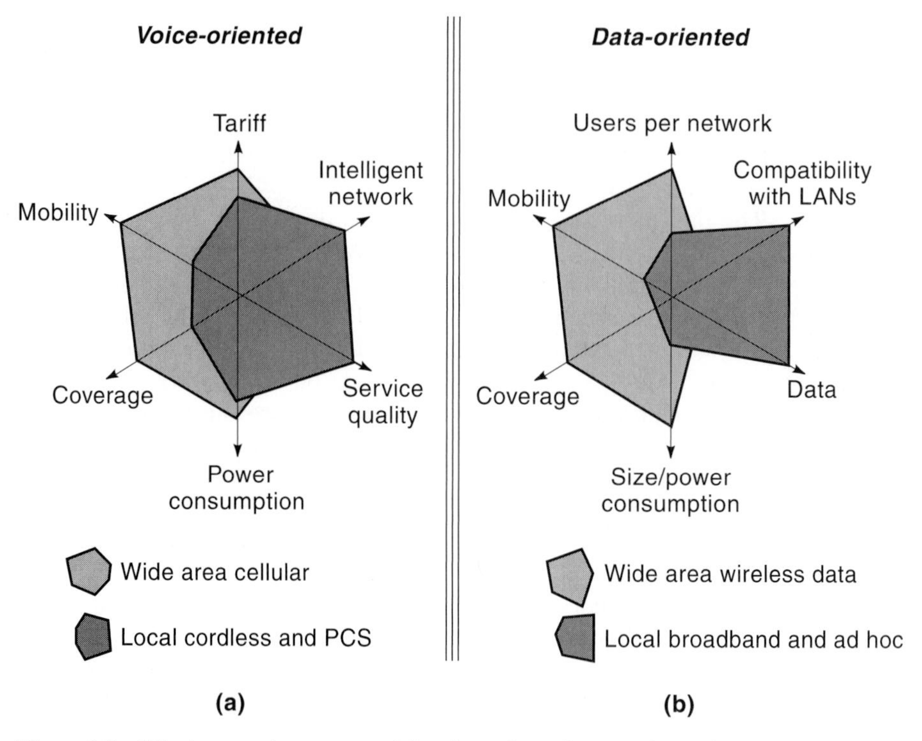

Figure 1.5 Wireless market sectors: (a) voice-oriented networks and (b) data-oriented networks.

1.1.4 Evolution of Voice-Oriented Wireless Networks

Table 1.1 shows a brief chronology of the evolution of voice-oriented wireless networks. The technology for frequency division multiple access (FDMA) analog cellular systems was developed at AT&T Bell Laboratories in the early 1970s. However, the first deployment of these systems took place in the Nordic countries as the Nordic Mobile Telephony (NMT) about a year earlier than the deployment of the Advanced Mobile Phone System (AMPS) in the United States. Because the United States is a large country, the frequency administration process was slower and it took a longer time for the deployment. The digital cellular networks started in Nordic countries with the formation of the Groupe Special Mobile standardization group that became the Global System for Mobile Communications (GSM). The GSM standard group was originally formed to address the international roaming, a serious problem for cellular operation in the European Union (EU) countries. The standardization group shortly decided to go for a new digital time division multiple access (TDMA) technology because it could allow integration of other services to expand the horizon of wireless applications [HAU94]. In the United States, however, the reason for migration to digital cellular was that the capacity of the analog systems in major metropolitan areas such as New York City and Los Angeles had reached its peak value, and there was a need for increasing it in the existing allocated bands. Although Nordic countries, led by Finland,

Table 1.1 History of Voice-Oriented Wireless Networks

Year	Event
Early 1970s	Exploration of first-generation mobile radio at Bell Labs
Late 1970s	First-generation cordless phones
1982	Exploration for second-generation digital cordless CT-2
1982	Deployment of first generation Nordic analog NMT
1983	Deployment of U.S. AMPS
1983	Exploration of the second-generation digital cellular GSM
1985	Exploration of wireless PBX, DECT
1988	Initiation for GSM development
1988	Initiation for IS-54 digital cellular
1988	Exploration of the QUALCOMM CDMA technology
1991	Deployment of GSM
1993	Deployment of PHS/PHP and DCS-1800
1993	Initiation for IS-95 standard for CDMA
1995	PCS band auction by FCC
1995	PACS finalized
1998	3G standardization started

have always maintained the highest rate of cellular telephone penetration, in the early days of this industry the United States was by far the largest market. By 1994, there were 41 million subscribers worldwide, 25 million of them in the United States. The need for higher capacity motivated the study of code division multiple access (CDMA) that was originally perceived to provide capacity that was orders of magnitude higher than other alternatives, such as analog band splitting or digital TDMA.

While the debate between TDMA and CDMA was in progress in the United States, deployment of GSM technology started in the EC in the early 1990s. At the same time, developing countries started their planning for cellular telephone networks, and most of them adopted the GSM digital cellular technology over the legacy analog cellular. Soon after, GSM had penetrated into more than 100 different countries. An interesting phenomenon in the evolution of the cellular telephone industry was the unexpected rapid expansion of this industry in developing countries. In these countries the growth of the infrastructure for wired POTS was slower than the growing demand for new subscriptions, and there was always a long waiting time to acquire a telephone line. As a result, in most of these countries telephone subscriptions were sold on the black market at a price several times the actual value. Penetration of cellular telephones in these countries was much easier because people were already prepared for a higher price for telephone subscriptions, and the expansion of cellular networks occurred much faster than POTS.

In the beginning of the race between TDMA and CDMA, CDMA technology was deployed in only a few countries. Besides, experimentation had shown that the capacity improvement factor of CDMA was smaller than expected. In the mid 1990s when the first deployment of CDMA technology started in the United States, most companies were subsidizing the cost to stay in race with TDMA and analog alternatives. However, from day one, the quality of voice using CDMA was superior to that of TDMA systems installed in the United States. As a result, CDMA service providers under banners like "you cannot believe your ears" started marketing this technology in the United States, which soon became very popular with the users. Meanwhile, with the huge success of digital cellular all manufacturers worldwide started working on the next generation IMT-2000 wireless networks. Most of these manufacturers adopted wideband CDMA (W-CDMA) as the technology of choice for the IMT-2000, assuming that W-CDMA eases integration of services, provides better quality of voice, and supports higher capacity.

The local voice-oriented wireless applications started with the introduction of the cordless telephone, which appeared in the market in the late 1970s. A cordless telephone provides a wireless connection to replace the wire between the handset and the telephone set. The technology for implementation of a cordless telephone is similar to the technology used in walkie-talkies which had existed since the Second World War. The important feature of the cordless telephone was that as soon as it was introduced to the market it became a major commercial success, selling tens of millions of units and generating an income exceeding several billion dollars. The success of the cordless telephone encouraged further developments in this field. The first digital cordless telephone was CT-2, a standard developed in the United Kingdom in the early 1980s. The next generation of cordless telephones was the wireless PBX using the digital European cordless telephone (DECT) standard. Both CT-2 and DECT had minimal network infrastructures to go beyond the simple cordless telephone and cover a larger area and multiple applications. However, in spite of the huge success of the cordless telephone, neither CT-2 nor DECT has yet been considered a very commercially successful system. These local systems soon evolved into PCS, which was a complete system with its own infrastructure, very similar to the cellular mobile telephone.

In the technical communities of the early 1990s, PCS systems were differentiated from the cellular systems as presented in Figure 1.5(b). A PCS service was considered the next generation cordless telephone designed for residential areas, providing a variety of services beyond the cordless telephone. The first real deployment of PCS systems was the personal handy phone (PHP), later renamed as personal handy system (PHS), introduced in Japan in 1993. At that time, the technical differences between PCS services and cellular systems were perceived to be smaller cell size, better quality of speech, lower tariff, lesser power consumption, and lower mobility. However, from the user's point of view the terminals and services for PCS and cellular looked very similar; the only significant differences were marketing strategy and the way that they were introduced to the market. For instance, around the same time in the United Kingdom DCS-1800 services were introduced as PCS services. The DCS-1800 was using GSM technology at a higher frequency of 1,800 MHz, but it employed a different marketing strategy. The last PCS standard was the personal access communications system (PACS) in the United States, finalized in 1995. All together, none of the PCS standards became a major commercial success or a competitor with cellular services.

In 1995, the Federal Communications Commission (FCC) in the United States auctioned the frequency bands around 2 GHz as the PCS bands, but PCS-specific standards were not adopted for these frequencies. Eventually the name PCS started to appear only as a marketing pitch by some service providers for digital cellular services, in some cases not even operating at PCS bands. Although the more advanced and complex PCS services evolving from simple cordless telephone application did not succeed and were merged into the cellular telephone industry, the simple cordless telephone industry itself still remains active. In more recent years, the frequency of operation of cordless telephone products has shifted into unlicensed industrial, scientific, and medical (ISM) bands rather than the licensed PCS bands. Cordless telephones in the ISM bands can provide a more reliable link using spread spectrum technology.

1.1.5 Evolution of Data-Oriented Wireless Networks

Table 1.2 provides the chronology of data-oriented wireless networks. As shown in Figure 1.5(b), data-oriented wireless networks are divided into the wide area wireless data and local broadband and ad hoc networks. Wireless local networks support higher data rates and ad hoc operation for a lower number of users. The broadband wireless local networks are usually referred to as WLANs and the ad hoc local networks as WPANs. The concept of WLANs was first introduced around 1980. However, the first WLAN products were completed about 10 years later. Today a key feature of the local broadband and ad hoc networks is operation in the unlicensed bands. The first unlicensed bands were the ISM bands released in the United States in 1985. Later in 1994 and then in 1997, unlicensed PCS and U-NII (Unlicensed National Information Infrastructure) bands were also released in the United States. The major WLAN standard is the IEEE 802.11 started in the late 1980s and completed in 1997. The IEEE 802.11 and 802.11b operate in the ISM bands and the IEEE 802.11a in the U-NII bands. The competing European standard for WLANs is the HIgh PErformance Radio LAN (HIPERLAN). The HIPERLAN-1 was completed in 1997, and the HIPERLAN-2 is currently under development. In 1996, the wireless ATM working group of the ATM (Asynchronous Transfer Mode) Forum was formed to merge ATM technology with wideband local access. More recently, after the announcement of Bluetooth technology in 1998, WPANs have attracted tremendous attention. The coverage of WPANs is

Table 1.2 History of Data-Oriented Wireless Networks

Date	Event
1979	Diffused infrared (IBM Rueschlikon Labs—Switzerland)
1980	Spread spectrum using SAW devices (HP Labs—California)
Early 1980s	Wireless modem (Data Radio)
1983	ARDIS (Motorola/IBM)
1985	SM bands for commercial spread spectrum applications
1986	Mobitex (Swedish Telcom and Ericsson)
1990	IEEE 802.11 for Wireless LAN standards
1990	Announcement of wireless LAN products
1991	RAM mobile (Mobitex)
1992	Formation of WINForum
1992	ETSI and HIPERLAN in Europe
1993	Release of 2.4, 5.2, and 17.1–17.3 GHz bands in EU
1993	CDPD (IBM and 9 operating companies)
1994	PCS licensed and unlicensed bands for PCS
1996	Wireless ATM Forum started
1997	U-NII bands released, IEEE 802.11 completed, GPRS started
1998	IEEE 802.11b and Bluetooth announcement
1999	IEEE 802.11a/ HIPERLAN-2 started

smaller than traditional WLANs, and they are intended for ad hoc environments to interconnect such personal equipment as the laptop, cell phone, and headset together. At the time of this writing, the IEEE 802.11 products generated around half a billion dollars per year. In the past couple of years, huge investments have been poured into WLAN and WPAN chip set developments all over the world. These investments expect sizable incomes from the possible incorporation of WLANs into the prosperous cellular industry and a large WPAN market for consumer products and home networking. A more complete history of WLANs and WPANs is provided in Chapter 10.

Mobile data services were first introduced with the ARDIS (now called DATATAC) project between Motorola and IBM in 1983. The purpose of this network was to allow IBM field crew to operate their portable computers wherever they want to deliver their services. In 1986, Ericsson introduced Mobitex technology, which was an open architecture implementation of the ARDIS. In 1993, IBM and nine operating companies in the United States started the CDPD (cellular digital packet data) project, expecting a huge market by the year 2000. In the late 1990s the GPRS (general packet radio service) data services that are integrated in the successful GSM systems and can support an order of magnitude higher data rates than previous technologies attracted considerable attention. These higher data rates are perceived to be essential for wireless Internet access, the most popular wireless data application. The third generation (3G) cellular systems are planning to provide up to 2 Mbps mobile data service that is substantially higher than the GPRS data rates. These data rates, however, would not have the comprehensive coverage of GPRS. The early mobile data networks, ARDIS and Mobitex, were independent networks owning their infrastructure. As time passed, CDPD overlaid its infrastructure over the AMPS systems, and GPRS was actually integrated within the GSM infrastructure. This gradual assimilation of the mobile data industry into the cellular telephone industry will be completed in the next-generation cellular systems.

With the integration of PCS and mobile data industries in the next-generation cellular systems we see the emergence of two industries: the next-generation traditional cellular systems operating in licensed bands and the local broadband and ad hoc networks operating in unlicensed bands.

1.2 DIFFERENT GENERATIONS OF WIRELESS NETWORKS

It is customary among the cellular telephone manufacturers and service providers to classify wireless communication systems into several generations. The first-generation (1G) systems are voice-oriented analog cellular and cordless telephones. The second-generation (2G) wireless networks are voice-oriented digital cellular and PCS systems and data-oriented wireless WANs and LANs. The third-generation (3G) networks integrate cellular and PCS voice services with a variety of packet-switched data services in a unified network. In parallel to the unified 3G standardization activities, broadband local and ad hoc networks attracted tremendous attention, and they developed their own standards. One of the major current differences between these two waves is that the 3G systems use licensed bands, and broadband and ad-hoc networks operate in unlicensed bands. The manner in which broadband local access in unlicensed bands and 3G standards in licensed bands may be integrated forms the core of the forthcoming generations of wireless networks.

1.2.1 1G Wireless Standards

Table 1.3 shows the worldwide 1G analog cellular systems. All these systems use two separate frequency bands for forward (from base station to mobile) and reverse (from mobile to base station) links. Such a system is referred to as a frequency division duplex (FDD) scheme. The typical allocated overall band in each direction, for example, for AMPS, TACS, and NMT-900, was 25 MHz in each direction. The dominant spectra of operation for these systems were the 800 and 900 MHz bands. In an ideal situation, all countries should use the same standard and the same frequency bands, however, in practice, as shown in Table 1.3, a variety of frequencies and standards were adopted all over the world. The reason for the different frequencies of operation is that the frequency administration agencies in each country have had previous frequency allocation rulings that restricted the assignment

Table 1.3 Existing 1G Analog Cellular Systems

Standard	Forward Band (MHz)	Reverse Band (MHz)	Channel Spacing (kHz)	Region	Comments
AMPS	824–849	869–894	30	United States	Also in Australia, southeast Asia, Africa
TACS	890–915	935–960	25	EU	Later, bands were allocated to GSM
E-TACS	872–905	917–950	25	UK	
NMT 450	453–457.5	463–467.5	25	EU	
NMT 900	890–915	935–960	12.5	EU	Freq. overlapping; also in Africa and southeast Asia
C-450	450–455.74	460–465.74	10	Germany, Portugal	
RMTS	450–455	460–465	25	Italy	
Radiocom 2000	192.5–199.5	200.5–207.5	12.5	France	
	215.5–233.5	207.5–215.5			
	165.2–168.4	169.8–173			
	414.8–418	424.8–428			
NTT	925–940	870–885	25/6.25	Japan	First band is nationwide, others regional
	915–918.5	860–863.5	6.25		
	922–925	867–870	6.25		
JTACS/ NTACS	915–925	860–870	25/12.5	Japan	All are regional
	898–901	843–846	25/12.5		
	918.5–922	863.5–867	12.5		

TACS: Total Access Communication System
E-TACS: Enhanced TACS
NTT: Nippon Telephone and Telegraph

choices. The reason for adopting different standards was that at that time cellular providers assumed services to be mainly used in one country, and they did not have a vision for a universal service. The channel spacing or bandwidth allocated to each user was either 30 kHz or 25 kHz or a fraction of either of them. The 25 kHz band was previously used for mobile satellite services and the 30 kHz band was something new used for cellular telephone applications. All the 1G cellular systems were using analog frequency modulation (FM) for which the transmission power requirement depends on the transmission bandwidth. On the other hand, power is also related to the coverage and size of the radios. Therefore, one can compensate for the reduction in transmission bandwidth per user by reducing the size of a cell in a cellular network. Reduction in size of the cell increases the number of cells and the cost of installation of the infrastructure.

Example 1.3: Infrastructure Density versus Channel Size in Analog Cellular Systems

The AMPS system in North America uses 30 kHz band whereas C-450 in Germany uses 10 kHz spacing that is ⅓ of the 30 kHz. Therefore, one expects a heavier infrastructure density for deployment of C-450.

Example 1.4: Cell Size and Channel Bandwidth in Analog Cellular Systems

The Japanese use 25 kHz, 12.5 kHz, and 6.25 kHz bands in their systems that support the entire band, ½, and ¼ of the band, respectively. The size of the cells for split-band operation is smaller than the size for full-band operation.

Band splitting obviously allows more subscribers at the expense of more investment in the infrastructure. In Chapter 5, where we address deployment of the cellular networks, we introduce a technique to improve the capacity of a cellular network with the same number of base stations using band splitting.

In the cellular industry, often 1G only refers to analog cellular systems because it is the only system implemented based on popular standards such as AMPS or NMT. However, we can generalize the 1G systems to other sectors of the wireless services. The analog cordless telephone that appeared on the market in the 1980s can be considered the 1G cordless telephone. Paging services that were deployed around the same time frame as analog cellular systems and cordless telephones can be referred to as 1G mobile data services providing one-way short data messages. In the early 1980s before the release of ISM bands and start of the WLAN industry, a couple of small companies in Canada and the United States developed low-speed connectionless WLANs using voiceband modem chip sets and commercially available walkie-talkies. These products were operating at the speed of voice-band modems (< 9,600 bps) but using medium-access control techniques used in data-oriented LANs. Because of their low speed they do not comply with the IEEE 802 definition of LANs, but one may refer to them as 1G WLAN products.

1.2.2 2G Wireless Systems

The 2G systems supported a complete set of standards for all four sectors of the wireless network industry. As we discussed in the history of voice-oriented and data-oriented networks, we have a number of digital cellular, PCS, mobile data, and WLAN standards and products that can be classified as 2G systems. In the remainder of this section, we cover each of these four sectors of 2G systems in a separate subsection.

1.2.2.1 2G Digital Cellular

Table 1.4 summarizes the major 2G digital cellular standards. There are four major standards in this category: GSM, the pan-European digital cellular, the North American Interim Standard (IS-54) that later on improved into IS-136 and Japanese digital cellular (JDC)—all of them using TDMA technology and IS-95 in North America, which uses CDMA technology. Like the 1G analog systems, the 2G systems are all FDD and operate in the 800–900 MHz bands. The carrier spacing of IS-54 and JDC is the same as the carrier spacing of 1G analog systems in their respective regions, but GSM and IS-95 use multiple analog channels to form one digital carrier. GSM supports eight users in a 200 kHz band; IS-54 and JDC support three users in 30 and 25 kHz bands, respectively. As we explain in Chapter 4, the number of users for CDMA depends on the acceptable quality of service; therefore, the number of users in the 1,250 kHz CDMA channels cannot be theoretically fixed. But this number is large enough to convince the standards organization to adopt CDMA technology for next generation 3G systems. By looking into these numbers, one may jump to the conclusion that GSM uses 25 kHz for each bearer, and IS-54 uses 10 kHz per user. Therefore GSM supports 2.5 times less the number of users in the given bandwidth. The reader should be warned that this is an illusive conclusion because when the network is deployed, the ultimate quality of voice also depends on the frequency reuse factor and signal-to-noise/interference

Table 1.4 2G Digital Cellular Standards

System	GSM	IS-54	JDC	IS-95
Region	Europe/Asia	United States	Japan	United States/Asia
Access method	TDMA/FDD	TDMA/FDD	TDMA/FDD	CDMA/FDD
Modulation scheme	GMSK	$\pi/4$-DQPSK	$\pi/4$-DQPSK	SQPSK/QPSK
Frequency band (MHz)	935–960 890–915	869–894 824–849	810–826 940–956 1,477–1,489 1,429–1,441 1,501–1,513 1,453–1,465	869–894 824–849
Carrier spacing (kHz)	200	30	25	1,250
Bearer channels/carrier	8	3	3	Variable
Channel bit rate (kbps)	270.833	48.6	42	1,228.8
Speech coding	13 kbps	8 kbps	8 kbps	1–8 kbps (variable)
Frame duration (ms)	4.615	40	20	20

requirements that will change these calculations significantly. These issues are addressed in Chapter 5.

The channel bit rate of GSM is 270 kbps whereas IS-54 and JDC use 48 and 42 kbps, respectively. Higher channel bit rates of a digital cellular system allow simple implementation of higher data rates for data services. By assigning several voice slots to one user in a single carrier, one can easily increase the maximum supportable data rate for a data service offered by the network. How the higher channel rate of GSM allows support of higher data rates is discussed in Chapter 9, when we discuss GPRS mobile data services. Using a similar argument, one may notice that the 1228.8 kcps channel chip rate of IS-95 provides a good ground for integration of higher data rates into IS-95. This fact has been exploited in IMT-2000 systems to support data rates up to 2 Mbps.

The speech coding technique of 2G systems are all around 10 kbps. As shown in Table 1.4, cellular standards assume large cell sizes and a large number of users per cell, which necessitates lower speech coding rates. On the other hand, cellular standards were assuming operation in cars for which power consumption and battery life were not issues. The peak transmission power of the mobile terminals in these standards can be between several hundreds of mW up to 1W [PAH95] and on the average they consume around 100 mW. All of these systems employ central power control, which reduces battery power consumption and helps in controlling the interference level. In digital communications, information is transmitted in packets or frames. The duration of a packet/frame in the air should be short enough so that the channel does not change significantly during the transmission and long so that the required time interval between packets is much smaller than the length of the packet. A frame length of around 5 to 40 ms is typically used in 2G networks.

1.2.2.2 2G PCS

As we discussed in the history of wireless voice-oriented networks, the 2G PCS standards evolved out of the 1G analog cordless telephone industry and merged into the 3G cellular systems. Figure 1.5(a) illustrates the difference between these two industries during the evolution of the 2G networks; before they

Table 1.5 Quantitative Comparison of PCS and Cellular Philosophies

System Aspects	PCS	Cellular
Cell size	5–500 m	0.5–30 km
Coverage	Zonal	Comprehensive
Antenna height	< 15 m	> 15 m
Vehicle speed	< 5 kmph	< 200 kmph
Handset complexity	Low	Moderate
Base station complexity	Low	High
Spectrum access	Shared	Exclusive
Average handset power	5–10 mW	100–600 mW
Speech coding	32 kb/s ADPCM	7–13 kb/s vocoder
Duplexing	Usually TDD	FDD
Detection	Non-coherent	Coherent

were integrated into one in 3G systems. Table 1.5 illustrates a more quantitative comparison of PCS and cellular industries that at the time was used to justify the existence of two separate voice-oriented standards. The basic philosophy was that PCS is for residential applications, the cell size is small, coverage is zonal, antennas are installed on existing posts (such as electricity or telephone posts), it is not designed to be used in the car, and the complexity of the handset and base station is low. These standards preferred 32 kbps speech coding to support wireline quality and shared the same spectrum in different zones; time-division-duplex (TDD) and noncoherent modulation techniques were mostly used to support simpler implementation.

Table 1.6 provides a summary of the specifications of the four major PCS standards. CT-2 and CT-2+ were the first digital cordless telephone standards; PHS, which later on became PHP, was the first and the only nationwide deployment of these systems; and PACS is the last standard developed with this philosophy. Except for CT-2+ all these standards were designed for 1.8 and 1.9 GHz frequency bands, which are commonly referred to as PCS bands; all systems use TDMA/TDD except PACS, which adopted FDD. To support wireline quality of voice, speech coding at 32 kbps is used in all of these standards. This rate is around three times higher than the speech coding rate of digital cellular systems. The carrier bit rate of 1,728 kbps in DECT is even higher than GSM, which had the highest carrier rate of all TDMA digital cellular systems. This carrier bit rate is even higher than the chip rate in IS-95, the 2G CDMA standard. This feature provides an edge to DECT in supporting high-speed data connections for Internet access. Perhaps this is the major reason why DECT is the only PCS standard that is still considered in new technologies like HomeRF. The power consumption of PCS standards is almost one order of magnitude less than that of the digital cellular standards because PCS systems are designed for smaller cells. If digital cellular systems were deployed with the same cell sizes, the average power consumption could be comparable to PCS systems. Modulation techniques used for PCS standards, GFSK and DQPSK, are less bandwidth efficient and more power efficient than modulation schemes used in digital cellular systems. These modulation techniques can be implemented with simpler noncoherent receivers reducing the size of the handset. The shorter propagation

Table 1.6 2G PCS Standards

System	CT-2 and CT-2(+)	DECT	PHS	PACS
Region	Europe/Canada	Europe	Japan	United States
Access Method	TDMA/TDD	TDMA/TDD	TDMA/TDD	TDMA/FDD
Frequency band (MHz)	864–868 944–948	1,880–1,900	1,895–1,918	1,850–1,910 1,930–1,990
Carrier spacing (kHz)	100	1,728	300	300
Bearer channels/carrier	1	12	4	8 per pair
Channel bit rate (kbps)	72	1,152	384	384
Modulation	GFSK	GFSK	$\pi/4$-DQPSK	$\pi/4$-DQPSK
Speech coding (kbps)	32	32	32	32
Average handset Tx power (mW)	5	10	10	25
Peak handset Tx power (mW)	10	250	80	200
Frame duration (ms)	2	10	5	2.5

time for the smaller-cell PCS standards allows shorter packet frames which help the quality of voice in spite of the wireless channel impairments.

1.2.2.3 Mobile Data Services

As shown in Figure 1.5(b), mobile data services provide moderate data rate and wide coverage area access to packet-switched data networks. The mobile data networks emerged after the success of the paging industry to provide a two-way connection for larger messages. Table 1.7 provides a comparison among a number of important mobile data services. ARDIS and Mobitex use their own frequency bands in 800–900 MHz, terrestrial European trunked radio (TETRA) uses its own band at 300 MHz, CDPD shares the AMPS bands and site infrastructure, GPRS shares the GSM's complete radio system, and Metricom uses the unlicensed ISM bands. The early systems, ARDIS, Mobitex, and CDPD, were developed before the popularity of the Internet, and the dominant design criteria was coverage and cost rather than data rate. These systems were a wireless replacement for voice-band modems operating at data rates up to 19.2 kbps, which was the rate of modems at that time. TETRA is designed for pan-European civil service application and has its own features for that purpose. Metricom and GPRS support data rates more suitable for Internet access. The advantage of GPRS is that it is incorporated in the popular GSM digital services with large numbers of terminals all over the world. Except for Metricom, the channel spacing of the rest of the mobile data services is based on the channel spacing of cellular telephone networks with 25 or 30 kHz bands or a fraction (12.5 kHz) or a multiple of them (200 kHz). These services are designed to use multiple carriers in an FDMA format and use different versions of random access techniques such as DSMA, BTMA, or ALOHA, which are explained in Chapter 4. Modulation techniques for these systems are like digital cellular and PCS systems explained in Chapter 3. A more detailed comparison of mobile data services is provided in Chapter 9.

1.2.2.4 WLANs

As shown in Figure 1.5(b), WLANs provide high data rates (minimum of 1 Mbps) in a local area (< 100 m) to provide access to wired LANs and the Internet. Today all successful WLANs operate in unlicensed bands that are free of charge

Table 1.7 Mobile Data Services

System	ARDIS	Mobitex	CDPD	TETRA	GPRS	Metricom
Frequency band (MHz)	800 bands 45 kHz sep	935–940 896–961	869–894 824–849	380–383 390–393	890–915 935–960	902–928 ISM bands
Channel bit rate (kbps)	19.2	8.0	19.2	36	200	100
RF channel spacing	25 kHz	12.5 kHz	30 kHz	25 kHz	200 kHz	160 kHz
Channel access/ Multiuser access	FDMA/ DSMA	FDMA/ Dynamic S-ALOHA	FDMA/ DSMA	FDMA/ DSMA	FDMA/ TDMA/ Reservn.	FHSS/ BTMA
Modulation technique	4-FSK	GMSK	GMSK	$\pi/4$-QPSK	GMSK	GMSK

Table 1.8 Wireless LAN Standards

Parameters	IEEE 802.11	IEEE 802.11b	IEEE 802.11a	HIPER-LAN/2	HIPER-LAN/1
Status	Approved, Products	Products	Approved, Products in development	Approved	Approved, No products
Freq. Band	2.4 GHz	2.4 GHz	5 GHz	5 GHz	5 GHz
PHY, modulation	DSSS: FHSS:	DSSS: CCK	OFDM	OFDM	GMSK
Data rate	1, 2 Mbps	1, 2, 5.5, 11 Mbps	6, 9, 12, 18, 24, 36, 54 Mbps		23.5 Mbps
Access method	Distributed control, CSMA/CA or RTS/CTS			Central control; reservation-based access	Active contention resolution; priority signaling

and rigorous regulations. Considering that PCS bands were auctioned at very high prices, in the past few years WLANs have attracted a renewed attention. Table 1.8 provides a summary of the IEEE 802.11 and HIPERLAN standards for WLANs. IEEE standards include 802.11 and 802.11b operating at 2.4 GHz and 802.11a, which operates at 5 GHz. Both HIPERLAN-1 and -2, developed under the European Telecommunication Standards Institute (ETSI), operate at 5 GHz. The 2.4 GHz products operate in ISM bands using spread spectrum technology to support data rates ranging from 1 to 11 Mbps. HIPERLAN-1 uses GMSK modulation with signal processing at the receiver that supports up to 23.5 Mbps. IEEE 802.11a and HIPERLAN-2 use the orthogonal frequency division multiplexing (OFDM) physical layer to support up to 54 Mbps. The access method for all 802.11 standards is the same and includes CSMA/CA, PCF, and RTS/CTS, which are described in Chapters 4 and 11. The access method of the HIPERLAN-1 is on the lines of the 802.11, but the access method for HIPERLAN-2 is a voice-oriented access technique that is suitable for integration of voice and data services. The details of these transmission techniques and access methods are described in Chapters 3 and 4. IEEE 802.11, IEEE 802.11b, and HIPERLAN-1 are completed standards, and IEEE 802.11 and 11.b are today's dominant products in the market. IEEE 802.11a and HIPERLAN-2 are still under development. The IEEE 802.11 and HIPERLAN-1 standards can be considered 2G wireless LANs. The OFDM wireless LANs are forming the next generation of these products. The last four chapters of this book are focused on wideband local access systems which describe these systems in further detail.

1.2.3 3G and Beyond

The purpose of migration to 3G networks was to develop an international standard that combines and gradually replaces 2G digital cellular, PCS, and mobile data services. At the same time, 3G systems were expected to increase the quality of the voice, capacity of the network, and data rate of the mobile data services. Among several radio transmission technology proposals submitted to the International Telecommunication Union (ITU), the dominant technology for 3G systems was W-CDMA, which is discussed in detail in Chapter 11. Outside the 3G standards, WLAN and WPAN standards are forming the future for the broadband and ad hoc wireless networks. Figure 1.6 illustrates the relative coverage and data rates of 2G, 3G, WLAN, and WPANs. WPANs, studied in Chapter 13, are formed under the IEEE 802.15 standard. This community has adopted Bluetooth technology as its first standard. Bluetooth is a new technology for ad hoc networking which was introduced in 1998. Like WLANs, ad hoc Bluetooth-type technologies operate in unlicensed bands. Bluetooth operates at lower data rates than WLANs but uses a voice-oriented wireless access method that provides a better environment for integration of voice and data services. The WPAN ad hoc networking technologies are designed to allow personal devices such as laptops, cellular phones, headsets, speakers, and printers to connect together without any wiring.

From the point of view of cellular service providers, 3G provides multimedia services to users everywhere, WLANs provide broadband services in hot spots where a short proximity is needed, and WPANs connect the personal devices together. The telecommunications industry is a multidisciplinary industry, and it has always been difficult to predict its future. However, there are certain current trends that one may perceive as important. In terms of frequency of operation, 3G systems use licensed bands whereas WLANs and WPANs use unlicensed bands. Unlicensed bands are wider and free of charge and rigorous rules, but there is no regulation to control the interference in these bands. Some researchers and vision-

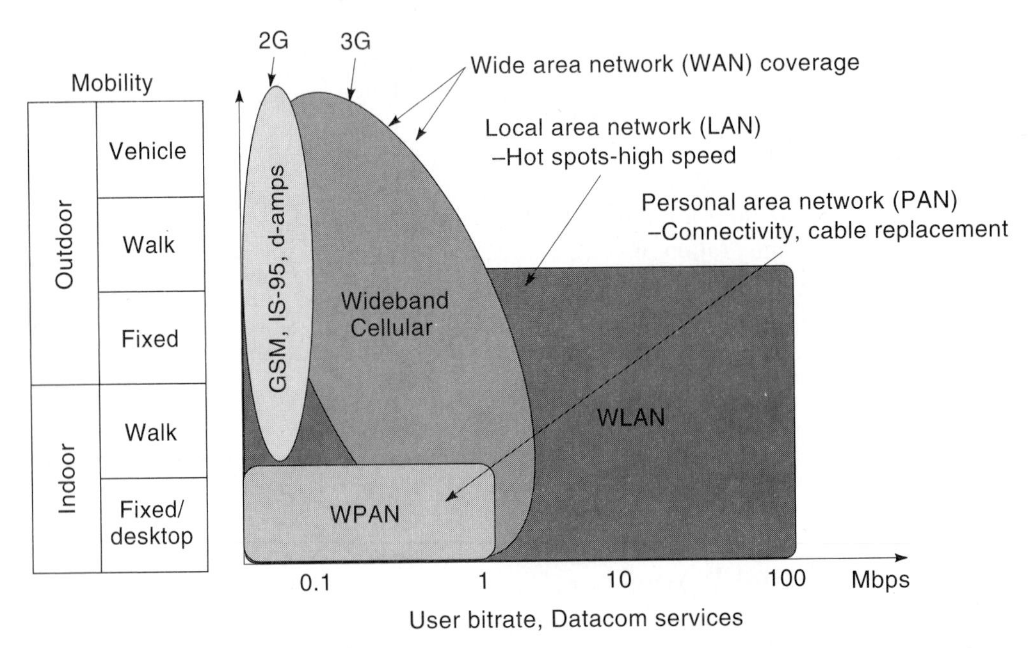

Figure 1.6 Relative coverage, mobility, and data rates of generations of cellular systems and local broadband and ad hoc networks.

aries have gone too far to state that the future is "everything unlicensed," but it is safer to say that recent years have witnessed an immense growth of hope for an increase in using unlicensed bands. Wideband CDMA is the dominant transmission technology for 3G systems and OFDM is becoming very popular in broadband WLANs operating in 5 GHz. Again some visionaries also state that the next-generation systems will be based on OFDM; however, it is safe to say that OFDM appears to play an increasing role in the future of broadband wireless access. Another important evolving technology is the ultra wide band (UWB) which is expected to support a myriad of users with noiselike impulses with a spectrum spreading over several GHz [SCH00]. We address this evolving technology in Chapter 3. There are some visionaries who predict that UWB may be the next step to OFDM, but it is safe to state that it is an evolving, promising technology. Another evolving technology is position finding, in particular in indoor areas, which is becoming an integral part of the wireless networks for the next generations. We address location finding in the last chapter of this book. The FCC in the United States has already mandated the integration of position location systems with cellular systems; however, the extent and method of integration is not yet clear.

1.3 STRUCTURE OF THE BOOK

Wireless networks are very complex multidisciplinary systems; to describe these networks, we need to divide the details into several categories to create a logical organization for presentation of important material. This book is organized in five sections: one introduction chapter and four parts. The introductory material presented in Chapter 1 defines the meaning and sketches the path of evolution of the wireless networks. This chapter also provides an overview of the important wireless systems and outlines the details of the rest of the book. The material presented in Chapter 1 identifies different sectors of the wireless market, familiarizes the reader with the forces behind the growth of these sectors, and provides an overview of the standards developed to address them. The material presented in Chapter 1 should motivate the reader to study the details provided in the remainder of the book.

Each of the four parts of the book consists of several chapters directed toward the description of certain aspects of wireless networks. Parts One and Two are devoted to the principles of wireless network operations. These parts provide the technical background needed for understanding wireless networks. The technical aspects are either related to the design of the air-interface or issues related to deployment and operation of the infrastructure. Part One consists of three chapters describing technical aspects of the air-interface. Part Two has two chapters devoted to the technical aspects of the wireless network infrastructure. Parts Three and Four are devoted to describing the details of typical wireless networks in a comparative manner. Part Three consists of three chapters describing both voice- and data-oriented wireless WANs. Part Four consists of four chapters on local broadband and ad hoc wireless networks, including WLAN, WPAN, and indoor positioning.

1.3.1 Part One: Principles of Air-Interface Design

Wired terminals connect to power lines, and wired access to information networks is reliable, fixed, and relatively simple. Wireless mobile terminals are battery operated, and wireless access is through the air, unreliable, and band limited. The design of the physical connection and access method and understanding of the behavior of the medium for wireless operation is far more complicated than wired operation scenarios. The design of air-interface for wireless connections needs a far deeper understanding of the behavior of the channel and a more complex physical and medium access control mechanism. The behavior of the wireless medium is more complex than the wired medium because in wireless channels the received signal strength suffers from extensive power fluctuations caused by temporal and spatial movements. Wireless transmission and access techniques are more complex because they have to be power and bandwidth efficient, and they need to employ techniques to mitigate the fluctuations caused by the medium. Part One of the book is devoted to the analysis of the behavior of the channel in Chapter 2, an overview of the applied wireless transmission techniques in Chapter 3, and a description of the medium access control techniques used in voice- and data-oriented networks in Chapter 4.

Chapter 2 describes modeling of the path-loss, fluctuation of the channel, and the multipath arrivals of the signal. Path-loss models describe the relation between the average received power in a mobile station and its distance with the base station. These models are used in deployment of the networks to determine the coverage of a base station. The received power at the receiver is not fixed, and it changes in time as the mobile moves or the environment changes. Models for the variations of the channel are used to design the adaptive elements of the receiver, such as synchronization circuits or equalizers, to cope with the variation of the channel. Models for multipath characteristics allow the design of a receiver that can handle the interference of the signals arriving from different paths to the receiver.

The second chapter related to air-interface is Chapter 3, which describes transmission techniques that are used for the physical (PHY) layer of wireless networks. The diversity and complexity of transmission techniques in wireless systems are far more complex than those in wired networks. This chapter provides a comprehensive coverage of all transmission techniques that are employed in voice-oriented cellular and PCS systems, as well as data-oriented mobile data and WLAN and WPAN systems. The transmission techniques are divided into pulse transmission techniques used in infrared (IR) networks and UWB systems, traditional modulation techniques mostly used in digital cellular and mobile data networks, and spread spectrum technologies used in CDMA cellular, as well as WLANs and WPANs. More advanced techniques such as CCK used in 802.11b and OFDM used in 802.11a and HIPERLAN-2 are also described in this chapter.

The third chapter related to air-interface is Chapter 4, which is devoted to multiple access alternatives applied to wireless networks. This chapter starts with a description and comparison of the voice-oriented FDMA, TDMA, and CDMA access methods. The second part of this chapter is devoted to CSMA- and ALOHA-based random access techniques used in data-oriented wireless networks. The last part analyzes the applied access methods for integration of the voice and data that has evolved to operate in the voice- and data-oriented networks.

1.3.2 Part Two: Principles of Wireless Network Operation

In Part One of the book, technical issues related to design of the air-interface are presented. In Part Two we address technical aspects of fixed infrastructure of the wireless networks. This part consists of Chapters 5 and 6, addressing deployment and operation, respectively. Service providers often start with minimal infrastructure and antenna sites to keep the initial investment low. As the number of subscribers grows, service providers expand the wireless infrastructure to increase the capacity and improve the quality. The technology related to the deployment and expansion of the cellular infrastructure is discussed in Chapter 5, which is the first chapter related to technical aspects of the network infrastructure. This chapter discusses different topologies for wireless networks, describes cellular infrastructure deployment, and addresses issues related to the expansion of the size and migration to new technologies.

Chapter 6 is devoted to functionalities of the fixed network infrastructure to support the mobile operation. These functionalities include mobility management, radio resource and power management, and security management. These issues are addressed in three separate parts of Chapter 6. The mobility management part of Chapter 6 describes how a mobile terminal registers with the network at different locations and how the network tracks the mobile as it changes its access to the network from one antenna to another. The radio resource and power management part of Chapter 6 is devoted to the technologies used for controlling the transmitted power of the terminals. Voice-oriented networks control the transmitted power of the mobile station to minimize the interference with other terminals using the same frequency and to maximize the life of the battery. Data-oriented networks also use the sleep mode to avoid unnecessary consumption of power. Explanation of the methodologies and examples of how to implement power control and sleeping modes are provided in this part of Chapter 6. The last section of Chapter 6 is devoted to security in wireless networks. Wireless connections are inherently vulnerable to fraudulent connections and eavesdropping and need security features. The security of wireless networks is provided by authentication and ciphering. When a wireless terminal connects to a network, an authentication process between the network and the terminal checks the authenticity of the terminal. When the connection is established, the transmitted bits are scrambled with a ciphering mechanism to prevent eavesdropping. Algorithms used for these purposes are discussed in the last part of Chapter 6.

1.3.3 Part Three: Wireless Wide Area Networks

After completion of the overview of the standards in Chapter 1 and study of the technical aspects in Parts One and Two, we start detailed descriptions of specific wireless networks. These detailed descriptions are divided into two parts addressing WANs and LANs. The third part of the book is devoted to the description of important voice- and data-oriented wireless wide area networks. In Chapter 7 we describe GSM as an example for TDMA systems, and in Chapter 8 we describe IS-95 and

IMT-2000 as examples of CDMA technology. The obvious reason for these selections is the current worldwide popularity of the GSM and emergence of the IMT-2000 and CDMA technologies as the choice of the emerging 3G systems. The last chapter in this part of the book is Chapter 9 which is devoted to mobile data networks.

Details of the architecture, mechanisms to support mobility, and layered protocols adopted by the GSM standard are described in Chapter 7. GSM is a complete standard that includes specification of the air-interface as well as fixed wired infrastructure to support the services. Other TDMA digital cellular standards, such as IS-136 or JDC, are very similar to GSM. Chapter 7 provides the most comprehensive coverage of a standard in this part of the book. First we describe all elements of the network architecture. Then we address mobility support mechanism with details of registration, call establishment, handoff, and security. The last part of this chapter provides details of how packets are formed and transmitted over the channel. The study of this chapter introduces the reader to the complexity and diversity of the issues involved in development of a wireless standard.

Chapter 8 is devoted to the CDMA and IMT-2000 technology that is the direction of the immediate future of 3G cellular systems. After a brief description of IS-41 and IS-634, which are the standards for communication between cellular switches and base stations in North America, details of the air-interface, IS-95, for cmdaONE that was developed by Qualcomm will be described. Then the 3G systems and how they differ from IS-95 are described. This chapter is completed with a summary of ITU's IMT-2000 standard. Because the wired backbone of the TDMA and CDMA systems are very similar, most of our attention is paid to the air-interface that is completely different from TDMA systems such as GSM. The details of GSM, IS-95, and IMT-2000 conclude our description of popular voice-oriented wide area systems.

The last topic in this part is the description of mobile data networks provided in Chapter 9. This is a comprehensive coverage of all mobile data services including short messaging and application programming for wireless data networks. Although a number of technologies have been examined for mobile data services, the market for this industry has not yet reached the market size of the popular cellular telephone industry. The delay in formation of the market has created a fragmented industry. To cover the details of this fragmented industry, we first provide an overview of all technologies and classify them into logical groups. Then we provide details of CDPD, a data-oriented network overlaid on AMPS infrastructure, and GPRS, which is embedded in the voice-oriented GSM system. In spite of small growth of the mobile data industry, recently short messaging systems (SMS) and application protocols for mobile data networks have attracted considerable market attention. As a result, we have devoted the last two sections of this chapter to SMS and application protocols, respectively.

1.3.4 Part Four: Local Broadband and Ad Hoc Networks

As we discussed earlier, 3G systems integrated cellular telephone, PCS, and mobile data services into one standard system operating in licensed bands. However, WLAN and WPAN have their own standards for broadband and ad hoc networking in unlicensed bands. At the time of this writing, most visionaries in wireless technology developments believe that the future is shaping with the merger of the broadband and ad hoc networking and 3G systems. In addition, there is a trend for integration of geolocation features into the next generation wireless networks. These networks support broadband wireless access to the backbone infrastructure, as well as ad hoc wireless networking. The last part of the book is devoted to these short-range local wireless networks. This part consists of five chapters, three of them devoted to WLAN, one to WPAN, and one to wireless geolocation principles.

The first of the three chapters on WLAN is Chapter 10 which provides an overview of the WLAN industry. This chapter analyzes the evolutionary path of the WLAN and explains how it started for office and manufacturing environments and is currently heading toward home area networking (HAN). Chapter 11, which describes details of the IEEE 802.11 standard for WLAN, follows the introduction to WLAN. Chapter 11 plays the same role as Chapter 7 on GSM was playing in Part Three; it provides details of the IEEE 802.11 to demonstrate all aspects of a data-oriented wireless standard operating in unlicensed bands. The medium access technology for the IEEE 802.11 is CSMA/CA which sets this standard as a connectionless, data-oriented standard. This feature eases the Internet access either by direct connection or connection through an existing wired LAN. The contents of this chapter describe the objective of the standard, explain specifications of the physical (PHY) and medium access control (MAC) layer alternatives supported by this standard, and provide the details of mobility support mechanisms such as registration, handoff, power management, and security. The next chapter, Chapter 12, is devoted to wireless ATM activities and the HIPERLAN standard. HIPERLAN-1 and IEEE 802.11 are considered data-oriented WLANs. Wireless ATM and HIPERLAN-2 are considering integration of voice and data from the point of view of the voice-oriented networks. The medium access for these standards is keen on supporting QoS for voice applications, which makes them suitable for integration into the existing prosperous cellular telephone networks. This chapter starts with an overview of the technical aspect of wireless ATM, then provides a short description of HIPERLAN-1, and at last describes the necessary details of HIPERLAN-2.

Chapter 13 describes WPANs with particular emphasis on details of Bluetooth technology. As discussed earlier, WPANs are ad hoc networks designed to connect personal equipment to one another. However, today's personal devices need to be connected to both voice- and data-oriented networks. The access method for these systems are designed to accommodate that need. Beyond that, WPANs are perceived to have lower power consumption which makes them compromise on the highest supportable data rate as compared with WLANs. This chapter starts by describing the IEEE 802.15 standards committee on WPAN and HomeRF activities under this standard. Then details of Bluetooth technology are described in some depth. The last part of this chapter analyzes the interference between IEEE 802.11 and Bluetooth, both operating in the 2.4 GHz ISM bands.

Chapter 14 is devoted to indoor geolocation and cellular positioning as emerging technologies to complement the local and wide local area wireless services. This chapter provides a generic architecture for the wireless geolocation services, describes alternative technologies for implementation of these systems, and gives examples of evolving location-based services. Location-based services provide a fertile environment for the emergence of E- and M-commerce applications.

APPENDIX 1A BACKBONE NETWORKS FOR WIRELESS ACCESS

Table 1A.1 provides a brief history of important developments in the telecommunications industry. The telecommunications industry started with the simple wired telegraph that used Morse code for digital data communication over long distance wires between the two neighboring cities of Washington, D.C., and Baltimore in 1839 [COU01]. It took 27 years, until 1866, for engineers to communicate over the ocean. In 1900, 34 years after the challenging task of deploying cables in the ocean and three years after the first trial of wireless telegraph, Marconi demonstrated wireless transoceanic telegraph as the first wireless data application. In 1867, Bell started the telephone industry, the first wired analog voice telecommunication service. It took 47 years for the telephone to become a transoceanic service in 1915, and it took almost 100 years for this industry to flourish into the wireless cellular telephone. The wireless telegraph was a point-to-point solution that eliminated the tedious task of laying very long wires in harsh environments. The wireless telephone was a network that had to support numerous mobile users. The challenge for wireless point-to-point communications is the design of a radio; the challenge for a wireless network is the design of a system that allows many mobile radios to work together. The telegraph was a manual SMS that needed a skilled worker to decode the transmitted message.

The first computer communication networks started after the Second World War by using voiceband modems operating over the PSTN infrastructure to exchange large amounts of data among computers located a far distance from one another [PAH88]. Approximately two decades after the era of circuit switched computer communication networks around 1970, wide area packet switched networks (DARPAnet) and wideband local area networks, which were tailored for bursty data applications, were invented.

By the end of the 20th century, multimedia wireless networks emerged to integrate all networks and provide wireless and mobile access to them. SMS services, E- and M-commerce are becoming very popular. It is interesting to note that SMS provides a similar service as the wireless telegraph using the telephone keypad as the terminal. Finally, after more than 100 years, when a terminal with easy user interface became available, the market for the same service started to explode. In the late 1990s, income from SMS in Finland, the current leader in development and consumption of wireless services, is 20% of the income from cellular telephones in that country. This phenomenon also reflects the trends in change of habits, as user-friendly terminals become available for an old application.

Connection to the wired infrastructure is a very important issue for implementation of a wireless network. In the rest of this appendix, we give a brief description of the evolution of the three major existing wired telecommunications network infrastructures: PSTN, the Internet, and cable TV.

1A.1 Evolution of PSTN and Cellular Telephony

The invention of the telegraph in 1834 started wired data communication, and the invention of telephone in 1876 was the start of analog telephone networking. At

Table 1A.1 A Brief History of Telecommunications

Year	Event
1834	Wired telegraph for manually digitalized data (Gauss & Weber)
1839	First demonstration of telegraph between Washington DC and Baltimore (Morse)
1858	First transoceanic cable for telegraph (second working version in 1866)
1867	Manually switched telephone for analog voice (Bell)
1897	Wireless telegram (Marconi)
1900	Transoceanic wireless telegraph (Marconi)
1905	Radio transmission (Fessenden)
1908	Idea of TV (Campbell-Swinton)
1915	Transcontinental telephone (Bell)
1920	Commercial radio broadcast (KDKA); also sampling in comm (Carson)
1926	TV demonstration (Baird, England, and Jenkins)
1933	FM modulation invented (Armstrong)
1941	TV broadcast starts in the United States
1946	First computer (University of Pennsylvania)
1950	Time division multiplexing (TDM), microwave radio, and voice band modems were used in PSTN
1953	Color TV and transoceanic telephone
1957	First satellite (Sputnik I)
1962	Transoceanic satellite TV (Telstar I)
1965	Videotape (Sony)
1968	Cable TV development
1968	ARPANET started (first node at UCLA)
1971	9600 bps voice band modems (Codex)
1972	Demonstration of cellular systems (Motorola)
1973	Ethernet was invented (Metcalfe); also international ARPANET
1980	Fiber optic systems were applied to the PSTN
1995	Netscape introduced and Internet industry started to evolve as the first popular data communications network competing with the legacy PSTN
Recently	Introducing broadband services (cable modems and xDSL), IP switching, home networking, pervasive networking, and incorporation of positioning systems into wireless networks

that time, operators were used to manually switch or route a session from one terminal to another. At the beginning of 20th century, the telecommunications industry had already been exposed to a number of important issues, which played different roles, culminating in the emergence of modern wireless networks. Among these important issues were analog versus digital, voice versus data, wireless versus wired, local versus long haul communications, and personal versus group services.

Example 1A.1: Cabling

It took 28 years for telegraphy to provide transoceanic services after the first on-land service. On the other hand, it took only three years for wireless telegraphy to become transoceanic after a local installation. This reflects the disadvantages of laying cables for wired communications compared with wireless services. To install a wireline in a town, one needs to get wiring permits which take extremely long processing times; conduct expensive, long, and laborious digging to lay the wire; and maintain a service organization for wiring maintenance. The PSTN has emerged as the expert of these details, and as we will later discuss, all networks that emerged use this expertise in one way or another for developing their own network infrastructure. The telephone service providers have the requisite knowledge for laying wires of three types: first, they know how to connect long-haul networks; second, they know how to provide a twisted pair line to homes and offices; and last, they know how to wire a home or office. Wiring of homes and offices is an expertise shared between telephone companies and electricians.

Example 1A.2: Personal Services

The telegraph infrastructure evolved as a private network not directly accessible to the public user. Each end terminal of a telegraph network supported a community whereas the telephone end terminals supported a home or an office with at most a few users. As a result, the number and the usage periods of the telephone terminal were orders of a magnitude larger than the telegraph network, resulting in a far more prosperous telephone industry. This fact reflects the importance of the extension of a service to the home or even further to the user himself/herself as a personal service. In order to develop an intuitive understanding of the size of the telephone market, it is noteworthy to remember that by the mid-1980s, before AT&T's divestiture, its annual budget was comparable to the budget of the fifth largest economy of the world.

By the 1950s, the PSTN had more than 10 million customers in the United States, and those interested in long-haul communication issues also needed PSTN services to solve their problems. Although end users are still mostly connected to the PSTN with twisted-pair analog lines, to provide flexibility and ease of maintenance and operation of the PSTN, the core network gradually changed to digital switches and digital wired lines connecting switches together. A hierarchy of digital lines (the T-carriers in the United States) evolved as trunks to connect switches of different sizes together.

Another advancement in the PSTN was the development of private branch exchanges (PBXs) as privately owned local telephone networks for large offices. A PBX is a voice-oriented local area network owned by the end organization itself, rather than the telephone service provider. This small switch allows the telephone company to reduce the number of wires that are needed to connect all the lines in an office to the local office of the PSTN. This way, the service provider reduces the number of wires to be laid to a small area where large offices with many subscribers are located. The end user also pays less to the telephone company. The organization thus has an opportunity to enhance services to the end users connected to the PBX.

In the 1920s, Bell Laboratories conducted studies to use the PSTN facilities for data communications. In this experiment the possibility of using analog telephone lines for transferring transoceanic telegrams was examined. Researchers involved in this project discovered several key issues that included the sampling theorem and effects of phase distortion on digital communications. However, these discoveries did not affect applications until after World War II when Bell Laboratories developed voice band modems for communication among air force computers in air bases that were geographically separated by large distances [PAH98]. These modems soon found their way into commercial airlines and banking industries, resulting in the associated private long-haul data networks. These pioneering computer communications networks consisted of a central computer and a bank of modems operating over four-wire commercial grade leased telephone lines to connect several terminals to the computer. In late 1960s, the highest data rate for commercial modems was 4,800 bps. By the early 1970s, with the invention of quadrature amplitude modulation (QAM), the data rate of four-wire voice band modems reached 9,600 bps. In the early 1980s, trellis-coded modulation (TCM) was invented which increased data rates to 19.2 kbps and beyond.

In parallel with the commercial four-wire modems used in early long haul computer networks, two-wire modems emerged for distance connection of computer terminals. The two-wire modems operate over standard two-wire telephone lines, and they are equipped with dialing procedures to initiate a call and establish a POTS line during the session. These modems started at data rates of 300 bps. By the early 1970s, they reached 1,200 bps, and by the mid 1980s, they were running at 9,600 bps. These two-wire voice band modems would allow users in the home and office to have access to a regular telephone to develop a data link connection with a distant modem also having access to the PSTN. Voice band modems using two-wire telephone connections soon found a large market in residential and small office remote computer access (telnet), and the technology soon spread to a number of popular applications such as operating a facsimile machine or credit card verification device. With the popularity of Internet access, a new gold rush for higher speed modems began, which resulted in 33.6 kbps full-duplex modems in 1995 and 56 kbps asymmetric modems by 1998. The 56 kbps modems use dialing procedures and operate within the 4 kHz voice band, but they directly connect to the core pulse code modulated (PCM) digital network of the PSTN that is similar to digital subscriber lines (DSLs). DSLs use the frequency band between 2.4 kHz and 1.1 MHz to support data rates up to 10 Mbps over two-wire telephone lines.

More recently cellular telephone services evolved. To connect a cellular telephone to the PSTN, the cellular operators developed their own infrastructure to support mobility. This infrastructure was connected to the PSTN to allow mobile-to-fixed telephone conversations. The addition of new services to the PSTN demanded increases in the intelligence of the core network to support these services. As this intelligence advanced, the telephone service provider added value features such as voice mail, autodialing through network operators, call forwarding, and caller identification to the basic POTS service traditionally supported. Figure 1A.1 shows a simplified representation of today's PSTN network.

1A.2 Emergence of Internet

Data networks that evolved around voice band modems connected a variety of applications in a semiprivate manner. The core of the network was still the PSTN, but the application was for specific corporate use and was not offered privately to individual users. These networks were private data networks designed for specific applications, and they did not have standard transport protocols to allow them to interconnect with one another. Another irony of this operation was that the digital data was first converted to analog to be transmitted over the telephone network; then within the telephone network, it was again converted to digital format for

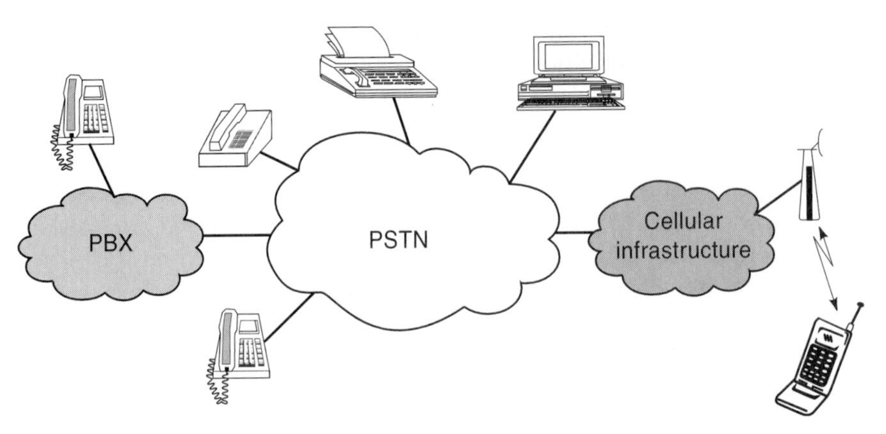

Figure 1A.1 Simplified representation of PSTN and surrounding voice networks.

transmission over long distances using the digital subcarrier system. To avoid this situation, starting in the mid 1970s, telephone companies started to introduce digital data services (DDS) which provided a 56 kbps digital service directly delivered to the end user. The idea was great because at that time the maximum data rate for voice band modems was 9,600 bps. However, like many other good and new ideas in telecommunications, this idea did not become popular. A large amount of capital was already invested in the existing voice band based data networks. It was not practical to replace them at once, and DDS services were not interoperable with the analog modems. The DDS services later emerged as integrated services digital network (ISDN) services providing two 64 kbps voice channels and a 16 kbps data channel to individual users. Penetration rates of ISDN services were not as expected, but laid a foundation for digital cellular services. Digital cellular systems can be viewed as a sort of wireless ISDN technology that integrates basic digital voice with a number of data services at the terminal.

The major cost for operation of a computer network over the four-wire lines was the cost of leasing lines from the telephone company. To reduce the operation cost, multiplexers were used to connect several lower speed modems and carry all of them at once over a higher speed modem operating over a long distance line. The next generation of multiplexers consisted of statistical multiplexers that multiplexed flows of data rather than multiplexing individual modem connections. Statistical multiplexer technology later evolved into router technology, which are generalized packet data switches.

In the early 1970s, the rapid increase in the number of terminals at the offices and manufacturing floors was the force behind the emergence of LANs. LANs provided high-speed connections (greater than 1 Mbps) among terminals facilitating the sharing of printers or mainframes from different locations. LANs provided a local medium specifically designed for data communication that was completely independent from the PSTN. By the mid 1980s, several successful LAN topologies and protocols were standardized, and LANs were installed in most large offices and manufacturing floors connecting their computing facilities. However, the income of the data communication industry, both LANs and public data networks (PDNs), was far below that of the PSTN still leaving the PSTN as the dominant economical force in the information networking industry.

Another important and innovative event in the 1970s was the implementation of ARPANET, the first packet-switched data network connecting 50 cities in the United States. This experimental network used routers rather than the PSTN switches to interconnect data terminals. The routers were originally connected via 56 kbps digital leased lines from the telephone company. This way, ARPANET interconnected several universities and government computers around a large geographic area. This network was the first packet-switched network supporting end-to-end digital services. This basic network later upgraded to higher speed lines and numerous additional networks. To facilitate a uniform communication protocol to interconnect these disparate networks, transmission control protocol/Internet protocol (TCP/IP) evolved that allowed LANs, as well as a number of other PDNs, to interconnect with one another and form the Internet. In the mid 1990s with the introduction of popular applications such as Telnet, FTP, email, and Web browsing, the Internet industry was created. Soon, the Internet penetrated the home market, and the number of Internet users became comparable with that of the PSTN, creating another economical power, namely, computer communications applications, which compete with the traditional PSTN. The IP-based Internet provides a cheaper solution than circuit switched operations, and today people are thinking of employing IP to capture a large share of the traditional telephony market served by the PSTN. The Internet provides a much lower-cost alternative to PSTN for support of multimedia applications. With the growth of the wireless industry in the past two decades, wireless LANs, wireless WANs, and wireless access to the Internet have become very popular.

In a manner similar to cellular telephony, wireless data network infrastructures have evolved around the existing wired data network infrastructures. Wireless LANs are designed mostly for in-building applications to cover a small area, and the network has a minimal infrastructure. Wireless LANs are usually connected to the existing wired LANs as an extension. Mobile data services are designed for low-speed wireless data applications with metropolitan, national, and, global coverage. These networks sometimes have their own infrastructure (e.g., ARDIS, Mobitex), sometimes use the existing cellular infrastructure but their own radio interface (e.g., CDPD), and sometimes use the infrastructure, as well as air-interface, of a cellular telephone service (e.g., GPRS). In all cases ultimately they connect to the Internet and run its popular applications. The 3G systems also provide competing packet-switched service. Figure 1A.2 provides a simplified sketch of the overall data networks surrounding the Internet. As compared with the PSTN, the Internet provides a cheaper and easier means to connect and expand networks. However, the telephone company owns the wires connecting the Internet and the telephone wires that can, among other alternatives, bring the Internet to the home or office.

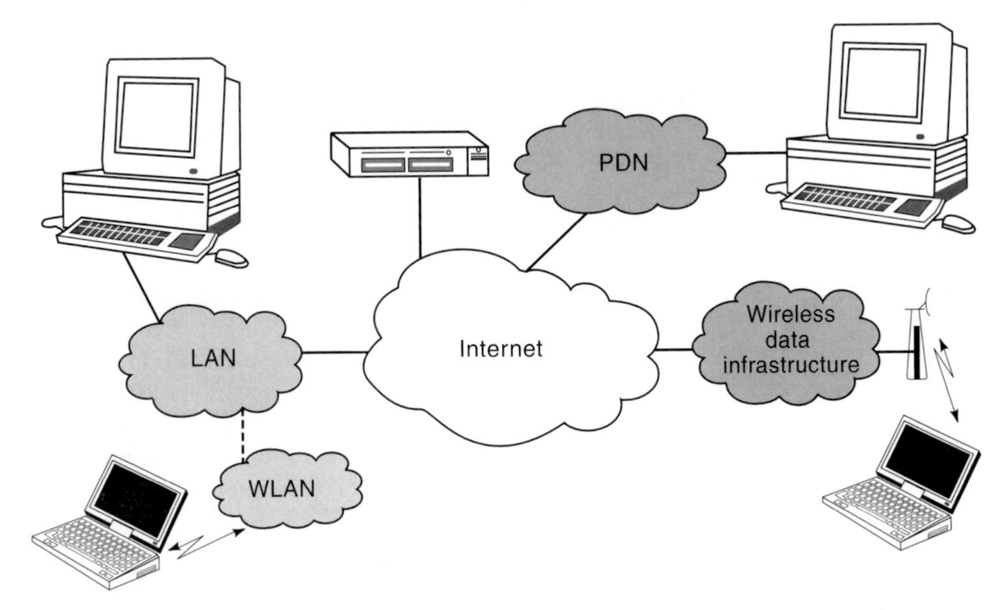

Figure 1A.2 Simplified representation of the Internet and surrounding data networks.

1A.3 The Cable TV Infrastructure

Another competing wired infrastructure that evolved in the last few decades of the 20th century was the cable television network. Installation of cable TV distribution networks in the United States started in 1968 and has penetrated more than 60 percent of the residential homes. This penetration rate is getting close to that of the PSTN. The cable TV network consists of three basic element: a regional hub, a distribution cable bus, and a fiber ring to connect the hubs to one another. Because of the hybrid usage of fiber and cable, this network is also referred to as the hybrid fiber coax (HFC) network. The signals containing all channels at the hub are distributed through the cable bus in a residential area, and each home taps the signal off the bus. This is radically different in many ways from home access through twisted pair wires provided by the PSTN. The bandwidth of the coaxial cable supports about 100 TV channels, each around 6 MHz, whereas the basic telephone channel is around 4 kHz. The extended telephone channel using DSL uses about 1 MHz of bandwidth. The cable access is via a long bus originally designed for a one-way multicast that has a number (up to 500) of taps, creating a less controllable medium. The twisted pair star access for the PSTN is designed for two-way operation and is easier to control. The HFC channel is noisier than the telephone channel, and despite its wider bandwidth, its current supported broadband data rate is at the same range as the digital subscriber line (DSL) services operating on telephone wiring.

Figure 1A.3 shows a general picture of a futuristic HFC and the way it connects to the PSTN and Internet. The cable TV network was also considered a backbone for wireless PCS systems, and it is considered the leading method for broadband home access to support the evolving home networks. Some of the cable TV providers in the United States also offer telephone services over this medium.

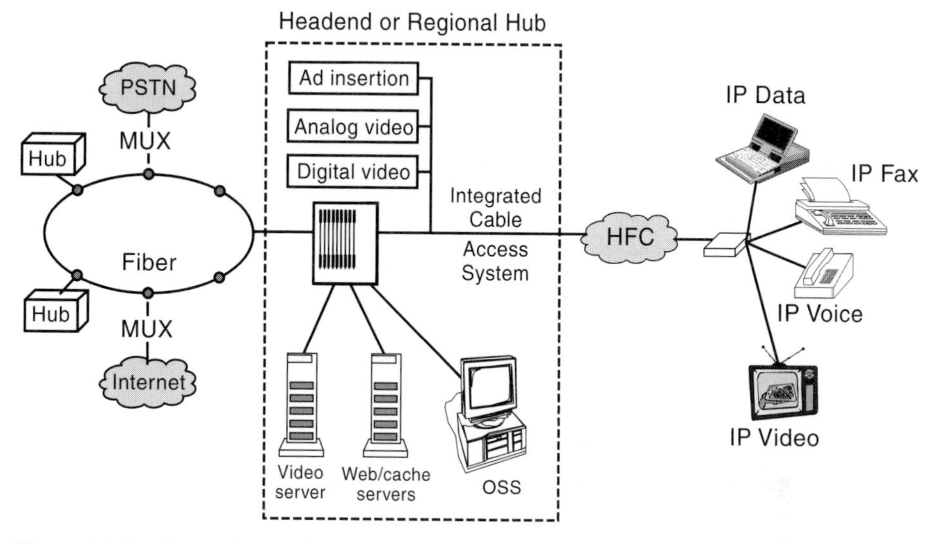

Figure 1A.3 General overview of HFC networks.

In the late 1990s, success of cable in broadband access encouraged some of the PSTN providers, such as AT&T, to acquire cable companies such as MediaOne.

APPENDIX 1B SUMMARY OF IMPORTANT STANDARDS ORGANIZATIONS

Figure 1B.1 provides an overview of the standardization process. The process starts in a group, for example, IEEE 802.11 or GSM, which implements the standard. The implemented standard is then moved for the approval of a regional organiza-

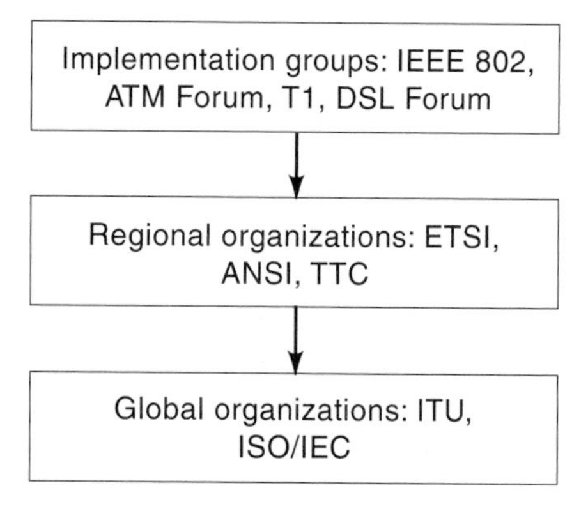

Figure 1B.1 Standard development process.

Table 1B.1 Summary of Important Standards

Standard / Organization
IEEE: Publishes 802 series standards for LANs and 802.11, 802.15, and 802.16 for wireless applications.
T1: Sponsored by Alliance for Telecommunications Information Solutions (ATIS) telecommunication standards body working on North American standards.
ATM (Asynchronous Transfer Mode) Forum: An industrial group working on a standard for ATM networks.
DSL (Digital Subscriber Loop) Forum: An industrial group working on xDSL services.
CableLab: Industrial alliance in America to certify DOCSIS compatible cable modems.
UNII (formerly SUPERNet): Industrial alliance for unlicensed bands for wireless LANs in 5.2 and 5.7 GHz.
IETF (Internet Engineering Task Force): Publishes Internet standards that include TCP/IP and SNMP. It is not an accredited standards organization.
FCC (Federal Communication Commission): The frequency administration authority in the United States.
EIA/TIA (Electronic/Telecommunication Industry Association): U.S. national standard for North American wireless systems.
ANSI: Accepted 802 series and forwarded to ISO. Also published FDDI, HIPPI, SCSI, and Fiber Channel. Developed JTC models for wireless channels.
ETSI: Published GSM, HIPERLAN-1, and UMTS.
CEPT (Committee of the European Post and Telecommunication): Standardization body of the European Posts Telegraph and Telephone (PTT) ministries. Co-published GSM with ETSI.
TTC (Telecommunication Technology Committee): Japanese national standard organization that approved JTC.
IEC (International Electrotechnical Commission): Publishes jointly with ISO.
ISO (International Standards Organization): Ultimate international authority for approval of standards.
ITU (International Telecommunication Union; formerly CCITT): International advisory committee under United Nations. The Telecommunication Sector, ITU-T, published ISDN and wide area ATM standards. Also works on IMT-2000.

tion such as the European Telecommunications Standards Institute (ETSI) or the American National Standards Institute (ANSI). The regional recommendation is finally submitted to world-level organizations, such as ITU or ISO/IEC, for final approval as an international standard. There are a number of standards organizations involved in wireless networking. Table 1B.1 provides a summary of the important standards playing major roles in shaping the wireless networking industry. This table helps the reader as a reference point while reading the text.

QUESTIONS

1.1 Draw a diagram showing the positioning of wireless networks vis a vis wired networks. Why is a wired network usually part of the wireless infrastructure?

1.2 Differentiate between portability, nomadicity, and mobility.

1.3 How is a wireless network different from a wired network? Explain at least five differences.

1.4 Distinguish between horizontal and vertical applications. Which of these two types of applications will reduce device and service costs? Why?

1.5 Name the four categories of 2G wireless networks and explain how they are related to 3G and WLAN systems.

1.6 What is the difference between the 3G cellular networks and WLANs in terms of frequency of operation, orientation of the application over the network, and supported rates for data services?

1.7 Name the three major telecommunication services that dominate today's commercial services and give the approximate time when they were first introduced.

1.8 Name five mobile data services, data rates supported by them, worldwide coverage, and frequency of operation.

1.9 Name the three major cellular standards in the United States and give the name of their wireless access technology.

1.10 What is the difference between the bandwidth per user of the analog cellular telephone systems in United States, European Union, and Japan? Which one has employed bandwidth-splitting techniques?

1.11 Name five WLAN standards and identify the transmission techniques and supported data rates of each of them.

1.12 Why do PCS standards such as CT-2 and DECT use 32 kbps ADPCM rather than the lower rate speech codes used in digital cellular standards such as GSM or IS-95?

1.13 What is WPAN? What is the difference between WPAN and WLAN? Name two example technologies for WPAN.

1.14 What is the difference between registration and call establishment in a cellular network?

1.15 When the ISM bands were released, what was new about them, and what are the available ISM bandwidths at 0.9, 2.4, and 5.7 GHz?

1.16 What is the difference between connectionless and connection-oriented packet data network protocols?

1.17 Create a kiviat chart like the one shown in Figure 1.5 comparing ARDIS and Metricom with respect to carrier spacing, data rate per carrier, number of channels, and ease of implementation.

PART ONE Principles of Air-Interface Design

Part One consists of three chapters providing the technical background for the design of the air-interface for wireless networks.

CHAPTER 2 CHARACTERISTICS OF THE WIRELESS MEDIUM

Learning about the behavior of the wireless medium is essential to understand the reasoning behind specific designs for wireless communication protocols. In particular, the design of the physical and medium access protocols are highly affected by the behavior of the channel that varies substantially in different indoor and outdoor areas. Chapter 2 analyzes the behavior of the wireless medium and provides a number of models for the prediction of the channel behavior in different environments.

CHAPTER 3 PHYSICAL LAYER ALTERNATIVES

The diversity and complexity of transmission techniques in wireless systems are far more involved than those of wired networks. Chapter 3 provides a comprehensive coverage of major transmission techniques that are employed in voice-oriented cellular and PCS systems as well as data-oriented mobile data, WLAN, and WPAN systems. These modulation techniques are logically divided into pulse modulation, carrier modulation, and spread spectrum techniques, and under these headings different practical examples are described.

CHAPTER 4 WIRELESS MEDIUM ACCESS METHODS

This chapter is devoted to multiple access alternatives applied to wireless networks. It starts with description and comparison of the voice-oriented FDMA, TDMA, and CDMA access methods. Then it addresses CSMA and ALOHA-based random access techniques used in data-oriented wireless networks. The last part of the chapter analyzes the applied access methods for integration of the voice and data that have evolved to operate with the voice- and data-oriented networks.

CHAPTER 2

CHARACTERISTICS OF THE WIRELESS MEDIUM

2.1 Introduction

2.1.1 Comparison of Wired and Wireless Media
2.1.2 Why Radio Propagation Studies?

2.2 Radio Propagation Mechanisms

2.3 Path-Loss Modeling and Signal Coverage

2.3.1 Free space propagation
2.3.2 Two-Ray Model for Mobile Radio Environments
2.3.3 Distance-Power Relationship and Shadow Fading
2.3.4 Path Loss Models for Megacellular Areas
2.3.5 Path Loss Models for Macrocellular Areas
2.3.6 Path Loss Models for Microcellular Areas
2.3.7 Path Loss Models for Picocellular Indoor Areas
2.3.8 Path Loss Models for Femtocellular Areas

2.4 Effects of Multipath and Doppler

2.4.1 Modeling of Multipath Fading
2.4.2 Doppler Spectrum
2.4.3 Multipath Delay Spread
2.4.4 Summary of Radio Channel Characteristics and Mitigation Methods
2.4.5 Emerging Channel Models

2.5 Channel Measurement and Modeling Techniques

2.6 Simulation of the Radio Channel

2.6.1 Software Simulation
2.6.2 Hardware Emulation

Appendix 2A What is dB?

Appendix 2B Wired Media

Appendix 2C Path Loss Models

Appendix 2D Wideband Channel Models

Questions

Problems

2.1 INTRODUCTION

In the past century, analysis, modeling, and simulation of radio propagation for a variety of applications have been studied in depth. In the 1970s, modeling of radio propagation for cellular networks operating from a mobile vehicle was investigated, and in the 1980s it was extended to include modeling of indoor radio propagation for cordless telephony and wireless LAN applications. References [JAK94], [LEE98], [RAP95], [BER00] provide details of these studies for mobile radio applications, and [PAH95] provides an overview with emphasis on indoor applications. This chapter discusses radio propagation modeling and simulation for wireless applications which we believe are necessary for a systems engineer to understand the principles of design and deployment of wireless networks. The details needed for development of an intuition and an understanding of how the wireless medium operates are discussed, and we give a number of models that can be used for simulation of the behavior of the wireless medium. We also point to current and emerging issues in radio channel modeling for wireless applications.

2.1.1 Comparison of Wired and Wireless Media

A wired medium provides a reliable, *guided* link that conducts an electric signal associated with the transmission of information from one fixed terminal to another. There are a number of alternatives for wired connection that include twisted pair (TP) telephone wiring for high-speed LANs, coaxial cables used for television distribution, and optical fiber used in the backbone of long-haul connections. Wires act as *filters* that limit the maximum transmitted data rate of the channel because of band limiting frequency response characteristics. The signal passing through a wire also radiates outside of the wire to some extent which can cause interference to close-by radios or other wired transmissions. These characteristics differ from one wired medium to another. Laying additional cables in general can duplicate the wired medium, and thereby increase the bandwidth.

Compared with wired media, the wireless medium is unreliable, has a low bandwidth, and is of broadcast nature; however, it supports mobility due to its tetherless nature. Different signals through wired media are physically conducted through different wires, but all wireless transmissions share the same medium—air. Thus it is the frequency of operation and the legality of access to the band that differentiates a variety of alternative for wireless networking. Wireless networks operate around 1 GHz (cellular), 2 GHz (PCS and WLANs), 5 GHz (WLANs), 28–60 GHz (local multipoint distribution service [LMDS] and point-to-point base station connections), and IR frequencies for optical communications. These bands are either licensed, like cellular and PCS bands, or unlicensed, like the ISM bands or U-NII bands. As the frequency of operation and data rates increase, the hardware implementation cost increases, and the ability of a radio signal to penetrate walls decreases. The electronic cost has become less significant with time, but in-building penetration and licensed versus unlicensed frequency bands have become an important differentiation. For frequencies up to a few GHz, the signal penetrates through the walls, allowing indoor applications with minimal wireless infrastructure inside a building. At higher frequencies a signal that is generated outdoors does not penetrate into buildings, and the signal generated indoors stays confined to a room. This phenomenon imposes restrictions on the selection of a suitable band for a wireless application.

Example 2.1: In-building Penetration of Signals

If one intends to bring a wireless Internet service to the rooftop of a residence and distribute that service inside the house, using other alternatives such as existing cable or TP wiring, that person may select LMDS equipment operating in licensed bands at several tens of GHz. If the intention is to penetrate the signal into the building for direct wireless connection to a computer terminal, the person may prefer equipment operating in the unlicensed ISM bands at 900 MHz or 2.4 GHz. The first approach is more expensive because it operates at licensed higher frequencies, where implementation and the electronics are more expensive and the service provider has paid to obtain the frequency bands. The second solution does not have any interference control mechanism because it operates in unlicensed bands.

Example 2.2: Licensed versus Unlicensed Bands

This example clarifies the differentiation between licensed and unlicensed bands with an analogous situation. Assume we equate radio transmission to barbequing: the interference caused by the transmission is like the smoke of the barbeque, and the frequency band of operation is similar to the property in which the barbeque grill is fixed. Then we have the affluent (cellular voice operators) that can afford a backyard (a licensed band) to operate their services with reasonable smoke (interference) from their neighbors occasionally. The less prosperous operators with larger barbeque grills (wideband data service providers) cannot afford to have huge private backyards (licensed bands) and have to use the public park (large unlicensed bands). In public parks, the space is provided on a first-come, first-serve basis. The only rule that the government can exercise is to restrict the overall smoke (interference) created by each barbeque (user) so as to allow peaceful coexistence. Although it may sound scary, both public parks as well as private backyards are used through natural selection. Because licensed bands are very expensive (the PCS bands in the United States were sold for around $20 billion), it is time-consuming to deploy a number of new applications rapidly at low costs. As such, new applications such as WLANs and Bluetooth are evolving in unlicensed bands.

Wired media provide us an easy means to increase capacity—we can lay more wires where required if it is affordable. With the wireless medium, we are restricted to a limited available band for operation, and we cannot obtain new bands or easily duplicate the medium to accommodate more users. As a result, researchers have developed a number of techniques to increase the capacity of wireless networks to support more users with a fixed bandwidth. The simplest method, comparable to laying new wires in wired networks, is to use a cellular architecture that reuses the frequency of operation when two cells are at an adequate distance from one another. Then, to further increase the capacity of the cellular network, as explained in Chapter 5, one may reduce the size of the cells. In a wired network, doubling the number of wired connections allows twice the number of users at the expense of twice the number of wired connections to the terminals. In a wireless network, reducing the size of the cells by half allows twice as many users as in one cell. Reduction of the size of the cell increases the cost and complexity of the infrastructure that interconnects the cells.

2.1.2 Why Radio Propagation Studies?

An understanding of radio propagation is essential for coming up with appropriate design, deployment, and management strategies for any wireless network. In effect, it is the nature of the *radio channel* that makes wireless networks far more complicated than their wired counterparts. Radio propagation is heavily site-specific and can vary significantly depending on the terrain, frequency of operation, velocity of the mobile terminal, interference sources, and other dynamic factors. Accurate characterization of the radio channel through key parameters and a mathematical model is important for predicting signal coverage, achievable data rates, specific performance attributes of alternative signaling and reception schemes, analysis of interference from different systems, and determining the optimum location for installing base station antennas.

In Chapter 5, we look at cellular hierarchy, where cells are classified into femto-, pico-, micro-, macro-, and megacells depending on their size. Radio propagation is different in each of these cell types. Radio propagation in open areas is much different from radio propagation in indoor and urban areas. In open areas across small distances or free space, the signal strength falls as the square of the distance. In other terrain, the signal strength often falls at a much higher rate as a function of distance depending on the environment and radio frequency. In urban areas the shortest direct path (the line-of-sight [LOS] path) between the transmitter and receiver is usually blocked by buildings and other terrain features outdoors. Similarly in indoor areas, walls, floors, and interior objects within buildings obstruct LOS communications. Such scenarios are called non-LOS (NLOS) or obstructed LOS (OLOS). This further complicates radio propagation in these areas, and the signal is usually carried by a multiplicity of indirect paths with various signal strengths. The signal strengths of these paths depends on the distance they have traveled, the obstacles they have reflected from or passed through, the architecture of the environment, and the location of objects around the transmitter and receiver. Because signals from the transmitter arrive at the receiver via a multiplicity of paths with each taking a different time to reach the receiver, the resulting channel has an associated *multipath delay spread* that affects the reception of data.

The radio frequency of operation also affects radio propagation characteristics and system design. At lower frequencies (less than 500 MHz) in the radio spectrum, the signal strength loss is much smaller at the first meter. However, bandwidth is less plentiful, and the antenna sizes required are quite prohibitive for wide-scale deployment. The separation of antennas also has to be much larger, and it is challenging to adopt diversity schemes for improving signal quality. On the other hand, ample bandwidth is available at much higher frequencies (greater than a few GHz). At such frequencies, it is still possible to use sufficiently low-power transmitters (of about 1 W) for providing adequate signal coverage over a few floors of a multistory building, or a few kilometers outdoors in LOS situations. The antenna sizes are also on the order of an inch, making transmitters and receivers quite compact and efficient. Diversity techniques can be employed to improve the quality of reception because antenna separations can be small. The downside of using higher frequencies is that they suffer a greater signal strength loss at the first meter, and also suffer larger signal strength losses while passing through obstacles such as walls. At a few tens of gigahertz, signals are usually confined within the walls of a room. From a security point of view, this is an attractive feature of these frequencies. At even higher frequencies (such as 60 GHz), atmospheric gases such as oxygen absorb signals, which results in a much larger attenuation of signal strength as a function of distance.

In the rest of this chapter, we present an overview of radio propagation characteristics and radio channel modeling techniques that are important in modern wireless networks. A more detailed discussion can be found in [PAH95]. The three most important radio propagation characteristics used in the design, analysis, and installation of wireless networks are the achievable signal coverage, the maximum data rate that can be supported by the channel, and the rate of fluctuations in the channel. The achievable signal coverage for a given transmission power determines the size of a cell in a cellular topology and the range of operation of a base station transmitter. This is usually obtained via empirical path-loss models obtained by measuring the received signal strength as a function of distance. Most of the path loss models are characterized by a distance-power or path-loss *gradient* and a random component that characterizes the fluctuations around the average path loss due to shadow fading and other reasons. For efficient data communications, the maximum data rate that can be supported over a channel becomes an important parameter. Data rate limitations are influenced by the multipath structure of the channel and the fading characteristics of the multipath components. This also influences the signaling scheme and receiver design. Another factor that is intimately related to the design of the adaptive parts of the receiver such as timing and carrier synchronization, phase recovery, and so on is the rate of fluctuations in the channel, usually caused by movement of the transmitter, receiver, or objects in between. This is characterized by the *Doppler spread* of the channel. We consider path-loss models in detail and provide a summary of the effects of multipath and Doppler spread in subsequent sections of this chapter.

Depending on the data rates that need to be supported by an application and the nature of the environment, certain characteristics are much more important than others. For example, signal coverage and slow fading are more important for low data-rate narrowband systems such as cordless telephones, low-speed data, and cellular voice telephony. The multipath delay spread also becomes important for high data rate wideband systems, especially those that employ spread spectrum such as CDMA, WLANs, and 3G cellular services. Other areas where the properties of the radio channel become important are in determining battery consumption, the design of transmitter and receivers, the design of medium access control protocols, the design of adaptive and smart antennas, link-level monitoring for higher layer protocol performance (e.g., number of retransmissions tries, window sizes, etc.), the design of wireless protocols (handoffs, power control, co-channel rejection via color codes), and system design.

2.2 RADIO PROPAGATION MECHANISMS

Radio signals with frequencies above 800 MHz, used in the wireless networks described in this book, have extremely small wavelengths compared with the dimensions of building features, so electromagnetic waves can be treated simply as rays [BER94]. This means that ray-optical methods can be used to describe the propagation within and even outside buildings by treating electromagnetic waves as traveling along localized ray paths. The fields associated with the ray paths change sequentially based on the features of the medium that the ray encounters.

In order to describe radio propagation with ray optics, three basic mechanisms [RAP95, PAH95] are generally considered while ignoring other complex mechanisms. These mechanisms are illustrated in Figures 2.1 and 2.2 for indoor and outdoor applications, respectively.

1. *Reflection and transmission.* Specular reflections and transmission occur when electromagnetic waves impinge on obstructions larger than the wavelength.

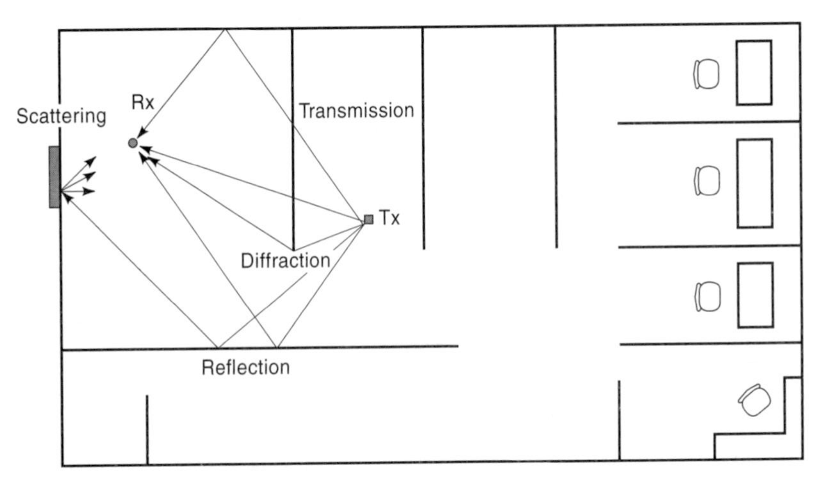

Figure 2.1 Radio propagation mechanisms in an indoor area.

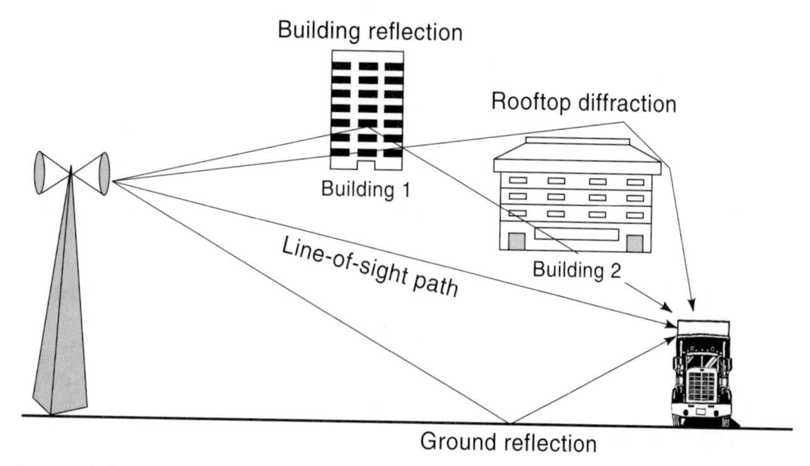

Figure 2.2 Radio propagation mechanisms in an outdoor area.

Usually rays incident upon the ground, walls of buildings, the ceiling, and the floor undergo specular reflection and transmission with the amplitude coefficients usually determined by plane wave analysis. Upon reflection or transmission, a ray attenuates by factors that depend on the frequency, the angle of incidence, and the nature of the medium (its material properties, thickness, homogeneity, etc.). These mechanisms often dominate radio propagation in indoor applications. In outdoor urban areas, this mechanism often loses its importance because it involves multiple transmissions that reduce the strength of the signal to negligible values.

2. *Diffraction.* Rays that are incident upon the edges of buildings, walls, and other large objects can be viewed as exciting the edges to act as a secondary line source. Diffracted fields are generated by this secondary wave source and propagate away from the diffracting edge as cylindrical waves. In effect, this results in propagation into *shadowed* regions because the diffracted field can reach a receiver, which is not in the line of sight of the transmitter. Because a secondary source is created, it suffers a loss much greater than that experienced via reflection or transmission. Consequently, diffraction is an important phenomenon outdoors (especially in microcellular areas) where signal transmission through buildings is virtually impossible. It is less consequential indoors where a diffracted signal is extremely weak compared to a reflected signal or a signal that is transmitted through a relatively thin wall.
3. *Scattering.* Irregular objects such as walls with rough surfaces and furniture (indoors) and vehicles, foliage, and the like (outdoors) scatter rays in all directions in the form of spherical waves. This particularly occurs when objects are of dimensions that are on the order of a wavelength or less of the electromagnetic wave. Propagation in many directions results in reduced power levels, especially far from the scatterer. As a result, this phenomenon is not that significant unless the receiver or transmitter is located in a highly cluttered environment. This mechanism dominates diffused IR propagations when the

wavelength of the signal is such that the roughness of the wall results in extensive scattering. In satellite and mobile radio applications, foliage often causes scattering.

2.3 PATH-LOSS MODELING AND SIGNAL COVERAGE

Calculation of signal coverage is essential for design and deployment of both narrowband and wideband wireless networks. Signal coverage is influenced by a variety of factors; most prominently the radio frequency of operation and the terrain. Often the region where a wireless network is providing service spans a variety of terrain. An operation scenario is defined by a set of operations for which a variety of distances and environments exist between the transmitter and the receiver. As a result, a unique channel model cannot describe radio propagation between the transmitter and the receiver, and we need several models for a variety of environments to enable system design. The core of the signal coverage calculations for any environment is a path-loss model which relates the loss of signal strength to distance between two terminals. Using path-loss models, radio engineers calculate the coverage area of wireless base stations and access points, as well as maximum distance between two terminals in an ad hoc network. In the following we consider path-loss models developed for several such environments that span different cell sizes and the terrain in the cellular hierarchy used for deployment of wireless networks.

2.3.1 Free Space Propagation

In most environments, it is observed that the radio signal strength falls as some power α of the distance, called the power-distance gradient or path-loss gradient. That is, if the transmitted power is P_t, after a distance d in meters, the signal strength will be proportional to $P_t d^{-\alpha}$. In its most simple case, the signal strength falls as the square of the distance in free space ($\alpha = 2$). When an antenna radiates a signal, the signal propagates in all directions. The signal strength density at a sphere of radius d is the total radiated signal strength divided by the area of the sphere, which is $4\pi d^2$. Depending on the radio frequency, there are additional losses, and in general the relationship between the transmitted power P_t and the received power P_r in free space is given by:

$$\frac{P_r}{P_t} = G_t G_r \left(\frac{\lambda}{4\pi d} \right)^2 \tag{2.1}$$

Here G_t and G_r are the transmitter and receiver antenna gains respectively in the direction from the transmitter to the receiver; d is the distance between the transmitter and receiver; $\lambda = c/f$ is the wavelength of the carrier; c is the speed of light in free space (3×10^8 m/s); and f is the frequency of the radio carrier. If we let $P_0 = P_t G_t G_r (\lambda / 4\pi)^2$ be the received signal strength at the first meter ($d = 1$ m), we can rewrite this equation as:

$$P_r = \frac{P_0}{d^2} \tag{2.2}$$

In decibels (dB), this equation takes the form

$$10\log(P_r) = 10\log(P_0) - 20\log(d) \tag{2.3}$$

where the logarithm is to the base 10. This means that there is a 20 dB per decade or 6 dB per octave loss in signal strength as a function of distance in free space. The transmission delay as a function of distance is given by $\tau = d/c = 3d$ ns or 3 ns per meter of distance.

Problem 1: Free Space Received Power and Path Loss

a) What is the received power (in dBm) in the free space of a signal whose transmit power is 1 W and carrier frequency is 2.4 GHz if the receiver is at a distance of 1 mile (1.6 km) from the transmitter? Assume that the transmitter and receiver antenna gains are 1.6.

b) What is the path loss in dB?

c) What is the transmission delay in ns?

Solution:

a) $10\log(P_t) = 30$ dBm because 1 W in dBm is 10 log (1 W/1 mW) = 30 dBm. Using Eq. (2.1) for $f = 2.4$ GHz, antenna gains of 1.6, distance of 1 meter, we have $10\log(P_0) = 30 - 40.046 = -10.046$ dBm and $P_r = -10.046 - 20\log(1600) = -74.128$ dBm.

b) The path loss is given by the difference between $10\log(P_t)$ and $10\log(P_r)$ (where both are in dBm). This is 104.128 dB.

c) The transmission delay is clearly $3 \times 1600 = 4800$ ns = 4.8 ms.

2.3.2 Two-Ray Model for Mobile Radio Environments

The distance-power relationship observed for free space does not hold for all environments. In free space, the signal travels from the transmitter to the receiver along a single path. In all realistic environments, the signal reaches the receiver through several different paths. The simple free space model of the previous section will not be valid for such scenarios and several complex models are required. We consider such models in the rest of this chapter.

We start with the two-path or two-ray model that is used for modeling the land mobile radio. The propagation environment and the two-ray model are shown in Figure 2.3. Here, the base station and the mobile terminal are both assumed to be at elevations above the earth, which is modeled as a flat surface in between the base station and the mobile terminal. Usually there is an LOS component that exists between the base station and the mobile terminal which carries the signal as in free space. There will also be another path over which the signal travels that consists of a reflection off the flat surface of the earth. The two paths travel different distances based on the height of the base station antenna, h_b, and the height of the mobile terminal antenna, h_m, and result in the addition of signals either constructively or destructively at the receiver.

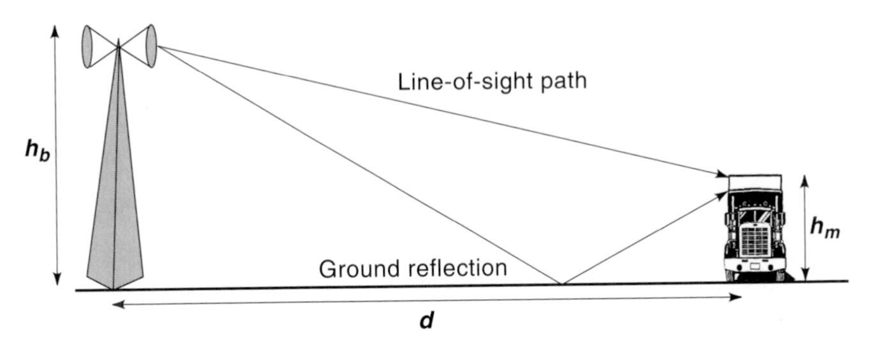

Figure 2.3 Two-ray model for mobile radio environments.

It can be shown that the relationship between the transmit power and the received power for the two-ray model can be approximated by [PAH95]:

$$P_r = P_t G_t G_r \frac{h_b^2 h_m^2}{d^4} \tag{2.4}$$

It is interesting to see that the signal strength falls as the fourth power of the distance between the transmitter and the receiver. In other words, there is a loss of 40 dB per decade or 12 dB per octave. The other interesting observation here is that the received signal strength can be increased by raising the heights of the transmit and receive antennas.

Problem 2: Comparison of Coverages

If a base station covers 1 km in a plain area modeled as a two-ray channel, what would be the coverage if it were used with a satellite?

Solution:

In the open area, the path loss gradient is 40 dB per decade of distance so the signal strength is reduced by 120 dB in covering 1 km. In free-space communication for satellites, the loss is 20 dB per decade of distance which allows 6 decades of distance or 1,000 km, for 120 dB loss of the signal.

2.3.3 Distance-Power Relationship and Shadow Fading

The simplest method of relating the received signal power to the distance is to state that the received signal power P_r is proportional to the distance between transmitter and receiver d, raised to a certain exponent α, which is referred to as the *distance-power gradient,* that is,

$$P_r = P_0 d^{-\alpha} \tag{2.5}$$

where P_0 is the received power at a reference distance (usually one meter) from the transmitter. For free-space, as already discussed, $\alpha = 2$, and for the simplified two-path model of an urban radio channel, $\alpha = 4$. For indoor and urban radio channels, the distance-power relationship will change with the building and street layouts, as well as with construction materials and density and height of the buildings in the area. Generally, variations in the value of the distance-power gradient in different outdoor areas are smaller than variations observed in indoor areas. The results of indoor radio propagation studies show values of α smaller than 2 for corridors or large open indoor areas and values as high as 6 for metallic buildings.

The distance-power relationship of Eq. (2.5) in decibels is given by

$$10\log(P_r) = 10\log(P_0) - 10\alpha\log(d) \tag{2.6}$$

where P_r and P_0 are the received signal strengths at d meters and at one meter, respectively. The last term in the right-hand side of the equation represents the power loss in dB with respect to the received power at one meter, and it indicates that for a one-decade increase in distance, the power loss is 10α dB and for a one-octave increase in distance, it is 3α dB. If we define the path loss in dB at a distance of one meter as $L_0 = 10\log_{10}(P_t) - 10\log_{10}(P_0)$, the total path loss L_p in dB is given by:

$$L_p = L_0 + 10\alpha\log(d) \tag{2.7}$$

This presents the total path-loss as the path-loss in the first meter plus the loss relative to the power received at one meter. The received power in dB is the transmitted power in dB minus the total path loss L_p. This normalized equation is occasionally used in the literature to represent the distance-power relationship.

As we will see in Chapter 5, the path-loss models of this form are extensively used for deployment of cellular networks. The coverage area of a radio transmitter depends on the power of the transmitted signal and the path loss. Each radio receiver has particular power sensitivity, for example, it can only detect and decode signals with a strength larger than this sensitivity. Because the signal strength falls with distance, using the transmitter power, the path-loss model, and the sensitivity of the receiver, one can calculate the coverage.

Problem 3: Coverage of a Base Station

What is the coverage of a base station that transmits a signal at 2 kW given that the receiver sensitivity is -100 dBm, the path loss at the first meter is 32 dB, and the path loss gradient is $\alpha = 4$?

Solution:

The transmit power in dB is $10\log(P_t) = 10\log(2000/.001) = 63$ dBm, and the receiver sensitivity $10\log(P_r)$ is -100 dBm. The total path loss that is allowed will thus be $10\log(P_t) - 10\log(P_r) = 63 - (-100) = 163$ dB. There is a loss of 32 dB at the first meter. Using Eq. (2.7) the loss due to distance can at most be 131 dB. Because the path loss gradient $\alpha = 4$, $10\alpha\log(d) = 131$ dB will imply that $d = 10^{(131/40)} = 1{,}883$ m $= 1.88$ km. So the coverage of the cell is 1.88 km. As we will see later, the path-loss models are far more complex than the simple model discussed here.

2.3.3.1 Measurement of the Distance Power Gradient

To measure the gradient of the distance-power relationship in a given area, the receiver is fixed at one location, and the transmitter is placed at a number of locations with different distances between the transmitter and the receiver. The received power or the path loss in dB is plotted against the distance on a logarithmic scale. The slope of the best-fit line through the measurements is taken as the gradient of the distance-power relationship. Simulations can also be used to arrive at similar results. Figure 2.4 shows a set of measured data taken in an indoor area at distances from 1 to 20 meters, together with the best-fit line through the measurements.

2.3.3.2 Shadow Fading

Depending on the environment and the surroundings, and the location of objects, the received signal strength *for the same distance* from the transmitter will be different. In effect, Equation (2.6) provides the mean value of the signal strength that can be expected if the distance between the transmitter and receiver is d. The actual received signal strength will vary around this mean value. This variation of the signal strength due to location is often referred to as *shadow fading* or *slow fading*. The reason for calling this shadow fading is that very often the fluctuations around the mean value are caused due to the signal being blocked from the receiver by buildings (in outdoor areas), walls (inside buildings), and other objects in the environment. It is called slow fading because the variations are much slower with distance than another fading phenomenon caused due to multipath that we discuss later. It is also found that shadow fading has less dependence on the frequency of operation than multipath fading or fast fading as discussed later. The path loss of Equation (2.7) will have to be modified to include this effect by adding a random component as follows:

$$L_p = L_0 + 10\alpha \log_{10}(d) + X \tag{2.8}$$

Here X is a random variable with a distribution that depends on the fading component. Several measurements and simulations indicate that this variation can be expressed as a log-normally distributed random variable. The log-normal probability density function is given by:

$$f_{LN}(x) = \frac{1}{\sqrt{2\pi}\,\sigma x} \exp\left(\frac{-(\ln x - \mu)^2}{2\sigma^2}\right) \tag{2.9}$$

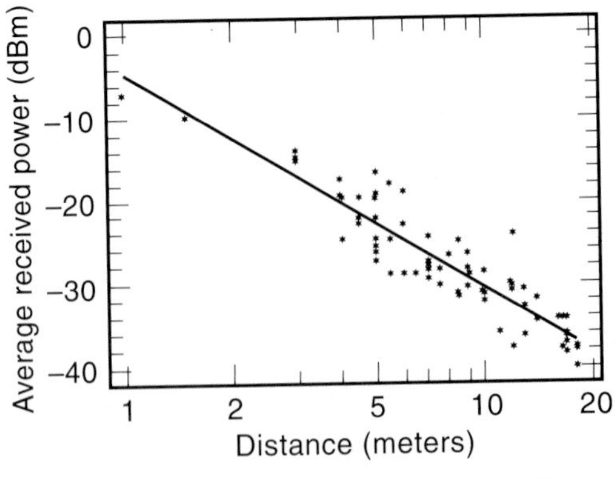

Figure 2.4 Measured received power and a linear regression fit to the data.

where μ is the mean received signal strength and σ is its standard deviation.

The problem caused by shadow fading is that all locations at a given distance may not receive sufficient signal strength for correctly detecting the information. In order to achieve sufficient signal coverage, the technique employed is to add a *fade margin* to the path loss or received signal strength. The fade margin is usually taken to be the additional signal power that can provide a certain fraction of the locations at the edge of a cell (or near the fringe areas) with the required signal strength. For computing the coverage, we thus employ the following equation:

$$L_p = L_0 + 10\alpha \log d + F_\sigma \tag{2.10}$$

where F_σ is the fade margin associated with the path loss to overcome the shadow fading effects.

The distribution of X in Eq. (2.8) is used to determine the appropriate fade margin. Note that a log-normal absolute fading component has a Gaussian distribution in dB. That is, X is a zero mean Gaussian random variable that corresponds to log-normal shadow fading in Eq. (2.8). At the fringe locations, the mean value of the shadow fading is zero dB. Fifty percent of the locations have a positive fading component, and 50 percent of the locations have a negative fading component. This will mean that the locations that have a positive fading component X will suffer a larger path loss resulting in unacceptable signal strength. To overcome this, a fading margin is employed to move most of these locations to within an acceptable received signal strength (RSS) value. This fading margin can be applied by increasing the transmit power and keeping the cell size the same, or reducing the cell size.

Problem 4: Computing the Fading Margin

A mobile system is to provide 95 percent successful communication at the fringe of coverage with a shadow fading component having a zero mean Gaussian distribution with standard deviation of 8 dB. What fade margin is required?

Solution:

Note that the location variability component X (in dB) in this case is a zero mean Gaussian random variable. In this example, the variance of X is 8 dB. We have to choose F_σ such that 95 percent of the locations will have a fading component smaller than the tolerable value. The distribution of the fading component X is Gaussian. So the fading margin depends on the tail of the Gaussian distribution that is described by the Q-function or the complementary error function erfc. Using the complementary error function and a software like Matlab, we can determine the value of F_σ as the solution to the equation $0.05 = 0.5\,\text{erfc}(F_\sigma\sqrt{2})$ i.e., 5% of the fringe areas have fading values that cannot be compensated by F_σ. For this example, the fade margin to be applied is 13.16 dB.[1]

---

[1]The function $Q(x) = \int_x^\infty f_x(x)dx = 0.05$ where $Q(x)$ is the probability that the normal random variable X has a value greater than x is tabulated or it can be determined using the complementary error function via the relation: $Q(x) = 0.5\,\text{erfc}(x/\sqrt{2})$. For a good discussion of the Q-function, see [SKL01].

So far, we have discussed achievable signal coverage in terms of the received signal strength and the path loss. In the following sections, we discuss parameters and path loss models for a variety of cellular environments. We also discuss, where relevant, the important factors that lead to these path loss models.

2.3.4 Path Loss Models for Megacellular Areas

Megacellular areas are those where the communication is over extremely large cells spanning hundreds of kilometers. Megacells are served mostly by mobile satellites (usually low-earth orbiting—LEO). The path loss is usually the same as that of free space, but the fading characteristics are somewhat different.

2.3.5 Path Loss Models for Macrocellular Areas

Macrocellular areas span a few kilometers to tens of kilometers, depending on the location. These are the traditional "cells" corresponding to the coverage area of a base station associated with traditional cellular telephony base stations. The frequency of operation is mostly around 900 MHz, though the emergence of PCS has resulted in frequencies around 1,800 to 1,900 MHz for such cells.

There have been extensive measurements in a number of cities and locations of the received signal strength in macrocellular areas that have been reported in the literature. The most popular of these measurements corresponds to those of Okumura who determined a set of path loss curves as a function of distance in 1968 for a range of frequencies between 100 MHz and 1,920 MHz. Okumura also identified the height of the base station antenna h_b and the height of the mobile antenna h_m as important parameters. Masaharu Hata [HAT80] created empirical models that provide a good fit to the measurements taken by Okumura for transmitter-receiver separations d of more than 1 km. The expressions for path loss developed by Hata are called the Okumura-Hata models or simply the Hata models. Table 2.1 provides these models.

Table 2.1 Okumura-Hata Models for Macro-Cellular Path Loss

General Formulation:

$$L_p = 69.55 + 26.16 \log f_c - 13.82 \log h_b - a(h_m) + [44.9 - 6.55 \log h_b] \log d \tag{2.11}$$

where f_c is in MHz, h_b and h_m are in meters, and d is in km.

Range of Values

Center frequency f_c in MHz			150–1,500 MHz
h_b, h_m in meters			30–200m, 1–10m
$a(h_m)$ in dB	Large City	$f_c \le 200$ MHz	$8.29 [\log (1.54 h_m)]^2 - 1.1$
		$f_c \ge 400$ MHz	$3.2 [\log (11.75 h_m)]^2 - 4.97$
	Medium–Small City	$150 \ge f_c \ge 1{,}500$ MHz	$1.1 [\log f_c - 0.7] h_m - (1.56 \log f_c - 0.8)$

Suburban Areas Formulation:
Use Eq. (2.11) and subtract a correction factor given by:

$$K_r\,(\text{dB}) = 2 [\log (f_c/28)]^2 + 5.4 \tag{2.12}$$

where f_c is in MHz.

Problem 5: Using the Okumura-Hata Model

Determine the path loss of a 900 MHz cellular system operating in a large city from a base station with the height of 100 m and mobile station installed in a vehicle with antenna height of 2 m. The distance between the mobile and the base station is 4 km.

Solution:

We calculate the terms in the Okumura-Hata model as follows:

$$a(h_m) = 3.2\,[\log(11.75\,h_m)]^2 - 4.97 = 1.045 \text{ dB}$$

$$L_p = 69.55 + 26.16 \log f_c - 13.82 \log h_b - a(h_m) + [44.9 - 6.55 \log h_b]\log d = 137.3 \text{ dB}$$

To extend the Okumura-Hata model for PCS applications operating at 1,800 to 2,000 MHz, the European Co-operative for Scientific and Technical Research (COST) came up with the COST-231 model for urban radio propagation at 1.900 MHz, which we provide in Table 2C.1 in Appendix 2C. In this table $a(h_m)$ is chosen from Table 2.1 for large cities.

In a similar way, the Joint Technical Committee (JTC) of the Telecommunications Industry Association (TIA) has come up with the JTC models for PCS applications at 1,800 MHz [PAH95].

2.3.6 Path Loss Models for Microcellular Areas

Microcells are cells that span hundreds of meters to a kilometer or so and are usually supported by below rooftop level base station antennas mounted on lampposts or utility poles. The shapes of microcells are also no longer circular (or close to circular) because they are deployed in streets in urban areas where tall buildings create *urban canyons*. There is little or no propagation of signals through buildings, and the shape of a microcell is more like a cross or a rectangle, depending on the placement of base station antennas at the intersection of streets or in between intersections. The propagation characteristics are quite complex with the propagation of signals affected by reflection from buildings and the ground and scattering from nearby vehicles. For obstructed paths, diffraction around building corners and rooftops become important. Many individual scenarios should be considered, unlike radio propagation in macrocells.

Bertoni and others [BER99] have developed empirical path-loss models based on signal strength measurements in the San Francisco Bay area which are similar to the Okumura-Hata models for a variety of situations. The corresponding path loss models are summarized in Table 2.2.

As usual, d is the distance between the mobile terminal and the transmitter in kilometers, h_b is the height of the base station in meters, h_m is the height of the mobile terminal antenna from the ground in meters, and f_c is now the center frequency of the carrier in GHz that can range between 0.9 and 2 GHz.

In addition, the following parameters are defined. The distance of the mobile terminal from the last rooftop (in meters) is denoted by r_h. A rooftop acts as a diffracting screen (see Fig. 2.5), and distance from the closest, such rooftop (around 250 m in many cases) becomes important in NLOS situations and introduces a correction factor. The height of the nearest building above the height of the receiver

Table 2.2 Path Loss Formulas for Microcells

Environment	Scenario	Path Loss Expression
Low Rise	NLOS	$L_p = [139.01 + 42.59 \log f_c] - [14.97 + 4.99 \log f_c] \operatorname{sgn}(\Delta h)\log(1 + \lvert\Delta h\rvert) + [40.67 - 4.57 \operatorname{sgn}(\Delta h) \log(1 + \lvert\Delta h\rvert) \log d + 20 \log (\Delta h_m/7.8) + 10 \log (20/r_h)$
High Rise $h_m = 1.6$m	Streets perpendicular to the LOS streets	$L_p = 135.41 + 12.49 \log f_c - 4.99 \log h_b + [46.84 - 2.34 \log h_b] \log d$
	Streets parallel to the LOS Streets	$L_p = 143.21 + 29.74 \log f_c - 0.99 \log h_b + [47.23 + 3.72 \log h_b] \log d$
Low Rise + High Rise	LOS	$L_p = 81.14 + 39.40 \log f_c - 0.09 \log h_b + [15.80 - 5.73 \log h_b] \log d$, for $d < d_{bk}$ $L_p = [48.38 - 32.1 \log d] + 45.7 \log f_c - (25.34 - 13.9 \log d) \log h_b + [32.10 + 13.90 \log h_b] \log d + 20 \log (1.6/h_m)$, for $d > d_{bk}$

antenna is denoted by Δh_m and introduces a correction factor similar to r_h. The average building height in the environment is an important parameter in microcellular environments. The relative height of the base station transmitter compared with the average height of buildings is denoted by Δh. Usually Δh ranges between -6 m and 8 m. In LOS situations, it is observed that there are two distinct slopes of the path loss curves, one in the near-end region and one in the far-out segment. A *breakpoint* distance d_{bk} is used to separate the two piecewise linear fits to the measured path loss. The breakpoint distance is dependent on the heights of the base station and mobile antennas, as well as the wavelength λ of the carrier (all in meters), and is given by $d_{bk} = 4\, h_b h_m/1000\lambda$.

Problem 6: Path-Loss Calculation in a Microcell

Determine the path loss between the base station (BS) and mobile station (MS) of a 1.8 GHz PCS system operating in a high-rise urban area. The MS is located in a perpendicular street to the location of the BS. The distances of the BS and MS to the corner of the street are 20 and 30 meters, respectively. The base station height is 20 m.

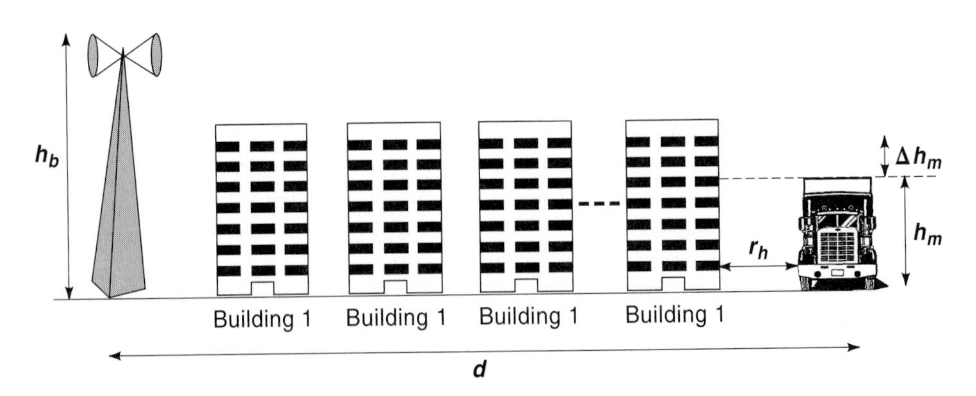

Figure 2.5 Geometry in a microcell; definitions of r_h and Δh_m.

Solution:

The distance of the mobile from the base station is $(20^2 + 30^2)^{1/2} = 36.05$ m. Using the appropriate equation from Table 2.2, we can write the path loss as:

$$L_p = 135.41 + 12.49 \log f_c - 4.99 \log h_b + [46.84 - 2.34 \log h_b] \log d = 68.89 \text{ dB}$$

In addition to the empirical models presented, there are theoretical models [BER94] that predict the path loss in microcellular environments which have been adopted by a variety of standard bodies. Another model available for the microcellular environments is the JTC model explained in [PAH95]. This model provides for PCS microcells in a manner similar to the COST model.

2.3.7 Path Loss Models for Picocellular Indoor Areas

Picocells correspond to radio cells covering a building or parts of buildings. The span of picocells is anywhere between 30 m and 100 m. Usually picocells are employed for WLANs, wireless PBX systems, and PCSs operating in indoor areas. One of the earliest statistical measurements of signal amplitude fluctuations in an office environment for a cordless telephone application is reported in [ALE82]. The measurements were made by fixing the transmitter while moving the receiver to various locations in a multiple-room office. Because of those earliest measurements, many researchers have performed narrowband measurements within buildings, primarily to determine the distance-power relationship and arrive at empirical path loss models for a variety of environments [PAH95].

2.3.7.1 Multifloor Attenuation Model

For describing the path loss in multistory buildings, signal attenuation by the floors in the building can be included as a constant independent of the distance [MOT88b]. The path loss in this case is given by

$$L_p = L_0 + nF + 10 \log (d) \tag{2.13}$$

Here F represents the signal attenuation provided by each floor; L_0 is the path loss at the first meter, shown in Equation (2.7); d is the distance between the transmitter and receiver in meters; and n is the number of floors through which the signal passes. The received power is plotted versus distance, and the best-fit line is determined for each different value of F. The value of F, which provides the minimum mean-square error between the line and the data, is taken as the value of F for the experiment. For indoor radio measurements at 900 MHz and 1.7 GHz, values of $F = 10$ dB and 16 dB respectively are reported in [MOT88a].

Interior objects such as furniture, equipment, and so on cause shadow fading as discussed earlier. Result of measurements on indoor [MOT88], [GAN91], [HOW91] radio channels show that a log-normal distribution provides the best fit to randomness introduced by shadow fading. In [MOT88b], the variations of the mean value of the signal were found to be log-normal with a standard deviation of 4 dB.

2.3.7.2 The JTC Model

In Equation (2.13) the relationship between the path loss and the number of floors is linear. However, results of measurements in [BER93] do not agree with this assumption. There a theoretical explanation indicates that diffraction out of windows becomes significant as the number of intervening floors increases. As such, an improvement to Equation (2.13) is to include a nonlinear function of the number of floors in the path loss model as follows:

$$L_p = A + L_f(n) + B\log(d) + X \tag{2.14}$$

Here $L_f(n)$ represents the function relating the power loss with the number of floors n, and X is a log-normally distributed random variable representing the shadow fading. Table 2.3 gives a set of suggested parameters in dB for the path loss calculation using Equation (2.14) at carrier frequencies of 1.8 GHz. The rows of the table provide the path loss in the first meter, the gradient of the distance-power relationship, the equation for calculation of multifloor path loss, and the standard deviation of the log-normal shadow fading parameter. It is assumed that the base and portable stations are inside the same building. The parameters are provided for three classes of indoor areas: residential, offices, and commercial buildings. This table is taken from a TIA recommendation for RF channel modeling for PCS applications [JTC94].

2.3.7.3 Path-Loss Models Using Building Material

Several other models for indoor radio propagation have been proposed in the literature. The *partition-dependent* model [RAP96] tries to improve upon standard models such as Equation (2.7) by fixing the value of the path-loss gradient α at 2 for free space and introducing losses for each partition that is encountered by a straight line connecting the transmitter and the receiver. The path loss is given by:

$$L_p = L_0 + 20\log d + \Sigma\, m_{type}\, w_{type} \tag{2.15}$$

Here m_{type} refers to the number of partitions of that type and w_{type} the loss in dB attributed to such a partition. The partition dependent model has been investigated in [SEI92]. Two partitions were considered here: soft partitions that have a loss of 1.4 dB and hard partitions that have a loss of 2.4 dB. Several other loss values (w_{type}) have been reported in [RAP96], which vary between 1 dB for dry plywood to 20 dB for concrete walls depending on the carrier frequency. Table 2.4 shows some dB loss values measured at Harris semiconductors at 2.4 GHz for different

Table 2.3 Parameters for Indoor Path-Loss Calculation (JTC model)

Environment	Residential	Office	Commercial
A (dB)	38	38	38
B	28	30	22
$L_f(n)$ (dB)	$4n$	$15 + 4(n-1)$ dB	$6 + 3(n-1)$ dB
Log Normal Shadowing (Std. Dev. dB)	8	10	10

Table 2.4 Partition Dependent Losses

Signal Attenuation of 2.4 GHz through	dB
Window in brick wall	2
Metal frame, glass wall into building	6
Office wall	6
Metal door in office wall	6
Cinder wall	4
Metal door in brick wall	12.4
Brick wall next to metal door	3

types of partitions. Once again, appropriate fading margins have to be included to account for the variability in path loss for the same distance *d*.

2.3.8 Path Loss Models for Femtocellular Areas

There have been few radio channel measurements or modeling work available for femtocells. Femtocells are expected to span from a few meters to a few tens of meters. Femtocells are probably going to exist in individual residences and use low-power devices employing Bluetooth chips or HomeRF equipment. Data rates are initially expected to be around 1 Mbps and increase with the availability of technology to operate at higher frequencies. Because femtocells are mostly deployed in residential environment, the JTC path loss model in the previous section for residential environments may be used to predict the coverage of a femtocell at 1.8 GHz. However, femtocells will use carrier frequencies in the unlicensed bands at 2.4 and 5 GHz. For these frequencies, path loss models are not readily available. Selected measurements [PRA92], [McD98], [GUE97] indicate indoor path loss models based on Equation (2.7) at these frequencies that are shown in Table 2.5.

Table 2.5 Path Loss Models at 2.4 GHz and 5 GHz for Femtocells

Center Frequency f_c	Environment	Scenario	Path Loss at the First Meter	Path Loss Gradient α
2.4 GHz	Indoor office	LOS	41.5 dB	1.9
		OLOS	37.7 dB	3.3
5.1 GHz	Meeting room	LOS	46.6 dB	2.22
		OLOS	61.6 dB	2.22
5.2 GHz	Suburban residences	LOS and same floor	47 dB	2 to 3
		OLOS and same floor		4 to 5
		OLOS and room in the higher floor directly above the Tx		4 to 6
		OLOS and room in the higher floor not directly above the Tx		6 to 7

2.4 EFFECTS OF MULTIPATH AND DOPPLER

So far we have considered achievable signal coverage in a variety of cell types based on the mean received signal strength or path loss suffered by a signal. Such a characterization of the received signal strength corresponds to *large-scale* average values. In reality, the received signal is rapidly fluctuating due to the mobility of the mobile terminal causing changes in multiple signal components arriving via different paths. This rapid fluctuation of the signal amplitude is referred to as *small-scale fading,* and it is the result of movement of the transmitter, the receiver, or objects surrounding them. Over a small area, the *average* value of the signal is employed to compute the received signal strength or path loss. But the characteristics of the instantaneous signal strength are also important in order to design receivers that can mitigate these effects. We briefly describe the features of small-scale fading in this section.

Two effects contribute to the rapid fluctuations of the signal amplitude. The first, caused by the movement of the mobile terminal toward or away from the base station transmitter, is called *Doppler.* The second, caused by the addition of signals arriving via different paths, is referred to as multipath fading.

2.4.1 Modeling of Multipath Fading

Multipath fading results in fluctuations of the signal amplitude because of the addition of signals arriving with different *phases.* This phase difference is caused due to the fact that signals have traveled different distances by traveling along different paths. Because the phase of the arriving paths are changing rapidly, the received signal amplitude undergoes rapid fluctuation that is often modeled as a random variable with a particular distribution.

To model these fluctuations, one can generate a histogram of the received signal strength in time. The density function formed by this histogram represents the distribution of the fluctuating values of the received signal strength. The most commonly used distribution for multipath fading is the Rayleigh distribution, whose probability density function (PDF) is given by:

$$f_{ray}(r) = \frac{r}{\sigma^2} \exp\left(-\frac{r^2}{2\sigma^2}\right), \quad r \geq 0 \tag{2.16}$$

Here it is assumed that all signals suffer nearly the same attenuation, but arrive with different phases. The random variable corresponding to the signal amplitude is r. Theoretical considerations indicate that the sum of such signals will result in the amplitude having the Rayleigh distribution of Eq. (2.16). This is also supported by measurements at various frequencies [PAH95]. When a strong LOS signal component also exists, the distribution is found to be Ricean, and the probability density function of such a distribution is given by:

$$f_{ric}(r) = \frac{r}{\sigma^2} \exp\left(\frac{-(r^2 + K^2)}{2\sigma^2}\right) I_0\left(\frac{Kr}{\sigma^2}\right), \quad r \geq 0, K \geq 0 \tag{2.17}$$

Here K is a factor that determines how strong the LOS component is relative to the rest of the multipath signals.

Equations (2.9), (2.16), and (2.17) are used to determine what fraction of time a signal is received such that the information it contains can be decoded or what fraction of area receives signals with the requisite strength. The remainder of the fraction is often referred to as outage.

Small-scale fading results in very high bit error rates. In order to overcome the effects of small-scale fading, it is not possible to simply increase the transmit power because this will require a huge increase in the transmit power. A variety of techniques are used to mitigate the effects of small-scale fading—in particular error control coding with interleaving, diversity schemes, and using directional antennas. These techniques are discussed in Chapter 3.

2.4.2 Doppler Spectrum

Equations (2.16) and (2.17) provide the distributions of the amplitude of a radio signal that is undergoing small-scale fading. In general, it is also important to know for what time a signal strength will be below a particular value (duration of fade) and how often it crosses a threshold value (frequency of transitions or fading rate). This is particularly important to design the coding schemes and interleaving sizes for efficient performance. We see that this is a second-order statistic, and it is obtained by what is known as the *Doppler spectrum* of the signal.

Doppler spectrum is the spectrum of the fluctuations of the received signal strength. Figure 2.6 [HOW90] demonstrates the results of measurements of amplitude fluctuations in a signal and its spectrum under different conditions. In Figure 2.6(a), the transmitter and receiver are kept stationary, and nothing is moving in their vicinity. The received signal has constant envelope, and its spectrum is only an impulse. In Figure 2.6(b), the transmitter is randomly moved, resulting in fluctuation of the received signal. The spectrum of this signal is now expanded to around 6 Hz, reflecting the rate of variations of the received signal strength. This spectrum is referred to as the Doppler spectrum.

In mobile radio applications, the Doppler spectrum for a Rayleigh fading channel is usually modeled by:

$$D(\lambda) = \frac{1}{2\pi f_m} \times [1 - (\lambda/f_m)^2]^{-1/2} \qquad \text{for } - f_m \leq \lambda \leq f_m \tag{2.18}$$

Here f_m is the maximum Doppler frequency possible and is related to the velocity of the mobile terminal via the expression $f_m = v_m/\lambda$, where v_m is the mobile velocity and λ is the wavelength of the radio signal. This spectrum, commonly used in mobile radio modeling, is also called the classical Doppler spectrum and is shown in Figure 2.7. Another popular model for the Doppler spectrum is the uniform distribution that is used for indoor applications [PAH95].

From the rms Doppler spread, it is possible to obtain the fade rate and the fade duration for a given mobile velocity [PAH95]. These values can then be used in the design of appropriate coding and interleaving techniques for mitigating the effects of fading. Diversity techniques are useful to overcome the effects of fast fading by providing multiple copies of the signal at the receiver. Because the probability

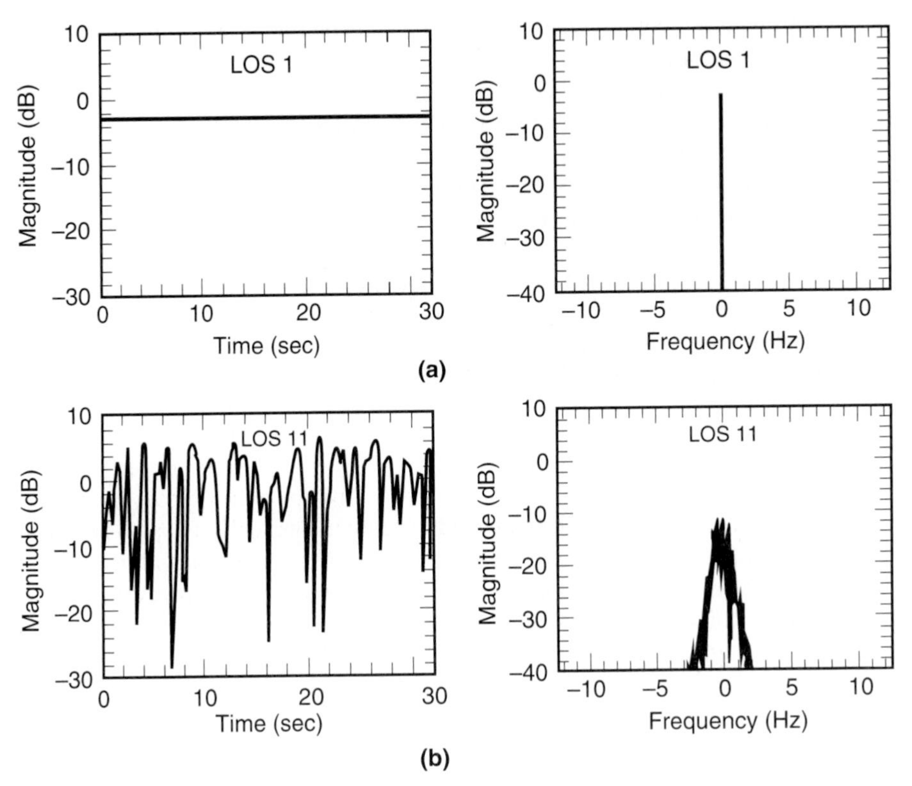

Figure 2.6 Measured values of the Doppler.

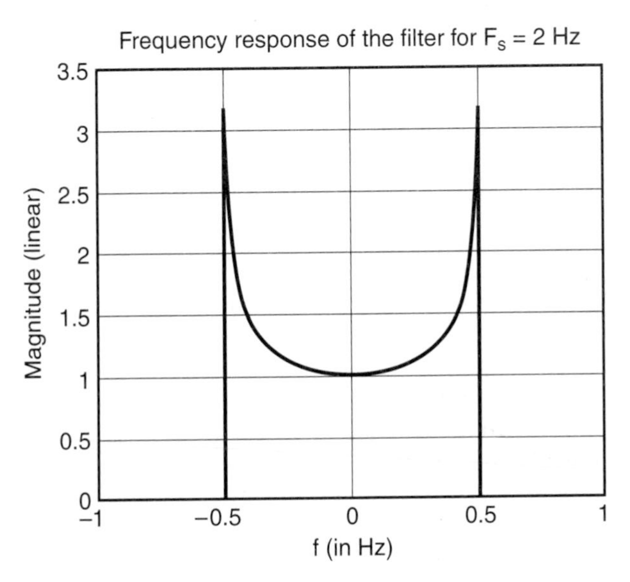

Figure 2.7 The classical Doppler spectrum.

that all these copies suffer fading is small, the receiver is able to correctly decode the received data. Frequency hopping is another technique that can be used to combat fast fading. Because all frequencies are not simultaneously under fade, transmitting data by hopping to different frequencies is an approach to combat fading. This is discussed in Chapter 3.

2.4.3 Multipath Delay Spread

Figure 2.8 shows a sample measured time and frequency response of a typical radio channel. In time domain, shown in Figure 2.8(a), a transmitted narrow pulse arrives as multiple paths with different strengths and arriving delay. In the frequency domain, shown in Figure 2.8(b), the response is not flat, and it suffers from deep frequency selective fades. From Figure 2.8(a) we observe that radio signals arrive at a receiver via a multiplicity of paths. One of the significant problems caused by this phenomenon along with fading is intersymbol interference (ISI). If the multipath delay spread is comparable to or larger than the symbol duration, the received waveform spreads into neighboring symbols and produces ISI. The ISI results in irreducible errors that are caused in the detected signal. This effect is modeled using a *wideband* multipath channel model, sometimes called the delay power spectrum, shown in Figure 2.9, which is usually given by the impulse response:

$$h(t) = \sum_{i=1}^{L} \alpha_i \delta(t - \tau_i) e^{j\varphi_i} \tag{2.19}$$

Here the model assumes that α_i is a Rayleigh distributed amplitude of the multipath with mean local strength $E\{\alpha_i^2\} = 2\sigma_i^2$ and the multipath arrives at a time delay τ_i with phase φ_i assumed to be uniformly distributed in $(0, 2\pi)$.

The delays and multipath interarrival times have various models [PAH95]. In many of the models, the time delays are assumed to be fixed and only the mean-square values of the amplitudes are provided. In Appendix 2D we list several wideband channel models with the multipath delays and mean square values of the amplitudes.

A measure of the data rate that can be supported over the channel without additional receiver techniques is determined by the RMS multipath delay spread given by:

$$\tau_{rms} = \sqrt{\frac{\sum_{k=1}^{N} \tau_k^2 \sigma_k^2}{\sum_{k=1}^{N} \sigma_k^2} - \left(\frac{\sum_{k=1}^{N} \tau_k \sigma_k^2}{\sum_{k=1}^{N} \sigma_k^2} \right)^2} \tag{2.20}$$

A rule of thumb is that it is possible to support data rates that are less than the *coherence bandwidth* of the channel that is approximately $1/5\tau_{rms.}$ The coherence bandwidth is the range of frequencies that are allowed to pass through the channel without distortion. The RMS delay spread varies depending on the type of environment. In indoor areas, it could be as small as 30 ns in residential areas or as large as 300 ns in factories [PAH95]. In urban macrocells, the RMS delay spread is on the order of a few microseconds. This means that the maximum data rates can

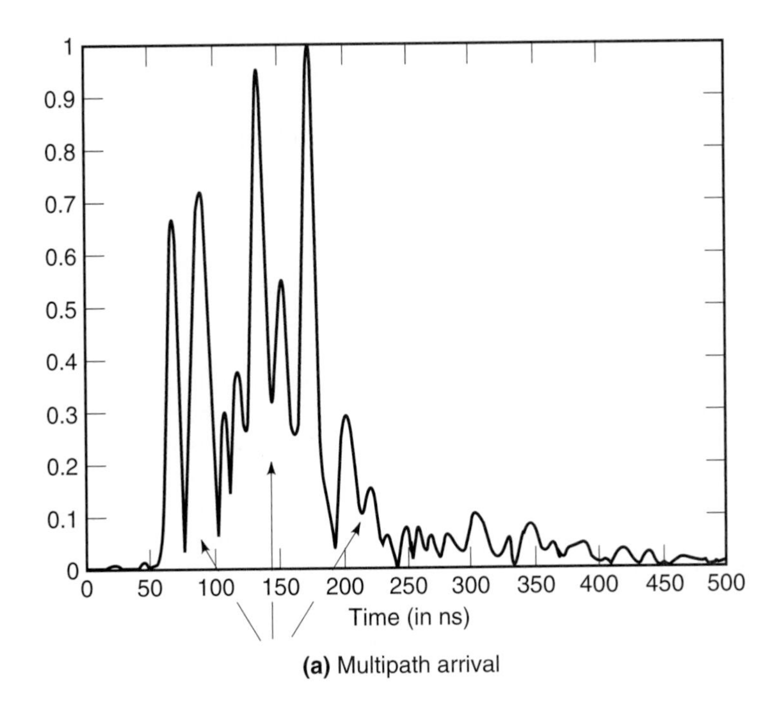

(a) Multipath arrival

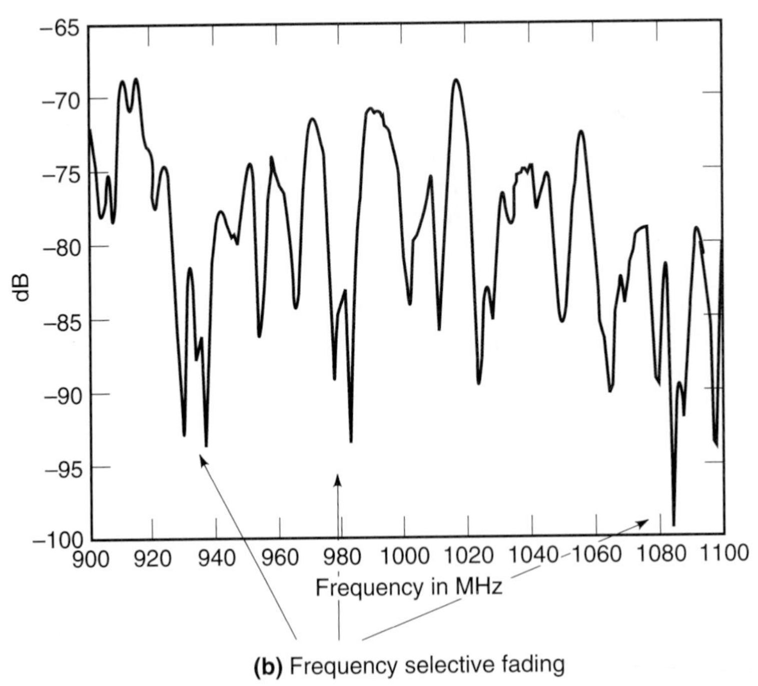

(b) Frequency selective fading

Figure 2.8 Time and frequency response of a typical radio channel over 200 MHz of bandwidth at 1 GHz.

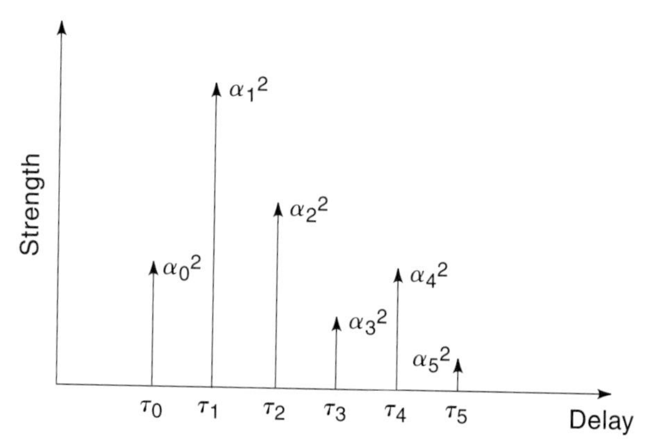

Figure 2.9 Wideband multipath channel model.

be supported in indoor areas is around 6.7 Mbps (at 30 ns) and 50 kbps in outdoor areas (at 4 μs).

In order to support higher data rates, different receiver techniques are necessary. Equalization is a method that tries to cancel the effects of multipath delay spread in the receiver. Direct sequence spread spectrum enables resolving the multipath components and using them to improve performance. OFDM uses multiple carriers, spaced closely in frequency, each carrying low data rates to avoid ISI. Directional antennas reduce the number of multipath components, thereby reducing the total delay spread itself. We discuss these topics in Chapter 3.

2.4.4 Summary of Radio Channel Characteristics and Mitigation Methods

In the previous sections, we have looked at various issues in radio propagation—signal coverage and shadow fading, multipath and Doppler fading leading to large bit error rates, and multipath delay spread that causes irreducible errors in the detected bits due to ISI. Table 2.6 summarizes these effects and techniques to combat them.

Table 2.6 Radio Channel Effects on Performance and Techniques to Improve Performance

Issue	Performance Affected	Mitigation Technique
Shadow fading	Received signal strength	Fade margin—Increase transmit power or decrease cell size
Fast fading	Bit error rate Packet error rate	Error control coding Interleaving Frequency hopping Diversity
Multipath delay spread	ISI and irreducible error rates	Equalization DS-spread spectrum OFDM Directional antennas

2.4.5 Emerging Channel Models

In this section we discuss some new radio channel models that are gaining importance for different applications. Position location is becoming important for emergency and location-aware applications (see Chapter 16), and models developed for communication systems are no longer sufficient to address the performance of geolocation schemes. The use of smart antennas and adaptive antenna arrays require knowledge of the *angle of arrival* (AOA) of the multipath components in order to steer antenna beams in the right directions.

2.4.5.1 Wideband Channel Models for Geolocation

With the advent of widespread wireless communications, the location of people, mobile terminals, pets, equipment, and the like by employing radio signals is gaining importance as well. Several new position location applications [COM00] are rapidly emerging in the market. Civilian applications include intelligent transportation systems (ITS), public safety (enhanced 911 or E-911 services) [NJW97], automated billing, fraud detection, cargo tracking, accident reporting, and so on. It is possible to employ position location for additional benefits such as cellular system design [COM98] and futuristic *intelligent office* environments [WAR97]. Most tactical military units on the other hand are also heavily reliant on wireless communications. Ad hoc connectivity among individual warfighters (for instance in the *Small Unit Operations* (SUO) [DAR97]) in restrictive RF propagation environments such as inside buildings, tunnels, and other urban structures, caves, mountainsides, and double canopy coverage in jungles and forests requires *situation awareness systems* (SAS) that enable the individual warfighters to determine their location and associated information. In either case, the position location service will have to operate within buildings where traditional geolocation techniques such as the global positioning system (GPS) fail due to a lack of sufficient signal power and the harsh multipath environment.

Although RF propagation studies in the past have focused on telecommunications applications, position location applications require a different characterization of the indoor radio channel [PAH98]. For position location applications, accurately detecting the *direct line-of-sight* (DLOS) path between the transmitter and receiver is extremely important. The DLOS path corresponds to the straight line connecting the transmitter and receiver even if there are obstructions like walls in between. Detecting the DLOS path is important because the time of arrival (TOA) or the AOA of the DLOS path corresponds to the distance between the transmitter and receiver (or to the direction between them). This information is used in conjunction with multiple such measurements to locate either the transmitter or the receiver as the case may be. This is in contrast to telecommunications applications where the emphasis is on how data bits can be sent over a link efficiently and without errors. Another issue in positioning systems is the relation between the bandwidth of the transmitted signal and the required accuracy in ranging. An error of 100 ns in estimating the delay of an arriving multipath component could result in an error of 30 meters in calculation of the distance between a transmitter and a receiver. Therefore, positioning systems using TOA often require wide bandwidths to resolve multipath components and detect the arrival of the first path.

In wideband indoor radio propagation studies for telecommunications applications often channel profiles measured in different locations of a building are divided into LOS and OLOS because the behavior of the channel in these two classes has substantially different impacts on the performance of a telecommunications system.

A logical way to classify channel profiles for geolocation applications is to divide them into three categories as shown in Figure 2.10. The first category is the dominant direct path (DDP) case in which the DLOS path is detected by the measurement system, and it is the strongest path in the channel profile. In this case, traditional GPS receivers [ENG94], [GET93], [KAP96] designed for outdoor applications where multipath components are significantly weaker than the DLOS path lock on to the DLOS path and detect its TOA accurately. The second category is the nondominant direct path (NDDP) case where the DLOS path is detected by the measurement system, but it is not the dominant path in the channel profile. For these profiles traditional GPS receivers, expected to lock to the strongest path, will make an erroneous decision on the TOA that leads to an error in position estimation. The amount of error made by a traditional receiver is the distance associated with the difference between the TOA of the strongest path and the TOA of the

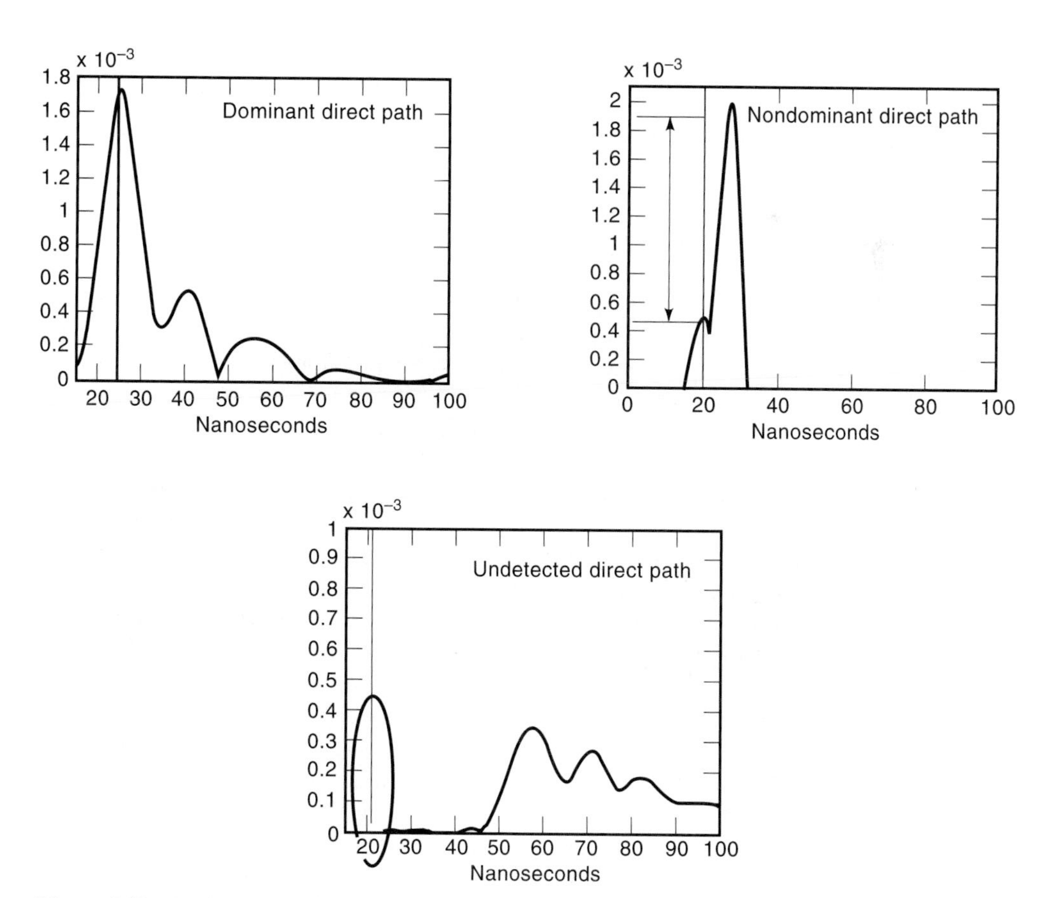

Figure 2.10 Multipath profiles for indoor geolocation.

DLOS path. For the second category locations with NDDP profiles, a more complex RAKE type receiver [PAH95a] can resolve the multipath and make an intelligent decision on the TOA of the DLOS path. The third category of channel profiles is the undetected direct path (UDP) profiles. In these profiles the measurement system cannot detect the DLOS path, and therefore traditional GPS or RAKE type receivers cannot detect the DLOS path. If we define the ratio of the power of the strongest path to the power of the weakest detectable path of a profile as the dynamic range of a receiver, then in NDDP profiles the strength of the DLOS path is within the dynamic range of the receiver and in UDP profiles it is not. If practical considerations regarding the dynamic range are neglected, one can argue that we have only two classes (DDP and NDDP) of profiles because the DLOS path always exists but sometimes we cannot detect it with a practical system. Figure 2.11 shows the results of ray tracing simulations of regions in the first floor of Atwater Kent Laboratories at Worcester Polytechnic Institute with different types of multipath profiles for a centrally located channel sounder.

In the same way that the bit error rate is the ultimate measure for comparing performance of different digital communication receivers, the error in the measurement of the TOA or AOA of the DLOS path is a measure of the performance

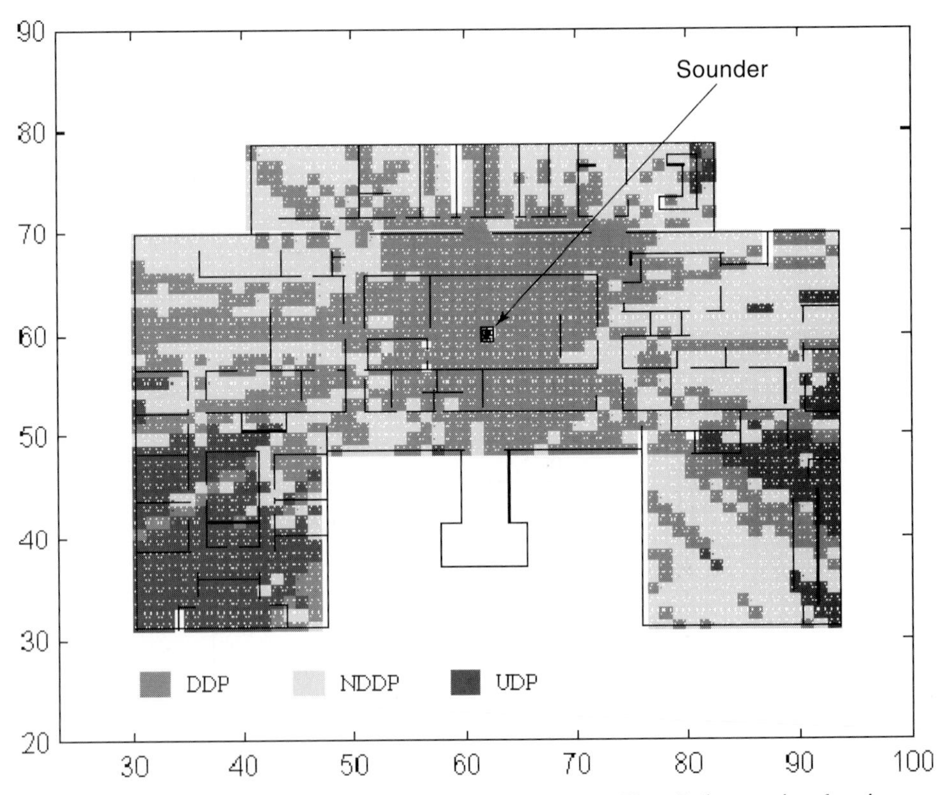

Figure 2.11 Simulated regions in the first floor of Atwater Kent Laboratories showing regions with different types of multipath profiles for a centrally located channel sounder. *Note:* Scale is in meters.

of the geolocation receivers. Traditional RF studies consider the path loss and τ_{rms} as mentioned earlier and these are not sufficient for the geolocation problem. The relative power and delay of the signal arriving via other paths, the channel noise, the signal bandwidth, and interference all influence the detection of the DLOS path and thus the error in estimating the range (distance) between the transmitter and receiver. Efforts to model the indoor radio channel for geolocation based on the error in the detection of the DLOS path are reported in [PAH98], [KRI99a,b,c]. In these efforts, parameters of importance and development of a model are based on measurements of the radio channel used as input to software simulations.

2.4.5.2 SIMO and MIMO Channel Models

Recently, there has been a lot of attention placed on *spatial wideband channel models,* that not only provide the delay-power spectrum discussed in Equation (2.19), but also the AOA of the multipath components. The advent of antenna array systems that are used for interference cancellation and position location applications has made it necessary to understand the spatial properties of the wireless communications channel.

A significant amount of research has been carried out in the area of *single-input multiple output* (SIMO) radio channel models [ERT98]. In these models, a typical cellular environment is considered where it is assumed that the mobile transmitters are relatively simple and the base station can have a complex receiver with adaptive smart antennas with M antenna elements. As shown in Figure 2.12, the multipath environment is such that up to L signals arrive at the base station from different mobile terminals (l) with different amplitudes (α) and phases (φ) at

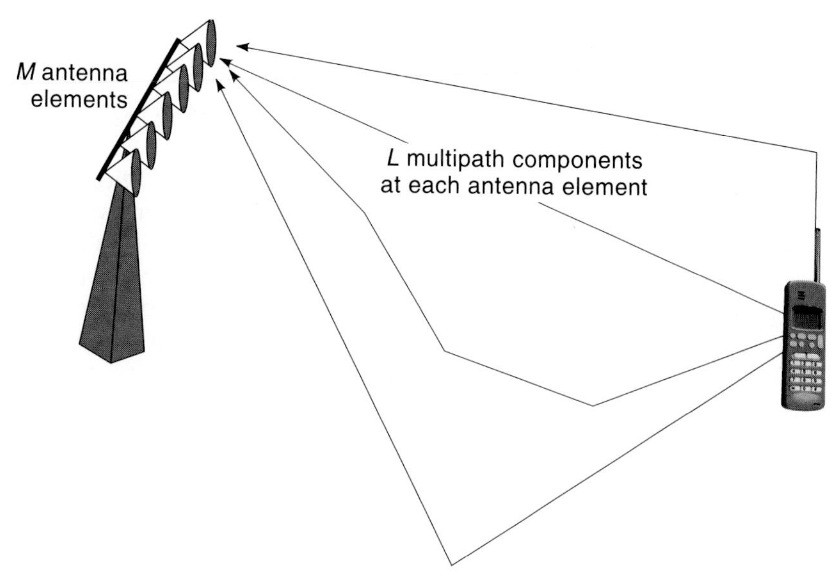

Figure 2.12 Illustration of SIMO radio channel.

different delays (τ) from different directions (θ). These are in general time-invariant, and, as a result, the channel impulse response is usually represented by:

$$\vec{h}(t) = \sum_{l=1}^{L(t)} \alpha_l(t) e^{j\varphi_l(t)} \delta(t - \tau_l(t)) \vec{a}(\theta_l(t)) \tag{2.21}$$

Note that the channel impulse response is now a *vector* rather than a scalar function of time. The quantity $\vec{a}(\theta_l(t))$ is called the array response vector and will have M components if there are M antenna array elements. Thus, there are M channel impulse responses each with L multipath components. A variety of models are available in [ERT98]. The amplitudes are usually assumed to be Rayleigh distributed although they are now dependent on the array response vector $\vec{a}(\theta_l(t))$ as well.

An extension of this model to the scenario where there are N mobile antenna elements and M base station antenna elements [PED00] is called a *multiple-input multiple-output* (MIMO) channel. In this case the channel impulse response is an $M \times N$ matrix that associates a *transmission coefficient* between each pair of antennas for each multipath component. Experimental results and models are considered in [PED00, KER00].

There appears to be tremendous potential for improving capacity using smart antenna systems. Capacity increases between 300 percent and 500 percent are possible in cellular environments as discussed in Chapter 5. In a microcellular environment, preliminary experimental results with a 4×4 antenna array system seem to indicate that over the MIMO channel, a spectral efficiency of 27.9 b/s/Hz is possible [PED00] compared with spectral efficiencies of 2 b/s/Hz in traditional radio systems.

2.5 CHANNEL MEASUREMENT AND MODELING TECHNIQUES

In order to study and model radio propagation characteristics, there is a need to perform different types of measurements and enhance them with simulations. It is extremely hard to analytically derive the expressions for how a radio signal may propagate through a complex environment. Figure 2.13 shows the general modeling procedure. Measurements of the radio channel are obtained at the site where the wireless network is required to be set up. These could be narrowband or wideband measurements or both as the case may be. The results of the measurements could be directly used or enhanced via simulations of the environment. All the data are then included in a database. Based on realistic assumptions about the nature of the channel, the data are used to construct empirical models. Path loss models are an example of such empirical models. Similarly, it is possible to generate wideband multipath models. These models are usually site-specific, dependent on the frequency and the type of measurement taken. For example, it is impossible to obtain time of arrival of multipath components or the RMS delay spread from narrow-

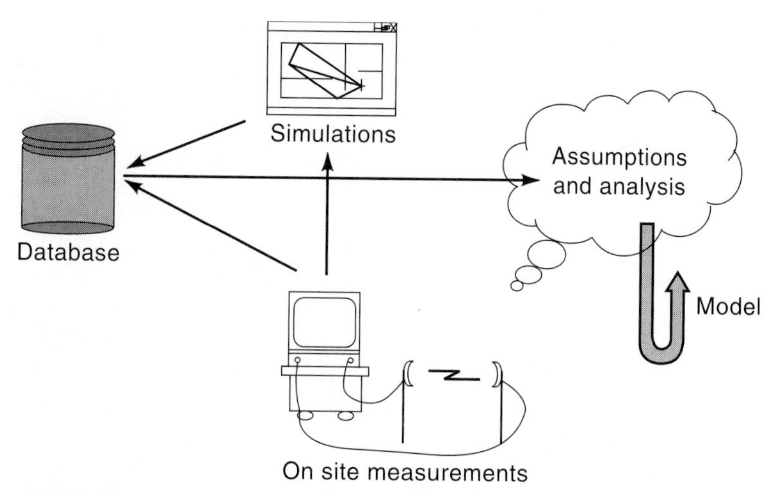

Figure 2.13 General procedure for modeling the radio channel.

band measurements. A detailed discussion of the various measurement and modeling techniques is available in [PAH95].

The procedure for narrowband radio channel measurements [PAH95, BER94], which are useful for determining the expressions for path loss, involves radiating a continuous wave signal from a radio transmitter. The receiver usually employs a vertical dipole antenna along with a position locator so that the distance from the transmitter is known for each value of the measurement. The effects of fast fading are averaged out, and the slow fading values are used in determining the path loss expression. Wideband channel measurements are more complicated [PAH95]. In order to obtain a multipath profile such as the one shown in Figure 2.7, *channel sounders* have to be used. The accuracy of the measurement depends on the channel sounder, the type of antennas used at the transmitter and receiver, the digital signal processing used, the accuracy of the synchronization between transmitter and receiver, and so on. Several channel sounding techniques have been used. Most of them employ one of three techniques: a spread spectrum signal with a matched filter receiver, transmission of a short RF pulse, or use of swept carrier techniques. Spread-spectrum channel sounders are complex to implement but provide large coverage areas. RF pulse systems require large peak power, and pulse amplifiers are expensive. Swept carrier techniques are not suitable for mobile wideband measurements but are quite useful in characterizing the RMS delay spread in indoor areas. Usually a network analyzer is employed for swept carrier wideband measurements. Figure 2.14 shows a picture of a network analyzer and a pair of wideband bicone antennas used for transmitting and receiving signals. Details of all of the measurement techniques can be found in [PAH95].

More recently 3D measurement and modeling of the radio channel characteristics, which includes the angle of arrival as well as delay of arrival of the paths, have become popular for the SIMO and MIMO types of applications. An indoor measurement system and statistical modeling of the angle of arrival of the paths is available in [TIN01]. Figure 2.15 shows a sample measurement of the 3D character-

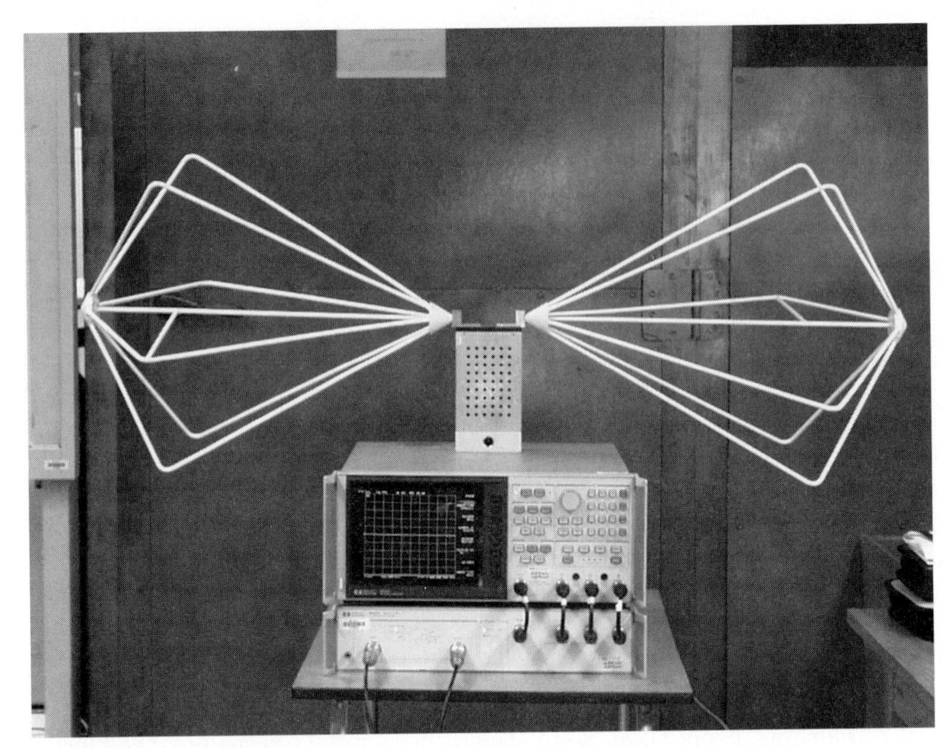

Figure 2.14 A network analyzer and wideband bicone antennas.

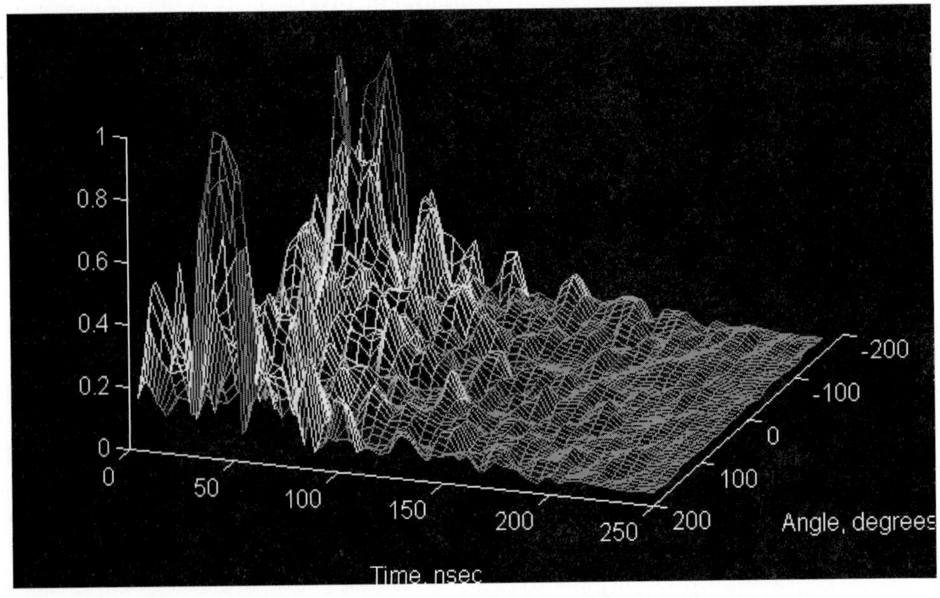

Figure 2.15 Measured time-space characteristics of a typical indoor environment.

istics using an eight-element antenna array. These measurements represent the delay-angle power spectrum that contains the angle of arrival, as well as strength of the signal at different arrival delays. The measurements can then be used for statistical modeling of the 3D characteristics that are useful in performance evaluations of the emerging communications systems exploiting time-space characteristics of the medium to improve the capacity and provide for positioning in E911 or indoor tracking systems [TIN00].

2.6 SIMULATION OF THE RADIO CHANNEL

With the basic understanding of the radio channel, system engineers can use simulations in software or hardware emulation of the radio channel in order to design, analyze, and deploy a variety of wireless communication systems. Of the many important areas where simulations play a role included are issues such as designing a cellular system where the number of base stations, frequency planning, capacity guarantees and so on can influence the economic viability and competitive edge for service providers. The radio channel models play a role in virtually all aspects of wireless systems, including performance evaluation, coverage calculation and receiver designs, and design of protocols for handoff and power control. Thus it is important to have some idea of how to simulate path loss, outage probabilities, and multipath models.

2.6.1 Software Simulation

Simulating path loss or received signal strength requires adding a random variable with the lognormal distribution to the path loss in Equation (2.7). It is fairly straightforward to simulate path-loss values based on the equations for the path loss and the shadow-fading component.

Radical optical methods can be employed to describe and model electromagnetic wave propagation within and outside buildings when the wavelengths of radio signals are smaller compared with the dimensions of objects that are in the environment [PAH95]. Such a simulation of the radio channel is called *ray tracing.* Figure 2.16 shows a graphical user interface (GUI) of a software that can do ray tracing in two dimensions.

In order to describe radio propagation with ray optics, the three basic mechanisms discussed previously are generally considered. Specular reflections and transmission occur when electromagnetic waves impinge on obstructions larger than the wavelength. Usually, rays that are incident upon the ground, walls of buildings, and so on undergo specular reflection and transmission with the amplitude coefficients determined by plane wave analysis or empirical data. Rays incident upon the edges of buildings, walls, or other large objects can be viewed as exciting the edges to act as a secondary line source. Diffracted waves cause propagation into shadowed regions that are not in clear LOS of the transmitter. Irregular objects such as rough walls, furniture, vehicles, and foliage scatter rays in all directions in the form of

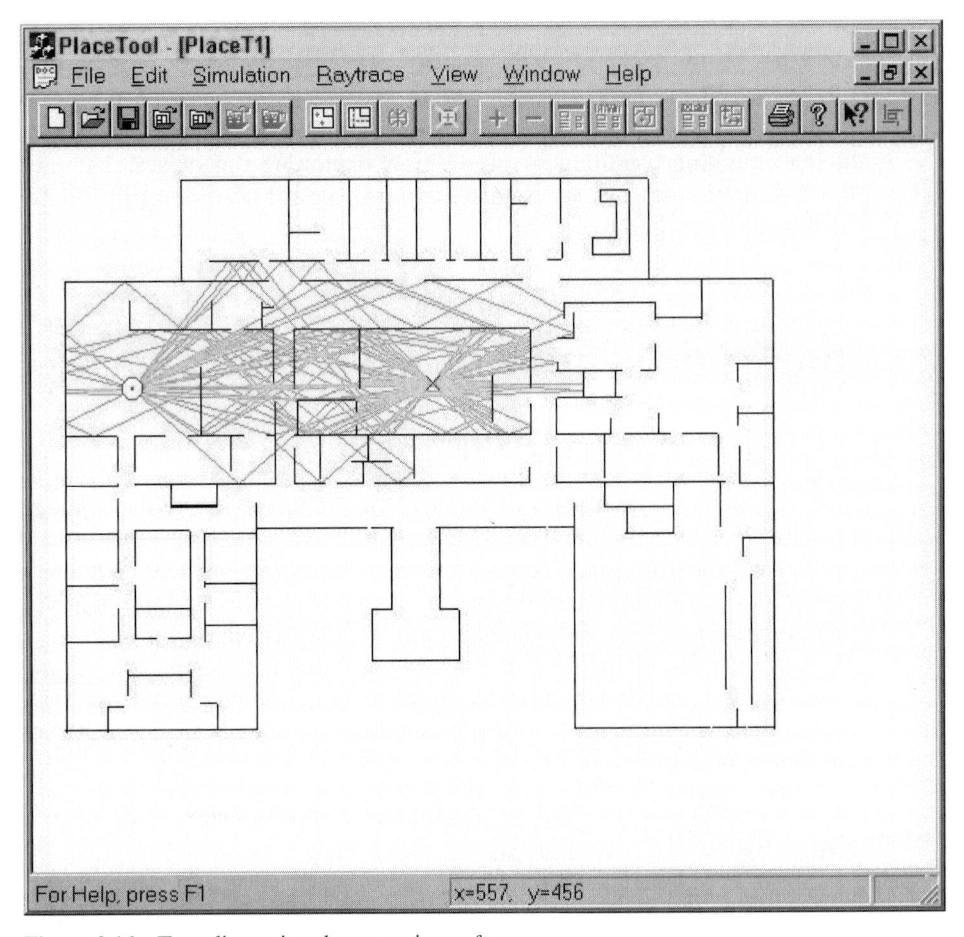

Figure 2.16 Two-dimensional ray tracing software.

spherical waves, especially when the dimensions of objects are on the order of a wavelength or less. The scattered waves are however very weak.

Specular reflection and transmission dominate in indoor environments and are extremely suitable for modeling with ray tracing. When a ray is incident upon a wall, voltage reflection coefficients (Γ) reduce the amplitude of the reflected ray. If there is no absorption, the transmission coefficient (T) will account for the rest of the energy, which will be manifest in the transmitted signal. On the other hand, there is usually some amount of diffuse scattering and absorption that reduces the transmitted energy, resulting in the sum of the reflected and transmitted energies being less than the incident energy. Detailed expressions for Γ and T are available in [PAH95].

The ray tracing approach for dealing with specular reflections and transmissions is as follows. Rays emanating from the transmitter reach the receiver after transmission through and reflection from walls. The unfolded path length determines the signal associated with a specular component. In order to determine all possible ray paths, two approaches exist in practice. In the ray shooting or ray launching approach, rays are launched from the transmitter at regular angular intervals on the order of a degree. Each ray is traced as it intersects with the first wall where it gives rise to a transmitted and reflected ray. Both new rays are traced to the next interaction and so on building a binary tree of rays that continues until some termination criterion is met. As discrete rays have a zero probability of intersecting a point, a region around the receiver is used to represent the receiver itself. The second approach called the image approach determines the exact ray paths between points by imaging the source in the plane of each wall. The former approach is easier for computer implementation. Having determined the required number of contributing rays that intersect the receiver when launched from the transmitter, the received power associated with the ith ray, assuming isotropic antennas is given by:

$$P_{r(i)} = P_0 \frac{1}{R_i^2} \prod_j |\Gamma_j(\phi_{ji})|^2 \prod_m |T_m(\phi_{mi})|^2 \tag{2.22}$$

where R_i is the total unfolded path length of the ith ray, $\Gamma_j(\phi_{ji})$ is the reflection coefficient encountered by the ith ray at the jth wall as a function of the incident angle $\phi_{ji,}$ $T_m(\phi_{mi})$ is the transmission coefficient encountered by the ith ray at the

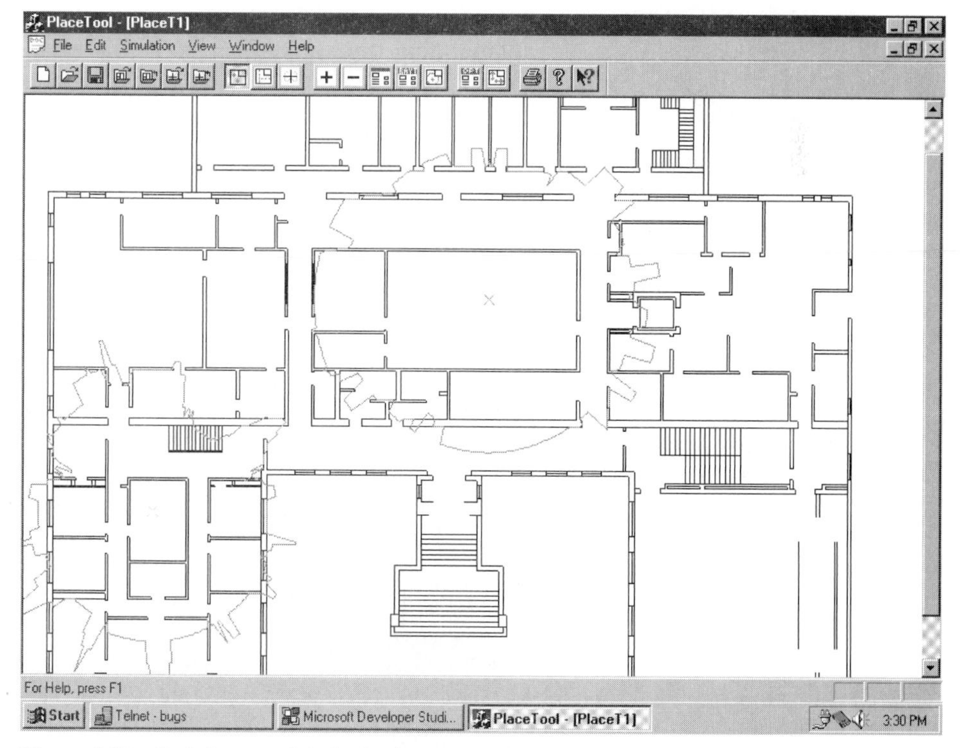

Figure 2.17 Path loss models in deployment software.

mth wall as a function of the incident angle ϕ_{mi} and P_0 is the received power at the first meter from the transmitter. The total received power is calculated by the sum of the individual received powers. The power delay profile is obtained by calculating the total delay from the unfolded path length R_i using the relation $\tau_i = R_i/c$, c being the speed of light. Ray tracing has been used successfully in predicting the signal coverage indoor, as well as in microcellular environments [PAH95].

In several commercial applications, deployment of radio transmitters (base stations and access points) becomes very important in terms of providing the required service, as well as consumer satisfaction. *Deployment tools* for macro-, micro-, and picocellular environments are used to design wireless networks. As discussed in later chapters, most wireless networks use a cellular topology where each base station or access points covers an area called a cell. Path loss models discussed earlier are used in many of the deployment tools to roughly determine the coverage area of base stations and to determine how many base stations are required for covering an area and where they should be placed. Figure 2.17 shows the GUI of one such deployment tool for indoor areas that employs the partition-dependent path loss model of Equation (2.15). Ray tracing may also be employed to estimate path loss via simulations.

2.6.2 Hardware Emulation

Real-time hardware RF channel simulators allow communication systems to be tested quickly and thoroughly under controlled, realistic, and repeatable channel conditions in the laboratory, eliminating the time and cost of lengthy field tests. There are currently several commercial real-time channel simulators available. These RF channel simulators, sometimes called RF channel emulators, are being

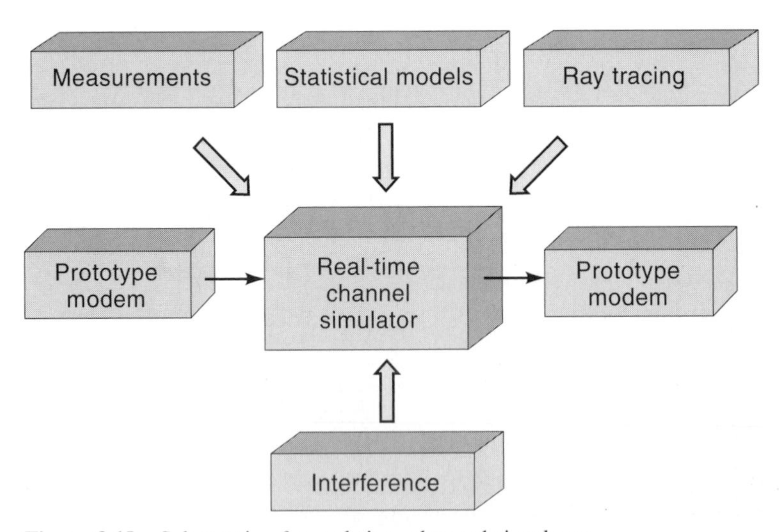

Figure 2.18 Schematic of a real-time channel simulator.

developed for testing modern wireless communications equipment (WCDMA, IS-136, IS-95, GSM, IEEE 802.11, and Bluetooth). They can emulate RF channel characteristics such as multipath fading, lognormal shadowing, path loss attenuation, delay spread, Doppler spread, and additive white Gaussian noise (AWGN). Recently these emulators are being augmented to include means for diversity testing and simulation of jamming. The emulators have embedded statistical channel models, which can use Rayleigh, Gaussian, Nakagami, constant, Ricean, and pure Doppler fading distributions (see Fig. 2.18). They can also be programmed to simulate custom channel models.

The technical specifications of an emulator depends strongly on wherever the communication system under test is for outdoor or indoor use. As an example, the specifications for the PROPSim radio channel simulator (see Fig. 2.19) from Elektrobit Ltd., which can be used for PCS and DECT indicates it has up to 312 taps, with tap spacing ranging from 25 to 3200 ns, a delay resolution as low as 2.5 ns, an RF bandwidth as high as 35 MHz and an RF frequency range from 100 to 400 MHz and from 600 to 2,650 MHz. This emulator uses DSP interpolation techniques to improve its delay resolution beyond that possible from the RF bandwidth limitations. It is possible to emulate the site-specific Ray Tracing–based channel models

Figure 2.19 The PROPSim hardware.

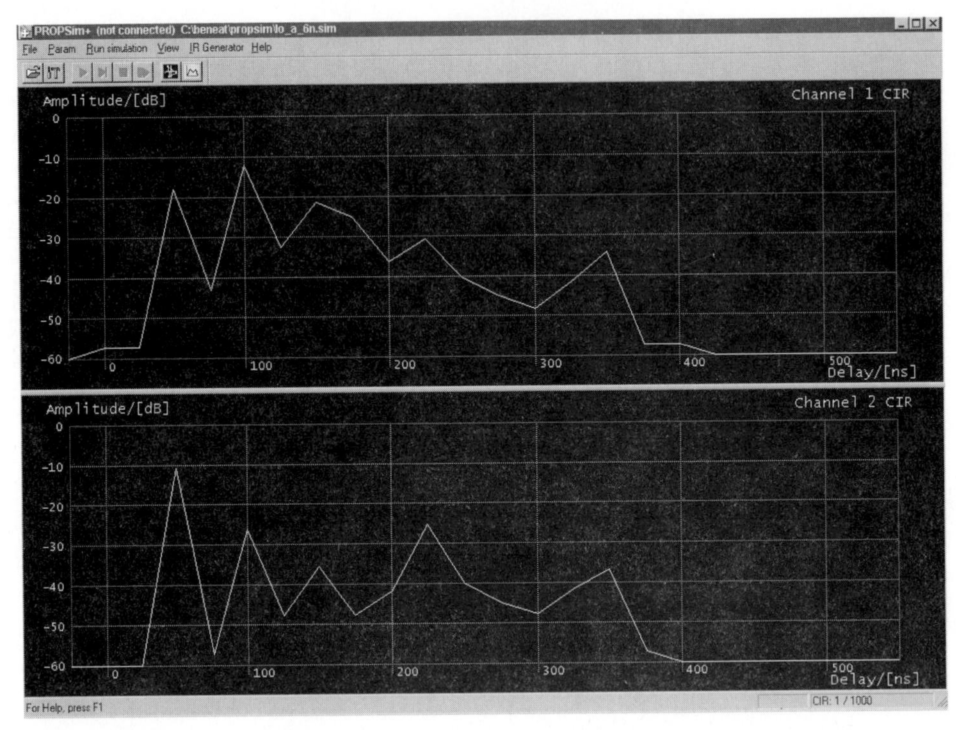

Figure 2.20 Snapshot of the GUI of PROPSim.

into the Elektrobit channel simulator. In this simulator the user interactively (see Fig. 2.20 for a snapshot of the GUI) identifies the location of transmitter and receiver in the users interface floor plan to obtain the channel model that reflects the identified communication link.

APPENDIX 2A WHAT IS dB?

Decibel or dB is usually the unit employed to compute the logarithmic measure of power and power ratios. The reason for using dB is that all computation reduces to addition and subtraction rather than multiplication and division. Every link, node,

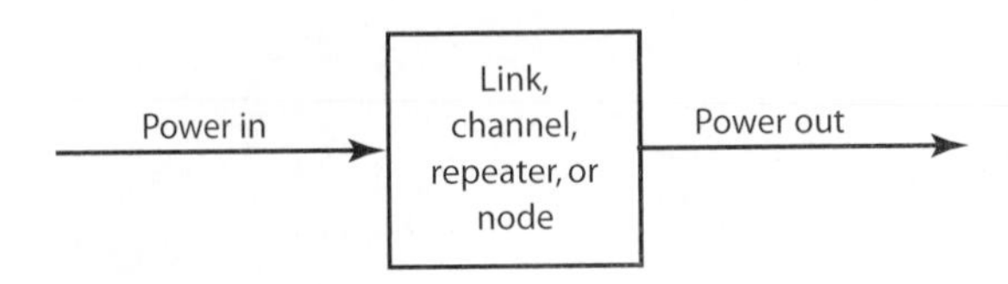

Figure 2A.1 Description of model for defining dB.

repeater, or channel can be treated as a *black box* (see Fig. 2A.1) with a particular decibel gain. The decibel gain of such a black box is given by

$$\text{db gain} = 10 \log\left(\frac{\text{power of output signal}}{\text{power of input signal}}\right) = 10 \log\left(\frac{P_{out}}{P_{in}}\right) \qquad (2A.1)$$

This corresponds to the *relative* output power with respect to the input power. The logarithm is always to the base 10. If the ratio in Eq. (2A.1) is negative, it is a decibel loss.

The decibel gain relative to an absolute power of 1 mW is denoted by dBm and that relative to 1 W is denoted by dBW. For example, if the input power is 50 mW, relative to 1 mW, the input power is 10 log (50 mW/ 1 mW) = 16.98 dBm. If this is followed by a link having a loss of 10 dB, the absolute power at the output of the link will be 16.98 − 10 = 6.98 dBm. Relative to 1 W, these values will be 10 log $(50 \times 10^{-3}/ 1) = -13$ dBW and −23 dBW, respectively.

Antenna gains are represented similarly with respect to an *isotropic* antenna (which radiates with a gain of unity in all directions) or a *dipole* antenna. The former gain is in units of dBi and the latter in units of dBd. The units in dBi are 2.15 dB larger than the units in dBd.

APPENDIX 2B WIRED MEDIA

The most common wired media for communications are TP wires, coaxial cables, optical fibers, and power line wires. TP wires, in shielded (STP) and unshielded (UTP) forms, is available in a variety of categories. It is commonly used for local voice and data communications in home and offices. The telephone companies use category 3 UTP to bring POTS to customer premises and distribute it in the premises. This wiring is also used for voice-band modem data communications, ISDN, DSL, and home phone networking (HPN). The wired media in local area networks are dominated by a variety of UTP and STP to support a range of local data services from several Mbps up to over a gigabit per second within 100 meters of distance.

Coaxial cable provides a wider band useful for a multichannel FDM operation, lower radiation, and longer coverage than TP, but it is less flexible in particular in indoor areas. The early LANs were operating on the so-called thick cable to cover up to 500 meters per segment. To reduce the cost of wiring, the thin cable, sometime referred to as Cheaper-LAN, that covers up to 200 meter per segment, replaced thick cables. Cabled LANs are not popular anymore, and TP wirings are taking over the LAN market. Another important application for cable is cable television which has a huge network to connect homes to the cable TV network. This network is also used for broadband access to the Internet. The cable-LANs were using baseband technology with data rates of 10 Mbps while broadband cable-modems provide comparable data rates over each of around hundred cable TV channels. Today, in addition to traditional cable TV, cable-modems are becoming a popular access method to the Internet, and some cable service providers are offering voice services over the cable connections.

Fiber optic lines provide extremely wide bandwidth, smaller size, lighter weight, less interference, and very long coverage (low attenuation). However, fiber lines are less flexible and more expensive to install, not suitable for FDM (though wavelength division multiplexing—WDM—is becoming common), and have expensive electronics for TDM operation. Because of the wide bandwidth and low attenuation, optical fiber lines are becoming the dominant wired medium to interconnect switches in all long-haul networks. In the LAN applications fiber lines mostly serve the backbone to interconnect servers and other high-speed elements of the local networks. Optical fibers have not found a considerable market for distributing any service to the home or office desktop.

Power-line wiring has also attracted some attention for low speed and high-speed home networks. The bandwidth for power lines is more restricted, the interference caused by appliances is significant, and the radiated interference in the frequency of operation of AM radio is high enough that frequency assignment agencies do not allow power line data networking in these frequencies. However, power lines have good distribution in exiting homes because power plugs are available in all rooms. In addition almost all appliances are connected to the power line, and with only one connection a terminal can connect to the network and the power supply. Existing power line networks either operate at low data rate in low frequencies, below frequency of operation of the AM radios, to interconnect evolving smart appliances or they operate at high data rates of up to 10 Mbps in frequencies above AM radio for home computing.

The wireless channels considered in this book are all for relatively short connections. For long haul transportations, these networks are complemented with wired connections. As we have discussed in this chapter, the average path loss in wireless channels is exponentially related to the distance, and we often describe the path loss in a fixed dB value per decade of distance (see Appendix 2A for a definition of dB). The path loss in wired environments is linearly related to the distance. Therefore wired connections are more power efficient for shorter distances and wireless for longer distances.

Example 2B.1: Path Loss in Wired and Wireless Media

The category 3 UTP wires used for Ethernet lose around 13 dB per 100 meters of distance [STA00]. Therefore, the path loss for 1 km is 130 dB. In free-space, the path loss of a 1 GHz radio in the first meter is around 30 dB (with an omnidirectional dipole antenna), and after that the path loss is 20 dB per decade of distance. Therefore, path loss at 100 meters is 70 dB and for 1 km it is 90 dB. Indeed for 130 dB path loss, the 1 GHz radio can cover 1,000 km. Therefore, for short distances, the path loss for wired transmission is smaller, but for long distances path loss for radio systems is smaller. This is the reason why in long haul wired transmissions, repeaters are commonly used. Also, for the same reason radio signals are able to provide for very long-distance satellite communications.

APPENDIX 2C PATH LOSS MODELS

Table 2C.1 tabulates the COST-231 path loss model. These formulas can be applied for rough calculations in terms of system dimensioning, power budget calculations, and so on, but site specific measurements are the only true way of determining what the radio propagation characteristics are in a particular area.

Table 2C.1 The COST-231 Model for PCS Applications in Urban Areas

General Formulation:

$$L_p = 46.3 + 33.9 \log f_c - 13.82 \log h_b - a(h_m) + [44.9 - 6.55 \log h_b] \log d + C_M$$

where f_c is in MHz, h_b and h_m are in meters and d is in km

Range of Values		
Center frequency f_c in MHz		1,500 – 2,000 MHz
h_b, h_m in meters		30–200m, 1–10m
d		1 – 20 km
C_M	Large City	0 dB
	Medium City/ Suburban areas	3 dB

APPENDIX 2D WIDEBAND CHANNEL MODELS

Table 2D.1 is a list of the JTC wideband multipath channel models in indoor areas. Each table has three channel models. A, B, and C associated with good, medium, and bad conditions. See [PAH95] for more details.

Table 2D.1 JTC Parameters of the Wideband Multipath Channel for Indoor Commercial Buildings

	Channel A		Channel B		Channel C		
Tap	***Rel Delay[1] (nSec)***	***Avg Power (dB)***	***Rel Delay[1] (nSec)***	***Avg Power (dB)***	***Rel Delay[1] (nSec)***	***Avg Power (dB)***	**Doppler Spectrum** $D(\lambda)$
1	0	0	0	0	0	0	FLAT
2	100	−5.9	100	−0.2	200	−4.9	FLAT
3	200	−14.6	200	−5.4	500	−3.8	FLAT
4			400	−6.9	700	−1.8	FLAT
5			500	−24.5	2100	−21.7	FLAT
6			700	−29.7	2700	−11.5	FLAT

[1]A ± 3 percent variation about the relative delay is allowed.

Table 2D.2 JTC Parameters of the Wideband Multipath Channel for Indoor Office Buildings

	Channel A		Channel B		Channel C		
Tap	***Rel Delay[1] (nSec)***	***Avg Power (dB)***	***Rel Delay[1] (nSec)***	***Avg Power (dB)***	***Rel Delay[1] (nSec)***	***Avg Power (dB)***	**Doppler Spectrum** $D(\lambda)$
1	0	0	0	0	0	0	FLAT
2	100	−8.5	100	−3.6	200	−1.4	FLAT
3			200	−7.2	500	−2.4	FLAT
4			300	−10.8	700	−4.8	FLAT
5			500	−18.0	1100	−1.0	FLAT
6			700	−25.2	2400	−16.3	FLAT

[1]A ± 3 percent variation about the relative delay is allowed.

Table 2D.3 JTC Parameters of the Wideband Multipath Channel for Indoor Residential Buildings

	Channel A		Channel B		Channel C		
Tap	***Rel Delay[1] (nSec)***	***Avg Power (dB)***	***Rel Delay[1] (nSec)***	***Avg Power (dB)***	***Rel Delay[1] (nSec)***	***Avg Power (dB)***	**Doppler Spectrum** $D(\lambda)$
1	0	0	0	0	0	0	FLAT
2	100	−13.8	100	−6.0	100	−0.2	FLAT
3			200	−11.9	200	−5.4	FLAT
4			300	−17.9	400	−6.9	FLAT
5					500	−24.5	FLAT
6					600	−29.7	FLAT

[1]A ± 3 percent variation about the relative delay is allowed.

QUESTIONS

2.1 What are the three important radio propagation phenomena at high frequencies? Which of them is predominant indoors?

2.2 Explain what path-loss gradient means. Give some typical values of the path-loss gradient in different environments.

2.3 Explain the meaning of the expression "a loss of 37 dB per decade of distance" in terms of the path loss gradient α.

2.4 Why does multipath in wireless channels limit the maximum symbols transmission rate? How can we overcome this limitation?

2.5 What is the Doppler spectrum and how can one measure it?

2.6 Differentiate between shadow fading and fast fading.

2.7 What distributions are used to model fast fading in LOS situations? In OLOS situations?

2.8 What are the differences between multipath, shadow, and frequency-selective fading? Give an example distribution function that is used to model multipath fading and an example that is used to model shadow fading.

2.9 What techniques can be used to combat the effects of frequency selective fading (multipath delay spread)?

2.10 For position location applications, how are wideband radio channels classified? How is this classification useful?

2.11 What is the difference between a SIMO and a MIMO radio channel?

2.12 Name three channel sounding techniques. Give the advantages and disadvantages of each.

2.13 Explain ray tracing.

2.14 How can we emulate a radio channel in hardware?

PROBLEMS

2.1 What is the received power (in dBm) in free space of a signal whose transmit power is 1 W and carrier frequency is 2.4 GHz if the receiver is at a distance of 1 mile (1.6 km) from the transmitter? What is the path loss is dB?

2.2 Use the Okumura-Hata and COST-231 models to determine the maximum radii of cells at 900 MHz and 1,900 MHz respectively having a maximum acceptable path loss of 130 dB. Use $a(h_m) = 3.2\,[\log(11.75\,h_m)]^2 - 4.97$ for both cases.

2.3 In a mobile communications network, the minimum required signal-to-noise ratio is 12 dB. The background noise at the frequency of operation is −115 dBm. If the transmit power is 10 W, transmitter antenna gain is 3 dBi, the receiver antenna gain is 2 dBi, the frequency of operation is 800 MHz, and the base station and mobile antenna heights are 100 m and 1.4 m respectively, determine the maximum in building penetration loss that is acceptable for a base station with a coverage of 5 km if the following path loss models are used.

a. Free space path loss model

b. Two-ray path loss model

c. Okumura-Hata model for a small city

2.4 A mobile system is to provide 95 percent successful communication at the fringe of coverage with a location variability having a zero mean Gaussian distribution with standard deviation of 8 dB. What fade margin is required?

2.5 Signal strength measurements for urban microcells in the San Francisco Bay area in a mixture of low-rise and high-rise buildings indicate that the path loss L_p in dB as a function of distance d is given by the following linear fits:

$$L_p = 81.14 + 39.40 \log f_c - 0.09 \log h_b + [15.80 - 5.73 \log h_b] \log d, \text{ for } d < d_{bk}$$

$$L_p = [48.38 - 32.1 \log d_{bk}] + 45.7 \log f_c + (25.34 - 13.9 \log d_{bk}) \log h_b + [32.10 + 13.90 \log h_b] \log d + 20 \log(1.6/h_m), \text{ for } d > d_{bk}$$

Here, d is in kilometers, the carrier frequency f_c is in GHz (that can range between 0.9 and 2 GHz), h_b is the height of the base station antenna in meters, and h_m is the height of the mobile terminal antenna from the ground in meters. The *breakpoint* distance d_{bk} is the distance at which two piecewise linear fits to the path loss model have been developed and it is given by $d_{bk} = 4h_b h_m / 1000\lambda$, where λ is the wavelength in meters. The shadow-fading component in dB is given by a zero mean Gaussian random variable with a standard deviation of 5 db.

a. If 90 percent of the locations at the cell edge need coverage, what should be the fading margin applied? What percent of locations would be covered if a fading margin of 5 dB is used?

b. What would be the radius of a cell covered by a base station (height 15 m) operating at 1.9 GHz and transmitting a power of 10 mW that employs a directional antenna of gain 5 dBi? The fading margin is 7.5 dB and the sensitivity of the mobile receiver is −110 dBm. Assume that $h_m = 1.2$ m. How would you increase the size of the cell?

Note that the path losses predicted by the two equations are very close, but not exactly the same at $d = d_{bk}$. You can use either value in your calculation.

2.6 The path loss in a building was discovered to have two factors adding to the free space loss: a factor directly proportional to the distance and a floor-attenuation factor. In other words, path loss = free space loss + βd + FAF. If the FAF is 24 dB, and the distance between transmitter and receiver is 30 m, determine what should be the value of β so that the path loss suffered is less than 110 dB.

2.7 Sketch the power-delay profile of the following wideband channel. Calculate the excess delay spread, the mean delay, and the RMS delay spread of the following multipath channel. A channel is considered "wideband" if its coherence bandwidth is smaller than the data rate of the system. Would the channel be considered a wideband channel for a binary data system at 25 kbps? Why?

Relative delay in microseconds	Average relative power in dB
0.0	−1.0
0.5	0.0
0.7	−3.0
1.5	−6.0
2.1	−7.0
4.7	−11.0

2.8 The modulation technique used in the existing American Mobile Phone (AMP) cellular radio systems in analog FM. The transmission bandwidth is 30 kHz per channel and the maximum transmitted power from a mobile user is 3 W. The acceptable quality of the input SNR is 18 dB and the background noise in the bandwidth of the system is −120 dBm (120 dB below the 1 mW reference power). In the cellular operation we may assume that the strength of the signal drops 30 dB for the first meter of distance from the transmitter antenna and 40 dB per decade of distance for distances beyond 1 meter.

a. What is the maximum distance between the mobile station and the base station at which we have an acceptable quality of signal?

b. Repeat (a) for digital cellular systems for which the acceptable SNR is 14 dB.

2.9 The modulation technique used in the exiting Advanced Mobile Phone System (AMPS) is analog FM. The transmission bandwidth is 30 kHz per channel and the maximum transmitted power from a mobile user is 3 W. The acceptable quality of the received SNR is 18 dB and the power of the background noise in the system is −120 dBm. Assuming that the height of the base and mobile station antennas are $h_b = 100$ m and $h_m = 3$ m respectively and the frequency operation of is $f = 900$ MHz, what is the maximum distance between the mobile station and the base station for an acceptable quality of communication?

a. Assume free space propagation with transmitter and receiver antenna gains of 2.

b. Use Hata's equations for Okumura's model in a large city.

c. Use JTC model in residential areas and assume that the mobile unit is used inside a building.

2.10 The Doppler spectrum of the indoor radio is often assumed to have a uniform distribution with a maximum Doppler shift of 10 Hz.

a. Determine the rms Doppler spread of the channel.

b. Determine the average number of fades per second and the average fade duration, assuming that the threshold for fading is chosen to be 10 dB below the average rms value of the signal.

2.11 Using specifications of Channel A in Table 2D.1 for the JTC model in the indoor office areas, determine the mean delay spread of the channel. Determine the rums delay spread of the channel. Give an approximated maximum data rate that can be supported over this channel.

2.12 Consider a transmitter located in a multistory building. The receiver is located three floors below. The floor loss is 5 dB per floor. There are two brick walls, one office wall, and a glass wall between the transmitter and receiver in addition to the floors. The distance between the transmitter and receiver is 16 m. Assuming a frequency of operation of 1 GHz and transmitter and receiver antenna gains of 1.6, calculate the path-loss using:

a. The partition-dependent model described in Section 2.3.7.2 that assumes a distance power gradient of 1.8, 2.0, and 4.2.

b. Using the JTC model for indoor office areas.

2.13 In Eq. (2.19) we looked at the impulse response of a multipath channel. This is a specific instance of the general characterization of a multipath channel called the *scattering function.* The uncorrelated scattering function of an indoor radio channel is defined as the product of a time function $Q(\tau)$ that represents the delay-power spectrum and a frequency function $D(\lambda)$ that represents the Doppler Spectrum (See [PAH95] for details), i.e.,

$$S(\tau;\lambda) = Q(\tau)\, D(\lambda)$$

The scattering function provides a better description of the radio channel. Suppose for τ (in ns) we have

$$Q(\tau) = 0.4\,\delta(\tau - 50) + 0.4\,\delta(\tau - 100) + 0.2\delta(\tau - 200)$$

and for λ (in Hz):

$$D(\lambda) = 0.1\,U(\lambda + 5) + 0.1\,U(\lambda - 5)$$

with $\delta(.)$ and $U(.)$ functions representing the impulse and unit step functions.

a. Determine the RMS delay spread of the channel assuming $Q(\tau)$ is similar to Eq. (2.19) and the mean square values are specified here.

b. Determine the RMS Doppler spread of the channel using an expression similar to Eq. (2.20) but with $D(\lambda)$ instead of $Q(\tau)$.

c. Give the coherence bandwidth of the channel.

2.14 Use a software tool like Matlab™ or Mathcad to generate 1,000 impulse responses of the JTC indoor residential radio channel (all three cases). Determine the RMS multipath delay spread for each sample and plot the cumulative distribution function.

2.15 The inter-arrival times of multipath components can be modeled as either a constant or as a random process. In many cases, the inter-arrival times are modeled as samples from an exponential distribution (path arrivals are from a Poisson process—see [PAH95]). If the variance of the exponential distribution used to model one such radio channel is 4 ns^2, what is the average arrival time of the process? Use a software tool to generate 100 samples of the inter-arrival times and plot arrival times of the multipath components. Compute the average for your simulations.

CHAPTER 3

PHYSICAL LAYER ALTERNATIVES FOR WIRELESS NETWORKS

3.1 Introduction

3.1.1 Wired Transmission Techniques
3.1.2 Considerations in the Design of Wireless Modems

3.2 Applied Wireless Transmission Techniques

3.3 Short Distance Baseband Transmission

3.4 UWB Pulse Transmission

3.5 Carrier Modulated Transmission

3.6 Traditional Digital Cellular Transmission

3.6.1 Digital Frequency Modulation and GMSK
3.6.2 Digital Phase Modulation and π/4-QPSK

3.7 Broadband Modems for Higher Speeds

3.7.1 Multicarrier, Multisymbol, Multirate OFDM Modulation

3.8 Spread Spectrum Transmissions

3.8.1 Frequency Hopping Spread Spectrum
3.8.2 Direct Sequence Spread Spectrum

3.9 High-Speed Modems for Spread Spectrum Technology

3.9.1 PPM-DSSS
3.9.2 CCK Modulation
3.9.3 Multicarrier CDMA

3.10 Diversity and Smart Receiving Techniques

3.10.1 Time Diversity Techniques
3.10.2 Frequency Diversity Techniques
3.10.3 Space Diversity Techniques

3.11 Comparison of Modulation Schemes

3.12 Coding Techniques for Wireless Communications

3.12.1 Error Control Coding
3.12.2 Speech Coding
3.12.3 Coding for Spread Spectrum Systems

3.13 A Brief Overview of Software Radio

Appendix 3A Performance of Communication Systems

Appendix 3B Coding and Correlation

Questions

Problems

3.1 INTRODUCTION

In this chapter we describe transmission technologies, which have been adopted in many of the developing standards and products for wireless networks. In principle these techniques are applicable to all wired and wireless modems because the basic design issues are common to both systems. In general we would like to transmit data with the highest achievable data rate with the minimum expenditure of signal power, channel bandwidth, and transmitter and receiver complexity. In other words we usually want to maximize both bandwidth efficiency and power efficiency and minimize the transmission system complexity. However, the emphasis on these three objectives varies according to the application requirement and medium for transmission, and there are certain details that are specific to particular applications and media of transmission. Also these design objectives are often conflicting and the trade-offs decide what factors are considered more important than others. We start this chapter with a brief description of specific characteristics of the wireless medium that affect the design of transmission techniques. Then we provide a comprehensive overview of applied wireless transmission techniques, followed by a review of coding techniques and a brief description of software implementation of these radios.

3.1.1 Wired Transmission Techniques

In most wired data applications, such as LANs, transmission schemes over TP, coaxial cable, or optical fiber are very simple. The received data from the higher layers are line coded (e.g., Manchester coded on Ethernet) and the voltages (or optical signals) are applied directly to the medium. These transmission techniques are often referred to as baseband transmission schemes. In voice-band modems, DSL, and coaxial cable modem applications, the transmitted signal is modulated over a *carrier*. The amplitude, frequency, phase of the carrier, or a combination of these is used to carry data. These digital modulation schemes are correspondingly called amplitude shift keying (ASK), frequency shift keying (FSK), phase shift keying (PSK), or quadrature amplitude modulation (QAM). In voice-band modems, this carrier is around 1,800 Hz which is the center of the passband of 300–3,300 Hz of the telephone channels. The purpose of modulation here is to eliminate the DC component from the transmission spectrum and to allow the usage of more bandwidth-efficient modulation to support higher data rates over the telephone channel. For DSL services, the spectrum that is utilized is shifted away from the lower frequencies used for voice applications. Discrete multitone transmission, a form of OFDM, is employed there. In cable modems, modulation is employed to shift the spectrum of the signal to a particular frequency channel and to improve the bandwidth efficiency of the channel to support higher data rates. In the data networking industry, cable modems are referred to as broadband modems because they provide a much higher data rate (broader band) than the voice-band modems. High bandwidth efficiency in the voice-band modems has a direct economic advantage to the user, as it can reduce connect time and avoids the necessity for leasing additional circuits to support the application at hand. The typical telephone channel is less hostile than a typical radio channel, providing a fertile environment for examining and employing complex modulation and coding techniques such as QAM and *trellis coded modulation* (TCM) and signal processing algorithms such as equalizers and echo cancellers. Specific impairments seen on telephone channels are amplitude and delay distortion, phase jitter, frequency offset, and effects of nonlinearity. Many of the practical design techniques of wired modems have been developed to efficiently deal with these categories of impairments.

3.1.2 Considerations in the Design of Wireless Modems

Radio channels are characterized by multipath fading and Doppler spread, and a key impediment in the radio environment is the relatively high levels of average signal power needed to overcome fading. However, there are other considerations that impact the selection of a modem technique for a wireless application. For example, in radio systems, bandwidth efficiency is an important consideration, because the radio spectrum is limited, and many operational bands are becoming increasingly crowded. There are a number of considerations that enter into the choice of a modulation technique for use in a wireless application, and here we briefly review the key requirements. These requirements can vary somewhat from one system to another, depending on the type of system, the requirements for delivered services, and the users' equipment constraints.

3.1.2.1 Bandwidth Efficiency

Most wireless networks that support mobile users have a need for bandwidth-efficient modulation, and this requirement steadily grows in importance each year. One of the major incentives of the cellular telephone industry for moving from analog to digital and then from TDMA to CDMA was to increase the bandwidth efficiency and consequently the number of users. A cellular carrier company is assigned a specified amount of licensed bandwidth in which to operate their system, and therefore an increase in system capacity leads directly to increased revenues.

This defines another clear need for modulation techniques that provide efficient utilization of available bandwidth.

An area of wireless communications network development where the modulation bandwidth efficiency is not yet as critical as that of the cellular industry is that of WLANs. Unlike cellular systems, which to date have been used to support circuit-switched services, WLANs are typically used for burst-mode traffic. Due to the bursty nature of the user data, the aggregate data traffic on a WLAN rarely approaches system capacity. Furthermore, almost all WLAN products operate in the unlicensed bands, where the same frequencies are reused again and again in relatively close geographic areas. It is for these reasons that the WLAN industry had placed relatively little emphasis on bandwidth-efficient modulation techniques.

3.1.2.2 Power Efficiency

Power efficiency is another parameter, which is not of major importance for wireless equipment using AC power sources, such as interLAN bridges, but is of crucial importance in battery-oriented applications such as handheld cellular or WLAN cards used in laptops. In these applications power consumption translates into battery size and recharging intervals, and even more important to the mobile user, into the size and weight of portable terminals. Figure 3.1 demonstrates how the power consumption is directly related to device size and weight. Power effi-

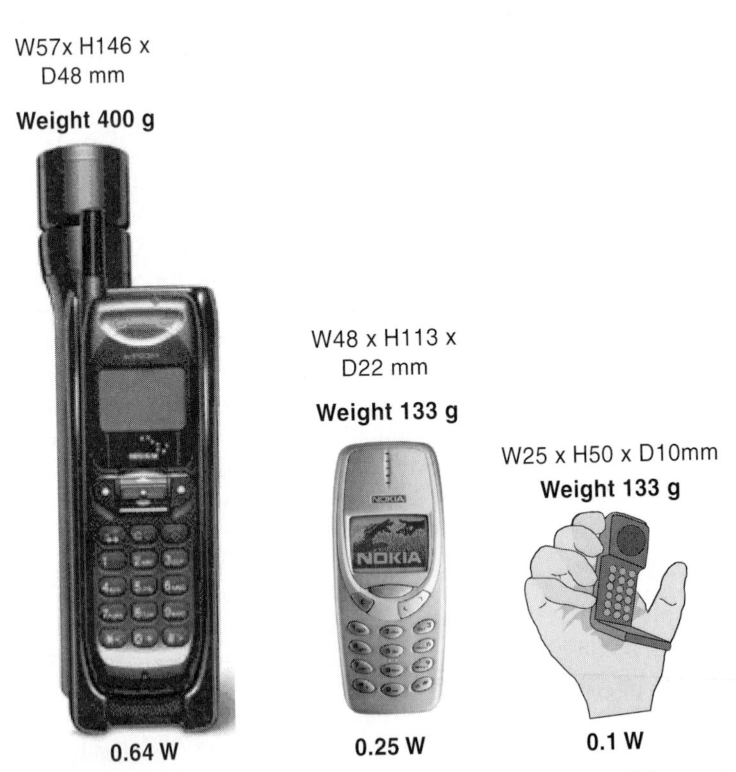

Figure 3.1 Power consumption and the size/weight of a mobile terminal.

ciency will become increasingly important as consumers become accustomed to the convenience of small, handheld communication devices.

There are two facets of the power requirement: One is the power needed to operate the electronics in the terminal; the other is the amount of power needed at the output of the power amplifier in order to radiate a given amount of signal power from the antenna. The radiated signal power, of course, translates directly into signal coverage, and it is a function of the data rate and the complexity of the receiver. Higher data rates require higher operating levels of SNR. More complex systems using computationally demanding coding techniques such as TCM or employing adaptive equalization require less transmission power. However, a more complex receiver design increases the power consumed by the electronics and consequently reduces battery life. In some applications a compromise has to be made between the complexity of the receiver and the electronic power consumption. For example, in handheld local communication devices some manufacturers avoid the use of complex speech coding techniques in order to reduce battery consumption. Also, in the design of high-speed data communication networks for laptop or handheld computers, some designers find it difficult to justify the additional electronic power consumption required for the inclusion of adaptive equalization algorithms.

In spread spectrum CDMA systems, power efficiency and overall system bandwith efficiency are closely related. The use of a more power-efficient modulation method allows a system to operate at lower SNR. The performance of a CDMA system is limited by the interference from other users on the system, and an improvement in power efficiency in turn increases the bandwidth efficiency of the system.

3.1.2.3 Out-of-Band Radiation

An important issue in the selection of a modulation technique for a radio modem is the amount of transmitted signal energy lying outside the main lobe of the signal spectrum. In cellular operation the performance is limited by adjacent-channel interference rather than additive noise. Figure 3.2 illustrates the situation: two users—one close to the BS antenna and the other at a larger distance—are operating in two adjacent channels. The out-of-band interference of the mobile closer to the antenna is a serious source of interference for the mobile located in a farther distance.

The adjacent-channel interference (ACI), which is the interference that a transmitting radio presents to the user channels immediately above and below the transmitting user's channel, is a major parameter in the design of cellular systems. The ACI determines the geographic area where mobile users can be served by a single base station. As shown in Figure 3.2, a low level of ACI will permit a distant mobile transmitter to reach the base station with a weak signal while another mobile much closer to the base station is transmitting in an adjacent channel. Thus ACI specifications will indirectly influence system capacity and cost. The characteristics of the transmit and receive channel filters, nonlinearity in the transmitter, and of course the height and roll-off characteristics of the skirts of the transmitted signal spectrum influence the level of ACI. Radio manufacturers strive to design radios that keep ACI below a specified level, typically −60 dB below the main lobe, and the out-of-band spectral power of the modulation scheme is the principal ingredient in achieving that goal. In contrast, the out-of-

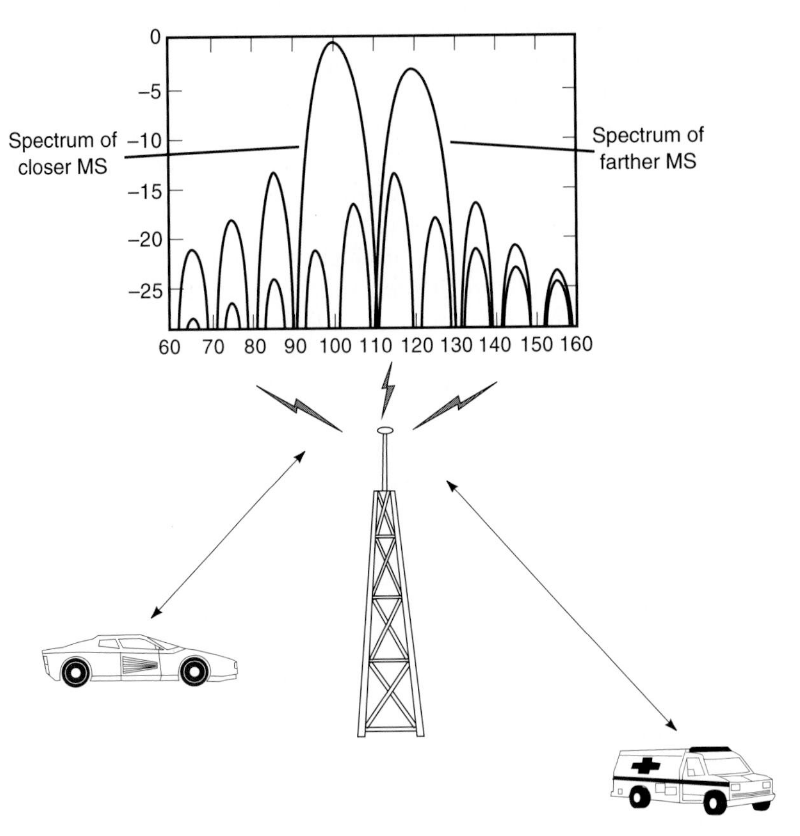

Figure 3.2 Illustration of adjacent channel interference.

band signal power in voice-band modems is not as critical and a voice-band modem manufacturer would be satisfied with an out-of-band power of around −40 dB below the main lobe.

3.1.2.4 Resistance to Multipath

Another important issue in the design of a radio modem is sensitivity to multipath. Various modulation techniques have different degrees of resistance to multipath. This was a major issue in the development of the digital cellular and PCS standards, where it was necessary that each standard be written to accommodate the worst-case multipath conditions likely to be encountered by users over the entire geographic region of usage for that standard.

3.1.2.5 Constant Envelope Modulation

Most mobile radio products are designed with Class-C power amplifiers, which provide the highest power efficiency among the common types of power amplifiers. However, Class-C amplifiers are highly nonlinear, and therefore it is necessary that the signal to be amplified has a constant-envelope or as nearly so as is practical. Though analog frequency modulation (FM) mobile radio systems were originally designed for analog voice, they have been extended to digital service simply by feeding baseband data streams to the frequency modulator. The FM modulation always provides a constant envelope for the transmitted signal, and the information is contained in the variations of the frequency of the transmitted signal. A popular modulation technique developed this way is Gaussian minimum shift keying (GMSK) which will be described in the next section. An FM signal by its very nature has a constant-envelope, but it is not spectrally efficient due to its large side lobes. Thus as the need for greater bandwidth efficiency had grown, efforts have been made to design modulation schemes that are less wasteful of bandwidth while preserving (or nearly so) the constant-envelope nature of FM. The $\pi/4$-quadrature phase shift keying (QPSK) modulation has struck a good balance, and it has emerged as another popular transmission technique in wireless networks. This modulation technique is also described in the next section.

3.2 APPLIED WIRELESS TRANSMISSION TECHNIQUES

As we saw in Chapter 1, 1G wireless cellular and cordless telephone systems used analog FM modulation. With the emergence of 2G wireless networks, digital techniques replaced analog modulation. To increase the capacity, analog voice was source coded into digital format at the mobile terminal for digital transmission, then POTS transmits the analog voice to the network where it is digitized for long haul transmission. Speech coding at the terminal also facilitates the integration of voice and data services in a single terminal. After the emergence of 2G systems, digital transmission has become the dominant choice for wireless communication networks. Therefore, in the rest of this chapter we describe digital transmission techniques applied to modern wireless networks.

Popular digital wireless transmission techniques can be divided into three categories according to their applications. The first category is pulse transmission techniques used mostly in IR applications and more recently applied to the so-called *impulse radio* or ultra wideband (UWB) transmission [SCH00]. The second category is basic modulation techniques widely used in TDMA cellular, as well as a number of mobile data networks. The third category is spread spectrum systems used in the CDMA, as well as wireless LANs operating in ISM bands. More recently, new transmission technologies are emerging to increase the data rate of all these systems to support broadband access for popular Internet access and other data-oriented applications. Variations of the spread spectrum technology and OFDM have been adopted by WLAN standard organizations and are considered for incorporation into the future voice-oriented cellular networks. In the following sections, we provide a detailed description of all these popular modulation techniques to provide the reader with an understanding of the applied wireless physical layer alternatives.

3.3 SHORT DISTANCE BASEBAND TRANSMISSION

In baseband transmission, the digital signal is transmitted without modulating with a carrier. Without a carrier signal, multiple channels cannot be accommodated in an FDM format, and consequently each user occupies the entire available bandwidth of the medium. With one user occupying the entire band, the designer does not need to pay attention to the out-of-band radiation. However, to arrange a multiuser environment, one has to resort to innovative techniques other than the simple, traditional FDM. There are two basic steps in baseband transmission: line coding and pulse modulation. In the first step, the digital data stream is line coded to facilitate synchronization at the receiver and avoid the DC offset during transmission. Baseband line coded signaling is commonly used in short distance wired as well as wireless applications. Wired and IR-based WLANs often only employ line-coded baseband transmission. In pulse modulation, the transmitted information is coded into amplitude, location, or duration of a pulse shape. Pulse-modulation baseband transmission is commonly used for low-speed IR data communications such as remote controls or connections between personal computers (PCs) and the printers or keyboards. More recently UWB pulse modulation is being considered for very low power short-range radio communications.

If the data stream produced by a computer is applied directly to the wires, the receivers will have difficulty in sychronizing with the transmitted symbols. To provide better synchronization at the receiver, the format of the incoming data stream is modified before transmission. This modification process is often referred to as *line coding.*

In wired applications, baseband signaling using differential Manchester line coding is used in the IEEE 802.3 Ethernet, the dominant standard for LANs, as well as IEEE 802.5 Token Ring, the competitor of Ethernet in the early days of the LAN industry. In wireless applications baseband transmission with line coding is popular in high-speed diffused and directed beam IR wireless LANs.

Example 3.1: Manchester Encoding in an IR Transmitter

Figure 3.3 shows the Manchester code implementation of an IR transmitter used in a number of IR-based WLANs. The received nonreturn to zero (NRZ) digital stream is first Manchester encoded. The line coded signal is then intensity modulated by the emitted IR light by simply turning the transmitted light to on and off positions. The receiver consists of a simple photosensitive diode detecting presence and lack of the light and either producing or not producing an electric signal that is then amplified and used as the received signal. The Manchester coded data double the transmission rate but provide one transition per bit. These transitions are important for the receiver because the receiver uses them to synchronize its clock with that of the transmitter clock. In the NRZ transmission if we have a long stream of "0"s of "1"s, the receiver loses the reference timing of the transmitter.

Base band pulse modulation techniques are usually divided into pulse-amplitude-modulation (PAM), pulse-position-modulation (PPM), and pulse-

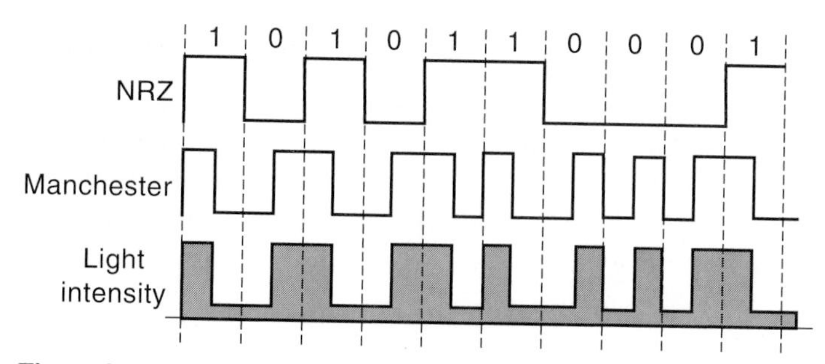

Figure 3.3 Manchester code implementation of an IR transceiver.

duration- or width-duration-modulation (PDM or PWM). As the names suggest, in PAM, PPM, and PDM the transmitted information is signaled through the amplitude, position (location), and duration of a basic pulse shape. Because wireless channels suffer from extensive amplitude fluctuations caused by fading and near-far problems, PAM is not popular in wireless operations, leaving variations of PPM and PDM as favorite choices for pulse modulation over wireless channels.

Example 3.2: Practical Implementation of PPM

A practical implementation of PPM and PDM used in applications such as connecting a keypad to a computer is shown in Fig. 3.4. The data stream is encoded to a pulse at the start of a bit to represent a "1" and a pulse in the middle to represent a "0" to generate the PPM signal. Rather than one single pulse, multiple narrow pulses are transmitted to code the transmission of digitized information. When detected by the photosensitive diode at the receiver, multiple narrow pulses will produce a single continuous pulse that is close to what would have been received if a single pulse was transmitted. However, with multiple narrow pulses, the required transmission power is smaller because the light is on for a shorter period of time. Therefore, multiple pulse transmission saves the life of the battery at the transmitter. In Figure 3.4 one of the options uses three narrow pulses and the other one uses nine narrow pulses. A disadvantage of using several pulses per symbol is the large bandwidth occupied by each pulse. However, this is not too important in IR applications.

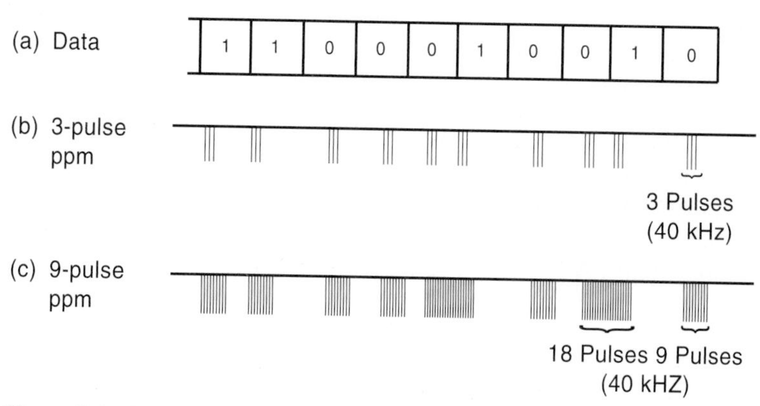

Figure 3.4 Practical implementation of PPM.

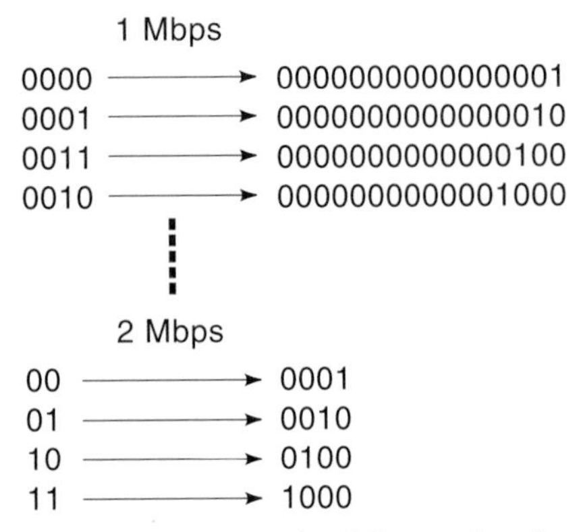

Figure 3.5 PPM using 250 ns pulses for IEEE 802.11.

Example 3.3: The IEEE 802.11 IR Standard

The IEEE 802.11 standard specifies a standard physical layer for high-speed diffused IR medium using PPM. The wavelength range specified by the standard is 850 nm–950 nm. The basic data rates are 1 and 2 Mbps which are consistent with the other two options using spread spectrum technology in ISM bands. As shown in Figure 3.5, the 1 Mbps physical layer uses 16-PPM, and the 2 Mbps uses 4-PPM. The width of each pulse for both cases is 250 ns. In 16-PPM, a 250 ns pulse occupies one of 16 positions, the duration of one symbol being $16 \times 250 =$ 4 microseconds. Each symbol corresponds to four bits (16 symbols => $\log_2 16 =$ 4 bits/symbol). The result is 4 bits being transmitted every 4 microseconds or a data rate of 1 Mbps. Similarly, in 4-PPM, two data bits are transmitted every 1 microsecond for a data rate of 2 Mbps.

3.4 UWB PULSE TRANSMISSION

Impulse radio [SCH00], [MIT01] has recently attracted considerable attention for short-range communications. In this technique, a very narrow width (on order of a few tenths of a nanosecond) and low power (high-duty cycle of several hundreds of nanoseconds) pulse are used for information transmission. The spectrum of this pulse obviously occupies a very wide band (several GHz), and for that reason this technology is sometimes referred to as ultra wide band (UWB). The spectral height of the UWB signal is very low because a small transmission power is spread over a large bandwidth. Therefore, UWB signals can be designed to coexist with existing radio systems. As we saw in Chapter 2, signal fading is caused by the overlap of the received signal from different paths. Because of its extremely wide bandwidth and large duty cycle, the UWB signal isolates (resolves) multipath components, resulting in a stable received power signal with minimal fading effects. Implementation of the transmitter does not involve modulation, and if carefully designed, it can be much simpler than traditional narrowband or wideband spread spectrum systems.

Designers of UWB systems chose a transmission waveform that (1) has such a high bandwidth and signal processing gain that interference from existing systems into the system is negligible; (2) its spectral height is comparable with background noise so that the FCC allows it to coexist with the existing systems; (3) its implementation is easy; and (4) the spectrum looks like a pass band transmission so that there is no DC component in the spectrum.

Example 3.4: UWB Pulse Shape

Recently, Time Domain Corporation (TDC) has developed a range of UWB systems [TDCweb] trademarked as PulsON technology. TDC's UWB transmitters emit ultra-short "Gaussian" monocycles with tightly controlled pulse-to-pulse intervals. TDC has been working with monocycle pulse widths of between 0.20 and 1.50 nanoseconds and pulse-to-pulse intervals of between 25 and 1,000 nanoseconds. These short monocycles are inherently UWB. The pulse shape used by TDC is mathematically given by:

$$v(t) = 6A\sqrt{\frac{e\pi}{3}}\frac{t}{\tau}e^{-6\pi\left(\frac{t}{\tau}\right)^2} \tag{3.1}$$

where A represents the peak amplitude of the pulse, τ is a constant determining the width of the pulse, and t is the time. The spectrum of the monocycle pulse in frequency domain is given by:

$$V(f) = -j\frac{2f}{3f_c^2}\sqrt{\frac{e\pi}{2}}e^{\frac{\pi}{6}\left(\frac{f}{f_c}\right)^2} \tag{3.2}$$

where $f_c = 1/\tau$ is the center frequency of the pulse. Figure 3.6 shows a typical graph of the pulse for $\tau = 0.5$ ns associated with a center frequency of $f_c = 2$ GHz. The half power (3-dB) bandwidth of the pulse occupies around 2 GHz of bandwidth. The applications suggested for this technology include precision geolocation and higher performance radar.

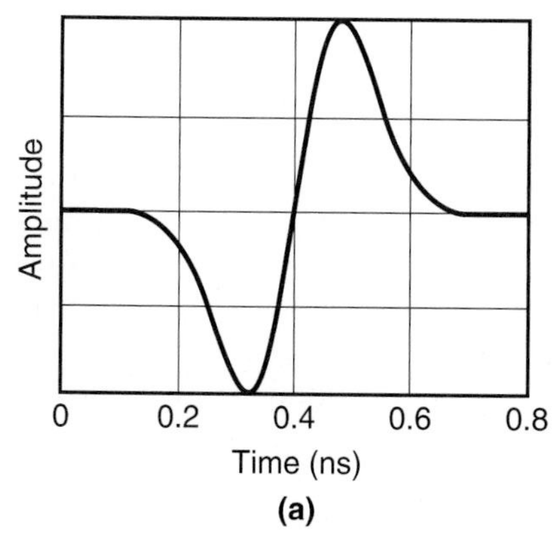

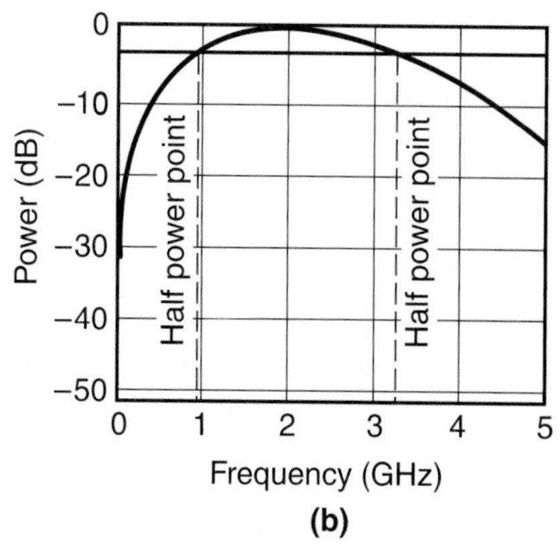

Figure 3.6 (a) The UWB pulse and (b) its spectrum.

3.5 CARRIER MODULATED TRANSMISSION

In broadband signaling, the message signal is mixed with a *carrier signal* at a higher frequency before transmission. The wired computer communication community is dominated by baseband systems, and the word broadband is used for all carrier-modulated systems. In the wireless community, however, the dominant technique is carrier modulation, and the word broadband is used when the transmission rate approaches a very high value. Carrier modulation shifts the spectrum of the transmitted signal to the location of the carrier in the spectrum allowing orderly coexistence of a number of transmissions via FDM. In wired networking, we can always add new wiring for new applications. We have telephone wiring for our telephones, cable for video distribution, and LAN wiring for data communications. In wireless networking, we have only one medium (air) that must be shared among a variety of applications using FDM. Our television, AM/FM radios, cordless and cellular phones, remote controls, and other wireless appliances share the same medium and are separated only by their carrier frequencies. Therefore, carrier modulation and FDM is the cornerstone of multiservice radio communications.

Carrier modulation also shifts the frequency operation to higher values providing better coverage and reducing the length of the antenna to a practical size. The size of the antenna is usually on the order of the transmission wavelength. As the carrier frequency is increased, the wavelength and, consequently, the size of the antenna, reduce. As the carrier frequency increases, there are wider frequency bands available to support higher data rates. However, with increasing carrier frequency, the design of RF circuits becomes more challenging, and also in-building penetration of the signal becomes smaller. Designers of wireless modems need to make a compromise among availability of frequency of operation, required bandwidth, coverage of indoor and outdoor areas, and the cost of implementation. There are two classes of carrier-modulated technologies that are used in a wireless network: traditional radio modem and spread spectrum modems. In the next four sections, we provide a description of traditional and spread spectrum carrier modulated techniques that are applied to voice- and data-oriented wireless networks.

3.6 TRADITIONAL DIGITAL CELLULAR TRANSMISSION

A number of alternatives for digital transmission over radio channel were evaluated for modern wireless communications in the past two decades. In their broadest form, carrier modulation techniques can be divided into three categories of amplitude-, frequency-, and phase-modulation techniques. Radio modems need low side lobes for operation using FDM, and they have to cope with extensive amplitude fluctuations caused by fading. As a result, amplitude modulation techniques are not desirable, and frequency and phase digital modulation techniques have emerged as the traditional techniques in this industry. The two most commonly used modulation techniques in traditional wireless networks are GMSK and $\pi/4$-QPSK modulation techniques. GMSK is adopted by GSM, today's most popular digital cellular standard, and a number of other wireless data networks such as CDPD and Mobitex. The $\pi/4$-QPSK modulation is adopted by the North American TDMA digital cellular standard, IS-136, and the Japanese digital cellular, JDC, as well as the European mobile data service called TETRA. In the next two subsections, we address evolution of digital frequency modulation techniques to GMSK and digital phase modulation techniques to $\pi/4$-QPSK.

3.6.1 Digital Frequency Modulation and GMSK

As we noted earlier in this chapter, FM is the predominant form of analog modulation used in the mobile radio industry. Digital FM modulation is referred to FSK which forms a simple and popular method for wireless communications. Figure 3.7(a) shows the basic concept behind binary FSK modulation. The binary baseband data stream is encoded into two different frequencies before transmission in the channel. To implement this modulation in it simplest form, as shown in Figure 3.7(b), one can input the binary data stream directly to a traditional analog FM transmitter and use an analog FM receiver to demodulate the signal at the receiver. In its ideal form, an analog FM transmitter linearly maps the instantaneous amplitude of the message signal to a constant amplitude sinusoid with varying frequency at the output of the transmitter. A binary input signal takes only two levels of amplitude, so the output would be a constant envelope signal with two frequencies associated with the two different levels of signal. If the baseband input signal has multiple levels, representing PAM symbols, the output would be still a constant envelope signal with as many frequencies as the number of levels in the PAM signal.

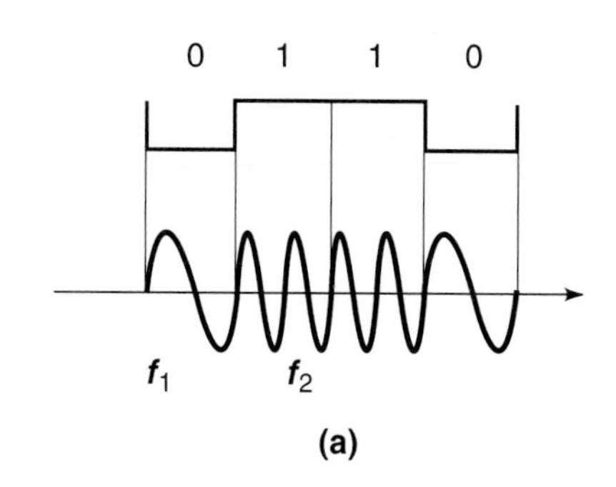

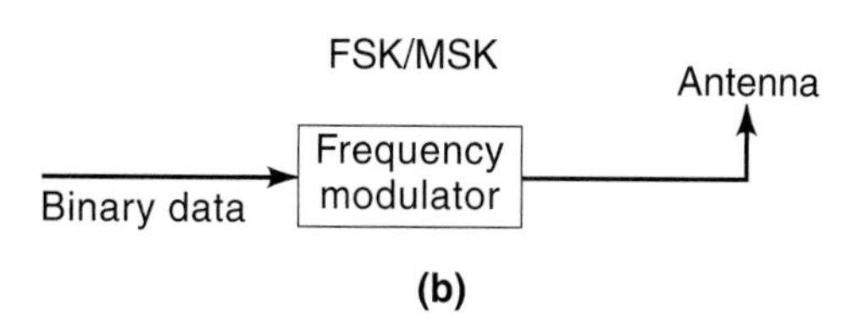

Figure 3.7 (a) Basic concept of FSK; (b) Implementation of FSK using an FM transceiver.

Example 3.5: Four-Level FSK

A 4-FSK-modulation technique was adopted for Altair, the first WLAN product that operated in the 18–19 GHz bands. This pioneering WLAN product operated at a bit rate of 10 Mbps and employed very advanced signal processing and fabrication techniques.

An important parameter in the design of FSK modems is the frequency spacing between the tones. This distance is representative of the occupied bandwidth of an FSK signal, and to maintain optimal detection at the receiver, it should take specific values that ensure orthogonality of the transmitted symbols. For noncoherent detection (when the receiver is not locked to the phase of the transmitted carrier), FSK modems use a minimum distance between the tones of $1/T$ where T is the duration of the transmitted data symbols. For coherent demodulation, the distance between the tones can be reduced to $1/2T$, that is, the minimum acceptable distance between the tones that ensures the orthogonality of the transmitted symbols. The FSK modulation with minimal tone distance of $1/2T$ is referred to as *minimum shift keying* (MSK), which is a very popular transmission technique in radio communications.

To make an MSK signal even more attractive for radio communications, as shown in Figure 3.8, the transmitted baseband signal is filtered before FM modulation to further reduce the side lobes. The most popular filters used for this implementation are Gaussian filters and associated modulation technique is referred to as Gaussian MSK or GMSK, which is one of the most popular modulation techniques in 2G wireless networks. The transmitted signal at the output of the FM modulator is still a constant envelope FM signal that avoids nonlinearities that can be introduced by the power amplifiers. In the time domain, the Gaussian filter smoothes sharp transitions of the voltage levels. As a result, rather than immediate changes of the tone frequency at the output of the FM modulator, we will have a smooth transition from one tone frequency to another that reduces the side lobes of the transmitted FM modulated signal. Because phase represents variations (the derivative) of the carrier frequency in time, modulation with gradual transition of the frequency is referred to as *continuous phase modulation* techniques. Desirable modulation techniques for radio channels are *constant-envelope continuous phase modulation* techniques that are only slightly affected by nonlinearity and have low side lobes. GMSK is an ideal example of a constant envelope continuous phase modulation technique.

An important factor that affects performance in GMSK is the time-bandwidth product B_bT. Here, B_b is the 3-dB bandwidth of the Gaussian filter and T is the symbol duration. Figure 3.9 shows the spectrum of the GMSK signal for va-

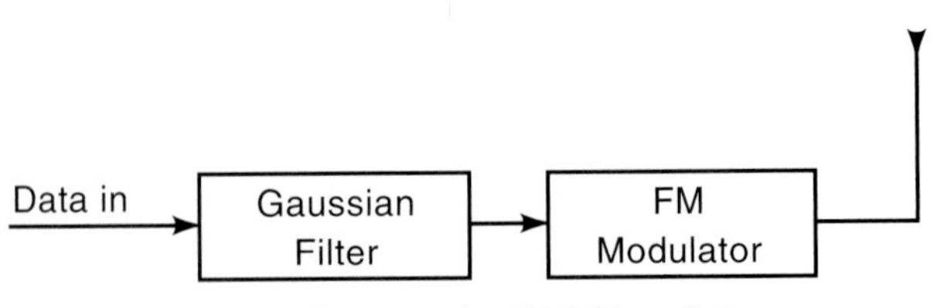

Figure 3.8 Block diagram of a GMSK modulator.

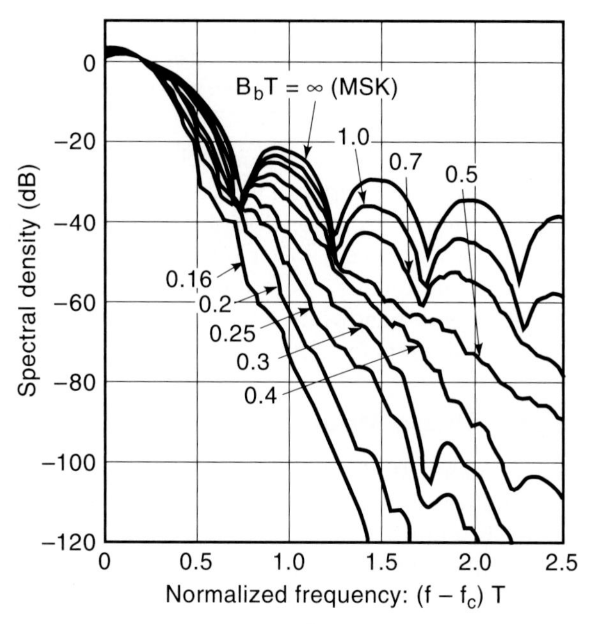

Figure 3.9 Spectra of GMSK signals for different B_bT products.

riety of B_bT, normalized 3-dB bandwidths, of the Gaussian filter. For $B_bT = \infty$, the bandwidth of the Gaussian filter is infinity (no filtering), and the system is indeed an MSK system. As the bandwidth of the filter becomes narrower, the power in the side lobes of the transmitted signal, and consequently adjacent channel interference, reduces. On the other hand, reduction of the bandwidth of the filter further smoothes the transition between levels in the time domain which increases the probability of erroneous detection at the receiver. The designer of the modem should decide on a compromise between the adjacent channel interference and detection error rate. The GSM voice-oriented standard recommends a B_bT value of 0.3, and the CDPD services recommend a B_bT value of 0.5.

Depending on the parameters of the filtering and the type of FSK-based systems, a number of modulation techniques with a variety of normalized bandwidth occupation are adopted by a variety of standards. The following example compares the bandwidth efficiency of these systems.

Example 3.6: Bandwidth Efficiency in Various Technologies

The ARDIS mobile data services use a 4-FSK modulation to support a data rate of 19.2 kbps in 25 KHz channels. The normalized bandwidth occupation or bandwidth efficiency of this system is 19.2/25 = 0.77 bits/sec/Hz. Mobitex mobile data services support 8 kbps in 12.5 kHz of bandwidth using GMSK modulation with a bandwidth efficiency of 0.64 bits/sec/Hz. CDPD uses GMSK to provide 19.2 kbps over a 30 kHz channel with a bandwidth efficiency of 0.64 bits/sec/Hz. The DECT system uses GFSK (Gaussian frequency shift keying) modulation to support 1.152 Mbps over a 1.728 MHz channel with a bandwidth efficiency of 0.67 bits/sec/Hz.

CT-2 cordless telephony uses GFSK to support 72 kbps over a 100 kHz channel with the bandwidth efficiency of 0.72 bits/sec/Hz. GSM uses GMSK with a data rate of 270.833 kbps in a 200 kHz band, giving an efficiency of 1.35 bits/sec/Hz.

Higher bandwidth efficiencies are achieved at the expense of more complex coherent implementations and lower uncoded bit error rate requirement. Data services often have more restrictions on the bit error rate than voice-oriented networks. Cellular phones accept lower qualities than cordless telephones that were designed for wireline quality operation. The diversity in requirements combined with the availability of bandwidth for a particular service has created this diversity in modulation parameters for different standards.

3.6.2 Digital Phase Modulation and $\pi/4$-QPSK

In digital phase modulation or PSK, the baseband information signal is encoded in the phase of the transmitted signal. Figure 3.10 illustrates the basic operation of a BPSK system. In Figure 3.10(a) two symbols used for binary communication are two sinusoids with 180-degree phase difference. The phase shift changes according to the voltage level of the baseband information signal.

It is customary to represent the magnitude and phase of the transmitted symbols in a complex coordinate system referred to as *signal constellation.* The signal constellation shows the unique characteristics of the modem necessary for calculation of the error rate in additive noise channels. The error rate is a function of the distance between the two points in the constellation and the level of the noise disturbing the channel.

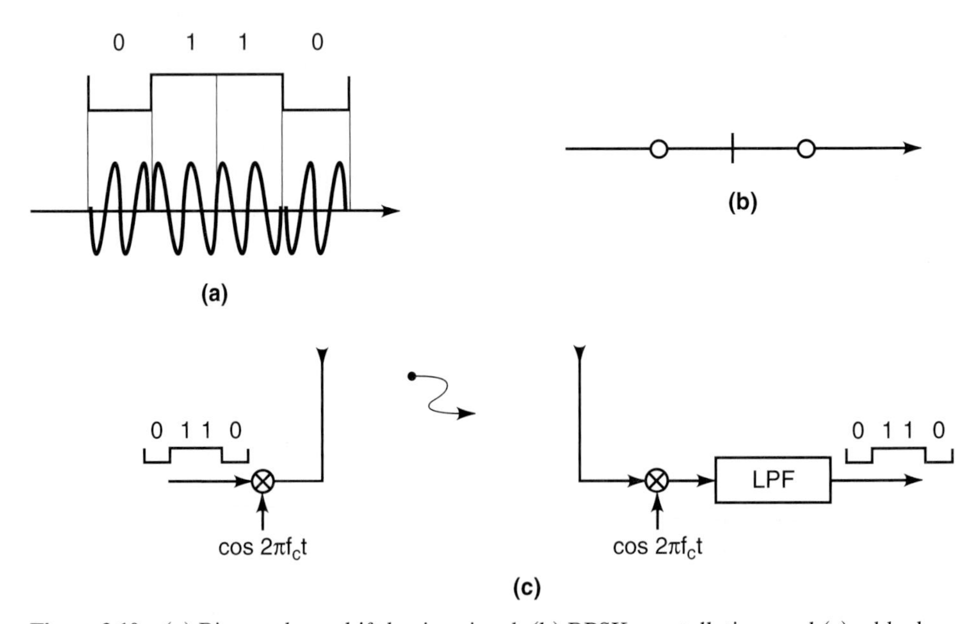

Figure 3.10 (a) Binary phase shift keying signal, (b) BPSK constellation, and (c) a block diagram of a BPSK system.

Figure 3.10(b) shows the signal constellation of the BPSK represented by two equal amplitude symbols with opposite polarity. Figure 3.10(c) represents a simple block diagram for implementation of a basic BPSK modem. The received baseband signal is simply multiplied (mixed) with a carrier frequency, and after filtering it is sent to the antenna. At the receiver the signal is multiplied with the carrier signal that is phase locked to the transmitter carrier (coherent demodulation) and then passed through a low pass filter to eliminate the unnecessary higher frequency components of the signal and discriminate the transmitted waveform. The resulting transmitted signal is a constant envelop signal shown in the lower part of the Figure 3.10(a).

Noncoherent detection of the PSK signal is also possible. The basic idea behind noncoherent detection of the PSK signal is to use the carrier in the current bit as the reference carrier of the following bit. To make this happen, the transmitted bits are differentially encoded. In differential encoding, rather than sending the actual bits the value of the exclusive-OR of the consequent bits are transmitted. Therefore the two phases that are transmitted represent the changes in the polarity of the current bit with respect to the previous bit. If this arrangement is made at the transmitter, the received signal can be detected noncoherently with the circuit shown in Figure 3.11. The value of the delay is equal to the duration of a transmitted bit. This way the carrier of the previous bit is used as the reference for detection of the current bit. The advantage of differential PSK (DPSK) over coherently detected BPSK is that the receiver circuitry does not need to recover the phase of the transmitted carrier. The disadvantage, as shown in Figure 3A.3, is that there is about 1 to 2 dB degradation in performance compared with coherent BPSK.

In a manner similar to multisymbol FSK modulation, one can design multiphase PSK modulation schemes. Four-phase PSK, often referred to as quadrature PSK (QPSK), is commonly used in radio modems. The two-dimensional signal constellation for QPSK is shown in Figure 3.12(a). Four transmitted symbols assume four different phase values of 0°, 90°, 180°, and 270°, each representing a block of two information bits. Figure 3.12(b) illustrates the basic principle of the 2-D modulation techniques used for transmission of the complex symbols. The basic structure is the same as the BPSK system except that here we have two branches of BPSK modems—one modulated over a sine wave and the other modulated over a cosine wave. Because the carriers in the two branches are orthogonal to one another at the receiver, the transmitted sine does not go through the cosine branch and vice versa. In other words, this structure has two independent modems operating over

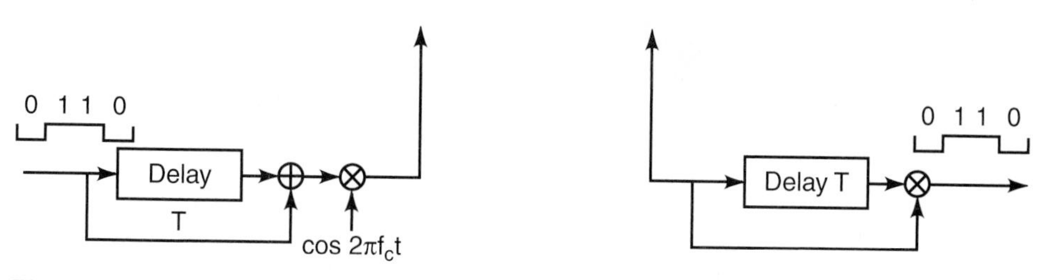

Figure 3.11 Differentially encoded BPSK.

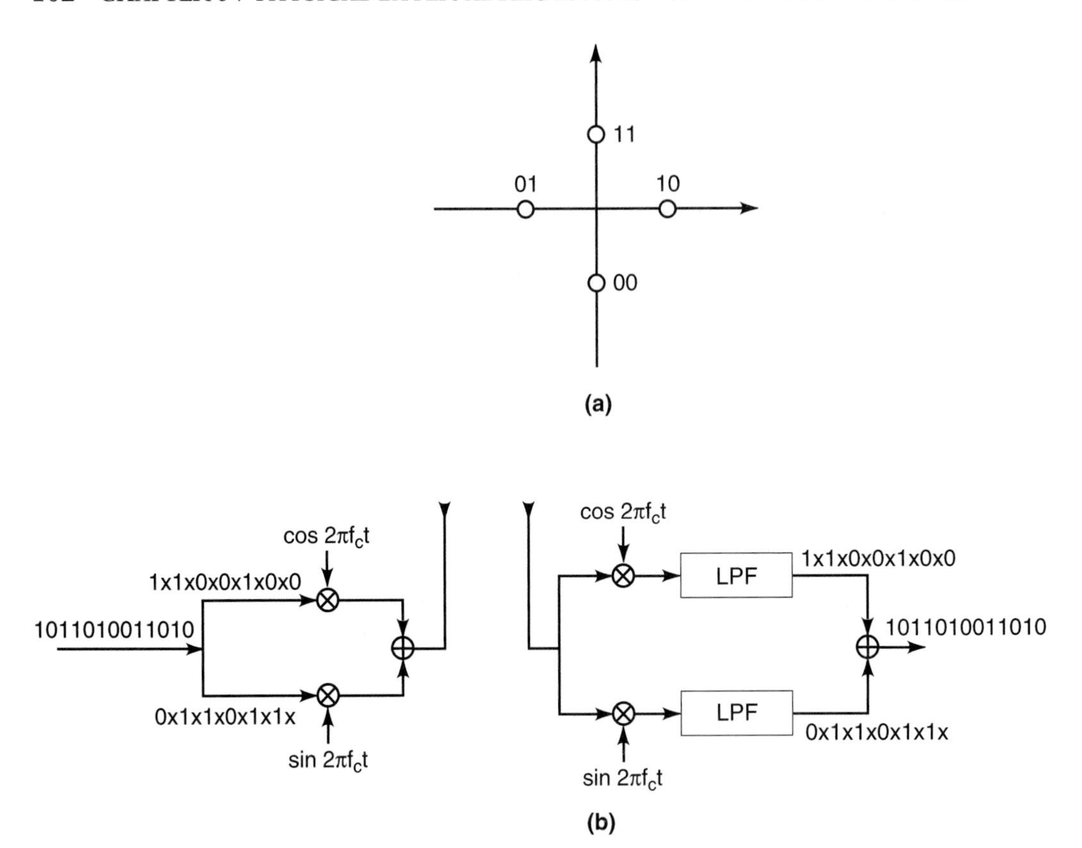

Figure 3.12 (a) Signal constellation for QPSK and (b) modulation scheme for transmission of QPSK.

the same bandwidth and separated using their orthogonal[1] carriers. With two independent channels, we can always form a complex transmission system that can implement phase modulation. The advantage of this implementation is that we have two channels sharing the same bandwidth, resulting in a system with twice the bandwidth efficiency. The amplitude of the real part of a symbol is coded in the amplitude of the transmitted pulse over the cosine branch, usually referred to as in-phase or I-channel, and the amplitude of the imaginary part of the same symbol is coded in the amplitude of the transmitted pulse over the sine branch, often referred to as the quadrature-phase or Q-channel.

Figure 3.13 shows the normalized frequency spectrum (the actual spectrum with the abscissa divided by the bit rate of the channel and shifted from the carrier frequency to zero) of the transmitted signal for BPSK, MSK, and QPSK. Two important parameters in evaluating the spectrum of a radio modem are the width of the main lobe (i.e., a measure of the occupied bandwidth) and the peak of the side-lobes, reflecting the level of adjacent channel interference. The bandwidth of

[1]Orthogonality is an important concept in digital communications. Essentially this means that two orthogonal signals cause zero interference *after* processing at the receiver. This processing at the receiver usually implies integration over a symbol period.

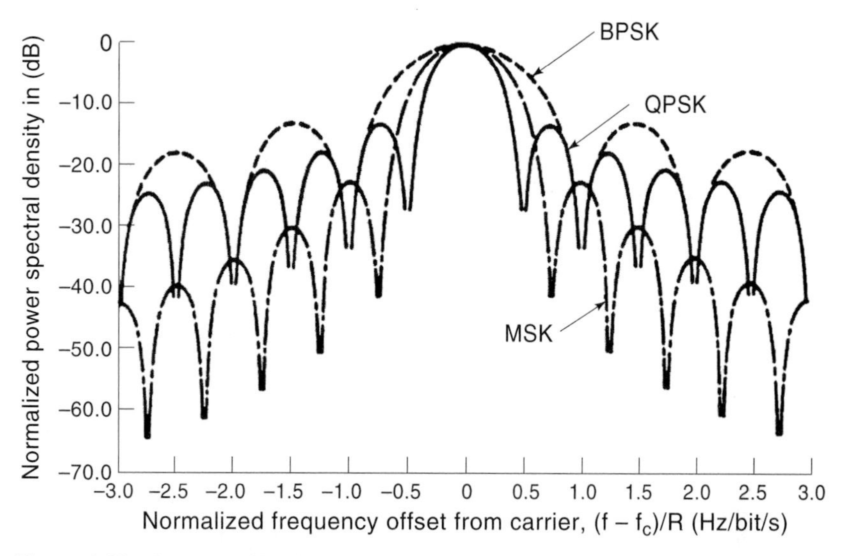

Figure 3.13 Spectra of BPSK, MSK, and QPSK.

QPSK is half of that of BPSK, and MSK lies somewhere in between. The bandwidths here are normalized to the data rate. The sidelobes of MSK, however, are more than 10 dB lower than those of BPSK and QPSK. The discussion provided in this section explains why QPSK and MSK are favored as the basis for radio modems. However, for reliable multiuser, multichannel wireless communications, we need to further reduce the height of the sidelobes. In the previous section, we introduced simple Gaussian filtering and continuous phase GMSK modulation as an improved version of the constant envelope MSK modulation. Filtering techniques can also be applied to QPSK systems to further control the sidelobes. As a matter of fact, if the baseband data stream is kept in its rectangular form, the sidelobes of the QPSK are very high (around 13 dB below the peak), which cannot attract attention for any serious radio application. All traditional QPSK and in general 2-D modulation systems use pulse-shaping filters (PSF) to control the sidelobes.

Figure 3.14 shows the basic implementation of a PSF in BPSK modem. The filter can be placed as a low-pass filter before mixing the signal with the carrier or as a band-pass filter after mixer; in either case the sidelobes of the transmission bandwidth can be controlled by this filter. Also in practice, a pair of identical filters at the receiver and transmitter is used for pulse shaping. Such a pair is referred to as a pair of matched filters. As shown in Figure 3.14, the baseband rectangular pulses are changed to pulses with smoother transitions to control the sidelobes. At the receiver side, the pulse shapes are sampled at their peak values, and then the rectangular transmitted pulses are reconstructed based on the value of the sample. Ideally the best PSF filter is an ideal low-pass filter that passes all the frequencies in the band with equal gain and eliminates all other frequencies and consequently has no sidelobes and adjacent channel interference. However, the time domain pulse associated with an ideal filter is a sinc pulse that has strong sidelobes in time domain

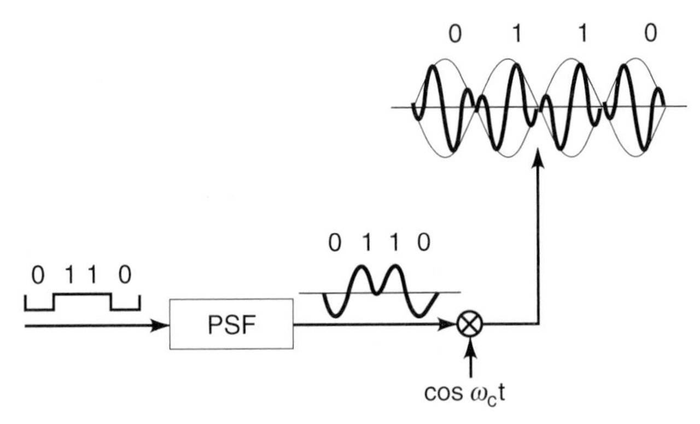

Figure 3.14 Basic implementation of PSFs for BPSK modems.

and is extremely difficult to implement. Sidelobes in time domain cause ISI, which acts as a source of noise to increase the bit error rate of the received signal. *Raised cosine* pulses provide a practical compromise between ISI and the occupied bandwidth.

Figure 3.15 shows the time domain pulse and the spectrum of the raised cosine pulses. The spectrum consists of two half-cosines at the two sides and a flat portion in the middle. The parameter β, referred to as the roll-off factor, takes values between zero, for ideal case, and one as the maximum where the flat part disappears. It controls the expansion of the spectrum. As the spectrum stretches further from $1/2T$ toward $1/T$ the sidelobes of the time domain pulse reduces, resulting in smaller values of ISI. In practical implementations, values of β between 0.2–0.5 are used in a variety of wired and wireless modems that bring the sidelobes 40–60 dB below the main lobe. Pulses shown in Figure 3.15 can be sent every $2T$ seconds where the main lobes of the received pulses do not overlap. For this case, only one sample per pulse is adequate for digital processing at the receiver side. If the receiver can manage several (typically eight) samples per main lobe, as shown in Figure 3.16, the raised cosine pulse can be sent in an overlapping manner every T seconds, reducing the transmission bandwidth by half. Overlapping and nonoverlapping pulses in the time domain in QPSK can be thought to be similar to the spectral overlap of FSK versus MSK. There is a twofold increase in bandwidth efficiency involved that is offset, however, by the complexity of the receiver design. Depending on the roll-off factor and overlapping or nonoverlapping pulses, a variety of QPSK modulation formats occupy different normalized bandwidths. An example will illustrate the situation.

Example 3.7: Bandwidth Efficiency in QPSK-Based Systems

The IS-95 standard uses QPSK modulation for its spread spectrum modulated baseband signal. The transmission bandwidth per carrier is 1.25 MHz, and the chip rate supported by the system is 1.2288 Mcps. The normalized bandwidth occupancy or efficiency of this QPSK system is 0.98 chips/sec/Hz. The IEEE 802.11 system occupies a bandwidth of 26 MHz for a spread spectrum QPSK system with a chip rate of 22 Mcps that yields a bandwidth efficiency of 0.85 chips/sec/Hz. In

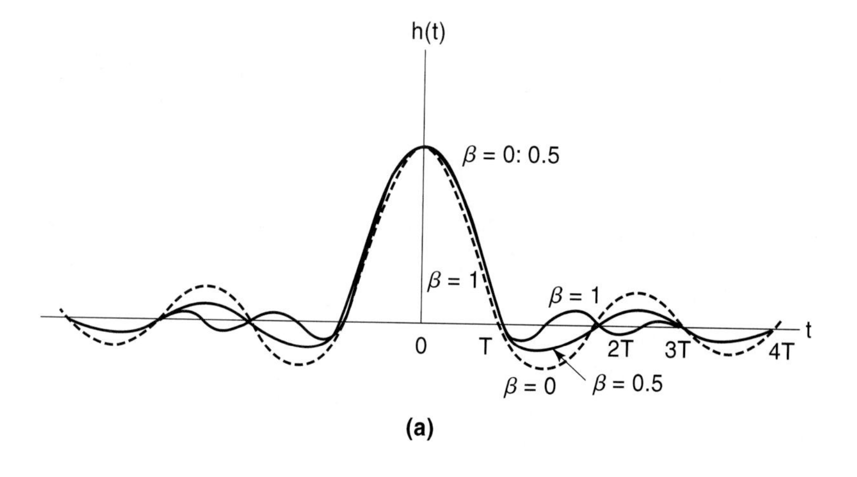

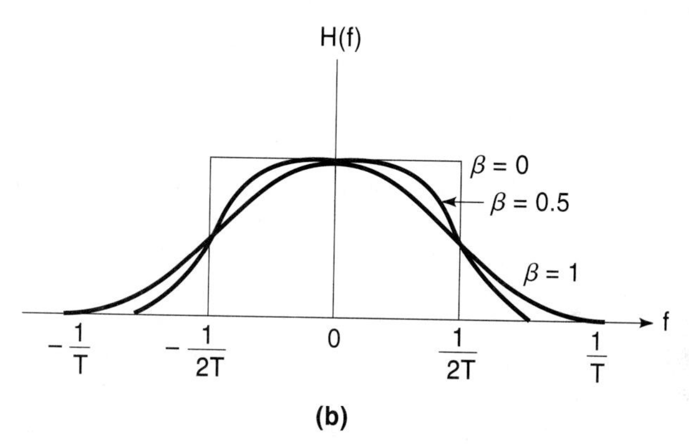

Figure 3.15 Time domain raised cosine pulse and its spectrum.

both of these spread spectrum systems, nonoverlapping pulses are used to implement the system. The overlapping pulse π/4-QPSK system in IS-136 TDMA standard uses a nominal bandwidth of 30 kHz to support a data rate of 48.6 kbps per channel that has the bandwidth efficiency of 1.62 bits/sec/Hz. The Japanese 2G digital cellular system uses overlapping pulses based on π/4-DQPSK to support a 42 kbps data rate over a 25 kHz channel resulting in a 1.68 bits/sec/Hz bandwidth efficiency. The PACS system uses π/4-QDPSK with a bandwidth efficiency of 500 kbps/350 kHz = 1.42 bits/sec/Hz. If we compare this example with Example 6, we can easily see that PSK-based modems are capable of supporting slightly higher bandwidth efficiencies compared with the equivalent FSK-based systems.

So far we have examined the similarities of the effects of filtering to reduce the sidelobes for both QPSK and GMSK modems. However, there is a major difference between the filtered QPSK and GMSK. In QPSK, when we use filters the transmitted signal is no longer a constant envelope signal. Also, QPSK has discontinuities of up to 180 degrees in the phase of the carrier. Therefore, filtered QPSK signals have neither a constant envelope nor a continuous phase. This gives a

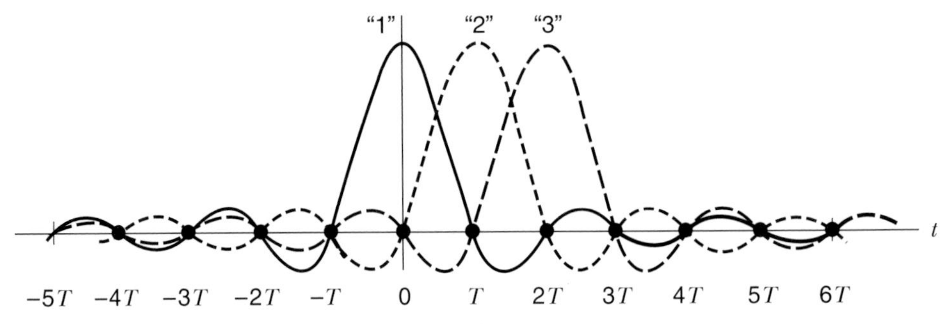

Figure 3.16 Transmission of raised cosine pulses.

practical edge for GMSK, compensating for its lower bandwidth efficiency. In the past few decades, a number of modifications to QPSK have been examined to improve its envelope and phase characteristics. The two most popular systems used in the wireless industry are offset or staggered QPSK (OQPSK or SQPSK) and π-/4-QPSK.

3.6.2.1 Further Improvements to QPSK

In an OQPSK modulator, instead of applying the source data bits to the two branches simultaneously every T seconds, as shown in Figure 3.12, the two branches are offset by $T/2$ seconds. The benefits of this scheme are twofold: (1) The envelopes of the in-phase and quadrature-phase signals overlap one another, resulting in fewer fluctuations of the amplitude of the transmitted signal (the peaks of the envelope in one branch occur in between the peaks of the other branch); and (2) the changes in the phase of the carrier in the two branches occurs every $T/2$ seconds rather than every T seconds. This way, every $T/2$ seconds, the phase is shifted by ± 90 degrees, and it avoids the ± 180-degree phase shift that was possible in standard QPSK. Therefore, OQPSK provides a better constancy of envelope and a better continuity in phase than QPSK which improves the power requirements and further reduces the sidelobes. A drawback of OQPSK modulation is that it is difficult to develop noncoherent receivers for this modulation technique, and it is more sensitive to multipath fading channels with large Doppler shifts [PAH85a], [FEH91]. The search for nonstaggered modulation schemes having low postfiltering amplitude variations led to work by Akaiwa and Nagata [AKA87] and others, including [LIU89], [GOO90], [LIU90], on π/4-QPSK modulation.

Simply described, π/4-QPSK is a form of QPSK modulation in which the QPSK signal constellation is shifted by 45 degrees each symbol interval T. This means that the phase transitions from one symbol to the next are restricted to ± 45 degrees and ± 135 degrees. By eliminating the 180-degree transitions of QPSK, the amplitude variations after filtering are significantly reduced. Figure 3.17(a) shows the eight possible phase states of the signal constellation of the π/4-QPSK modulation. The eight phases represent the four-phase QPSK constellation in its two shifted positions. The four dotted lines radiating from each point on the circle indicate the allowed phase transitions. In the implementation depicted, the modulation is implemented using a sine wave pulse shaping [FEH91]. Figure 3.17(b) shows the spectra measured for two versions of π/4-QPSK, nonlinearly amplified. The upper trace is strict-sense π/4-QPSK, while the lower trace is the spectrum for a sine wave pulse-shaped π/4-QPSK.

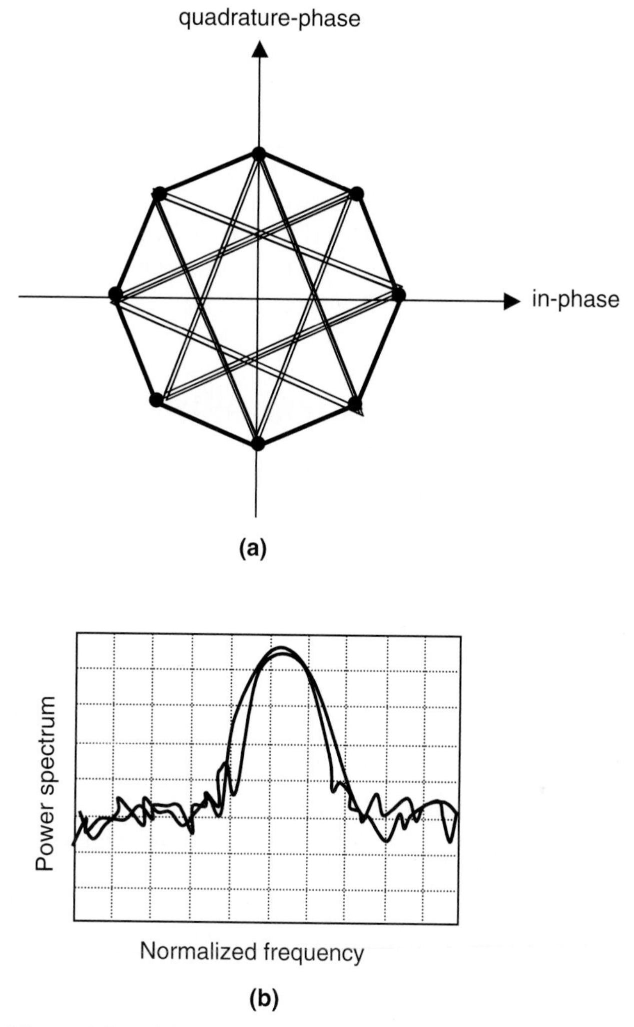

Figure 3.17 (a) Signal constellation and phase transitions of $\pi/4$-QPSK and (b) spectra for two different pulse shapes.

Thus $\pi/4$-QPSK modulation provides the bandwidth efficiency of QPSK together with a diminished range of amplitude fluctuations. Furthermore, the $\pi/4$-QPSK modulation has the advantage that it can be implemented with coherent, differentially coherent, or discriminator detection [LIU90]. These multiple advantages of $\pi/4$-QPSK led to its adoption for the North American Digital Cellular TDMA standard, IS-136, as well as the Japanese Digital Cellular standard [NAK90] and the standards for TETRA [HAI92].

Although it is not essential, $\pi/4$-QPSK modulation is frequently implemented with differential encoding, because this permits the use of differential detection in the receiver, though coherent detection may also be used to achieve optimum performance. The use of differential detection avoids the complexity required to reliably extract a coherent carrier reference under multipath fading conditions. This scheme is termed $\pi/4$-differential QPSK, denoted simply as $\pi/4$-DQPSK.

In summary, GMSK and $\pi/4$-QPSK are the most popular traditional modem technologies used in wireless networks. Overall, considering the bandwidth efficiency, adjacent channel interference, and the ease of implementation, these two techniques are very comparable. It can be shown that it is possible to implement GMSK using the 2-D branches of QPSK modems [PAH95] which reveals a great similarity between the two schemes.

3.7 BROADBAND MODEMS FOR HIGHER SPEEDS

In the previous section, we described popular modulation techniques used in digital cellular networks. These networks are designed for voice or other relatively low-speed modulation techniques for which comprehensive coverage and mobility are the dominant design concerns. In WLANs and point-to-point fixed wireless communications, coverage and mobility are restricted, but achievable data rate is the prime concern. In the past decade, a number of modulation techniques have been examined to support higher speed wireless networks for indoor wireless LANs and outdoor broadband fixed wireless networks, such as local multipoint distribution service (LMDS), used as trunks, providing the last mile access to the PSTN or the Internet. In the late 1990s OFDM emerged as the most popular modulation technique for these applications and is adopted in the IEEE 802.11a and HIPERLAN-2 next-generation WLANs, as well as LMDS and digital audio broadcast (DAB) systems.

3.7.1 Multicarrier, Multisymbol, Multirate OFDM Modulation

What is known as OFDM today combines three transmission principles: multirate, multisymbol, and multicarrier modulation (MCM). However, its name comes from the method of implementation of the MCM, using orthogonality of the adjacent carriers. An MCM system is indeed an FDM system for which a single user uses all the FDM channels together. OFDM is an implementation of MCM that takes advantage of the orthogonality of the channels and develops a computationally efficient implementation based on the Fast Fourier Transform (FFT) algorithm.

3.7.1.1 Multicarrier Modulation

MCM was first evaluated for high-speed voice band modems [HOL63], and it was augmented with FFT implementation for the same application in the early 1980s [PAH88]. It found its way into wireless modems in the early 1990s [BIN90]. The concept here is very simple. Instead of modulating a single carrier at a rate of R_s symbols/sec, we use N carriers spaced by about R_s/N Hz and modulate each of the carriers at the rate R_s/N symbols/sec. The advantage of this scheme is that on a multipath channel the multipath is in effect reduced relative to a symbol interval by the ratio of $1/N$ and thus imposes less distortion in each demodulated symbol. If the symbols are made sufficiently long relative to the multipath spread, reliable demodulation performance can be achieved without the need for any antimultipath signal processing technique. Therefore, in frequency selective multipath fading channels, MCM provides a simpler alternative to single-carrier modulation that needs complex signal processing algorithms at the receiver. A further advantage of MCM is that in frequency-selective fading, the subchannels provide a form of frequency diversity, which can be exploited by applying error-control coding *across symbols* in different subchannels. In the OFDM implementation, this latter technique is referred to as coded OFDM or COFDM. To further improve the performance of an MCM system, one may measure the received signal power in different subchannels and using a feedback channel may change the modulation and/or coding of the transmitted subcarriers to optimize performance. One may adjust the transmitted power per channel to compensate for the frequency selective fading affecting different channels differently. With these features MCM is an ideal solution for broadband communications because increasing the data rate is simply a matter of increasing the number of carriers. The limitations are complexity of implementation and the limitation on the transmitted power. To avoid overlap between consecutive symbols, a time guard is enforced between transmissions of two OFDM pulses that will reduce the effective data rate. Also some of the carriers are dedicated to the synchronization signal, and some are reserved for redundancy.

Example 3.8: MCM in Wireless LANs

Figure 3.18 shows the 64 subchannel implementation of MCM for the IEEE 802.11a and HIPERLAN-2 physical layer specifications. Each channel carries a symbol rate of 250 kilo symbols per second (ksps). We have 48 subcarriers devoted to information transmission, four subcarriers for pilot tones used for synchroniza-

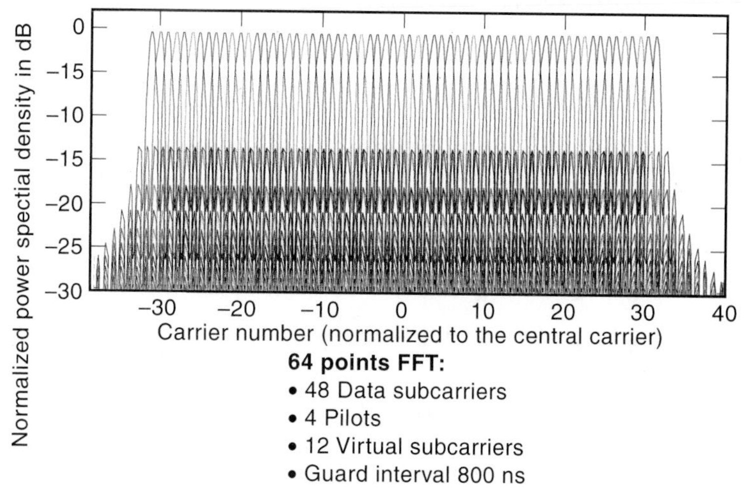

Figure 3.18 OFDM implementation of IEEE 802.11a and HIPERLAN-2 physical layer.

tion, and 12 reserved for other purposes. The occupied bandwidth is 20 MHz, providing a channel occupancy of 20 MHz/64 kHz = 312.5 kHz per subchannel. Therefore, the modulation efficiency is 250 ksps/312.5 kHz = 0.8 symbols/sec/Hz, and the user symbol transmission rate is 48 × 250 ksps = 12 Msps. The bit transmission rate depends on the number of bits per symbol. The guard time between two transmitted symbols is 800 ns compared with the symbol duration of 1/250 ksps = 4000 ns with a time utilization efficiency of 4000/4800 = 83 percent.

3.7.1.2 Multisymbol Modulation

Multisymbol modulation uses multiamplitude and multiphase modulation and coding techniques for increasing the data rate. As we saw in the previous section, traditional radio modems, such as QPSK, are four-symbol systems encoding two bits in one of four transmitted symbols. They have a signal constellation with four points representing the amplitudes and phases of the four distinct symbols. The advantage of these systems is that with the same symbol transmission rate (the same occupied bandwidth) they could double the bit transmission rate compared with BPSK. Multiamplitude, multiphase modulation techniques extend this concept by increasing the number of symbols in the constellation, allowing more encoding of the number of bits per symbols. These symbols are then modulated over a QAM modem that transmits the real and imaginary part of the encoded symbols in the in-phase and quadrature-phase channels of a modem similar to the one shown in Figure 3.12. The number of bits per symbol of a signal constellation represents the increase of the data rate over binary communication systems using a one bit per symbol scheme.

In radio channels, often symbols are encoded with a coding technique with a certain rate that represents the ratio of the actual information rate to the encoded data stream. The coding rate, number of points in the signal constellation, and the symbol rate are used to find the actual information transmission rate of a system.

Example 3.9: Data Rate in QAM Systems

Figure 3.19 shows the 16-QAM (4-bit per symbol) and 64-QAM (6-bit per symbol) signal constellations. If the symbol transmission rate for these constellations are 250 ksps, the bit rate of the 16-QAM is 4 × 250 ksps = 1 Mbps, and the bit

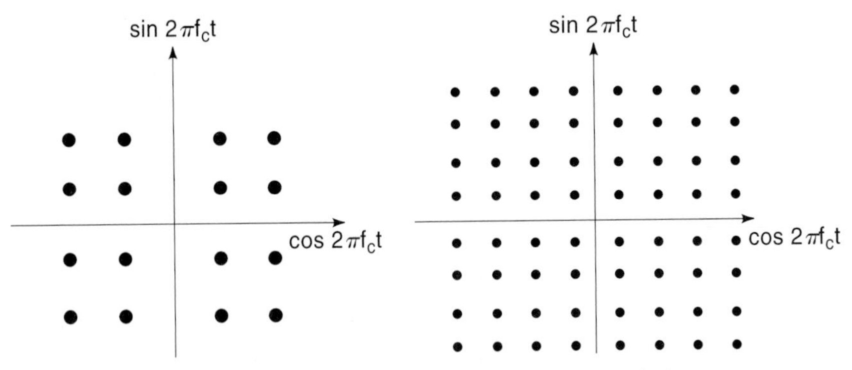

Figure 3.19 QAM constellations for four and six bits per symbol.

rate of the 64-QAM is 6 × 250 ksps = 1.5 Mbps. If the information bits were convolutional encoded with a rate of 3/4, the actual data transmission rates would be 750 kbps and 1.125 Mbps, respectively. These system specifications are used in the IEEE 802.11a/HIPERLAN-2 standards to provide data rates of 36 Mbps and 54 Mbps over the multicarrier structure described in our previous example.

3.7.1.3 Multirate Transmission

Yet another approach to increasing data rate is to use a multirate modem. A multirate modem provides one or more "fallback" modes of operation for increased reliability of communication under degraded channel conditions. The idea here is as follows: If the modulation efficiency is increased (the number of bits per symbol is increased), the required signal-to-noise ratio at the receiver also increases. In voice band modems if the modem is connected to a line with poor characteristics, the data rate is reduced. In the radio modems, as the user moves away from a base or an access point, the received signal-to-noise ratio reduces, and the modem falls to a lower rate, providing reasonable error rates at lower values of the signal to noise ratio. Readers interested in quantitative performance improvements due to multirate transmission in radio channels can refer to [WIN85], [ZHA90], and [PAH95]. Most wireless LAN products and standards have adopted multirate transmission.

Example 3.10: Multirate Transmission in Wireless LANS

The IEEE 802.11/HIPERLAN-2 standard using OFDM modulation that was described in Examples 3.8 and 3.9 uses a number of fallback options. As the distance between the transmitter and the receiver is increased, the data rate is reduced by adjusting the coding rate and the symbol transmission rate (size of the constellation). The fallback data rates are 54 Mbps to 36, 27, 18, 12, 9, and 6 Mbps to cover distances up to 100 meters.

3.8 SPREAD SPECTRUM TRANSMISSIONS

The main difference between the spread spectrum transmission and traditional radio modem technologies is that the transmitted signal in spread spectrum systems occupies a much larger bandwidth than the traditional radio modems where the transmitted signal has a bandwidth of the same order as the information signal at baseband [PIC91]. Compared with UWB, however, the occupied bandwidth by spread spectrum is still restricted enough so that the spread spectrum radio can share the medium with other spread spectrum and traditional radios in an FDM format. There are two basic methods for spread spectrum transmission: direct sequence spread spectrum (DSSS) and frequency hopping spread spectrum (FHSS). Spread spectrum technology was first invented during the Second World War, and it has dominated military communication applications, where it is attractive because of its resistance to interference and interception, as well as its amenability to high-resolution ranging. In the 1980s, commercial applications of spread spectrum technology were investigated, and today it is the transmission choice for emerging 3G cellular as well as IEEE 802.11 wireless LAN standards. The voice-oriented digital cellular and PCS industries have selected spread spectrum technology to support CDMA networks as an alternative to TDMA/FDMA networks. This increases system capacity, provides a more reliable service, and supports soft handoff of cellular connections that are discussed in more detail in subsequent chapters. In the WLAN industry, spread spectrum technology was adopted primarily because the first unlicensed frequency bands suitable for high-speed radio communication were ISM bands, which were released by the FCC under the condition that the devices operating in these bands use spread spectrum. The FCC ruling on the ISM bands was to protect users from interfering with one another [MAR85].

The principle advantages of spread spectrum transmission are as follows:

1. Spread-spectrum signals can be overlaid onto bands where other systems are already operating, with minimal performance impact to both systems.
2. Spread spectrum is a wideband signal that has a superior performance over traditional radios on frequency selective fading multipath channel. Spread spectrum provides a robust and reliable transmission in urban and indoor environments where wireless transmission suffers from heavy multipath conditions.
3. The anti-interference characteristics of spread spectrum are important in some applications, such as networks operating on manufacturing floors, where the signal interference environment can be harsh.
4. Cellular systems designed with CDMA spread spectrum technology offer greater operational flexibility and overall system capacity than systems built on FDMA or TDMA access methods.
5. The convenience of unlicensed spread-spectrum operation in ISM bands in the United States is attractive to manufacturers and users alike.

Many of these features are also shared with UWB transmission discussed earlier in this chapter. UWB can be compared with DSSS transmission; however, from an application point of view, the major difference is that the UWB is for very short-range communications whereas spread spectrum technology can cover wide areas.

3.8.1 Frequency Hopping Spread Spectrum

The FHSS technique, invented by the Austrian-born movie star Hedy Lamarr to protect guided torpedoes from jamming, is a relatively simple concept. In order to avoid a jammer, the transmitter shifts the center frequency of the transmitted signal. The shifts in frequency, or *frequency hops,* occur according to a random pattern that is known only to the transmitter and the receiver. If we move the center frequency randomly among 100 different frequencies, then the required transmission bandwidth is 100 times more than the original transmission bandwidth. We call this new technique a *spread spectrum technique* because the spectrum is spread

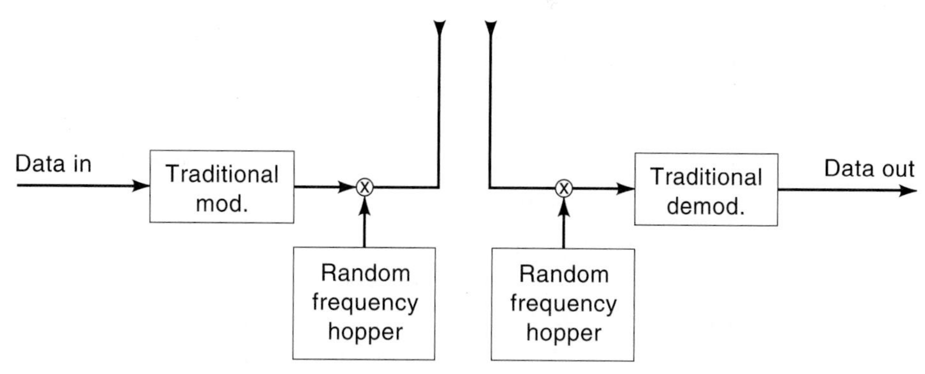

Figure 3.20 The frequency-hopping concept.

over a band that is 100 times larger than original traditional radio. FHSS can be applied to both analog and digital communications, but it has been applied primarily for digital transmissions. Figure 3.20 shows a simple diagram to describe the basic concept of FHSS transmission.

In the first stage, the input data is modulated with a traditional modulator, and in the second stage, the center frequency is changed according to a random hopping pattern generated by a random number generator. Ideally, the random pattern or spreading code is designed so that the occurrence of frequencies is statistically independent of one another. Because the sequenced pattern is coded to only *appear* random, the sequences are referred to as pseudorandom sequences or codes. At the receiver, first a dehopper, synchronized to the transmitter, repeats the hopping pattern of the transmitted signal, and then a traditional demodulator detects the received data. In digital implementation of this system, the sampling rate is the same as the sampling rate of the traditional system, leaving the complexity of the implementation in the same range as traditional modems. As we will see later, DSSS needs much higher sampling rates and consequently a more complex hardware implementation.

Example 3.11: Hopping Sequence of an FHSS System

Figure 3.21 shows the hopping pattern and associated frequencies for a frequency hopping system transfering data packets over the air. Each packet is transmitted using a different frequency. The sequence of frequencies is $f_3, f_5, f_6, f_1, f_4, f_8, f_2, f_7$ before returning to the first frequency, f_3. This is a slow frequency hopping system where a long packet with a number of data bits are transferred at each hop using the same frequency. In a fast frequency hopping system, the frequency hops occur much more rapidly and in each hop a very short packet is transmitted. In fast frequency hopping systems used in military applications, sometimes the same bit is transferred using several frequencies.

In the FHSS, the hopping of the carrier frequency does not affect the performance in additive noise because the noise level in each hop remains the same as the noise level of the traditional modems. Therefore, the performance of the FHSS

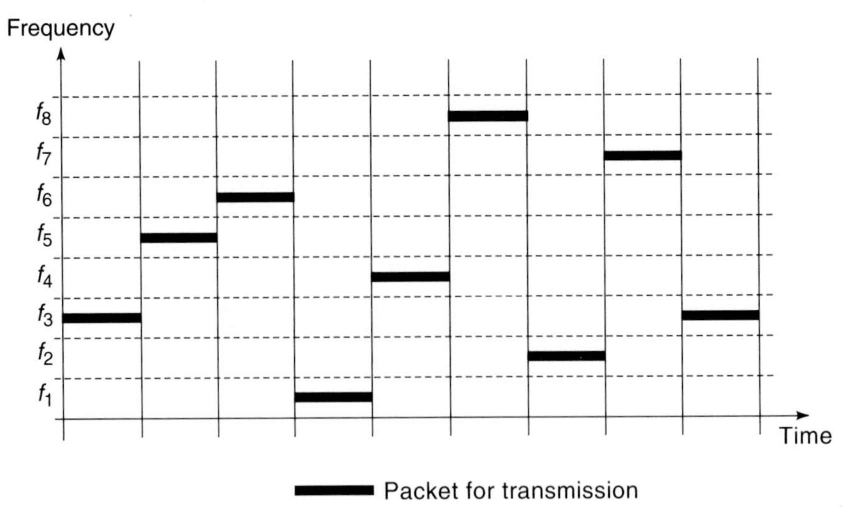

Figure 3.21 Example of an FHSS system.

systems in noninterfering environments remains exactly the same as the performance of the traditional systems without frequency hopping. In the presence of a narrowband interference, the signal-to-interference ratio of a traditional modem operating at the frequency of the interferer becomes very low, corrupting the integrity of the received digital information. The same situation happens in frequency selective fading channels when the center frequency of a traditional system coincides with a deep frequency selective fade. In an FHSS system, because the carrier frequency is constantly changing, the interference or frequency selective fading corrupts only a fraction of the transmitted information, and transmission in the rest of the center frequencies remains unaffected. This feature of the FHSS is exploited in the design of wireless networks to provide a reliable transmission in the presence of interfering signals or when a system works over a frequency selective fading channel.

Example 3.12: FHSS and Retransmissions

In a wireless packet data network, often an ACK mechanism (either at MAC, LLC, or higher layers) is in place to ensure retransmission of corrupted or lost packets. In a traditional system, if the channel is distorted with interference, all the transmissions will be corrupted, and a retransmission mechanism of any sort would not help. In a system using FHSS, the packets transmitted during the hop that coincides with the interference frequency would be corrupted. If we design the system so that we send one or a few packets per hub, when these packets are retransmitted, the hop frequency is changed and the retransmission mechanism works successfully. In IEEE 802.11, the maximum time of each hop is specified as 400 ms, and the maximum length of the packet is around 30 ms. Therefore, a packet corrupted by narrowband interference can be retransmitted during the next hop, around 400 ms later, which is a reasonable delay compared to the maximum duration of the packet.

Example 3.13: Frequency Selective Fading Channels and FHSS

The same feature of the FHSS that was described in the previous example also helps successful transmission in frequency selective fading channels. As we discussed in Chapter 2, when a frequency selective fading occurs, traditional systems operating over center frequencies coinciding with the faded frequencies cannot operate properly. An FHSS system can be designed so that the deep fades in the environment only corrupt a few hops, leaving the rest of the hops for successful retransmission. In an indoor environment, the width of the fade is around several MHz, and the FHSS system used in IEEE 802.11 uses hops that are 1 MHz apart from one another. Therefore, if a hop occurs in a deep fade and the data transmitted in that hop is not reliable, the retransmitted data after around 400 ns (as in the last example) will be successful.

Example 3.14: FHSS and GSM

In voice-oriented networks, often there is no retransmission mechanism. Corrupted packets are either discarded or are retained with a wrong value, in both cases causing distortions in the voice signal at the receiver. In TDMA systems, if the channel coincides with a deep frequency selective fading or when the cochannel interference (CCI) from another cell using the same frequency is excessive, the distortion in the received voice signal will persist until the terminal moves adequately and the frequency selective fading pattern is changed or the CCI is reduced. One method to reduce the duration of the frequency selective fade or excessive CCI situations is to provide for a slow frequency hopping pattern that forces a restriction on the duration of the frequency selective fading or CCI effects. This option is exercised in the GSM system that supports an optional frequency-hopping pattern of 217.6 hops per second.

Frequency hopping spread spectrum allows the coexistence of several transmissions in the same frequency band using CDMA. To implement CDMA we simply assign a different random hopping pattern to each terminal. Then multiuser interference occurs when two different users transmit on the same hop frequency. If the codes are random and independent from one another, the "hits" will occur with some calculable probability. If the codes are synchronized and the hopping patterns are selected so that two users never hop to the same frequency at the same time, multiple-user interference is eliminated. The number of frequency slots in this case limits the number of users.

Example 3.15: Multiple Access-Points in IEEE 802.11 Using FHSS

The IEEE 802.11 FHSS WLAN specifies 78 hopping channels each separated by 1 MHz. These frequencies are divided into three patterns of 26 hops each corresponding to channel numbers (0, 3, 6, 9, . . . , 75), (1, 4, 7, 10, . . . , 76), (2, 5, 8, 11, . . . , 77). These choices are available for three different systems to coexist without any hop collision or "hit." This mechanism allows installation of three APs in the same area in an overlapping format that results in a threefold increase in the capacity of the cell.

3.8.2 Direct Sequence Spread Spectrum

In a manner similar to FHSS, DSSS can be thought of as a two-stage modulation technique, shown in Figure 3.22. In the first stage, each transmitted information bit is spread (mapped) into N smaller pulses referred to as chips. In the second stage, the chips are transmitted over a traditional digital modulator. At the receiver, the transmitted chips are first demodulated and then passed through a correlator that despreads the signal. The despreader correlates the received signal with the duplicated transmitted spreading signal (chip sequence). The peak of the autocorrelation function is used to detect the transmitted bit. The autocorrelation[2] function of a good random code has a very high peak-to-sidelobe ratio that approximately equals N which is usually referred to as the *processing gain* [GAR00] of the receiver.

Example 3.16: DSSS in IEEE 802.11

A Barker code of length 11, used in the IEEE 802.11 as the spreading signal for the DSSS physical layer, is given by $[1, 1, 1, -1, -1, -1, 1, -1, -1, 1, -1]$. Figure 3.23 shows a data bit "1" in a binary communication system, the transmitted Barker code for the data bit, and the autocorrelation function at the receiver with its high peak and low sidelobes.

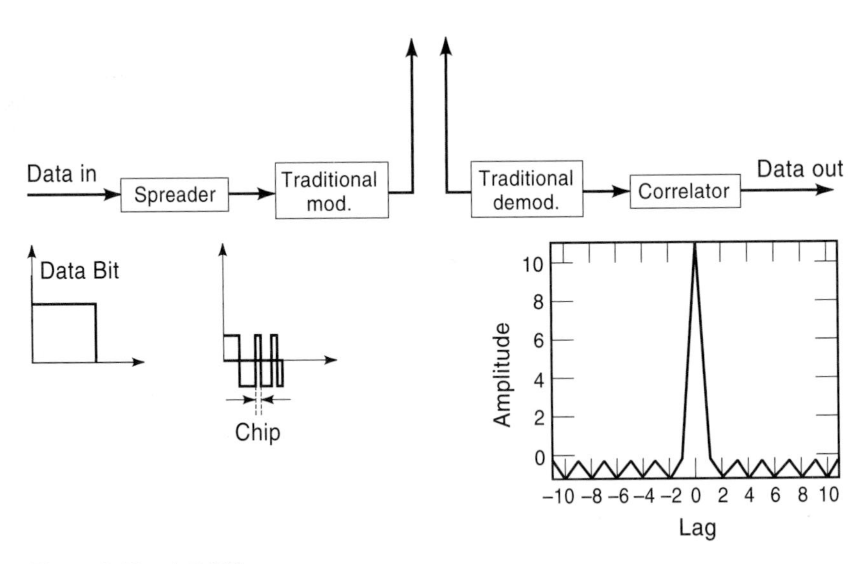

Figure 3.22 A DSSS system.

[2]Autocorrelation corresponds to correlating the pulse with itself, and this involves multiplying the pulse with a delayed version of itself and integrating the product over the duration of the pulse. For more details, see Appendix 3B.

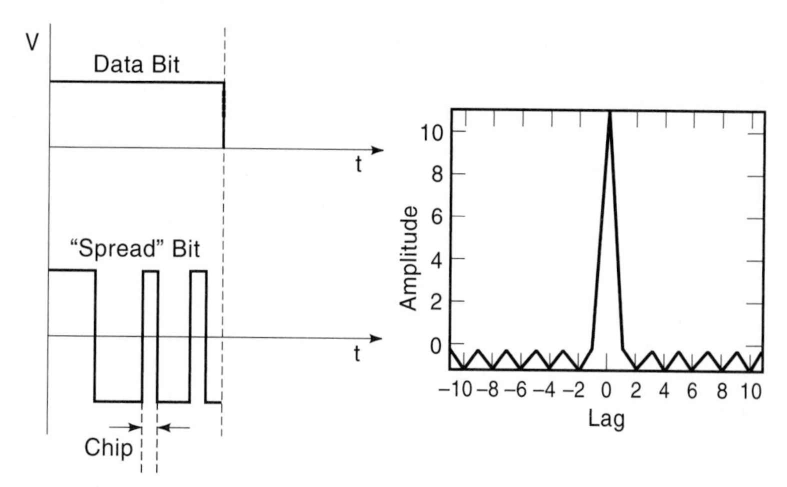

Figure 3.23 Barker code modulated DSSS signal in IEEE 802.11 and its autocorrelation.

The bandwidth of any digital system is inversely proportional to the duration of the transmitted pulse or symbol. Because the transmitted chips are N times narrower than data bits, the bandwidth of the transmitted DSSS signal is N times larger than a traditional system without spreading. As a result, N is also referred to as the *bandwidth expansion factor.* In a manner similar to FHSS, DSSS is also anti-interference and resistant to frequency selective fading. The transmission bandwidth of the DSSS is always wide, whereas the FHSS is a narrowband system hopping over a number of frequencies in a wide spectrum. As a result there is some distinction between the two methods. The DSSS systems provide a robust signal with better coverage area than FHSS. The FHSS system can be implemented with much slower sampling rates saving in the implementation costs and power consumption of the mobile units.

The DSSS can also be employed for code division multiple access. In a multiuser DS-CDMA environment, different codes are assigned to different users. In other words, each user has its own unique "key" code which is used to spread and despread only that user's messages. The codes assigned to other users are selected so that during the despreading process at the receiver, they produce very small signal levels (like noise) that are on the order of the sidelobes of the autocorrelation function. Consequently, they don't interfere with the detection of the peak of the autocorrelation function of the target receiver. In this manner each user is a source of noise for the detection of other users' signals. As the number of users increases, the multiuser interference increases for all of the users. This phenomenon continues up to a point that mutual interference among all terminals stops the proper operation for all of them. We discuss this topic in more detail in Chapter 4 when we compare the capacity of CDMA with TDMA and FDMA systems.

3.9 HIGH-SPEED MODEMS FOR SPREAD SPECTRUM TECHNOLOGY

3.9.1 PPM-DSSS

If we neglect the inner modulation in the DSSS systems and we look at the input and output of Figure 3.23, we see a signaling system in which every transmitted bit is received as an N times narrower pulse. Therefore, from the receiver's point of view it is very similar to a pulse transmission technique that sends a narrow pulse every N slots to signal the transmitted information. However, during the transmission the total energy of the pulse is spread over the entire N chips. In a way, each chip carries a piece of the information related to the received pulse, and the receiver knows the pattern (the spreading code) to put all these pieces together and construct the associated pulse. This feature can be used to implement a pulse transmission system without actually transmitting a narrow pulse. The advantage of this approach is that it refrains from transmitting high-power, narrow, short-time pulses and instead spreads the transmitted energy over time. Pulse transmission techniques of this sort are very popular in measuring the multipath characteristics of the radio channels [PAH95]. In these systems, in order to measure the impulse response of the channel, rather than sending a direct narrow pulse, a long DSSS waveform is sent periodically. After correlation at the receiver, a number of pulses associated with the signal arriving from different paths reveals the multipath characteristics of the medium. To explain this phenomenon, sometimes it is said that spread spectrum systems isolate or resolve the transmission paths. These isolated paths can also be used to provide the so-called RAKE-based time diversity in the received signal. As we will see in the next section, a smart receiver can take advantage of the time diversity of the received spread spectrum signal to improve the performance of a DSSS system.

Because we can think of DSSS as a pulse transmission technique, we might as well consider implementing a pulse modulation technique using the DSSS. The idea of using PPM with DSSS has been used for the design of high-speed wireless LANs. Let's assume that we have no serious multipath and consider a DSSS system similar to Figure 3.23. If at the transmitter we move the position of the transmitted pulse, the peak of the received pulse is also shifted. Therefore we can code our information in the position of the pulse similar to direct PPM techniques.

3.9.2 CCK Modulation

Orthogonal codes provide another method of increasing the bandwidth efficiency of a spread spectrum system [PAH87]. With this approach, each user has a set of orthogonal sequences representing a set of symbols for transmission. The set of orthogonal sequences, or *orthogonal codes,* will typically be a *Hadamard* or *Walsh* code, which are discussed later in this chapter. A stream of information bits is segmented into groups, and each group represents a nonbinary information symbol, which is associated with a particular transmitted code sequence. If there are N bits per group, one of a set of 2^N sequences is transmitted in each symbol interval. The received signal is correlated with a set of 2^N matched codes, each matched to the code sequence of one symbol. The correlator outputs are compared, and the symbol associated with the largest output is declared as the transmitted symbol.

Example 17: Orthogonal Sequences in IEEE 802.11

Recently, while the IEEE 802.11 committee was evaluating a variety of proposals for increasing the data rate of the first standard beyond 2 Mbps, they adopted an orthogonal coding technique called complementary code keying (CCK). A simplified block diagram of the basic principles of the CCK, adopted by IEEE 802.11b for 11 Mbps operation in 2.4 GHz ISM bands, is shown in Figure 3.24. The input data stream is grouped into 256 eight-bit symbols at 11 Mbps/8 bpS = 1.375 MSps. The encoder maps each eight-bit symbol into 8 four-phase coded symbols. The 256 coded symbols are selected from the 4^8 = 65,536 available alternatives, so that they are orthogonal to one another. The elements of the blocks of 8 four-phase symbols are transmitted serially using a QPSK four-phase modulator. At the receiver, each eight received complex waveforms from the QPSK demodulator are grouped in a block and sent to the decoder to find the closest eight-bit symbol associated with the demodulated 8 four-phase signals. The chip rate and occupied bandwidth of this system is the same as that of the original IEEE 802.11 standard, but the data rate is increased to 11 Mbps. The following equation gives the mapping rule for generation of the codes:

$$c = \{e^{j(\varphi 1 + \varphi 2 + \varphi 3 + \varphi 4)}, e^{j(\varphi 1 + \varphi 3 + \varphi 4)}, e^{j(\varphi 1 + \varphi 2 + \varphi 4)}, \\ -e^{j(\varphi 1 + \varphi 4)}, e^{j(\varphi 1 + \varphi 2 + \varphi 3)}, e^{j(\varphi 1 + \varphi 3)}, -e^{j(\varphi 1 + \varphi 2)}, e^{j\varphi 1}\}$$

The eight input bits are further grouped into 4 two-bit complex four-phase symbols. The resulting four phases $\varphi 1$, $\varphi 2$, $\varphi 3$, and $\varphi 4$ are inserted in the equation to find the eight complex-coded waveform that is serially modulated by the QPSK modem.

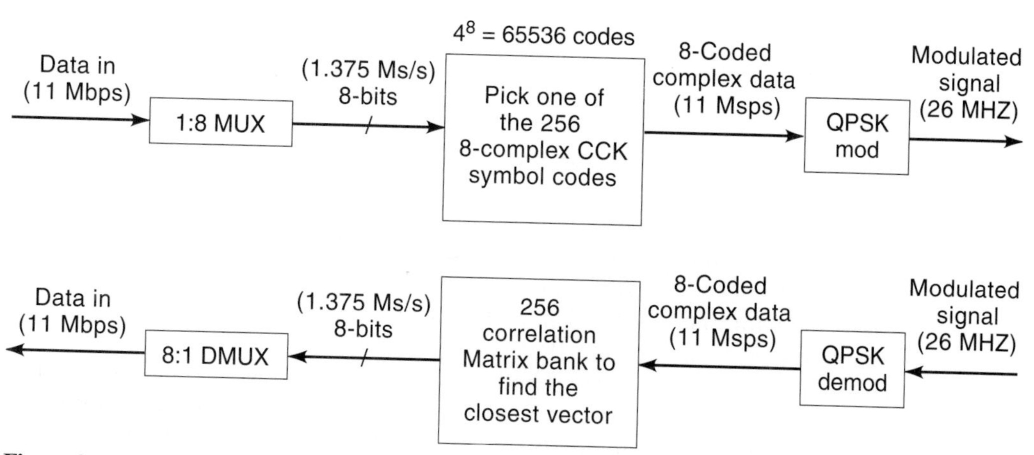

Figure 3.24 Principle of operation of CCK modulation used in IEEE 802.11b.

3.9.3 Multicarrier CDMA

Multicarrier modulation is also applied to 3G voice-oriented cellular networks using CDMA technology. The incentive of using multicarrier modulation in the CDMA systems is to provide for higher data rates for data application, wider bandwidth and consequently better quality for the voice users, and backward compatibility with the existing systems.

Example 3.18: Multicarrier in cdma2000

The cdma2000 standard for 3G cellular systems is based on the IS-95 or cdmaOne standard, which uses 1.25 Mcps for each user to support voice or data applications. The cdma2000 employs a multicarrier operation in which a user is allowed to use 1, 3, 6, or 9 of the cdmaOne channels to support a more reliable voice and variety of data channels.

3.10 DIVERSITY AND SMART RECEIVING TECHNIQUES

As we described earlier in this chapter and in Chapter 2, the main difference between the behavior of the wireless and wired media is the extensive fluctuations of the received signal that is caused by multipath and shadow fading. In particular, during short periods of time, the channel goes into deep fades, causing a significant number of errors that virtually dominate the overall average error rate of the system. In order to compensate for the effects of fading when operating with a fixed-power transmitter, the power must typically be increased by several orders of magnitude relative to the nonfading operation. This increase of power protects the system during the short intervals of time when the channel is deeply faded. The most effective method of counteracting the effects of fading is to use diversity techniques in the transmission and reception of the signal. The concept here is to provide multiple received signals whose fading patterns are different. With the use of diversity, the probability that all the received signals are in a fade at the same time reduces significantly, which in turn can yield a large reduction in the average error rate of the system.

A variety of techniques are available for reception of the diversity signals. With *selection diversity,* one signal is chosen from the set of diversity branches, usually on the basis of received signal strength. With *linear combining,* as the name suggests, the diversity branches are simply summed together before demodulation. In the optimum method of combining, called *maximal-ratio combining,* the diversity branches are weighted prior to summing them, each weight being proportional to the received branch signal strength. It can be shown that maximal-ratio combining provides the optimum performance among all diversity combining techniques [PAH95].

As the data rate of a modem increases beyond fractions (about 20%) of the inverse of the multipath spread, the channel becomes frequency selective for which a deep null may occur in the passband of the channel. In a frequency selective fading multipath transmission, the data rate is large enough, so that the bit duration is on the order of the multipath spread of the channel, resulting in performance degradation due to ISI. The performance degradation caused by ISI forces the performance curves into flat areas where any increase in the transmitted power does not improve the bit error rate performance of the modem. As a result, in a frequency selective fading channel, increase in the transmitted power level is not effective, and diversity techniques remain as the only remedy for implementation of a reliable system over these channels.

Diversity can be provided spatially by using multiple antennas, in frequency by providing signal replicas at different carrier frequencies, or in time by providing signal replicas with different arrival times. In the past several decades, a number of spatial-, frequency-, and time-diversity techniques have emerged for the implementation of advanced wireless modems that are now adopted in many standards for wireless communications.

3.10.1 Time Diversity Techniques

As we discussed earlier, in the wireless medium, due to reflection, diffraction, and scattering the received signal arrives from different paths, and because the length of these paths are not the same, signals arrive at different time delays. Therefore, in the wireless medium we have to deal with the multipath arrival problem. We also explained that for traditional receivers, the multipath arrival of the signal has the harmful effect of causing ISI. Looking into this issue conceptually, the signals arriving from different paths are exposed to different fading patterns, and if a receiver isolates these paths, it can provide a source of diversity for the arriving signal. If the paths are isolated at the receiver, the multipath signals can be regarded as a form of diversity, and a smart receiver can use this diversity to improve its performance.

In this section we describe several applied receivers that take advantage of the time diversity of the multipath arrival in wireless medium. Historically speaking the first receiver taking advantage of the in-band time diversity was the RAKE receiver, which was invented in the 1950s in MIT's Lincoln Laboratory [PRI58]. Today, to provide a robust reception, RAKE receivers are commonly used in DSSS receivers in CDMA cellular telephones. The second class of receivers exploiting time diversity operates with traditional modems. These modems either adaptively estimate the channel multipath characteristic to unfold the effect of multipath, or they use an adaptive filter with the inverse of the characteristics of the channel. These two approaches serve the same purpose of equalizing the effects of the channel, and for that reason they are referred to as *equalization* techniques.

3.10.1.1 DSSS and the RAKE Receiver

The original RAKE system was designed for wireless teletype communications with a symbol rate of 90 cps operating over an ionospheric channel. The envelope of the transmitted binary FSK signal was a 10 kHz PN-sequence. The received signal was passed through two tapped delay lines for mark and space frequencies, and the outputs were compared for decision making. The tap gains of the delay lines were adaptively adjusted by cross-correlating the received signal with both mark and space reference signals at the receiver. Because the data rate was considerably smaller than the transmission bandwidth, the ISI was negligible. More recently, other versions of RAKE receivers have been examined for urban radio [KAM81], HF radio [BEL88], and indoor radio [PAH90a] channels.

As we described in Section 3.9, from the receiver point of view, the operation of a DSSS system is similar to a pulse modulation system. After correlation in the receiver (see Fig. 3.23), the signal is a narrow pulse with a width twice the chip duration that occurs every bit interval. In a multipath environment, the output of the receiver (matched filter) will have several peaks associated with the signal arriving at different path delays. Figure 3.25 illustrates the output of the receiver with three multipath components. By comparing this figure with Figure 3.23, can one observe the effects of multipath on the receiver output in a DSSS system. Figure 3.23 represents the received signal in a single path channel, whereas Figure 3.25 represents the received signal in a three-path multipath channel. In Figure 3.25, the interarrival delay among the multipath signals is greater than the width of the base of the autocorrelation function, and the delay spread is less than the information bit interval. However, these conditions do not apply in every situation. For instance, if the delay spread of the channel is greater than bit duration, we will have ISI. To avoid interference between detected information symbols, the bit duration must be kept larger than the multipath spread of the channel. In other words, the symbol transmission rate should be kept below the coherence bandwidth of the channel. When consecutive signal paths arrive with delay differences greater than the width of the autocorrelation function (i.e., on the order of the chip duration), the receiver output will exhibit separate peaks, as shown in Figure 3.25. If the delay between two consecutive paths is significantly less than the chip duration, the two paths will merge and appear as one path equivalent to the phasor sum of the two actual paths. Thus, as the transmission bandwidth is made smaller, the chip duration becomes correspondingly longer, and fewer isolated paths can be resolved at the receiver output. Of course, as paths merge together, the fluctuations in their amplitudes and phases produce an overall fluctuation in the phasor sum, which we observe as fading.

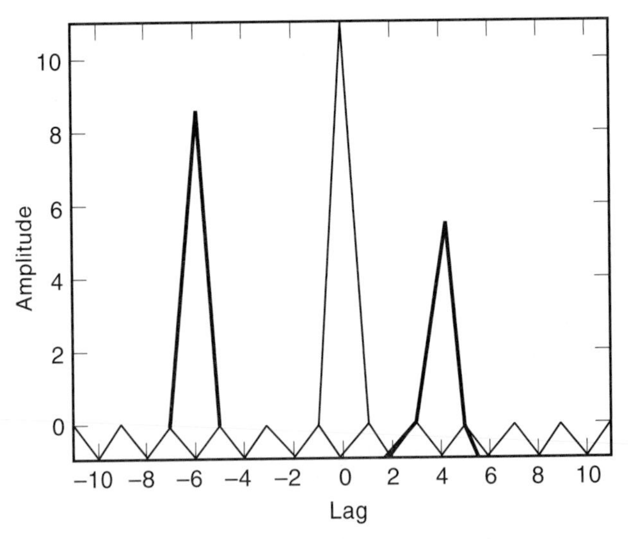

Figure 3.25 The output of a matched filter with signals arriving via multiple paths at the input.

Example 3.19: Multipath Reception in IS-95 CDMA

The IS-95 CDMA system uses a chip rate of 1.25 Mcps and a symbol transmission rate of 4,800 Sps (equivalent to 9,600 bps at two-bits per symbol in QPSK modulation). Therefore, it can resolve multipath components on the order of 1/1.25 Mcps = 800 ns apart, and a multipath spread of up to 1/4800 bps = 2.08 ms cannot cause ISI in the system. The multipath spread in the outdoor microcellular environment is on the order of several tens of microseconds, and in indoor picocell areas, it is on the order of several hundreds of nanoseconds. Therefore, IS-95 does not suffer from ISI in any of the indoor or outdoor environments. However, in an indoor picocellular environment, it is unlikely that the system resolves the multipath components. Some 3G systems offer similar bit rates with chip rates that are up to an order of magnitude shorter (10 times larger bandwidth). With a pulse resolution of around 80 ns for systems, one expects resolving several multipath components even in the indoor picocellular environment.

Example 3.20: Multipath Reception for IEEE 802.11

The IEEE 802.11-based systems have a chip rate of 11 Mcps and a symbol transmission rate of 1 Mbps (2 Mbps for QPSK modem). The resolution of the channel is on the order of 1/11 Mcps = 90 ns, and multipath spread of up to 1/1 MSps = 1000 ns does not cause ISI in the system. Wireless LANs are designed for indoor applications; therefore the IEEE 802.11 receiver can isolate several paths and it does not suffer from ISI.

If we operate with chip durations short enough to resolve individual paths, we can design a smart receiver to take advantage of the multiple paths to provide diversity and enhance the reliability of the decision on each received information symbol. In a DSSS system, a receiver that optimally combines the multipath components as part of the decision process is referred to as a RAKE receiver. A typical RAKE receiver structure for a DSSS system is shown in Figure 3.26. The received signal is passed through a tapped-delay line, and the signal at each tap is passed through a correlator similar to the one used for standard DSSS receivers. The outputs of the correlators are then brought together in a diversity combiner whose output is the estimate of the transmitted information symbol.

In the original RAKE receiver [PRI58], the delays between the consecutive taps or "fingers" of the RAKE receiver were fixed at half of the chip duration to provide two samples of the overall correlation function for each chip period. Using this method for a rectangular chip pulse with triangular correlation function, we will have four samples of each triangle in the correlation function. Because the peaks in general are not aligned precisely at multiples of the sampling rate, it is not possible to capture all the major peaks of the correlation function. But a RAKE receiver implemented with a sufficiently large number of taps will provide a good approximation of all major peaks. Modern digitally implemented receivers typically have only a few RAKE taps and the capability to adjust the tap locations. An algorithm is used to search for a few dominant peaks of the correlation function and then position the taps accordingly.

A RAKE receiver can combine the arriving signal paths using any standard diversity combiner such as a selective, equal-gain, square-law, or maximal ratio

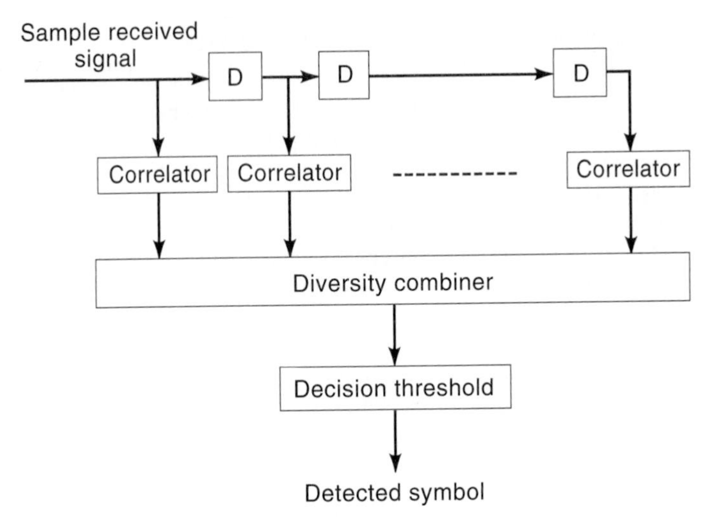

Figure 3.26 The RAKE receiver structure.

combiner [SKL01]. As we discussed earlier in this section, the optimum diversity combiner is the maximal ratio combiner, which weights the received signal from each branch by the signal-to-noise ratio at that branch. A maximal ratio combining RAKE receiver that resolves all the paths and does not introduce ISI provides optimum system performance in the presence of time diversity.

Example 3.21: Rake Reception in IS-95

The QUALCOMM original receiver uses three moving fingers to implement the RAKE receiver. The original WaveLAN that was the model for implementation of transmission system for the IEEE 802.11 standard was not using a RAKE receiver. Later on other manufacturers started implementing RAKE receivers in wireless LANs to support a more robust transmission.

In summary, in the operation of a DSSS system, multipath does not cause ISI unless the information symbol transmission rate approaches the coherence bandwidth of the channel. Also, it is possible to design a receiver which takes advantage of the isolated arriving paths to improve or even optimize system performance. The wider the transmission bandwidth, the greater is the order of implicit time diversity that can be utilized. The isolated paths will provide a source of implicit diversity to a DSSS receiver, which improves the performance of the system. On the other hand, if not utilized in some diversity combining receiver, the signal arriving from each path is a wideband interference to the signals arriving from other paths, which degrades the performance of the DSSS system.

3.10.1.2 Traditional Modems and Equalizers

In the past few decades, other adaptive techniques exploiting time diversity have emerged in the development of radio modems for fading multipath channels that operate in conjunction with traditional modems. Time-gating of the transmitted pulse to avoid ISI, with adaptive matched filtering of each received pulse, was the approach taken for a family of military troposcatter radios [PAH80], [CON78]. The adaptive decision feedback equalizer was another approach investigated for application to troposcatter [GRZ75], [MON84]; microwave LOS, [BEL84], [PAH85]; HF [FAL85]; and more indoor radio [SEX89a] channels. Finally, an adaptive version of the maximum likelihood sequence estimation (MLSE) technique [FOR72] was investigated for troposcatter in [BEL69] and for HF in [CHA75]. More recently, these techniques have found their way into wireless standards and products.

Example 3.22: Equalization in Wireless Networks

The GSM standard recommends using equalization, and to support proper operation of the equalizer, as we will see later, provides a 26-bit training sequence in each transmitted packet. The HIPERLAN-1 wireless LAN standard specification recommends using equalization in indoor radio channels to achieve data rates of up to 23 Mbps.

A standard usually does not specify the design of a receiver, but it provides a training sequence that can be used for the equalization. Therefore, manufacturers have different options for implementation of an equalizer. There are two methods that can be used to take advantage of the training sequence and implement an equalizer. One is to use the training sequence to estimate the channel multipath characteristics and then use the estimate of the channel to eliminate the effects of ISI. This technique was originally named MLSE after the name of the algorithm that uses channel estimates to eliminate the effects of ISI [FOR72]. The second technique is to use the training sequence to design an adaptive filter at the receiver that inverses the channel distortions. In the early literature in this field, this method was referred to as equalization [MON77]. More recently, both techniques are referred to as adaptive equalization. This is partially because of the fact that both techniques use the training sequence reserved for equalization, and partially due to the fact that though they are two different approaches to equalize the distortions caused by the channel, they are both adaptive in nature.

The MLSE receiver is the optimum receiver in the presence of ISI. Given the estimates of the discrete channel impulse response, an MLSE receiver uses a trellis diagram with the Viterbi algorithm to obtain maximum-likelihood estimates of the transmitted symbols. The adaptive MLSE receiver is shown in Figure 3.27. It consists of two parts: the adaptive channel estimator and the MLSE algorithm. The sampled channel impulse response is measured with the adaptive channel estimator, which operates in a way similar to the top part of the RAKE receiver described in the previous section. Using the reference sequence (a PN sequence similar to those used in DSSS systems), the top part of the figure estimates the sampled channel impulse response. In the RAKE receiver, these samples were taken at half the symbol duration, whereas in the MLSE receiver they are taken at symbol intervals. The lower part of the receiver uses the samples of channel impulse response to compare the sequence of the sampled received signal with all possible received sequences and determine the most likely transmitted sequence of symbols. The MLSE procedure [FOR72]

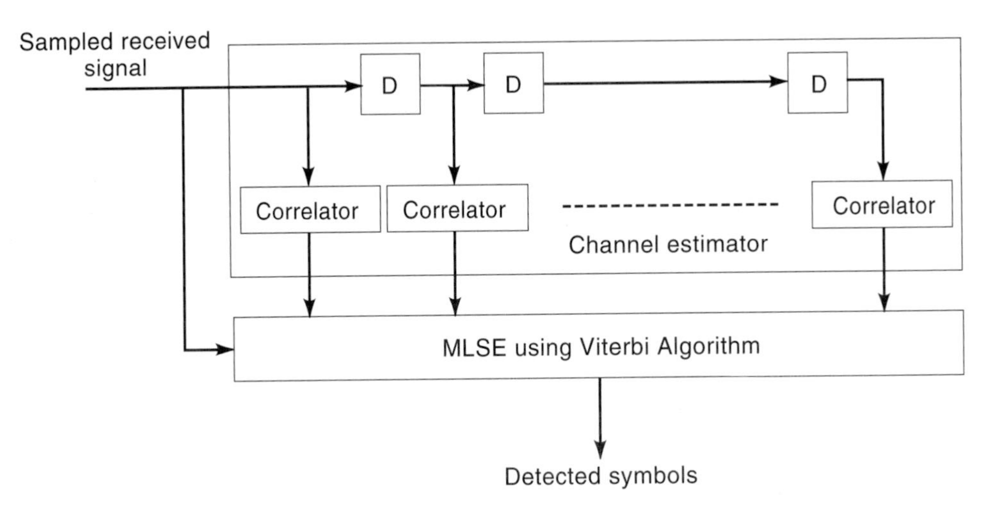

Figure 3.27 The adaptive MLSE receiver.

uses the Viterbi algorithm [VIT67] to minimize the computational complexity of the maximum likelihood selection among all possible transmitted sequences.

The MLSE is the optimal method of canceling the ISI; however, the complexity of this receiver grows exponentially with the length of the channel impulse response, whereas, as we discuss later, the complexity of the other equalizer grows only linearly with the length of the impulse response. For this reason, MLSE is an attractive option for channels with short impulse responses, but for longer impulse responses, equalizers are more practical. In the radio communication literature, MLSE is usually compared with decision feedback equalization (DFE), which is the equalizer of choice for frequency selective fading multipath channels. Comparisons of MLSE versus DFE performance are given for telephone line modems in [FAL76b,c], for HF radio in [FAL85], and for troposcatter radio links in [MON77], [CHA75].

Here we provide a short description of the principles of operation of DFE. For more detailed treatments of equalization techniques, the reader can refer to [PAH95], [PRO00], [QUR85], and [MES87]. To describe the DFE, it is easier to start with the linear transversal equalizer (LTE). Figure 3.28 shows the principal elements of an LTE. In an LTE, similar to the RAKE and channel estimator, the received signal is passed through a tapped delay line. The delayed signal is then multiplied with different tap gains, and the weighted delayed signals are added together to form a sample that is measured against some threshold to detect the value of the transmitted symbol. The weight of the tap gains is determined by variety of algorithms based on the difference between the detected symbol and the output of the adder that was the estimate of the received value of the transmitted symbol. The algorithm indeed determines the best tap weights that somehow minimize the error in the system. The LTE equalizer is a discrete-time filter intended to compensate for the amplitude and phase distortions of the channel. One can see intuitively that for an infinite-tap equalizer, the sampled frequency response of the equalizer should be the inverse of the frequency response of the channel.

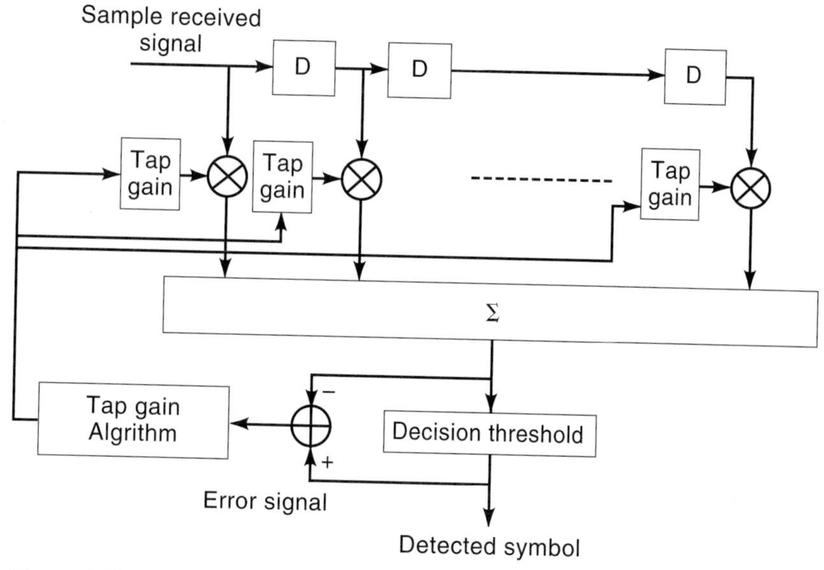

Figure 3.28 The linear transversal equalizer.

Linear equalizers are unable to properly equalize channels having deep nulls in the pass band. For these channels the linear equalizer applies a high gain in its frequency response to compensate for the null, which in turn causes noise enhancement. However, the backward filter of a DFE does not suffer from the noise enhancement problem, because it estimates the channel rather than its inverse. As a result, for channels with deep nulls in the pass band, DFEs are superior to linear equalizers.

In frequency selective fading radio channels, channels occasionally experience deep nulls in the pass band, resulting in an unsatisfactory performance for linear equalizers. On telephone channels, the most significant amplitude and delay distortion is found at the edges of the pass band. As a result a DFE, which is more effective against nulls in the middle of the pass band, has little to offer over a linear equalizer with a large number of taps. Thus linear equalization is still the predominant design choice for voice-band data modems operating over telephone channels.

DFEs [BEL79], [SAL73], shown in Figure 3.29, consist of two TDL filters, referred to as the forward and the backward equalizers. The input to the forward equalizer is the received signal, and it operates similarly to the linear equalizers discussed earlier. The input to the backward equalizer section is the stream of detected symbols. The tap gains of this section are the estimates of the channel sampled impulse response, including the forward equalizer, and this section cancels the ISI due to past samples.

3.10.2 Frequency Diversity Techniques

As we discussed in Chapter 2, a fading multipath channel displays frequency selectivity. In other words, if we sweep a large spectrum of frequencies with the same transmitted power at different frequencies, the received power fluctuates signifi-

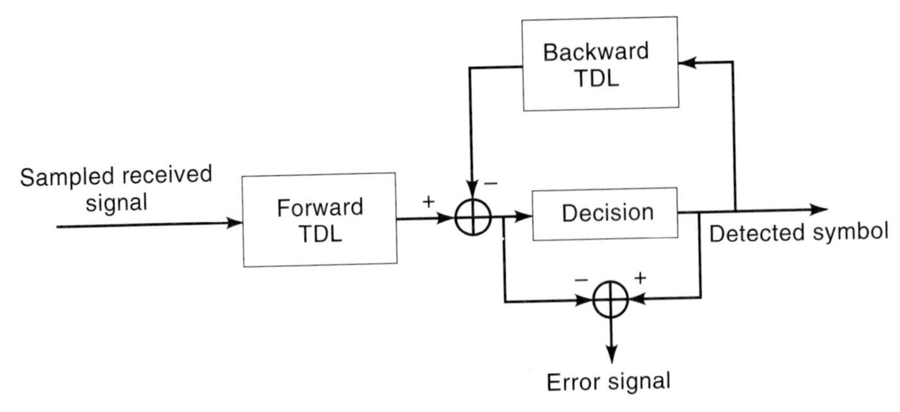

Figure 3.29 Structure of a decision feedback equalizer.

cantly, and in particular certain frequencies go to deep fades of 30–40 dB lower than the average received power at other frequencies. The width of the fades is proportional to the delay spread of the multipath arrival of the signal. In the radio channel modeling literature, this phenomenon is referred to as frequency selective fading.

Example 3.23: Frequency Selective Fading Channel

Figure 3.30 represents a sample measured impulse response and frequency response of an indoor radio channel in the 900–1,100 MHz range. We see a number of fades and, in particular, one very deep fade close to the center of the band. The multipath spread in this experiment is around several hundred ns, resulting in fade durations of around several MHz. The difference between the peak received power and the deepest fade is around 35 dB.

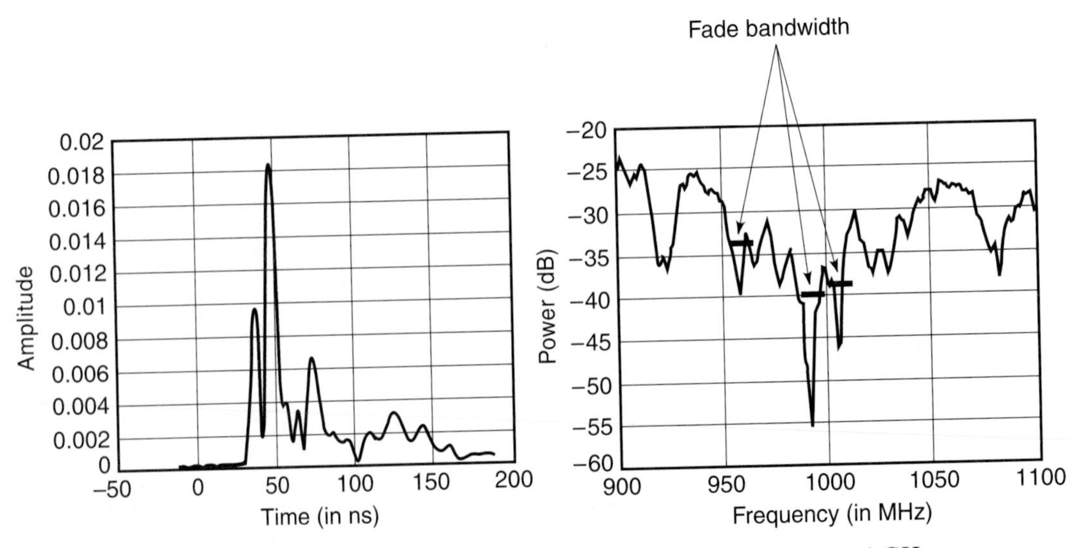

Figure 3.30 Frequency response of the channel over 200 MHz frequency band at 1 GHz.

As a terminal moves or people move around the terminal, the multipath profile and the frequency selectivity pattern both change continually. With no movement of the terminal or movement of the objects close to the terminals, the channel behavior will remain quasi-stationary. This inequality of the received signal at different frequencies provides a source of *frequency diversity* in the received signal that a smart system can use to improve its performance.

As we also saw before, if the symbol transmission rate of a signal is much smaller than coherence bandwidth, the entire signal bandwidth is similarly affected. The good part of this situation is that there is no ISI, and the bad part is that when the signal is hit by fade the received signal strength becomes very low because all the spectrum of the signal goes into a fade. In time diversity techniques, we increased the data rate so that the entire signal did not fade at the same time, and then we used an adaptive technique to take advantage of multipath arrivals. Receivers taking advantage of the frequency diversity intend to send the signal at different frequencies, so that only a portion of the transmitted signal is damaged due to the frequency selective fading. The first class of systems taking advantage of the frequency diversity are the FHSS systems, which are naturally hopping over different random frequencies.

Example 3.24: Frequency Hopping and Bluetooth

The Bluetooth system designed for personal area networking uses a symbol transmission rate of 1 Msps that does not need time diversity receivers to operate in indoor areas. However, it uses a fast frequency hopping technique (1,600 hops per second) that transmits one packet per fixed hop slot of 625 μsec hopping over 79 MHz of bandwidth. If the hop frequency hits the deep fade, the packet is lost, and it will be retransmitted in the next hop whose frequency is selected in a random manner. Bluetooth is a simple system with low power consumption and reliable transmission.

Example 3.25: Frequency Hopping and IEEE 802.11

The IEEE 802.11 FHSS system operates at 1 and 2 Mbps with a transmission bandwidth of 1 MHz using MSK modulation. This system has a variable packet duration of up to 20 ms, and the frequency hopping recommended by this standard is 2.5 hops per second. In a fade, a packet has a chance to be successfully received in a few attempts, and if not successful in the next hop, it will go through the system. The maximum delay jitter for the user would be around 400 ms, which is associated with a deep and long persisting frequency selective fading. FHSS again provides a low power simple implementation as it is compared with the DSSS alternative capable of taking advantage of time diversity.

Example 3.26: Frequency Hopping and GSM

The GSM standard has provisions for the implementation of equalization techniques using the reference (training) signal. The GSM standard also recommends slow frequency hopping transmission to introduce frequency diversity. If a mobile terminal is in a persisting fading situation, the transmitted power controlled by the BS is maximized, and the equalizer must be operating too in order to consume the life of the battery very quickly. Frequency hopping will allow the hop to equalize the performance and battery consumption among all terminals.

Another class of systems taking advantage of frequency diversity is the class of multicarrier systems. If the bandwidth of each carrier is smaller than the coherence bandwidth of the channel and the number of carriers is large enough so that the bandwidth is much larger than the fade durations, we will always have a number of carriers that are not affected by deep frequency selective fading.

Example 3.27: Frequency Diversity and MCM in IEEE 802.11a/HIPERLAN-2

Figure 3.31 shows a fading pattern and the band used by IEEE 802.11a/HIPERLAN-2 standards. As shown in the figure, we always have a number of channels that are not affected by deep null and can provide a reliable error rate.

A smart receiver using frequency diversity of this form has several choices. One is to measure the received power in all subcarriers and improve the performance of the faded subcarrier by reducing the transmission rate of that subcarrier. As a rule of thumb, the reduction of each bit per symbol will reduce the power requirement by about 3 dB.

Example 3.28: SNR Requirements versus Bit Rate

IEEE 802.11a and HIPERLAN-2 have transmission choices from 6 bits per symbol (64-QAM) to 1 bit per symbol (BPSK), which provide a range of 15 dB adjustments for frequency selective fading effects by changing from highest to lowest data rate.

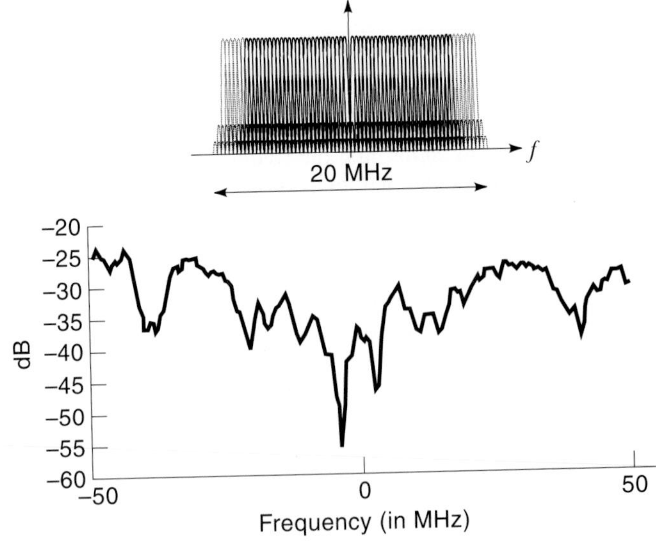

Figure 3.31 Spectrum of an OFDM carrier and the corresponding frequency response.

Similar results can be obtained if the transmit power in a subcarrier is increased as the signal in that subcarrier goes further into fade. These approaches require a feedback channel and channel measurements at the receiver that can be implemented for data applications preferring reliability to delay. Yet another approach is to use strong error correcting codes for the transmitted bits in parallel subcarriers. This way there is no need for a feedback channel and channel measurements, and the low error rate bits in the normal channels are used to correct the high error rate data from subcarriers affected by deep fades. Application of any of these frequency diversity techniques does not prevent the use of others in the same system, but the complexity of the system increases as more features are included. Coding techniques are similarly useful for FHSS systems. It means that the bits over several hops can be encoded to recover the bits lost during the fade. In many FHSS applications, the transmitted data stream is scrambled over several hops so that if a few bits are lost due to the fade, the strong coding techniques help in the recovery of these bits by providing reliable communication. Scramblers will prevent the occurrence of a string of errors in data that is more difficult to correct.

3.10.3 Space Diversity Techniques

In a fading multipath channel, the received signal at the antenna of a receiver is composed of a number of signals arriving through different paths from different spatial angles. Each path is formed after a series of reflections, transmissions, diffractions, and scattering patterns that are unique to that path. As a result, signal strength, polarization of the waveform, delay, and the angle of arrival of each path is quite different from other paths. In addition, as the location of the antenna is changed by about the wavelength of the transmitted signal, all the path structures and its associated parameters will change as well. Therefore, besides time and frequency diversity of the received signal, discussed in the past two sections, there is a significant amount of diversity in the special behavior of the signal. In the same way as time- and frequency-diversity techniques, a number of space-diversity techniques have been developed that allow a smart receiver to take advantage of the diversity in the arriving signal in the space.

Figure 3.32 illustrates the basic concepts behind four approaches to take advantage of the spatial diversity. Spatial diversity can be implemented using multiple antennas located in different locations (Fig. 3.32[a]), by using multiple antennas with different polarization located in the same location (Fig. 3.32[b]), by a sectored antenna limiting the angle of arrival of the signal (Fig. 3.32[c]) or by an antenna array that changes its antenna pattern adaptively (Fig. 3.32[d]). Although all four techniques are smart methods to take advantage of the diversity of the signal in the space, only the fourth technique is referred to as a smart antenna. Using spatially separated antennas is very common in fixed wireless systems. However, installing multiple antennas for mobile terminals is challenging in practice. Polarization diversity, however, has found its way into mobile terminals for cellular voice and wireless LAN applications.

Example 3.29: Array Antennas for Spatial Diversity

The first DFE modem [MON76] designed for fixed troposcatter communications used a quadruple antenna array with antennas spatially separated in the order of

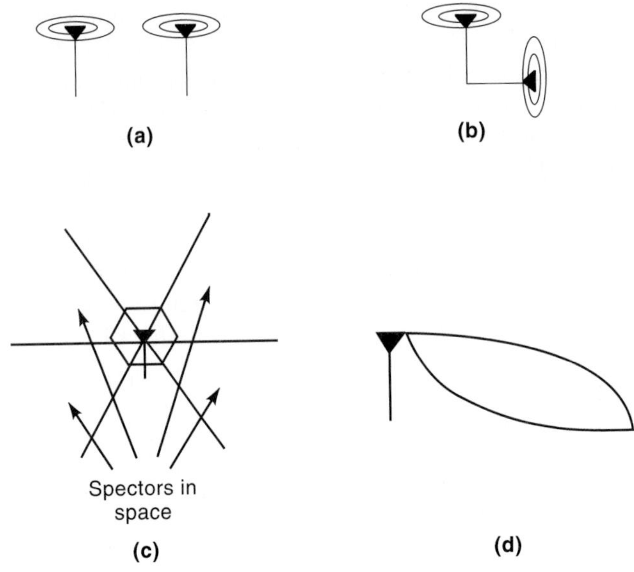

Figure 3.32 Spatial diversity schemes using (a) multiple antennas, (b) polarization diversity, (c) sectored antennas and angle diversity, and (d) adaptive angle diversity.

wavelength of the channel. WaveLAN, the first successful wireless LAN, used polarized receiver antennas for its mobile terminals [TUC91].

A sectored antenna has several sectors each selecting only the signal arriving in their field of view. In other words, the sectored antenna divides the space into several noninterfering zones. The use of sectored antennas has several advantages relative to the first two techniques. A sectored antenna reduces the interference from other users operating in the same band because it restricts the spatial angle of the arriving interference signals. A sectored antenna reduces the multipath delay spread because it only accepts a fraction of the arriving paths that fall in the sector of the antenna pattern. The reduction of multipath delay spread allows increasing the maximum data rate achievable on the channel. Using a sectored antenna, an effective diversity reception can be provided without requiring wide physical separation between antennas, making compact product packaging feasible. Sectored antennas are used in cellular systems to restrict the interference and increase the capacity of the system. In wireless LAN applications, sectored antennas have been used to increase the maximum supportable data rate by controlling the multipath arrival.

Example 3.30: Trisectored Antennas for Capacity in CDMA

Three sectored antennas are commonly used in digital cellular industry, as we will see in Chapter 5. These antennas improve the quality of transmission and increase the capacity of the network. In the next chapter when we calculate the capacity of the cellular CDMA systems, we will quantitatively show the significance of the sectored antennas in increasing the capacity of a network.

Example 3.31: Switched Sector Antennas in Motorola Altair

The Altair wireless LAN product, the first wireless LAN operating at 18–19 GHz licensed bands, used 2 six-sector antennas, one each at the transmitter and receiver, which provided a total of 36 effective sector-pairs. With such a design, the array of all multiple-reflected signal paths arriving at the receiver is divided into 36 subsets, and with appropriate signal processing (say selection diversity), the receiver can extract the subset producing the least signal degradation, discarding all the other signal arrivals. With this approach, Altair was able to support data transmission rates of up to 15 Mbps using simple four-level FSK modulation scheme.

Smart antenna systems use an adaptive antenna array that can form a beam toward a target. These antennas have found a number of applications in the cellular telephone industry that range from interference cancellation to control of eavesdropping. Because the antenna is highly directive, the gain is substantially increased, enabling better building penetration and longer ranges, thereby reducing initial deployment costs. By employing smart receiver antennas, signals from mobile terminals are isolated, reducing the near-far effect discussed earlier.

Example 3.32: Smart Antennas and CDMA

In CDMA, the multiple access interference is the primary cause for capacity limitations. Smart antennas reduce the multiple access interference by focusing the signals in one direction only, thereby increasing the system capacity.

3.11 COMPARISON OF MODULATION SCHEMES

In this section, following [FAL96], we compare the bandwidth and power requirements of various transmission techniques operating in a indoor test area, as shown in Figure 3.33. An indoor area is considered because broadband communications for applications such as WLANs are becoming more important these days. However, the overall conclusions can be extended to any other wireless fading multipath channel. The test area consists of seven rooms in the second floor of the Atwater Kent Laboratories at the Worcester Polytechnic Institute (WPI). Results of more than 600 wideband channel measurements in these rooms [HOW90] are used to calibrate a ray-tracing algorithm [YAN94] that is then used to generate several hundreds of thousands of channel profiles in the area. These profiles are then used for performance evaluation of different modem design technologies operating in this area. The basic modulation for all techniques was QPSK, and acceptable performance for a modem was considered to be a bit error rate that is better than 10^{-5} in 99 percent of location in the area. The purpose of this exercise was to examine the relationship between bandwidth and power requirements to the maximum data rate of each technique.

First we assume a fixed bandwidth of 10 MHz that is appropriate for the unlicensed operation in the PCS bands, and we compare the minimum required radia-

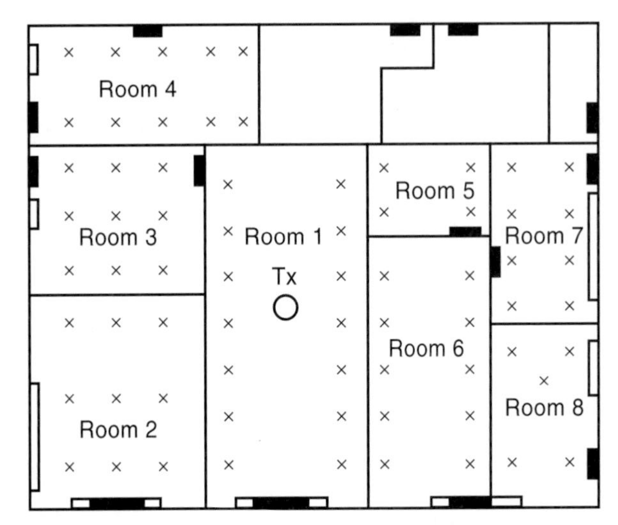

Figure 3.33 Indoor test area in WPI's Atwater Kent Laboratories.

tion power for each transmission technique to cover the test area. This section addresses the important issue of power consumption for battery-operated mobile terminals. It was found that with a maximum transmission power of 100 mW all the transmission techniques discussed here are able to cover the test area. If we assume the transmission power is maintained at 100 mW and there is no restriction on the bandwidth, we can determine the maximum data rate that can be achieved with any of the transmission techniques in the test area. This exercise addresses the transmission technologies in the context of demand for higher data rate, which has been an extremely important factor in the evolution of the LAN industry.

Figure 3.34 shows the minimum radiation power required to achieve a data rate for a 10 MHz channel operating in the test area for DSSS with a four-tap

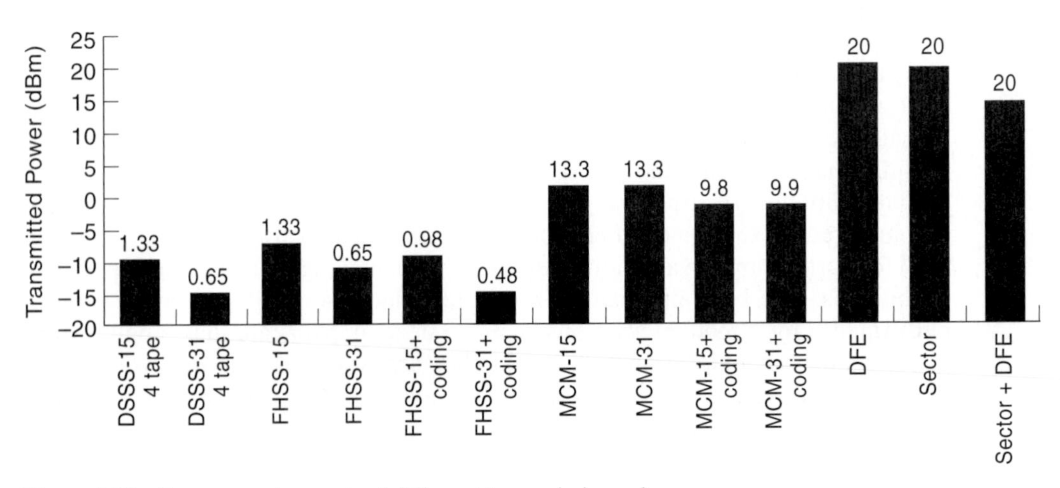

Figure 3.34 Power requirements of different transmission schemes.

RAKE receiver and processing gains of 15 and 31, as well as FHSS and MCM with and without Reed-Solomon coding (COFDM), using 15 and 31 carriers. Also presented are the power requirements to achieve the same outage probability using DFE with three forward and three feedback taps, six sectored antenna systems (SAS), and a DFE/SAS system. The number on top of each bar in the chart represents the data rate supported with that technology. Therefore, using this figure we can find the power requirements and supported data rate for a variety of modem design technologies in a fixed bandwidth (10 MHz) environment.

Example 3.33: Comparison between DFE and SAS Systems

From Figure 3.34, one can find out that in the typical area shown in Figure 3.33, with 10 MHz bandwidth, a maximum supportable data rate of 20 Mbps would be provided by DFE, SAS, or DFE/SAS systems. The DFE, a time-diversity system, and SAS, a space-diversity technique both require approximately 20 dBm (100 mW) to cover 99 percent of the area with a reasonable bit error rate. The DFE/SAS system, taking advantage of both time- and space-diversity, would need around 6 dB (four times) less transmit power at the expense of a more complex receiver that, however, will consume more electronic power for operation.

Example 3.34: Comparison between DSSS, FHSS, and DFE

A DSSS system with a processing gain of 15 (comparable to the value of 11 in IEEE 802.11) and a four-tap RAKE receiver, using time diversity, needs −10 dBm to cover the area with a data rate of 1.33 Mbps. Compared with DFE or SAS operating in the same 10 MHz bandwidth, the DSSS technique consumes about 30 dB (1,000 times) less power and supports a data rate that is about 15 times smaller. A FHSS system with 15 hopping frequencies and Reed-Solmon coding, taking advantage of frequency diversity, provides a 1 Mbps system with almost the same power consumption as DSSS-15 system.

These two examples lead us to the conclusion that for fixed bandwidth channels, DFE and SAS provide the highest data rates at the expense of considerably higher power consumption. The spread spectrum systems provide a better coverage at the expense of lowering the operating data rate.

The quantitative study of radiated power consumption is important for two reasons: (1) Given the restriction on the maximum radiated power by the FCC, we need to know how the coverage of various transmission techniques compare with one another; (2) considering the increasing expansion of the market for battery-operated wireless LANs, we need to examine the total power consumption of a modem. The results of Figure 3.34 can be used directly to analyze the coverage of various transmission techniques in a typical indoor area. This allows a designer or a standards regulator to justify the selection of a particular transmission technique for a specified system description. Because the power consumption varies substantially with the design and fabrication technology used in the implementation of a WLAN, the designer can use the results of Figure 3.34 with their estimation of the electronic power consumption for their own implementation of the system.

Figure 3.35 presents the maximum attainable data rate for different transmission techniques in the test area. The transmission power is maintained a 100 mW, and there is no constraint on the bandwidth—a situation similar to U-NII bands where several hundreds of MHz of bandwidth are available for implementation of a broadband service. In Figure 3.35, the required bandwidth for each technology is indicated on top of the bar graph representing the highest supportable data rate with that technology. In a manner similar to Figure 3.34, a number of interesting practical conclusions can be drawn out of this figure as well.

Example 3.35: Data Rates with MCM

An MCM modem (OFDM) using frequency diversity with 15 carriers can achieve a data rate close to 40 Mbps, and this data rate doubled when coding (COFDM) is added to the system. Coding is a smart method to combine the frequency diversity of the received signal in an OFDM modem. With seven carriers, the data rate drops to slightly more than half of the data rate for a 15 carriers system, but the coding is not as effective as before. The significant difference between the seven carriers and 15 carriers performance reflects the fact that the bandwidth of the seven-carrier modem is not wide enough to take advantage of the frequency diversity in the received signal.

From Example 3.35 one can observe that by appropriate selection of the bandwidth of the individual carrier and by applying the correct coding technique one may completely eliminate the effects of multipath fading for the system. As a result, if there is no power or bandwidth constraint, one can achieve any data rate with multicarrier modems. This property is not shared by other techniques where the increase of data rate finally reaches to a point that the effects of frequency selectivity of the channel are dominant and an increase in the transmit power is no longer effective. The restriction in MCM is implementation complexity that increases with an increase in the number of carriers. In the practical band- and power-limited applications, this

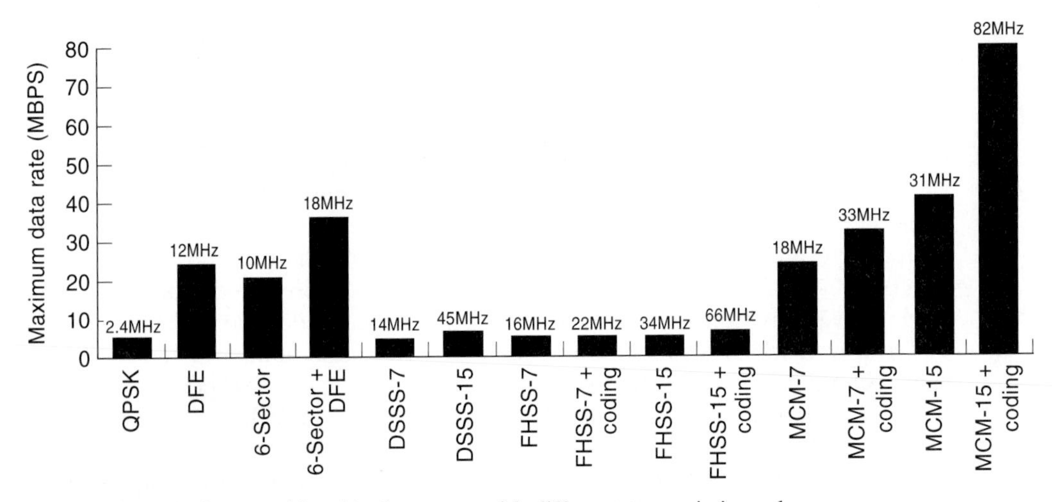

Figure 3.35 Maximum achievable data rates with different transmission schemes.

performance is expected to improve even further if, similar to IEEE 802.11a/HIPERLAN-2, a smart frequency diversity receiver uses multirate transmission or exploits measurement of the channel characteristics to adjust the power in different carriers.

The DFE modem has the advantage of having a single channel, and it can achieve data rates on the order of 20 Mbps that is considered for HIPERLAN-1. Similar results are obtained for sectored antenna systems. However, the complexity of the system for mobile applications has been the drawback for ETSI's RES-10 to adopt SAS for HIPERLAN-1 [WIL95]. Higher data rates on the order of 30 Mbps are obtained using DFE/SAS at the expense of a very complex implementation.

Obviously spread spectrum provides lower data rates. However, bandwidth efficiency is only one side of the equation; the processing gain that spread spectrum provides can serve to increase the fade margin, reducing the transmitter's power requirement. This is an important consideration if the wireless network is being used to connect portable, battery-operated computers. Besides, the results presented here are based on a single-code spread spectrum that is considered by IEEE 802.11. As we saw earlier in this chapter, by using M-ary orthogonal coding the data rate of DSSS used for 802.11 can be increased significantly to support 11 Mbps (the 802.11b standard). Then each spread spectrum code is analogous to a carrier in MCM. In a manner similar to MCM, a CDMA modem of this form has limitations caused by the hardware complexity.

3.12 CODING TECHNIQUES FOR WIRELESS COMMUNICATIONS

In this section, we discuss coding techniques for wireless communications. Coding of bits is a common technique that is employed for a variety of reasons. The most popular reason for coding is error control. Codes are also employed to convert voice into bits. The perceived quality of voice often depends on the speech coding employed, and there is a trade-off between this quality and the bandwidth requirement for digitized voice. In code division multiple access, several different codes are employed to differentiate between multiple users on the same frequency-time channel, between transmissions in various cells, and for error control. Complementary codes employed over regular spread spectrum transmissions can be used to increase the data rate by encoding data symbols with orthogonal codes.

3.12.1 Error Control Coding

In Section 3.9 we discussed a variety of diversity techniques that essentially employ *redundancy* in time, frequency, or space to improve the reliability of reception of transmitted data. In some sense, error control coding is also a diversity scheme because it introduces redundancy in the transmitted bits to correct errors that may be introduced by a channel and if correction is not possible, to provide the capability to detect the occurrence of errors.

Error control coding, as the name suggests, is a technique to *code* the transmitted bits to control the error rate. This becomes increasingly important in radio communications because of the harsh channel conditions. Errors in wireless channels usually occur in bursts. That is, a string of data bits is subject to fading or other harsh impairments such as interference, resulting in several consecutive bits (up to 50 percent of the bits in the burst) arriving at the receiver in error. This is in contrast to wired communications where errors usually occur at random, a bit at a time. Consequently, error control coding schemes are different for wired and wireless channels.

Error control coding is also dependent upon the application under consideration. Voice packets can usually tolerate error rates as high as 1 in 100 bits or 10^{-2}. Such error rates are generally unacceptable for data packets and messaging systems that require error rates as low as 10^{-5}. In some cases, it is impossible to achieve such low error rates, and in such cases, it is reasonable to *retransmit* the data packets that are lost. Such schemes are referred to as *automatic repeat request* (ARQ) schemes. In order to determine whether a packet (or a block of data bits) has been received in error, *block-coding* schemes are employed, and the process is called error detection. Block codes can also be used to correct errors, and this is called *forward error correction* (FEC). Another coding scheme that can be used for FEC is convolutional coding, which employs some memory of previously transmitted bits to determine a *most likely* sequence of transmitted bits (similar to the MLSE receiver in Section 3.10.1.2). These are discussed briefly in Appendix 3B. An interested reader may investigate details of coding schemes in references such as [SKL01], [LIN83], and [MIC85].

3.12.2 Speech Coding

Encoding analog voice into digital format has received much attention as it influences not only the quality of voice, but also the performance and capacity of the system. The important parameters of a speech code are the transmitted bit rate, the speech quality, the robustness in the presence of transmission errors, and the complexity of implementation. Speech quality is usually subjective and is determined by the mean-opinion-score [JAY84]. A low rate speech coder will essentially require less bandwidth for transmission (which is beneficial in wireless environments) but usually compromises on the quality of speech. Table 3.1 includes the details of some important speech coding techniques employed in wireless networks.

Table 3.1 Speech Coders Employed in Some Wireless Systems

System	Application	Voice Coder	Uncoded Rate	Overall Rate
GSM	European digital cellular (2G)	RPE-LTP	13 kbps	22.8 kbps
IS-136	U.S. digital cellular (2G)	VSELP	8 kbps	13 kbps
JDC	Japanese digital cellular (2G)	VSELP	8 kbps	13 kbps
IS-95	U.S. and other digital cellular CDMA	QCELP	9.6, 4.8, 2.4 and 1.2 kbps	28.8 or 19.2 kbps (FEC + repetition)
DCS-1800	PCS in the United States	RPE-LTP	13 kbps	22.8 kbps
PHS	Personal handiphone in Japan	ADPCM	32 kbps	32 kbps
CT-2	European cordless	ADPCM	32 kbps	32 kbps
DECT	Cordless and WPBX	ADPCM	32 kbps	32 kbps

There are two bit rates associated with a speech coder—the uncoded or raw bit rate and the encoded bit rate to account for error correction. Where the bandwidth efficiency is not the most important criterion and the voice quality is, a higher rate encoder such as an adaptive differential pulse code modulation (ADPCM) scheme is employed for speech coding. Resistance to channel errors is also an important issue. Voice codes may give a poor performance when the error rates are as large as 10^{-2}. This is the reason why some error control coding is applied with low-rate voice coders. The voice compression scheme removes redundancies in digitized voice, and the error-coding scheme introduces some structured redundancy to provide better performance. Block interleaving techniques (see Appendix 3B) are sometimes used to improve performance.

The important speech coding techniques are waveform encoding techniques such as pulse code modulation (PCM) and ADPCM; model-based speech coders such as linear predictive coders (LPCs), regular-pulse excitation (RPE), and code-excite linear predictive (CELP) techniques; and hybrid schemes. The complexity of implementation is low in the case of waveform encoding schemes, and the quality of voice is extremely high. The bit rates are also correspondingly larger, making them unattractive for wireless applications. LPC techniques can provide good voice quality at bit rates as low as 2,400 bps compared with the 64 kbps rate of PCM at the expense of burdensome computation. GSM uses a version of RPE, called RPE-LTP, that has an acceptable implementation complexity and delay. It operates at 13 kbps and utilizes a speech frame that lasts for 20 ms. For still further low bit rates, the quality of RPE-LTP coded voice is not adequate, and CELP techniques are preferred. Vector-sum excitation linear predication (VSELP) and QUALCOMM's CELP (QCELP) are employed in the North American digital cellular TDMA and CDMA standards, respectively, to achieve voice coding rates of around 8 kbps.

3.12.3 Coding for Spread Spectrum Systems

Spread spectrum systems and, in particular, CDMA systems are heavily dependent on coding schemes for good performance. The literature in this field is exhaustive, and the interested reader is referred to it [PRO00]. Coding is employed for a variety of functions in CDMA systems. These include codes for separation of channels over the same frequency bands, identification of base stations covering an area, error control coding, and so on. We briefly consider these in the following sections.

3.12.3.1 PN Spreading Codes

Pseudonoise (PN) codes are also called pseudorandom sequences, and they are used as codes for spreading bits in a spread spectrum system. Even though they are called pseudorandom, the sequences are not random, but appear random because they contain almost equal numbers of zeros and ones. The most common PN sequences are the maximal-length sequences or M-sequences that are created using maximal-length linear feedback shift registers (LFSRs) of length m. The name arises from the fact that the M-sequences are the longest sequences that can be generated by an m-stage LFSR. The contents of the LFSRs repeat after a cycle of $2^m - 1$ shifts. The length of the PN sequence before it repeats itself is also $2^m - 1$. The maximal length shift registers are represented by polynomials that denote the connections that are active in the LFSR. For example, an LFSR of length $m = 3$ with feedback connections from the first (exponent 0) and second (exponent 1) shift registers (but not the third) is represented by $1 + x + x^3$. The corresponding LFSR is shown in Figure 3.36. The states of the LFSR and the outputs are also shown in this figure.

Example 3.36: PN Sequences in IS-95

On the reverse channel, IS-95 employs two PN sequences, one for the in-phase and the other for the quadrature channel of the QPSK modulations scheme. The two sequences are each of length $2^{15} - 1 = 32{,}767$. The corresponding polynomials are given by:

$$G_I(x^3) = x^{15} + x^{13} + x^9 + x^8 + x^7 + x^5 + 1$$
$$G_Q(x) = x^{15} + x^{12} + x^{11} + x^{10} + x^6 + x^5 + x^4 + x^3 + 1$$

PN sequences are widely employed because of their nice properties. Some of the important properties include the fact that they have nearly equal numbers of zeros and ones and their autocorrelation exhibits a strong peak with low sidelobes. The periodic autocorrelation has a peak value of $2^m - 1$ for zero lag and a value of -1 for all other lags. Consequently it is possible to differentiate between users by computing the correlation between their allocated sequences and a replica of the sequence of the user under consideration. Figure 3.37 shows the periodic autocorrelation of an M-sequence of length 7 generated by the LFSR of Figure 3.36.

Several other sequences exhibit good correlation properties. These include Kasami sequences [DIN98] and Barker sequences. Barker sequences of length 11 are employed in IEEE 802.11 for simply spreading the codes. The autocorrelation of this sequence is shown in Figure 3.23.

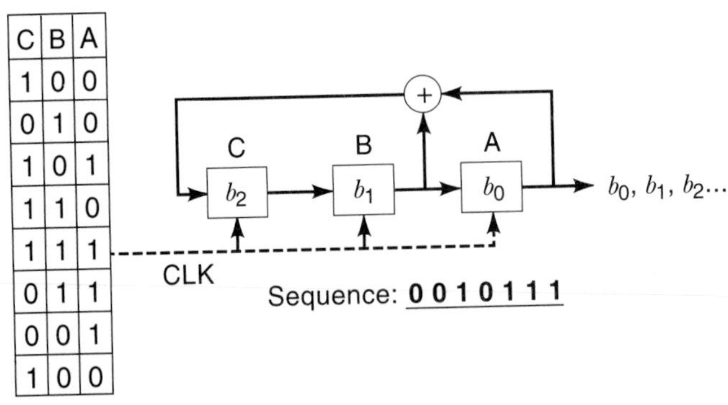

Figure 3.36 An LFSR of length $m = 3$.

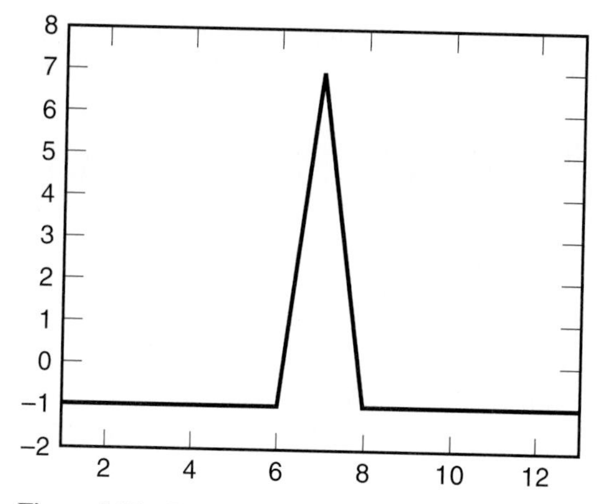

Figure 3.37 Periodic autocorrelation of an M-sequence of length 7.

3.12.3.2 Orthogonal Codes

PN sequences have good correlation properties, but the sidelobes are nonzero and cause interference to other users operating over the same channel. In order to increase capacity, CDMA systems employ orthogonal sequences. The cross-correlation between two orthogonal sequences is zero when synchronized. If two users employ orthogonal sequences for spreading at the receiver, it is possible to completely separate the two signals by correlating them with the signal replicas, without interference. The problem with orthogonal sequences is that the users must be synchronized because orthogonal sequences do not have good correlation properties outside of the zero-lag case. Consequently orthogonal sequences are employed on the downlink where the base station is able to synchronize transmissions. PN sequences are preferred for spreading on the uplink.

Example 37: Orthogonal Sequences in IS-95

IS-95 employs Walsh sequences that are generated by the rows of a Hadamard matrix. A Hadamard matrix is a square matrix of order $n \times n$ where all pairs of rows are orthogonal. The following matrix is an example of a Hadamard matrix.

$$\begin{bmatrix} 0 & 0 & 0 & 0 \\ 0 & 1 & 0 & 1 \\ 0 & 0 & 1 & 1 \\ 0 & 1 & 1 & 0 \end{bmatrix}$$

The row [0 0 0 0] corresponds to the all-zero Walsh code. Each row of the Hadamard matrix corresponds to a unique Walsh code. It is easy to construct Hadamard matrices of order $2^m \times 2^m$. In IS-95, Walsh codes of order 64 generated by a Hadamard matrix of order 64×64 are employed on the downlink for spreading the signal.

Complementary codes are also orthogonal codes employed in the IEEE 802.11b standard. Example 3.17 discusses complementary codes.

3.13 A BRIEF OVERVIEW OF SOFTWARE RADIO

The receiver techniques discussed so far have considered traditional ways of implementation. Recently, with the immense popularity of cellular services and the emergence of a large number of standards, software implementation of the mobile terminal that can dynamically adapt itself with time to the radio environment in which it is located is becoming an attractive solution. This concept is generally referred to as *software radio* [MIT00], [BUR00]. Software radio provides the impetus for fast rollout of new services; mix-and-match services offered by a variety of standards provide choice to the customer and increase the hardware lifetime of mobile terminals. In the literature, software radio has several definitions, including (1) a software-controllable and flexible transmitter/receiver architecture, (2) replacement of radio functionality by signal processing as much as possible, (3) the ability to download an air-interface architecture and dynamically reconfigure the user terminal, (4) multimode or multistandard support, and (5) a transceiver that can define the frequency bands, the modulation and coding schemes, radio resource and mobility management, as well as user applications to be used in software. DSP technology and reconfigurable hardware technology is driving efforts toward actual implementation of software radio.

From a mobile terminal's point of view, software radio needs to have limited circuit complexity, low cost, low power consumption, and a small form factor. Ideally the analog components of a software radio are limited, and most of the radio functionality should be implemented digitally to enable software reconfigurability. However, analog-to-digital conversion at RF is extremely difficult, and instead a programmable down converter appears promising. However the limited bandwidth of the converters, the jitter introduced by the digital operations, and intermodulation products are problems with the sampled signal that are yet to be completely addressed. The processing power becomes an issue of significance in mobile terminals operating with battery power. Special-purpose DSPs, which are required for real-time computation, are expensive and complex.

For adaptive reconfigurability several solutions have been considered. Each manufacturer could have proprietary software for a variety of hardware platforms. This provides the ability to differentiate products in the market, but creates a problem for network operators, especially when mobile terminals are required to "download" the air-interface. A standard hardware platform would eliminate the numerous proprietary solutions, but would restrict the differentiability of products. A third solution proposed includes a real-time compiler that would compile a common source code into solutions for different hardware platforms. In [BUR00], Java programming language is suggested as a language for implementing the third option. This is because Java already possesses the ability to have a uniform "bytecode"[3] for all hardware platforms, requiring only an interpreter for the bytecode for each platform.

---

[3] A bytecode is a compiled format for Java programs that run on any Java virtual machine.

Connecting the terminal to a PC, by using smart cards or over the air, could do the download of the air-interface. Using a PC is not a feasible solution, especially while the user is on the move. Potentially, smart cards could provide a fast solution for changing the air-interface, but there are technology limitations as of today. Over-the-air downloads are preferable and require no effort by the user, and intelligent updates are possible. In this case, there is a suggestion of a *universal control channel* for accessing the radio personalities over the air. The problems with such a solution are the security of such a download, the possibility of the radio channel introducing errors during the download, delay and slowness of the download procedure, and the need for protocols, resources, and bandwidth to assist the procedure. The interested reader is referred to the software radio forum [SDRweb] for details.

APPENDIX 3A PERFORMANCE OF COMMUNICATION SYSTEMS

3A.1 Signal-to-Noise Ratio

Signal-to-noise ratio (SNR) is an important measure of performance in communication systems. Broadly speaking, the SNR is the ratio of the average signal power to the average noise power, either at the input or output of a component of the communication system. The SNR determines quantitatively the *quality* of a signal that is corrupted by noise. In digital communications, where the information is contained in the transmitted bits, the ratio of the signal energy per bit (E_b) to average noise power spectral density (N_0), commonly written as E_b/N_0 or γ_b, is important. A plot of the bit error rate versus γ_b provides the tradeoff between the transmit power requirements of modulation schemes and receiver techniques and the performance in terms of the bit error rate as discussed in the chapter. As we see there, a larger data rate can be supported over the same bandwidth using higher level modulation schemes at the expense of larger transmit powers.

Detailed discussion and calculation of the signal power or energy or the noise power is beyond the scope of this textbook and the reader is referred to [HAY00] or [SKL01]. We do however illustrate a simple case of determining the signal energy per bit E_b. In digital modulation schemes, the information-carrying signal usually consists of a train of symbols selected from a set of M symbols, each of duration T. Since there are M possible symbols, each symbol carries $\log_2 M$ bits. If there are only two symbols, we have the binary case ($M = 2$). In binary phase shift keying (BPSK), there are two symbols: $s_1(t) = \cos(2\pi f_c t)$ and $s_2(t) = \cos(2\pi f_c t + 180°)$, each lasting for a duration T seconds. Suppose $T = n/f_c$ where n is an integer. The energy in one symbol is given by:

$$E_b = \int_0^T s_1^2(t)dt = \int_0^T \cos^2(2\pi f_c t)dt = \frac{1}{2}\left(\int_0^T [1 + \cos(4\pi f_c t)]dt\right) = T/2 \quad (3A.1)$$

As another example, if $M = 8$, each symbol carries 3 bits. Phase shift keying with eight phases (8-PSK) is an example of this case. Each symbol consists of a truncated sinusoid of duration T seconds. There are eight possible truncated sinusoids (each having the same frequency but a different phase). As noted in the chapter, information can be carried either in the phase, frequency, or amplitude of the sinusoid.

The average noise power depends on the type of noise under consideration. Usually, noise arising from thermal effects on electrons is considered additive and its statistical properties are described by a Gaussian uncorrelated random process. In cellular topologies, sometimes the co-channel interference is approximated as additive, white, Gaussian noise (AWGN). In the case of CDMA, the interference from other users in the same cell and neighboring cells is also noise. While this is not Gaussian, sometimes this approximation is made. There have been proposals of schemes that use knowledge of the interference to subtract it from a given signal and thus increase capacity in CDMA systems [MOS96].

3A.2 Performance over Wired Channels

In a band-limited wired or wireless channel, the transmitted waveform is distorted because of the effects of filtering and additive noise. Figure 3A.1 shows a simple square pulse and the received pulse if the effects of filtering and additive noise are considered. In digital communications, the additive noise causes erroneous deci-

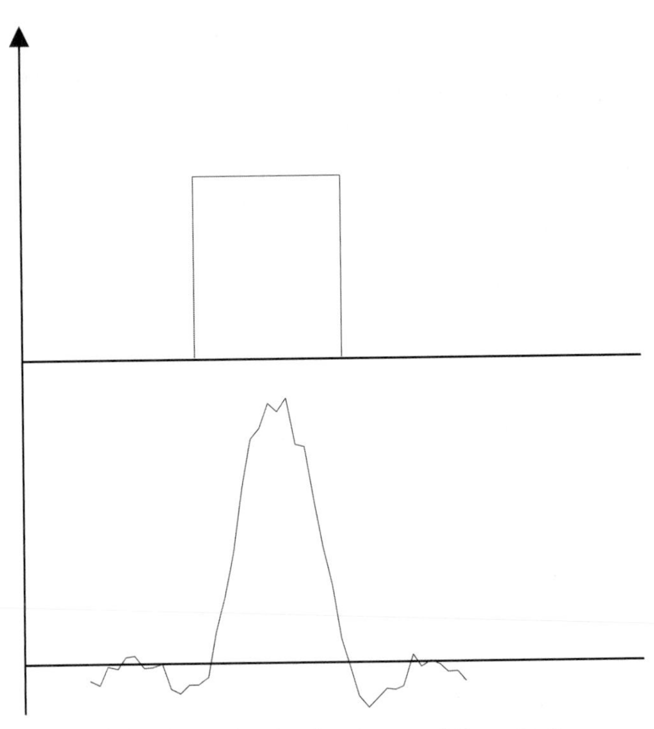

Figure 3A.1 Transmitted pulse shape and the output (received pulse shape) of a bandlimited noisy channel.

sions in detecting the transmitted bits at the receiver. To measure the performance of these transmission techniques, the bit error rate (BER) or probability of bit error (P_e) for transmission of the signal is often plotted against the signal to noise ratio (SNR) in dB for particular implementations of transceivers. The SNR can have several interpretations [PAH95], but the most common interpretation is the ratio of the signal energy *per bit* to the background noise power (γ_b). This is a measure of how much received power is required to detect information in a signal correctly with a given probability of error.

Example 3A.1: Probability of Error in AWGN

The graph in Figure 3A.2 shows the probability of bit error versus SNR for an implementation of BPSK[4] modulation technique The modem designer uses these curves to translate the customer's error rate requirement to the SNR. From the figure, if the customer demands a BER of 10^{-5}, the designer should implement the system so that the ratio of the received signal power to the background noise power at the receiver is higher than 10 dB.

Figure 3A.3 shows the graphs relating the bit error performance to γ_b in AWGN channels representing basic phase and frequency shift modulation techniques. The differential BPSK system (DBPSK) and noncoherent FSK (NC-FSK) do not need a reference of the carrier phase at the receiver, so they can be implemented easier. The purpose of the plots is to demonstrate how much performance degradation is resulted when we design the simpler receivers.

Example 3A.2: Power Requirements of Different Modulation Techniques for a Given Error Rate

As shown in Figure 3A.3, coherent FSK (CFSK) for an error rate 10^{-5} needs roughly 13 dB of γ_b. For the same error rate, an NC-FSK requires around 14 dB. Therefore, a CFSK needs a dB less power to maintain the same error rate as NC-FSK. This means that in practice if the modem is designed so that the main consumption of the battery is due to the transmission power, the lifetime of the battery in a mobile terminal using CFSK modulation is 1.25 times the life of an NC-FSK terminal. As shown in Figure 3A.3, BPSK needs around 10 dB of γ_b for an error rate 10^{-5}. For the same error rate, CFSK requires 13 dB of γ_b. Therefore, a coherent FSK needs 3 dB more power to maintain the same error rate as a BPSK system. Similarly DPSK with simpler implementation needs one more dB transmission power to match the performance of a BPSK.

3A.3 Performance over Wireless Channels

The two main characteristics of the wireless medium affecting the performance of a modem are large fluctuations of the received power level (called fading) and arrival of the received signal via delayed multiple paths referred to as multipath propagation. In the rest of this section, we give an overview of the performance of

---

[4]BPSK uses the phase of the carrier to carry one information bit per symbol.

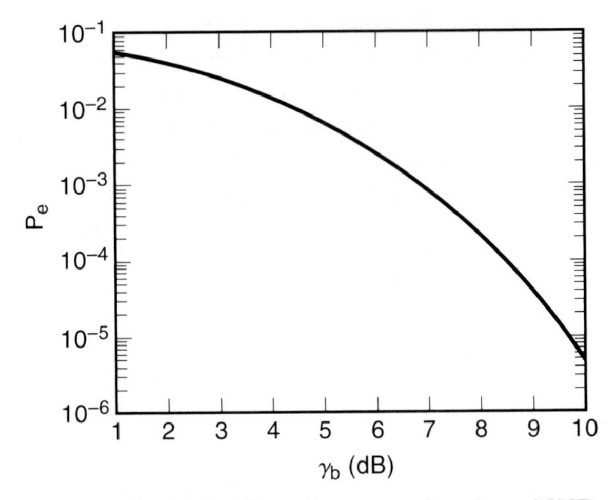

Figure 3A.2 Probability of error as a function of SNR for BPSK signals in an AWGN channel.

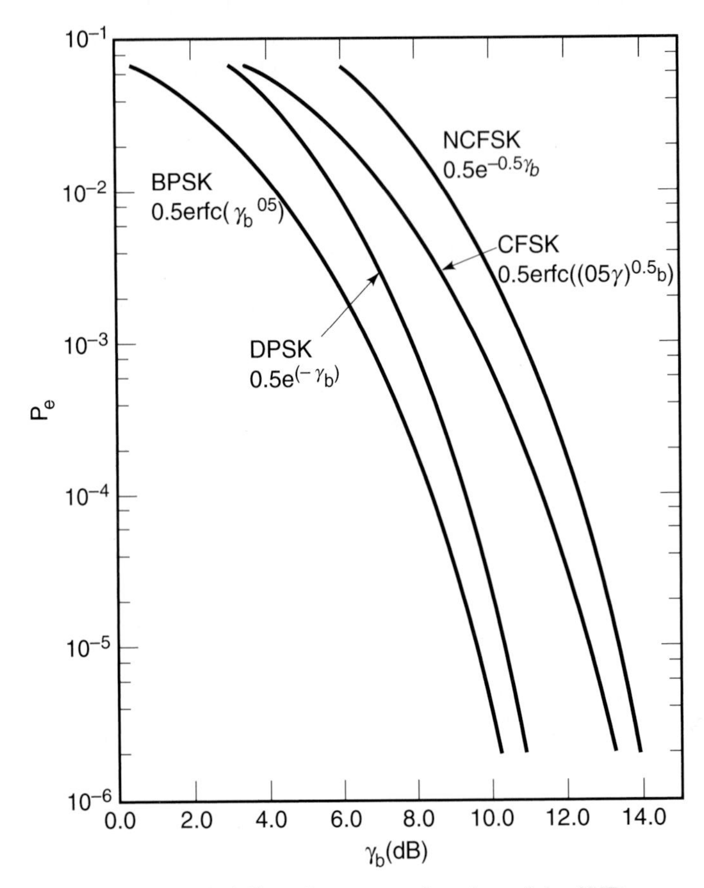

Figure 3A.3 Probability of error as a function of the SNR per bit for various modulation schemes.

basic amplitude, phase, and frequency modulation techniques, and we analyze the performance as Rayleigh fading and simple multipath conditions affect it. This will provide the reader with an intuitive understanding of the measures that are used to evaluate the performance of a modulation technique and how this measure is affected when we increase the power level or when we get exposed to fading or multipath conditions.

3A.3.1 Effects of Fading

As opposed to wired channels, the received signal from wireless channels suffers from strong amplitude fluctuations (of the order of 30–40 dB) which cause fading in the received signal. Figure 3A.4 shows a simple diagram describing the effects of fading on the BER. During periods of signal fading, the error rate of the transmission system increases, and when the system is out of fade, the error rate becomes negligible. To evaluate the performance over a fading radio channel, the *average* BER versus *average* received SNR-γ_b is used.

Example 3A.3: Probability of Error of BPSK in Rayleigh Fading Channel

The graph in Figure 3A.4 shows the average probability of error versus average received SNR over a flat Rayleigh fading channel for the same implementation of BPSK as in Example 3A.1. This time, for an average BER of 10^{-5}, we need an average received SNR of around 40 dB, which is 30 dB more than the required SNR for a nonfading wired channel.

To compensate for the additional SNR requirement and to maintain the average BER during fading, designers of wireless modems increase the SNR re-

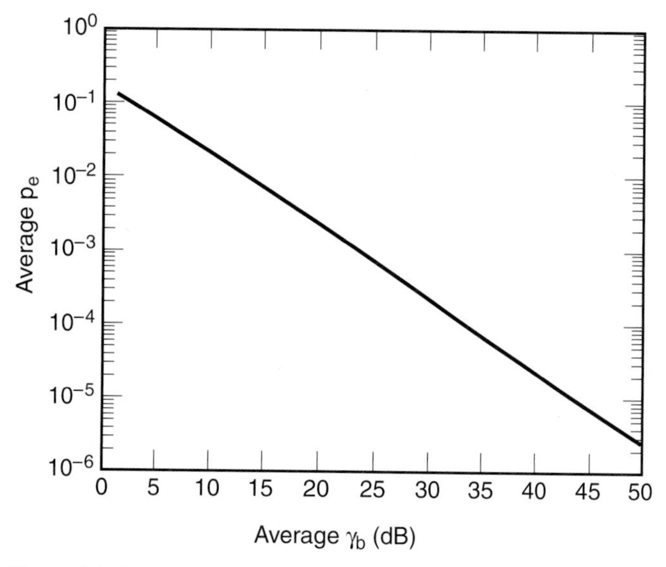

Figure 3A.4 Probability of error of binary phase shift keying in a flat Rayleigh fading channel with additive white Gaussian noise.

quirement with a margin referred to as the *fading margin.* Addition of the fading margin to the SNR requirement maintains the average performance at the required BER.

Table 3A.1 gives the BER versus SNR over a nonfading and flat Rayleigh fading channel with AWGN for basic modulation techniques. In this table, M is the number of symbols, $2^m = M$, and erfc is the complementary error function.

3A.3.2 ISI Effects Due to Multipath

One of the main differences between wireless and wired channels is that the wireless channel suffers from multipath propagation. The shape of the received pulse and its time duration of the signaling pulse are both changed due to multipath arrivals. The difference between the first and the last arriving pulses is the *delay spread* of the channel. If the symbol duration is much larger than the multipath spread, this means that the data rate is much smaller than the coherence bandwidth of the channel, and all pulses received via different paths arrive roughly on top of one another, causing amplitude fluctuations and fading. If the ratio of the delay spread to the pulse duration becomes considerable, the received pulse shape is severely distorted, and it also interferes with neighboring symbols, causing ISI. In addition to SNR fluctuations due to fading effects, the interference power also degrades the performance. However, the ISI effect of multipath degrades the performance in a different manner than fading. The effects of fading can be compensated via an increase in the transmit power by a fading margin. Increase of the transmit power cannot compensate the effects of ISI. This is because an increase in the transmit power increases the signal, as

Table 3A.1 Probabilities of Bit Error for Common Modulation Schemes

Modulation Scheme	BER in AWGN Channels	Average BER in Flat Rayleigh Fading Channels
On-Off Keying or Amplitude Shift Keying	$\frac{1}{2}\operatorname{erfc}\sqrt{\frac{\gamma_b}{2}}$	Hard to use in fading channels because of amplitude dependence
Binary Phase Shift Keying	$\frac{1}{2}\operatorname{erfc}\sqrt{\gamma_b}$	$\frac{1}{2}\left(1-\sqrt{\frac{\gamma_b}{(1+\gamma_b)}}\right)$
Differential Phase Shift Keying	$\frac{1}{2}e^{-\gamma_b}$	$\frac{1}{2(1+\gamma_b)}$
Frequency Shift Keying (Coherent Detection)	$\frac{1}{2}e^{-\frac{\gamma_b}{2}}$	$\frac{1}{2}\left(1-\sqrt{\frac{\gamma_b}{(2+\gamma_b)}}\right)$
M-ary Phase Shift Keying	$\approx\frac{2^{m-1}}{M-1}\operatorname{erfc}\sqrt{\sin^2\left(\frac{\pi}{M}\right)m\gamma_b}$	$\approx\frac{2^{m-1}}{M-1}\left(1-\sqrt{\frac{\sin^2\left(\frac{\pi}{m}\right)m\gamma_b}{\left(1+\sin^2\left(\frac{\pi}{m}\right)m\gamma_b\right)}}\right)$
M-ary QAM	$\approx\frac{2^{m-1}}{M-1}M\operatorname{erfc}\sqrt{\frac{3}{2M-1}m\gamma_b}$	Hard to use in fading channels because of amplitude dependence

well as the ISI interference power keeping the signal to interference ratio at the same level. Example 3A.4 will clarify the situation.

Example 3A.4: ISI Effects of Multipath

Figure 3A.5(a) represents a simple two-path channel model. The delay between the arrival of the paths is t and the duration of the symbol is T. Figure 3A.5(b) shows the probability of error for a BPSK modem versus the normalized delay spread t/T for different relative powers of the second path to the power of the first path and SNR of 15 dB. In the leftmost corner, we have $t/T = 0$ (no multipath), and the error rate of the system is 10^{-8}. Until a value of t/T of around 10 percent the effects of multipath are negligible. For higher normalized values of the multipath spread, the lowest curve belongs to the case where the ratio of the strength of the second path amplitude to the main path is 10 percent, the next is for a ratio of 50 percent, and the top path belongs to the case where both paths have the same strength. For a 10 percent multipath strength, the error rate remains reasonable even for high ratios of t/T. For 50 percent and 100 percent multipath strengths, the BER degrades drastically to around 10^{-2}.

In practical situations, the multipath delay spread is measured by the RMS multipath delay spread, τ_{rms}. The RMS delay spread τ_{rms}, is the square root of the second central moment of the multipath profile of a channel that is a good measure of both distribution of amplitude strength and arrival times of the multiple signal paths. We referred to the inverse of the RMS multipath spread as the coherence bandwidth of the channel. Usually, the coherence bandwidth of the channel is known and designers try to increase the data rate by reducing the transmitted pulse duration T, while avoiding the effects of multipath arrivals. Experimental work in this field has shown that as long as the ratio τ_{rms}/T is less than 20 percent, the ISI effects of the multipath

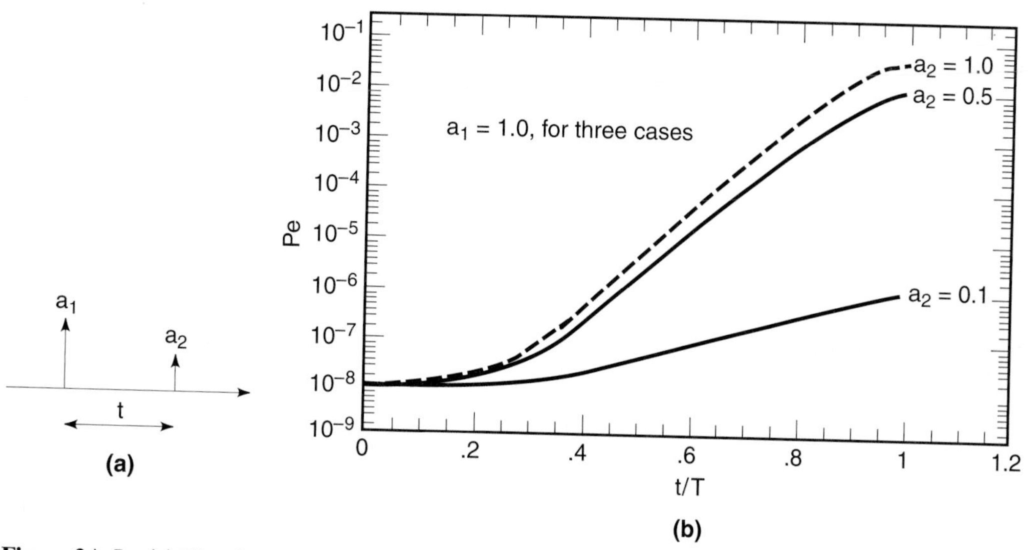

Figure 3A.5 (a) The simple two-path model and its parameters and (b) performance of BPSK in a two-path channel.

are negligible. In other words, radio design engineers try to keep the data rate to below 20 percent of the coherence bandwidth of the channel to avoid ISI caused by multipath. A number of signal processing techniques such as equalization have been developed that allow higher data rates over wireless channels.

APPENDIX 3B CODING AND CORRELATION

3B.1 Block Codes

Block coding, as the name suggests, involves encoding a *block* of bits into another block of bits, with some redundancy to combat errors. Block coding in its simplest form consists of a *parity check* bit. An extra bit is added to each block of k bits, and the extra bit is selected so that each new block of $k + 1$ bits has either an even number of ones or an odd number of ones. The extra bit is called the parity check bit. The result is that if the channel introduces a single bit error in a block of $k + 1$ bits, the number of ones in the block will no longer be even (or odd), and the receiver can *detect* the error. The simplicity of this code is clear because if there is an even number of errors in the block the errors cannot be detected because number of ones is maintained even or odd.

Using a variety of algebraic techniques, efficient encoding rules have been obtained that calculate a set of $n - k$ parity check bits which apply parity checks to a group of bits in the block of k bits. Together with these parity check bits, the size of the encoded block is $k+(n-k) = n$ bits. The block code is called an (n, k) block code, and the code rate is $C = k/n$. This means that if the raw data rate is R bps, only kR/n bps corresponds to actual data. The factor k/n is called the code rate. The rest of the bits do not contain useful information and are included only for error control purposes.

Example 3B.1: Code Rate for the GSM Control Channel

In GSM a block of 184 bits is encoded into 224 bits of codeword on the control channel before it is sent to a convolutional encoder. The number of parity check bits is 40. The code rate of this block encoder is 184/224 = 0.82.

In multilevel modulation schemes, it is possible to encode *symbols* (nonbinary alphabets) in a similar manner. Bose-Chaudhuri-Hocquenghem (BCH) codes are a popular class of nonbinary codes. Reed-Solomon (RS) codes, a subset of the BCH codes, are a good example of nonbinary block codes employed in a variety of wireless systems. The symbols are commonly drawn from a set of 2^m alphabets (each alphabet represents m bits). An RS codeword has length $n = 2^m - 1$. The number of symbols being encodes can vary from 1 to $n - 1$. Depending on the number of symbols encoded, the minimum distance between the codes (see next section), which in the case of RS codes is $n - k + 1$, changes.

Example 3B.2: Nonbinary Block Codes in CDPD

CDPD uses a (63, 47) RS encoder. Instead of encoding bits, CDPD encodes alphabets from a set of 64 symbols, each representing six bits. This means that a block of

47 symbols is encoded into a block of 63 symbols for a symbol code rate of 47/63 = 0.746. The minimum distance between the codes is $d_{min} = 63 - 47 + 1 = 17$. It can thus correct up to eight symbol errors.

3B.1.1 Operation of Block Codes

Block codes use finite-field arithmetic (modern algebraic techniques) properties to encode and decode blocks of bit or symbols. Most operations are based on linear feedback shift registers that are easy to implement and inexpensive. Most block codes are created in *systematic* form where the k data bit are retained as is and the $n-k$ parity check bits are either prepended or appended to them (see Fig. 3B.1). The parity check bit are generated via a generator matrix or generator polynomial. The encoded block of n bits is called the *codeword,* and this is transmitted over the channel. Codes generated by a polynomial are called cyclic codes, and block codes of this nature are called *cyclic redundancy check* (CRC) codes and are employed in a variety of data transmission schemes for error correction and detection.

The received word may be identical to the codeword in the case of error-free transmission and may have been modified due to channel errors. The modifications may result in another valid codeword, in which case, it is not possible either to detect or correct the errors. The probability of such a false detection is upper bounded [WOL82] by:

$$P_{FD} \leq 2^{-(n-k)} \tag{3B.1}$$

The idea behind the design of block codes is to thus have a large *distance* between any pair of codewords. This distance is measured in terms of the number of positions (bits or symbols) in which the codewords differ, and is called the Hamming distance. The *minimum* Hamming distance between the set of all codewords of a block code determines its error detection or correction capability. A block code with a minimum distance of d_{min} can detect blocks of errors that have a "weight"[5] of less than d_{min} and can correct blocks of errors that have a weight up to t_{max} where

$$t_{max} = \left\lfloor \frac{d_{min} - 1}{2} \right\rfloor \tag{3B.2}$$

Here, $\lfloor x \rfloor$ refers to the largest integer less than or equal to x. An error block is also represented in a manner similar to a block of data bits. Those bits that are not changed are represented by zeros and those that are represented by ones. The weight of the error block corresponds to the number of ones in the block, or the number of bits that are changed and thus in error. Intuitively, we can see why a block code with a minimum distance of d_{min} can correct up to t_{max} errors. Given two codewords in the set, the distance between them is greater than or equal to d_{min}. An error block modifies a codeword into the received word. If its distance from the correct codeword is *less* than half the distance between the correct codeword and

[5]"Weight" here refers to the number of ones in the codeword.

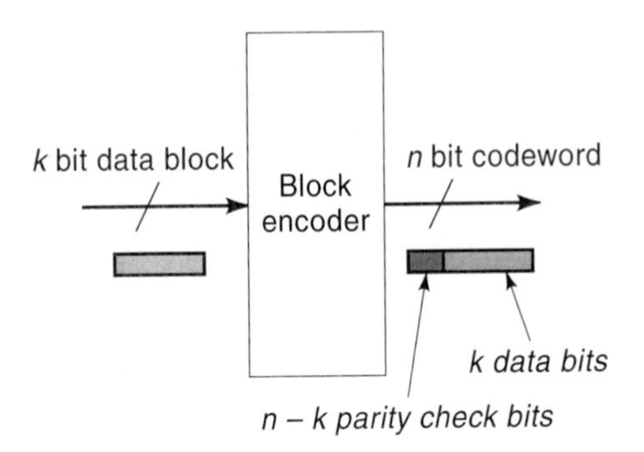

Figure 3B.1 Operation of a block code.

any other codeword, we can associate the received word as being *closest* to the original codeword, and correct it accordingly. If, however, the error block modifies the received word to make it closer to some other codeword, the error correction procedure will not work.

For sensitive data transfer, it is still possible to *detect* errors with weights larger than t_{max}. The detection of errors is performed by determining whether the received word is a valid codeword. This can be done by computing the parity check bits again from the k data bits and comparing it with the parity check bits received over the channel. It could be possible that the data bits were received correctly but the parity checks bits were in error, but the two cases are indistinguishable.

3B.2 Convolutional Codes

Unlike block codes convolutional codes do not map individual blocks of bits into blocks of codewords. Instead they accept a continuous stream of bits and map them into an output stream introducing redundancies in the process. Usually a code rate can be defined for convolutional codes as well. If there are k bits per second input to the convolutional encoder and the output is n bits per second, the code rate is once again k/n. The redundancy is however dependent on not only the incoming k bits, but also several of the preceding k bits. The number of the preceding k bits used in the encoding process is called the constraint length m that is similar to the *memory* in the system. Typically, the values of k, n, and m are 1–2, 2–3, and 4–7 in commonly employed convolutional codes.

Example 3B.3: Convolutional Coding in Wireless Systems

In the IS-95 CDMA standard, a convolutional encoder is employed in both the forward and reverse links. In the forward link, a rate 1/2 convolutional encoder is used that has a constraint length of $m = 9$. On the reverse link, a rate 1/3 convolutional encoder is employed with the same constraint length.

GSM also employs a convolutional encoder. Digitized voice is broken up into 182 class-I bits and 78 class-II bits. The most significant 50 bits of the class-I bits is enhanced with a block code that adds three parity bits. The sum total of 182 class-I bits plus the three parity bits, plus four tail bits (189 bits in all) are passed through a rate 1/2 convolutional coder to produce 378 bits. The 78 class-II bits are added to these 378 bits to produce 456 bits of encoded data.

In general, convolutional codes are more powerful than block codes in terms of FEC, but are not useful for error detection or ARQ schemes. At the receiver, FEC is performed using a maximum-likelihood decoding algorithm that determines what sequence was most likely transmitted given the received sequence of bits. The Viterbi algorithm is the most common algorithm of all, and several VLSI implementations of this algorithm are available.

3B.2.1 Hard Decision versus Soft Decision

Most receivers simply decide whether a received bit is a zero or a one and send this information to the channel decoder. The decoder employs its knowledge of the coding scheme to either detect or correct errors at this stage. Such a procedure is called *hard decision.* In soft decision decoding schemes, the receiver will convert the received signal into one of several *levels* of output. Usually there are Q quantized levels, where Q is larger than the number of alphabets. For example, in a binary system, there may be eight levels of quantized demodulator outputs instead of the usual two levels with hard decision. The decoder will use the additional information now available in order to make a decision on the received block. The Viterbi algorithm can be used for soft decisions in both convolutional encoding and block coding techniques.

3B.3 Automatic Repeat Request Schemes

In voice networks, if a packet is received in error, it is either dropped or replaced with an attenuated version of the previous packet to provide a semblance of continuity, and transmissions of damaged packets are not done because of the sensitivity of voice to delay. ARQ schemes essentially are used in data networks where reliability of received information is of paramount importance and delay is less of a problem compared with real-time multimedia applications. If a block of data is received in error, the receiver requests retransmission of the block of data. This request may be explicit or built into several protocols already operational in the system for flow control or other purposes. Usually an acknowledgment packet is employed to indicate correct reception of one or more transmitted packets. If an acknowledgment is not received in a certain time frame, or a negative acknowledgment is received, the transmitter will retransmit the packet. Such mechanisms are commonly employed with the random-access protocols discussed in Chapter 4.

There are three basic ARQ schemes. The stop-and-wait ARQ scheme waits for an acknowledgment for each individual packet before sending the next one. This is especially inefficient if the round-trip times are large because the transmitter spends a lot of time waiting for the acknowledgment. In order to improve upon this scheme, the Go-back-N ARQ scheme transmit up to N packets at a time and waits for acknowledgments. Multiple packets can be acknowledged with one response. Depending on the receipt of acknowledgments, the transmitter will back up to the last correctly received packet and retransmit the following ones (N or less packets). It is possible that some of the subsequent packets are received correctly, but they will be discarded. In order to eliminate this inefficiency, a selective-repeat ARQ scheme can be employed. Here, only those packets that are received in error are retransmitted.

Example 3B.4: Acknowledgments and Retransmissions in IEEE 802.11

In IEEE 802.11, every transmitted packet is acknowledged because the channel is unreliable. It is similar to a stop-and-wait protocol in that sense. Because round-trip times are small, and both the AP and the mobile stations share the channel, the inefficiency is limited. It is often possible to piggyback the acknowledgments over data packets transmitted in the other direction.

3B.4 Block Interleaving

Block interleaving is a technique used in wireless systems to spread the errors out over a large number of codewords. For instance, consider a Hamming code that can correct single bit errors over codewords of seven bit size. This means that if there is a single error over a block of seven bits, the coding scheme can correct it. On the other hand, a burst of five errors cannot be corrected by this code. If we can, however, spread these errors over five codewords so that each codeword "sees" only one error, it is possible to correct each of the errors. The way this works is shown in Fig. 3B.2. Codewords are arranged one below the other, and bits are transmitted vertically. At the receiver, the codewords are reconstructed, and the bits are decoded horizontally. Because the burst of errors affects the serially transmitted vertical bits that are spread over several codewords, the errors can be corrected. Block interleaving introduces delay because several codewords have to be first received before the voice packet can be reconstructed. There is only so much delay that is acceptable for normal voice conversations, and the interleaving process should not create an unacceptable value of delay in the process.

Example 3B.5: Block Interleaving in GSM

In GSM, the output of the convolutional encoder consists of 456 bits for each input of 228 bits. The 456 bits are split into eight blocks of 57 bits each. The 57 bits are spread over eight frames so that even if one frame out of five is lost, the voice quality is not affected. The delay in reconstructing the codewords corresponds to

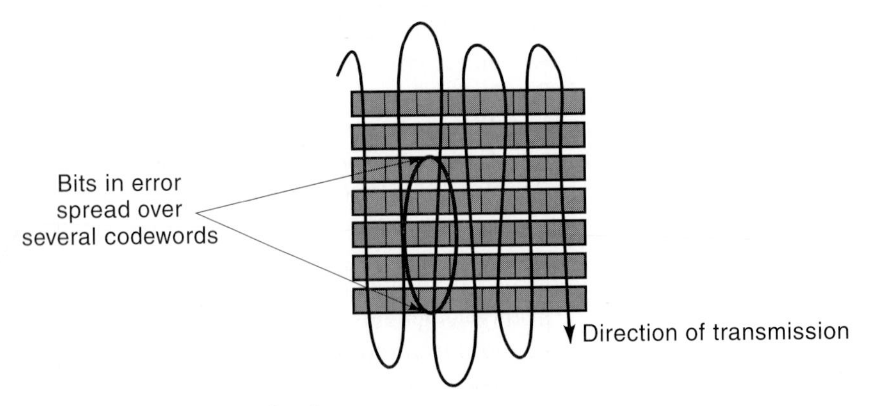

Figure 3B.2 Block interleaving.

the reception of eight frames, which takes 37 ms. A delay of 50 ms is usually tolerable for voice conversations.

3B.5 Correlation

Correlation is a measure of how similar two quantities are. When we talk of correlation between signals, we are measuring their similarity as a function of time or time lag. This essentially provides an idea of how similar a given signal $s_1(t)$ is to another signal $s_2(t)$ or a time shifted version of it [such as $s_2(t + \tau)$]. If the two signals being considered for comparison are the same, we call this the *autocorrelation.* If the two signals being considered are different, we call it the *cross-correlation.* Detailed calculation of autocorrelation and cross-correlation belong to a signals and systems course and such details are readily available in [HAY00]. We only discuss the discrete version of these and briefly.

Consider samples of two signals $a(t)$ and $b(t)$ given by two sequences $\boldsymbol{a} = [a_0, a_1, a_2, a_3, \ldots a_n]$ and $\boldsymbol{b} = [b_0, b_1, b_2, b_3, \ldots b_n]$. The periodic cross-correlation between the two sequences is given by:

$$R_{ab}(l) = \sum_{j=-n}^{n} a_j b_{j-l} \tag{3B.3}$$

where the values a_i and b_i are zero for $i < 1$ and $i > n$.

Example 3B.6: Autocorrelation and Cross-Correlation

For instance, suppose $a = [1, 1, 1, -1]$ and $b = [-1, 1, -1, 1]$, the cross-correlation is: $[1, 0, 1, -2, 1, -2, 1]$. The periodic cross-correlation does the same, but with the sequences wrapping around themselves so that each term in the correlation will be a scalar product of sequence a and a cyclically shifted version of b, i.e., the periodic autocorrelation of a and b will have the following form:

$$R_{per,ab} = [a_0b_0 + a_1b_1 + \ldots + a_nb_n, a_0b_1 + a_1b_2 + \ldots + a_nb_0, a_0b_2 + a_1b_3 + \ldots + a_nb_1, \ldots a_0b_n + a_1b_{n-1} + \ldots + a_nb_0]$$

For the particular example shown here, the periodic autocorrelation is $[-2, 2, -2, 2]$.

Autocorrelation and cross-correlation play an important role in spread spectrum where sequences with good correlation properties need to be used for robustness under interference and multipath.

QUESTIONS

3.1 Name the two most popular modulation techniques used in digital cellular modems and give one example standard that uses each of them.

3.2 What is the difference between GMSK and FSK modulation techniques? What is the difference between $\pi/4$-QPSK and QPSK modulation techniques?

3.3 For a fixed given bandwidth, which transmission technique among DFE, sectored antenna, MCM, DSSS, and FHSS provides the highest data rate and which one consumes the minimum power?

3.4 For a fixed transmission power and unlimited available bandwidth, which transmission technique among DFE, sectored antenna, MCM, DSSS, and FHSS provides the highest data rate.

3.5 In an OFDM modem with 48 channels, each channel uses 16-QAM modulation. If the overall transmission rate is 10 Mbps, what is the symbol transmission rate per channel?

3.6 Explain inter-symbol interference. What are the sources of ISI? What techniques can be used to combat ISI?

3.7 Name five design considerations in selecting a modulation scheme for a wireless network.

3.8 Why is out-of-band radiation an important issue in designing modulation schemes? How is GMSK a good solution for this?

3.9 Why is PPM used with infrared communications instead of PAM?

3.10 Differentiate between frequency hopping and direct-sequence spread spectrum.

3.11 Name four diversity techniques.

3.12 Explain how spread spectrum receivers can exploit multipath diversity using RAKE receivers.

3.13 What are sectored antennas? How are they useful in combatting multipath?

3.14 Differentiate between block codes and convolutional codes.

3.15 What is block interleaving? How is it useful in combating the effects of fast fading?

PROBLEMS

3.1 Use the results of data rate versus power consumption for the central part of the Atwater Kent building (see Figures 3.34 and 3.35) to answer the following questions:

a. For 10MHz of bandwidth, what is the power requirement and maximum data rate supported by DSSS-15, MCM-15, and DFE modulation?

b. What is the maximum data rate and required bandwidth for DSSS-15, MCM-15, and MCM modulations to cover this area with 100 mW power?

c. Name three standards using DSSS, DFE, and MCM for implementation of wireless LANs.

3.2 For a 64-QAM modem:

a. Give the SNR at which the error rate over a telephone line is 10^{-5}.

b. Give the average SNR at which the average error rate over a flat Rayleigh fading radio channel is 10^{-5}.

c. Give the outage rate from the threshold error rate of 10^{-5} if the system operates in a Rayleigh fading radio channel and the receiver uses a single antenna. Assume that the average received SNR per symbol is 14 dB.

3.3 The IS-136 digital cellular replaces the AMPS analog cellular. The modulation technique for the IS-136 is $\pi/4$-QPSK.

a. What is the minimum required average SNR for the IS-136 modems if the minimum acceptable average error rate is 10^{-3} and the channel is assumed to be flat Rayleigh fading?

b. What is the threshold SNR if the acceptable error is 10^{-3}?

c. With the average SNR of part (a) and the threshold SNR of part (b), what is the outage rate of the system?

3.4 In the following differential encoded Manchester coded signal

a. Show the beginning and the end of each bit.

b. Identify all the bits in the data sequence.

c. Identify the bits if it was non-differential Manchester coded.

3.5 Consider Table 3A.1 that gives the expressions for the probability of error of various modulation schemes in an AWGN channel in terms of the complementary error function.

a. Plot the error rate curves as a function of the SNR per bit. Use $M = 16$ and 64 for M-ary phase shift keying and M-ary QAM. Note that $2^m = M$ and that the SNR per bit should be expressed in absolute values in these expressions. However, plot the error curves as a function of the SNR per bit in dB.

b. Use the Matlab erfinv function (or other software tools you may prefer) to determine the SNR per bit required to obtain an error rate of 10^{-5} for each of the above modulation schemes. Comment on the results. Which modulation scheme(s) is (are) power efficient? Why?

c. Repeat (a) and (b) for Rayleigh fading channels.

3.6 This problem illustrates the concept of using orthogonal waveforms in CDMA. Figure 3.38 shows the waveforms used by three users—A, B, and C—to send messages from a wireless text transmitter device to a base station. The chip duration is 100 ns and the bit duration is 400 ns. Each of them is transmitting two bits each, corresponding to 00, 01, and 11, respectively. They transmit their waveform if the bit is a zero, and the negative of the waveform if it is a one.

a. Draw the signals corresponding to the bits transmitted by each of them.

b. Draw the composite transmitted signal that will be the sum of the individual signals if they are transmitting at the same time and are synchronized perfectly.

c. Let us refer to the transmitted signal in (b). We shall call the part of the signal during the durations 0–400 ns, 400–800 ns, and so on, as "symbols." Compute the cross-correlation at zero lag of each of the symbols in the transmitted waveform with the waveforms of A, B, and C. Comment on the results.

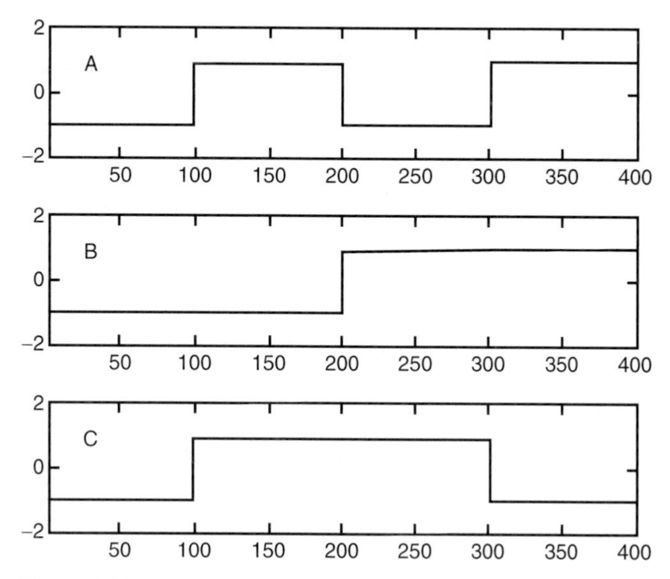

Figure 3.38 Orthogonal waveforms transmitted by three users.

3.7 Assume that instead of the waveforms shown in Figure 3.38, A, B, and C use the following waveforms:

A: $\cos(\pi t/T)$, $0 \leq t \leq T$; B: $\cos(2\pi t/T)$, $0 \leq t \leq T$; C: $\cos(3\pi t/T)$, $0 \leq t \leq T$

Draw these three waveforms for $T = 1$ microsecond. Show that they are orthogonal. Also draw the spectrum of the three waveforms. Note that these waveforms are at baseband. If a single user is transmitting all three waveforms on a single carrier, we have OFDM with three sub-carriers.

3.8 In Matlab, the function conv performs the convolution of two vectors. Samples of a signal can be represented as vectors (as in the case of spread spectrum pulses). Suppose that the M-sequence of Problem 3.1 is used as the basic waveform for transmitting a zero and its negative is used to transmit a one. The matched filter will have the flipped version of the M-sequence as its impulse response, i.e., [1, −1, −1, −1, −1, 1, 1, 1, −1, 1, 1, 1, −1, −1, 1]. You convolve the input to the matched filter with this vector to get the output. Let us suppose we are transmitting four bits: 0, 1, 1, 0. Assume also that the channel is a three-path channel with interpath delays of $5T_c$ and $8T_c$, respectively. Plot the output of the matched filter. What will be the output if you are using NRZ with a matched filter? (Note that in this case, you need to replace the M-sequence by all ones. What will the MF impulse response be?)

3.9 Show that the sequences shown in Example 3.37 are orthogonal to one another. (*Hint:* Represent the zeros by −1s in the sequences. Orthogonality is demonstrated as follows: Multiply the sequences element-wise and sum the resulting elements. If the sum is zero, the sequences are orthogonal.)

3.10 Show the steps to generate the periodic M-sequence of period 7 from the linear feedback shift register shown in Figure 3.36.

3.11 Draw the linear feedback shift register used to generate the PN sequences in the CDMA standard for digital cellular systems. Use the polynomials given in Example 3.36.

3.12 Using Reed-Solomon codes as an example, plot a graph showing the relationship between the code rate n/k and the minimum distance d_{min}. Discuss the tradeoff between reduced data rates and improved performance as the error correcting code is made more powerful.

3.13 Suppose the maximum fade duration over a radio channel is 0.001 ms. Assume that all the bits are in error when a signal encounters a fade. What is the maximum number of consecutive bits that are in error for a transmission through this channel if the data rate is 10 kbps? If the data rate is 11 Mbps?

3.14 Block interleaving is a solution to enable simple error correcting codes to correct long bursts of errors. For both the situations in Problem 3.13, determine the number of codewords over which interleaving has to be performed if the length of the codeword is 7 bits and a single bit error can be corrected.

3.15 If codewords have to be received in sequence for message delivery, determine the delay encountered by the block interleaving scheme of Problem 3.14. How does this impact voice transmission?

CHAPTER 4

WIRELESS MEDIUM ACCESS ALTERNATIVES

4.1 Introduction

4.2 Fixed-Assignment Access for Voice-Oriented Networks

4.2.1 Frequency Division Multiple Access (FDMA)
4.2.2 Time Division Multiple Access (TDMA)
4.2.3 Code-Division Multiple Access (CDMA)
4.2.4 Comparison of CDMA, TDMA, and FDMA
4.2.5 Performance of Fixed-Assignment Access Methods

4.3 Random Access for Data-Oriented Networks

4.3.1 Random Access Methods for Mobile Data Services
4.3.2 Access Methods for Wireless LANs
4.3.3 Performance of Random Access Methods

4.4 Integration of Voice and Data Traffic

4.4.1 Access Methods for Integrated Services
4.4.2 Data Integration in Voice-Oriented Networks
4.4.3 Voice Integration into Data-Oriented Networks

Questions

Problems

4.1 INTRODUCTION

This chapter presents an overview of the access methods commonly used in wireless networks. Access methods form a part of Layer 2 of the OSI protocol stack and Layer 3 of the IEEE 802 standard for LANs that is responsible for interacting with the medium to coordinate the successful operation of multiple terminals over the wireless channel. Most multiple access methods were originally developed for wired networks and later on adopted to the wireless medium. However, requirements on the wired and wireless media are different, thereby demanding modifications in the original protocols to make them suitable for the wireless medium. Today the main differences between wireless and wired channels are availability of bandwidth and reliability of transmissions. The wired medium is moving toward optical media with enormous bandwidth and very reliable transmission. Bandwidth in wireless systems is always limited because the medium (air) cannot be duplicated, and the medium is shared between all wireless systems, including multichannel broadcast television and a number of other bandwidth demanding applications and services. In the case of wired operation, we can always lay additional cable to increase the capacity as needed even if it is an expensive proposition. In a wireless environment, we can reduce the size of *cells* to increase capacity as discussed in Chapter 5. With the reduction of the size of the cells, the number of cells increases, and the need for improvements in the wired infrastructure to connect these cells increases. Also the complexity of the network for handling additional handoffs and mobility management increases posing a practical limitation upon the maximum capacity of the network. As far as transmission reliability is concerned, as we saw in Chapter 2, the wireless medium always suffers from multipath and fading, which causes a serious threat to reliable data transmission over the communication link. Because the wireless channel is so unreliable, as discussed in Chapter 3, people have developed a number of signal processing techniques to improve transmission reliability over the wireless channel. In spite of these techniques, the reliability of the wireless medium is far below that of the wired medium used as the backbone of the wireless networks.

Although in practice we prefer to have the same access method and the same frame structure for wired backbone and the wireless access, wireless networks often use different packet sizes and a modified access method to optimize the performance to the specifics of the unreliable wireless medium.

Example 4.1: IEEE 802.3 and IEEE 802.11

The IEEE 802.3 standard (based on Ethernet) is the successful and dominant standard for wired LANs. Consequently, the IEEE 802.11 WLAN standard, in ideal situations, desired using the same access method as previously established with IEEE 802.3. Carrier sense multiple-access with collision detection (CSMA/CD) is the protocol used in the Ethernet. However, collision detection in wireless channels is extremely challenging, and IEEE 802.11 had to resort to carrier sense multiple-access with collision avoidance (CSMA/CA) that can be viewed as a wireless adaptation of IEEE 802.3.

Example 4.2: ATM and Wireless ATM

In the 1990s, ATM was perceived to be the transmission scheme for all future networking. In the mid 1990s when wireless solutions were considered, a wireless ATM working group was formed to extend the ATM short packet solution with QoS for wireless access. The group had to make significant compromises as discussed in Chapter 12 because ATM was designed for broadband and reliable transmission over optical channels.

To avoid substantial overlap with existing literature, we describe access methods used in wireless networks with justification of why and how they are employed in different wireless networks.

As we explained in previous chapters, wireless networks have evolved around voice or data applications, and as a result we can divide them into voice- and data-oriented networks. The access methods adopted by voice- and data-oriented networks are quite different. Voice-oriented networks are designed for relatively long telephone conversations as the main application. Each communication session for this application exchanges several megabytes of information in both directions. A signaling channel that exchanges short messages between two calling components sets up the call by obtaining resources (such as the link, switches, etc.) in the telephone network at the beginning of the conversation and terminates these arrangements by releasing the resources at the end of the call. The wireless access methods evolved for interaction with these networks assigns a slot of time, a portion of frequency, or a specific code to a user preferably for the entire length of the conversation. We refer to these techniques as fixed-assignment channel access methods or channel partitioning techniques. Data networks were originally designed for bursts of data for which the supporting network does not have a separate signaling channel. In packet communications each packet carries some "signaling information" related to the address of the destination and the source. We refer to the access methods used in these networks as random-access methods accommodating randomly arriving packets of data. Certain local area data networks also *take turns* in accessing the medium as in the case of token passing and polling schemes. In some other cases, the random access mechanisms are used to temporarily *reserve* the medium for transmitting the packet. In the next two sections of this chapter, we provide a short description of the fixed-assignment and random access methods used in voice- and data-oriented wireless networks, respectively.

4.2 FIXED-ASSIGNMENT ACCESS FOR VOICE-ORIENTED NETWORKS

All existing voice-oriented wireless networks such as cellular telephony or PCS services use fixed-assignment channel access or channel partitioning techniques. In the fixed-assignment access method, a fixed allocation of channel resources, frequency, time, or a spread spectrum code are made available on a predetermined basis to a single user for the duration of the communication session. The three basic fixed-assignment multiple-access methods are FDMA, TDMA, and CDMA. The choice of an access method will have a great impact on the capacity and QoS provided by a network. The impact of multiple access schemes is so important that we commonly refer to various voice-oriented wireless systems by their channel access method, which is only a part of the layer two specification of the air interface of the network.

Example 4.3: Common Terminology for Digital Cellular Systems

The GSM and the North American IS-136 digital cellular standards are commonly referred to as digital TDMA cellular systems and the IS-95/IMT-2000 standards are called digital CDMA cellular systems.

In reality, different systems use different modulation techniques as well. However, as we will see in the rest of this book, the impact of the choice of access method on the capacity and overall performance of the network is much more profound. Consequently, the system is really distinguished by its access method. As we will see in our examples of cellular networks, a network that is identified with an access technique often uses other random or fixed assignment techniques as a part of its overall operation. However, it is identified by the access techniques employed for transferring the main information source for which the network is designed to carry.

Example 4.4: Random Access Techniques in Cellular Networks

GSM uses slotted ALOHA (a random access method) to establish a link between the mobile terminal and the base station. It also has an optional frequency-hopping pattern that improves the system performance when there is fading of the radio signal. However, the GSM network is built for voice communications, and each session uses TDMA as the access method.

Another important design parameter related to the access method is the differentiation between the carrier frequencies of the forward (downlink—communication between the base station and mobile terminals) and reverse (uplink—communication between the mobile terminal and the base station) channels. If both forward and reverse channels use the same frequency band for communications, but the forward and reverse channels employ alternating time slots, the system is referred to as employing TDD. If the forward and reverse channels use different carrier frequencies that are sufficiently separated, the duplexing scheme is referred to as FDD. With TDD, because only one frequency carrier is needed for a duplex operation, we can share more of the RF circuitry between the forward and the reverse channels. The reciprocity of the channel in TDD allows for exact open-loop power control and simultaneous synchronization of the forward and reverse channels. TDD techniques are used in systems intended for low-power local area communications where interference must be carefully controlled and where low complexity and low-power consumption are very important. Thus TDD systems are often used in local area pico- or microcellular systems deployed by PCS networks. FDD is mostly used in macrocellular systems designed for coverage of several tens of kilometers where implementation of TDD is more challenging (see Fig. 4.1).

4.2.1 Frequency Division Multiple Access (FDMA)

In an FDMA environment, all users can transmit signals simultaneously, and they are separated from one another by their frequency of operation. The FDMA technique is built upon FDM. FDM is the oldest and still a commonly used multiplexing technique in the trunks connecting switches in the PSTN. It is also the choice of radio and TV broadcast, as well as cable TV distribution. FDM is more suitable for analog technology because it is easier to implement. When FDM is used for channel access, it is referred to as FDMA.

Example 4.5: FDMA in AMPS with FDD

Figure 4.1(a) shows the FDMA/FDD system commonly used in 1G analog cellular systems such as AMPS and a number of early cordless telephones. In FDMA/FDD systems, forward and reverse channels use different carrier frequencies, and a fixed subchannel pair is assigned to a user terminal during the communication session. At the receiving end, the mobile terminal filters the designated channel out of the composite signal. As shown in Figure 4.2(a), the AMPS system allocates 30 kHz of bandwidth for each forward and reverse channel. There are a total of 421 channels in 25 MHz of spectrum assigned to each direction; 395 of these channels are used for the voice traffic and the rest for signaling.

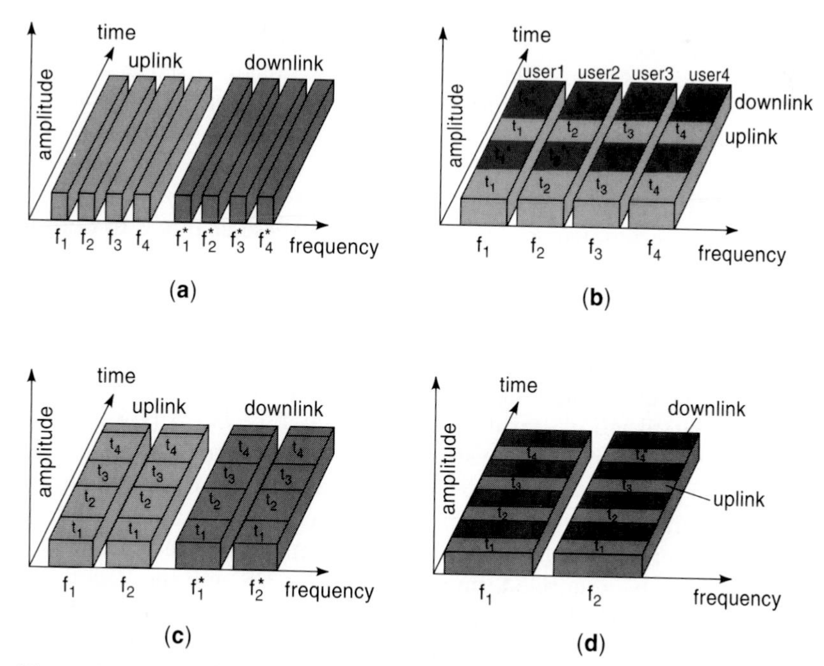

Figure 4.1 (a) FDMA/FDD, (b) FDMA/TDD, (c) TDMA/FDD with multiple carriers, (d) TDMA/TDD with multiple carriers.

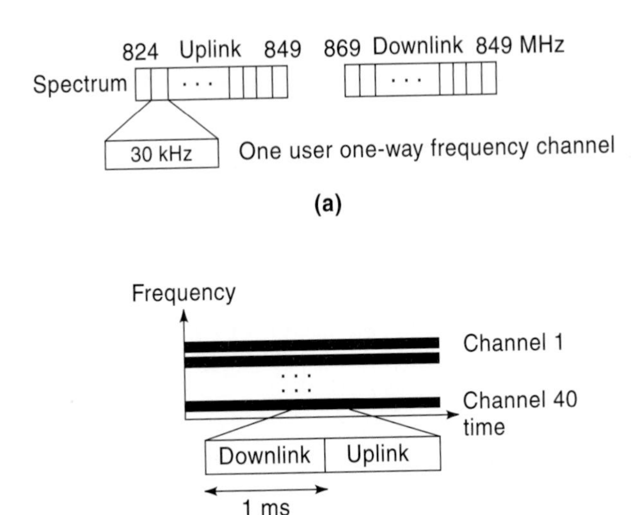

Figure 4.2 (a) FDMA/FDD in AMPS and (b) FDMA/TDD in CT-2.

Example 4.6: FDMA in CT-2 with TDD

Figure 4.1(b) shows an FDMA/TDD system used in the CT-2 digital cordless telephony standard. Each user employs a single carrier frequency for all communications. The forward and reverse transmissions take turns via alternating time slots. This system was designed for distances of up to 100 meters, and a voice conversation is based on 32 kbps ADPCM voice coding. As shown in Figure 4.2(b) the total allocated bandwidth for CT-2 is 4 MHz, supporting 40 carriers each using 100 kHz of bandwidth.

The designer of an FDMA system must pay special attention to adjacent channel interference, in particular in the reverse channel. In both forward and reverse channels, the signal transmitted must be kept confined within its assigned band, at least to the extent that the out-of-band energy causes negligible interference to the users employing adjacent channels. Operation of the forward channel in wireless FDMA networks is very similar to wired FDM networks. In forward wireless channels, in a manner similar to that of wired FDM systems, the signal received by all mobile terminals has the same received power, and interference is controlled by adjusting the sharpness of the transmitter and receiver filters for the separate carrier frequencies. The problem of adjacent channel interference is much more challenging on the reverse channel. On the reverse channel, mobile terminals will be operating at different distances from the BS. The RSS at the BS, of a signal transmitted by a mobile terminal close to the BS, and the RSS at the BS of a transmission by a mobile terminal at edges of the cell are often substantially different, causing problems in detecting the weaker signal. This problem is usually referred to as the near-far problem. If the out-of-band emissions are large, they may swamp the actual information-carrying signal.

Problem 1: Near-Far Problem

a) What is the difference between the received signal strength of two terminals located in 10 m and 1 km from a base station in an open area?

b) Explain the effects of shadow fading on the difference in the RSS.

c) What would be the impact if the two terminals were operating in two adjacent channels? Assume out-of-band radiation that is 40 dB below the main lobe.

Solution:

a) As we saw in Chapter 2, the received signal strength falls by around 40 dB per decade of distance in open areas. Therefore, the received powers from a mobile terminal that is 10 meters from a BS and another, that is at a distance of 1 km, are 80 dB apart.

b) In addition to the fall of the RSS with distance, we also discussed the issues of multipath and shadow fading in radio channels that cause power fluctuations on the order of several tens of dBs. Therefore, the difference in the received powers due to the near-far problem may exceed even 100 dB.

c) If the out-of-band emission is only 40 dB below that of the transmitted power, it may exceed the strength of the information-bearing signal by almost 60 dB.

To handle the near-far problem, FDMA cellular systems adopt two different measures. First, when frequencies are assigned to a cell, they are grouped such that the frequencies in each cell are as far apart as possible. The second measure employed is power control that is discussed in Chapter 6. In addition, whenever FDMA is employed, *guard bands* are included in the frequency channel to further reduce adjacent channel interference. This, however, has the effect of reducing the overall spectrum efficiency.

4.2.2 Time Division Multiple Access (TDMA)

In TDMA systems, a number of users share the same frequency band by taking assigned turns in using the channel. The TDMA technique is built upon the TDM scheme commonly used in the trunks for the telephones systems. The major advantage of the TDMA over FDMA is its format flexibility. Because of the fully digital format and the flexibility of buffering and multiplexing functions, time-slot assignments among multiple users are readily adjustable to provide different access rates for different users. This feature is particularly adopted in the PSTN, and the TDM scheme forms the backbone of all digital connections in the heart of the PSTN. The hierarchy of digital transmission trunks used in North America is the so-called T-carrier system that has an equivalent European system (the E-carriers) approved by the ITU. In the hierarchy of digital transmission rates standardized throughout North America, the basic building block is the 1.544 Mbps link known as T-1 carrier. A T-1 transmission frame is formed by TDD 24 PCM-encoded voice channels, each carrying 64 kbps of users data. Service providers often lease T-carriers to interconnect their own switches and routers and for forming their own networks.

Example 4.7: The Use of T-carriers in Cellular Networks

Cellular networks often lease T-carriers from the long-haul telephone companies to interconnect their own switches referred to as mobile switching centers

(MSCs). The difference between the MSC and a regular switch in the PSTN is that the MSC can support mobility of the terminal. The details of these differences are discussed in later chapters when we provide examples of cellular networks. The end-user subscribes to the cellular service provider.

Example 4.8: The Use of T-1 Lines in the Internet

The routers in the Internet are sometimes connected through leased T-carrier telephone lines to form part of the Internet. The difference between a router and a PSTN switch is that the router can handle packet switching whereas the PSTN switch uses circuit switching. The end-user subscribes to an Internet service provider (ISP) in this case.

With TDMA, a transmit controller assigns time slots to users, and an assigned time slot is held by a user until the user releases it. At the receiving end, a receiver station synchronizes to the TDMA signal frame, and extracts the time slot designated for that user. The heart of this operation is synchronization that was not needed for FDMA systems. The TDMA concept was developed in the 1960s for use in digital satellite communication systems and first became operational commercially in telephone networks in the mid-1970s [PAH95].

In cellular and cordless systems, the migration to TDMA from FDMA took place in the 2G systems. The first cellular standard adopting TDMA was GSM. The GSM standard was initiated to support international roaming among Scandinavian countries in particular and the rest of Europe in general. The digital voice adoption in TDMA format facilitated the network implementation, resulted in improvements in the quality of the voice, and provided a flexible format to integrate data services in the cellular network. The FDMA systems in the United States very quickly observed a capacity crunch in major cities, and among the options for increasing capacity, TDMA was adopted initially through the IS-54 system that was later on replaced by IS-136. TDMA was adopted in 2G cordless telephones such as DECT to provide format flexibility and to allow more compact and low-power terminals.

Example 4.9: TDMA in GSM

Figure 4.1(c) shows an FDMA/TDMA/FDD channel used in 2G digital cellular networks. Figure 4.3 shows a particular example of the 8-slot TDMA scheme used in the GSM system. Forward and reverse channels use separate carrier frequencies (FDD). Each carrier can support up to eight simultaneous users via TDMA, each using a 13 kbps encoded digital speech, within a 200 MHz carrier bandwidth. A total of 124 frequency carriers (FDMA) are available in the 25 MHz allocated band in each direction. One hundred kHz of band is allocated as a guard band at each edge of the overall allocated band.

Example 4.10: TDMA in DECT

Figure 4.1(d) shows an FDMA/TDMA/TDD system used in the Pan-European digital PCS standard DECT. Because distances are short, a TDD format allows

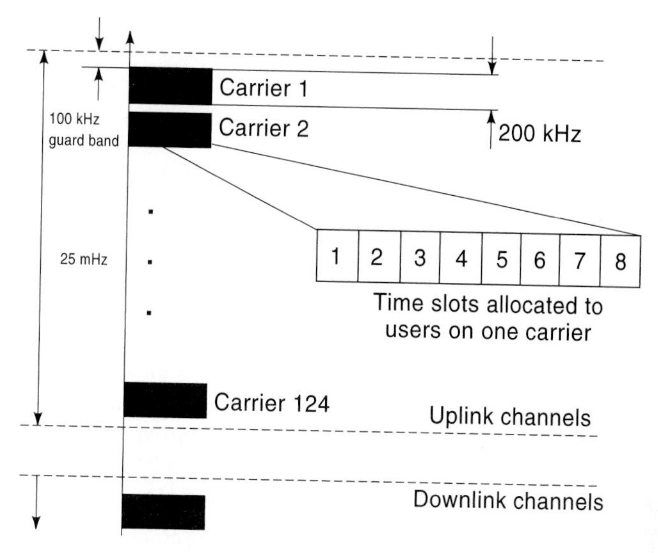

Figure 4.3 FDMA/TDMA/FDD in GSM.

using the same frequency for forward and reverse operations. The bandwidth per carrier is 1.728 MHz which can support up to 12 ADPCM coded speech channels via TDMA. The total allocated band in the EC is 10 MHz that can support five carriers (FDMA). Figure 4.4 shows the details of the TDMA/TDD time slots use in the DECT system. The frame duration is 10 ms, with 5 ms for portable-to-fixed station and 5 ms for fixed-to-portable. The transmitter transfers information in signal bursts which it transmits in slots of duration 10/24 = 0.417 ms. With 480 bits per slot (including a 64-bit guard time), the total bit rate is 1.152 Mbps. Each slot contains 64 bits for system control (C, P, Q, and M channels) and 320 bits for user information (I channel).

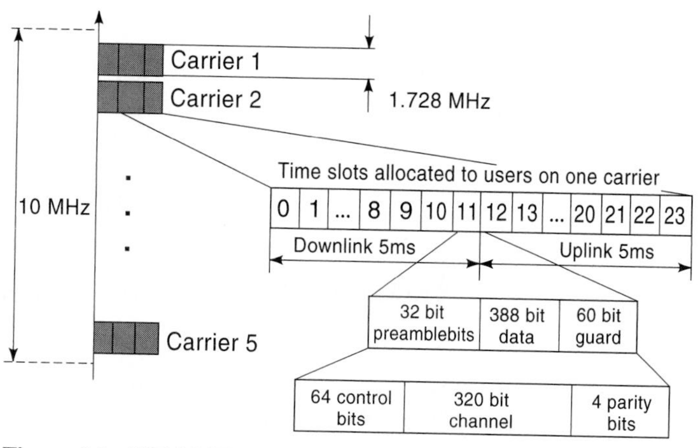

Figure 4.4 FDMA/TDMA/TDD in DECT.

Example 4.11: TDMA in IS-136

Figure 4.5 shows the frame format for the TDMA/FDD with six slots considered for IS-136 both for the forward (base to mobile) and reverse (mobile to base) channels. In IS-136 each 30 kHz digital channel has a channel transmission rate of 48.6 kbps. The 48.6 kbps stream is divided into six TDMA channels of 8.1 kbps each. The IS-136 slot and frame format, shown in Figure 4.5, is much simpler than that of the GSM standard. The 40-ms frame is composed of six 6.67-ms time slots. Each slot contains 324 bits, including 260 bits of user data, and 12 bits of system control information in a slow associated control channel (SACCH). There is also a 28-bit synchronization sequence, and a 12-bit digital verification color code (DVCC) used to identify the frequency channel to which the mobile terminal is tuned. In the mobile-to-base direction, the slot also contains a guard time interval of a six-bit duration when no signal is transmitted, and a six-bit ramp interval to allow the transmitter to reach its full output power level.

Due to the near-far problem, the received signal on the reverse channel from a user occupying a time slot can be much larger than the received power from the terminal using the adjacent time slot. In such a case, the receiver will have difficulties in distinguishing the weaker signal from the background noise. In a manner similar to FDMA systems, TDMA systems also use power control to handle this near-far problem.

4.2.3 Code-Division Multiple Access (CDMA)

With the growing interest in the integration of voice, data, and video traffic in telecommunication networks, CDMA appears increasingly attractive as the wireless access method of choice. Fundamentally, integration of various types of traffic is readily accomplished in a CDMA environment as coexistence in such an

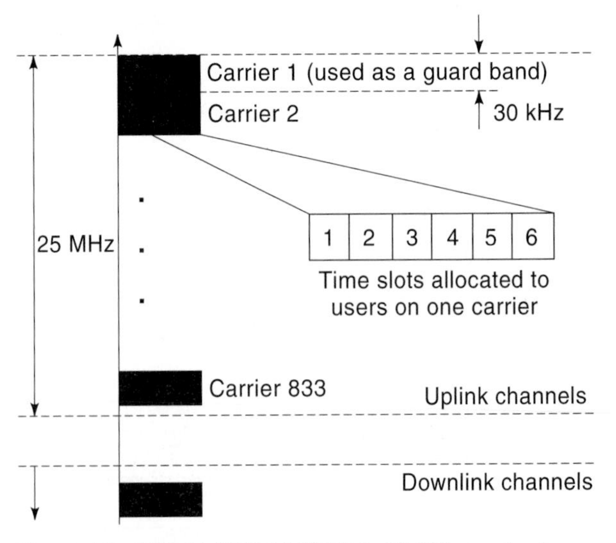

Figure 4.5 FDMA/TDMA/FDD in IS-136 standard.

environment does not require any specific coordination among user terminals. In principle, CDMA can accommodate various wireless users with different bandwidth requirements, switching methods and technical characteristics without any need for coordination. Of course, because each user signal contributes to the interference seen by other users, power control techniques are essential in the efficient operation of a CDMA system.

To illustrate CDMA and how it is related to FDMA and TDMA, it is useful to think of the available band and time as resources we use to share among multiple users. In FDMA, the frequency band is divided into slots, and each user occupies that frequency throughout the communication session. In TDMA, a larger frequency band is shared among the terminals, and each user uses a slot of time during the communication session. As shown in Figure 4.6, in a CDMA environment multiple users use the same band at the same time, and the user is differentiated by a code that acts as the key to identify that user. These codes are selected so that when they are used at the same time in the same band a receiver knowing the code of a particular user can detect that user among all the received signals. In the CDMA/FDD [Figure 4.7(a)] that is used in IS-95 and IMT-2000, the forward and reverse channels use different carrier frequencies. If both transmitter and receiver use the same carrier frequency [Figure 4.7(b)], the system is CDMA/TDD.

In CDMA, each user is a source of noise to the receiver of other users, and if we increase the number of users beyond a certain value, the entire system collapses because the signal received in each specific receiver will be buried under the noise caused by many other users. An important question is, how many users can simultaneously use a CDMA system before the system collapses?

4.2.3.1 Capacity of CDMA

CDMA systems are implemented based on the spread spectrum technology that is presented in Chapter 3. In its most simplified form, a spread spectrum transmitter spreads the signal power over a spectrum N times wider than the spectrum

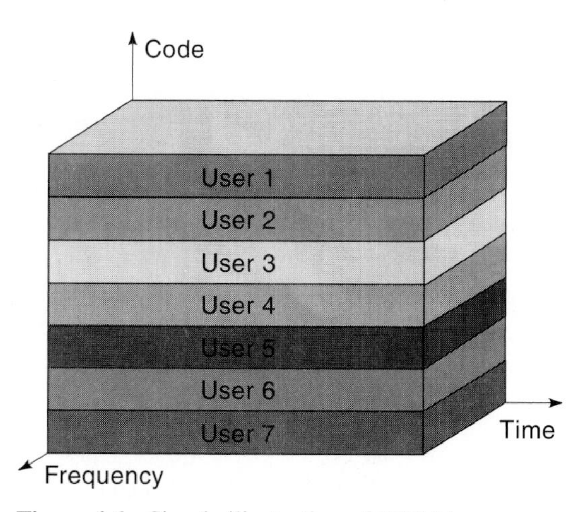

Figure 4.6 Simple illustration of CDMA.

of the message signal. In other words, an information bandwidth of R occupies a transmission bandwidth of W, where:

$$W = NR \tag{4.1}$$

The spread spectrum receiver processes the received signal with a *processing gain* of N. This means that during the processing at the receiver, the power of the received signal having the code of that particular receiver will be increased N times beyond the value before processing.

Let us consider the situation of a single cell in a cellular system employing CDMA. Assume that we have M simultaneous users on the reverse channel of a CDMA network. Further let us assume that we have an ideal power control enforced on the channel so that the received power of signals from all terminals has the same value P. Then, the received power from the target user after processing at the receiver is NP, and the received interference from $M - 1$ other terminals is $(M - 1)P$. If we also assume that a cellular system is interference limited and the background noise is dominated by the interference noise from other users, the received signal-to-interference ratio for the target receiver will be:

$$S_r = \frac{NP}{(M-1)P} = \frac{N}{M-1} \tag{4.2}$$

All users always have a requirement for the acceptable error rate of the received data stream. For a given modulation and coding specification of the system, that error rate requirement will be supported by a minimum S_r requirement that can be used in Eq. (4.2) to solve for the number of simultaneous users. Then, solving Eqs. (4.1) and (4.2) for M, we will have:

$$M = \frac{W}{R}\frac{1}{S_r} + 1 \cong \frac{W}{R}\frac{1}{S_r} \tag{4.3}$$

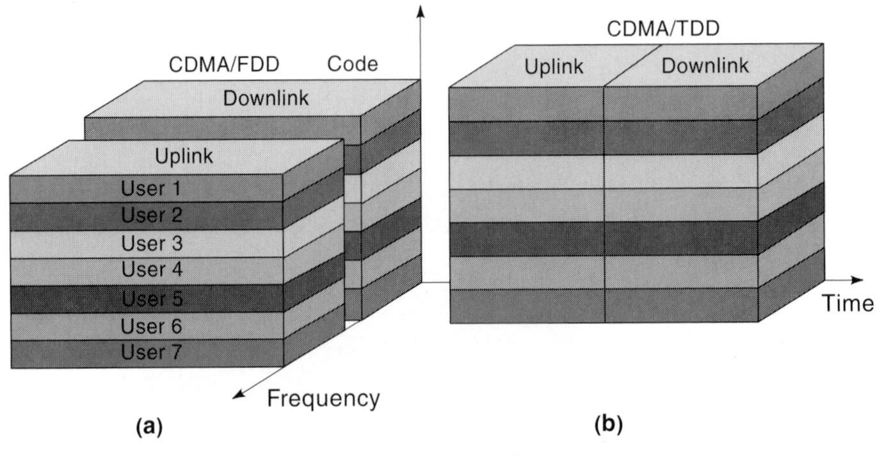

Figure 4.7 (a) CDMA/FDD and (b) CDMA/TDD.

Problem 2: Capacity of One Carrier in a Single-Cell CDMA System

Using QPSK modulation and convolutional coding, the IS-95 digital cellular systems require 3 dB $< S_r <$ 9 dB. The bandwidth of the channel is 1.25 MHz, and the transmission rate is R = 9600 bps. Find the capacity of a single IS-95 cell.

Solution:

Using Equation (4.3) we can support from up to

$$M = \frac{1.25\text{ MHz}}{9600\text{ bps}}\frac{1}{8} \approx 16 \text{ to } M = \frac{1.25\text{ MHz}}{9600\text{ bps}}\frac{1}{2} \approx 65 \text{ users.}$$

4.2.3.2 Practical Considerations

In the practical design of digital cellular systems, three other parameters affect the number of users that can be supported by the system, as well as the bandwidth efficiency of the system. These are the number of sectors in each base station antenna, the voice activity factor, and the interference increase factor. These parameters are quantified as factors used in the calculation of the number of simultaneous users that the CDMA system can support. The use of sectored antennas is an important factor in maximizing bandwidth efficiency. Cell sectorization using directional antennas reduces the overall interference, increasing the allowable number of simultaneous users by a *sectorization gain factor,* which we denote by G_A. With ideal sectorization the users in one sector of a base station antenna do not interfere with the users operating in other sectors, and $G_A = N_{sec}$ where N_{sec} is the number of sectors in the cell. In practice antenna patterns cannot be designed to have ideal characteristics, and due to multipath reflections, users in general communicate with more than one sector. Three-sector base station antennas are commonly used in cellular systems, and a typical value of the sectorization gain factor is assumed to be G_A = 2.5 (4 dB). The voice activity interference reduction factor G_v is the ratio of the total connection time to the active talkspurt time. On the average, in a two-way conversation, each user talks roughly 50 percent of the time. The short pauses in the flow of natural speech reduce the activity factor further to about 40 percent of the connection time in each direction. As a result, the typical number used for G_v is 2.5 (4 dB). The interference increase factor H_0 accounts for users in other cells in the CDMA system. Because all neighboring cells in a CDMA cellular network operate at the same frequency, they will cause additional interference. This interference is relatively small due to the processing gain of the system and the distances involved; a value of H_0 = 1.6 (2 dB) is commonly used in the industry.

Incorporating these three factors as a correction to Equation (4.3), the number of simultaneous users that can be supported in a CDMA cell can be approximated by

$$M = \frac{W}{R}\frac{1}{S_r} + 1 \cong \frac{W}{R}\frac{1}{S_r}\frac{G_A G_v}{H_0} \tag{4.4}$$

If we define the *performance improvement factor* in a digital cellular system as

$$K = \frac{G_A G_v}{H_0} \tag{4.5}$$

assuming the typical parameter values given earlier, the performance improvement factor is $K = 4$ (6 dB).

Problem 3: Capacity of One Carrier in a Multi-Cell CDMA System with Correction Factors

Determine the multicell IS-95 CDMA capacity with correction for sectorization and voice activity. Use the numbers from Problem 2.

Solution:

If we continue the previous example with the new correction factor included, the range for the number of simultaneous users becomes $64 < M < 260$.

4.2.4 Comparison of CDMA, TDMA, and FDMA

With the success of IS-95 CDMA systems in its challenge to conventional IS-136 TDMA systems in the United States and the adoption of W-CDMA as the primary choice for the 3G cellular networks, one wonders why CDMA has become the favorite choice for wireless access in voice-oriented networks. Spread spectrum technology became the favorite technology for military applications because of its capability to provide a low probability of interception and strong resistance to interference from jamming. In the cellular industry, CDMA was introduced as an alternative to TDMA to improve the capacity of 2G cellular systems in the United States. As a result, much of the early debates in this area were focused on calculation of the capacity of CDMA as it is compared with TDMA. However, capacity is not the only reason for the success of the CDMA technology. As a matter of fact, calculation of the capacity of CDMA using the simple approach provided earlier is *not* very conclusive and is subject to a number of assumptions such as perfect power control that cannot be practically met. The first CDMA service providers in the United States were using slogans such as "you cannot believe your ears!" to address the superior quality of voice for the CDMA. However, the superiority of voice is partially dependent on the speech coder, and it is not a CDMA versus TDMA issue. In order to provide a good explanation for the success of a complex and multidisciplinary technology, such as a cellular network, addressing consumer market issues has always been very important. Those of us involved in this debate for the past decade have seen the discussion of the ups and downs of CDMA in variety of forums. One of the most interesting events that the principal author remembers was in 1997 in a major wireless conference in Taipei where one the most famous figures in this debate in his keynote speech at the opening of the conference declared that "we have seen in the past that the VHS which was not a better technology defeated BETA." In his perception, at that time, CDMA was similar to BETA. In less than a year or so after that, CDMA was selected by a number of different communities around the world as the technology of choice for 3G and IMT-2000.

In the rest of this section, we bring out a number of issues that may enlighten the reader toward a deeper understanding of the technical aspects of CDMA systems as they are compared with TDMA and FDMA networks. We hope that this may lead the reader to her/his own conclusion about the success of CDMA.

Format Flexibility. Telephone voice was the dominant source of income for the telecommunication industry up to the end of the past century. In the new millennium, the strong emergence of Internet and cable TV industries has created a case for other popular multimedia applications. The cellular phones that were designed for telephony applications are now being used for other applications and need support for multimedia applications. To support a variety of data rates with different requirements, a network needs format flexibility. As we discussed earlier, one of the reasons for migrating from analog FDMA to digital TDMA was that TDMA provides a more flexible environment for integration of voice and data. The time slots of a TDMA network designed for voice transmission can be used individually or in a group format to transmit data from users and to support different data rates. However, all these users should be time synchronized and the quality of the transmission channel is the same for all of them. The chief advantage of CDMA relative to TDMA is its flexibility in timing and the quality of transmission. In CDMA users are separated by their codes, unaffected by the transmission time relative to other users. The power of the user can also be adjusted with respect to others to support a certain quality of transmission. In CDMA each user is far more liberated from the other users, allowing a fertile setting to accommodate different service requirements to support a variety of transmission rates with different qualities of transmission to support multimedia or any other emerging application.

Performance in Multipath Fading. As we saw in Chapter 2, multipath in wireless channels causes frequency selective fading. In frequency selective fading, when the transmission band of a narrowband system coincides with the location of the fade, no useful signal is received. As we increase the transmission bandwidth, fading will occupy only a portion of the transmission band, providing an opportunity for a wideband receiver to take advantage of the portion of the transmission band not under fade and a more reliable communication link. In Chapter 3 we introduced DFE, OFDM, sectored antennas, and spread spectrum as technologies that can be employed in wideband systems to handle frequency selective fading. The wider the bandwidth, the better is the opportunity for averaging out the faded frequency.

These technologies are not used in the 1G analog cellular FDMA systems because they were analog systems and these techniques are digital. The Pan European GSM digital cellular system uses 200 kHz of band, and the standard recommends using DFE. The North American digital cellular system, IS-136, uses digital transmission over the same analog band of 30 kHz of the North American AMPS system and does not recommend equalization because the bandwidth is not very large. An equalizer needs additional circuitry, and some power budget at the receiver that was one of the drawbacks considered in IS-136. The bandwidth of the IS-95 CDMA system is 1.25MHz and W-CDMA systems for 3G networks use bandwidths that are as high 10 MHz. RAKE receivers are used to increase the benefits of wideband transmission by taking advantage of the so-called in-band or time diversity of the wideband signal. This is one of the reasons for having a better quality of voice in CDMA systems. As we mentioned earlier, quality of voice is also effected by the robustness of the speech-coding algorithm, coverage of service, methods to handle interference, handoffs, and power control as well.

System Capacity. Comparison of the capacity depends on a number of issues, including the frequency reuse factor, speech coding rate, and the type of antenna. Therefore a fair comparison would be difficult unless we go to practical systems. The following simple example compares the capacity of FDMA (AMPS), TDMA (IS-136), and CDMA (IS-95) used in debates to evaluate alternatives for the 2G North American digital cellular systems to replace the 1G analog.

Problem 4: Comparison of the Capacity of Different 2G Systems

Compare the capacity of IS-95 CDMA with AMPS FDMA and IS-136 TDMA systems. For the CDMA system, assume an acceptable signal to interference ratio of 6 dB, data rate of 9600 bps, voice duty cycle of 50 percent, effective antenna separation factor of 2.75 (close to ideal 3-sector antenna), and neighboring cell interference factor of 1.67.

Solution:

For the IS-95 CDMA using Equation (4.4) for each carrier with $W = 1.25$ MHz, $R = 9600$ bps, $S_r = 4$ (6dB), $G_v = 2$ (50 percent voice activity), $G_A = 2.75$, and $H_0 = 1.67$ we have:

$$M = \frac{W}{R}\frac{1}{S_r}\frac{G_A G_v}{H_0} = 108 \text{ users per cell}$$

For the IS-136 with a carrier bandwidth of $W_c = 30$ kHz, the number of users per carrier of $N_u = 3$, and frequency reuse factor of $K = 4$ (commonly used in these systems), each $W = 1.25$ MHz of bandwidth provides for

$$M = \frac{W}{W_c}\frac{N_u}{K} = 31.25 \text{ users per cell}$$

For the AMPS analog system with carrier bandwidth of $W_c = 30$ kHz, and frequency reuse factor of $K = 7$ (commonly used in these systems), each $W = 1.25$ MHz of bandwidth provides for

$$M = \frac{W}{W_c}\frac{1}{K} = 6 \text{ users per channel}$$

Another example of this form is instructive to compare these systems with the GSM.

Problem 5: Comparison of NA Systems with GSM

Determine the capacity of GSM for $K = 3$.

Solution:

For the GSM system with a carrier bandwidth of $W_c = 200$ kHz, the number of users per carrier of $N_u = 8$, and frequency reuse factor of $K = 3$ (commonly used in these systems), each $W = 1.25$MHz of bandwidth provides for

$$M = \frac{W}{W_c} \frac{N_u}{K} = 16.7 \text{ users per cell}$$

Handoff. As we discuss in Chapter 6, handoff occurs when a received signal in an MS becomes weak and another BS can provide a stronger signal to the MS. The 1G FDMA cellular systems often used the so-called hard-decision handoff in which the base station controller monitors the received signal from the BS and at the appropriate time switches the connection from one BS to another. TDMA systems use the so-called *mobile-assisted handoff* in which the mobile station monitors the received signal from available BSs and reports it to the base station controller which then makes a decision on the handoff. Because adjacent cells in both FDMA and TDMA use different frequencies, the MS has to disconnect from and reconnect to the network that will appear as a click to the user. Handoffs occur at the edge of the cells when the received signals from both BSs are weak. The signals also fluctuate anyway because they are arriving over radio channels. As a result, decision making for the handoff time is often complex, and the user experiences a period of poor signal quality and possibly several clicks during the completion of the handoff process. Because adjacent cells in a CDMA network use the same frequency, a mobile moving from one cell to another can make "seamless" handoff by the use of signal combining. When the mobile station approaches the boundary between cells, it communicates with both cells. A controller combines the signals from both links to form a better communication link. When a reliable link has been established with the new base station, the mobile stops communicating with the previous base station, and communication is fully established with the new base station. This technique is referred to as soft handoff. Soft handoff provides a dual diversity for the received signal from two links which improves the quality of reception and eliminates clicking as well as the ping-pong problem. Handoff is an important issue that has many more details and we will discuss these details in Chapter 6.

Power Control. As we discussed earlier in this chapter, power control is necessary for FDMA and TDMA systems to control adjacent channel interference and mitigate the unexpected interference caused by the near-far problem. In FDMA and TDMA systems, some sort of power control is needed to improve the quality of the voice delivered to the user. In CDMA, however, the capacity of the system depends *directly* on the power control, and an accurate power control mechanism is needed for proper operation of the network. With CDMA, power control is the key ingredient in maximizing the number of users that can operate simultaneously in the system. As a result, CDMA systems adjust the transmitted power more often and with smaller adjustment steps to support a more refined control of power. Better power control also saves on the transmission power of the MS, which increases the life of the battery. The more refined power control in CDMA systems also helps in power management of the MS, which is an extremely important practical issue for users of the mobile terminals. These issues are further discussed in Chapters 6 and 8.

Implementation Complexity. Spread spectrum is a two-layer modulation technique requiring greater circuit complexity than conventional modulation schemes.

This in turn will lead to higher electronic power consumption and larger weight and cost for mobile terminals. Gradual improvements in battery and integrated circuit technologies, however, have made this issue transparent to the user.

4.2.5 Performance of Fixed-Assignment Access Methods

Fixed assignment access methods are used with circuit switched cellular and PCS telephone networks. In these networks, in a manner similar to the wired multichannel environments, the performance of the network is measured by the blockage rate of an initiated call. A call does not go through for two reasons: (1) when the calling number is not available, and (2) when the telephone company is out of resources to provide a line for the communication session. In POTS, for both cases the user hears a busy tone signal and cannot distinguish between the two types of blockage. In most cellular systems, however, type (1) blockage results in a response that is a busy tone and type (2) with a message such as "All the circuits are busy at this time please try your call later." In the rest of this book, we refer to blockage rate only as a type (2) blockage rate. The statistical properties of the traffic offered to the network are also a function of time. The telephone service providers often design their networks so that the blockage rate at peak traffic is always below a certain percentage. Cellular operators often try to keep this average blockage rate below 2 percent.

The blockage rate is a function of the number of subscribers, number of initiated calls, and the length of the conversations. In telephone networks, the Erlang equations are used to relate the probability of blockage to the average rate of the arriving calls and the average length of a call. In wired networks, the number of lines or subscribers that can connect to a multichannel switch is a fixed number. The telephone company monitors the statistics of the calls over a long period of time and upgrades the switches with the growth of subscribers so that the blockage rate during peak traffic times remains below the objective value. In cellular telephony and PCS networks, the number of subscribers operating in a cell is also a function of time. In the downtown areas, everyone uses their cellular telephones during the day, and in the evenings they use them in their residential area which is covered by a different cell. Therefore, traffic fluctuations in cellular telephone networks are much more than the traffic fluctuations in POTS. In addition, telephone companies can easily increase the capacity of their networks by increasing their investment on the number of transmission lines and quality of switches supporting network connections. In wireless networks, the overall number of available channels for communications is ultimately limited by the availability of the frequency bands assigned for network operation. To respond to the fluctuations of the traffic and cope with the bandwidth limitations, cellular operators use complex frequency assignment strategies to share the available resources in an optimal manner. Some of these issues are discussed in Chapter 5.

4.2.5.1 Traffic Engineering Using the Erlang Equations

The Erlang equations are the core of the traffic engineering for telephony applications. The two basic equations used for traffic engineering are Erlang B and Erlang C equations. The Erlang B equation relates the probability of blockage

$B(N,\rho)$ to the number of channels N and the normalized call density in units of channels ρ. The Erlang B formula is:

$$B(N, \rho) = \frac{\rho^N / N!}{\sum_i^N (\rho^i / i!)} \tag{4.6}$$

where $\rho = \lambda / \mu$, λ is the call arrival rate and μ is the service rate of the calls.[1]

Problem 6: Call Blocking Using Erlang B Formula

We want to provide a wireless public phone service with five lines to a ferry crossing between Helsinki and Stockholm carrying 100 passengers where on the average each passenger makes a three-minute telephone call every two hours. What is the probability of a passenger approaching the telephones and none of the four lines are available?

Solution:

In practice, often the probability of call blockage is given, and we need to calculate the number of subscribers. Here we need an inverse function for the Erlang equation that is not available. As a result, a number of tables and graphs are available for this inverse mapping. Figure 4.8 shows a graph relating the probability of blockage $B(N,\rho)$ to the number of channels N and the normalized traffic per available channels ρ. From this graph, we can estimate the blocking probability. The traffic load is 100 users $\times$ 1 call/user $\times$ 3 minutes/call per 120 minutes = 2.5 Erlangs. Because there are five lines available and the traffic is 2.5 Erlangs, the blocking probability is roughly 0.07.

Problem 7: Capacity Using Erlang B Formula

An IS-136 cellular phone provider owns 50 cell sites and 19 traffic carriers per cell each with a bandwidth of 30 kHz. Assuming each user makes three calls per hour and the average holding time per call of five minutes, determine the total number of subscribers that the service provider can support with a blocking rate of less than 2 percent.

Solution:

The total number of channels is $N = 19 \times 3 = 57$ per cell. For $B(N,\rho) = 0.02$ and $N = 57$ Figure 4.8 shows that the $\rho = 45$ Erlangs. With an average of five calls per minute, the service rate is $\mu = 1/5$ minutes, and the acceptable arrival rate of the calls is $\lambda = \rho \times \mu = 1/5\ (\text{min}^{-1}) \times 45$ (Erlang) = 9 (Erlang/min). With an average of 3 calls per hour, the system can accept 9 (Erlang/min) / 3 (Erlang) / 60 (min) = 180 subscribers per cell. Therefore the total number of subscribers are 180 (subscribers/cell) $\times$ 50 (cells) = 8,000 subscribers.

The Erlang C formula relates the waiting time in a queue if a call does not go through, but it is buffered until a channel is available. These equations start with

[1] The equation assumes that the arrivals are Poisson, and the service rate is exponential. For details, see [BER87].

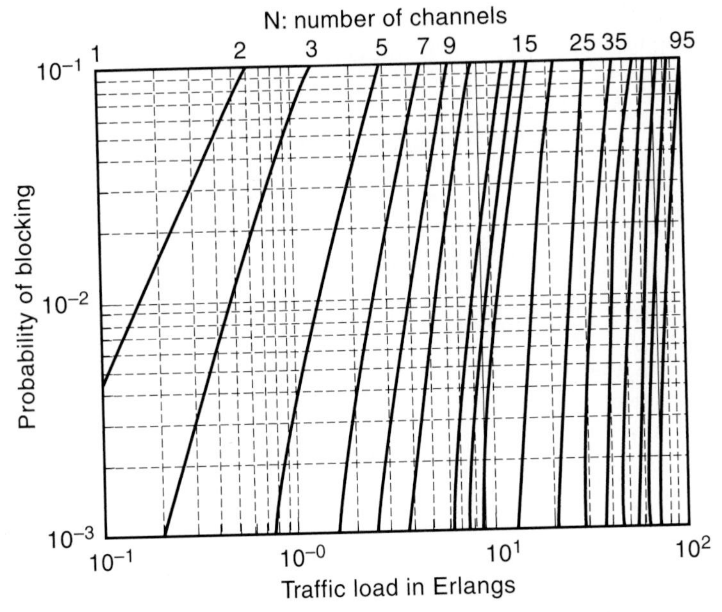

Figure 4.8 Erlang B chart showing the blocking probability as a function of offered traffic and number of channels.

the probability that a call does not get processed immediately and gets delayed. The probability that a call gets delayed is given by:

$$P(\text{delay} > 0) = \frac{\rho^N}{\rho^N + N!\left(1 - \frac{\rho}{N}\right)\sum_{k=0}^{N-1}\frac{\rho^k}{k!}} \tag{4.7}$$

Because of the complexity of the calculation, tables or graphs are again used to provide values for this probability based on normalized values of ρ. Figure 4.9 illustrates the relationship between probability of delay, number of channels N, and the normalized traffic per available channel ρ. The probability of having a delay that is more than a time t is given by:

$$P[delay > t] = P[delay > 0]e^{-(N-\rho)\mu t} \tag{4.8}$$

This indicates the exponential distribution of the delay time. The average delay is then given by the average of the exponential distribution:

$$D = P[delay > 0]\frac{1}{\mu(N - \rho)} \tag{4.9}$$

Problem 8: Call Delay Using Erlang C Formula

For the ferry described in Problem 6 answer the following questions:

a) What is the average delay for a passenger to get access to the telephone?

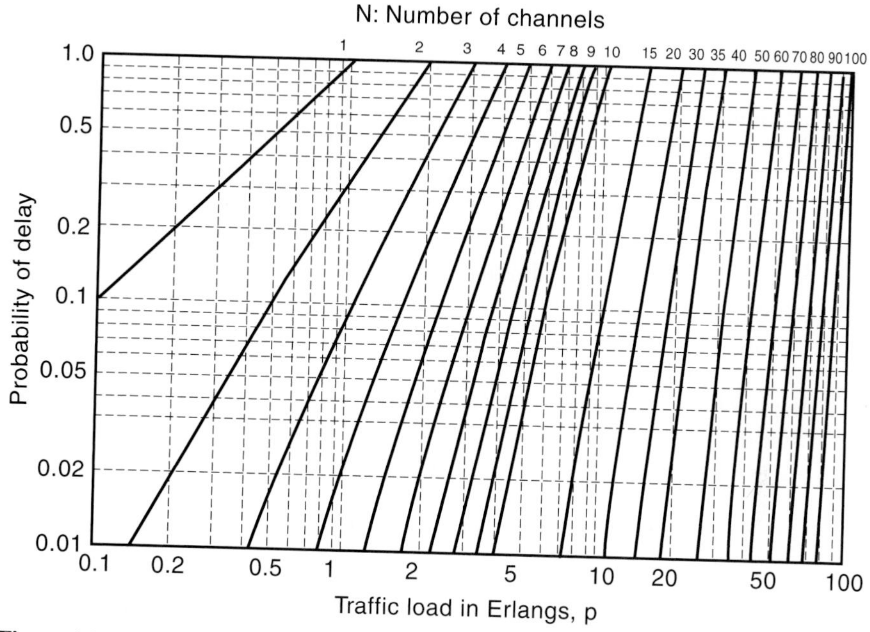

Figure 4.9 Erlang C chart relating the offered traffic to the number of channels and the probability of queuing.

b) What is the probability of having a passenger waiting more than a minute for the access to the telephone?

Solution:

a) Using Equation (4.7) for $N = 5$ and $\rho = 2.5$ we have $P[\text{Delay} > 0] = 0.13$. Using Equation (4.9), the average delay is $0.13/(5 - 2.5)/3 = 0.17$ minutes.

b) Using Equation (4.8) $P[\text{delay} > 1\text{min}] = 0.13 \exp[-(5 - 2.5)1/3] = 0.13 \exp(-0.83) = 0.0565$.

4.3 RANDOM ACCESS FOR DATA-ORIENTED NETWORKS

Random access methods have evolved around bursty data applications for computer communications. In our discussion of fixed-assignment access methods, we noted that such methods make relatively efficient use of communications resources when each user has a steady flow of information to be transmitted. This would be the case, for example, with digitized voiced traffic, data file transfer, or facsimile transmission. However, if the information to be transmitted is intermittent or bursty in nature, fixed-assignment access methods can result in communication resources being wasted for much of the duration of the connection. Furthermore, in wireless networks, where subscribers pay for service as a function of channel connection time, fixed-assignment access can be an expensive means of transmitting short messages and will also involve large call setup times. *Random access* methods provide a more flexible and efficient way of managing channel access for communicating short bursty messages. In contrast to fixed-assignment access schemes, random access schemes provide each user station with varying degrees of freedom in gaining access to the network whenever information is to be sent. A natural consequence of randomness of user access is that there is contention among the users of the network for access to a channel, and this is manifested in collisions of contending transmissions. Therefore these access schemes are sometimes called contention-based schemes or simply *contention schemes*.

Random access techniques are widely used in wired LANs, and the literature in computer networking provides adequate description of these techniques. When applied to wireless applications, these techniques often are modified from their original wired version [CHA00]. The objective of the rest of this section is to describe the evolution of random access techniques that are used in wireless networks. We first discuss the random access methods used in wireless data networks, and then we provide some details of the access methods used in WLAN applications.

4.3.1 Random Access Methods for Mobile Data Services

The random access methods used in mobile data networks can be divided into two groups. The first group consists of ALOHA-based access methods for which the mobile terminals transmit their contention packet without any coordination between them. The second group is the carrier-sense based random access techniques for which the terminal senses the availability of the channel before it transmits its packets.

4.3.1.1 ALOHA-Based Wireless Random Access Techniques

The original *ALOHA protocol* is sometimes called *pure ALOHA* to distinguish it from subsequent enhancements of the original protocol. This protocol derives its name from the ALOHA system, a communications network developed by Norman Abramson and his colleagues at the University of Hawaii and first put into operation in 1971 [ABR70]. The initial system used ground-based UHF radios to connect computers on several of the island campuses with the university's main computer center on Oahu, by use of a random access protocol which has since been known as the ALOHA protocol. The word ALOHA means "hello" in Hawaiian.

The basic concept of ALOHA protocol is very simple. A mobile terminal transmits an information packet when the packet arrives from the upper layers of the protocol stack. Simply put, mobile terminals say "hello" to the air interface as the packet arrives. Each packet is encoded with an error-detection code. The BS checks the parity of the received packet. If the parity checks properly, the BS sends a short acknowledgment packet to the MS. Of course, because the MS packets are transmitted at arbitrary times, there will be collisions between packets whenever packet transmissions overlap by any amount of time, as indicated in Figure 4.10(a). Thus, after sending a packet, the user waits a length of time more than the round-trip delay for an acknowledgment from the receiver. If no acknowledgment is received, the packet is assumed lost in a collision, and it is transmitted again with a randomly selected delay to avoid repeated collisions.

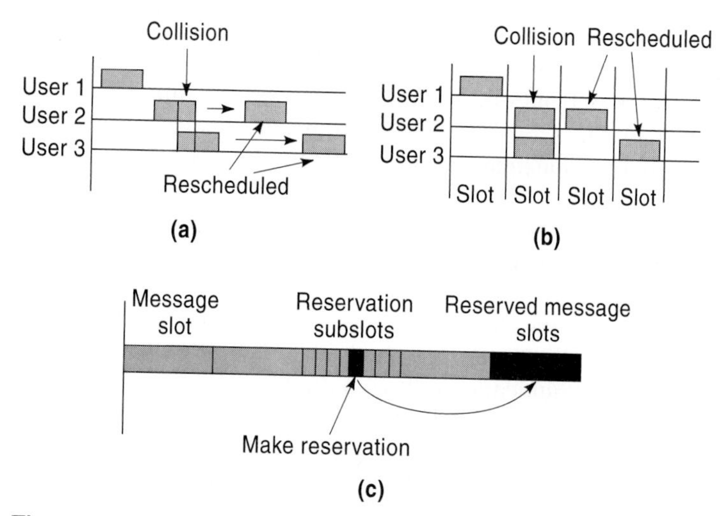

Figure 4.10 (a) Pure ALOHA protocol, (b) slotted ALOHA protocol, and (c) reservation ALOHA.

The advantage of ALOHA protocol is that it is very simple, and it does not impose any synchronization between mobile terminals. The terminals transmit their packets as they become ready for transmission, and if there is a collision, they simply retransmit. The disadvantage of the ALOHA protocol is its low throughput under heavy load conditions. If we assume that packets arrive randomly, they have the same length, and are generated from a large population of terminals. The maximum throughput of the pure ALOHA is 18 percent.

Problem 9: Throughput of Pure ALOHA

a) What is the maximum throughput of a pure ALOHA network with a large number of users and a transmission rate of 1 Mbps?

b) What is the throughput of a TDMA network with the same transmission rate?

c) What is the throughput of the ALOHA network if only one user was effective?

Solution:

a) For a large number of mobile terminals, each using a transmission rate of 1 Mbps to access a BS using ALOHA protocol, the maximum data rate that successfully passes through to the BS is 180 kbps.

b) If we have a TDMA system with negligible overhead (long packets), the throughput defined this way is nearly 100 percent and the BS receives data at a maximum rate of 1 Mbps.

c) The 1 Mbps can be attained in an ALOHA system only if we have one user (no-collision) who transmits all the time.

In wireless channels where bandwidth limitations often impose serious concerns for data communications applications, this technique is often changed to its synchronized version referred to as slotted ALOHA. The maximum throughput of a slotted ALOHA system under the conditions mentioned earlier is 36 percent which is double the throughput of pure ALOHA.

In slotted ALOHA protocol, shown in Figure 4.10(b), the transmission time is divided into time slots. The BS transmits a beacon signal for timing, and all MSs synchronize their time slots to this beacon signal. When a user terminal generates a packet of data, the packet is buffered and transmitted at the start of the next time slot. With this scheme we eliminate partial packet collision. Assuming equal length packets, either we have a complete collision or we have no collisions. This doubles the throughput of the network. The report on collision and retransmission mechanisms remains the same as in pure ALOHA. Because of its simplicity, the slotted ALOHA protocol is commonly used in the early stages of registration of an MS to initiate a communication link with the BS.

Example 4.12: Slotted ALOHA in GSM

In the GSM system, the initial contact between the MS and the BS to establish a traffic channel for TDMA voice communications is performed through a random access channel using slotted ALOHA protocol. Other voice-oriented cellular systems adopt similar approaches as the first step in the registration process of an MS.

Throughput of slotted ALOHA protocol is still very low for wireless data applications. This technique is sometimes combined with TDMA systems to form the so-called reservation-ALOHA (R-ALOHA) protocol, shown in Figure 4.10(c) In R-ALOHA, time slots are divided into contention periods and contention free periods. During the contention interval, an MS uses very short packets to contend for the upcoming contention free intervals that will be used for transmission of long information packets. The R-ALOHA protocol was used in the Altair WLANs that were developed in the early 1990s to operate in licensed frequency bands around 18–19 GHz. The detailed implementation of R-ALOHA can take a variety of forms, and for that reason sometimes it is used under different names. The follow-

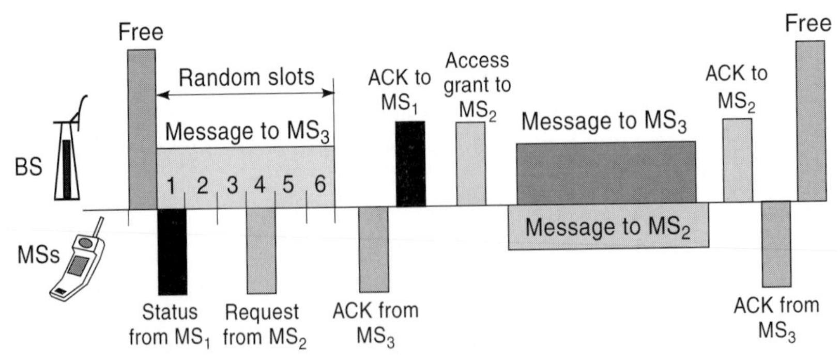

Figure 4.11 Dynamic slotted ALOHA used in Mobitex.

ing example provides some details of the so-called dynamic slotted ALOHA protocol that is used in the Mobitex mobile data networks.

Example 4.13: Dynamic Slotted ALOHA

Mobitex has a full-duplex communication capability (simultaneous transmissions on the uplink and downlink) and employs a *dynamic* slotted ALOHA protocol. Suppose that there are three mobile stations MS_1, MS_2, and MS_3 in a cell. The situation is such that the BS has two messages to send to MS_3, MS_1 has a short status update that requires one slot, MS_2 has a long message to send, and MS_3 has nothing to transmit. An MS can transmit only during certain "free" cycles consisting of several slots of equal length that are periodically initiated by the BS using a FREE frame on the downlink. In this example, shown in Figure 4.11, the BS indicates that there are six free slots for contention, each of a certain length. This can change depending on the traffic; hence the term "dynamic." Also note that MSs cannot transmit whenever they want as in slotted ALOHA. The MSs with traffic to send, such as, MS_1 and MS_2, select one of the six slots at random. In this case, MS_1 selects slot 1 and MS_2 selects slot 4. Hence there is no collision. MS_1 is able to transmit its short status update in slot 1 after which it ceases transmission. MS_2 transmits in slot 4 requesting access to the channel. Simultaneously, the BS would have transmitted its message to MS_3. Upon receipt of the message, MS_3 acknowledges it. The free slots are designed to be of the duration of the downlink message to MS_3 so that the acknowledgment from MS_3 can be received without contention. The BS also acknowledges the status report from MS_1 and sends an access grant to MS_2. As MS_2 transmits its long message on the uplink, the BS can simultaneously send the second message to MS_3. After proper ACKs are transmitted and received, a new FREE cycle is started.

Example 4.14: Packet Reservation Multiple Access (PRMA)

An example of a system that uses reservation for integrating voice and data services is the work done by David Goodman and his colleagues in developing the concept of packet reservation multiple access (PRMA) [GOO89], [GOO91]. PRMA is a method for transmitting, in a wireless environment, a variable mixture of voice packets and data packets. The PRMA system is closely related to R-ALOHA, in that it merges characteristics of slotted ALOHA and TDMA protocols. PRMA has been developed for use in centralized networks operating over short-range radio channels. Short propagation times are an important ingredient in providing acceptable delay characteristics for voice service.

The transmission format in PRMA is organized into frames, each containing a fixed number of time slots. The frame rate is identical to the arrival rate of speech packets. The terminals identify each slot as either "reserved" or "available" in accordance with a feedback message received from the base station at the end of the slot. In the next frame, only the user terminal that reserved the slot can use a reserved slot. Any terminal not holding a reservation that has information to transmit can use an available slot.

Terminals can send two types of information, referred to as periodic and random. Speech packets are always periodic. Data packets can be random, if they are isolated, or periodic if they are contained in a long unbroken stream of information. One bit in the packet header specifies the type of information in the packet. A terminal having periodic information to send starts transmitting in contention for the next available time slot. Upon successfully detecting the first packet in the information burst, the base station grants the sending terminal a reservation for exclusive use of the same time slot in the next frame. The terminal in effect "owns" that time slot in all succeeding frames as long as it has an unbroken stream of packets to send. After the end of the information burst, the terminal sends nothing in its reserved slot. This in turn causes the base station to transmit a negative acknowledgment feedback message indicating that the slot is once again available.

To transmit a packet, a terminal must verify two conditions. The current time slot must be available, and the terminal must have permission to transmit. Permission is granted according to the state of a pseudorandom number generator, permissions at different terminals being statistically independent. The terminal attempts to transmit the initial packet of a burst until the base station acknowledges successful reception of the packet or until the terminal discards the packet, because it has been held too long. The maximum holding time, D_{max} seconds, is determined by delay constraints on speech communication, and is a design parameter of the PRMA system. If the terminal drops the first packet of a speech burst, it continues to contend for a reservation to send subsequent packets. It drops additional speech packets as their holding times exceed the limit D_{max}. Terminals with data packets store packets indefinitely while they contend for slot reservations (equivalent to setting D_{max} to infinity). Thus as a PRMA system becomes congested, both the speech packet dropping rate and the data packet delay increase.

In [GOO91] Goodman and Wei analyze PRMA efficiency, which they quantify as the maximum number of conversations per channel that the system can support within a chosen constraint on packet-dropping probability. In their work they adopted a constraint of $P_{drop} < 0.01$. They used a speech source rate of 32 kbps and a header length of 64 bits in each packet. Using computer simulation methods, they investigated the effects of six system variables on PRMA efficiency: (1) channel rate, (2) frame duration, (3) speech activity detector, (4) maximum delay, (5) permission probability, and (6) number of conversations. Over the range of conditions examined, they found many PRMA configurations capable of supporting about 1.6 conversations per channel and found that this level of efficiency could be maintained over a wide range of conditions.

Example 4.15: Reservation in GPRS

A single 200 kHz carrier in GSM has eight time slots, each capable of carrying data at 9.6 kbps (standard), 14.4 kbps (enhanced), or 21.4 kbps (if forward error correction is completely omitted). The raw data rate can thus be as high as $8 \times 21.4 = 171.2$ kbps. The same time slots can be reserved for data access using slotted ALOHA. Medium access is based on a slotted ALOHA reservation protocol. In the contention phase, a slotted ALOHA random access technique is used to transmit reservation requests, the BS then transmits a notification to the MS indicating the channel allocation for an uplink transmission, and finally the MS can transfer data on the allocated slots without contention. On the downlink, the BS transmits a notification to the MS, indicating the channel allocation for downlink transmission of data to the MS. The MS will monitor the indicated channels, and the transfer occurs without contention.

4.3.1.2 CSMA-Based Wireless Random Access Techniques

The main drawback of ALOHA based contention protocols is the lack of efficiency caused by the collision and retransmission process. In ALOHA, users do not take into account what other users are doing when they attempt to transmit data packets and there are no mechanisms to avoid collisions. A simple method to avoid collisions is to sense the channel before transmission of a packet. If there is another user transmitting on the channel, it is obvious that a terminal should delay the transmission of the packet. Protocols employing this concept are referred to as CSMA or listen-before-talk (LBT) protocols. Figure 4.12 shows the basic concept of the CSMA protocol. Terminal "1" will sense the channel first and then sends a packet. This is followed by a sensing and packet transmission by terminal "1" again. During the second transmission time of terminal "1," terminal "2" senses the channel and discovers that another terminal is using the medium. It then delays its transmission for a later time using a back-off algorithm. The CSMA protocol reduces the packet collision probability significantly compared with ALOHA protocol. However, it cannot eliminate the collisions entirely. Sometimes, as shown in Figure 4.12, two terminals sense the channel busy and reschedule their packets for a later time but their transmission time overlap with each other causing a collision. Such situations do not cause a significant operational problem because the collisions can be handled in the same way as they were handled in ALOHA. However, if the propagation time between the terminals is very long, such situations happen more frequently, thereby reducing the effectiveness of carrier sensing in preventing collisions. As a result, several variations of CSMA have been employed in local area applications, whereas ALOHA protocols are preferred in wide area applications.

Example 4.16: Examples of Wireless Networks that Employ ALOHA and CSMA

As described earlier, Mobitex uses a variation of ALOHA protocol, and the IEEE 802.11 standard for WLANs employs a version of CSMA protocol. ALOHA protocol is also used in the random access logical channels in cellular telephone and satellite communication applications.

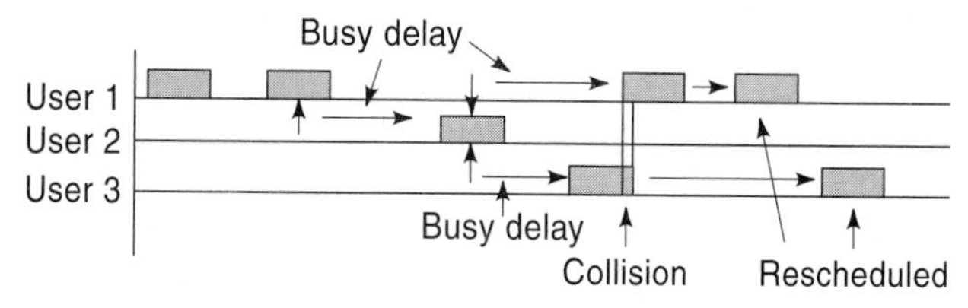

Figure 4.12 Basic operation of CSMA protocol.

A number of strategies are used for the sensing procedure and retransmission mechanisms that have resulted in a number of variations of the CSMA protocol for a variety of wired and wireless data networks. Figure 4.13 depicts the key elements of distinction among these protocols. If after sensing the channel, the terminal attempts another sensing only after a random waiting period the carrier-sensing mechanism is called "nonpersistent." After sensing a busy channel, if the terminal continues sensing the channel until the channel becomes free, the protocol is referred to as a "persistent." In persistent operation, after the channel becomes free, if the terminal transmits its packet immediately, it is referred to as "1-persistent" CSMA, and if it runs a random number generator and based on the outcome transmit its packet with a probability "p," the protocol is called p-persistent CSMA.

In a wireless network, due to multipath and shadow fading as well as the mobility of terminals, sensing the availability of the channel is not as simple as in the case of wired channels. Typically in a wireless network, two terminals can each be within range of some intended third terminal but out of range of each other, because they are separated by excessive distance or by some physical obstacle that makes direct communication between the two terminals impossible. This situation, where the two terminals cannot sense the transmission of each other, but a third terminal can sense both of them, is referred to as the *hidden terminal problem*. This is a more likely situation in cases of radio networks covering wider geographic areas in which hilly terrain blocks some groups of user terminals from sensing other groups. In this situation the CSMA protocol will successfully prevent collisions among the users of one group but will fail to prevent collisions between users in groups hidden from one another.

To resolve the hidden terminal problem, we need to facilitate the sensing procedure. In multihop ad hoc networks, where there is no centralized station or infrastructure, a protocol called *busy-tone multiple-access* (BTMA) has been used in packet radio for military applications. A brief summary of BTMA is given in [TOB80], where a number of packet communication protocols are discussed and compared. In the BTMA scheme, the system bandwidth is divided into two channels, a *message channel* and a *busy-tone channel*. Whenever a station sends signal

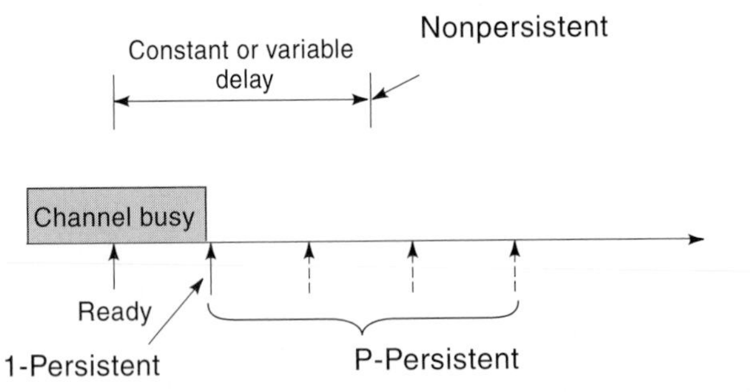

Figure 4.13 Retransmission alternatives for CSMA [STA00].

energy on the message channel, it transmits a simple busy-tone signal (e.g., a sinusoid) on its busy-tone channel. When any other terminal senses a busy-tone signal, it turns on its own busy tone. In other words, as a terminal detects that some user is on the message channel, it sounds the alarm on the busy tone channel in an attempt to inform every user, including those hidden to the transmitting terminal. A user station with a packet ready to send first senses the busy-tone channel to determine if the network is occupied.

Most cellular mobile data networks use different frequencies for forward (downlink) and reverse (uplink channels). The messages in the forward channel are transmitted from the mobile data base station that is designed and deployed to provide a comprehensive and reliable coverage. In another words, the base stations are not hidden to the mobile terminals, whereas the mobile terminals may be hidden from one another. In this situation one may use the forward channel to announce the availability of the channel for the mobile terminals. This concept is used in a protocol referred to as *digital* or *data sense multiple-access* (DSMA). DSMA is very popular in mobile data networks, and it is used in CDPD, ARDIS, and TETRA. In DSMA, the forward channel broadcasts a periodic busy-idle bit announcing availability of reverse channel for data transmission. A mobile terminal checks the busy-idle bit prior to transmission of its packet. As soon as the mobile station starts its transmission, the base station will change the busy-idle bit to the busy state to prevent other mobile terminals from transmission. Because the sensing process is performed after demodulation of data from the digital information, it is referred to digital or data sense, rather than carrier sense, multiple access.

4.3.2 Access Methods for Wireless LANs

Compared with a WAN, a LAN operates over shorter distances with smaller propagation delays and consequently a transmission medium that is well suited for variations of the CSMA protocol. Low-speed WANs are developed for communicating shorter messages while local area networks are designed to facilitate large file transfers at high data rates. As a result, the length of the packets in LANs is much larger than the length of the packets in low-speed mobile data networks. When the length of the packets is long, it would be very useful to pay further attention to packet collisions. LANs often employ variations of the CSMA protocol that either stop transmission as soon as a packet collision is detected or add additional features to the avoid collision.

Problem 10: Packet Sizes in Wireless Data Networks

a) Determine the transfer time of a 20 kB file with a mobile data network with a transmission rate of 10 kbps.

b) Repeat for an 802.11 WLAN operating at 2 Mbps.

c) What is the length of the file that the WLAN of part (b) can carry in the time that mobile data service of part (a) carries its 20 kB file?

Solution:

a) The early mobile data networks, such as ARDIS and Mobitex, limited the length of a file to around 20 kB. For a data rate of around 10 kbps it would take 20 (kB) × 8 (B/b) × 10 Mbps = 16 seconds to transfer such a file.

b) An IEEE 802.11 network operating at 2 Mbps would transfer this file in 80 ms.

c) In a 16-second time interval, the same WLAN transfers a 4 MB file.

The most popular version of the CSMA for wired LANs is CSMA with collision detection (CSMA/CD) adopted in the IEEE 802.3 (Ethernet) standard, the dominant standard for wired LANs supporting data rates that can be up to several gigabits per second. The basic operation of CSMA/CD is the same as CSMA implementations discussed earlier. The defining feature of CSMA/CD is that it provides for detection of a collision shortly after its onset, and each transmitter involved in the collision stops transmission as soon as it senses a collision. In this way colliding packets can be aborted promptly, minimizing the wastage of channel occupancy time by transmissions destined to be unsuccessful. Unlike CSMA, which requires an acknowledgment (or lack of an acknowledgment) to learn the status of a packet collision, CSMA/CD requires no such feedback information, because the collision-detection mechanism is built into the transmitter. When a collision is detected, the transmission is immediately aborted, a jamming signal is transmitted, and a retransmission back-off procedure is initiated, just as in CSMA [PAH95]. As is the case with any random access scheme, proper design of the back-off algorithm is an important element in assuring stable operation of the network.

Example 4.17: Binary Exponential Back-off

The back-off algorithm recommended by IEEE 802.3 Ethernet is referred to as the binary exponential back-off algorithm that is combined with 1-persistence CSMA protocol with collision detection. When a terminal senses a transmission, it continues sensing (persistent) until the transmission is completed. After the channel becomes free, the terminal sends its own packet. If another terminal was also waiting, a collision occurs because of the 1-persistence, and the two terminals reattempt transmission with a probability of ½ after a time slot that spans twice the maximum propagation delay allowed between the two terminals. A time slot that spans twice the maximum propagation delay is selected to ensure that in the worse case scenario the terminal will be able to detect the collision. If a second collision occurs, the terminals reattempt with the probability of ¼ that is half of the previous retransmission probability. If collision persists, the terminal continues reducing its retransmission probability by half up to 10 times, and after that it continues with the same probability six more times. If no transmission is possible after 16 attempts, the MAC layer reports to the higher layers that the network is congested and transmission shall be stopped. This procedure exponentially increases the back-off time and gives the back-off strategy its name. The disadvantage of this procedure is that the packets arriving later have a higher chance to survive the collision that results in an unfair first-come, last-serve environment. It can be shown that the average waiting time for the exponential back-off algorithm is $5.4T$ where T is the time slot used for waiting [TAN97], [STA00].

The CSMA/CD scheme is used in many IR-based LANs, where both transmission and reception are inherently directional. In such an environment, a transmitting station can always compare the received signal from other terminals with its own transmitted signal to detect a collision. Radio propagation is not directional, posing a serious problem in determining other transmissions during your own transmission. As a result, collision detection mechanisms are not well suited for radio LANs. However, compatibility is very important for WLANs, and, therefore, designers of these networks have had to consider CSMA/CD for compatibility with the Ethernet backbone LANs that dominate the wired-LAN industry.

Although collision detection is easily performed on a wired network, simply by sensing voltage levels against a threshold, such a simple scheme is not readily applicable to radio channels because of fading and other radio channel characteristics. The one approach that can be adopted for detecting collisions is to have the transmitting station demodulate the channel signal and compare the resulting information with its own transmitted information. Disagreements can be taken as an indicator of collisions, and the packet can be immediately aborted. However, on a wireless channel the transmitting terminal's own signal dominates all other signals received in its vicinity, and thus the receiver may fail to recognize the collision and simply retrieve its own signal. To avoid this situation, the station's transmitting antenna pattern should be different from its receiving pattern. Arranging this situation is not convenient in radio terminals because it requires directional antennas and expensive front-end amplifiers for both transmitters and receivers.

The approach called CSMA/CA, shown in Figure 4.14, is actually adopted by the IEEE 802.11 wireless LAN standard. The elements of CSMA/CA used in the IEEE 802.11 are interframe spacing (IFS), contention window (CW), and a back-off counter. The CW intervals are used for contention and transmission of the packet frames. The IFS is used as an interval between two CW intervals. The back-off counter is used to organize the back-off procedure for transmission of packets. The method of operation is best described by an example.

Example 4.18: Operation of Collision Avoidance in IEEE 802.11

Figure 4.15 provides an example for the operation of the CSMA/CA mechanism used in the IEEE 802.11 standard. Stations A, B, C, D, and E are engaged in contention for transmission of their packet frames. Station A has a frame in the air when stations B, C, and D sense the channel and find it busy. Each of the three stations will run its random number generator to get a back-off time by random. Station C followed by D and B draws the smallest number. All three terminals

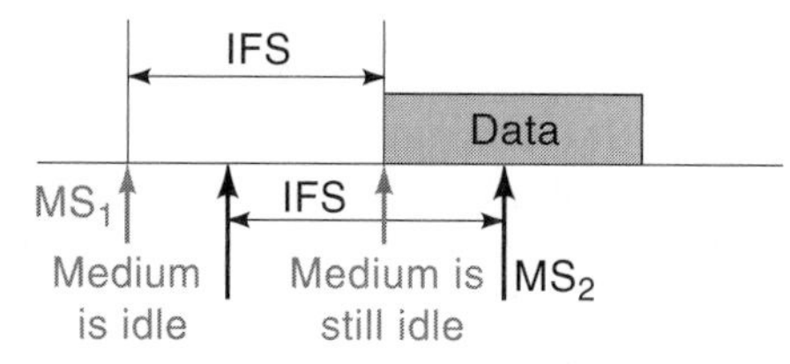

Figure 4.14 CSMA/CA adopted by the IEEE 802.11.

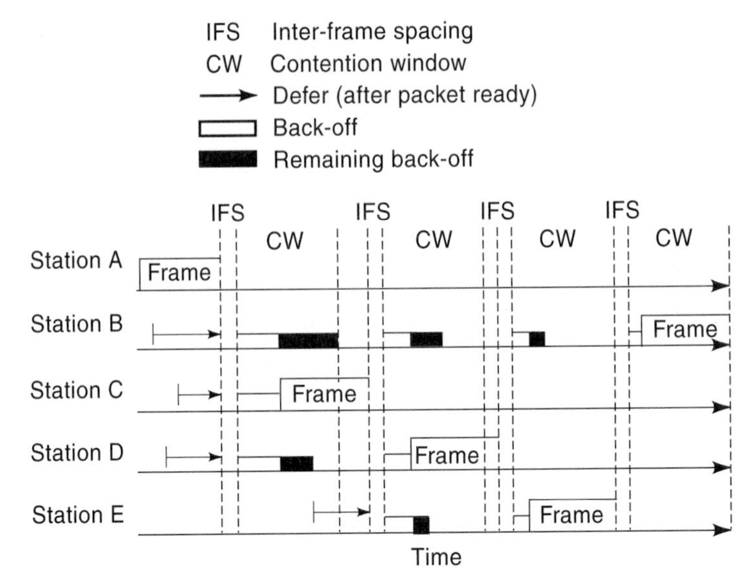

Figure 4.15 Illustration of CSMA/CA.

persist on sensing the channel and defer their transmission until the transmission of the frame from terminal A is completed. After completion, all three terminals wait for the IFS period and start their counters immediately after completion of this period. As soon as the first terminal, station C in this example, finishes counting its waiting time, it starts transmission of its frame. The other two terminals, B and D, freeze their counter to the value that they have reached at the start of transmission for terminal C. During transmission of the frame from station C, station E senses the channel, runs its own random number generator that in this case ends up with a number larger than the remainder of D and smaller than the remainder of B, and defers its transmission for after the completion of station C's frame. In the same manner as the previous instance, all terminals wait for IFS and start their counter. Station D runs out of its random waiting time earlier and transmits its own packet. Stations B and E freeze their counters and wait for the completion of the frame transmission from terminal D and the IFS period after that before they start running down their counters. The counter for terminal E runs down to zero earlier, and this terminal sends its frame while B freezes its counter. After the IFS period following completion of the frame from station E, the counter in station B counts down to zero before it sends its own frame. The advantage of this back-off strategy over the exponential back-off used in IEEE 802.3 is that the collision detection procedure is eliminated and the waiting time is fairly distributed in a way that on average enforces a first-come, first-serve policy.

Another related technique considered for collision avoidance in wireless LANs is the *combing* method [WIL95]. As shown in Figure 4.16, the time is divided into comb and data transmission intervals. During the comb period, each station alternates between transmission and listening periods according to a code assigned to the station. All stations will continue advancing in their code until they sense a carrier during their listening period. If they do not sense a carrier at the end of the

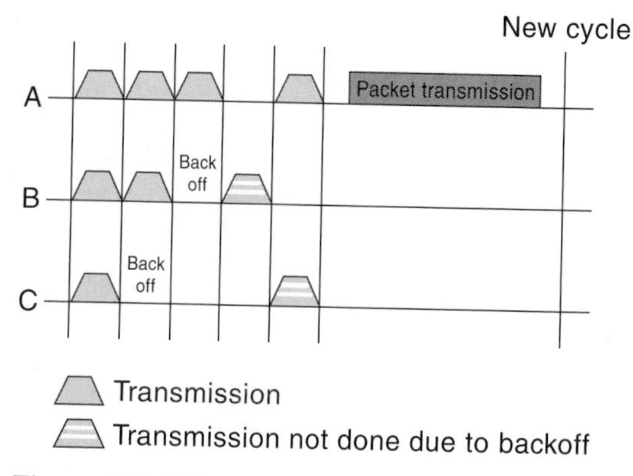

Figure 4.16 Illustration of combing.

code, they transmit their packet. If they sense a carrier, they postpone their transmission until the next comb interval. A simple example will further clarify this method.

Example 4.19: Combing for Collision Avoidance

Figure 4.16 shows three stations with five digit codes 11101 (terminal A), 11010 (terminal B), and 10011 (terminal C). All three terminals transmit their carrier during the first slot, because all codes have 1 in that slot. In the second slot, terminal C will listen and after sensing the carriers of other two terminals withdraws from contention and waits for the next combing. In the third period station, B goes to listening state, and after sensing the carrier of terminal A defers its transmission until the next cycle. Terminal A continues its sequence of alternating transmissions and listening until the end of the comb period when it transmits its data packet (as it heard no other terminal). After completion of the data transmission from station A, the other two terminals will wait for an interpacket spacing for a new contention after which station B transmits its packet. Station C will transmit after the second transmission cycle.

In CSMA/CA, as we will see later, priority is assigned by dividing the IFS into several different sized intervals associated with different priority levels. In combing, priority can be arranged by assigning different classes of numbers to the codes. The lower priority packets will receive earlier zero codes, and higher priority packets will have a zero in their codes in the later intervals.

Another access method used in wireless LANs is the request-to-send/ clear-to-send (RTS/CTS) mechanism shown in Figure 4.17. A terminal ready for transmission sends a short RTS packet identifying the source address, destination address, and the length of the data to be transmitted. The destination station will respond with a CTS packet. The source terminal will send its packet with no contention. After acknowledgment from the destination terminal, the channel will be available for other usage. IEEE 802.11 supports this feature as well as CSMA/CA. This method provides a unique access right to a terminal to transmit without any contention.

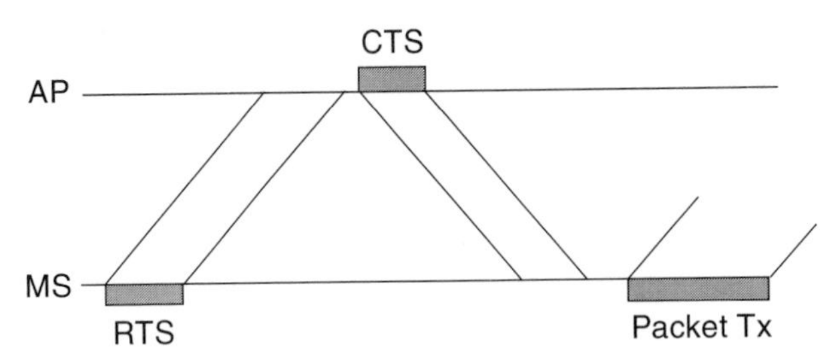

Figure 4.17 RTS/CTS in IEEE 802.11.

4.3.3 Performance of Random Access Methods

In voice-oriented circuit switched networks, performance is measured by the probability of blockage (blockage rate) of initiating a call. If the call is not blocked, a fixed rate full-duplex channel is allocated to the user for the entire communication session. In other words, interaction between the user and the network takes place in two steps. First, during the call establishment procedure, the user negotiates the availability of a line with the network, and if successful (not blocked), the network guarantees a connection with a certain QoS (data rate, delay, error rates) to the user. For real-time interactive applications such as telephone conversations or video conferencing, if the user does not talk, the resource allocated to the user is wasted. If these facilities, originally designed for two-way voice application, are used for data application then (1) for bursty data file transfers during the idle times between transmission of two packet bursts allocated resources are wasted, and (2) large file transfers suffer a long delay or waiting time for the transfer because resources allocated to each user is more restricted.

Users of packet switched networks are always connected, and there is neither an initiation (negotiation) procedure to be blocked nor a fixed QoS to be allocated. In this situation, analysis of the performance for real-time interactive applications such as telephone conversations is complicated and will be addressed later. Performance of packet switched networks for data applications is often measured by the average throughput, S, and average delay, D, versus the total offered traffic, G. The *channel throughput, S,* is the average number of successful packet transmissions per time interval T_p. The offered traffic, G, is the number of packet transmission attempts per packet time slot T_p that includes new arriving packets, as well as retransmissions of old packets. The average delay D is the average waiting time before successful transmission, normalized to the packet duration T_p. The standard unit of traffic flow is Erlang, which can be thought of as the number of the packets per packet duration time T_p. The throughput is always between zero and one Erlang, whereas the offered traffic, G, may exceed one Erlang.

The analyses of the relationships between S, G, and D for a variety of medium access protocols have been a subject of research for a few decades. This analysis depends on the assumptions on the statistical behavior of the traffic, number of terminals, relative duration of the packets, and the details of the implementation. Assuming a large number of terminals generating fixed length packets with a Poisson distribution,[2] Table 4.1 summarizes the throughput expressions for

[2]Poisson distribution assumes packets are generated independent from one another, and the interarrival time between the packets forms an exponentially distributed random variable.

Table 4.1 Throughput of Various Random Access Protocols

Protocol	Throughput
Pure ALOHA	$S = Ge^{-2G}$
Slotted ALOHA	$S = Ge^{-G}$
Unslotted 1-persistent CSMA	$S = \dfrac{G[1 + G + aG(1 + G + aG/2)]e^{-G(1+2a)}}{G(1 + 2a) - (1 - e^{-aG}) + (1 + aG)e^{-G(1+a)}}$
Slotted 1-persistent CSMA	$S = \dfrac{G[1 + a - e^{-aG}]e^{-G(1+a)}}{(1 + a)(1 - e^{-ag}) + ae^{-G(1+a)}}$
Unslotted nonpersistent CSMA	$S = \dfrac{Ge^{-aG}}{G(1 + 2a) + e^{-aG}}$
Slotted nonpersistent CSMA	$S = \dfrac{aGe^{-aG}}{1 - e^{-aG} + a}$

ALOHA, and 1-persistent and nonpersistent CSMA protocols, including the slotted and unslotted versions of each. The expressions for *p*-persistent protocols are very involved and are not included here. The interested reader should refer to [KLE75b], [TOB75], and [TAK85], where the derivations of the other CSMA expressions can also be found. The expressions in the table are also derived in [HAM86] and [KEI89]. The parameter a in this table corresponds to the normalized propagation delay defined as $a = \tau/T_p$, where τ is the maximum propagation delay for the signal to go from one end of the network to the other end.

Problem 11: Calculation of the Normalized Propagation Delay

Determine the parameter a in IEEE 802.3 (Ethernet) 10 Mbps LANs and IEEE 802.11 2 Mbps LANs.

Solution:

The IEEE 802.3 standard for star LANs allows a maximum length of 200 m between two terminals. The propagation speed in the cables is usually approximated by 200,000 km/s, resulting in $\tau = 1\mu s$. The IEEE 802.11 allows maximum distance of 100 meters between the AP and the MS. The radio propagation is at the rate of 300,000Km/s, resulting in $\tau = 0.33\ \mu s$. For a star LAN operating at 10 Mbps with 1,000 bit packets, the value is $a = 0.01$. For an IEEE 802.11 operating at 2 Mbps with the same packet size $a = 0.00066$.

Figure 4.18 shows plots of throughput S versus offered traffic load G for the six protocols listed in Table 4.1, with a normalized propagation delay of $a = 0.01$. All curves follow the same pattern. Initially as the offered traffic G increases the throughput, S also increases up to a point where it reaches a maximum S_{max}. After the throughput reaches its maximum value, an increase in the offered traffic actually reduces the throughput. The first region depicts the stable operation of the network in which an increase in aggregating traffic G, which includes arriving traffic as well as retransmissions due to collisions, increases the total successful transmissions and thus S. The second region represents unstable operation where an increase in G actually reduces the throughput S because of congestion and the eventually halting of the operation. In

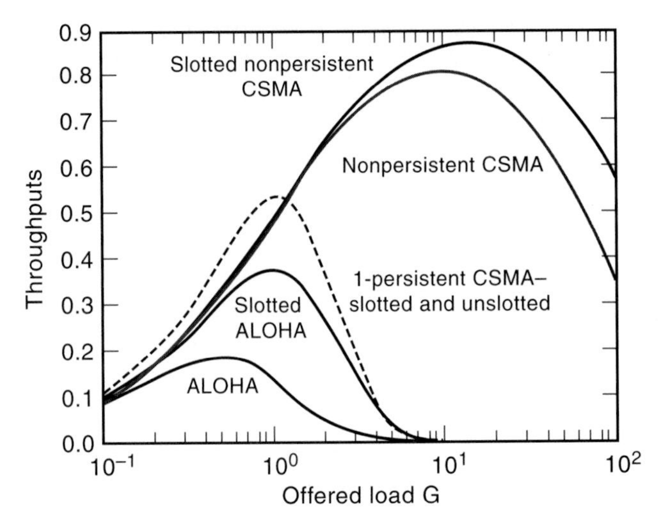

Figure 4.18 Throughput S versus offered traffic load G for various random access protocols.

practice, as we saw in the last section, retransmission techniques adopted for the real implementation include back-off mechanisms to prevent operation in unstable regions.

The throughput curves for the slotted and unslotted versions of 1-persistent CSMA are essentially indistinguishable. It can be seen from the figure that for low levels of offered traffic, the 1-persistent protocols provide the best throughput, but at higher load levels the nonpersistent protocols are by far the best. It can also be seen that the slotted nonpersistent CSMA protocol has a peak throughput almost twice that of persistent CSMA schemes.

The equations in Table 4.1 can also be used to calculate capacity, which is defined as the peak value S_{max} of throughput over the entire range of offered traffic load G [HAM86]. An example is helpful to show how to relate the curve to a particular system.

Problem 12: Relating Throughput and Offered Traffic to Data Rates

To relate throughput and offered traffic to data rates, assume that we have a centralized network that supports a maximum data rate of 10 Mbps and serves a large set of user terminals with the pure ALOHA protocol.

a) What is the maximum throughput of the network?

b) What is the offered traffic in the medium and how is it composed?

Solution:

a) Because the peak value of the throughput is $S = 18\%$ the terminals contending for access to the central module can altogether succeed in getting at most 1.8 Mbps of information through the network.

b) At that peak the total traffic from the terminals is 5 Mbps (because the peak occurs at $G = 0.5$), which is composed of 1.8 Mbps of successfully delivered packets (some mixture of new and old packets) and 3.2 Mbps of packets doomed to collide with one another.

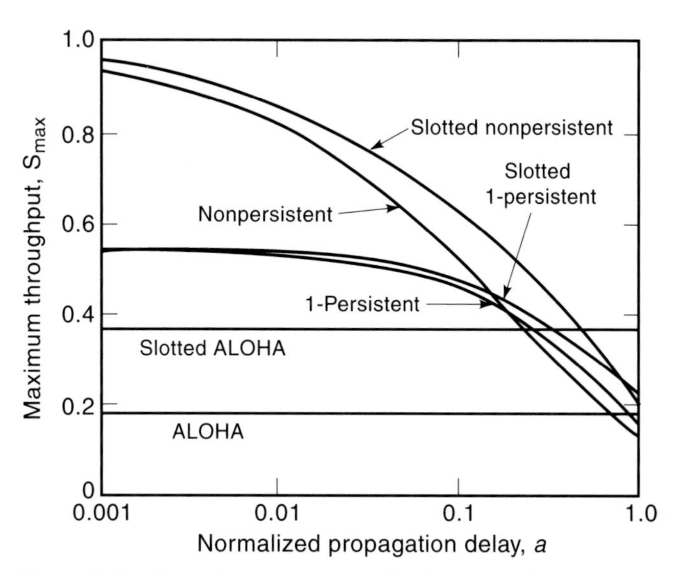

Figure 4.19 Capacity versus normalized propagation delay for various random access protocols.

Plots of capacity versus normalized propagation delay are plotted in Figure 4.19 for the same set of ALOHA and CSMA schemes. The curves show that for each type of protocol the capacity has a distinctive behavior as a function of normalized propagation delay a. For the ALOHA protocols, capacity is independent of a, and is the largest of all the protocols (compared when a is large). As we discussed earlier, this is the case where the area of coverage is large and propagation delays are comparable to the length of packets. The plots in Figure 4.19 also show that the capacity of 1-persistent CSMA is less sensitive to the normalized propagation delay for small a, than is nonpersistent CSMA. However, for small a, nonpersistent CSMA yields a larger capacity than does 1-persistent CSMA, though the situation reverses as a approaches the range of 0.3 to 0.5 [HAM86].

Another important performance measure for packet data communications is the delay characteristics of the transmitted packets. For real-time applications and voice conversations, if the delay is more than a certain value (several hundred milliseconds), the packet is not useful, and it is dropped. Therefore, we need to analyze the delay characteristics of the channel to determine the capacity of the access method. In the data transfer applications, the delay characteristic is usually related to the throughput of the medium, and it usually follows a hockey-stick shape. At low traffic when a small fraction of the maximum throughput is utilized, the delay often remains the same as the transmission delay. As the throughput increases, the number of retransmitted packets increases, resulting in higher average delay for the packets. Around the maximum throughput, the delay retransmissions grow rapidly, pushing the network toward unstable condition where the channel is dominated with retransmissions, and the packet delays grow extremely large. Figure 4.20 shows the delay-throughput behavior of the ALOHA, slotted ALOHA, and CSMA protocols [TAN00].

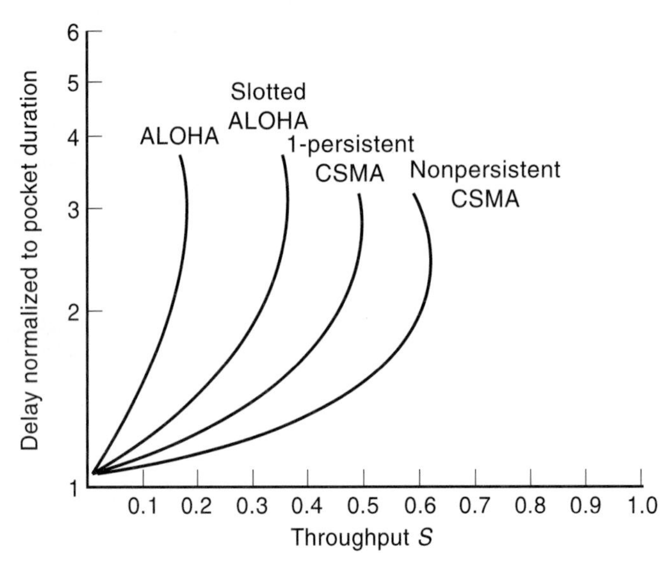

Figure 4.20 Delay-throughout behavior of random access protocols.

4.3.3.1 Practical Considerations

The analysis provided in the previous chapter is abstract and is used to provide an intuitive framework for the operation of different classes of access methods. In practice implementations deviate considerably from the abstract, and the performance is evaluated by analysis or simulation of case-by-case situations. Examples of this type of analysis for the CSMA/CD with exponential back-off algorithm used in the IEEE 802.3 Ethernet and performance of token ring access method used in IEEE 802.5 are available in the last chapter of [STA00].

4.3.3.2 Complications Caused by Wireless Channel

Three factors that are effective in throughput analysis in wired environment are propagation delay, users' idle period (not transmitting), and packet collisions. In a wireless environment, analysis of the real throughput of a protocol is much more complicated because it involves hidden terminal and capture effects. To analyze these effects, let's assume we have a centralized AP with a number of terminals connected to it, which communicate via a random access method.

Figure 4.21 demonstrates the basic concept behind the hidden terminal problem. The two terminals contending to communicate with the AP are both in the coverage area of the AP, but they are out of the coverage area of each other. Limited antenna range and shadowing are two major causes for hidden terminal degradation. The hidden terminal problem does not effect the performance of the ALOHA type protocols, but it degrades the performance of CSMA protocols. In a CSMA environment effected by the hidden terminal problem, some terminals cannot sense the carrier of the transmitting terminal and their transmitted packets have a higher probability of colliding and degrading the overall throughput.

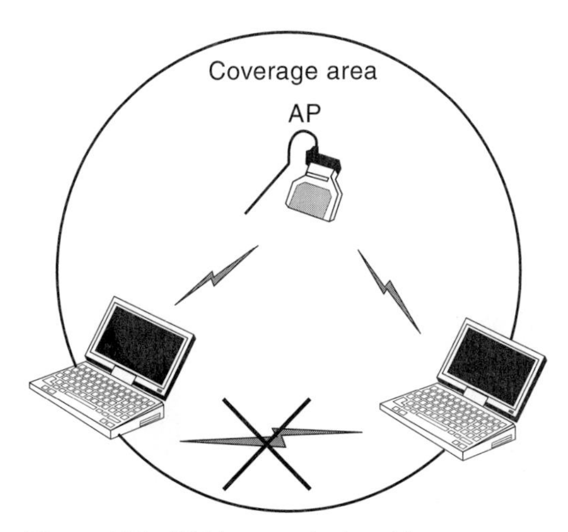

Figure 4.21 Hidden terminal problem.

In real installations, the coverage area of the AP is usually larger than that the mobile terminals, because the AP is installed in a selected location to optimize the coverage (high on the walls or on the ceiling), which will increase the negative impacts of the hidden terminal problem. Assuming that the coverage area of the AP and the mobile terminals are the same, there is still no guarantee that all the terminals in the coverage area of the AP can hear one another. This is because two terminals at the maximum distance L from the AP could be as far as $2L$ apart. Therefore, the hidden terminal problem is unavoidable and natural to the operation of the centralized access systems using CSMA protocol that are common in WLAN operations.

Another phenomenon impacting the throughput of a radio network is capture. In radio channels, sometimes collision of two packets may not destroy both packets. Because of signal fading or the near-far effect, packets from different transmitting stations can arrive with different power levels, and the strongest packet may survive a collision. Figure 4.22 shows the basic concept of the capture phenomenon. The received power from the terminal closer to the access point is much larger than the received power from the terminal located at a distance. If two packets collide in time, the packet with the weaker signal will appear as a background noise, and the AP captures (detects) the packet from the closer terminal successfully. The capture effect increases the throughput of the radio network because in calculating the throughput, we always assume that the colliding packets are destroyed (not detected).

The hidden terminal effects were first analyzed for different types of CSMA protocols used in rapidly moving packet radio networks for military applications, and busy tone signaling was suggested for eliminating the hidden terminal problem. More recently, there have been efforts to analyze the effects of capture and hidden terminals in a WLAN environment using various assumptions [ZHA92], [ZAH97].

In reality, the capture of a packet is a random process, which is a function of the modulation technique used for transmission, received signal-to-noise ratio, and the length of the packet.

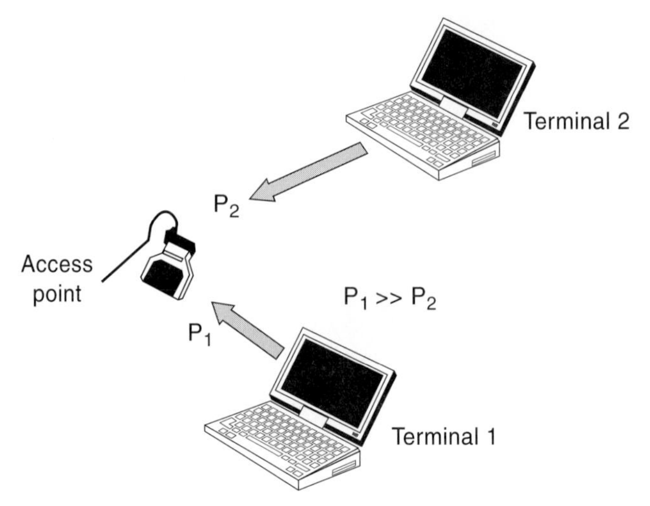

Figure 4.22 The capture effect.

Example 4.20: Capture Effect and Throughput

Figure 4.23 [ZHA92] shows the effects of capture on the throughput of the conventional slotted ALOHA and the CSMA systems for a variety of packet lengths of 16, 64, and 640 bits. Also shown for comparison are the curves for conventional nonpersistent CSMA and slotted ALOHA without capture. With capture, the maximum throughput of CSMA with packet length 16 bits is 0.88 Erlang, which is 0.065 Erlang more than the case without capture. The maximum throughput for

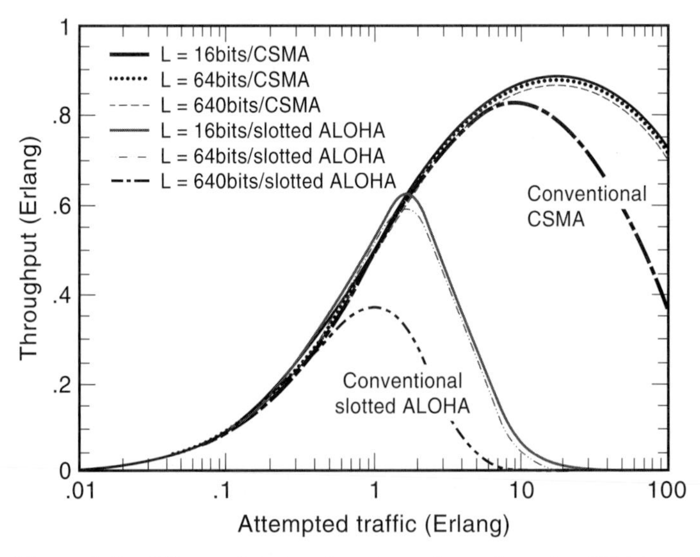

Figure 4.23 Effects of packet length on throughput for CSMA and slotted ALOHA with capture. The modulation is BPSK, and the SNR = 20 dB.

slotted ALOHA with the same packet length is 0.591 Erlang, which is 0.231 Erlang higher than the case without capture.

In slow fading channels, if the terminal generating the test packet is in a "good" location, the interference from other packets is small, and all the bits of the test packet survive the collision. In contrast, for a test packet originating from a terminal in a "bad" location, all the bits are subject to high probability of error, and the packet does not survive the collision. As a result, the system shows minimal sensitivity to the choice of packet length, which is consistent with our assumption of slow fading. Figure 4.24 [ZHA92] shows the delay-throughput for the CSMA protocol with and without capture for a 640-bit packet network. The packet delay is normalized to the length of the packet. For both cases as the throughput reaches around its maximum value system becomes rapidly unstable, causing unacceptable delays for the delivery of the packets. When the capture effects are included, maximum throughput is increased, and the instability occurs at a slightly higher value of the throughput.

Example 4.21: Capture and Hidden Terminals in WLANs

Figure 4.25 [ZAH97] shows the throughput versus offered traffic curves for a WLAN access point using CSMA protocol and surround by a large number of terminals uniformly distributed within the AP's coverage area. In this scenario, as shown in Figure 4.26 each terminal senses a group of terminals within its coverage area (area I in the figure) and cannot sense those that are out of its coverage area but are still within the coverage area of the AP (area II in the figure). The

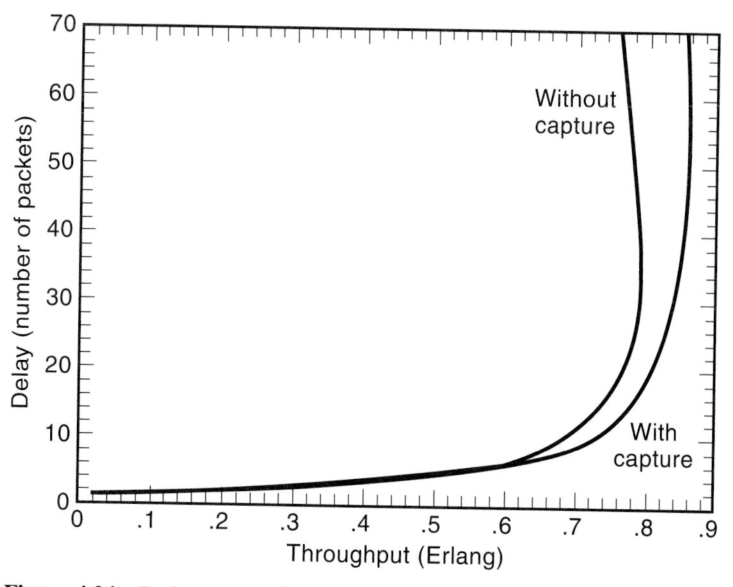

Figure 4.24 Delay versus throughput of CSMA for BPSK modulation and SNR = 20 dB, with and without capture.

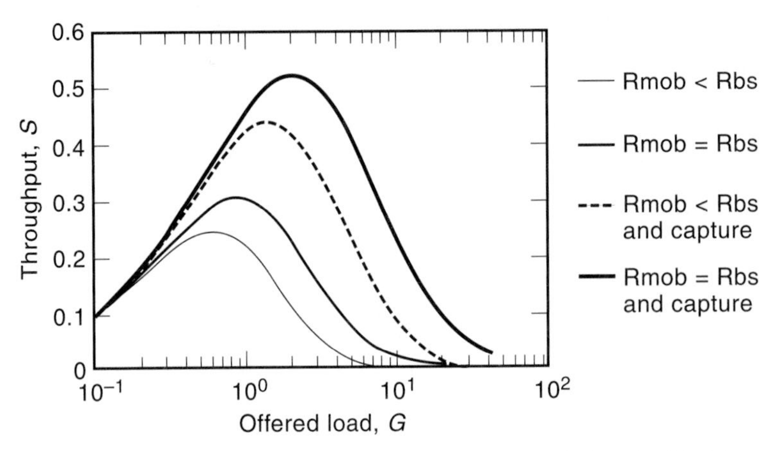

Figure 4.25 Throughput versus offered traffic for a wireless LAN with a large number of terminals.

throughput of the target terminal with respect to terminals in area I is the same as the throughput of a CSMA system. However, the throughput of the target terminal with respect to terminals in area II is the same as the throughput of ALOHA networks because carrier sensing does not work and the terminals transmit their packets without knowledge of the transmission from terminals in area II. Using these facts in the throughput at each point in the area of the coverage of the AP is calculated [ZHA96]. Then it is averaged over the entire coverage area of the AP for different coverage areas for the mobile terminals. Obviously, this average throughput will always remain between the throughput of CSMA and that of ALOHA. The lowest curve in Figure 4.25 shows the throughput when the hidden terminal problem is considered, and the coverage (radius R_{mob}) of each terminal is 70 percent of the coverage (radius *rbs*) of the AP. Because a number of termi-

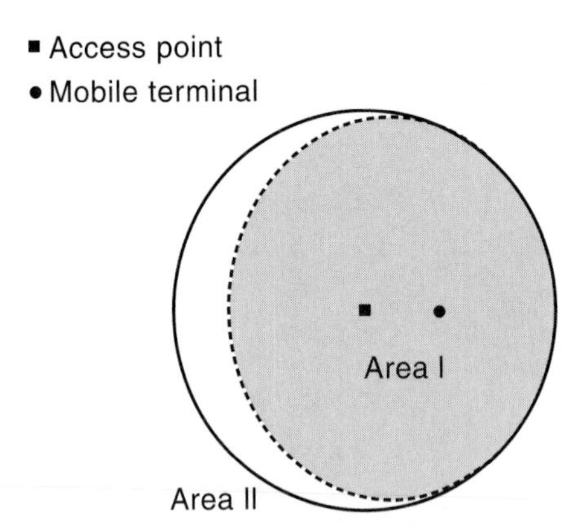

Figure 4.26 Coverage areas of an AP and a tagged mobile terminal in a WLAN.

nals cannot sense the transmission of others, the peak throughput has declined to less than 25 percent which is slightly higher than the 18 percent maximum throughput of the ALOHA and far below the maximum of CSMA. The second curve from below, with a peak value of around 30 percent, represents the same results where the coverage of the AP and the mobile terminals are the same. The third curve depicts the performance when both effects of the hidden terminal and capture are considered, and the coverage of the mobile terminals is smaller than that of the AP. The capture effect increases the throughput to more than 40 percent. The top curve is the same as the third curve where the coverage of the AP and the mobile terminals are the same. This situation has increased the throughput by another 10 percent to above 50 percent, which is getting closer to the performance of conventional CSMA.

4.4 INTEGRATION OF VOICE AND DATA TRAFFIC

4.4.1 Access Methods for Integrated Services

As the wireless communications industry moves toward 3G and 4G networks, one of the important objectives is the use of a single wireless system for multimedia applications to support a variety of communications services, including voice, data, and voice in various forms and combinations. A key technical problem to be dealt with in such integrated systems is that of multiuser access. As we saw earlier in this chapter, an access method that efficiently supports one category of service may be unsuitable for another category of service. In the 1G and 2G networks, as we saw in Chapter 1, the wireless industry evolved around two separate paths for voice- and data-oriented applications. If a data service can be efficiently integrated with a voice service, transmission resources that are otherwise wasted (because there is no voice transmission) can be used for data, which typically do not have stringent delay requirements. First, the voice-oriented networks evolved into supporting data. More recently with the popularity of voice over IP over the Internet and PSTN, supporting voice in data-oriented WLANs has become attractive as well.

As we saw earlier, in a packet communication environment, voice and data have different requirements. Voice packets can tolerate errors and even packet losses (a loss of 1 to 2 percent of voice packets has insignificant effect on the perceived quality of reconstructed voice [KUM74]), whereas data packets are sensitive to loss and errors but can generally tolerate delays. Also, the rate at which information is transmitted is constant in the case of voice, thereby making circuit-switching a viable and efficient approach, whereas information generated for data transmission is very bursty. As a result, voice- and data-oriented networks use different multiple access methods. In a wireless environment, the simplest approach is to assign different frequency bands to isochronous (voice) and asynchronous (data) packets. However, integration in one frequency band will result in a more efficient use of the bandwidth, a simpler radio interface, and an environment that provides a better control for synchronizing voice and video (e.g., lip-synch).

4.4.2 Data Integration in Voice-Oriented Networks

Fixed access methods such as FDMA, TDMA, and CDMA were basically designed for access to the circuit-switched voice-oriented networks. Later on, as we saw in Chapter 1, several data services evolved around these systems. The economical incentive for using this medium for mobile data services is to take advantage, either partially or fully, of the existing infrastructure, the terminals, and the frequency bands designed for the voice-oriented networks. This way the mobile data service provider saves in the major costs of deployment, which includes the cost of real estates and the installation of the antenna, and there no longer is a need to obtain new frequency bands for operation of the data service. If possible, using the same terminal for voice and data will reduce the cost and facilitates marketing of the service.

Example 4.22: Mobile Data over FDMA Analog Cellular

The CDPD packet data system introduced in the early 1990s uses available frequency channels in the existing analog FDMA cellular telephone network (AMPS) to provide an overlaid packet data service supporting data rates similar to voice-band modems (up to 19.2 kbps). In its present form, this system does not exploit the pauses between talk spurts but simply takes advantage of the frequency bands temporarily unused by mobile telephone users in each cell area. CDPD uses the unused AMPS channel to develop a communication link between a mobile data unit and a mobile data base station. Ideally, a CDPD terminal can use the RF and antenna of an AMPS terminal to communicate packet data bursts. However, the most important issue for the CDPD network is that it can use the same antenna site and the antenna towers and the frequency bands of an existing AMPS network. Because real estate, installation of the antenna post, and the frequency bands are perhaps the most expensive parts for implementation of a network, CDPD was perceived to provide a cost-efficient solution for a mobile data service with a comprehensive coverage. The air interface protocol and modulation technique used in the CDPD are, however, different from the AMPS system.

Example 4.23: Mobile Data over TDMA Systems

The GPRS packet data network, introduced in late 1990s uses the air interface and the infrastructure of the GSM network to provide a mobile packet data service that can support data rates of up to a couple of hundred kilobits per second. GPRS uses the same physical packet format and modulation technique as GSM. The logical channels used in GPRS do not use the dialing procedure used in the GSM. In a manner similar to CDPD, through the wired infrastructure of the network, the packets of data in GPRS are routed to the packet switched data networks rather than being switched through the PSTN. GPRS is designed to take advantage of the unused time slots of a TDMA voice-oriented GSM network.

Example 4.24: Mobile Data over CDMA Networks

In TDMA systems and FDMA systems, the data users may use free time slots and free channels, respectively, as they become available. In a CDMA system, the situation is somewhat different. The structure of CDMA is such that all active users use the entire bandwidth-time space simultaneously. The resource to be managed is signal power. With the application of efficient power control algorithms, the signal levels transmitted by mobile stations and the base station are continually adjusted in response to the changing locations of mobiles and the number of users on the system at any given time. In a CDMA network, the integration of data calls with voice calls is straightforward in principle, because various numbers of both categories of calls are readily mixed together, with each call accessing the channel with its unique user signal code. Therefore, in a CDMA system no modification need be made to the channel access scheme to accommodate integration of voice and data "channels," and the information rate for voice or data traffic in any one channel can in principle be varied by a variable-rate scheme such as that used for voice service in the IS-95 standard. The integration of voice and data services in a single user channel is not necessarily straightforward.

From a technical point of view, there are two incentives for integration of data into voice-oriented fixed-assignment access methods:

1. The fixed-assignment access methods used in voice-oriented networks are designed to support a certain number of simultaneous users. When the number of active users falls below that number, some portion of the transmission resources is wasted.
2. A typical two-way conversation does not make full use of the call connection time, because only one of the parties talks at a time. Furthermore the flow of natural speech is because actually composed of *talk spurts* with intervening short pauses. It is generally estimated that in a two-way voice connection, the average *voice activity factor* for each party is in the vicinity of 40 percent and thus about 60 percent of available transmission time remains unused.

Assume that we have N voice channels available for a given area (e.g., the coverage area of a sectored antenna in a cellular deployment) to accommodate newly originated calls as well as calls handed off from other areas. Further assume that the overall calls in the area are generated according to a Poisson process with a normalized rate of $\rho = \lambda/\mu$ calls/unit channel and length of call holding is generally distributed with a unit mean. The number of idle channels, N_{idle}, according to renewal theory [BUD97] is given by

$$N_{idle} = N - \rho(1 - B(N,\rho)), \tag{4.10}$$

where $B(N,\rho)$ represents the blocking probability calculated from the Erlang B equation given by:

$$B(N,\rho) = \frac{\rho^N / N!}{\sum_{i}^{N} (\rho^i / i!)} \tag{4.11}$$

Figure 4.27 shows the average number of the idle voice channels per area N_{idle} versus the number of available channels N for a variety of call blocking rates. At call blocking rates of around 2 percent that is desirable for most cellular systems, a number of free channels are available for data communications. As the network

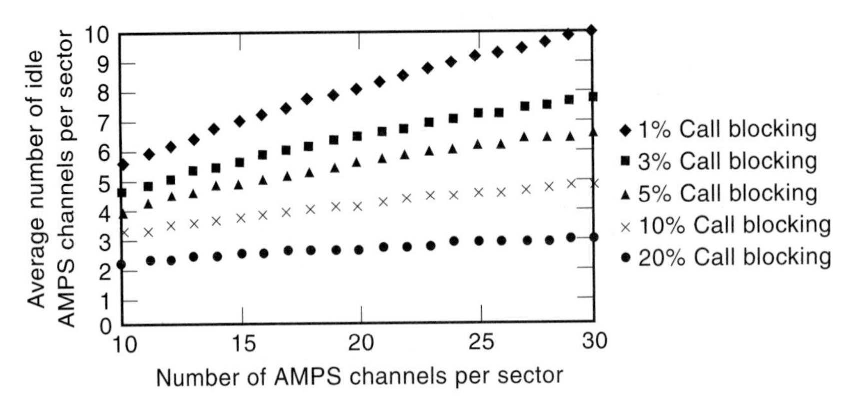

Figure 4.27 Average number of idle channels per area as a function of the number of available channels for a given call blocking rate [BUD 97].

accepts larger values of blocking rates, because most of the time all channels are in use, regardless of the number of channels, only a few channels will be available for data. If the integrated system uses the idle channels for data transmission, on the average, the maximum throughput available to the data user is N_{idle} times the encoding rate of the voice channel. If the system can take advantage of the silence periods in the two-way telephone conversations, as we indicated earlier, an additional throughput of up to 60 percent is available for data applications.

Another important parameter is the idle time for a voice channel. The idle time is the period of time that a voice channel is not occupied by a voice user. In an N channel voice-oriented network, assuming that each channel receives an equal fraction of the call load, one may calculate T, the average length of time a channel is idle, by [BUD97]:

$$T = \frac{N - \rho(1 - B(N,\rho))}{\rho(1 - B(N,\rho))} \tag{4.12}$$

Figure 4.28 shows the normalized average idle period per channel T versus the number of available voice channels for different blocking rates. For a typical holding time of around a couple of minutes and a blocking rate of around 2 percent, the average idle periods are fairly long implying that a data network has a reasonable time to detect the availability and send its bursts of data.

There are periods of time for which all the voice channels are occupied for the voice users, and there is no channel available to the data service. If these periods are short and infrequent, some data applications may accept the situation. Otherwise, specific channels should be allocated for data only usage. In these situations the data users have their own channel, as well as unused portions of the voice channels.

Assuming that we accept the block-out periods and assign no dedicated channel resources to the data application, we will have periodic operation between the available and blocked-out periods. Assuming that the holding time for the telephone conversations is exponentially distributed, it can be shown [BUD97] that the average active period where a channel is available for data, T_a, is given by:

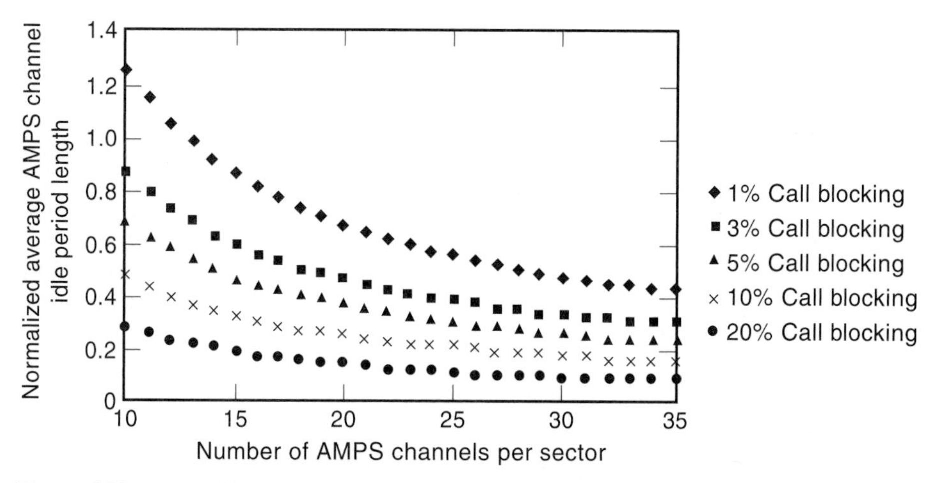

Figure 4.28 Normalized average idle period per channel as a function of the number of channels and call blocking probability [BUD97].

$$T_a = \frac{1 - B(N,\rho)}{NB(N,\rho)} \tag{4.13}$$

The mean length of the blackout period, T_b, for this case is independent of the call load and hence blockage rate and it is given by:

$$T_b = \frac{1}{N} \tag{4.14}$$

Figure 4.29 shows the normalized mean length of the active period T_a versus the number of channels. As the call blocking rate increases, the active period shortens, leaving the system in more blackout periods. For a blocking rate of 5 percent or less and with fewer than 25 channels, the system has the equivalent of one dedicated channel for data applications. At higher blocking rates and data applications that cannot tolerate blackout periods, the data service must be deployed with at least one dedicated channel. Some practical examples are helpful at this stage.

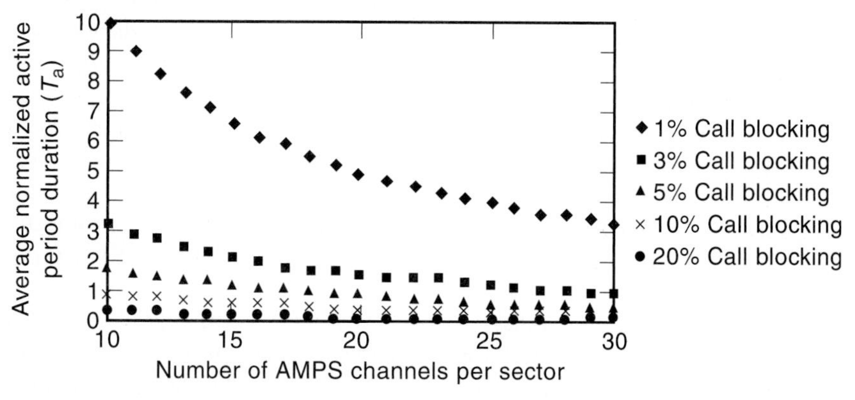

Figure 4.29 Mean length of available time for data as a function of the number of voice channels [BUD97].

Example 4.25: Data Overlay in FDMA Systems—CDPD

The above was actually developed for CDPD, and all the analyses and discussions are directly applicable to it. As we saw in Chapter 1, CDPD operates over analog FDMA cellular networks at the rate of 19.2 kbps per channel, which corresponds to digital transmission using GMSK modulation over 30 kHz AMPS channels. CDPD supports a channel hopping feature that allows a mobile data terminal to move to another channel during a communication session releasing the current channel for voice telephone conversation. This feature helps maintain the blockage probability at its nominal value and allows continual operation during the handoffs. The weakness of CDPD data overlay is that it does not assign several voice channels for one data user to support higher data rates. In general, in an FDMA system, the assignment of multiple voice channels to a single data user involves simultaneous operation of several RF channels by one terminal that is not practically attractive. For the same practical reason, data overlay in FDMA systems encounters difficulties in taking advantage of the silence periods during telephone conversations.

Example 4.26: Data Overlay in TDMA Systems—GPRS

An example of TDMA data overlay is GPRS. All the earlier analysis is applicable to GPRS applications as well. However, the format flexibility of TDMA allows multislot assignment to support higher data rates. GPRS also does not take advantage of silence periods in two-way telephone conversation.

An efficient method of integrating voice and data packets is a *movable boundary TDMA scheme with silence detection*. This method has been applied in the time-assignment speech interpolation (TASI) system used in T1-carrier telephone networks [FIS80] to maximize the number of voice users carried and to integrate data transmission into the channel. Using this basic idea, it is possible to design a *TDMA/framed-polling* protocol to integrate voice and data packets in a WLAN. This system consists of a number of voice and data terminals and a central station, which coordinates all the transmissions [ZHA90]. The protocol for integration of voice and data packets is a movable boundary TDMA scheme shown in Figure 4.30. A frame is divided into two regions with a boundary between them. The first region is used for both voice and data traffic where the voice traffic has priority. If not, voice packets occupy all the slots in this region, and the remaining slots are used for data traffic. The second region is reserved exclusively for data traffic. The boundary between the voice and data regions moves in accordance with the number of active voice packets in each frame. The maximum number of voice packets per frames is N_1, which is assigned an appropriate value to ensure some minimum data traffic capacity and to keep the blockage of voice packets below a selected value (2 percent in [ZHA90]).

The result of extensive analysis and simulation in [ZHA90] provides a simple experimental relation between the capacity that can be allocated for voice and data applications: $D = R_T - 0.032 \times N_v - 0.29$ where D is the data rate in Mbps available for data applications, N_v is the number of active 64 kbps PCM-encoded telephone conversations and R_T is the transmission rate available on the medium. We can

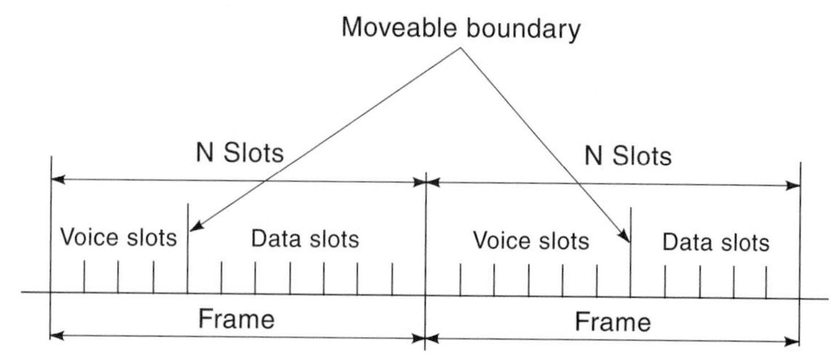

Figure 4.30 Frame structure in a moveable boundary frame-polling system.

apply this equation to the TASI system that uses this protocol to accommodate 30 voice users with some additional data in a traditional 24-voice channels system (T1 carrier). The transmission rate is $R_T = 0.064 \times 24 = 1.536$ Mbps to support 24 voice users at 64 kbps. Using the equation for $N_v = 30$ active users, the data throughput with maximum delay of 10 ms (used in the simulation) is 286 kbps. The new protocol supports 6 more voice users plus 280 kbps data. With the same 24-voice users, 487 kbps would be available for data applications that are around 30 percent of the overall transmission rate.

Example 4.27: Data Overlay on CDMA

As we saw earlier in this chapter, integration of bursty data with voice in the CDMA system is very simple, and CDMA systems already take advantage of the voice activity factor. Therefore, with the same infrastructure and terminals, data services can be overlaid on CDMA. If higher data rates are needed, one can either reduce the processing gain of the data channel or assign several parallel channels for one data link. Indeed the natural flexibility of CDMA to accommodate a variety of data services is one of the major reasons behind selection of the CDMA for 3G systems. Qualcomm has suggested *high data rate* (HDR) where asymmetric uplink and downlink data rates can be supported by simply using multiple carriers on the downlink for higher data rates.

4.4.3 Voice Integration into Data-Oriented Networks

Integration of voice and data has been discussed extensively in the literature. Most of these studies are concerned with protocols with explicit synchronization between the receiver and the transmitter. These approaches use assignment-based protocols for integration of voice and data, which allocate a fixed reference time such as a slot for transmission of packets. Synchronous systems provide more control on delay for voice traffic but less flexibility for bursty data traffic. Another approach that does not need explicit synchronization between the receiver and the transmitter is the asynchronous approach. The asynchronous packet approach mostly uses protocols extended from packet data networks, which are more suited for bursty data traffic. The voice traffic in this approach requires relatively complicated handling to limit the delay.

Contention-based packet communications protocols such as ALOHA and CSMA are used for data-oriented wireless networks. They are especially well suited to networks comprised of many user stations, each with low-average data rate and potentially high-peak rates. These protocols can operate with little or no centralized control, and can generally accommodate variable numbers of users in the network. However, contention-based schemes can become very inefficient in sharing the communications resources when the traffic load is heavy, as the system throughput degrades and the transmission delays increase. The unpredictability of throughput and time delays make these access methods unattractive for a voice-dominated communication service, where a minimum throughput and delay is essential for user acceptance of the QoS. Up to recent times (the Internet and wireless age), wired telephone services and the PSTN were producing the dominant source of income for the telecommunications industry. In the past century, telephone users have accepted the quality of the PSTN wired voice services as a normal standard for telephone conversations.

4.4.3.1 QoS in Voice Services

In the language of digital packet communications, the QoS of the PSTN voice user is specified by a guaranteed 64 kbps PCM (or 32Kbps ADPCM) coded data rate and a maximum delay of around 100 ms. We refer to this QoS as *wire-line-voice quality*. With the introduction of cellular telephone services in the late twentieth century, users accepted a lower QoS that suffered from the effects of fading due to the radio channel and dropped calls due to handoff, lack of coverage, or other reasons. As we saw in Chapter 1, cordless telephony and PCS services were aimed at bringing their QoS close to that of wire-line quality and 3G cellular, that is, a merger of cellular telephony and PCS service follows the same pattern. If the quality of voice in a cordless telephone was far below that of wireline quality, users would have rejected the service. This is because they have the choice to receive a better QoS (no drops or fading effects) with their wired telephone that is also available at home or in the office. However, users had to accept lower QoS with cellular telephony because there was no other alternative service provision for vehicles and other mobile applications.

Another recent event deviating from the wireline QoS is the emergence of voice over Internet or, as most people refer to it, the voice over IP, phenomenon. The popularity of the Internet, its penetration into the home market, its capability to support multimedia, and the most important advantage, its uniform cost for local- and long-haul communications, encouraged development of Internet telephony. Operating on a packet switched environment with contention-access, the QoS of these services is not guaranteed at all, and in its present stage of technology, the quality of voice calls is well below that of wire-line quality. However, free international calls through an Internet connection have been an incentive for some users to try this option as well.

In wireless networks, voice-over IP (VoIP) does not make sense for mobile data applications because these services provide low data rates, and after all they are evolving as an auxiliary network over already existing voice-oriented networks. However, VoIP can be considered for WLAN environments. Imagine a WLAN installation in a stock market hall supporting wireless terminals for the users working in the hall. It would be useful and beneficial if they had a VoIP service at the same terminal to use it for their telephone conversations as well. This incentive has initiated preliminary work on VoIP in a WLAN environment using contention-based access methods. The work in [ZAN00, FEI00] determines the number of supported voice terminals in a WLAN environment under a variety of conditions.

4.4.3.2 Capacity of a WLAN with Voice and Data

WLANs are becoming very popular in indoor applications such as in stock exchange halls where mobile users demand high-speed wireless data access to the network and voice capabilities for telephone conversations. To deploy such a network, a mathematical framework is very helpful to compare the capacity performance of WLANs with voice and data services in different scenarios. Therefore, for an asynchronous WLAN using the TCP/IP protocol suite, we need to find an answer to two questions:

1. What is the number of network telephone calls that can be carried with a given amount of data traffic?
2. What is the maximum data traffic per user for a given number of voice users?

A mathematical framework to answer these two questions is provided in [ZAH00] where the integration of voice and data with TCP/IP protocol that operates in an asynchronous CSMA access environment is analyzed.

To integrate voice packets in a TCP/IP environment, the first step is to select a speech coder. A variety of speech coding algorithms exist with different rates as discussed in Chapter 3. To reduce the load on the network generated by voice traffic, [ZAH00] adopted IMBE that is a popular low-data-rate vocoder (4.8 kbps) with an acceptable QoS. This vocoder has been used in INMARSAT-M and AUSSAT mobile satellite communication systems and is proposed for APCO-25 standard for narrow-band digital land mobile radio.

Using IP will provide two options for sending packets in the network: the transmission control protocol—TCP and the user datagram protocol—UDP. As a streaming protocol, UDP has no support for error correction, acknowledgment, sequencing, and flow control. Under high traffic conditions, the lack of flow control in UDP may cause bandwidth saturation in the Internet that should be prevented by the application program. In contrast, TCP has an error-correcting mechanism, uses acknowledgment, and guarantees in-order packet delivery. These requirements demand additional overhead that increases the average delay and reduces the overall throughput of the network. In general, data packets can tolerate delay but cannot tolerate packet loss, whereas the voice packets can accept packet loss of the order of 1 percent but cannot afford delays of more than 200 ms between consecutive packets. In the system described in [ZAH00], TCP is used for the data packets to guarantee accuracy of information and UDP for voice packets to handle the delay requirement. This approach is adopted in several products available in the market, whereas other products use TCP for both voice and data. Although in this analysis TCP is selected for data transmission, there are some products, that use UDP for data transmission. In this case the upper layers are responsible for reliable delivery of data.

Figure 4.31 represents a general overview of the system where several voice and data terminals communicate with an AP of a WLAN. Because the human ear is sensitive to time delays larger than 200 ms (T_{th}) in a voice conversation, wireless terminals should provide some facilities to minimize voice time delay. Allocating higher priority for transmitting voice over the data traffic is one way to decrease the voice packet time delay. Therefore, voice and data packets should be stored in a queue and wait for transmission as shown in Figure 4.32. The total delay for each packet transmission consists of a *queuing delay* at the terminal and *channel transfer delay*. The work in [ZAH00] uses an *M/G/1* queue[3] with two priorities and a single server for node modeling. The arrival is modeled by a Poisson random variable. Figure 4.33 shows the system capacity and delay for T_{th} = 100 ms and 50 ms with a 1 and a 2 Mbps channel bandwidth. The maximum number of the voice users declines with reducing T_{th}.

Problem 13: Capacity of a WLAN with Voice and Data Users

Using Figure 4.33, find the number of users for T_{th} = 100 ms and T_{th} = 50 ms when the channel bandwidth is 1 Mbps.

Solution:

From the figure a maximum of 18 voice users are supported with T_{th} = 100 ms, and decreasing T_{th} to 50 ms will reduce the maximum voice users to 14. In this example the data traffic is less than 10 kbps.

4.4.3.3 IP Telephony Using WLANs

The previous section provided an analysis of how many voice channels can be supported using UDP protocol over a CSMA channel in a wireless LAN environment where data transmission is coexisting. In practice, there are a number of VoIP software packages and services, such as Speakfreely, Net2Phone, DialPad, and so on, that can be used to implement a real-time testbed to analyze the behavior of voice in a WLAN environment. The purpose of setting such experimental testbeds or simulations is to determine the number of voice users that can be supported and relate it to the design parameters. In this section, we provide a practical framework for the analysis of VoIP in a WLAN environment. The principles of operation of voice over contention-based packet-switched networks and assigned access circuit-switched networks are very different. In a circuit-switched network, a fixed connection between the two terminals is established during the call establishment procedure. This connection supports end-to-end communications with a fixed data rate and a controlled delay dominated by propagation delay and a negligible delay jitter. In packet voice communications with contention access and packet-switched

[3]An M/G/1 queue implies that packet arrivals are Poisson and the service rate of the queue has a general distribution with known mean and variance (see [BER87]).

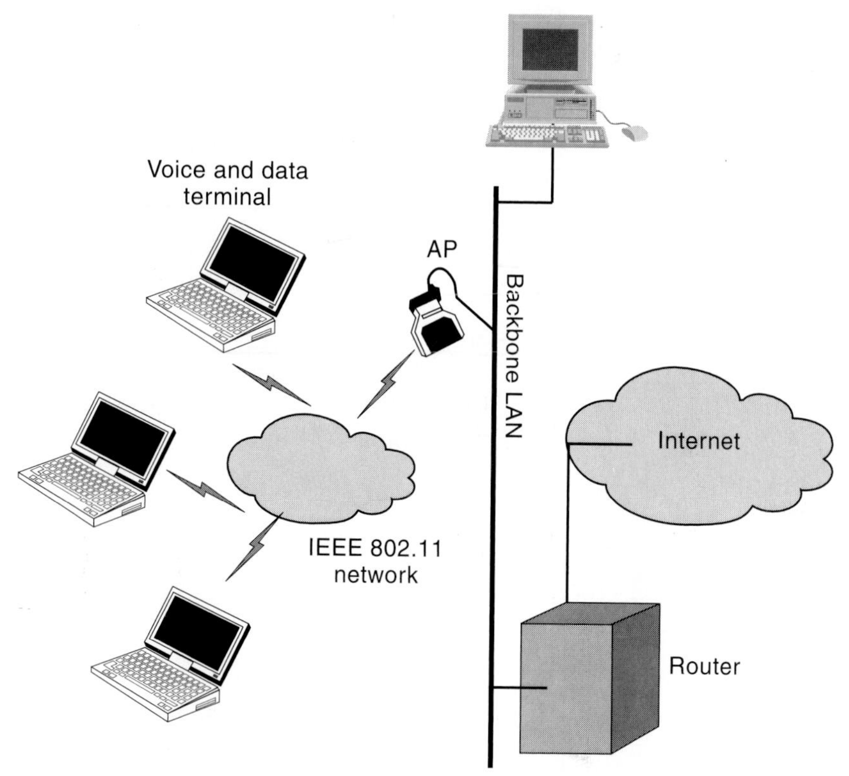

Figure 4.31 Schematic of a system employing voice and data terminals in a WLAN.

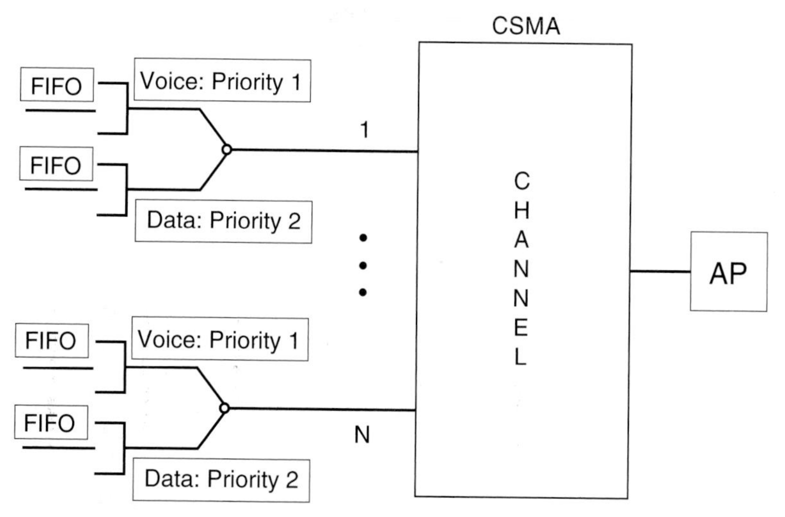

Figure 4.32 Queuing model for prioritizing voice traffic.

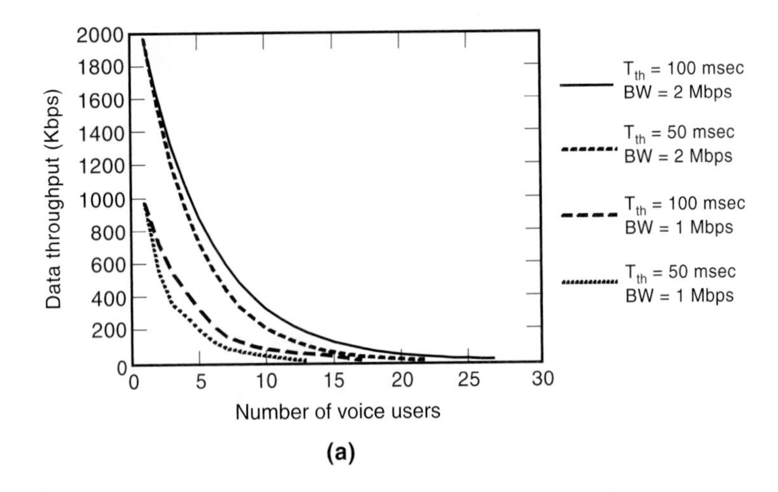

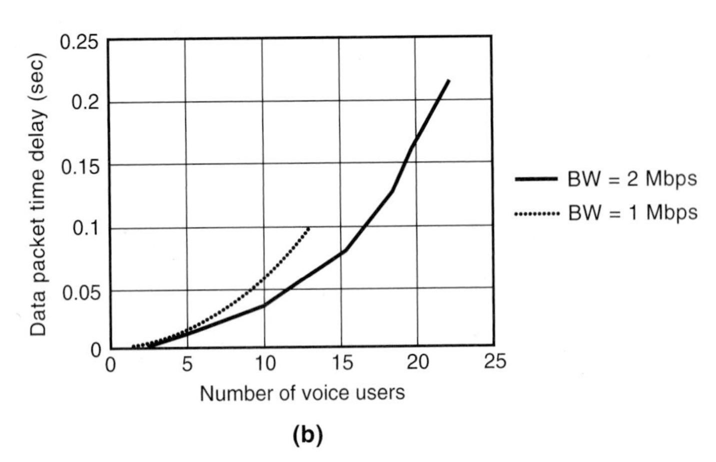

Figure 4.33 (a) Throughput of data versus number of voice users for variety of thresholds for acceptable delay in voice packets and (b) Data packet time delay versus number of voice users.

networks, delay and jitter are the dominant sources of the performance degradation. To regulate jitter, the receiver has a buffer to store the received packets at different delays but pump them to the user at constant intervals to reconstruct real-time voice. The performance of the system is then related to the size of the buffer at the receiver. This section provides a summary of the experimental work presented in [FEI00] to relate the throughput to the buffer size.

The first step is to describe the overall scenario in a practical situation for the implementation of the VoIP in a WLAN environment. Figure 4.34 describes the arrival of packets. Because of random access and packet-switched networking, the packets sent at fixed intervals arrive at a variety of delays.

The overall delay is the minimum network delay plus the individual jitter per packet that will be regulated with the jitter compensation buffer at the receiver.

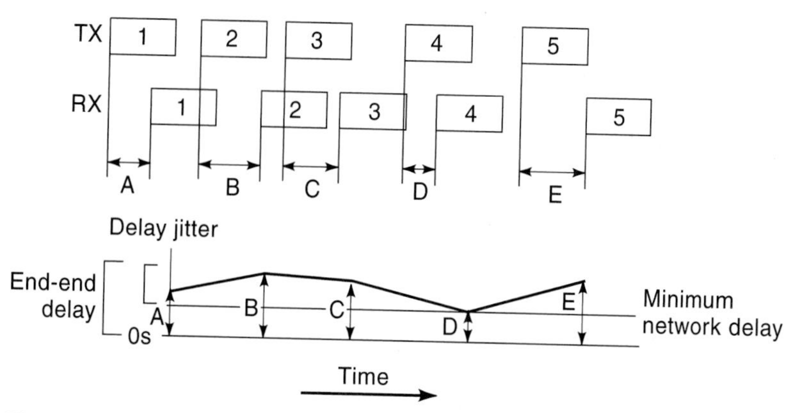

Figure 4.34 Illustration of the arrival of voice packets transmitted at a constant rate.

The packets arriving at different delays are stored in a buffer, and the application at the receiver reads this buffer periodically. When the packet with the right sequence number is available, the receiver reads the packet and plays it through the speaker. When the packet is not available, the application software at the receiver skips that packet. A simple example further clarifies the operation.

Example 4.28: Jitter in VoIP on WLANs

Figure 4.35 illustrates the details of the operation of the receiver and the relationship between the jitter compensation buffer, arrival, and playing time of the packets. When the first packet arrives, it is delayed in the receiver's jitter compensation buffer and after the maximum allowed delay, it is delivered to the user application to be played in the first slot. The second packet is discarded because it has arrived after the deadline for playing. The third packet arrives normally before the deadline, and it is delivered at its appropriate time to the speaker. The fourth and the fifth packets arrive off-sequence. The fourth one is late (arriving after its deadline), and it is discarded. The fifth packet has arrived before the deadline of the fourth packet, and it is shifted to its own time slot.

As we have discussed before, the user can accept a packet drop rate of around 1 percent. To reach the goal of 1 percent packet drop, the receiver has the choice of increasing the length of the jitter compensation buffer at the expense of additional overall delay at the receiver. Therefore, the length of the jitter compensation buffer is an important parameter in VoIP applications. This parameter is adjusted by changing the length of the buffer at the receiver. The delay observed by the user is the minimum network delay plus the jitter compensation buffering delay. Example 4.28 also illustrates that in VoIP applications. In addition to transmission packet losses occurring in the network, we have packet losses due to late arrival at the receiver (that is a function of the length of the jitter compensation buffer). Therefore, the length of the jitter compensation buffer and packet loss are interrelated.

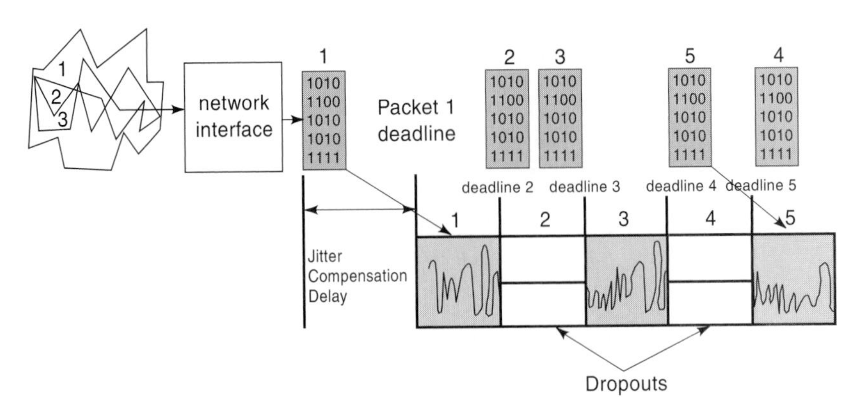

Figure 4.35 Reception of voice packets and buffering to maintain appearance of no jitter.

To determine the relationship between the jitter compensation delay and the packet loss rate, a testbed was developed in [FEI00] to implement the scenario shown in Figure 4.31. In this test bed, an infrastructure for wireless LAN operation using an AP and a number of laptops is used for measurement of the statistics of the delay jitter in a VoIP application. Figure 4.36(a) shows the statistics of the delay jitter for 1,800 packets. The measurement system transmits, time stamps, and stores the packet at the transmitting and the receiving laptops. The stored files are then post processed to eliminate the effects of differences between the clocks of the transmitter and the receiver laptops, and we extract the refined delay jitter measurements. Figure 4.36(b) shows the accuracy of the system (that is measured by comparing the results obtained from two separate laptops connected to the same point and receiving the same message from a receiver). The measurement error (difference of measurements in two identical laptops) has a mean of around .01 ms, whereas the mean of the measurements is around 1 ms, restricting the measurement error to around 1 percent. Using a delay jitter distribution, one can simply find the relationship between the packet loss and the jitter compensation buffer length. For any given jitter compensation buffer length, the probability of packet loss is the same as the probability of having a delay jitter larger than the jitter compensation buffer length. Figure 4.37 shows the experimental results for up to five stations operating in the wireless LAN test bed. If we fix the acceptable packet loss rate to 1 percent, the minimum buffer length increases from 0.5 ms to 7 ms when we increase the number of users from one to seven. The details of algorithms for the implementation of the test bed and the results of OPNET simulation for large number of voice users are available in [FEI99].

QUESTIONS

4.1 Name two duplexing methods and one example standard that uses each of these technologies.

4.2 What are the popular access schemes for data networks? Classify them.

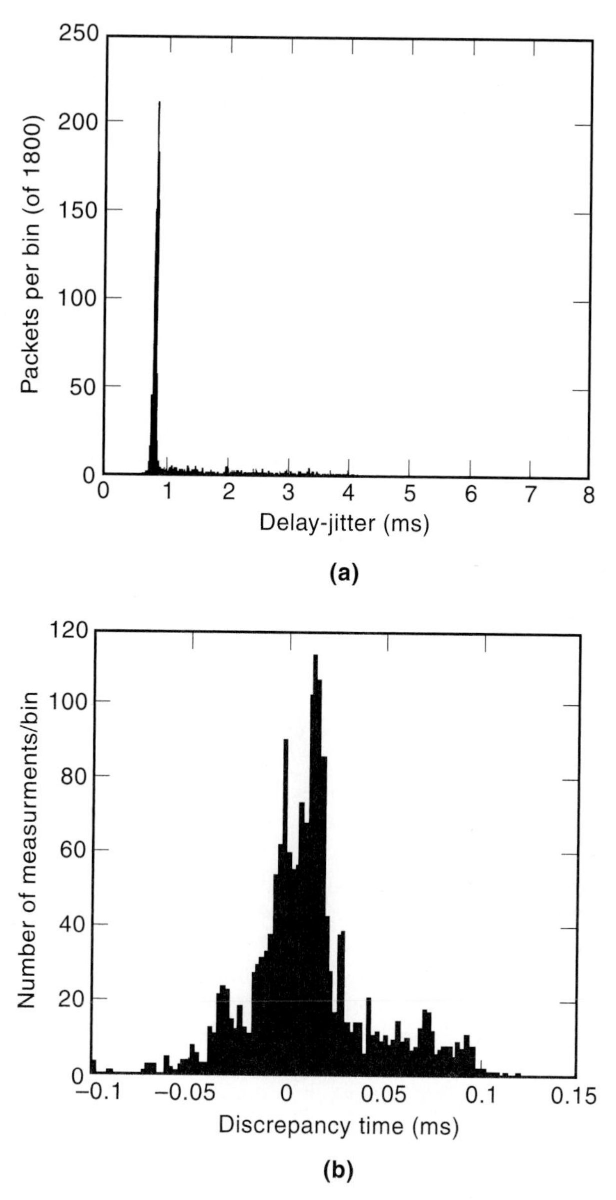

Figure 4.36 (a) Measured delay jitter in the wireless LAN testbed and (b) accuracy of the measurements.

4.3 Name a cellular telephony standard that employs FDMA.

4.4 What is binary exponential back-off algorithm, which standard uses it, and what is the purpose of using it? What is its weakness?

4.5 What is the purpose of the IEEE 802 standard committee? What are the steps taken to make its recommendations an international standard?

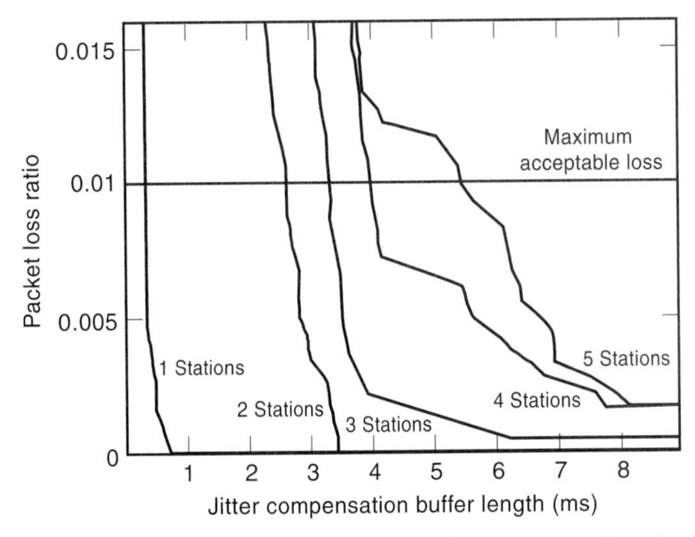

Figure 4.37 Packet loss versus jitter compensation buffer length.

4.6 What is the difference between the access techniques of IEEE 802.3 and IEEE 802.11?

4.7 Why do most PCS standards use TDD, while most cellular standards use FDD?

4.8 Why in the PSTN backbone hierarchy has FDM multiplexing lost its popularity to TDM multiplexing?

4.9 Why did the 2G cellular systems shift from analog FDMA to digital TDMA and CDMA?

4.10 Name three standards using TDMA/TDD as their access method.

4.11 What are the advantages of the CDMA access techniques?

4.12 What is the difference between performance evaluation of voice-oriented fixed assignment and data-oriented random access methods.

4.13 Explain the difference between the effects of power-control on the capacity of TDMA and CDMA systems.

4.14 In a radio ALOHA network, how does a terminal learn that its packet is collided?

4.15 What is the difference between the maximum throughputs of ALOHA and Slotted ALOHA networks? What causes this difference?

4.16 What is the difficulty of implementing CSMA/CD in a wireless environment?

4.17 Explain the difference between carrier-sensing mechanisms in the wired and wireless channels.

4.18 What is the hidden terminal problem and how does it impact the performance of a CSMA-based access method?

4.19 What is the capture effect and how does it impact the performance of the random access methods?

4.20 Explain the differences between integration of data into a voice-oriented network and integration of voice into a data-oriented network.

4.21 Explain the relation between the receiver buffer size and packet error rate in voice-over IP applications.

PROBLEMS

4.1 To provide public telephone access to commercial ferries a telephone company installs a multi-channel wireless telephone system in a ferry. This wireless radio system connects to a base station on the shore through the air. The base station is connected to the PSTN using wires.

a. If the telephone company installs a four-channel system, what is the probability of having a person come to the telephone and finding none of the lines are available? Assume that the average length of a telephone call is 3 minutes and each of the 150 passengers of the ferry make on the average one call per hour.

b. What is the average delay for accessing the telephones?

c. How many channels are needed to keep the blockage probability below 2 percent?

4.2 a. Neglecting the frequency spectrum used for control channels, what is the maximum number of two-way voice channels that can fit inside the frequencies allocated to the AMPS system.

b. What is the number of channels in each cell? Note that $K = 7$ was originally used in the AMPS.

c. Repeat (b) for IS-136 in which $K = 4$ and the number of slots per TDMA channel is three.

d. Repeat (b) for IS-95 CDMA, assuming the minimum required E_b/N_0 is 6 dB. Include the effects of antenna sectorization, voice activity, and extra CDMA interference.

e. Repeat (d) for broadband CDMA where 5 MHz bands are used in each direction.

4.3 a. Sketch the throughput versus offered traffic G for a mobile data network using slotted nonpersistent CSMA protocol. The packets are 20 ms long and the radius of coverage of each BS is 10 km. Assume the radio propagation speed is 300,000 km/second and use the worst delay for calculation of the a parameter.

b. Repeat (a) for slotted ALOHA protocol.

c. Repeat (a) for 1-persistent CSMA protocol.

d. Repeat (a) for a WLAN with access point coverage of 100 m.

e. Repeat (a) for a satellite link with a distance of 20,000 km from the earth.

4.4 Use the equations from Section 4.2.2 to reproduce Figures 4.27 and 4.29.

4.5 A cellular carrier has established 100 cell sites using AMPS with 395 channels and $K = 7$.

a. Use Figure 4.8 to calculate the total number of subscribers for a blocking probability of 0.02, average of 2 calls per hour, and average telephone conversation of 5 min.

b. Use computer software, such as MathLAB, or MathCAD to calculate the same values using the Erlong B equation directly.

c. Determine (either from Fig. 4.9 or calculation) the average delay for a call.

d. Repeat (a) for a blocking probability of 0.01.

e. Repeat (a) if IS-136 was used with $K = 7$.

f. Repeat (a) if IS-136 was used with $K = 4$.

4.6 We want to use a GSM system with sectored antennas ($K = 4$) to replace the existing AMPS system ($K = 7$) with the same cell sites.

a. Determine the number of voice channels per cell for the AMPS system.

b. Determine the number of voice channels per cell for the GSM system.

c. Repeat (b) if we were using a W-CDMA system with the bandwidth of 12.5 MHz for each direction. Assume a signal-to-noise ratio requirement of 4 (6 dB) and include the effects of antenna sectorization (2.75), voice activity (2), and extra CDMA interference (1.67).

4.7 We provide a wireless public phone with six lines to a ferry crossing between Helsinki and Stockholm carrying 100 passengers where, on the average, each passenger makes a 3-min telephone call every 2 hours.

a. What is the probability of a passenger approaching the telephones and none of the four lines are available?

b. What is the average delay for a passenger to get access to the telephone?

c. What is the probability of a passenger waiting more than 3 minutes for access to the telephone?

d. What would be the average delay if the ferry had 200 passengers?

4.8 A WLAN hop accommodates 50 terminals running the same application. The transmission rate is 2 Mbps and the terminals are using slotted ALOHA protocol. The commutative traffic produced by the terminals is assumed to form a Poisson process.

a. Give the throughput versus offered traffic equation for the system and determine the maximum throughput in Erlong.

b. What is the maximum throughput in bits per second?

c. What is the maximum throughput in bits per second for each terminal?

4.9 A local 3-hour tour boat with 50 passengers has one AMPS radio phone to connect to the shore. On the average, each user places one call per tour and the average holding time for the calls is 3 minutes.

a. What is the probability that a person attempts to use the phone and he/she finds it occupied?

b. Repeat (a) if the AMPS phone is replaced by three IS-136 phones using the three slots of the existing IS-136 TDMA over the same band.

c. Repeat (a) if this phone is replaced by six upgraded IS-136 phones using 6-slot upgraded IS-136 TDMA over the same band.

4.10 In a datagram packet switched network with

P: packet size in bits

N: number of hops between two given systems

B: data rate in bps on all links

H: overhead (header in bits per packet)

T: end-to-end delay

N_p: number of packets

L: message length in bits

D: propagation delay per hop

a. Give N_p in terms of L, P, and H.

b. Give T in terms of L, P, H, N, B, and D.

c. What value of P, as a function of N, B, and H, results in minimum end-to-end delay T? Assume that the message length is much larger than the packet size and propagation delay is negligible ($D = 0$).

4.11 An ad hoc 2 Mbps WLAN using ALOHA protocol connects two stations with a distance of 100 m from one another—each, on the average, generating 10 packets per second. If one of the terminals transmits a 100-bit packet, what is the probability of successful transmission of this packet? Assume that the propagation velocity is 300,000 km/sec and the packets are produced according to the Poisson distribution.

PART TWO Principle of Wireless Network Operation

Part Two of the book is devoted to technical aspects of fixed infrastructure of the wireless networks. This part consists of two chapters addressing deployment and operation of wireless networks, respectively.

CHAPTER 5 NETWORK PLANNING

Wireless service providers often start with a minimal infrastructure and antenna sites. As the number of subscribers grows, the service provider expands the wireless infrastructure and migrates to more advanced technologies to increase the capacity and improve the quality. Chapter 5 presents the technologies related to the initial deployment and later expansion of the infrastructure for wireless networks. Different topologies, channel allocation techniques, and architectural methods used for expansion of the network and issues related to migration to CDMA technology are explained in this chapter.

CHAPTER 6 WIRELESS NETWORK OPERATION

Chapter 6 is devoted to functionalities of the fixed network infrastructure that are needed to support mobile operation. These functionalities are divided into three categories: mobility management, radio resource and power management, and security management. Mobility management defines how a mobile terminal registers with the network at different locations and how network tracks the mobile as it changes its access to the network from one antenna to another. Radio resource and power management is the technology used for controlling the transmitted power of the terminals. Security management of wireless networks is implemented by authentication and ciphering. Authentication is a process between the network and the terminal checking the authenticity of the terminal, and ciphering scrambles the transmitted bits to prevent eavesdropping.

CHAPTER 5

NETWORK PLANNING

5.1 Introduction

5.2 Wireless Network Topologies

5.2.1 Infrastructure Network Topology
5.2.2 Ad Hoc Network Topology
5.2.3 Comparison of Ad Hoc and Infrastructure Network Topologies

5.3 Cellular Topology

5.3.1 The Cellular Concept
5.3.2 Cellular Hierarchy

5.4 Cell Fundamentals

5.5 Signal-to-Interference Ratio Calculation

5.6 Capacity Expansion Techniques

5.6.1 Architectural Methods for Capacity Expansion
5.6.2 Channel Allocation Techniques and Capacity Expansion
5.6.3 Migration to Digital Systems

5.7 Network Planning for CDMA Systems

5.7.1 Issues in CDMA Network Planning
5.7.2 Migration from AMPS to IS-95 Systems

Questions

Problems

5.1 INTRODUCTION

One can envision a *telecommunications network* as a kind of medium interconnecting a collection of devices that is equipped to exchange information between them. The most common information sources are voice, data, and video. Traditional communications networks have evolved around applications using one of these specific information sources. The future of communications networks, however, is directed toward multimedia services supporting applications using a combination of all these information sources. The entire communications network consists of two types of elements, the communicating devices and the network infrastructure. The communicating devices provide an interface between the user application and the network infrastructure. These devices are usually referred to as communication *stations, terminals,* or *hosts.* The network infrastructure is a collection of point-to-point wired or wireless links and a number of switches and routers interconnecting several of these communication terminals in geographically separated locations. Traditional communication devices are connected to the communications network through a fixed connection point. The geographic location of the terminal and its connection to the infrastructure remains fixed. Figure 5.1 shows a simple traditional wired network with examples of its elements.

The future direction of communications networks is toward wireless mobile connections to the infrastructure that shall support continual connection though the geographical location of the terminal is constantly changing due to the mobility of the terminal. In order to modify the traditional fixed network infrastructures to support wireless connections, a new wireless infrastructure is needed as an interface between the backbone wired network infrastructure and the mobile communication terminals.

The mobile communication terminals need to be equipped with wireless front-ends to communicate with the wired backbone through the new wireless in-

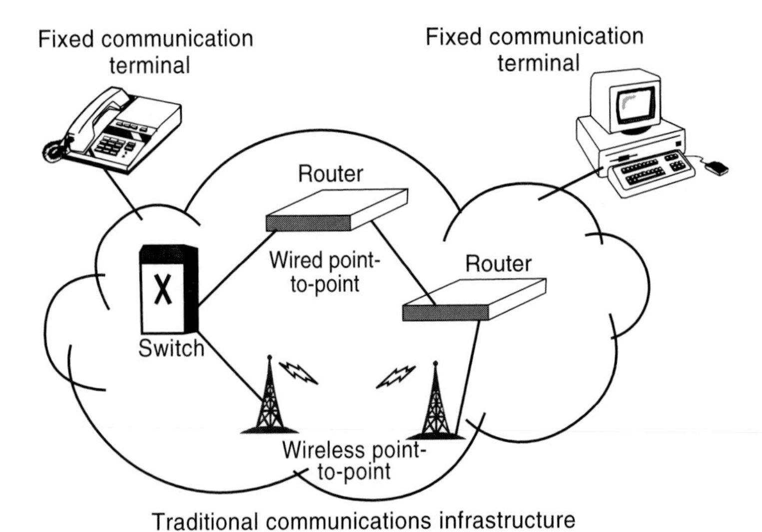

Figure 5.1 Traditional wired networks.

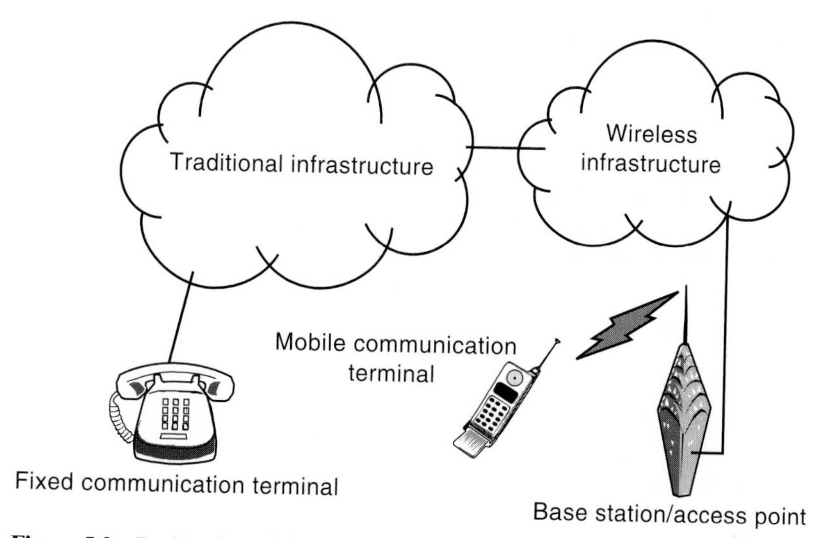

Figure 5.2 Positioning of the wireless network infrastructure in relation to the wired network infrastructure.

frastructure. Figure 5.2 shows the positioning of the wireless infrastructure in relation to the wired infrastructure.

In addition to switches, routers, and point-to-point links, the wireless network infrastructure also needs wireless transceivers to communicate with the wireless communication terminals and act as points of access to the fixed part of the wireless network infrastructure. These transceivers are referred to as *base stations* (BSs) or *access points* (APs). Any wireless base station has a limited coverage area. If the coverage area is less than the desirable coverage area for the wireless service, we need multiple base stations to cover the service area. In the case of multiple base stations operating in an area, the wireless network infrastructure needs to coordinate the continuity of a wireless connection as the mobile communication terminal moves through the coverage areas of different base stations.

5.2 WIRELESS NETWORK TOPOLOGIES

We refer to *wireless network topology* as the configuration in which a mobile terminal communicates with another. There are two fundamental types of topologies used in wireless networks. They are infrastructure, centralized, or hub-and-spoke topology, and the ad hoc or distributed topology.

5.2.1 Infrastructure Network Topology

In the infrastructure topology, there is a fixed (wired) infrastructure that supports communication between mobile terminals and between mobile and fixed terminals. The infrastructure networks are often designed for large coverage areas and multiple base station or access point operations. Most of the discussion in this section is around this type of operation. Figure 5.3 shows the basic operation of an infrastructure network with a single BS/AP. The BS/AP serves as the hub of the network, and the mobile terminals are located at the ends of the spokes. Any communication from one wireless user station to another, that is, between peers, has to be sent through the BS/AP. The hub station usually controls the mobile stations and monitors what each station is transmitting. Thus the hub station is involved in managing user access to the network.

Example 5.1: Systems That Employ Infrastructure Network Topology

All standardized cellular mobile telephone and wireless data systems use an infrastructure network topology to serve mobile terminals operating within the coverage area of any BS. The IEEE 802.11 standard and most of the wireless LAN products support infrastructure operation.

5.2.2 Ad Hoc Network Topology

Ad hoc or distributed network topology applies to reconfigurable networks that can operate without the need for a fixed infrastructure. These networks are primarily used by the military and also in a few commercial applications for voice and data transmission. Such a topology is suitable for rapid deployment of a wireless network in a mobile or fixed environment. Figure 5.4 shows two variations of the ad hoc network topology. Figure 5.4(a) is a single-hop ad hoc network where, as

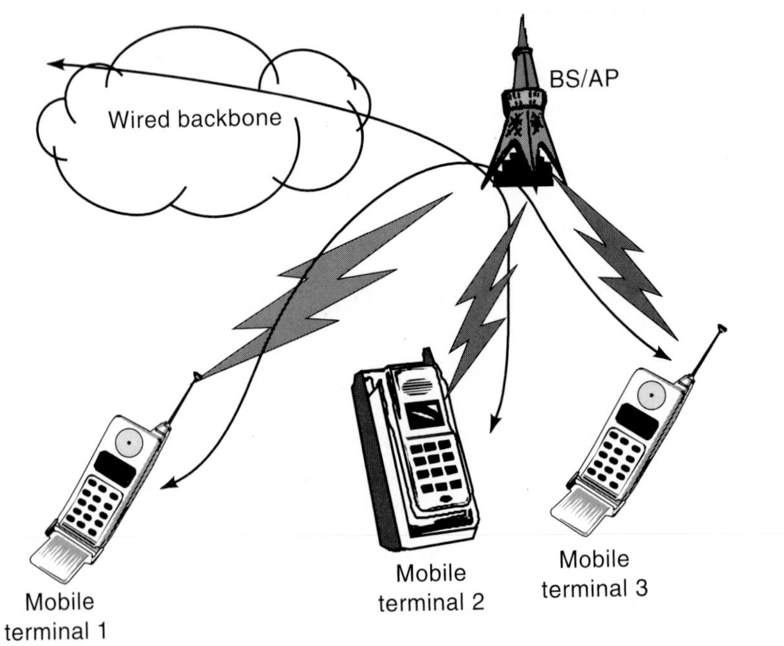

Figure 5.3 Basic operation of an infrastructure network topology.

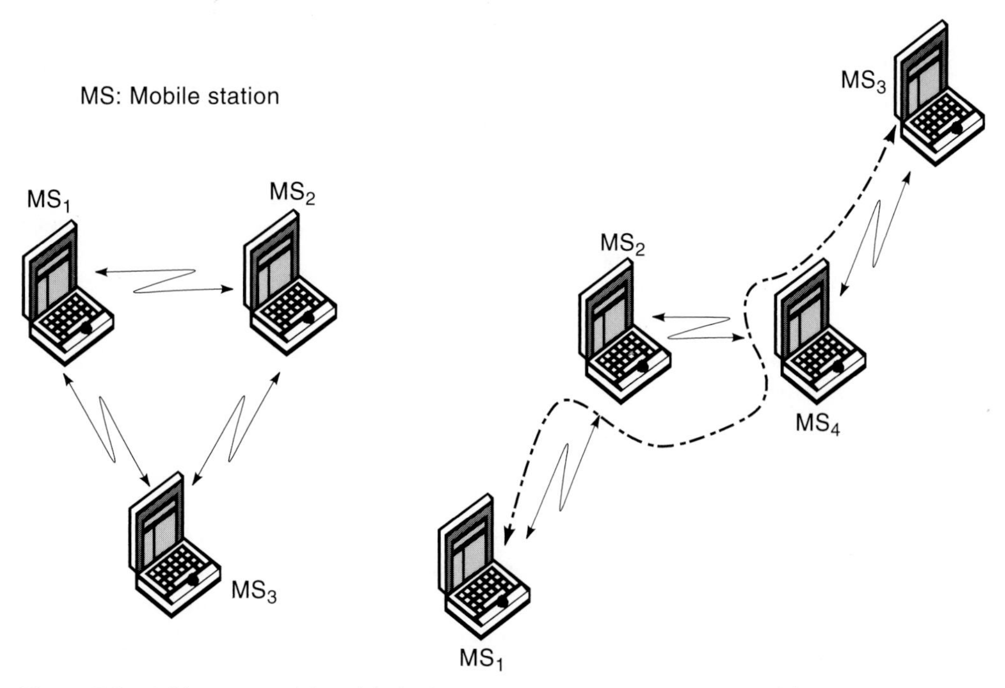

Figure 5.4 Ad hoc networking: (a) single-hop peer-to-peer topology and (b) multi-hop ad hoc networking topology.

the name implies, every user terminal has the functional capability of communicating directly with any of the other user terminals.

Example 5.2: Systems That Support Single-Hop Ad Hoc Network Topology

The IEEE 802.11 wireless LAN standard supports single-hop peer-to-peer topology for ad hoc networking. When a terminal is turned on, it first searches for a beacon signal from an AP or another terminal announcing the existence of an ad hoc network. If a beacon is not detected, the terminal takes the responsibility of announcing the existence of an ad hoc network. Several PCS services, such as PHS and NEXTEL satellite, support peer-to-peer walkie-talkie type communication among voice terminals.

In some ad hoc networking applications, where users may be distributed over a wide area, a given user terminal may be able to reach only a portion of the other users in the network due to transmitter signal power limitations. In this situation, user terminals will have to cooperate in carrying messages across the network between widely separated stations. Networks designed to function this way are called multihop ad hoc networks and are illustrated in Figure 5.4(b). In an ad hoc multihop network, each terminal should be aware of the neighboring terminals in its coverage range. The multihop network configuration was originally used in military tactical networks, where providing reliable communications under unpredictable propagation conditions and over widely varying geographic areas was important.

Example 5.3: Systems That Support Multihop Ad Hoc Network Topology

The early packet radio networks studied for military applications in 1970s were employing multihop ad hoc network topology. ETSI BRAN's HIPERLAN standard for wireless LANs that was developed during the mid-1990s, supports multihop ad hoc networking for commercial applications.

5.2.3 Comparison of Ad Hoc and Infrastructure Network Topologies

A number of attributes can be used to compare infrastructure and ad hoc network topologies.

Scalability. In peer-to-peer single-hop networks, expansion is always limited to the coverage of the radio transmitter and receiver, and there is no simple way to scale up the network coverage or capacity (wireless traffic) that can be supported by the network. In multihop ad hoc networks, as the number of terminals increases the potential coverage of the network is increased. However, the traffic handling capacity of the network remains the same. To connect an ad hoc network to the backbone wired network, one needs to use a proxy server with a wireless connection as a member of the ad hoc network. In practice all terminals supporting ad hoc networking operate in a dual mode that also supports infrastructure operation. Wireless infrastructure networks are inherently scalable. To scale up a wireless infrastructure network, the number of BSs or APs is increased to expand the coverage area or to increase the capacity while using the same available spectrum. Therefore, for wide area coverage and for applications with variable traffic loads, infrastructure networks are always used. We look at how the available capacity can be expanded in an infrastructure topology later on in this chapter.

Flexibility. Operation of infrastructure networks requires deployment of a network infrastructure which is very often time-consuming and expensive. Ad hoc networks are inherently flexible and can be set up instantly. Therefore, ad hoc networks are always used for temporary applications where flexibility is of prime importance.

Controllability. To coordinate proper operation of a radio network, we need to centrally control certain features such as time synchronization, transmitted power of the mobile stations operating in a certain area, and so on. In an infrastructure network, all these features are naturally implemented in the BS or AP. In an ad hoc network, implementation of these features requires more complicated structures demanding changes in all terminals.

Routing Complexity. In multihop peer-to-peer networks, each terminal should be able to route messages to other terminals. This capability requires each terminal to monitor the existence of other terminals and be able to connect to those available in the immediate neighborhood. For this, there is a need for a routing algorithm that directs information to the next appropriate terminal. Implementation of these features adds to the complexity of the terminal and the network operation. In infrastructure and peer-to-peer single hop ad hoc networks, this problem does not exist.

Coverage. In WLANs, coverage of the network is an issue of concern because it has an effect on the selection of the topology. In peer-to-peer single hop network topology, the maximum distance between two terminals is the range of coverage of the wireless interface used in the terminal. In an infrastructure network, two wireless terminals communicate through an AP or a BS. The maximum distance between two terminals is thus twice the range of coverage of a single wireless modem because the communicating terminals may be located at the edge of the coverage area of the BS or AP. In practice often APs or BSs are fixed in opportunistic locations using elevated mountings that increase the coverage of the wireless modem. This usually results in a maximum coverage distance between two terminals that is greater than twice the coverage distance of the same modem in an ad hoc configuration.

Reliability. Another issue of concern in small-scale WLAN operation and military applications in battlefields is resistance to failure. Infrastructure networks are "single failure point" networks. If the AP or BS fails, the entire communications network is destroyed. This problem does not exist in ad hoc peer-to-peer configurations.

Store and Forward Delay and Media Usage Efficiency. In peer-to-peer single-hop networks, information is transmitted only once, and there is no store and forward procedure. In the infrastructure topology, we have transmission of data twice, once from the source to the BS/AP and once from the BS/AP to the destination. The BS/AP also should store the message and forward it later. This adds to the delay encountered by the data packets. Multihop ad hoc networks may have several transmissions and several store and forward delays that depend on the instantaneous topology and number of hops required to send the data from the source to the destination.

5.3 CELLULAR TOPOLOGY

Cellular topology is a special case of an infrastructure multi-BS network configuration that exploits the *frequency reuse* concept. Radio spectrum is one of the scarcest resources available, and every effort has to be made to find ways of utilizing the spectrum efficiently and to employ architectures that can support as many users as theoretically possible with the available spectrum. This is extremely important especially today in light of the huge demand for capacity. Spatially reusing the available spectrum so that the same spectrum can support multiple users separated by a distance is the primary approach for efficiently using the spectrum. This is called *frequency reuse.* Employing frequency reuse is a technique that has its foundations in the attenuation of the signal strength of electromagnetic waves with distance. For instance, in vacuum or free space, the signal strength falls as the square of the distance. This means that the same frequency spectrum may be employed without any interference for communications or other purposes, provided the distance separating the transmitters is sufficiently large and their transmit powers are reasonably small (depending on the reuse distance). This technique has been used, for example, in commercial radio and television broadcast where the transmitting stations have a constraint on the maximum power they can transmit so that the same frequencies can be used elsewhere. The cellular concept is an intelligent means of employing frequency reuse. Cellular topology is the dominant topology used in all large-scale terrestrial and satellite wireless networks. The concept of cellular communications was first developed at Bell Laboratories in the 1970s to accommodate a large number of users with a limited bandwidth [MAC79].

5.3.1 The Cellular Concept

By cellular radio, we mean deploying a large number of low-power base stations for transmission, each having a limited coverage area. In this fashion, the available capacity is multiplied each time a new base station or transmitter is set up because the same spectrum is being *reused* several times in a given area. The fundamental principle of the cellular concept is to divide the coverage area into a number of contiguous smaller areas which are each served by its own radio base station. Radio channels are allocated to these smaller areas in an intelligent way so as to minimize the interference, provide an adequate performance, and cater to the traffic loads in these areas. Each of these smaller areas is called a *cell.* Cells are grouped into *clusters.* Each cluster utilizes the entire available radio spectrum. The reason for clustering is that adjacent cells cannot use the same frequency spectrum because of interference. So the frequency bands have to be split into chunks and distributed among the cells of a cluster. The spatial distribution of chunks of radio spectrum (which are called subbands) within a cluster has to be done in a manner such that the desired performance can be obtained. This forms an important part of network planning in cellular radio. The number of cells in a cluster is called cluster size or frequency reuse factor.

Two types of interference are important in such a cellular architecture. The interference due to using the same frequencies in cells of different clusters is referred to as *cochannel interference.* The cells that use the same set of frequencies or channels are called *cochannel cells.* The interference from different frequency channels used within a cluster whose side-lobes overlap is called *adjacent channel interference.* The allocation of channels within the cluster and between clusters must be done so as to minimize both of these.

The cellular concept can increase the number of customers that can be supported in the available frequency spectrum as illustrated by the following examples by deploying several low power radio transmitters.

Example 5.4: Cellular Concept

Consider a single high-power transmitter (see Fig. 5.5) that can support 35 voice channels over an area of 100 square km with the available spectrum. If seven lower power transmitters are used so that they support 30 percent of the channels over

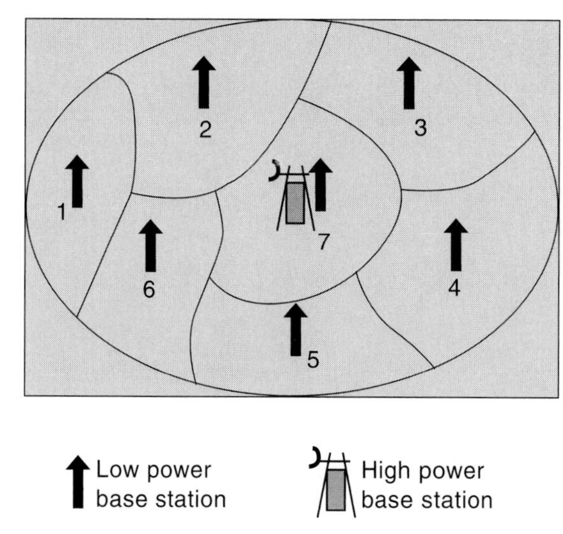

Figure 5.5 The cellular concept.

an area of 14.3 square km each, a total of ≈80 voice channels are now available in this area instead of 35. In reality, channels will have to be allocated to base stations in such a way as to prevent interference between one base station and another. In Figure 5.5, base stations 1 and 4 could use the same channels, as their coverage areas are sufficiently far apart and so also are base stations 3 and 6. Suppose the cells labeled 1, 2, 5, 6, and 7 use disjoint frequency bands and the channels used in 1 and 6 are reused in 3 and 4. The set of cells {1, 2, 5, 6, 7} forms a cluster. Cells 3 and 4 form part of another cluster. In the limiting case, the density of base stations can be made so large that the capacity is infinite. However, in practice this is impossible for several reasons that include drastic increases in the network and signaling load, number and rate of handoffs, and cost of infrastructure and network planning.

Example 5.5: Importance of Cellular Topology

We want to provide a radio communication service to a city. The total bandwidth available is 25 MHz, and each user requires 30 KHz of bandwidth for voice communication. If we use one antenna to cover the entire town, we can only support 25 MHz/30 KHz = 833 simultaneous users. Now let us employ a cellular topology where 20 lower power antennas are opportunistically located to minimize both kinds of interference. We divide our frequency band into four sets and assign one set to each cell. Each cell has a spectrum of 25 MHz/4 = 6.25 MHz allocated to it. We have a *cluster* of four cells in this example. The number of simultaneous users supported per cell is 6.25 MHz/30 KHz = 208. The number of users per cluster is $4 \times 208 = 832$. The total number of simultaneous users is now $832 \times 5 = 4{,}160$ because we have five clusters of four cells each. The new capacity is roughly five times the capacity with a single antenna.

Examples 5.4 and 5.5 illustrate the main benefits and elements of a cellular network planning by relating the bandwidth, number of cells, frequency reuse factor, and capacity of the network. If W is the total available spectrum, B is the bandwidth needed per user, N is the frequency reuse factor, and m is the number of cells required to cover an area, the number of simultaneous users is given by:

$$n = \frac{m(W / N)}{B} \tag{5.1}$$

In particular we observe that the capacity of the network can be increased by (a) increasing m, such as the number of cells (by reducing the size of each cell) and (b) decreasing the frequency reuse factor. A major remaining question at this point is how to assign the groups of subbands to individual cells so that interference between different users using the same subbands is acceptable. We address this issue in subsequent sections. Let us now consider some other important issues related to a cellular topology.

A cellular topology reduces the coverage requirements of both the mobile terminal and the BS. The reduction of the size of coverage lowers the required transmitted power by the mobile terminal because mobile terminals are located closer to the base stations and they require less power to communicate with the network. This increases the battery lifetime and reduces the size of a terminal. These issues are extremely important to the user of a handheld terminal. Therefore, the larger the number of cells, the larger the capacity and the smaller the size of the handheld terminal. However, we need a fixed network infrastructure to interconnect the cells and ensure that the entire system works in a coordinated manner. As we increase the number of cells, the cost and the time for deploying the network increases. In addition, as the cell size becomes smaller, the number of handoffs increases. Therefore, a reduction in the size of the cells increases the complexity of the design and deployment of the network, as well as the signaling load in the fixed part of the infrastructure. The art of designing a cellular topology involves striking a balance between all these elements, and this is the subject of details that follow in this chapter.

Another important factor in deployment of wireless cellular networks is provision for expansion. The main investment of a wireless service provider is toward the cost of the fixed infrastructure, which includes the BS and connections between them. When a service provider starts an operation, that person needs to minimize the cost of the infrastructure while continuously increasing the number of subscribers. As the number of subscribers increases, new income is generated, and the service provider can afford to expand the network by increasing the complexity of its infrastructure to support a further larger population of subscribers. Therefore, there is a need for a plan to take into account the growth of the subscriber base and thus the entire wireless network.

In summary, we need to address the following technical issues for planning a cellular network:

- Selection of a frequency reuse pattern for different radio transmission techniques
- Physical deployment and radio coverage modeling
- Plans to account for the growth of the network
- Analysis of the relationship between the capacity, cell size, and the cost of the infrastructure

5.3.2 Cellular Hierarchy

There are three reasons to use a hierarchical cellular infrastructure supporting cells of different sizes. One is to extend the coverage to the areas that are difficult to cover by a large cell. For example, cells designed to cover suburban areas have antennas on tall towers and cover a large area. Signals from these antennas, however, cannot propagate sufficiently into urban canyons or indoor environments. For urban canyons we need to install antennas at lower heights, and in indoor areas we may mount the antennas on walls to provide a comprehensive coverage. Antennas mounted in these locations are of low power and cover a smaller area, resulting in the creation of a smaller sized cell. The second reason to have a cellular hierarchy is to increase the capacity of the network for those areas that have a higher density of users. Imagine the number of cellular phone users in the downtown area of a large city and compare it with the number of mobile users on an interstate highway. To support the larger subscriber demand and higher traffic in smaller areas, we need to increase the number of cells by reducing their sizes. The third reason is that sometime an application needs certain coverage. Consider the increasing number of wireless devices that we are carrying in our bags these days and the increasing need for communication between these devices. This necessitates extremely small-sized cells that provide a wireless network for connecting laptops or notepads to cellular phones.

In a modern deployment of a cellular network, a number of cell sizes are used to provide a comprehensive coverage supporting traffic fluctuations in different geographic areas and supporting a variety of applications. One way of dividing the cells into a hierarchy is to define the following cell sizes:

- *Femtocells:* These are the smallest unit of the cellular hierarchy used for connection of personal equipment such as laptops, notepads, and cellular telephones. These cells need to cover only a few meters where all these devices are in physical range of the user.
- *Picocells:* These are small cells inside a building that support local indoor networks such as wireless LANs. The size of these networks is in the range of a few tens of meters.
- *Microcells:* These cells cover the inside of streets with antennas mounted at heights lower than the rooftop of the buildings along the streets. They cover a range of hundreds of meters and are used in urban areas to support PCS.
- *Macrocells:* Macrocells cover metropolitan areas, and they are the traditional cells installed during the early phases of the cellular telephony. These cells cover areas on the order of several kilometers, and their antennas are mounted above the rooftop of typical buildings in the coverage area.
- *Megacells:* Megacells cover nationwide areas with ranges of hundreds of kilometers and are mainly used with satellites.

Figure 5.6 illustrates the relationship between different cells with example applications. An ideal network has a hierarchy of these cells to cover airplane travelers with megacells, car drivers in suburban areas with macrocells, pedestrians in the

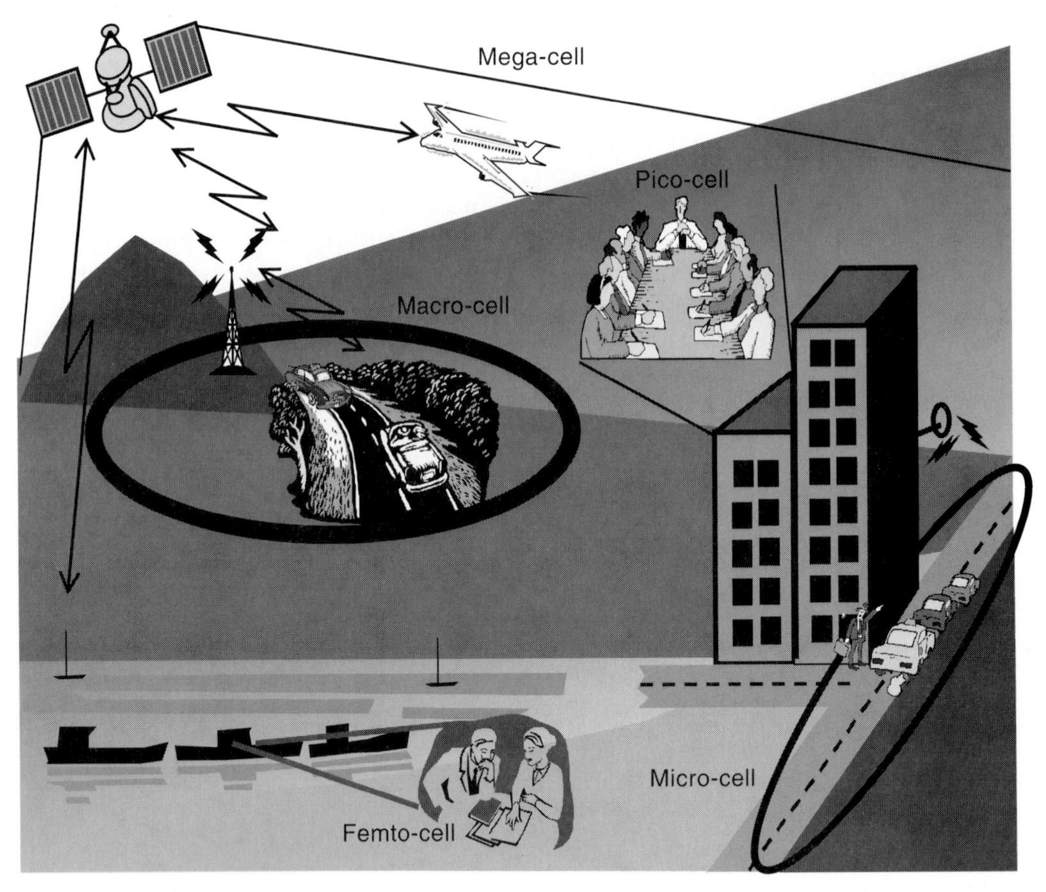

Figure 5.6 Cellular hierarchy.

streets via microcells, indoor users with picocells, and connect personal equipment with femtocells.

5.4 CELL FUNDAMENTALS

Having looked at the cellular topology and the concept of employing a cellular architecture to increase the communications capacity and to cater to a large subscriber demand in hotspots, we now consider quantitative means to characterize the interference in a cellular topology. This in turn leads to quantitative means for determining the best cluster size and simple techniques for allocating the subbands of spectrum within a cluster.

Even though in practice cells are of arbitrary shape (close to a circle) because of the randomness inherent in radio propagation, it is easier to obtain insight and

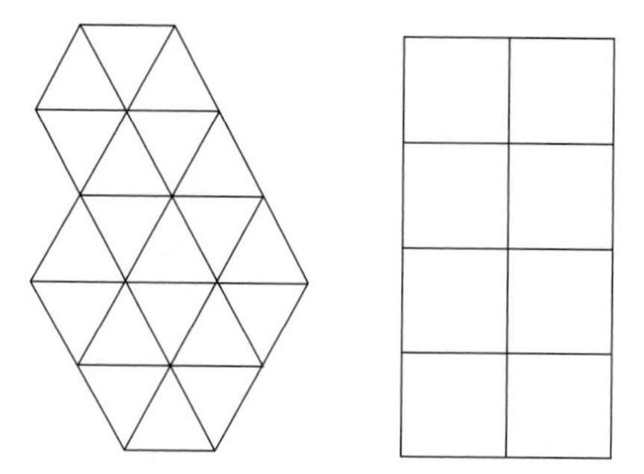

Figure 5.7 Triangular and rectangular cells.

understanding for system design by visualizing all cells as having the same shape. Also it is easier to mathematically analyze a cellular topology by assuming a uniform cell size for all cells. Once some insight is obtained as to what the effects of interference are, measurements, simulation, and a combination of these can be employed in actually determining the planning of a network.

For cells of the same shape to form a tessellation so that there are no ambiguous areas that belong to multiple cells or to no cell, the cell shape can be of only three types of regular polygons: equilateral triangle, square, or regular hexagon as shown in Figure 5.7.

A hexagonal cell is the closest approximation to a circle of these three and has been used traditionally for system design (see Figure 5.8). The argument for a hexagonal shape comes from the fact that among the three shapes mentioned, for a given radius (largest possible distance between the polygon center and its edge), the hexagon has the largest area.

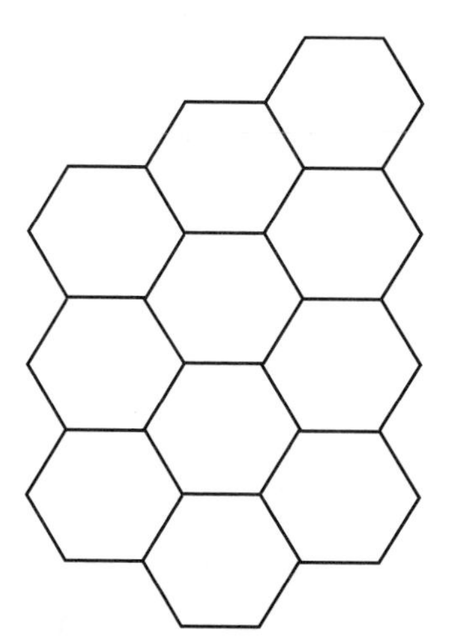

Figure 5.8 Arranging regular hexagons that can cover a given area without creating ambiguous regions.

In most of the literature and in the back of the envelope design, the hexagonal cell shape is chosen as the default cell shape. In particular cases that consider continuous distributions of traffic load and interference between different transmission schemes, a circular cell shape is employed for tractable calculation.

In order to investigate the effects of interference, which changes with distance, there is a need to come up with an elegant way of determining distances and identifying cells. Fortunately, it is possible to do this easily in the case of hexagonal cells [MAC79]. In order to maximize the capacity, cochannel cells must be placed as far apart as possible for a given cluster size. It can be shown that there are *only* six cochannel cells for a given reference cell at this distance. The relationship among the distance between cochannel cells, D_L, the cluster size, N, and the cell radius, R_L, is given by:

$$\frac{D_L}{R_L} = \sqrt{3N} \tag{5.2}$$

This quantity is also referred to as the *cochannel reuse ratio.* Values for N can only take on values of the form $i^2 + ij + j^2$ where i and j are integers.

Example 5.6: Cluster Size of $N = 7$

As described earlier, i and j can only take integer values. If we take $i = 2$ and $j = 1$, we see that $N = 4 + 2 + 1 = 7$. Selecting a cell A, we can determine its cochannel cell by moving two units along one face of the hexagon and one unit in a direction 60° or 120° to this direction. Clusters of size $N = 7$ can be created as shown in Figure 5.9. A value of $N = 7$ is employed in the United States in the advanced mobile phone service (AMPS).

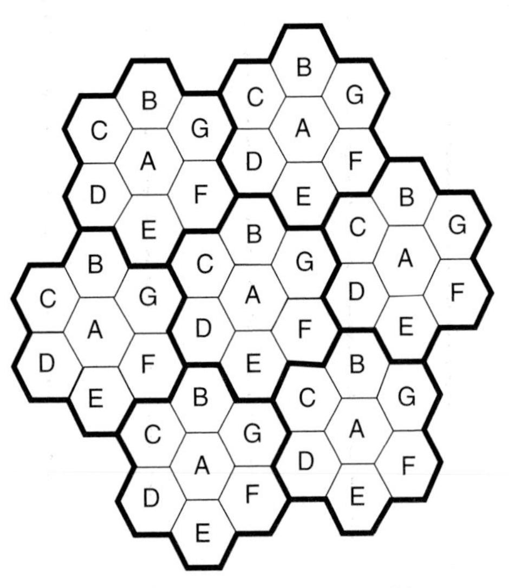

Figure 5.9 Hexagonal cellular architecture with a cluster size of $N = 7$.

The number of cells in a cluster N determines the amount of cochannel interference and also the number of frequency channels available per cell. Suppose there are N_c channels available for the entire system. Each cluster uses all the N_c channels. With fixed channel allocation, each cell is allocated N_c/N channels. It is desirable to maximize the number of channels allocated to a cell. This means that N should be made as small as possible. However, reducing N increases the signal-to-interference ratio (as discussed in the following section). There is thus a trade-off between the system capacity and performance. The reader should note that the development frequency of reuse using hexagonal cells is for mathematical tractability only. In reality, this approximation is useful only to obtain insight into designing cellular systems. Actual deployment is far more complicated because of irregular differing cell sizes and propagation mechanisms.

5.5 SIGNAL-TO-INTERFERENCE RATIO CALCULATION

In Section 5.1, we mentioned that a cellular architecture was essential in order to reuse the available spectrum while reducing interference caused by reusing the frequency spectrum. In this section, we look in detail at the performance measures that are useful in system design, in particular the signal-to-interference ratio and its relationship with the path loss, and the grade of service.

In general, the signal-to-interference ratio can be written as follows:

$$S_r = \frac{P_{desired}}{\sum_i P_{interference,\,i}} \tag{5.3}$$

Here $P_{desired}$ is the signal strength from the desired BS and $P_{interference,i}$ is the signal strength from the ith interfering BS. The signal strength falls as some power of the distance α called the power-distance gradient or path-loss gradient. That is, if the transmitted power is P_t, after a distance d in meters, the signal strength of a radio signal will be proportional to $P_t\, d^{-\alpha}$. In its most simple case, the signal strength falls as the square of the distance in free space ($\alpha = 2$). Suppose there are two base station transmitters BS_1 and BS_2 located in an area with the same transmit power P_t and a mobile terminal is at a distance of d_1 from the first and d_2 from the second. If the mobile terminal is trying to communicate with the first BS, the signal from the second BS is interference. The *signal-to-interference* ratio for this mobile terminal will be:

$$S_r = \frac{KP_t d_1^{-\alpha}}{KP_t d_2^{-\alpha}} = \left(\frac{d_2}{d_1}\right)^{\alpha} \tag{5.4}$$

The larger the ratio d_1/d_2 is, the greater is S_r and the better is the performance. The objective in a cellular radio system is to allocate frequencies or channels to cells within a cluster so that the distance between interfering cells (cochannel or adjacent channel) is as large as possible. For urban land mobile radio, the distance power gradient increases from two (in the case of free space) to roughly four so that the received signal strength falls as the fourth power of the distance. This further improves the signal-to-interference ratio. If there are J_s interfering BS surrounding a given BSs, the general form of the signal-to-noise ratio will be:

$$S_r = \frac{d_0^{\alpha}}{\sum_{n=1}^{J_s} d_n^{\alpha}} \tag{5.5}$$

where the distance of the mobile from the given base station is d_0 and its distance from the nth base station is d_n.

Example 5.7: S_r in a Hexagonal Cellular Architecture

Recalling that there are exactly six cochannel cells with a hexagonal cellular structure, it is clear that they will all cause similar levels of interference to a mobile terminal in the given cell. So $J_s = 6$ here. Also, the distance at which the cochannel cells are located depends on the size of the cluster from Eq. (5.2). The farthest distance a mobile terminal can be from the base station of a given cell is the cell radius R. The approximate distance of the mobile terminal from the base stations of each of the cochannel cells is D_L. For land mobile radio, if only the six cochannel cells that make up the first tier of interferers are considered, $J_s = 6$, and the signal-to-interference ratio can be approximated as:

$$S_r \approx \frac{R^{-4}}{J_s D_L^{-4}} = \frac{R^{-4}}{6 D_L^{-4}} = \frac{1}{6}\left(\frac{D_L}{R}\right)^4 = \frac{3}{2} N^2 \tag{5.6}$$

In terms of dB, we can write the signal-to-interference ratio as:

$$S_r = -7.78 + 40 \log(D_L / R_L) = 1.76 + 20 \log N \tag{5.7}$$

Figure 5.10 shows how the signal-to-interference ratio given by (5.7) varies with the cluster size N. Equation (5.7) is commonly used to determine the cluster size for an adequate performance. Note that the signal-to-interference ratio is influenced by the cochannel reuse ratio D_L/R_L in that a given D_L/R_L has to be maintained for a particular S_r. However, it is an approximation because different base stations may employ different transmit powers, and the path loss model may not be as simple as the d^{-4} model used here. The S_r calculation will be different for the uplink (mobile terminal to base station communication) compared with the downlink (base station to mobile terminal communication).

We have so far assumed that the received signal strength falls as the fourth power of the distance for land mobile radio. In dB, this translates to a path loss model of the form:

$$P_r(d)\,(\text{dB}) = P_t(\text{dB}) - 40 \log d + 10 \log K \tag{5.8}$$

The factor 10 log K usually corresponds to the path loss at the first meter, or kilometer as the case may be, and d is in the same units. This path loss model may not be appropriate, especially because measurements of the received signal strength indicate that the path loss is dependent not only on the distance between the base

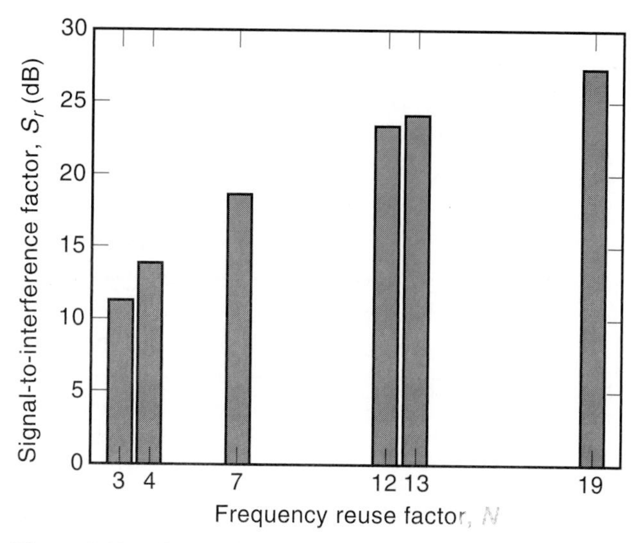

Figure 5.10 Signal-to-interference ratio, S_r, as a function of frequency reuse factor, N.

station and the mobile, but also on the radio frequency of operation and the antenna heights. The path loss is also dependent on the scenario, whether the cellular architecture corresponds to land mobile radio or to a microcellular PCS application. However, this simple model is appropriate for first-cut approximations in system design.

Let us consider an example of a real cellular system that tries to bring together many of the concepts that have been considered so far in this chapter.

Example 5.8: Cellular Architecture of AMPS

As an example of cellular architecture, we consider the very first cellular radio telephone system in the United States, called *AMPS* based on an analog FM modulation scheme. Each voice channel in AMPS occupies 30 kHz of bandwidth and uses FM. Figure 5.11 shows the spectrum allocations for AMPS.

A bandwidth of 25 MHz is allocated for both the uplink and downlink so that transmission is full duplex. The 25 MHz of spectrum is divided into two blocks of 12.5 MHz each. Block A is allocated to carriers who are not traditional telephone service providers. Block B is allocated to traditional telephone service providers. Each 12.5 MHz of spectrum can support 416 channels each of which is 30 kHz wide. Of these, 395 are dedicated channels for voice, and 21 are dedicated for call control.

Based on subjective voice quality tests, it was determined that a signal-to-interference ratio of 18 dB can be tolerated while providing a good voice quality to the user. From Eq. (5.7), this means that the cluster size has to be $N = 7$. Figure 5.9 shows the cellular architecture with this cluster size. Cells with the same label use the same frequency spectrum. They are separated by a distance $D_L = 4.58\ R_L$ in this case which ensures that the signal-to-interference ratio is around 18 dB.

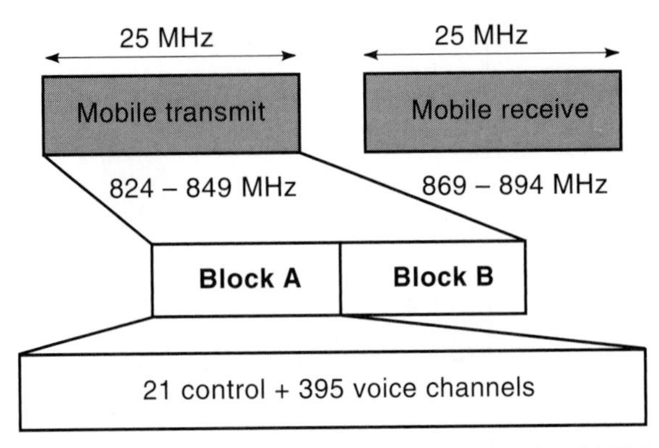

Figure 5.11 Frequency spectrum allocation for AMPS.

Let the 395 voice channels available for a service provider be numbered from 1 to 395. For example, on the downlink, 869–869.030 corresponds to channel 1, 869.030–869.060 to channel 2, and so on. Channels 1, 8, 15, . . . are allocated to cells labeled A. Channels labeled 2, 9, 16, . . . are allocated to cells labeled B and so on. This ensures that there is a sufficient separation between channels used within a cell so that adjacent channel interference is minimized. In practice, the numbering scheme is different because the entire 25 MHz of bandwidth was not initially available for AMPS. However, a separation of seven adjacent channels is maintained between channels used within a cell. It was also found in some cases that, because cells actually do not have a hexagonal shape and because the assumptions made in coming up with the value for N are not satisfied, a cluster size of $N = 12$ has to be employed for good voice quality. The reader is referred to [APP85] for another example.

5.6 CAPACITY EXPANSION TECHNIQUES

In the 1990s, the dominant source of income for the wireless telecommunication industry has been the cellular telephone service. During this period, the industry grew exponentially. Numerous companies are in fierce competition to gain a portion of the income of this profitable and prosperous industry. The main investment in deploying a cellular network is the cost of the infrastructure that includes the cost of BS and switching equipment, property (land for setting up the cell sites), installation, and links connecting the BSs. This cost is proportional to the number of BS sites. The income of the service is directly proportional to the number of subscribers. The number of subscribers grows with time and a cellular service provider has to develop a reasonable deployment plan that has a sound financial structure to account for many of these aspects. All service providers start their operation with the minimum number of cell sites to cover a service area that requires the least initial investment. As the number of subscribers increases, it generates a source of income for the service provider. At such a point of time, they can increase the investment on the infrastructure to improve service and to increase the capacity of the network to support additional subscribers. Therefore, a number of methodologies have evolved to facilitate the expansion of cellular telephone networks.

There are basically four methods to expand the capacity of a cellular network. The simplest method is to obtain additional spectrum for new subscribers. This is a very simple but expensive approach. The so-called PCS bands were sold in the United States for around $20 billion. If we assume that each new subscriber generates a profit of approximately $1,000 per year, we will still need 20 million additional subscribers to recover this amount in a year. With the fierce competition to provide the lowest cost to the customer, this has proved to be fatal. A case in example is that the top three companies that purchased the PCS bands have already filed for bankruptcy. The reader should not, however, conclude that this is not an acceptable method. With our pessimistic scenario, we are accentuating the vital importance of the need for other alternatives to expand capacity in addition to this simple approach of getting additional spectrum.

The second method to expand the capacity of a cellular network is to change the cellular architecture. Architectural approaches include cell splitting, cell sectoring using directional antennas, Lee's microcell zone technique [LEE91], and using multiple reuse factors [HAL83] (called reuse partitioning). These techniques, described in detail in the rest of this section, change the size and shape of the coverage of the cells by adding cell sites or modifying the nature of antennas to increase the capacity. These techniques do not need additional spectrum or any major changes in the wireless modem or access technique of the system that will require the user to purchase a new terminal. These features of architectural approaches distinguish them as one of the more practical and less expensive solutions to expand the network capacity.

The third method for capacity expansion is to change the frequency allocation methodology. Rather than distributing existing channels equally among all cells, it is possible to use a nonuniform distribution of the frequency bands among different cells according to their traffic need. The traffic load of each cell is dynamically changed by the geography of the service area and with time depending on the traffic load. In most downtown areas, we have the largest traffic loads during rush hours and a relatively light traffic load in the evening hours and weekends. This situation is reversed in residential areas. If we allocate channels dynamically to different cells, we can increase the overall capacity of the network. These techniques do not need any change in the terminal or physical architecture of the system, and they are implemented somewhere inside the computational devices used for network control and management.

The fourth and the most effective method to expand the network capacity is to change the modem and access technology. The cellular industry started with analog technology using FM modulation and has now evolved toward TDMA and then a CDMA air interface using digital modems. Digital technology increases the network capacity and also provides a fertile environment for integration of voice and data services. However, this migration requires the user to purchase new terminals and the service provider to install new components in the infrastructure.

5.6.1 Architectural Methods for Capacity Expansion

5.6.1.1 Cell Splitting

As the number of subscribers increase within a given area, the number of channels allocated to a cell is no longer sufficient for supporting the subscriber demand. It then becomes necessary to allocate more channels to the area that is being covered by this cell. This can be done by *splitting* cells into smaller cells and allowing additional channels in the smaller cells.

Consider Figure 5.12. In this figure, we have a cellular architecture where a cluster size of seven is employed. When the traffic load increases, a smaller cell is introduced such that it has half the area of the larger cells. This will ultimately increase the capacity fourfold (because area is proportional to the square of the radius). However, in practice, only a single small cell will be introduced such that it is midway between two cochannel cells. In this case, these are the larger cells labeled A. It is logical to thus reuse the channels allocated to these cells in the smaller cell to minimize the interference.

This approach gives rise to some problems. Let us suppose that the radius of the smaller split cell (labeled a) is $R/2$. Let the transmit power of the base station of the small cell be the same as the transmit powers of the larger cells. As far as the smaller cell is concerned, the signal-to-interference ratio is maintained because the maximum distance the mobile can be from the BS in this cell is $R/2$. So though the distance between this cell and the cochannel cells A is reduced by half, the value of S_r remains the same. On the other hand, this is not the case for the cells labeled A because the cochannel reuse ratio for these cells is now $D_L/2R$ with respect to the smaller split cell. In order to maintain the same level of interference, the transmit power of the base station in the smaller cell should be reduced. But this will increase the interference observed by the mobiles in the smaller cell. The other alternative is to divide the channels allocated to cells labeled A into two parts: those used by a and those not used by a. The channels used by a will be used in the larger cells only within a radius of $R/2$ from the center of the cell so that the cochannel reuse ratio will be maintained as far as these channels are concerned. This is called the *overlaid cell concept* where a larger *macrocell* coexists with a smaller *microcell.*

The downside of this approach is that the capacity of the larger cells is reduced which will ultimately lead to introducing split cells in their area, until such

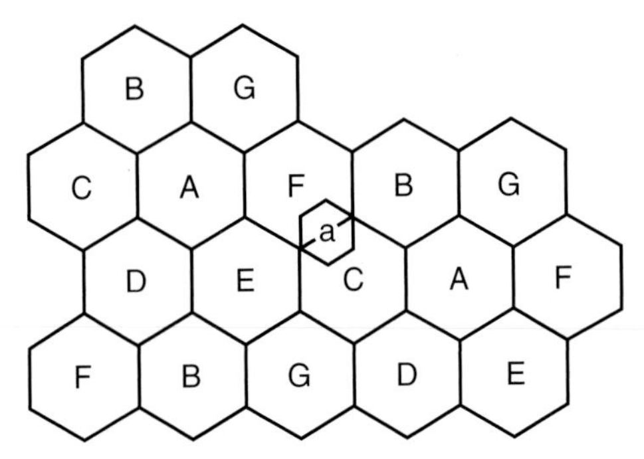

Figure 5.12 Illustrations of cell splitting for capacity expansion.

time as a chain reaction will result in the entire area being served by cells of a smaller radius. Also the BS in cells labeled A will becomc more complex, and there will be a need for handoffs between the overlays.

5.6.1.2 Using Directional Antennas for Cell Sectoring

The simplest and the most popular scheme for expanding the capacity of cellular systems is cell sectoring using directional antennas. This technique attempts to reduce the signal-to-interference ratio and thus reduce the cluster size, thereby increasing the capacity. The idea behind using directional antennas is the reduction in cochannel interference that results by focusing the radio propagation in only the direction where it is required. In order to achieve this, the coverage of a base station antenna is restricted to part of a cell called a *sector* by making the antenna directional. In implementing this technique, cell site locations remain unchanged, and only the antennas used in the site will be changed. The main objective here is to increase the signal to interference ratio to a level that enables us to use a lower frequency reuse factor. A lower frequency reuse factor allows a larger number of channels per cell, increasing the overall capacity of the cellular network.

As we discussed earlier [see Eq. (5.6)] the signal to interference ratio is given by:

$$S_r = \frac{1}{J_s}\left(\frac{D_L}{R}\right)^4 = \frac{9}{J_s}N^2 \tag{5.9}$$

where J_s is the number of interfering cell sites. Using a sector antenna reduces the factor J_s, resulting in the interference and an increase in S_r. The most popular directional antennas employed in cellular systems are 120° directional antennas. In some cases 60° directional antennas are also employed. In the following two examples, we evaluate the impact of these antennas which enables the reuse factor to be reduced from $N = 7$ to $N = 4$ and $N = 3$, respectively. In some cases, even though the reuse factor is reduced, we see that sectorization yields a better S_r from the required 18 dB value.

Example 5.9: Three-Sector Cells and a Reuse Factor of $N = 7$

Consider a seven-cell cluster scheme with 120° directional antennas as shown in Figure 5.13. Channels allocated to a cell are further divided into three parts, each used in one sector of a cell. As shown in the figure, the number of cochannel interfering cells is reduced from six to two. Thus, there is an improvement in the signal-to-interference ratio. For omnidirectional antennas (see Examples 5.7 and 5.8), the value of S_r for a cluster size of $N = 7$ is 18.66 dB. In this case, in a manner similar to Eq. (5.6), the signal-to-interference ratio is given by:

$$S_r \approx \frac{R^{-4}}{J_s D_L^{-4}} = \frac{R^{-4}}{6D_L^{-4}} = \frac{1}{2}\left(\frac{D_L}{R}\right)^4 = \frac{9}{2}N^2 \tag{5.10}$$

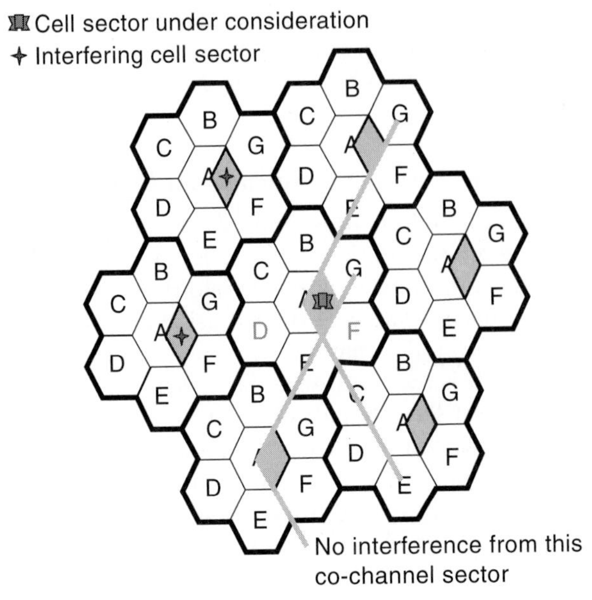

Figure 5.13 Seven-cell reuse with 120° directional antennas (3-sector cells)

For $N = 7$, this will give us $S_r = 23.43$ dB. To see the importance of this gain, note that the required signal-to-noise ratio for AMPS systems is 18 dB which suggests $N = 7$. However, a larger S_r is required because of nonideal situations.

Example 5.10: Three-Sector Cells and a Reuse Factor of $N = 4$

Equation (5.10) remains unchanged in this case as there are only two interfering cells again (see Figure 5.14). With omnidirectional antennas, $J_s = 6$ and for $N = 4$, we end up with $S_r = 13.8$ dB, which is inadequate for AMPS.

It can be seen that the signal-to-interference ratio with three-sector cells is substantially better compared with omnidirectional antennas and no cell sectoring. With $N = 4$, the signal to interference ratio is 19.9 dB. This value is larger than the requirement of 18 dB based on subjective mean-opinion-score tests of voice quality.

Example 5.11: Six-Sector Cells and a Reuse Factor of $N = 4$ and $N = 3$

With 60-degree directional antennas, we have six sectors within a cell. The number of interfering cochannel cells reduces to one, and the signal-to-interference ratio can be written as

$$S_r \approx \left(\frac{D_L}{R}\right)^4 = 9N^2 \tag{5.11}$$

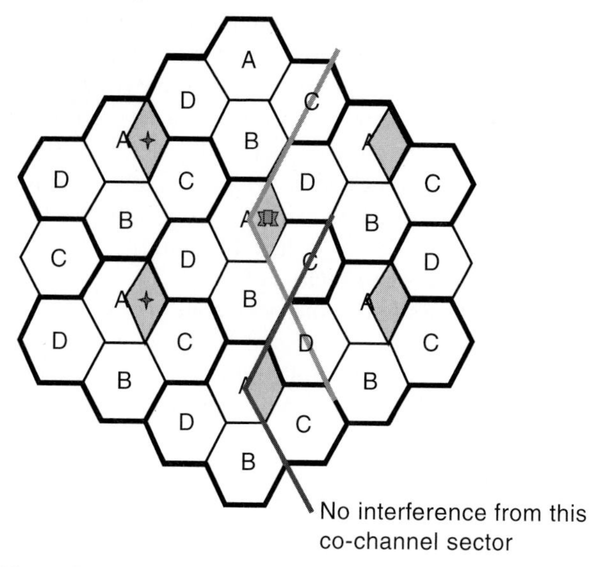

Figure 5.14 Four-cell reuse with 120° directional antennas (3-sector cells).

It is possible to employ a cluster size of four or three with six-sector cells because the signal-to-interference ratio will be 21.58 dB or 19.1 dB, respectively, which has a sufficient margin for AMPS. The cellular layout and relation to sectors in this case is left as an exercise for the readers.

In practice we cannot ideally sector a cell because ideal antenna patterns cannot be implemented. Therefore, the numbers obtained in the examples for ideal cell sectors are optimistic. However, our conclusion from these examples is that the use of sectoring increases the signal to interference ratio at the terminal. We should emphasize that in the particular examples we could reduce the frequency reuse factor from $N = 7$ to $N = 4$ or even $N = 3$ by using three- and six-sector cells, respectively. This reduction in frequency reuse from seven to four or even three would result in a capacity increase of 1.67 and 2.3, respectively, allowing an equal increase in the number of subscribers and consequently income of the service provider. The service provider needs to add these antennas to the BSs in the desired area. Compared with the cell-splitting method, using directional antennas is less effective in increasing capacity, but it can be significantly less expensive. The cost of additional cell sites, needed in cell splitting, includes costs of the property and installing the antenna mounting tower which are usually far expensive compared with deploying directional antennas. Cell splitting also requires additional planning efforts to maintain interference levels in the smaller cells. If directional antennas are used without reduction in the frequency reuse factor, the average required transmitted signal power from the mobile stations will be reduced which can potentially result in longer battery life for the user.

5.6.1.3 Lee's Microcell Method

The disadvantage of using sectors is that each sector is nothing but a new cell with a different shape, because channels have to be partitioned between the different sectors of a cell. The network load is substantially increased because a handoff has to be made each time a mobile terminal moves from one sector of a cell to another. Also, in all the discussion in the previous sections, it has been assumed that the BS antenna is located at the center of a cell, whether directional antennas are employed. In practice, employing directional BS antennas at the corners of cells can reduce the number of BSs [MAC79]. Lee's microcell zone technique [LEE91] exploits corner excited BSs to reduce the number of handoffs and eliminate partitioning of channels between sectors of a cell.

Figure 5.15 shows Lee's microcell zones concept. In this case, there is one BS per cell, but there are three "zone-sites" located at the corners of a cell. Directional antennas that span 135° are employed at these zone-sites. All three zone-sites act as receivers for signals transmitted by a mobile terminal. The BS determines which of the zone-sites has the best reception from the mobile and uses that zone-site to transmit the signal on the downlink. The zone-sites are connected to the BS by high-speed fiber links to avoid congestion and delay. Because only a single zone-site is active at a time, the interference faced by a mobile terminal from a cochannel zone-site is smaller compared with what would be the interference with an omnidirectional antenna. Consequently, the cluster size can be reduced to three, and a capacity gain of 2.33 is obtained over a seven-cell cluster scheme. Consider the following example:

Example 5.12: Lee's Microcell Zone Technique

Show that the cochannel reuse ratio D_Z/R_Z for the zones in Lee's microcell zone concept is larger than 4.6 if a cluster size of three is employed. Use Figure 5.16 for your calculations.

In this example, a cluster size of $N = 3$ is employed. Each "cell" is divided into three "zones." On the downlink, only one of the zones is active. Because the zone-sites are directional, they cause interference only in corresponding multi-zone sites in another cluster. The cochannel reuse ratio D_z/R_z in Figure 5.16 is clearly $6 \times \sqrt{3}/2 = 5.196$, which is larger than 4.6. Even if all six cochannel zones cause interference, the capacity is still larger by a factor of 2.33 because the cluster size is now $N = 3$ as compared to the usual value of $N = 7$.

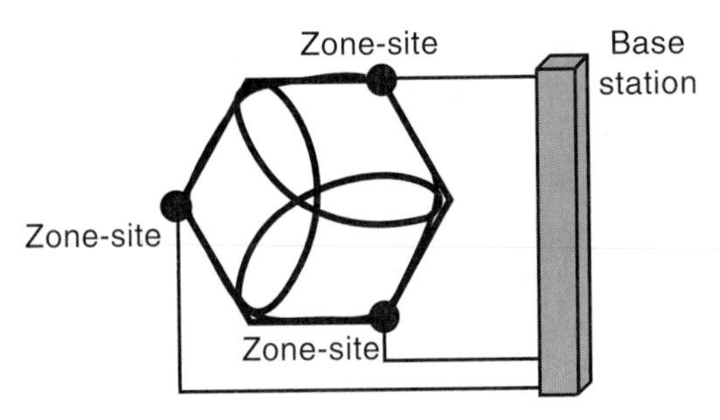

Figure 5.15 Lee's microcell zone concept.

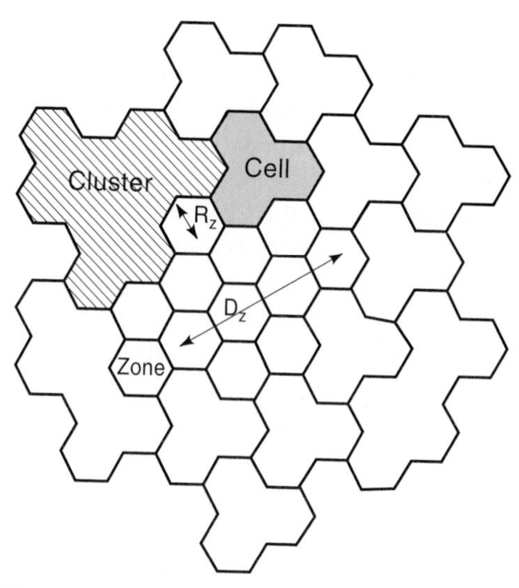

Figure 5.16 Example of Lee's microcell zone concept.

5.7.1.4 Using Overlaid Cells

The *overlaid cell concept* introduced in the section on cell splitting can be used to increase the capacity of a cellular network. Here, channels are divided among a larger macrocell that coexists with a smaller microcell contained entirely within the macrocell. The same BS serves both the macro- and microcells. Figure 5.17 illustrates the basic concept for overlaid cell concept. There are four parameters (R_1 and D_1) representing the radius of coverage and distance among cochannel cells for the macrocells; R_2 and D_2 denote the radius of coverage and the distance among cochannel cells for the microcells. The design is made such that D_2/R_2 is larger than D_1/R_1 and from Eq. (5.4) the signal-to-interference ratio (S_r) for the microcells will be substantially greater than that of the macrocells. There are two methods to exploit this situation to increase

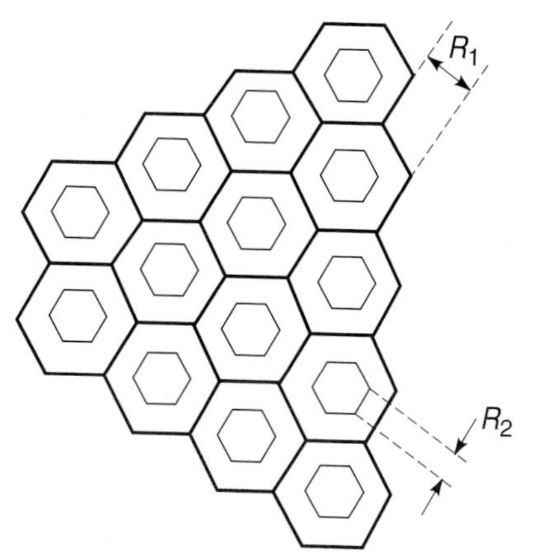

Figure 5.17 Reuse partitioning.

the capacity of the network: using split-band analog systems and reuse partitioning. Often the microcells are said to belong to an *overlay* network that is overlaid on top of an underlying macrocellular network referred to as the *underlay* network.

Split-Band Analog Systems. The split-band analog systems use a more bandwidth efficient modulation within the overlay cells. This technique is applied in analog cellular systems using FM. We have considered several examples of analog cellular systems in Chapter 1. In FM, the signal to noise requirement is inversely proportional to the square of the bandwidth. If we reduce the bandwidth to half the original value, the signal to noise ratio requirement will be increased four times (by 6 dB). If we arrange R_2 and D_2 to have a cochannel reuse ratio that is four times larger than usual, we end up with a signal to interference ratio (from Eq. (5.4)) that remains unchanged. The overlay system can then use FM with half the bandwidth of the underlay system, doubling the capacity within the overlay part of the network. An example will further clarify this situation.

Example 5.13: Band-Splitting in AMPS

The AMPS system uses a 30 kHz band for FM signals used for communication between the MS and the BS. As discussed earlier, the minimum required signal to interference ratio for this system is 18 dB. If we develop an overlay system with a 15 kHz bandwidth, the required S_r is 24 dB, which is 6 dB more than the system employing a bandwidth of 30 kHz. From Eq. (5.4) we have

$$10 \log \frac{\left(\frac{D_2}{R_2}\right)^4}{\left(\frac{D_1}{R_1}\right)^4} = 6\ dB$$

If we employ the same frequency reuse factor of $N = 7$ for the overlay and underlay networks, $D_1 = D_2$ and solving for the above equation we have $R_2 = 0.7079\ R_1$. Because the area covered by each cell is proportional to the square of the cell radius, the area of the overlay cell, A_2, will be half of the area of the underlay cell, A_1. The overlay is responsible for terminals within the smaller hexagon, while the underlay system supports users in the layer between the boundary of the overlay cell and the boundary of the underlay cell. These two areas are the same in our example. Therefore, the number of channels available to the overlay and underlay cells remain the same. If we represent this number by M then the total bandwidth used by the system is $M(15+30)$ kHz.

In the original AMPS network each service provider has 12.5 MHz of bandwidth that is divided into 416 channels, from which 395 channels are used for voice and 21 channels for control signaling. Therefore, 395 $\times$ 30 kHz of bandwidth was used for actual traffic. If we replace that system with a split-band underlay-overlay network we have

$$M(15 + 30) = 395 \times 30 \Rightarrow \mathrm{M} = 263$$

The total number of channels $M = 263$ for each of the overlay and underlay cells and $263 \times 2 = 526$ will be the total number of channels available. This number is 1.34 times larger than the original system, improving the capacity of the system by 34 percent.

Compared with cell splitting or using sectored cells, this technique provides a smaller improvement in capacity. However, it does not need any change in the hardware infrastructure. However, the MS and BS need minor changes to cope up with multiple bandwidths. To further improve the capacity of this technique, it is possible to use another layer of overlay system using even smaller cells. As we saw in Chapter 1, the Japanese analog systems use 25 kHz per user for underlay networks and 12.5 kHz (and even 6.25 kHz) for the overlay networks. The downside of underlay-overlay networks is the increased complexity at the BS for keeping track of which channel belongs to which overlay and increased number of handoffs when a mobile moves from one overlay to the next (or from a microcell to a macrocell). This requires additional complexity at the BS and handoffs when a mobile terminal moves from a microcell to a macrocell.

Reuse Partitioning. The overlaid cell concept described above can be used to increase the capacity of a cellular network through what is called the *reuse partitioning* concept [HAL83]. Here channels are divided among a larger macrocell and a smaller microcell contained entirely within the macrocell. The bandwidth in both cells remains the same. Because the radius of the microcell is smaller, the S_r for the overlay is larger, and it is able to employ a smaller cochannel reuse distance compared with the underlay or macrocell. The channels allocated to the microcell, for instance, may be reused in every third or fourth microcell, whereas the channels allocated to the macrocells can be reused in only every seventh or twelfth cell as the case may be. This requires additional complexity at the BS and handoffs when a mobile terminal moves from a microcell to a macrocell. To explain this situation, consider the following example:

Example 5.14: Reuse Partitioning of 7 and 3

Assume that in Figure 5.17 the radius of the underlay macrocells is R_1 and the radius of the microcells of the overlay is R_2. If we have an AMPS network operating on this infrastructure, the required S_r for both networks is 18 dB. From Equation (5.4), both underlay and overlay networks should have $D_1/R_1 = D_2/R_2 = 4.6$. Because R_2 is smaller than R_1, D_2 can be made smaller than D_1 by a factor equal to the ratio of R_1 to R_2. The improvement in cochannel reuse ratio comes from the fact that the microcells in the overlay are *not* contiguous to one another.

Suppose the cochannel reuse ratio without reuse partitioning was $D_L/R = Q$. The cluster size N in this case is $Q^2/3$ [from Eq. (5.4)], and the number of channels available per cell is $N_c/N = 3N_c/Q^2$. With reuse partitioning, let the ratio of the macrocell radius to the microcell radius be $\kappa = R_1/R_2$. From Example 5.14, the cluster size for the microcells can be reduced by a factor of κ^2 because the microcells are noncontiguous.

Example 5.15: Channel Allocation to Underlay and Overlay Cells

Consider Figure 5.18. Here we are using a cluster size of $N_1 = 7$ for the underlay macrocells to ensure that $D_1/R_1 = 4.6$ provides a suitable S_r for AMPS. We now overlay microcells with a radius R_2 such that the cluster size of the microcells is $N_2 = 3$. If $N_2 = 3$, we can see from Figure 5.18 that $D_2 = 3R_1$. Clearly $3R_1/R_2 = 4.6$

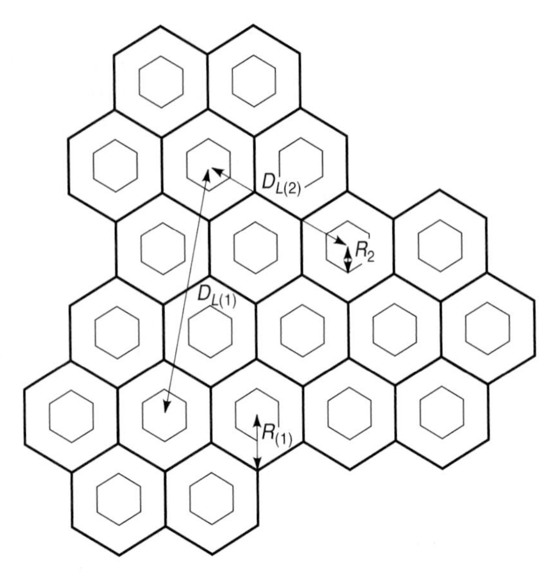

Figure 5.18 Reuse partitioning with a cluster size of seven for the macrocells and three for the microcells.

or $R_2 = 0.652R_1$. One way of allocating channels to the microcells and macrocells is to distribute them by area occupied. This may not be the best case. However, for this example, we employ this technique. The area of a cell is proportional to the square of its radius. We see that the area of a microcell is 0.652^2 times the area of a macrocell or $0.425 \times$ area of macrocell. Let the total number of channels available be N_c. If channels are distributed according to area and there are L channels available per cell, let us assume that $0.425L$ channels are allocated to the microcell and $0.575L$ channels are allocated to the macrocell.

Since the cluster sizes are 7 and 3, respectively, we have:

$$N_c = 7 \times 0.575L + 3 \times 0.425L \Rightarrow L = N_c/5.3$$

The total number of available channels for an AMPS operator is 395. Therefore, $L = 75$. The inner overlay uses approximately 32 channels, and the underlay uses 43 channels. Originally we had 395 channels with $N = 7$, providing approximately 56 channels per cell. The increase in the capacity is $75/56 = 1.34$, a 34 percent increase in capacity. In reality, a larger capacity can be expected because the channels allocated to the macrocells may also be used within the microcells.

Multiple overlays can provide an additional increase in capacity. As compared with the other expansion techniques, the advantages and disadvantages of the frequency reuse partitioning are very similar to frequency splitting. However, reuse partitioning does not need modification in the BS or MS radio equipment, and it can be easily applied to other modulation techniques. The derivation of S_r for frequency splitting was highly correlated with how FM works and, it cannot be extended to digital systems in a straightforward way.

5.6.1.5 Using Smart Antennas

Recently, using smart antennas for capacity expansion has attracted attention [LEH99]. Traditionally, frequency division, time division, and code division multiple access has been employed for cellular communications. Using smart antennas, users in the same cells can use the *same* physical communication channel, as long as they are not located in the same angular region with respect to a BS. Such a multiple access scheme, referred to as space division multiple access (SDMA), can be achieved by the BS *directing* a narrow antenna beam toward a mobile communicating with it. In addition to SDMA, interference between cochannel cells is greatly reduced because the antenna patterns are extremely narrow. In Section 5.7.1.2, we saw that by using sectored cells the reuse factor can be vastly reduced. Even larger advantages can be obtained with smart antennas. Simulations on a frequency-hopped GSM system have reported a capacity increase of 300 percent. A fivefold (500 percent) increase in capacity has been reported for CDMA [LEH99].

5.6.2 Channel Allocation Techniques and Capacity Expansion

In the previous section, we associated each cell with a group of channels that is assigned by the service provider to that cell. In the analysis of capacity, we assumed that all channels in a cell will be used, and we found methods to increase the number of available channels per cell in a given geographic area. We examined a variety of architectural methods to increase the number of available channels per cell. In all these schemes, the number of channels in cells of equal size were assumed to be the same. This assumption would be valid if the distribution of the users in the area was stationary and uniform over all cells. In practice, during the day, a cell in downtown areas of a city carries a high traffic that peaks during rush hour. But this very same cell does not carry a lot of traffic during late evenings or weekends. A cell in a residential area may have traffic characteristics that are the opposite of that of the cell in the downtown area. Clearly, the number of terminals in a cell changes in time, depending on the location of the cell, and this means there is a need for a more complex methodology to allocate channels to the cells dynamically based on the traffic load at a given time. A number of channel allocation strategies have been developed to address this issue [KAT96].

To address channel assignment or allocation techniques in cellular networks, we look at this problem from the point of view of a user. For the user it is not important how many channels are available or how they are allocated. A circuit switched (voice) user will dial a number, and if a channel is available the user is happy. If the call is blocked because a free channel is not available, the user will be dissatisfied with the service provider. A measure of whether channels will be available for a user when that person attempts a call is the probability of call blockage. It depends on the number of available channels and the traffic load. This is thus the quantity that represents user satisfaction. Service providers believe that a probability of call blockage of around 2 percent will keep customers happy, and they aim at this number. However, the probability of call blockage changes its value as mobile terminals move in and out of the boundaries of a cell. A service provider should maintain the number of subscribers under a particular value so that it is possible to accommodate fluctuations in the probability of call blockage over time across all the cells.

The main objective of channel allocation techniques is to stabilize the fluctuations in the probability of call blockage over the entire coverage area of a network over time. Reduction of the fluctuations in probability of call blockage allows service providers to accept a higher number of subscribers over the coverage area. This can be considered equivalent to expansion of the network through additional channels. In other words, as the number of subscribers increase, one way of accommodating such an expansion is to use a more efficient channel allocation technique to cope with the situation. Service providers use a variety of proprietary algorithms for channel allocations. These techniques can be divided into three main categories: fixed channel allocation (FCA), dynamic channel allocation (DCA), and hybrid channel allocation (HCA) techniques. We study these techniques in the following sections.

5.6.2.1 Fixed Channel Allocation (FCA)

In cellular telephony, the number of chunks of frequency spectrum available for the service provider and the bandwidth required per user governs the number of available channels. Fixed channel allocation techniques, in their simplest form, divide the available spectrum by the cluster size to determine the number of radio channels per cell. That is, if the available spectrum is W Hz and each channel needs B Hz, the total number of channels is $N_c = W/B$. If the cluster size is N, the number of channels per cell is $C_c = N_c/N$. These C_c available radio channels are then distributed in the cells in a manner so as to minimize adjacent channel interference. One obvious distribution pattern for channels among the cells is to assign adjacent radio frequency bands to different cells. In analog cellular systems, each radio channel corresponds to one user (one voice channel), whereas in digital TDMA or CDMA networks, each radio frequency channel carries several time slots or codes associated with voice channels.

Example 5.16: Fixed Channel Allocation in GSM

In the GSM cellular system, we have a pair of 25 MHz bands allocated to the downlink or forward channel and uplink or reverse channel. Each radio carrier uses 200 kHz of bandwidth, and each carrier contains eight time slots capable of supporting eight voice users. Potentially we have 125 carriers, but in practice only 124 of them are used. Let us number the channels as 1, 2, 3, . . . , 124. If the cluster size is $N = 4$, the simple FCA technique results in four sets of frequencies given by:

{1, 5, ... , 120} for the first set of cells
{2, 6, ... , 121} for the second set of cells
{3, 7, ... , 122} for the third set of cells
{4, 8, ... , 123) for the fourth set of cells

Example 5.17: AMPS and IS-136

In the U.S. cellular systems discussed in Example 5.8, each service provider has 12.5 MHz of spectrum available. In the downlink or forward channel, service providers use half of the 869–894 MHz band and in the uplink or reverse channel, half of the 824–849 MHz band. These bands are divided into 416 pairs of radio frequency bands, each having 30 kHz chunks for forward and reverse channels. In AMPS each frequency band carries one voice user, whereas in the IS-136 digital TDMA system, three users per 30 kHz of radio channel are supported via time slots. Of the 416 radio frequency channels, 21 are used for control channels and 395 for voice traffic. With FCA and a frequency reuse factor of seven, we can create the following seven sets of frequencies to minimize the interference.

{1, 8, ... , 390} for the first set of cells

{2, 9, ... , 391} for the second set of cells

...

{7, 14, ... , 396} for the seventh set of cells

The FCA strategy described earlier is simple to implement, and if the traffic in the network is uniform, so that the number of active users in each cell is the same, and it remains constant with time, this is also an optimum channel allocation strategy. However, in practice, traffic in each cell changes with time due to the movement of mobile terminals from one cell to another. This results in higher probabilities of call blocking in some cells and lower values in others, which results in poor utilization of the available bandwidth. To equalize the utilization of channels in all cells, the obvious solution is that the cells with higher traffic load should somehow use the free channels available in low traffic cells. This is possible by a nonuniform allocation of channels to cells in the first place. When we assume that the traffic density in all the cells is the same, as illustrated by Examples 5.16 and 5.17, the channel assignment algorithm is very straightforward. We simply divide the total number of available channels by the cluster size of the system and allocate this number of channels to each cell. Using the traffic density and the number of channels in one cell, we can determine the call-blockage probability in that cell. The probability of call blockage in all other cells and consequently the average probability of call blockage in the entire network will be the same as that of one cell. Here the channel assignment algorithm and the calculation of probability of call blockage are performed in two independent steps. With a nonuniform channel allocation technique, we need to include the call blockage probability as a criterion for the channel allocation algorithm. Because the relation between the number of channels and the call-blockage probability is a complex function, this algorithm becomes significantly more complex. The following example helps in the understanding of the complexity involved.

Example 5.18: Probability of Call Blockage with Nonuniform Traffic Distribution

Assume that we have only four cells and one cluster. We simply divide our available channels N_c by four. That is, we assign $N_c/4$ channels per cell. Also assume that for a uniform distribution of the traffic, the calculated probability of blockage for each cell is the desired number (2 percent). If the traffic becomes nonuniform, the blockage rate of all four cells will be changed, and the overall average of them does not remain at 2 percent anymore. The general idea is to increase the number of channels in cells with the higher traffic load and decrease it in cells with a lower traffic load so that the overall blockage rate of the network is minimized. There are four variables, namely, the number of channels per cell and the cost function to be minimized is the probability of call blocking. This has a very complex expression involving these variables. The minimization process is far more complex than the uniformly distributed traffic case for which we know that the same number of channels per cell would provide the minimum blockage rate. This is only one dimension of the complexity of the problem. The other dimension of complexity emerges when we consider cochannel cells having different traffic densities. The optimization problem is now a function of N_c variables, and, in addition, the regular frequency reuse pattern studied earlier does not work any more. The relatively simple frequency reuse pattern strategies discussed earlier were based on the assumption that the channels are fixed and thus calculating the cochannel interference on the basis of having the same number of channels per cell. These patterns become much more complex as we start thinking of unequal number of channels per cell.

Algorithms that distribute channels among the cells according to their traffic load have been investigated. For example, a *nonuniform compact channel allocation algorithm* is discussed in [ZHA91]. This algorithm first defines a set of patterns for nonuniform distribution of channels; then it selects the pattern that minimizes the average call blocking probability in the system. Results of example simulations in [ZHA91] show that nonuniform distribution of channels adopted by this algorithm provide better call blocking probabilities in the system. The reduction in call blocking probabilities allows an average of 10 percent and a maximum of 22 percent traffic to be added to the system while maintaining the same call blocking probability as that of uniform channel allocation. Note that channels are still permanently allocated to cells here, and this still corresponds to fixed channel allocation.

5.6.2.2 Channel Borrowing Techniques

Nonuniform channel allocation is quite complicated. A simpler scheme enables high-traffic cells to *borrow* channel frequencies from low traffic cells and maintain them until significant changes in traffic pattern are measured or predicted. (In other words the high-traffic cells borrow channels from low traffic cells). These techniques are usually referred to as *channel borrowing techniques* [KAT96]. The technical issues are how can we relate the traffic distribution to the channel allocation? And which cell should we borrow the channels from? There are two methods to borrow channels: *temporary channel borrowing* and *static channel borrowing*. In *temporary channel borrowing,* high-traffic cells return the borrowed channels after the call is completed. In *static channel borrowing,* channels are nonuniformly distributed among cells according to the available statistics of the traffic and changed in a predictive manner.

Temporary channel borrowing deals with short-term allocation of borrowed channels to cells. Once a call associated with the borrowed channel is completed, the channel is returned to the cell from which it was borrowed.

Example 5.19: Temporary Channel Borrowing

Let us suppose that we had 49 cells in a region. Let the cluster size be seven. In the case of uniform traffic density, we divide the total number of available channels, N_c, into seven groups and, using prescribed reuse patterns studied in earlier sections, we assign channel groups to different cells (see Figure 5.19). The calculation of the probability of call blockage will remain the same as before and let this value be 2 percent. If we make a pool of the N_c channels and allocate them purely on the basis of demand, we have N_c different channels each with a different characteristic in terms of the traffic on them.

Some channels may not be in use, and others may be continually in use. Depending on how often a channel is used and where it is being used, it may cause a high or low interference to its cochannel elsewhere. Suppose the cell A in the central cluster in Figure 5.19 borrows a channel from the solid-shaded cell F within its cluster to accommodate extra traffic load. This means that the corresponding channel in three cells labeled F and cross-hatched in neighboring clusters are locked until this channel is released by the cell A. This is because the reuse distance for the borrowed channel has decreased since it has been moved from the solidly shaded cell F, to the cell A.

A number of methods for selecting free channels from a lightly loaded cell for borrowing by another cell are summarized in [KAT96]. These methods either make all channels available for borrowing (this is called simple borrowing schemes) or they partition the channels into borrowable and nonborrowable (such schemes are referred to as hybrid borrowing schemes). Simple borrowing schemes are found to be better under light or moderate traffic loads. A quantitative comparison of some

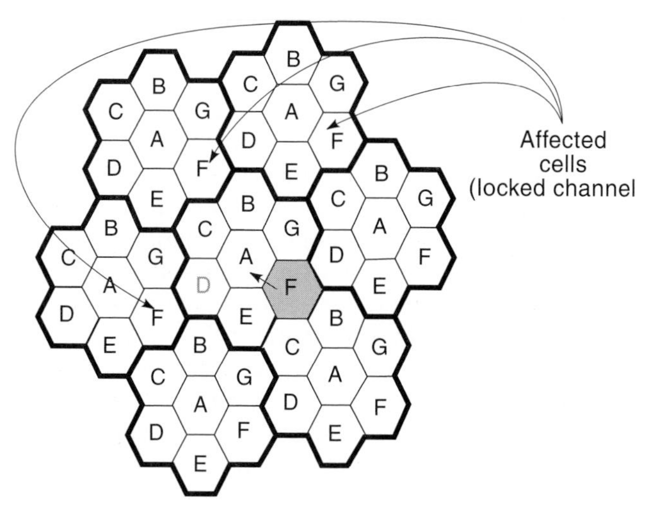

Figure 5.19 Temporary channel borrowing.

of these techniques is provided in [KUE92]. In specific examples addressed in this book, some of the suggested borrowing techniques are found to be capable of supporting up to 35 percent more traffic than uniformly distributed FCA. Channel borrowing schemes, however, require additional computational complexity, and frequent switching of channels. They may also affect handoff strategies.

5.6.2.3 Dynamic Channel Allocation (DCA)

Many researchers have studied the shortcoming of FCA techniques to accommodate temporal and spatial traffic variations and have suggested various DCA techniques in the past two decades. In DCA all channels are placed in a pool, and they are assigned to a new call according to the overall signal to interference pattern in all cells. Each channel can be used in any cell as long as it satisfies the signal to interference ratio requirements for the system. The channel is returned to the pool after the cellular call is terminated. This technique adapts well to the temporal and spatial changes in traffic load. In fact, according to [COX99], the capacity is maximized when the S_r of every set of cochannel users (users in cochannel cells using the same frequency bands) is balanced around some level that is no larger than strictly necessary. The downside is that DCA is extremely complex and inefficient under high-traffic load conditions. Although many claims have been made about the relative performance of each DCA scheme, the trade-offs and the range of achievable capacity gains are still unclear, and a number of questions remain unanswered [KAT96]. Microcellular systems of high-density personal communication networks have been shown to benefit the most out of the DCA, and the results of simulations shows a near ideal performance with DCA algorithms [KAT96].

The basic idea of the DCA is straightforward. However, a number of DCA schemes have been studied. The question then arises as to how these schemes differ from one another. In DCA, often several choices are available for assigning channels to a requesting cell. To devise a selection policy, we have to define a cost function to determine the appropriateness of the channel to be selected. This cost function quantitatively ranks the available channels based on the overall interference, average probability of call-blockage, or parameters that somehow relate to these quantities. The difference between DCA techniques lies in the selection and optimization of this cost function.

In reference [KAT96] a number of DCA schemes are introduced, and they are divided into two categories: centralized and distributed schemes. In centralized schemes, a central pool of all channels exists. The various schemes differ in the way the cost function handles the priorities in selecting a candidate channel and returning it to the central pool upon termination of the call. Distributed DCA techniques are considered for microcellular systems where channel propagation is less predictable and traffic is denser. In these schemes, the BS decides on the frequency assignment locally. Distributed DCA techniques are further divided into two classes: cell-based and interference-based techniques. Cell-based techniques require each BS to maintain a table of available channels in its vicinity and based on this table, the BS decides and assigns the channel to the users in its cell. This technique is very efficient, but the expense incurred is additional inter-BS communication traffic, which increases with the traffic in a cell. To avoid this situation, another subclass of distributed DCA schemes has evolved. Here each BS makes the channel assignment based on the RSS of the mobiles in its vicinity. In such schemes all channels are available to the BS, and the BS make its decision based on the local information without any need to communicate with other BSs. These schemes are self-organizing, simple, efficient, and fast, but they suffer from additional, unwanted cochannel interference which may result in channel interruption and network instability.

Example 5.20: Centralized DCA

In [ZHA89], a locally optimized dynamic assignment strategy is discussed. Here the particular cell allocating channels considers candidate channels based on whether they are being used in the first, second, or third tiers of cochannel cells and so on until the n-th tier. It then assigns the channel with minimum cost to the requested call from the mobile. Simulations of the locally optimized dynamic assignment scheme with 49 cells indicate that the call-blocking probability can be reduced by 40 percent for light loads compared with fixed channel assignment.

Example 5.21: Distributed DCA

Examples of a signal strength–based distributed dynamic channel assignment strategies are simulated in [GOO93]. The distributed DCA schemes are implemented for microcellular environments or one-dimensional cellular systems. Similar schemes are implemented in DECT [KAT96]. When a mobile requests a channel from the BS, the BS measures the interfering signal power on all channels not already assigned to mobiles in its cell. The channel with the maximum S_r is assigned to the mobile. For the same mean traffic, the distributed DCA strategies provide a much lower probability (around 30 to 50 percent lower) of call blocking compared with fixed channel assignment schemes. The exact increase in capacity depends on the number of mobile stations and the offered traffic load in the given cell.

Table 5.1 compares all three classes of algorithms for DCA. It also briefly summarizes the important characteristics of DCA strategies. Centralized DCA strategies do provide optimal or near optimal channel allocation. However, they require a lot of computational and signaling effort because the centralized location has to be aware of the available channel and the necessary parameters required to make an optimum decision on which channel to allocate to an incoming call. These parameters could be how channels are allocated in cochannel cells, what are the S_r values for the channel under consideration, what is the expected traffic load in and around the region, and so on. Also, because the decision mechanism is centralized, it is not robust, and a failure here could lead to an entire systemwide shutdown. In that sense, distributed DCA strategies are better. Cell-based distributed DCA strategies also can allocate channels optimally because BSs can communicate with each other to obtain knowledge of the entire system. Although the BSs make the decisions to allocate channels, the need for frequent communication and update between BSs, is a disadvantage. Signal-strength or measurement-based DCA

Table 5.1 Comparison of Three Classes of Algorithms for DCA [KAT96]

Algorithm	Centralized DCA	Distributed DCA	
		Cell-Based	*Measurement-Based*
Advantages	Near Optimum Channel Allocation	Near Optimum Channel Allocation	Suboptimal Channel Allocation Simple Assignment Algorithm Use of Local Information Minimum Communication between BSs System Capacity Increases, Efficiency, Radio Coverage Fast Real-Time Processing Adaptive to Traffic Changes
Disadvantages	High-Centralized Overhead	Extensive Communication between BSs	Increased Cochannel Interference Increased Cochannel Interference Deadlock Probability Instability

strategies do not provide optimal allocation of channels. Such schemes do, however, provide some capacity increases and are implemented in digital cordless systems like DECT.

5.6.2.4 Comparison of FCA, DCA, and HCA

Overall, DCA techniques have shown 30 to 40 percent performance improvements over the simple FCA techniques [GOO93]. Table 5.2 compares fixed and dynamic channel allocation strategies.

Under low-to-moderate traffic loads, DCA strategies perform far better than FCA techniques. Because DCA is based on random arrivals of mobiles and random allocation of channels to them, unless maximizing the "packing" of channels is an optimization criterion, it is likely that distances larger than what is required may separate cochannels. This will prevent channels from being reused as often as possible, resulting in less capacity at larger loads. DCA, however, reduces the fluctuations in the call blocking probabilities, as well as forced call terminations. FCA strategies require a lot of "offline" effort in frequency planning. DCA strategies need plenty of effort in real time for channel allocation. A unified framework for comparing all kinds of DCA strategies versus FCA is not available [KAT96], and it is hard to say which of these schemes is actually beneficial. In addition, DCA schemes that jointly optimize power control and handoff strategies have also been proposed.

Because DCA is better at lower traffic loads and FCA is better at higher traffic loads, the natural question of whether the two channel allocation techniques can be combined to provide both advantages arises. Indeed, *hybrid* channel allocation strategies have also been investigated. The total number of channels is partitioned into *fixed* and *dynamic* sets. The ratio of fixed to dynamic channels becomes important in the performance of the system. HCA schemes have been shown to perform better than FCA schemes for load increases up to 50 percent [KAT96].

Table 5.2 Comparison between FCA and DCA [KAT96]

Attribute	Fixed Channel Allocation	Dynamic Channel Allocation
Traffic load	Better under heavy traffic load	Better under light/moderate traffic load
Flexibility in channel allocation	Low	High
Reusability of channels	Maximum possible	Limited
Temporal and spatial changes	Very sensitive	Insensitive
Grade of service	Fluctuating	Stable
Forced call termination	Large probability	Low/moderate probability
Suitability of cell size	Macrocellular	Microcellular
Radio equipment	Covers only the channels allocated to the cell	Has to cover all possible channels that could be assigned to the cell
Computational effort	Low	High
Call setup delay	Low	Moderate/high
Implementation complexity	Low	Moderate/high
Frequency planning	Laborious and complex	None
Signaling load	Low	Moderate/high
Control	Centralized	Centralized, decentralized, or distributed

5.6.3 Migration to Digital Systems

Analog cellular systems had inherent capacity limitations and also had the corresponding disadvantages of analog communication systems. Migration to digital cellular systems during the early 1990s resulted in benefits for both the service providers and end users by providing additional capacity and feature flexibility. In digital AMPS or IS-136, the same 30 kHz channels employed with AMPS were deployed in TDMA format with six time slots per 30 kHz channel increasing the capacity by a factor of up to six. With full-rate voice coding, each user is allowed access to two time slots so that three users can be accommodated on each AMPS carrier. The frequency planning discussed earlier is equally applicable to such TDMA systems because they use exactly the same carriers as the AMPS systems. However, it has been found that the interference that can be tolerated by digital TDMA systems is much larger than AMPS so that a much tighter reuse ratio could be employed. For instance, for the IS-136 systems, an S_r of 12 dB is sufficient as against 18 dB for AMPS. This further increases the capacity because a reuse factor of $N = 4$ is possible. In GSM, an S_r of 9 dB is sufficient with slow frequency hopping and this enables employing a reuse factor of $N = 3$.

Digital CDMA systems can provide an even larger capacity increase because of the various interference combating capabilities of CDMA. With CDMA, the *same frequencies* can be employed in adjacent cells, thereby increasing the reuse factor to one.

5.7 NETWORK PLANNING FOR CDMA SYSTEMS

CDMA presents some unique features that are not present in traditional TDMA and FDMA systems. In TDMA and FDMA systems, the users operating in one channel are completely isolated from the users operating in other channels. The only interference comes from the fact that the same frequency bands are employed in spatially separated cells, and this interference is the cochannel interference. Of course leakage of signal from adjacent bands, causing adjacent channel interference, is also a factor, but intelligent design can reduce this effect greatly. However, in the case of CDMA, all users are operating on the same frequency channel at the same time, resulting in everyone causing cochannel interference. This problem is reduced on the downlink by employing time synchronized orthogonal codes. On the uplink, a combination of convolutional coding, spreading, and orthogonal modulation are employed to combat the effects of this interference. Network planning in the case of CDMA is far more complicated than in the case of TDMA/FDMA in that sense, but at the same time, using CDMA completely eliminates the concept of conventional frequency reuse because the same frequencies can be deployed in all cells.

Instead of defining an acceptable signal-to-interference ratio, in CDMA it is necessary to define the *quality of the signal* [HAL96]. Usually this is expressed in terms of the acceptable energy per bit to total noise ratio E_b/N_t which results in roughly a 1 percent data frame error rate. The reason for selecting this as a measure is that this frame error rate results in acceptable speech quality at the voice encoder output. The value of E_b/N_t is usually around 6 dB and depends on the speed of the mobile terminal, propagation conditions, the number of multipath signals that can be used for diversity, and so on. The value of N_t depends on the number of interfering signals and the transmit powers of the interfering users. Consequently, power control and thresholds play a very important role in the coverage of a CDMA cell and the soft handoff process associated with it. Details of power control and soft handoff in CDMA are discussed in Chapter 8.

5.7.1 Issues in CDMA Network Planning

Many of the principles that apply to TDMA/FDMA systems also apply to CDMA systems, but there are important differences. For example, the path loss is very similar to TDMA systems in that the signal strength drops roughly as the fourth power of the distance in macrocells and is quite site specific and terrain dependent. The design issues that differ are described next.

5.7.1.1 Managing the Noise Floor

In CDMA, managing the noise floor is very important. If the number of users in a particular area increases beyond that dictated by Eq. (5.12), the system is interference limited, and increasing the transmit power will not benefit any user or set of users as the total interference also increases. It is quite possible that interference from many cells can raise the noise floor to such a level that *holes* may be created in the region where the coding/spreading gain is not sufficient to overcome the interference levels. This is illustrated in Figure 5.20 [HAL96]. If there is an isolated three-sector cell, most of the cell has an E_b/N_t larger than 7 dB, and in the regions where there is soft handoff (where the mobile terminal can connect to more than one BS), the E_b/N_t value from each BS is around 3 dB providing sufficient diversity gain to allow communication. If too many cells are deployed as shown in Figure 5.20, there may be some regions where the noise level is so high that it is impossible to communicate. It is often possible to cover the same area with fewer cells to reduce the total interference levels, and it is usually not a very good idea to cover an area by more than three cells or cell sectors. The problem becomes more severe when terrain plays a role, and in addition to site selection, it will be important to use the down tilt of antennas and the use of minimum radiated power levels to manage the noise floor.

5.7.1.2 Cell Breathing

In CDMA, the boundary of a cell is not fixed and depends on where the E_b/N_t value is reached. For example, consider the uplink E_b/N_t value that is observed at a BS. As the number of traffic channels on the uplink is increased, this value also increases and it is clear from Equation (5.12) that the handoff boundary (where the mobile terminal has to move from one BS to another) shifts closer to the BS. This effect is called cell breathing. In order to ensure that a correct handoff is performed, the transmit power of the pilot channel of the BS (see Chapter 8 for details) must also be reduced so that the forward link handoff boundary is also maintained at the same level as the reverse link boundary. In some cases, cell breathing can have a harmful impact on the system performance, and this should be taken into account while planning the system, either by deploying more cells or offloading capacity to other carriers.

5.7.2 Migration from AMPS to IS-95 Systems

IS-95 CDMA systems typically operate on the same frequency bands allocated to AMPS and digital TDMA systems. A single CDMA channel requires removing

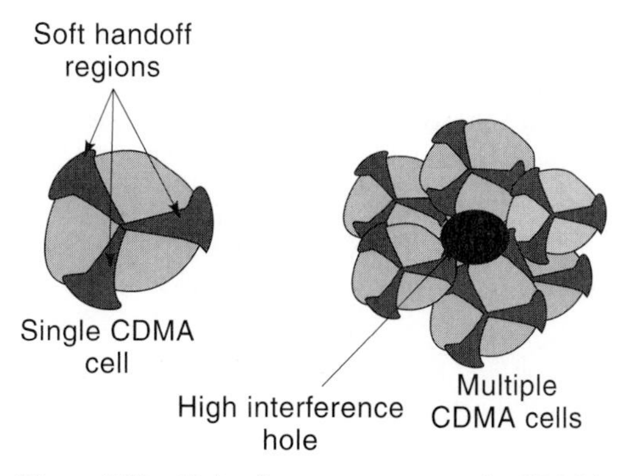

Figure 5.20 Noise floor management in CDMA.

41 contiguous AMPS channels. Because the CDMA carrier requires a different set of signal-to-interference constraints, when service providers are migrating from AMPS to IS-95 systems, care should be taken to minimize the interference between these systems. There are three possible approaches that a service provider may employ to migrate from AMPS to CDMA systems [GAR00]. They are as follows:

1. Two independent systems: one based on AMPS and the other on CDMA
2. An integrated AMPS/CDMA system
3. A partially integrated system with AMPS providing coverage at the fringes and CDMA in the core along with AMPS

The characteristics, advantages, and disadvantages of these approaches are summarized in Table 5.3.

Table 5.3 Approaches to Migrating from AMPS to IS-95

Approach	Characteristics	Advantages	Disadvantages
Independent systems	CDMA uses a separate subset of the spectrum; CDMA may cover a larger area (because of greater capacity)	• Allows independent operation and vendor independence • All digital service everywhere • Smaller number of CDMA BSs can be deployed, reducing capital costs	• Capacity loss due to spectrum segmentation • If there is dumping of analog terminals, the blocking rate for analog subscribers may increase • Operational complexity
Integrated systems	The same service provider provides both CDMA and AMPS over the entire service area	• All digital service everywhere • High-spectral efficiency • Operational simplicity • No need for dual mode phones	• Requires deployment of digital BSs everywhere and may underutilize the available capacity
Partially integrated systems	Part of the system is converted to support both AMPS and CDMA	• CDMA capacity advantage is placed only where it is required • Simpler than the independent approach in terms of operation	• Needs a buffer zone where adjacent AMPS channels are removed and cannot be operated • Digital service will not be available everywhere • Needs dual-mode phones • Voice quality changes are perceived when a handoff is made from CDMA to AMPS

QUESTIONS

5.1 Name any three advantages of an infrastructure topology over an ad hoc topology.

5.2 Compare peer-to-peer and multihop ad hoc topologies.

5.3 Name the five different cell types in the cellular hierarchy and compare them in terms of coverage area and antenna site.

5.4 Why is hexagonal cell shape preferred over square or triangular cell shapes to represent the cellular architecture?

5.5 What are the most popular frequency reuse factors for AMPS, GSM, and IS-95?

5.6 Of the following, what values are possible for a cluster size in a cellular topology? Why? Assume a hexagonal geometry: 8, 21, 23, 30, 47, 61, 75.

5.7 Name five architectural methods that are used to increase the capacity of an analog cellular system without increasing the number of antenna sites.

5.8 Explain why band splitting is not used in 2G cellular networks.

5.9 Explain why reuse-partitioning can be used for both 1G and 2G cellular networks.

5.10 What is the difference between band-splitting and underlay-overlay techniques for increasing the capacity of cellular networks? What is the effectiveness of each in improving the capacity? How do they differ from one another?

5.11 Explain how smart antennas can improve the capacity of a cellular network.

5.12 Explain why in fixed channel allocation techniques neighboring frequency channels are assigned to different cells.

5.13 How are the high interference holes in CDMA deployment created?

5.14 Compare FCA and DCA frequency assignment techniques.

PROBLEMS

5.1 Assume that you have six-sector cells in a hexagonal geometry. Draw the hexagonal grid corresponding to this case. Compute S_r for reuse factors of 7, 4, and 3. Comment on your results.

5.2 Assume that we wanted to deploy an analog FM AMPS system with half band of 15 kHz rather than the existing 30 kHz. Also assume that in analog FM, the carrier-to-interference ratio (C/I) requirement is inversely proportional to the square of the bandwidth (4 time increase in C/I for dividing the band into two).

a. What is the required C/I in dB for the 15 MHz channel if the required C/I for the 30 kHz systems is 18 dB?

b. Determine the frequency reuse factor N needed for the implementation of this 15 kHz per user analog cellular system.

c. If a service provider had a 12.5 MHz band in each direction (up-link and down-link) and it would install 30 antenna sites to provide its service, what would be the maximum number of simultaneous users (capacity) that the system could support in all cells? Neglect the channels that are used for control signaling.

d. If we use the same antenna sites but a 30kHz per channel system with $N = 7$ (instead of the 15 kHz system) what would be the capacity of the new system?

5.3 We have an installed cellular system with 100 sites, a frequency reuse factor of $N = 7$, and 500 overall two-way channels:

a. Give the number of channels per cell, total number of channels available to the service provider, and the minimum carrier-to-interference ratio (C/I) of the system in dB.

b. To expand the network, we decide to create an underlay-overlay system where the new system uses a frequency reuse factor of $K = 3$. Give the number of cells assigned to inner and outer cells to keep a uniform traffic density over the entire coverage area.

5.4 Repeat Problem 5.3 with $N = 12$.

5.5 **a.** What is the number of RF channels per cell in the GSM network described in Chapter 7? The frequency reuse factor of the GSM is $K = 4$.

b. What is the maximum number of simultaneous users per cell in this system?

c. Assume that we want to replace this GSM system with an IS-95 spread spectrum system in the same frequency bands. What is the maximum number of users per cell? Assume an ideal power control and use the practical considerations for the IS-95 system.

5.6 **a.** Determine the carrier-to-interference ratio, in dB, of a cellular system with a frequency reuse factor of $K = 7$.

b. Repeat (a) for $K = 4$.

c. If we consider multi-symbol QAM modulation for the digital transmission of the information, how many more bits per symbol can be transmitted with $K = 4$ as compared with $K = 7$ architecture?

CHAPTER 6

WIRELESS NETWORK OPERATION

6.1 Introduction

6.2 Mobility Management

6.2.1 Location Management
6.2.2 Handoff Management
6.2.3 Mobile IP

6.3 Radio Resources and Power Management

6.3.1 Power Control
6.3.2 Power Saving Mechanisms in Wireless Networks
6.3.3 Energy Efficient Designs
6.3.4 Energy Efficient Software Approaches
6.3.5 Implementation of Radio Resource and Power Management: A Protocol Stack Perspective

6.4 Security in Wireless Networks

6.4.1 Security Requirements for Wireless Networks
6.4.2 An Overview of Network Security
6.4.3 Identification Schemes

Appendix 6A The Diffie-Hellman (DH) Key Exchange Protocol

Appendix 6B Nonrepudiation and Digital Signatures

Questions

Problems

6.1 INTRODUCTION

In the previous chapters, we provided the principles of radio propagation, wireless modem design, wireless access methods, and deployment of cellular systems. These issues were all related to the air interface design and physical characteristics of the wireless medium which are needed to connect a MS to a BS or access point that then connects to the backbone wired networks. To support mobile operation, the backbone network has to add several new functionalities that do not exist for the wired terminal operations, because they are not usually necessary. These functionalities include mobility and location management, radio resource and power management, and security. The very nature of mobile communications implies that the MS is constantly changing locations, warranting a need for tracking the mobile and restructuring existing connections as it moves. Mobility and location management handle the operations required for these purposes. As we have discussed earlier, bandwidth is a scarce resource as also is the battery power of the MS. Also we have seen that a consequence of employing a cellular topology to "multiply" the bandwidth results in wireless networks that are limited by interference. Radio resource and power management schemes are used to address operational aspects related to reducing interference, improving battery life and handling the scarce radio resources. Wireless communications are inherently vulnerable to eavesdropping and need security features that are often not very important in wired networks because of the physical lack of access to the medium. A number of algorithms and methodologies have been implemented in different wireless networks to implement these three features. In this chapter we provide an overview of the techniques that are used for implementation of these three features in wireless networks.

6.2 MOBILITY MANAGEMENT

The primary advantage of wireless communications is the ability to support tetherless access to a variety of services, whether voice oriented as in the case of cellular radio and PCS or data services and access to the Internet as in the case of mobile data networks and wireless LANs. Tetherless access implies that the user has the ability to move around while connected to the network and continuously possesses the ability to access the services provided by the system to which the user is attached. This leads to a variety of issues because of the way in which most communications networks operate. First, in order for any message to reach a particular destination, there must be some knowledge of where the destination is (location) and how to reach the destination (route). In static networks, where the end terminals are fixed, the physical connection (wire or cable) is sufficient to indicate the destination. In wireless networks, where the terminal may be anywhere, there must be a mechanism to locate the terminal in order to deliver the communication to it. *Location management* refers to the activities a wireless network should perform in order to keep track of where the MS is. As discussed in Chapter 5, the most common wireless topology uses multiple cells to provide coverage over a larger area.

The location of the MS must be determined such that there is a knowledge of which point of access (BS or AP) is serving the cell in which the MS is located. Second, once the destination is determined, it is not enough to assume that the destination will remain at the same location with time. When a MS moves away from a BS, the signal level from the current BS degrades, and there is a need to switch communications to another BS. *Handoff* is the mechanism by which an ongoing connection between a MS and a correspondent terminal is transferred from one point of access to the fixed network to another. *Handoff management* handles the messages required to make the changes in the fixed network to handle this change in the location during an ongoing communication. Location and handoff management together are commonly referred to as *mobility management* [AKY98].

6.2.1 Location Management

Location management involves tracking of the location of the MS, as it moves, for delivery of voice or data communication to it. In the case of voice networks, when a call is made to a mobile number, a dedicated channel has to be set up from the calling party to the called party for the conversation to proceed. For this, a circuit has to be set up over the fixed part of the network, and a pair of radio channels have to be allocated to the MS for the voice conversation. The MS has to be located to set up this dedicated channel. Note that this is before the actual conversation takes place. If the MS moves during the course of a conversation, the steps taken to handle the continuity of conversation is called handoff and handoff management. In the case of data networks, packets are addressed to a destination terminal. Routers, within the data network, will use the destination address to deliver the packet. The address information is usually hierarchical and fixed, which means that the address points toward a physical location. If the terminal is fixed, the packet is routed appropriately to the physical location of the terminal. In the case of a MS, some steps are required to determine where it is before the packet is routed to it. Another important functionality of location management is to determine the status of the MS. If the MS is switched off, the network should be aware that it is unreachable so that appropriate action may be taken depending on the service requested. For example, short messages may be stored on a server for later delivery.

Location management in general has three parts to it: location updates, paging, and location information dissemination. *Location updates* are messages sent by the MS regarding its changing points of access to the fixed network. These updates may have varying granularity and frequency. Each time the MS makes an update to its location, a database in the fixed part of the network has to be updated to reflect the new location of the MS. Whether there is a change in the location, the update message will be transmitted over the air and over the part of the fixed network. Because the updates are periodic, there will be some uncertainty in the location of the MS to something around a group of cells. In order to deliver an incoming message to the MS, the network will have to *page* the MS in such a group of cells. The paged terminal will respond through the point of access that is providing coverage in its cell. The response will enable the network to locate the terminal to within the accuracy of the cell in which it is located. Procedures can then be initiated to either deliver the packet or set up a dedicated communications channel for voice conversation. In order to initiate paging however, the calling party or the incoming message should trigger a location request from some fixed network entity. The fixed network entity will then access some kind of database that will contain the most current location information related to the particular MS and use this information to generate the paging request, as well as deliver the message or set up a channel for the voice call. *Location information dissemination* refers to the procedures that are required to store and distribute the location information related to the MSs serviced by the network.

The basic issue in location management is the trade-off between the cost of the nature, number and frequency of location updates, and the cost of paging [WON00]. If the location updates are too frequent and the incoming messages few, the load on the network becomes an unnecessary cost, both in terms of the usage of the scarce spectrum, as well as network resources for updating and processing of the location updates. If the location updates are few and infrequent, a larger area and thus a larger number of cells will have to be paged in order to locate the MS. Paging in all the cells where the mobile is not located is a waste of resources. Also depending on the way paging is performed, there may be a delay in the response of the MS because the paging in the cell in which it is located might be performed much later than the cell in which it last performed its location update. For applications such as voice calls, this will result in unnecessary call dropping because the MS did not respond in a reasonable time. In the case of data networks, depending on the type of mobility management scheme implemented, packets might simply be dropped if the MS is not located correctly.

As we discussed earlier in this section, location management consists of three activities—location updates, paging, and location information dissemination. There are different types of location management schemes that employ a variety of location update mechanisms, paging schemes, and dissemination architectures. We discuss these in the following sections.

6.2.1.1 Location Update Algorithms

Location update algorithms are usually of two types—static and dynamic [WON00]. In static location updates, the topology of the cellular network decides when the location update needs to be initiated. In dynamic location updates, the mobility of the user, as well as the call patterns, is used in initiating location updates.

In the most common form of static location updates, which is the case in most cellular networks, a group of cells is assigned a *location area* (LA) identifier, as shown in Figure 6.1. Each BS in the LA broadcasts this identification number periodically over some control channel. An MS is required to continually listen to the control channel for the LA identifier. When the identifier changes, the MS will make an update to the location by transmitting a message with the new identifier to the databases containing the location information. If there is an incoming message, paging is performed in the group of cells corresponding to the location identifier stored in the database. The MS usually responds (unless the location area identifier has changed in the meanwhile), and the communication can be delivered successfully.

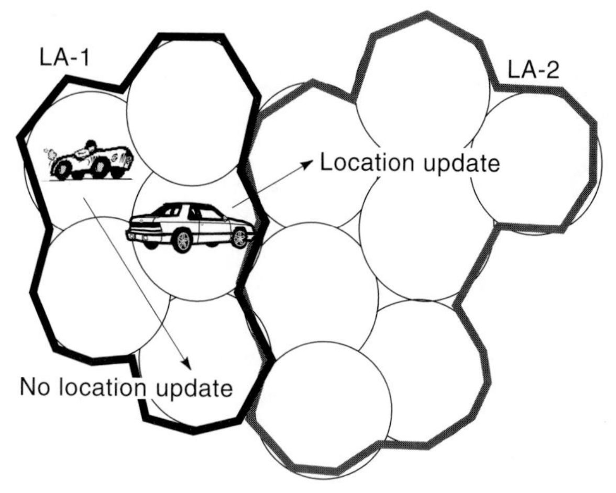

Figure 6.1 Location area (LA) based location updates.

Example 6.1: Location Update Mechanism in GSM

In GSM, an LA identity, also called a paging area, is used for location updates. An LA usually consists of a group of cells controlled by a base station controller (BSC). An MS will perform a location update under three circumstances: (1) Upon powering up, it compares the location area identity it previously had recorded with the one currently being broadcast. If the two location area identities are different, a location update is performed. (2) When the MS crosses the boundary of a location area, it performs a location update. (3) After a period of time predetermined by the network, a location update is performed to ensure that the MS is available. In case (2), the MS detects a change in LA because the BS broadcasts the location area identity, which the MS is required to monitor and compare with the stored value. In case (3), the update mechanism might be costly if the MS does not leave an LA for long periods of time.

The primary problem with the static LA identifier approach is that if an MS is frequently crossing the boundary of two LAs as shown in Figure 6.2, there will be a *Ping-Pong* effect of continually switching between two LAs. A solution to this problem is to employ a dwell timer that persists without a location update for sometime to ensure that the location update is worthwhile. Similar problems and associated algorithms to resolve the problem are encountered during handoff which we will see later on in this chapter.

A variety of other static location update schemes is possible. This includes distance-based—where the location update is performed after crossing a certain number of cells; timer-based—where a location update is performed after a certain time elapses; and variations on these two schemes which take into account signaling load on control channels, and the location and velocity of the MS [WON00].

Examples of dynamic location update schemes are the *state-based* and *user-profile-based* location updates. In the state-based location update scheme, the MS makes a decision on when to perform an update based on its current state information. The state information can include several metrics that include the time

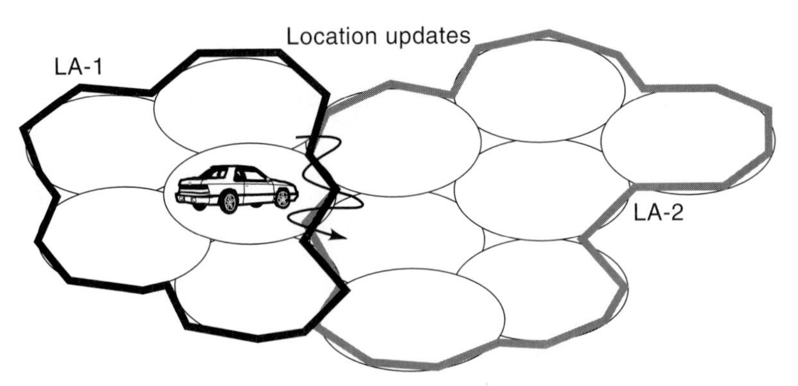

Figure 6.2 Location area ping-pong effect.

elapsed, the distance traveled, the number of LAs crossed, and the number of calls received that could be changed based on the user's mobility and call patterns. The user-profile-based location update schemes maintain a sequential list of LAs that the MS may be located in at different points of time based on the history of the MS. A detailed, comparative evaluation of several of these schemes has been discussed in [WON00].

6.2.1.2 Paging Schemes

Paging is broadcasting a message in a cell or a group of cells to elicit a response from the MS for which a call or message is incoming. Transmitting the page in only the cell in which the mobile terminal is located, which makes the most accurate location estimate, can reduce the cost of paging. The problem with paging in the most accurate cell location is that it is quite impossible to determine location accurately, especially if the location update cost has to be kept low.

Example 6.2: Blanket Paging in GSM

Blanket paging refers to paging the MS in all the cells within an LA simultaneously. This means that, if the LA update is correct, in the very first paging cycle, the MS will receive a paging request and respond to it. The advantage of this system, employed in GSM, is that the delay of the response to paging is kept at a minimum. The disadvantage is that paging has to be done in several cells—all of which have the same LA identity.

Another strategy for paging is to use "closest-cells first" approach. Here the cell where the MS was last seen is paged first followed by subsequent *rings* of cells that are equidistant from this cell in each paging cycle. If there are delay constraints, as in the case of voice calls, several rings may be polled simultaneously in a paging cycle. In general, if the first location estimate is not correct, the next page should be performed so that the probability of locating the mobile is the next largest and so on. The paging is performed in the area corresponding to the last location update, and then subsequent pages are performed in most likely locations based on parameters that may include past history and distance. A timer is used to declare the MS as unreachable in a particular paging cycle. This is sometimes called *sequential paging*. Results indicate that blanket polling provides the lowest delay at small load, whereas sequential paging can sustain a higher paging request rate, especially when there are several incoming calls to a certain area. The behavior pattern of mobile terminals, as well as the user profile, may also be employed in paging algorithms in a manner similar to location update mechanisms.

6.2.1.3 Location Information Dissemination

When there is an incoming packet to the MS, there is the need for at least one fixed network entity, whose location and address is known, which can be reached to obtain information about the MS. In general, this is often referred to as an *anchor*. The anchor has some information regarding the location and routing information of the MS. If a single anchor is used for all MSs, not only is the load on this entity increased making it a bottleneck for communications, but also it makes a failure point that can result in the collapse of the network. Usually multiple anchor points are employed. What we describe further is in general terms as to how network entities and databases are employed for location, and it can be also extended to other applications such as handoff management. Specific implementations are different.

Every MS is associated with a *home network* and a *home database*. The home database keeps track of the profile of the MS—such as the mobile identification, authentication keys, subscriber profile, accounting, and location. The location of the mobile is maintained in terms of a *visiting network,* where the MS is located, and a *visiting database,* which keeps track of the MSs in its service area. The home and visiting databases communicate with each other to authenticate and update each other about the MS. We will see more of this in the section on handoff management.

Example 6.3: Location Information Dissemination in GSM

In GSM, the home and visiting databases are called *home location register* (HLR) and *visiting location register* (VLR), respectively. When the MS observes a change in the LA identity, it transmits a location update message through the BS to a MSC. The MSC contacts its VLR with the location update. The VLR does nothing if it serves both the old and new LA. If the VLR has no information about the mobile terminal, it contacts the HLR of the MS via a location registration message. The HLR authenticates and acknowledges the location registration, updates its own database, and sends a message to the old VLR to cancel the registration there.

Example 6.4: Call Delivery in GSM

When a call is made to a mobile telephone number, the *anchor* entity contacts the MSC associated with the HLR of the mobile terminal. The HLR contacts the VLR that is associated with the MS and enables call setup. A detailed description of the call delivery in GSM is presented in Chapter 7.

6.2.1.4 Emerging Issues in Location Management

Location management has several potential issues associated with it [AKY98], [WON00]. Primarily, there is a lot of research going on in database architectures for next-generation wireless networks. Access to the database and management of queries is very important in order to reduce delay and maintain quality. To reduce the load on a centralized database (such as an HLR), local caches of the mobile terminal information can be maintained. Similar strategies are being considered in Mobile IP, as we will see later on. Alternative location update strategies and paging algorithms are being investigated. An important factor that influences the performance of all these techniques is traffic modeling, which can accurately represent the nature of incoming calls, paging requests, and movement of the MS.

6.2.2 Handoff Management

Handoff management involves the entire gamut of issues and actions that are required to handle an ongoing connection when a mobile terminal moves from the coverage of one point of access to another. Handoff [POL96], [TRI97], [TRI98] is extremely important in any mobile network because of the default cellular architecture employed to maximize spectrum utilization. Handoff, in the case of cellular telephony, involves the transfer of a voice call from one BS to another. In the case of WLANs, it involves transferring the connection from one AP to another. In hybrid networks, it will involve the transfer of a connection from a BS to another, from an AP to another, between a BS and an AP, or vice versa.

For a voice user, handoff results in an audible click interrupting the conversation for each handoff [POL96], and because of handoff, data users may lose packets and unnecessary congestion control measures may come into play [CAC95]. Degradation of the signal level is, however, a random process, and simple decision mechanism such as those based on signal strength measurements result in the *Ping-Pong effect.* The Ping-Pong effect refers to several handoffs that occur back and forth between two BSs. This has a severe toll on both the user's quality perception and the network load. A way of eliminating the Ping-Pong effect is to persist with a BS for as long as possible. However, if handoff is delayed, weak signal reception persists unnecessarily, resulting in a lower voice quality, increasing the probability of call drops and/or degradation of QoS. Consequently, more complex algorithms are needed to decide on the optimal time for handoff. Handoff also involves a sequence of events in the backbone network that include rerouting the connection and reregistering to the new point of access, which are additional loads on the network traffic. Handoff has an impact on traffic matching and traffic density for individual BSs (because the load on the air-interface is transferred from one point of access to another). In the case of random access techniques employed to access the air interface, or in the case of CDMA, moving from one cell to another impacts QoS in both cells because throughput and interference depend on the number of terminals competing for the available bandwidth.

Although significant work has been done on handoff mechanism in circuit switched mobile networks [POL96], [TRI98], there is not much of literature available for packet switched mobile networks. Performance measures such as call blocking and call dropping probabilities are applicable only to real-time traffic and may not be suitable for bursty traffic which exists in client-server type of applications. When a voice call is in progress, allowed latency is very limited and resource allocation has to be guaranteed, and although occasionally some packets may be dropped and moderate error rates are permissible, retransmissions are not possible, and connectivity has to be maintained continuously. On the other hand, bursty data traffic by definition needs only intermittent connectivity and can tolerate greater latencies and employ retransmission of lost packets. In such networks handoff is warranted only when the terminal moves out of coverage of the current point of attachment or the traffic load is so high that a handoff may result in greater throughput and utilization.

There are a variety of issues related to handoff. In particular we can consider handoff as consisting of two different steps as shown in Figure 6.3. In the first step, the handoff management process determines that a handoff is required (handoff decision and initiation). In the second step, the rest of the network is made aware of the handoff, and the connection is restructured to reflect the new location of the MS. Note that there is an *anchor* in the fixed part of the network that must be involved in the handoff management process in a manner similar to location management. Several issues arise during the handoff management process.

As shown in Figure 6.4, these issues are divided into two categories: architectural issues and handoff decision time algorithms. Architectural issues are those related to the methodology, control, and software/hardware elements involved in rerouting the connection. Issues related to the handoff decision time algorithms are the types of algorithms, metrics used by the algorithms, and performance evaluation methodologies.

6.2.2.1 Architectural Issues in Handoff

Handoff procedures involve a set of protocols to notify all the related entities of a particular connection that a handoff has been executed and that the connection has to be redefined. In data networks, the MS is usually registered with a particular point of attachment. In voice networks, an idle MS would have selected a

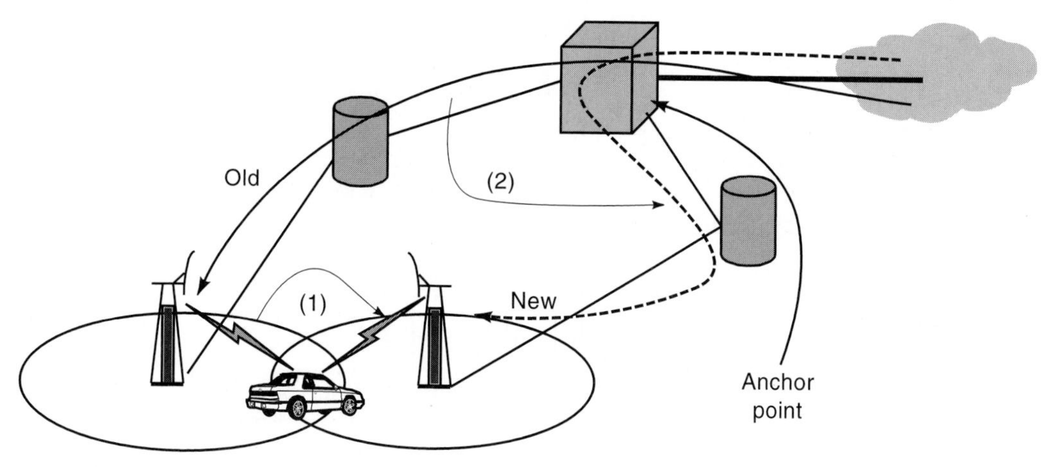

Figure 6.3 Two basic actions during handoff.

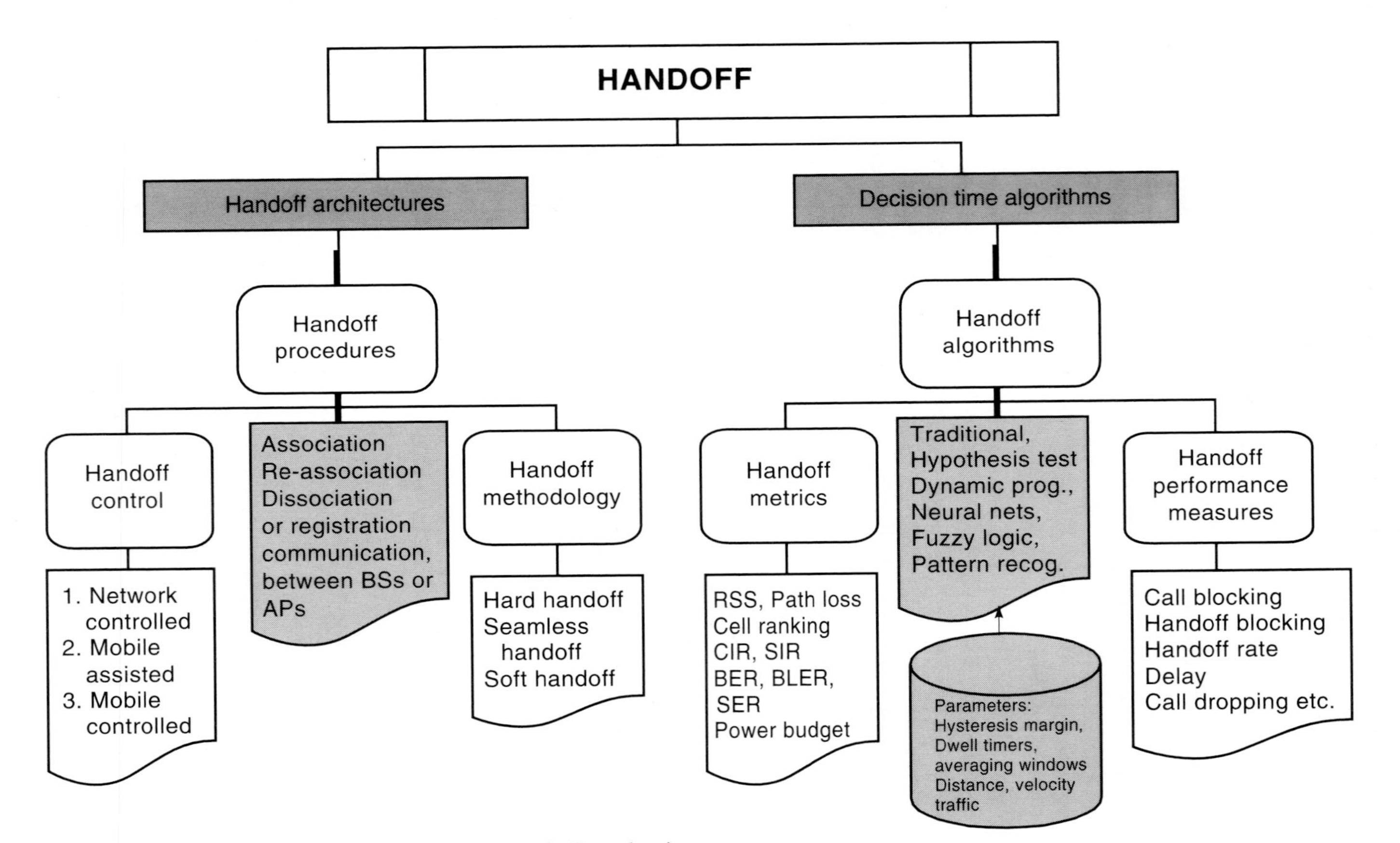

Figure 6.4 Important issues involved in the handoff mechanism.

particular BS that is serving the cell in which it is located. This is for the purpose of routing incoming data packets or voice calls appropriately. When the MS moves and executes a handoff from one point of attachment to another, the old serving point of attachment has to be informed about the change. This is usually called dissociation. The MS will also have to reassociate itself with the new point of access to the fixed network. Other network entities that are involved in routing data packets to the MS or in switching voice calls have to be aware of the handoff in order to seamlessly continue the ongoing connection or call.

Example 6.5: Registration in CDPD

The mobile terminal in CDPD is called the *mobile end system* (M-ES). It announces its presence to the network by sending a message called End_System_Hello to a mobile data intermediate system—MD-IS (which is the network entity controlling a group of cells). Registration is completed when the MD-IS responds with a confirmation. The MD-IS serving the cell has to contact the *home* MD-IS (which controls the home database) to authenticate the M-ES and also update it about the location of the M-ES.

Depending on whether a new connection is created before breaking the old one, handoffs are classified into a hard and seamless handoff. Hard handoff occurs when the mobile terminal completely breaks connection with the old BS before connecting to the new BS and synchronizing itself to it. Seamless handoff refers to the case where the mobile terminal sets up a traffic channel with the new BS before breaking off from the old BS. However, communication is possible only through one BS at a time. In CDMA, the existence of two simultaneous connections during handoff results in soft handoff [TEK91]. Soft handoff is discussed in more detail in Chapter 8.

The decision mechanism or *handoff control* could be located in a network entity (as in cellular voice) or in the mobile terminal (as in mobile data and WLANs) itself. These cases are called network controlled handoff (NCHO) and mobile controlled handoff (MCHO), respectively. In GPRS, information sent by the mobile terminal can be employed by the network entity in making the handoff decision. This is called the MAHO.

Example 6.6: Handoff Control in Different Systems

In AMPS, the analog 1G standard, handoff decision is NCHO. The mobile telephone switching office uses the RSS measurements from an MS at different BSs to initiate handoff. In the case of IEEE 802.11 LANs, the mobile station controls handoff decision (MCHO). It monitors the *beacon* of several APs to decide which AP to connect to. The network has no role in deciding when to make a handoff.

In any case, the entity that decides on the handoff uses some metrics, algorithms, and performance measures in making the decision. The measurement and handoff decision are usually part of the radio resource management procedures. However, we look at handoff decision mechanisms in this section to keep the procedures for handoff together.

6.2.2.2 Handoff Decision Time Algorithms

Several algorithms are being employed or investigated to make the correct decision to handoff [POL96], [TRI98]. Traditional algorithms employ thresholds to compare the values of *metrics* from different points of attachment and then decide on when to make the handoff. A variety of metrics have been employed in mobile voice and data networks to decide on a handoff.

Primarily, the RSS measurements from the serving point of attachment and neighbouring points of attachment are used in most of these networks. Alternatively or in conjunction, the path loss, carrier-to-interference ratio (CIR), signal-to-interference ratio (SIR), BER, block error rate (BLER), symbol error rate (SER), power budgets, and cell ranking have been employed as metrics in certain mobile voice and data networks. In order to avoid the Ping-Pong effect, additional parameters are employed by the algorithms such as hysteresis margin, dwell timers, and averaging windows. Additional parameters (when available) may be employed to make more intelligent decisions. Some of these parameters also include the distance between the MH and the point of attachment, the velocity of the MH, traffic characteristics in the serving cell, and so on.

Traditional handoff algorithms are all based on the RSS or received power P. Some of the traditional algorithms [POL96] are as follows:

1. *RSS:* The BS whose signal is being received with the largest strength is selected (choose BS B_{new} if $P_{new} > P_{old}$).
2. *RSS plus Threshold:* A handoff is made if the RSS of a new BS exceeds that of the current one and the signal strength of the current BS is below a threshold T (choose B_{new} if $P_{new} > P_{old}$ and $P_{old} < T$).
3. *RSS plus Hysteresis:* A handoff is made if the RSS of a new BS is greater than that of the old BS by a hysteresis margin H (choose B_{new} if $P_{new} > P_{old} + H$).
4. *RSS, Hysteresis, and Threshold:* A handoff is made if the received signal strength of a new BS exceeds that of the current one by a hysteresis margin H and the signal strength of the current BS is below a threshold T (choose B_{new} if $P_{new} > P_{old} + H$ and $P_{old} < T$).
5. *Algorithm plus Dwell Timer:* Sometimes a dwell timer is used with the other algorithms. A timer is started at the instant when the condition in the algorithm is true. If the condition continues to be true until the timer expires, a handoff is performed.

Figure 6.5 illustrates these algorithms in the case of a mobile terminal traveling between two BSs along a straight line. Note that the RSS is not smooth as shown in this figure but more random as illustrated in Figure 6.6.

Recently other techniques are emerging such as hypothesis testing [LIO94], dynamic programming [REZ95], and pattern recognition techniques [COX96] based on neural networks or fuzzy logic systems [TRI97] (for an excellent survey of various algorithms, see [POL96], [TRI97]). These complicated algorithms are necessitated by the complexity of the handoff problem, especially in hybrid data or voice networks. The mobile terminal has to monitor the air for wireless data services that may be available for attachment. As an example, consider an MS that

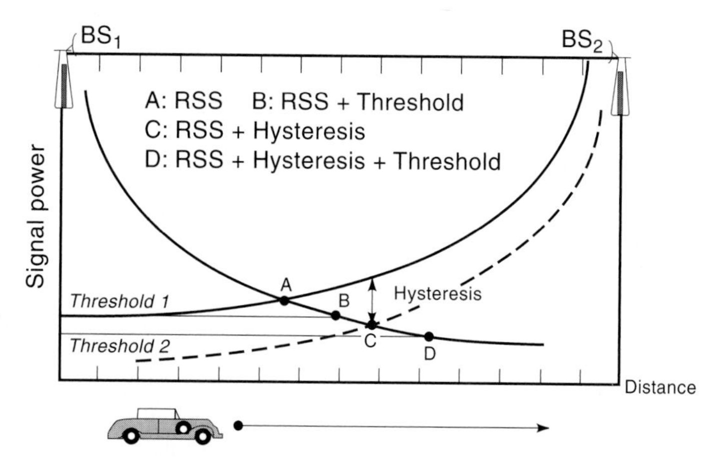

Figure 6.5 Traditional handoff algorithms using RSS thresholds and hysteresis.

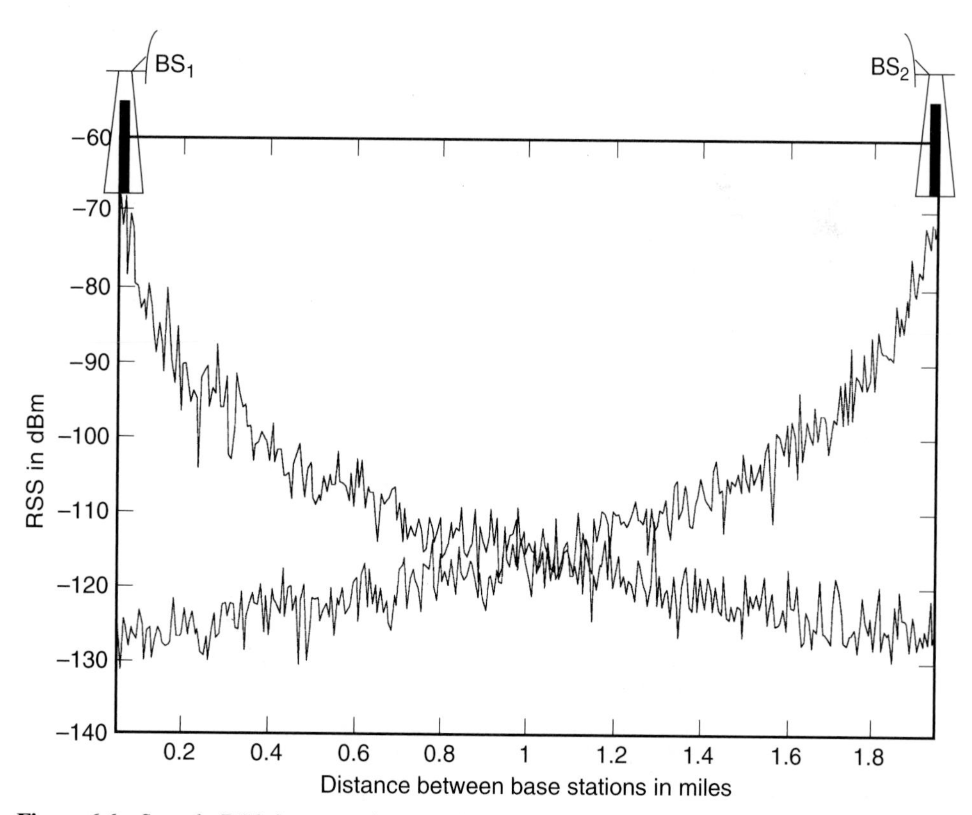

Figure 6.6 Sample RSS from two base stations as seen by an MS traveling in a straight line between them.

could connect to either an 802.11 WLAN AP connected to a LAN or a GPRS BS subsystem (BSS) connected to a backbone GPRS network. There must be a mechanism or algorithm within the mobile terminal that will enable it to choose the best available service and switch to this service as soon as it is available. For example, the mobile terminal must be able to switch from the GPRS service to the WLAN AP as soon as it detects the availability of a connection to an access point. Most of the emergent algorithms are in their nascent stages, and they have been analyzed or simulated only for voice networks and only for extremely simple scenarios.

The performance of handoff algorithms is determined by their effect on certain performance measures. Most of the performance measures that have been considered such as call blocking probability, handoff blocking probability, delay between handoff request and execution, call dropping probability, and so on are related to voice connections. Handoff rate (number of handoffs per unit time) is related to the ping-pong effect, and algorithms are usually designed to minimize the number of unnecessary handoffs. Although minimizing the handoff rate is important in mobile data networks, other issues include throughput maximization and maintaining QoS guarantees during and after handoff. However, these issues have not received sufficient attention in the literature.

6.2.2.3 Generic Handoff Management Process

In this section, we show the different messages and processes that are required for handoff management in a generic wireless network. As in the case of location management, the specific implementations will be different. We consider some specifics in later chapters.

In Figure 6.7, a generic architecture is shown for the handoff management process. There are two types of databases in the network; the home database that also acts as the anchor and the visiting database. Every mobile terminal is registered with a home database that keeps track of the profile of the mobile terminal.

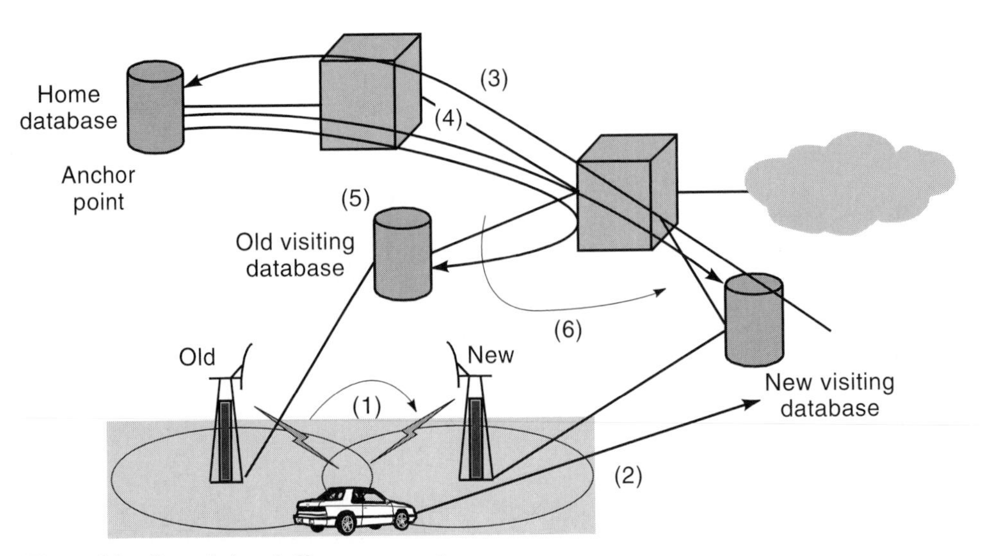

Figure 6.7 Generic handoff management process.

The visiting database keeps track of the mobile terminals in its service area. The home and visiting databases communicate with each other during the handoff management process.

1. In the first step, a decision is made to handoff, and handoff is initiated. This decision, as discussed earlier, may be made in the network by some entity with or without the help of the mobile terminal, or at the mobile terminal. For this purpose the decision time algorithms are employed.
2. The mobile terminal registers with the "new" visiting database via a handoff announcement message. This is the first information to a network entity in the case of a mobile-controlled handoff. In the case of a network controlled or mobile assisted handoff, the new visiting database may already be aware or expecting this message.
3. The new visiting database communicates with the home database to obtain subscriber profile and for authentication. This is the first information exchange between network entities about the changed location of the mobile in MCHO. In the case of MAHO or NCHO, these entities may already be in communication.
4. The home database responds to the new visiting database with the authentication of the mobile. If the mobile is authenticated, in the case of circuit-switched connections, a pair of traffic channels that might be kept ready is allocated to the mobile terminal for continuing the conversation. In the case of packet data traffic, no such dedicated channels are required because the traffic is bursty. The two databases are updated for delivering new messages that may arrive to the mobile terminal. The new visiting database includes the mobile terminal in its list of terminals that are being serviced by it.
5. The home database sends a message to the old visiting database to flush packets intended for and registration information related to the mobile terminal. This is because packets that may have been routed to the old visited network while the mobile was making a handoff need to be dropped or redirected, and the old visiting database needs to clear resources it had maintained for the mobile terminal because they are no longer required.
6. The old visiting database flushes or redirects packets to the new visiting database and removes the mobile terminal from its list.

Each of these steps is important in order to correctly, securely, and efficiently implement handoff and release resources that are otherwise not used in the system.

Mobility management procedures have details that are specific to the respective systems. They need some description of the network entities because the nomenclature for the databases and the controlling entities are different with different functionality. Descriptions of mobility management procedures are provided in subsequent chapters where individual technologies are described. There are other architectural issues such as handoff between channels in a cell (intracell handoff), handoff between two BSs associated with the same database (intercell intradomain handoff), and different databases (intercell interdomain handoff), and so on. These issues are also discussed when we describe specific technologies later on. Also procedures are adopted when an MS returns to a cell from which it had been handed off to simplify the connections in the backbone.

In the following section, we consider mobile IP as a specific handoff management scheme that has two drawbacks—it does not specify either the first step (handoff decision and initiation) or the last step (flushing and redirecting data). These are technology specific as far as mobile IP is considered.

6.2.3 Mobile IP

IP, which is the most popular network layer protocol for data networks, was not designed with wireless or mobile networks in view. Mobile IP tries to address this issue by creating an "anchor" for a mobile host that takes care of packet forwarding and location management. Mobile IP [PER97] is simplified because IP packets do not need mechanisms to set up dedicated bandwidth or channels as in the case of circuit switched connections. However, it solves a different kind of problem that IP created when the terminals were mobile. The IP address is used for dual purposes—for routing packets through the Internet and also as an endpoint identifier for applications in end-hosts. The connections in an IP network use *sockets* to communicate between clients and servers. A socket consists of the following tuple: `<source IP address, source port, destination IP address, destination port>`. A TCP connection cannot survive any address change because it relies on the socket to determine a connection. However, when a terminal moves from one network to another, its address changes. This is because the Internet uses domain names that are converted to an IP addresses. A packet addressed to one IP address gets routed to the *same place* always because the IP address also points to the location of a physical network.

An IP-mobility working group of the Internet Engineering Task Force (IETF) is in charge of activities related to Mobile IP. Several standards and requests for comments (RFCs) related to mobile IP are available [IETweb]. The basic design criteria for mobile IP were (a) compatibility with existing network protocols, (b) transparency to higher layers (TCP through application) and to the user, (c) scalability and efficiency in terms of not requiring a great deal of additional traffic or network elements, and (d) security due to changing locations of the mobile node (MN).

An MN is a terminal that can change its location and thus its point of attachment. The partner for communication is called the *correspondent node* (CN) that can be either a fixed or a mobile node. The IP network where the MN resides is called the *home network,* and the IP network where the MN is visiting is called the *foreign network*. The *home address* of an MN is a long-term IP address assigned to the MN that is part of the home IP network. It remains unchanged regardless of where the MN is, and it is used for domain name system (DNS) determination of the MN's IP address. The *care-of address* (COA) is an IP address in the foreign network that is the reference pointer to the MN when it is visiting the foreign network. The *home agent* (HA) is the anchor in the home network for the MN. All packets addressed to the MN reach the HA first unless the MN is already in its home network. A *foreign agent* (FA) (only in the case of IPv4) acts as the reference point in the foreign network for the MN. The COA is usually the IP address of the FA. The MN can act as its own FA, in which case it is called a *colocated* COA.

6.2.3.1 Location Management in Mobile IP

Location management in Mobile IP is achieved via a registration process and the so-called agent advertisement. Foreign agents and home agents periodically "advertise" their presence using *agent advertisement* messages. The same agent may act as both an HA and an FA mobility extension to ICMP messages which are used for agent advertisements. The messages contain information about the COA associated with the FA, whether the agent is busy, whether minimal encapsulation is permitted, whether registration is mandatory, and so on. The agent advertisement packet is a broadcast message on the link. If the MN gets an advertisement from its HA, it *must* deregister its COAs and enable a gratuitous address resolution protocol (ARP). If an MN does not "hear" any advertisement, it must solicit an agent advertisement using ICMP. The entire connection search flow is shown in Figure 6.8.

Once an agent is discovered, the MN performs either a registration or deregistration with the HA, depending on whether the discovered agent is an HA or an FA. The MN sends a *registration request,* using UDP to the HA through the FA (or directly if it is a colocated COA). The HA creates a *mobility binding* between the MN's home address and the current COA that has a fixed lifetime. The MN should reregister before the expiration of the binding. A *registration reply* indicates whether the registration is successful. A rejection is possible by either the HA or FA for such reasons as insufficient resources, the HA is unreachable, there are too many simultaneous bindings, for failed authentication, and so on. If an MN does

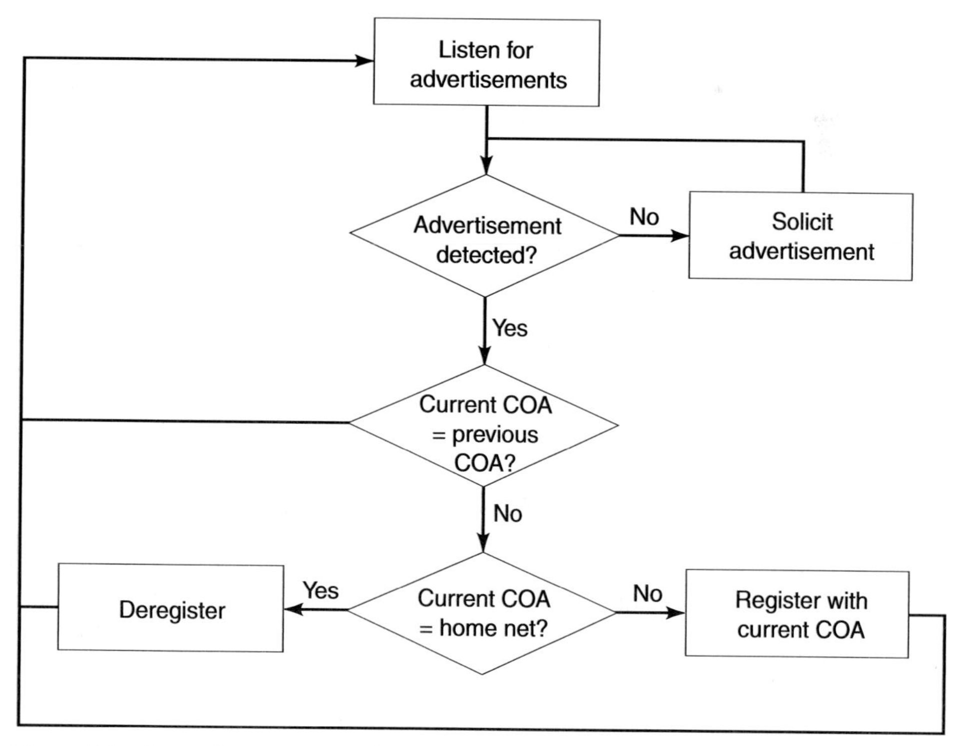

Figure 6.8 Agent discovery procedure.

not know the HA address, it will send a broadcast registration request to its home network called a *directed broadcast*. The response to this request is a reject by every valid HA. The MN uses one of the HA addresses in the reject message to make a valid registration request. The HA and FA maintain lists of MNs in what we can relate to as the home and visiting databases. Upon a valid registration, the HA creates an entry for an MN that has the MN's care of address, an identification field, and the remaining lifetime of the registration. Each foreign agent maintains a visitor list containing the following information: link layer address of the MN, MN's home IP address, UDP registration request source port, HA IP address, an identification field, the registration lifetime, and the remaining lifetime of pending or current registration.

6.2.3.2 Handoff Management in Mobile IP

Mobile IP enables datagrams addressed to the MN at the home address to be delivered wherever the MN is. As shown in Figure 6.9, the CN transmits a datagram to the MN that is routed to MH's home network as usual in step (1). The HA intercepts the packet, encapsulates and tunnels it to FA in step (2). The FA decapsulates and forwards the packet to the MN in step (3). Packets from the MN to the CN are sent as usual (4). This procedure is called triangle routing.

In order to intercept packets addressed to the MN, the HA performs a proxy ARP on behalf of the MN when it is away. The way ARP works is as follows. An ARP request is a broadcast message seeking the MAC (physical) address of a terminal given its IP address. When a packet arrives to the MN, an ARP request is made to obtain its MAC address on the home network. If the MN is away, the HA will respond with its own MAC address. When it returns to the home network, the MN will perform a gratuitous ARP that is an unsolicited ARP reply broadcast to each node on the home network clearing the ARP caches. Forwarding packets is achieved by encapsulation (tunneling). A virtual pipe between tunnel entry point (HA) and tunnel termination point (FA) is created through a datagram that includes the packet from the CN as its payload. The mandatory implementation for Mobile IP is IP-in-IP encapsulation as shown in Figure 6.10 though more efficient implementations (called minimal encapsulation) are optional. As far as the IP

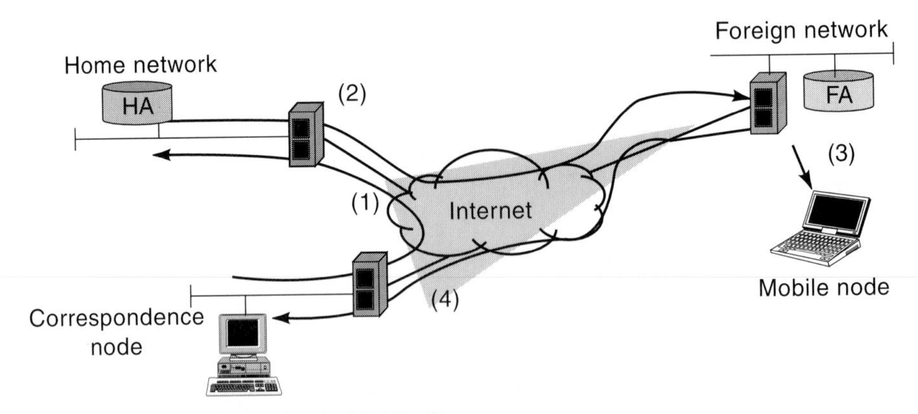

Figure 6.9 Triangle routing in Mobile IP.

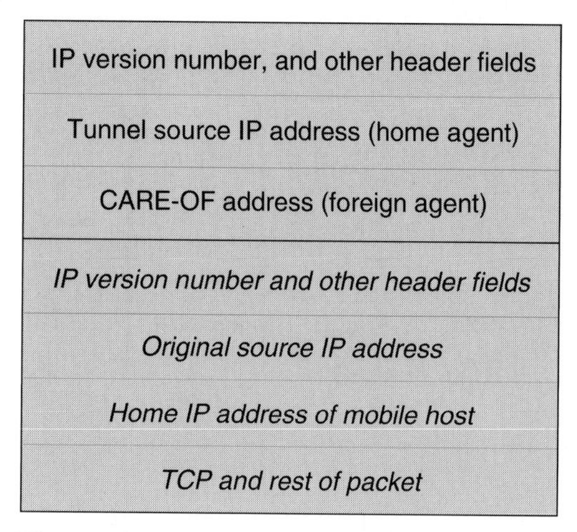

Figure 6.10 IP-in-IP encapsulation.

packet from the correspondent node is concerned, it looks like a single hop within the Internet.

Figure 6.11(a) shows the sequence of events when an MN moves from the home network to a foreign network, and Figure 6.11(b) shows the sequence of events when it returns to the home network.

6.2.3.3 Other Issues in Mobile IP

There are several issues in Mobile IP that are under consideration. Because the HA has to tunnel packets, it could be a potential bottleneck in the case of heavy traffic. Triangle routing is inefficient, especially if packets are routed to the home network, only to be tunneled back to a point close to the CN. A solution for both these problems is to enable routers in the Internet to cache the mobility binding and route packets accordingly. The packets addressed to the MN can be detected when they are being routed as packets that need tunneling to a new address and can be routed as such. This, however, leads to issues related to security and the need to change the way in which routers operate.

Suppose an MN changes its foreign network and while a new registration request is in progress, data are being tunneled to the old FA. These data have to be resent by the CN, as the old FA will drop the packets addressed to the MN. After the CN sends the data, the retransmitted data have to be tunneled again. If the old FA can tunnel packets it receives to the new FA, this can reduce delay and congestion. Such a procedure is called *smooth handoff*. It is also possible that the old FA retunnels the packet back to the HA, in what is called a "special tunnel." This enables the HA to detect a "loop" if a new registration request has not been enabled.

Sometimes packets will have to be tunneled through the HA. Two common reasons for this are that firewalls drop outgoing packets that have an IP address which corresponds to another network, so the packet cannot be directly sent from the MN to the CN. Also packets addressed by the MN to hosts on the home network usually have a small time-to-live because they are supposed to be on the same network. A small time-to-live implies that the packet needs to sense the Internet as

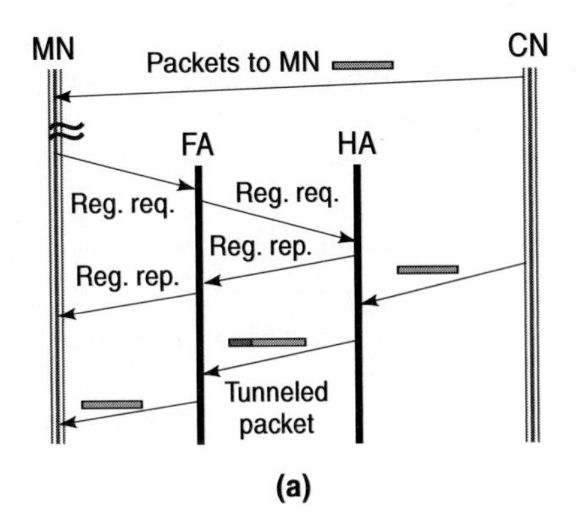

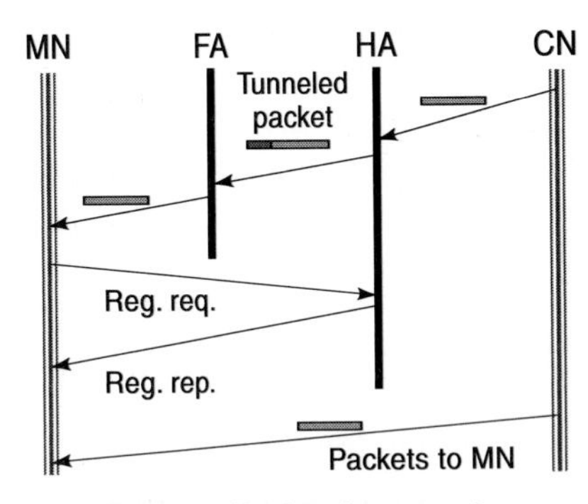

Figure 6.11 Sequence of events: (a) when the MN moves to a foreign network and (b) when the MN returns to the home network.

one hop. In both cases, the MN can tunnel the packet to the home agent for retransmission; this procedure is called *reverse tunneling*.

6.3 RADIO RESOURCES AND POWER MANAGEMENT

Radio resource and power management is an important part of any wireless network because of several reasons as explained in the following. The scarcity of radio spectrum has resulted in frequency reuse as discussed in Chapter 5 to increase capacity in a serving area. The installation of multiple BSs to provide service results in certain phenomena that need to be correctly addressed for proper operation of the wireless network. First, signals from MSs operating in the coverage area of one BS cause interference to the signals of MSs operating in the coverage area of another BS on the uplink. There is a need to reduce such an interference by properly controlling the transmit powers of MSs. Similarly, signals transmitted by one BS will interfere with the signals transmitted by another BS on the downlink. The transmit powers of BSs on interfering channels need to be controlled to minimize this interference. Second, correctly controlling the transmit powers of MSs can enhance their battery life and make mobile terminals lighter and handier to use. Because wireless terminals are mobile, they run on battery power which needs to be conserved as long as possible to avoid the inconvenience of requiring a fixed power outlet for recharging. Most of this power is consumed during transmission of signals. Consequently, the transmit power of the MS must be made as small as possible. In turn, this requires reducing the coverage area of a point of access—whether a BS or an AP, so that the received signals are of adequate quality. Also, as mobile terminals move, the ability to communicate with the current BS degrades, and they will need to switch their connection to a neighboring BS. At some point during the movement, a decision has to be made to handoff from one BS to another. This decision will have to be made based on the expected future signal characteristics from several BSs that may be potential candidates for handoff, the capacities and available radio resources of such BSs, and interference considerations. For example, if an MS continues to communicate with a BS when it is deep into the coverage area of another BS, it will cause significant interference in some other cell that employs the same channel. Third, then, there is a need for the wireless network to keep track of the radio resources, signal strengths, and other associated information related to communication between an MS and the current and neighboring BSs. All these tasks are not undertaken by a single entity. We have already discussed the last issue in the section on handoff decision algorithms. As we discuss in this chapter, several schemes and technologies are employed for the first two issues of radio resources and power management in wireless networks.

We distinguish between power control, power saving mechanisms, energy efficiency, and radio resource management. By *power control*, we shall mean the algorithms, protocols, and techniques that are employed in a wireless network to dynamically adjust the transmit power of either the MS or BS for reducing cochannel interference, near-far interference in the case of CDMA, or other reasons. *Power-saving mechanisms* are employed to save the battery life of a mobile terminal by explicitly making the MS enter a suspended or semisuspended mode of operation with limited capabilities. This is, however, done in cooperation with the network, so as to *not* disrupt normal communications or provide the user with a perception of such a disruption even if there was one. *Energy efficient design* is a new area of research that is investigating approaches to save the battery life, of an MS in fundamental ways such as in protocol design, via coding and modulation schemes, and in software. *Radio resource management* refers to the control signaling and associated protocols employed to keep track of relationships between signal strength, available radio channels, and so on in a system so as to enable an MS or the network to optimally select the best radio resources for communications.

6.3.1 Power Control

In this section we discuss basic power control mechanisms and why they are important through example implementations in cellular networks.

6.3.1.1 Basic Idea in Cellular Networks

Power control has been an issue of importance since the very first deployment of analog cellular systems. As discussed in Chapter 5, cochannel interference limits the capacity of a cellular network. Cochannel interference also causes the quality of a voice signal to deteriorate, and an attempt has to be always made to ensure that cochannel interference is minimized. This translates into forcing a mobile terminal or a BS to operate at the *lowest possible SIR* so that the voice or communications quality is acceptable. This appears to be a paradox because one would expect that it is important to maintain a high SIR for good communications quality. Although this is true in ordinary communications systems, in wireless communications with a cellular topology, operating at a high SIR implies that the transmit power of an MS or a BS is large. A large transmit power in one frequency channel in one cell results in a large cochannel interference in all the closest cochannel cells that employ the same frequency channel, albeit at a sufficient distance away from the given BS. This will reduce the communications quality all around and is not desirable.

Example 6.7: Minimum S_r Operation in AMPS on the Reverse Link

Consider an AMPS network. As discussed in Chapter 5, usually a cluster of seven cells uses the entire spectrum allocated to an operator. The spectrum is then reused in neighboring clusters. The approximate distance between the centers of cochannel cells D_L is $4.58R$ where R is the radius of a cell. Consider an MS in one of the cells of the cluster. Suppose it is located at a distance of $R/2$ from its own BS. Without power control, it would transmit at some power P_t, which is independent of distance. If radio propagation in the cell has a distance-power gradient of four, in order for it to operate such that the BS receives the signal at the lowest possible S_r, the transmit power must be reduced to $P_t/16$ because the BS would otherwise receive a signal that is 16 times stronger. This will in turn reduce the interference it causes to cochannel cells as well as adjacent channel cells. By reducing the transmit power, the mobile terminal will also save on its battery life.

Example 6.8: Effect of Large Transmit Power on the Forward Link in AMPS

Consider an AMPS network with a reuse factor of $N = 7$. Assume that channels 1, 8, 15, and so on are allocated to cells labeled A in Figure 6.12 (see Example 5.8). Channel 1 corresponds to the frequency band 869.0–869.030 MHz. Let the BS in the shaded cell transmit signal on Channel 1 at a transmit power six times as large as the other BSs of its co-channel cells. This is an increase in the transmit power by less than 8 dB. The effect on the SIR observed by the mobile terminal in Figure 6.12 is as follows:

$$S_r \approx \frac{P_t R^{-4}}{5P_t D_L^{-4} + 6P_t D_L^{-4}} = \frac{1}{11}\left(\frac{D_L}{R}\right)^4 \tag{6.1}$$

From Chapter 5, we know that the ratio D_L/R is 4.58 for the case where $N = 7$. Here D_L is approximately the distance of the mobile terminal from its cochannel cells. The SIR given by (1) is around 16 dB that is 2 dB lower than the required value for good communications quality. If all the BSs in the area are erratic, the signals received by the mobile terminals will all be of poor quality.

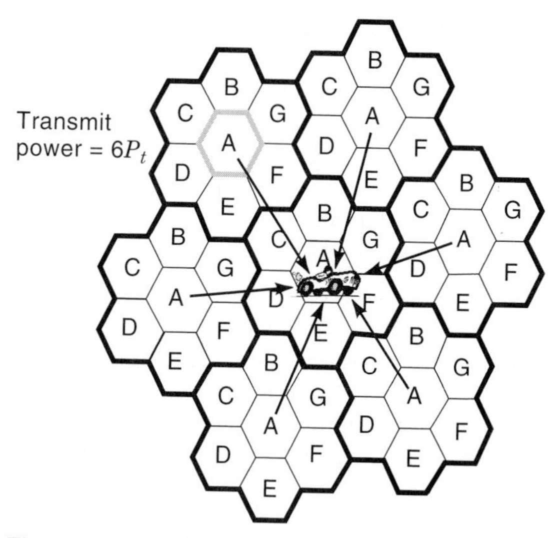

Figure 6.12 Effect of large transmit power.

From these examples, we can see that controlling the transmit power of both MSs and the BSs is important. When power control is applied properly, it can improve the quality of communications by increasing the SIR. An alternative way to view this is in terms of increase in system capacity. If the SIR can be increased, this will imply that a lower frequency reuse can be employed. As discussed in Chapter 5, this will increase the capacity accordingly.

6.3.1.2 Open and Closed Loop Power Control

The transmit powers of the BS and the MS must be dynamically changed because a variety of phenomena affect the quality of the signal. These include fading, velocity of the mobile, distance from the BS and so on. Also, there must be mechanisms by which the MS and BS can determine how the transmit power should be adjusted. This is achieved as follows.

Open loop power control is usually implemented on the reverse link. In open loop power control, the MS measures the quality of a reference channel from the base station. There may be a variety of measures such as RSS or frame or BER. If the RSS or BERs are above certain thresholds, the mobile terminal will automatically reduce its transmit power. If the signal quality is not good, the MS will increase it's transmit power. Clearly this is not a good mechanism for a variety of reasons. First, the decision on reducing or increasing the transmit power on the reverse channel is based on the measurement of the signal quality on the forward channel. These channels are not usually correlated, and a good signal reception on the forward channel does not necessarily mean the same on the reverse channel. The MS has no means of determining whether it has achieved the goal of minimizing the transmit power. Also, depending on the system, there may be significant delay in implementing this power control. In TDMA systems, the MS reception and transmission times are different, and there will be a lag time in implementing open loop power control.

Closed loop power control eliminates the disadvantages of open-loop power control by implementing a feedback mechanism between the BS and the MS. The BS measures the quality of the signal received from the MS and indicates what actions the MS should take via control signaling on the forward channel. Closed loop power control can also be used to control the transmit power at the BS. This is usually less important because the BS is not limited by battery power. However, adjusting the BS transmit power can benefit the system in terms of reducing the overall cochannel interference.

Example 6.9: Open Loop Power Control in IS-95 CDMA

In IS-95 (CDMA) standard, power control is extremely important because of the near-far effect described in Chapter 5. Because all the voice channels occupy the same time and frequency slots, the received signals from multiple users must all have the same RSS for each one to be detected correctly. An MS that transmits at a large power unnecessarily will essentially jam the signals of all the other MSs.

The open-loop power control scheme in IS-95 operates on the following principle. A mobile terminal that is closer to a BS should transmit less power because its signal suffers a smaller path-loss. Mobile terminals that are in deep fade or far away from a BS should transmit at a larger transmit power to overcome the loss in signal strength. Upon powering up, the MS adjusts its transmit power based on the *total* received power from all BSs on the forward channels [GAR00]. The BSs are not involved in the power control mechanism. The reference channel on the forward path used to determine the transmit power of the MS is the *pilot channel*. If the pilot channel is received very strongly, the MS transmits a weak signal to the BS. Else it transmits a strong signal to the BS.

Example 6.10: Closed Loop Power Control in GSM

The closed loop power control on the reverse link of GSM operates as follows [GAR99]. The MS measures the RSS and the signal quality of up to six neighboring BSs and reports its measurements to the base transceiver subsystem (BTS). The BTS also measures the RSS, signal quality, and distance to each of the mobile terminals in its serving area. From these measured values, it determines the minimum required transmit power and informs the MS of this value via a five-bit field in the slow associated control channel. The power control is performed in steps of 2 dB.

6.3.1.3 Centralized and Distributed Power Control

The open loop and closed loop power control mechanisms discussed earlier try to dynamically adjust the *transmit power* of the MSs or BSs based on the thresholds for SIR or BER set in the network. The goal of any power control scheme should be to uniformly render the SIR of all users to a value, which is usually the *maximum possible SIR* in the system. In terms of how such an optimization can be done are two approaches—centralized and distributed.

In a centralized power control (CPC) scheme, a central controller in the BS controller or mobile switching center has knowledge of *all* the radio links in the system. That is, the transmit powers, received powers, SIRs, and BERs for all mobile terminal–BS combinations are known to this centralized controller. Assuming that the system is interference limited, an optimization algorithm can be implemented to maximize the minimum SIR in the system and minimize the maximum SIR in the system, thereby equalizing the SIR of all radio links [GRA93]. Although this provides an optimum solution, this scheme is extremely hard to implement because the centralized controller has to dynamically keep track of all the links in the system and compute the transmit powers for each mobile terminal.

In distributed power control (DPC) [GRA94], the mobile terminals adjust their transmit powers in discrete steps. This is similar to what is actually done in practice. For theoretical simplicity, it is often assumed that the MSs adjust their transmit powers synchronously. The power adjustments made by the MSs result in the transmit powers iteratively converging to the optimum power control solution. Ideally this should result in all the mobile terminals having the same SIR (which should be the maximum possible in the system) as in the CPC scheme after a number of adjustments. In practice, the adjustment to the power levels is also discrete (in steps of a few dB).

Example 6.11: Power Adjustment Levels in Example Wireless Networks

In GSM, each mobile terminal is required to either increase or decrease its power level by 2 dB depending on the message sent by the BTS. In CDMA, mobile terminals can change their power levels in steps of 1 dB. The IS-136 standard requires that a mobile terminal be able to change its transmit power by 4 dB in response to a command from the BS in 20 ms. In CDPD, the transmit power is set based on the signal power received and *not* the SIR. Consequently, power control is not based on signal quality. It is based only on the absolute signal strength.

6.3.2 Power Saving Mechanisms in Wireless Networks

In addition to dynamically changing the transmit power, there are several other mechanisms built into the operation of most wireless networks for saving the battery power of the MS. A variety of measurements have indicated that the most battery power is consumed during transmission of signals. A significant amount of power is consumed during active reception of signals (although this is lesser than the power consumed during transmission). A third mode of operation, called *standby* mode, consumes nearly an order of magnitude less power than either during transmission or reception of signals [AGR98].

Example 6.12: Power Consumption in Lucent WaveLAN

Lucent's WaveLAN (now Orinoco) is a WLAN product based on the IEEE 802.11 standard. It operates in the 2.4 GHz ISM bands. The power consumed by a 15 dBm WaveLAN radio is 1.825 W in the transmit mode, 1.8 W in the receive mode, and 0.18 W in the standby mode [AGR98]. Clearly, the standby mode operates with very little power and operating in this mode can save the battery life.

The operation of wireless networks is often designed to ensure that the mobile terminal spends as much time as possible in either a standby or sleep mode in order to conserve power. Several techniques are employed in wireless networks to reduce the amount of time spent in transmitting or receiving signals. In the case of laptops and other data terminals, a *sleep* mode of operation is preferred where the radio transceiver is shut off to conserve power. In voice networks such as GSM and IS-95, the *voice activity* factor is used to reduce either the transmit power or completely stop transmission when there is no speech activity. We discuss these techniques in the following sections.

6.3.2.1 Discontinuous Transmission and Repetition at Lower Transmit Powers

A particularly attractive option for saving the battery power of a mobile terminal is not to send information unnecessarily. With real-time communications, it is often assumed that there is a constant stream of data to be transmitted and also such communications are sensitive to delay and jitter. Usually, this is not the case, especially in a two-way conversation where one of the users is listening for some duration of time. Data communications are less affected by delay and jitter; it is possible to buffer data and transmit it at a later time. Discontinuous transmission is mostly employed in cellular telephone networks where additional hardware and algorithms are used to detect the presence or absence of voice. In ancient mobile telephones, some amount of information would be transmitted all the time, whether the person was actually speaking or not. With the use of *voice activity detection* (VAD), it is now possible for an MS to behave differently when no voice activity is detected. One of the possibilities (assuming ideal voice activity detection) is to not transmit any signal when the user is not speaking. A second alternative is to repeat data, but at a far lower signal power than usual. This will ensure that data is transmitted all the time, but the total power consumed corresponds to only that data which was actually generated by speech. VAD has its associated problems. In high-noise situations, the mobile terminal must be able to distinguish between the presence of useful signals in high noise and simply noise. Also it should be able to detect low-level voice activity. If VAD is not implemented correctly, there may be clipping of speech or an additional hangover after a talk spurt. Also, if there is absolutely no transmission, subjective tests indicate that silent gaps are extremely annoying. Consequently, systems usually insert a very low power *comfort noise* signal during silent gaps.

Example 6.13: Discontinuous Transmission in GSM

GSM transmits a comfort noise signal when there is no speech activity. When a VAD determines that there is no speech activity, the MS enters a hangover state to prevent clipping of speech due to very short silence periods. If there is no speech activity after the hangover period has elapsed, a silence identifier frame is transmitted at larger intervals than voice frames. The receiver will insert comfort noise when it detects the presence of the silence identifier frame.

Example 6.14: Discontinuous Transmission in IS-95

In IS-95, speech coders operate at different rates depending on the voice activity. The number of bits generated per frame will be different depending on the rate of the speech coder. The data stream corresponds to 9,600, 4,800, 2,400 or 1,200 bps. If traffic is generated at 9,600 bps, bits are transmitted at 100 percent of the transmit power. At lower data rates, the bits are repeated and then transmitted at half, one-fourth, or one-eighth of the transmit power on the forward channel. On the reverse channel, discontinuous transmission is employed.

6.3.2.2 Sleep Modes

A common approach for saving battery power is to allow the MS to enter into a *sleep mode* during periods of inactivity. The idea here is based on earlier discussion where we mentioned that the most power is consumed during signal transmission and significant power is consumed during signal reception as well. By entirely shutting off the RF hardware, it is possible to further reduce the battery consumption. However, there are problems associated with shutting off the RF hardware completely. What happens if a call or a packet arrives for a mobile terminal when it is shut off? The network should be able to make provisions for handling calls or packets that arrive for an MS that is in sleep mode.

Example 6.15: Sleep Mode in IS-136

In IS-136, the *standby time* is defined as the time for which an MS can be powered on and is available for service on a control channel before it needs recharging. The operation has been designed to allow the MS to enter a *sleep mode* for long periods of time when it is on standby. An MS is required to monitor the forward link only for a few time slots in order to determine whether there is a call addressed to it. The network may, however, require the mobile to monitor channels more frequently. The MS also has to monitor neighboring channels for handoff and monitor broadcast information. These will affect the time for which it can enter the sleep mode. For the rest of the time slots, the MS can enter a sleep mode.

Example 6.16: Sleep Mode in IEEE 802.11

In IEEE 802.11 WLANs, an MS can enter the sleep mode and inform an AP of its decision. Because this is a LAN and handoffs are less frequent than in cellular systems, handoff is less of a problem. Because of the bursty nature of data traffic, the arrival of packets addressed to the MS is a bigger concern. The AP buffers packets addressed to the sleeping mobile in 802.11. A beacon signal is transmitted periodically that contains information about buffered packets intended for sleeping mobiles. The MS wakes up at times when it expects the beacon and determines whether it should reenter the sleep mode or awaken completely to receive packets.

6.3.3 Energy Efficient Designs

The most common technique for conserving energy in a mobile terminal is to use advanced hardware design. Low power digital CMOS, mobile CPU microprocessors, and other hardware design approaches that consume very little power are usually employed in laptops and handheld computers. Beyond the actual hardware design, there are approaches at other layers that can enable savings on power consumption, thereby improving the lifetime of a battery. There are three approaches for improving battery life. The first approach is to tune the protocols employed in a wireless network to reduce power consumption. The second approach is to investigate power efficient modulation and coding techniques (see Chapter 3 for more details on modulation and coding). The final approach is in software design for mobile terminals.

6.3.3.1 Energy Efficient Protocols

Most protocols in data and voice networks have some relationship with the OSI protocol model even though they do not exactly match the seven layers. In wireless networks, the more important layers are the physical layer that handles the actual transmission of symbols and the link-layer that handles transmission of data packets or voice packets in link-layer frames and also controls access to the wireless medium. With the emergence of TCP and IP as the most popular transport and network layer protocols, these two are usually seen in wireless networks, although they are more restricted to data networks currently. As VoIP becomes popular, IP will make its presence in voice networks as well. As discussed earlier, the important power conservation principle in mobile terminals is to *minimize* the amount of time spent in transmitting signals. This principle can be applied to the design of different protocols in different layers.

6.3.3.2 Link Layer and MAC Design

At the link layer and in the design of medium access control techniques, power conservation factors should be taken into consideration. Two areas of design have been addressed in the literature. The first area is MAC design where the design goal is to eliminate unnecessary collisions and to employ better protocols for sleep mode and broadcast operation, as well as eliminating unnecessary processing at the MAC layer. The second area of design involves the link layer where ARQ error control schemes are employed for retransmission of lost or damaged packets.

Design at the MAC layer has focused on the following issues. Techniques to avoid retransmission due to collision as far as possible have been incorporated. Collision is not an issue in cellular telephony, where a channel is dedicated to the voice call for the duration of the call. However, in most data networks, such as IEEE 802.11, HIPERLAN, and CDPD, collision is an issue. Even in reservation-based schemes such as GPRS, the access to the network is achieved by a contention-based protocol.

Example 6.17: Collision Avoidance Mechanisms in Wireless Data Networks

Collision avoidance mechanisms have been incorporated in both IEEE 802.11 and HIPERLAN based on carrier sensing. In IEEE 802.11, there are two forms of carrier sensing—at the physical layer and at the MAC layer. At the MAC layer, the length of transmission of a signal is detected via a field in the MAC frame and a *net allocation vector* (NAV) is set so that no signal transmission or physical carrier sensing is attempted in this period, thereby reducing the chance of a collision and also reducing the energy spent in monitoring the channel. In HIPERLAN, multiple contention phases eliminate the chance of collision to a great extent. In CDPD, the downlink carries a "status" flag that indicates whether the uplink is busy or idle. This once again reduces the possibility of collision.

Outside of collision avoidance, it is possible to use some intelligent techniques to further reduce unnecessary battery consumption. In WLANs, an MS will receive *all* packets regardless of whether they are addressed to it. If the packet is not addressed to the MS, it is discarded. This results in an unnecessary waste of battery resources. One possibility for improving this situation is to simply look at only the header information and continue receiving the signal only if the packet is addressed to the MS.

Example 6.18: Intelligent Processing in HIPERLAN

In HIPERLAN, some header information is transmitted via a low bit rate transmission scheme to reduce battery consumption [WOE98]. The reason for this is as follows. As the data rate increases, the effects of multipath delay spread require the use of equalization techniques as discussed in Chapter 3. Equalization schemes consume a lot of battery power. At 23 Mbps, which is the data rate supported by HIPERLAN, equalization becomes very important. In order to reduce battery power consumption, the header information is transmitted at a lower data rate (1.4706 Mbps). The entire header is not transmitted at a low data rate. Only a 34 bit value of the destination address is sent at this low data rate. The MS determines whether the received hash value matches its own hash value. If the hash value does not match, the rest of the packet is not received. If the hash value matches, the equalization circuitry is switched on, and the rest of the packet is decoded. Of course, there is a possibility that the packet still may not be intended for the MS, because the hash values are not unique. However, the chances of this happening are small. By not receiving signals or using the equalization circuitry unnecessarily, HIPERLAN terminals can save battery power significantly.

Other possibilities for intelligent packet reception are possible. Because the downlink is controlled in infrastructure networks by a BS or AP, this can schedule the broadcast of packets intended for different mobile terminals. The MS will then have to decode only those packets that arrive in the vicinity of its schedules reception time [AGR98].

ARQ schemes are employed at the link layer to retransmit data packets that are lost. ARQ schemes are not useful for real-time traffic such as voice. Packet losses can occur due to several reasons—collisions, interference, fading, and multipath delay spread. The collision avoidance mechanisms discussed earlier try to eliminate collisions to the extent possible. However, collisions cannot be entirely avoided as discussed in Chapter 4. Also, interference, fading, and other radio channel effects can result in errors in the received packet. Retransmission techniques are incorporated at the link layer based on error detection schemes. It is possible to reduce retransmissions if error recovery can be performed at the receiver via forward error correction. In fact several wireless systems include *block interleaving* as discussed in Chapter 3 to reduce the effects of burst errors and enable forward error correction. However, if channels conditions are very bad, none of these techniques can recover from errors, and retransmission of packets will be necessary.

Retransmitting packets will not be useful if the channel conditions continue to be harsh. In fact this will result in unnecessary transmissions of packets that are bound to be lost or damaged. In [ZOR97], the energy efficiency of an error control protocol is defined as follows:

$$\lambda = \frac{\text{total amount of data delivered}}{\text{total energy consumed}} \tag{6.2}$$

Zorzi and Rao argue that this metric, which corresponds to the average number of packets delivered correctly during the lifetime of a battery, will influence the choice of ARQ protocols and that suboptimal protocols may in fact be better as far as energy consumption is concerned.

Example 6.19: An Adaptive Energy Efficient Go-Back-N ARQ Protocol

A classic Go-back-N ARQ scheme will transmit up to M packets and wait for acknowledgments from the receiver. The receiver will only accept packets in order and will send a negative acknowledgment (NAK) if a packet is not received. The receiver may also acknowledge several packets received correctly and in sequence with a single acknowledgment for the last correctly received packet. If the sender times out while waiting for an acknowledgment for packet N in the set of M packets or receives a NAK for the packet N, it will retransmit all the packets starting from N until M once again. All these packets may be lost if the channel conditions continue to be bad. Instead, the adaptive protocol suggested in [ZOR97] operates as follows. A probe packet is transmitted when there is a NAK or a timeout. The probe packet is a small packet that has minimum payload or simply a header so that the mobile terminal does not waste resources transmitting large packets. Only if a positive acknowledgment is received for the probe packet will the sender resume normal transmission of packets. Under slow fading conditions and small energy consumption for probing packets, this scheme can increase the energy efficiency by three times. Such a scheme can be worse than the regular scheme if the channel conditions are varying rapidly because the channel may have degraded immediately after an acknowledgment is received for the probe.

6.3.3.3 Transport Layer Design

The most common protocol employed at the transport layer in data networks is the TCP. As discussed in the previous section, as long as the channel conditions are bad, it is wasteful to transmit packets because they will not be delivered correctly. TCP has in-built mechanisms to back off when it detects packet losses. This backing-off of transmissions is initiated not because channel conditions are suspected to be bad, but because TCP assumes that there is congestion in the network, and transmissions should be reduced to ease congestion. However, it is possible that indirectly this may also aid in reducing unnecessary energy consumption in wireless networks, especially when there are correlated errors on the wireless channel. In [ZOR99], the energy efficiency [defined in Eq. (6.2)] of TCP is investigated with this point of view. Analysis there indicates that depending on the nature of the channel, and the type of TCP implementation, TCP parameters can be tuned to increase energy efficiency significantly. In certain cases, it is possible to increase the energy efficiency by almost three times.

If parameters of TCP are not set correctly, the congestion avoidance mechanisms can degrade throughput and increase energy inefficiency. Split approaches for TCP [AGR98] introduce intermediate hosts in the network that keep track of missing packets and acknowledgments and handle TCP congestion avoidance mechanisms appropriately. These approaches can also increase the energy efficiency of the system.

6.3.4 Energy Efficient Software Approaches

Significant reduction of battery consumption can be obtained if the MS can be made to operate intelligently to reduce power consumption. Battery is consumed in mobile devices due to accessing of hard disks, operation of the CPU, and power consumption in the display in addition to the wireless communication unit and the communication protocols that we have considered so far. Energy management strategies are appropriate for each of these components, and this is usually achieved by the operating system (OS). These components usually have low power modes of operation and in many cases multiple modes of operation. A thorough discussion of these issues is provided in [LOR98]. The operating system will have to decide which mode of operation is appropriate at what time and when there should be a switch from one mode to another (*transition*). It should decide how a component's functionality can be modified to move it into low power modes as often as possible (*load change*) and also how software can be employed to permit novel power-saving use of such components (*adaptation*). Important factors in deciding strategies for any of these decisions involve what effect a strategy may have on the *overall* power savings because saving power in one component may affect the performance or power consumption of other components in addition to introducing unnecessary overhead. In certain cases, the lifetime of a battery will not be the sole issue, but how much of productivity can be obtained from a mobile terminal.

Table 6.1 provides a summary of results from [LOR98] that considers energy management issues related to secondary storage (hard disk, etc.), the processing unit, and the mobile terminal display units. A variety of power-saving strategies are possible as shown in the Table 6.1. The improvement in power saving varies depending on the type of mobile device, secondary storage device, processor, and display.

Significant power is consumed solely by access to hard disk, and it is suggested that it may be better to offload some of the storage onto a network file server. This assumes that the power consumed with the wireless transceiver will be smaller than the power consumed by disk access. Because transmission and reception does not involve mechanically moving parts, this may be a viable option, especially if energy efficient protocols are employed. Flash memory is nonvolatile

Table 6.1 Energy Management in Software in Mobile Terminals

Component	Secondary Storage	Processor	Display
Power-Saving Features	Five power modes—active, idle, standby, sleep and off	Clock slowdown, shut-off functional units, shut-off processor	Color to monochrome, reduce update frequency, turn display off
Issues	Sleep and standby modes consume far less power, but moving from standby to idle mode consumes power	Clock speed reduction can increase task time, thereby increasing power consumption; shutting off processor works best	Affects readability Can be annoying if there are flashes for updates
Transition Strategies	Enter sleep mode when inactive for some fixed threshold of time (several seconds) Use dynamic thresholds based on previous samples of disk access Predictive spin-ups of the disk (has not worked very well)	Process scheduler knows whether processes are ready to run or running. When processes are blocked, the CPU can be turned off (UNIX and Windows) Predict the number of busy CPU cycles and set the CPU clock speed	Turn off or turn down the display if there is no user input after a period of time
Load Change Strategies	Increase number of disk accesses—increase cache size Prefetch data based on prediction of usage before spinning disk down Reduce paging and improve memory access locality	Use lower power instructions (energy efficient compilers) Reduce the time taken by low-level tasks and pipeline tasks to units that can be turned off Reduce unnecessary tasks (block instead of busy-wait)	No formal load change strategies
Adaptation Strategies	Switch to flash memory for cache and storage Offload access to the network Reduce disk speeds in low power modes, improve energy consumption of disks	Design motherboards that automatically power down all components when the processor does not present them with a load	Use sensors to determine if a user is looking at a display or not Use lighter colors for display Provide ability to display only active windows and dim the rest of the display

and consumes less power. Against 0.9 W consumed only in idle mode for a hard disk, flash memory consumes 0.15–0.47 W for reading and writing. However, it is about 20 times more expensive than hard disks. Simulations indicate that flash memory used as secondary storage can reduce power consumption by 60 to 90 percent. Turning the processor off and reducing the voltage and clock speed of a processor can result in power savings that can be as large as 70 percent if the strategies are implemented correctly. The use of low-power states can further reduce power consumed by display units. Other components that can benefit from the OS employing low-power modes include sound cards, modems, and main memory [LOR98].

6.3.5 Implementation of Radio Resource and Power Management: A Protocol Stack Perspective

The purposes of the radio resources management (RRM) layer are threefold. First, it has to implement power control in the mobile terminals to reduce interference in the system. As a by-product this will increase the battery lifetime of the mobile terminal. Second, it has to assist the network and the mobile terminal to be connected on the best possible radio channel available to it within the cell. Last, it has to function to enable the mobile terminal and the network to handoff the mobile terminal's connection from one cell to another when the mobile moves across the boundary of a cell. All these tasks are performed by RRM entities that reside in the mobile terminal and sometimes in the network side of the wireless system.

The radio resource and power management functionality is usually handled by a management entity that interfaces across the lower three layers of the OSI protocol stack (as in the case of CDPD). Alternatively, it may be viewed as an application layer on top of the lower three layers (as in the case of GSM). In any case, handling of this functionality requires knowledge of exactly how signals are behaving at a particular point of time. This automatically requires feedback from the lower layers of the protocol stack.

In a manner similar to mobility management, there are two approaches to RRM. As we discuss in Chapter 9, in the case of GSM, the BS controller and the mobile switching center need to communicate with the mobile to obtain information about the state of the radio channel and provide the instructions for power control and selection of channels. A *bidirectional* logical channel is required for the communication between the mobile terminal and the network. In data networks, such as CDPD (discussed in Chapter 9), the mobile station has autonomous operation and decides for itself what should be the appropriate action. Power control is much harder to implement in this case. However, selection of radio channels and entering sleep mode of operation is done at the mobile station. In this case a *unidirectional* channel is sufficient. In CDPD, for example, the mobile–database station provides the mobile terminal information about available channels, transmit powers, and thresholds, but it does not expect any feedback from the mobile terminal in return. The mobile terminal is in charge of independently selecting channels and choosing what actions to perform. The protocol architecture and location of the RRM layers are described in subsequent chapters.

6.4 SECURITY IN WIRELESS NETWORKS

Wireless access to the Internet is becoming pervasive with diverse mobile devices being able to access the Internet in recent times. Current deployment is small, and the security risks are low as of today in these emerging technologies. However, the widely varying features and capabilities of wireless communication devices introduce several security concerns. The broadcast nature of wireless communications renders it very susceptible to malicious interception and wanton or unintentional interference. At least minimal security features are essential to prevent casual hacking into wireless networks. Since the advent of analog telephony, wireless service providers have suffered several billion dollars of losses due to fraud. In this section, we address security issues in general. We provide an overview of network security services and mechanisms and describe some specific wireless examples.

6.4.1 Security Requirements for Wireless Networks

Security requirements for wireless communications are very similar to the wired counterparts but are treated differently because of the applications involved and potential for fraud. Different parts of the wireless network need security. Over the air security is usually associated with privacy of voice conversations. This is changing with the increasing use of wireless data services. Message authentication, identification, authorization, and so on also become issues here. Wireless networks are inherently insecure compared to their wired counterparts. The broadcast nature of the channel makes it easier to be tapped. Analog telephones are extremely easy to tap, and conversations can be eavesdropped using an RF scanner. Digital systems such as TDMA and CDMA are much harder to tap, and RF scanners cannot do the trick anymore, but as the circuitry and chips are freely available, it is not hard for someone to break into the system. Very little work has been done on optimizing security services for wireless systems, and patchwork solutions have made wireless networks not very secure. As long as the deployment is sparse and potential for harm to the consumer small, such measures can be sufficient. As more people use wireless access to the Internet and use wireless networks for e-commerce, credit-card transactions, and so on the potential for harm increases. In this section, we try to address some security requirements that have been identified for wireless voice networks and some that are emerging for wireless data networks.

6.4.1.1 Privacy Requirements of Wireless Networks

Privacy requirements are twofold in wireless networks. As discussed in Chapter 1, along with the air-interface, there is also a fixed infrastructure for handling the registration of mobiles, billing, mobility, power control, and other issues. There are privacy requirements for the air-interface and others for the messages transmitted over the wired infrastructure.

A variety of control information is transmitted over the air in addition to the actual voice or data. These include call setup information, user location, user ID (or telephone number) of both parties, and so on. These should all be kept secure because there is potential for misusing such information. Calling patterns (traffic analysis) can yield valuable information under certain circumstances. A flurry of calls between the CEOs of two major companies may indicate certain trends if it was discovered, even if the actual information in the calls was secure. Hiding such information is also important. In [WIL95], various levels of privacy are defined for voice communications.

At the bare minimum, it is desirable to have wireline equivalent privacy for all voice conversations. We commonly assume that all telephone conversations are secure. Although this is not true, it is possible to detect a tap on a wireline telephone. It is impossible to detect taps over a wireless link. To provide privacy that is equivalent to that of a wired telephone, for routine conversations it is sufficient to employ some sort of an encryption that will take more than simple scanning and decoding to decrypt. Using DES with a 56-bit key would seem to be adequate for this purpose. In order to alert wireline callers about the insecure nature of a wireless call that is *not at all* encrypted, a "lack-of-privacy" indicator may be employed.

Wilkes [WIL95] calls these two levels of security as levels zero and one, respectively. Level-0 privacy is when there is no encryption employed over the air so that anyone can tap into the signal. Level-1 privacy provides privacy equivalent to that of a wireline telephone call, one possibility being encrypting the over-the-air signal. For commercial applications, a much stronger encryption scheme would be required that would keep the information safe for more than several years. Secret key algorithms with key sizes larger than 80 bits are appropriate for this purpose. This is referred to as Level-2 privacy. Encryption schemes that will keep the information secret for several hundreds of years are required for military communications and fall under Level-3 privacy.

For wireless data networks, a bare minimum level would be to keep the information secure for several years. The primary reason for this is that wireless electronic transactions are becoming common. Credit card information, dates of birth, social security numbers, email addresses, and so on can be misused (fraud) or abused (junk messages, for example). Consequently, such information should never be revealed easily. A Level-2 privacy will be absolutely essential for wireless data networks. In certain cases, a Level-3 privacy is required. Examples are wireless banking, stock trading, mass purchasing, and so on.

6.4.1.2 Other Security Requirements in Wireless Networks

Although privacy and confidentiality continue to be the important issue in wireless networks, other security requirements are becoming significant in recent times. There has been widespread fraud and impersonation of analog cellular telephones in the past. Although this is more difficult with digital systems, it is not impossible. There is thus a need to correctly *identify* and *authenticate* a mobile terminal. This becomes more important for private wireless data networks. For example, an organization that has installed a wireless LAN within its premises discovers that the coverage area extends into a neighboring street. With the reducing costs of wireless LAN PC-cards, anyone could buy a PC-card and access the organization's WLAN from the street. Privacy would assure that the person would not be able to read information being transmitted over the organization's communication network. However, the person could perhaps access the Web, thereby occupying the organization's bandwidth, or obtain an IP address from a DHCP server on the LAN if adequate authentication procedures are not employed. Similar situations exist with residential wireless networks, where it is important to keep security measures adequate, and yet not cumbersome to the layman.

6.4.1.3 Miscellaneous Issues

Even though traditional wired security measures are being put in place for wireless networks, wireless specific issues have been largely neglected. In addition to traditional security services such as privacy, authentication, message integrity, nonrepudiation, access-control, and availability, some of the wireless devices need certain intermediate security services such as authorization, identification, and varying degrees or classes of security and privacy as discussed earlier. The potential security implications and interaction between security requirements and wireless network/device limitations are unclear. Wireless communication devices are expected to be mobile and have the additional requirement that they must consume as little power as possible while performing computations for encrypting or decrypting data to conserve battery power. This is a significant issue because cryptographic algorithms are computationally intensive and may drain the battery of a mobile terminal quickly. Because the spectrum is scarce, cryptographic protocols should also not waste resources by requiring several handshakes between the mobile terminal and the fixed network. This requirement is usually contrary to security services as they are implemented in wired networks. The wireless channel is error prone and may also result in messages being lost, duplicated, or damaged. It is also not clear what effects this may have on the overall performance of security protocols. Interference, fading, disconnections, handoff and other mobility-related procedures, and other peculiarities of the wireless network require robust security services that are at the same time resource efficient. These issues are yet to be addressed in wireless security.

6.4.2 An Overview of Network Security

Network security has made tremendous strides with the explosive growth of the Internet, and several textbooks have been written on this topic [STI95, STA98, SCH94]. It is beyond the scope of this section to present all relevant concepts in detail. The goal here is to present sufficient detail to enable the reader to appreciate and understand the subject material that follows in later sections.

6.4.2.1 Security Services

The primary method of understanding and designing solutions to a particular problem is to first develop a set of requirements that must be satisfied. In network security, classifying and defining various features that must be available to a network to keep information secure accomplishes this. Such various features are commonly referred to as security services. Although there are several ways of defining security services, usually we identify them as specific measures employing *security mechanisms* that combat security attacks on a network. Based on [STA98], we present a list of security services as follows.

Confidentiality or *privacy* is a security service that provides resistance to the security attack known as *interception*. This is the most intuitive form of security service where two communicating parties do not wish to reveal the contents of their transactions to a third party. In more rigid cases, the existence of the communication itself must not be revealed to unauthorized entities. Encrypting the messages and the identities of the two communicating parties is the most popular method of providing confidentiality. *Message authentication* is a term that is used for a security service that provides *integrity* of the message and also provides a guarantee that the sender is who he or she claims to be (*sender authentication*). The corresponding security attacks are *modification* of the message and *impersonation* of the sender's identity. Message authentication can be provided by attaching a digest of the message, which is encrypted by a key known only to the communicating parties.

Nonrepudiation corresponds to a security service against denial by either party of creating or acknowledging a message. This service is similar in nature to a signature by the creator of a document and *digital signatures* based on public key encryption schemes discussed below are employed to provide this service. The corresponding security attack in this case is *fabrication*. *Access control* enables only authorized entities to access resources. *Masquerading* is the security attack corresponding to this service. *Availability* ensures that resources or communications are not prevented from access or transmission by malicious entities. *Denial of service* is the attack corresponding to the security service of availability.

Although the security services discussed are the most prominent ones, other security services also play a role in certain applications. *Authorization* is sufficient in certain cases instead of a sender authentication. Authorization allows a user or communication command to execute certain operations. Such a security service would be sufficient, for example, to instruct a coffee machine to execute certain operations. *Identification* is another security service often used in transactions such as automatic teller machines. Such a security service would be useful in other applications such as credit card transactions over a WAP-enabled microbrowser.

6.4.2.2 Security Mechanisms

There are several ways in which security mechanisms can be provided. However, there is no single technique that can provide all required security services. *Encryption* is a technique that is employed both in wired and wireless networks for providing several security services. Encryption, when employed in clever ways, can provide confidentiality, message authentication, nonrepudiation, access control, and identification. Availability cannot be guaranteed by encryption because encryption is powerless against attacks such as cutting wires or jamming signals. In succeeding subsections, we shall see how encryption can be used to provide security services.

6.4.2.3 Confidentiality or Privacy

In order to keep a message confidential, the easiest technique is to scramble the message using a *key* so that if the message falls into wrong hands, the adversary will not be able to understand or descramble the message. The scrambling technique is usually called *encryption*. The message is referred to as *plaintext* or *cleartext,* and the encrypted version of it is referred to as *ciphertext*. It is common to denote two communicating parties as Alice and Bob and the adversary or opponent as Oscar. Mathematically, the encryption of a plaintext x into a ciphertext y using a key k is written as:

$$y = e_k(x) \tag{6.3}$$

The corresponding decryption is written as:

$$x = d_k(y) \tag{6.4}$$

Ideally, we would like the encryption scheme to be such that it cannot be broken at all. Because there are no practical methods of achieving such an unconditional security, encryption schemes are designed to be *computationally secure*. The encryption scheme has to be powerful, in that given significant computational resources, an adversary must not be able to either find the key or decrypt the message in a reasonable time. Alternatively, if either the key or the plaintext can be determined in a short time, it should cost the adversary much more than what the value of the secret information would be to him. Usually, it is assumed that Oscar has knowledge of how the algorithm works, but not the key. Also, because of the standard formats of data packets and control messages in voice networks, Oscar usually has access to a limited number of plaintext-ciphertext pairs that he can use to perform a *known plaintext* attack to recover the key. Once the key is recovered, all subsequent ciphertext can be decrypted easily.

So far, we have discussed security services, and we have said that encryption can provide some of these services. What we have not discussed are the encryption algorithms that are employed or can be employed within these mechanisms. The details of these algorithms are beyond the scope of this book, and the subject matter forms a wide area of interest in itself. In this section, we briefly mention some of the algorithms that are in wide usage today. We also discuss the key sizes that are required to make these algorithms secure.

6.4.2.4 Secret-Key and Public-Key Algorithms

Encryption schemes have been available through the ages and have all been what are known as *secret-key* algorithms. Here, the communicating parties (Alice and Bob in Figure 6.13) share a secret key that they use to encrypt any communication between themselves. Usually, the encryption and decryption algorithms use

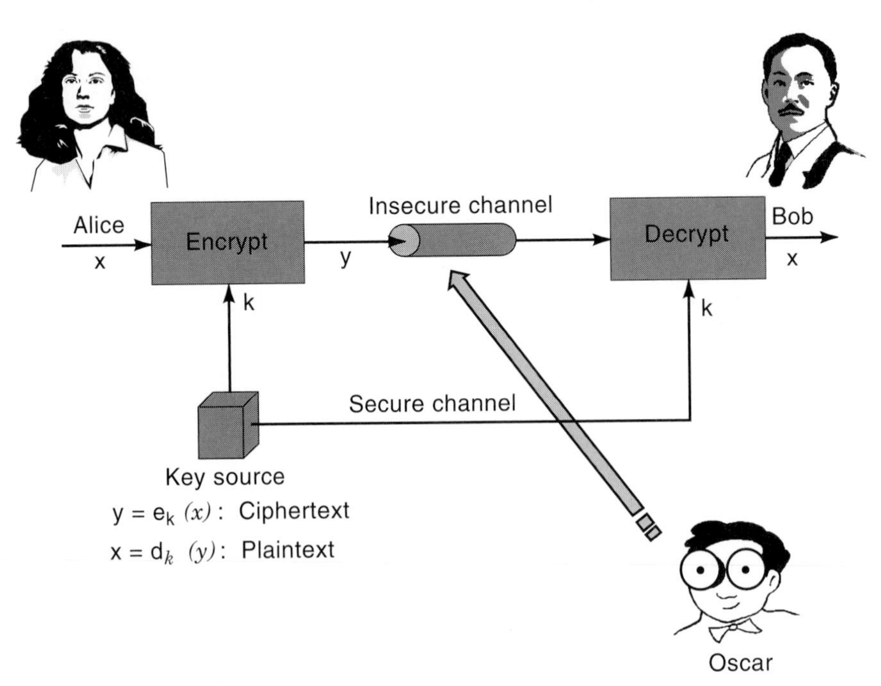

Figure 6.13 Conventional encryption model.

the *same* key, and, hence, such algorithms are also called *symmetric key* algorithms. Block ciphers such as the data encryption standard (DES) also fall under this category. Figure 6.13 illustrates a schematic of a conventional encryption scheme. The opponent Oscar has access to the insecure channel and thus the ciphertext. However, he has no knowledge of the secret key k shared by Alice and Bob.

Secret-key algorithms such as DES are based on two principles: confusion and diffusion. The former introduces a layer of scrambling that creates confusion as to what exactly might be the transmitted message. The latter creates a randomness whereby the effect of changing a small part of the plaintext message will result in changing half of the encrypted ciphertext. This eliminates matching patterns or frequencies of occurrence of messages. Most secret-key algorithms are thus unbreakable except by brute force [SIL99]. If the length of the key of a secret-key algorithm is n bits, at least 2^{n-1} steps are required to break the encryption. Today, a key length of 80 bits is considered to be sufficiently safe from brute force attacks even though a key size of 128 bits is usually recommended.

Example 6.20: Security of DES against Brute Force Attacks

DES is a block cipher that encrypts plaintext messages in blocks of 64 bits using keys that are 56 bits long. The total number of keys is 2^{56}. On average, half of them will have to be examined to determine the right key if a known plaintext-ciphertext pair is available. If a 500 MHz chip is employed for this attack, and one decryption (or encryption) can be performed in one clock cycle, to test 2^{55} keys, it will take $2^{55}/(500 \times 10^6)$ seconds = 834 days to break the encryption. This is not very secure if 834 chips are used in parallel, because the key can be obtained in a single day. The total cost will be about $16,680 if each chip costs $20!

Example 6.21: Security vs Advances in Chip Speeds

DES was broken in less than a day in January 1999 at a cost of $500,000. Today it is virtually impossible to break a well-designed block cipher with key sizes of more than 80 bits (which translates into examining around 280 keys by brute force). However, a common assumption (called Moore's Law) is that processor or chip speeds double every 18 months, thereby weakening any encryption scheme with time. For example, using a speed of 500 MHz for today, in 100 years, an encryption scheme that employs key sizes twice that of DES (i.e., 112 bits) can be broken in a day.

Example 6.22: Key Sizes in Wireless Systems

The key sizes used in current wireless systems are not sufficiently large enough for good security. In IEEE 802.11, a 40-bit key is used in the encryption algorithm (called wired-equivalent privacy or WEP) that is not secure by today's standards. IS-136 uses a 64-bit A-key that is more secure, but still considered weak.

The primary advantage of secret-key algorithms is that they are fast, and at the huge data rates that are being supported by today's networks, it is virtually impossible to employ *public-key* algorithms. However, because every pair of users has to have a key, for a communication system with N users, at least $N(N-1)/2$ keys need to be created and distributed. This is not a trivial exercise and has its own weaknesses.

Example 6.23: Number of Keys with Symmetric Key Encryption Algorithms

Assume a small corporate network with 500 computers. A total of 124,750 keys are required (one between each pair of computers). Each computer needs to store 499 keys associated with the remaining computers. Suppose an employee gets a new handheld personal computing device. Not only will this handheld device need to load 500 new keys, but the remaining 500 old computers also need to be each updated with a key for the handheld computer.

There are techniques of key distribution for symmetric key algorithms such as the Needham-Schroeder key distribution scheme and Kerberos [STA98]. All these schemes need several handshaking steps and also an initial configuration of computers with *master keys*. Such master keys can be distributed physically in a secure manner. However, key distribution is still a potential weakness in the system. Another way of generating fresh keys for each communicating session is to use the master key and a one-time random number (called nonce) as inputs to a one-way hash function to generate a key. Alternatively, the nonce can be encrypted using the master key.

Example 6.24: Generation of Secret Keys in Wireless Networks

Most wireless networks (including cellular networks) and several wired networks employ identification schemes that depend on a *shared master key* followed by secure forms of hash algorithms to generate "fresh" keys that can then be used with a secret key algorithm to provide various security services. In general hashing a random number concatenated with a secret identifying parameter known only to the communicating parties can securely generate keys. In most cases (like IEEE 802.11), the size of the identification parameter (PIN number or master key) and the algorithms employing it provide loopholes and vulnerability in the protocol. It is generally suggested that the shared secret (whether a password or a PIN) should be at least as long as 80 to 128 bits (and the hash algorithm employed should have an output of at least 160 bits) because of a square-root attack called the "birthday attack" [STA98].

Block ciphers such as DES and the *advanced encryption standard* (AES) encrypt blocks of data at a time. Stream ciphers encrypt bits or bytes of data [STI95]. The advantage of stream ciphers is that there is no need for buffering data up to the block size and for padding. Stream ciphers may also be more suitable for a jitter sensitive voice conversations. The disadvantage is that these have to be used carefully because encryption with stream ciphers uses simple XOR operation.

Example 6.25: Use of Stream Ciphers in IEEE 802.11

In IEEE 802.11, the encryption algorithm RC-4 [STA98] is used to generate a pseudorandom key stream using a 40-bit master key and an initial vector (IV). The data are then simply XOR-ed with the key stream to create the ciphertext.

DES was the secret-key encryption standard for over 20 years. The National Institute for Standards and Technology (NIST) examined proposals for an AES in 1998. Of the five candidate algorithms, NIST selected Rijndael as the algorithm for AES in October 2000. AES is being considered for use in IEEE 802.11. A variety of factors were considered by NIST to determine the suitability of the algorithm for a standard. Security—resistance to cryptanalysis, mathematical soundness, randomness of the algorithm output; cost—licensing requirements, computational efficiency on various platforms, memory requirements; and algorithm implementation characteristics—ability to handle variable key sizes and block lengths, implementation as stream ciphers and hash functions, hardware and software implementations, and algorithm simplicity are three categories used for evaluation of these algorithms.

In addition to these standards, several freeware and other secret-key algorithms are available such as IDEA, RC-4, and Blowfish [STA98]. RC-4 in particular has been widely employed in Web browsers, as well as in wireless networks such as IEEE 802.11. Many hash algorithms and MAC algorithms exist that are based on secret-key encryption schemes. The secure hash algorithm (SHA) and the hashed MAC (HMAC) are used widely over the Internet for message authentication. The encryption algorithms employed in GSM and the North American digital cellular standards are proprietary.

Public key encryption is a radical shift in the way data is encrypted. Diffie and Hellman introduced the concept in 1977. With secret key algorithms, we have a situation that is similar to having a locked mailbox for *each pair of users*. Both users associated with a mailbox share a key that can unlock or lock the mailbox. Consider Figure 6.14. Here, if Alice desires to communicate with Dan, she unlocks the mailbox shared between her and Dan, deposits the message, and locks the mailbox again. The message is now accessible only to Alice and Dan who also has an identical key.

Clearly the number of mailboxes required for N users like Alice and Dan is $N(N-1)/2$. For example, we have six mailboxes for four users as shown in Figure 6.14.

The situation described is not the natural way in which we employ mailboxes. Mailboxes are *associated with individuals* and not pairs of communicating parties. The natural way to employ a mailbox is as described in the following example. Alice owns a mailbox. Only she has a key to lock or unlock the mailbox (i.e., only Alice has complete control over the mailbox). *Any other person* who wishes to communicate with Alice will deposit the message through a *slot* in the mailbox. Once the message is deposited in the slot, *only Alice has access to it*. Even the originator of the message cannot retrieve it, although he or she may regenerate the message from knowledge of the contents.

Public-key algorithms are similar to this example. Each individual has a pair of keys—the public key and the private key. As the name suggests, everyone knows the public key. So anyone can employ the public key to encrypt a message intended for the owner of the key. The public key is like the slot in the mailbox. Only the owner knows the private key. As a result, once the message is encrypted using the public key of the owner, only the owner can decrypt the message. Not even the originator of the message can decrypt it once the message has been encrypted.

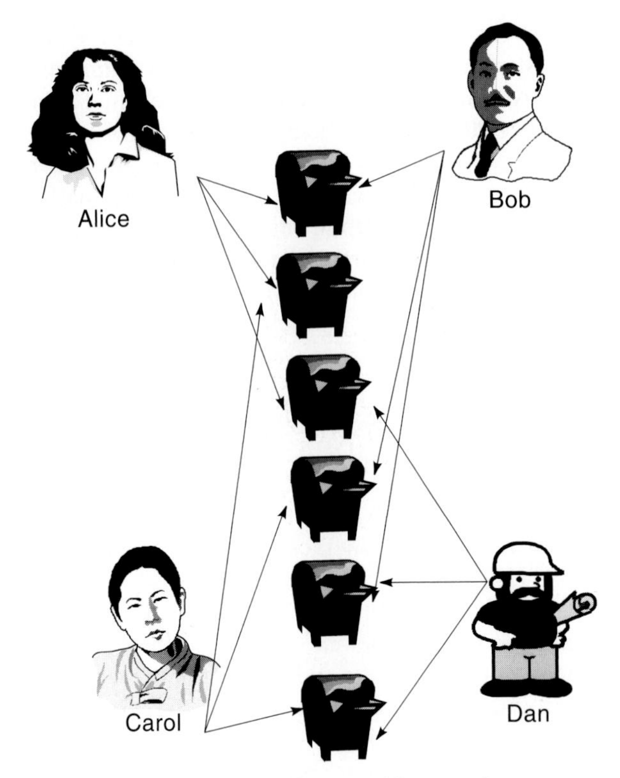

Figure 6.14 Multiple mailboxes with secret-key encryption.

Figure 6.15 shows the schematic of a public-key encryption scheme. Note that there is no need for secure transfer of the key anymore. Alice encrypts a message intended for Bob with Bob's public key $K_{pub,bob}$. The ciphertext is decrypted by Bob via his private key. The design criterion for public-key algorithms is as follows. Given a function $f(k,x)$, the following properties always hold.

- It is extremely easy to compute $y = f(kpub,x)$.
- Given *kpub,* and *y,* it is computationally not feasible to determine $x = f^{-1}(kpub,y)$.
- With a knowledge of *kprv* that is related to *kpub,* it is easy to determine $x = f^{1}(kprv,y)$.

Example 6.26: Trapdoor One-Way Functions

Functions that have these properties are called trapdoor one-way functions. Examples are the factorization problem and the discrete logarithm (DL) problem. The former is based on the fact that it is easy to multiply prime factors to arrive at a composite number (e.g., it is easy to find $7 \times 17 \times 109 \times 151 = 195{,}821$, but it is quite a hard task to split 30,616,693 into its prime number factors). The latter is based on the fact that it is easy to determine what 2^{23} mod 109 is (the answer

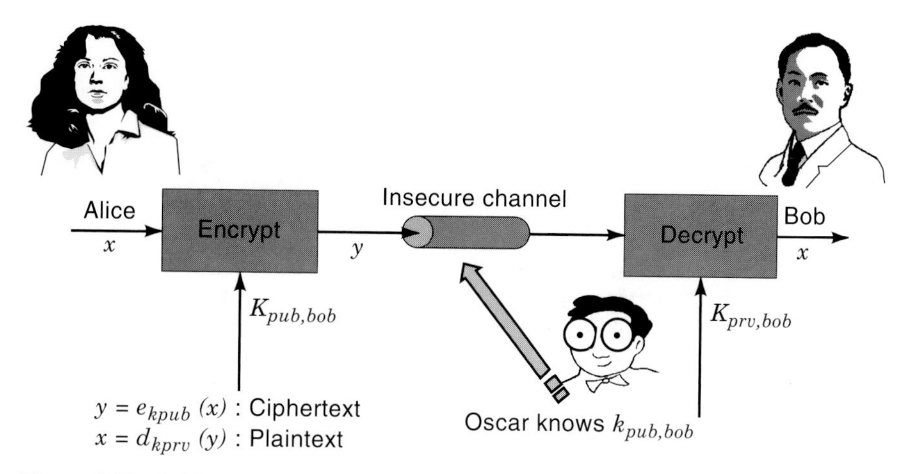

Figure 6.15 Public-key encryption scheme.

is 77). It is quite hard to find out what u is given $2^u \bmod 109 = 68$. Note that with real number arithmetic, it would have been trivial to determine u as $u = \log_2 68$. The modulo function which reduces the operations to be one set of numbers that are nonnegative integers less than 109 makes this problem very hard to solve. Integer factorization is employed in RSA, and the DL problem is used in the Diffie-Hellman Key Exchange Protocol and Digital Signatures.

RSA has been the most popular public-key algorithm. It employs integer factorization. The Diffie-Hellman key exchange protocol based on discrete logarithms is also very popular in wireless networks. This protocol is described in Appendix 6A and is commonly employed for key exchange for Web transactions, e-commerce, and IP security. The digital signature standard (DSS) is also based on discrete logarithms. Signature schemes based on RSA are also widely employed.

However, in the case of public-key algorithms, Oscar, the opponent, is aware of Bob's public key, and this adds an additional parameter to the problem. Because public-key algorithms are based on mathematical structures, for small key sizes, there are well-known results or tables that can be employed to break the encryption. As such, the key sizes are extremely large compared with secret-key algorithms. Today, for good security, public-key algorithms need keys that are three to 15 times larger than their secret-key counterparts. Because the mathematical bases on which public-key algorithms work are well known, they are susceptible to analytical attacks and require much larger key sizes compared with secret key algorithms. The mathematics of elliptic curves are also being employed in encryption schemes because they need smaller key lengths compared with RSA. Table 6.2 presents the key lengths and the time required to break some of the well-known public and secret-key algorithms [SIL99]. The values in this table are based on the assumption that $10 million is available for computer hardware. The key sizes in each row are equivalent.

The mathematical operations for public-key algorithms are also quite computationally intensive. Consequently, the encryption rates are quite small and public-

Table 6.2 Cost Equivalent Key Lengths (in Bits) of Various Encryption Schemes

Secret-key Algorithm	Elliptic Curve	RSA	Time to Break	Memory
56	112	430	Less than 5 minutes	Trivial
80	160	760	600 months	4 Gb
96	192	1,020	3 million years	170 Gb
128	256	1,620	10^{16} years	120 Tb

key algorithms are rarely used for bulk data transfer. Instead, they are employed to exchange a *session key* between a pair of communicating entities who will then use the session key with secret-key algorithms for the duration of that communication alone. This ensures that a new session key is employed each time a communication is initiated, thereby reducing the possibility of an adversary breaking the encryption scheme.

6.4.2.5 Message Authentication

Message authentication is a security service that provides two functions: sender authentication and message integrity. By sender authentication, what we mean is that the receiver can be assured that the message has been originated from the person who claims to have sent the message. Message integrity assures a receiver that no one has modified the message in transit. Both these functions can be accomplished by adding a *message digest* (MD) or MAC to a message. The MAC here should not be confused with the medium access control layer discussed in Chapter 4.

Example 6.27: Message Authentication in IEEE 802.11

In IEEE 802.11, as we saw before, the packet is encrypted using a stream cipher. If the key stream is not correct, upon decryption, the CRC (cyclic redundancy check) in the packet will fail and the access point can discard the packet. However, this implementation can be easily broken. Additionally, it is possible to filter packets based on the 48-bit 802.11 MAC address. Once again, it is not too hard to spoof the physical address of the device.

Secret-key algorithms can be employed for this purpose. The way a MAC is used to provide message authentication is as follows. It creates a fixed length sequence of bits that depend on the message itself and a secret key shared between the communicating parties. Irrespective of whether the message is a few kilobytes long or hundreds of megabytes, the MAC creates a sequence of bits of fixed length that directly depends on the message and the key. This sequence of bits is appended to the message, and then the result is transmitted over the insecure channel. Note that the message could be sent in plaintext form if confidentiality is not an issue. It is computationally infeasible to create a replica of the MAC without the message and key. If the message is modified in transit, the receiver can discover this fact by creating a MAC from the received message and comparing it with the transmitted MAC. Because the secret key is shared only between the communicating parties, it also assures the receiver of the origin of the sender.

An MD operates in a slightly different manner. The MD depends only on the message and not the key. Hash functions are used to create message digests. The message is appended with the MD, and the result is encrypted using a session key shared between the communicating parties. This way, both the message and the MD, which verifies it, are kept secure. The MD has to be sufficiently long to prevent what is known as the "birthday attack." Given a message digest of length b bits, with a good probability, a fake message with the same MD can be generated in $2^{b/2}$ trials. This result is due to the fact that good probabilities of finding two people with the same date of birth exist in a group with roughly the square root of the number of days in a year. That is, in a group of 20 people, it is quite likely that any two would have the same birthday.

Figure 6.16 shows a schematic of message authentication with hash functions. On the left-hand side, Alice concatenates the message x and its hash value $h(x)$ together before encryption the result with the secret key k. The ciphertext $y = e_k(x||h(x))$ is transmitted over an insecure channel. Bob decrypts the ciphertext y and expects to find a message and its hash value concatenated together. He separates the message x from the hash value, computes a new hash value, and compares the two together. If the ciphertext is modified or replaced in between, Bob is able to discover this fact easily. No one can impersonate Alice because it is computationally impossible to create a ciphertext that decrypts into a message and its hash value without knowledge of the key k. Thus both sender authentication and message integrity is assured. The interested reader is referred to [STA98] for other schemes for message authentication. Using the hash function is generally preferred because of its speed.

6.4.3 Identification Schemes

Encryption schemes and hash functions are widely employed in password protection schemes and access control lists that are used for access control—the ability to allow or deny people access to certain resources based on their identification.

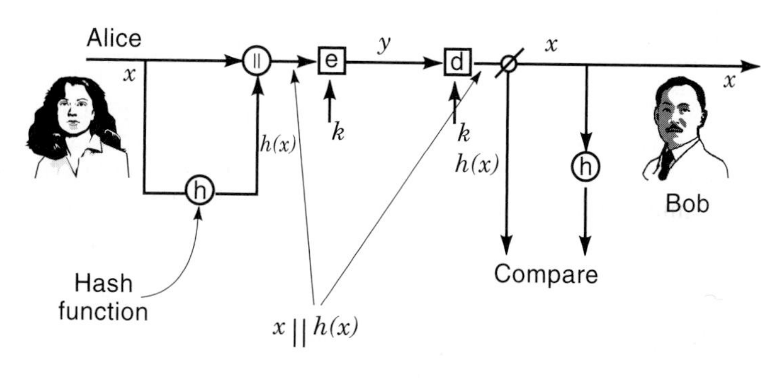

Figure 6.16 Message authentication with hash functions.

Identification by itself is an important security service that needs to be provided for a variety of applications. Access to an automatic teller machine, logging on to a computer, identifying the user of a cellular telephone to the network, etc. involve identification schemes. Note that there is a difference between identification and authentication. When we talk about authentication, there is usually some information-containing message that is exchanged between the parties and one or both parties need to be authenticated. Identification schemes (sometimes referred to as *entity authentication*) involve real-time verification of a party's identity and *need not* involve exchanging information-bearing messages.

Weak identification schemes are based on passwords or pin numbers that are time invariant. Usually the password or pin value is compared with a securely stored hash value. Such schemes are easily susceptible to replay attacks, especially if the password or pin is transmitted over the air in an insecure manner. *Challenge-response* identification or *strong identification* schemes are usually employed in wireless networks. Here, Alice proves her identity to Bob by demonstrating knowledge of a secret, rather than presenting the secret itself. For this purpose, a quantity called the "nonce" is used. A nonce is a value employed no more than once for the same purpose and eliminates replay attacks. Random numbers, time stamps, sequence numbers, and so on are used as nonce in practice. One example of a challenge-response protocol is as follows:

1. Alice is registered with Bob via a password and user name.
2. Bob sends Alice a random number (challenge).
3. Alice replies with an encrypted value of the random number where the encryption is done by using her password as the key (response).
4. Bob verifies that Alice indeed possesses the key (the password).

An eavesdropper Oscar cannot replay the response because the challenge is different if he tries to contact Bob. Oscar also cannot determine the password because the encryption scheme is sufficiently strong and the password is *never* revealed.

Example 6.28: Challenge Response Schemes in Cellular Networks

Figure 6.17 shows the architecture of the IS-136 digital TDMA standard. Details of this architecture are discussed in Chapter 8. The lower half of the figure shows part of the challenge response mechanism implemented in the IS-41 standard for network operations in IS-136. The network (Bob) generates a random number RANDU and sends it over the air to the mobile terminal (Alice). The mobile terminal computes a value AUTHU using the encryption algorithm called CAVE (Cellular Authentication and Voice Encryption) algorithm. The value AUTHU is transmitted over the air. The network computes its version of AUTHU and compares the two values. If the values match, the mobile terminal is identified (authenticated in the terminology of IS-41).

In order to look at the security mechanisms in specific technologies, details of the architectures of these technologies need to be described. We consider such details in subsequent chapters.

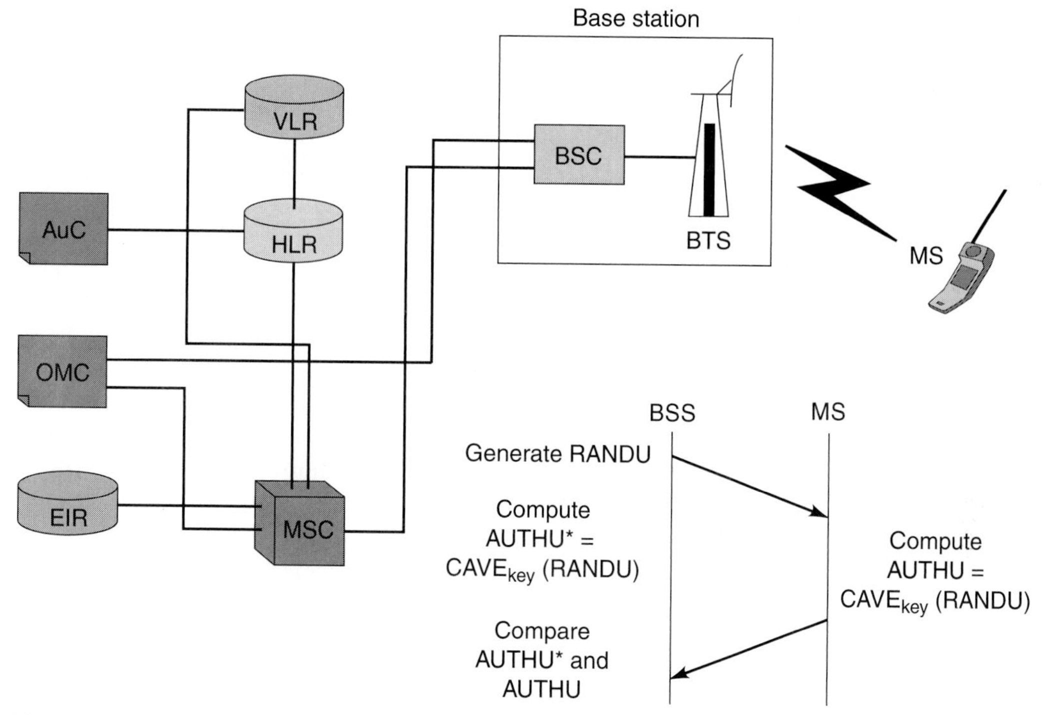

Figure 6.17 Challenge-response mechanism in IS-41.

APPENDIX 6A THE DIFFIE-HELLMAN (DH) KEY EXCHANGE PROTOCOL

The DH key exchange protocol is based on the DL problem discussed in Example 6.4. Let us suppose that Alice wishes to exchange a session key with Bob without sharing any secret with him. Alice chooses a base α and a large prime number p that are publicly known. She only chooses a random private number a. She computes $k_{pubA} = \alpha^a \bmod p$ which she sends to Bob. Note that given k_{pubA}, α, and p, it is computationally impossible to determine a. Similarly, Bob chooses a private random number b and computes $k_{pubB} = \alpha^b \bmod p$ which he transmits to Alice. Once again, it is extremely difficult to determine b. After obtaining the public keys of each other, Alice and Bob raise these public keys to the exponent corresponding to their private numbers respectively. That is, Alice will compute

$$k_s = {k_{pubB}}^a \bmod p = \alpha^{ab} \bmod p \tag{6A.1}$$

Bob computes

$$k_s = {k_{pubA}}^b \bmod p = \alpha^{ab} \bmod p \tag{6A.2}$$

This way both Alice and Bob have generated a common session key. An adversary Oscar cannot determine this key without solving the DL problem. At least, there is no known solution for obtaining the session key other than by solving the DL problem.

APPENDIX 6B NONREPUDIATION AND DIGITAL SIGNATURES

We considered sender authentication and message integrity in this chapter. This does not, however, assure nonrepudiation. For instance, let us suppose that Alice is a consumer and Bob an e-commerce service provider. Bob claims that Alice placed an order with him for purchasing books worth $350, and Alice denies the transaction. Alice claims that she had requested books only worth $100. Both of them are able to produce ciphertexts and messages purportedly used in the transaction. Because both parties know the shared session key, it is impossible to verify who is being truthful and who is not. Public-key algorithms and digital signatures can be employed to resolve such situations.

We know that *only* the owner of the key knows the private key part of a public key algorithm. Consequently, this information can be used to bind the owner to a message transmitted by him. Popular public-key algorithms operate such that it is possible to encrypt a message using a private-key as well. We can compare this with the following scenario. Only the owner of a mailbox can slip a message through the slot because only he has access to the private key that opens the mailbox. No one other than Alice can encrypt the message using her private key (or produce a meaningful ciphertext that can be decrypted with her public key). The problem with this encryption is that *anyone* will be able to decrypt the message because the public key dual is available to everyone. But this is exactly the concept of a signature. If Alice were to sign a document, this means that anyone should be able to verify her signature. However, no one should be able to forge her signature. This is indeed the case here when a message is encrypted using the private key.

Digital signatures take the concept a step further. The entire document *need* not be encrypted. As already discussed, this process would be extremely slow. Instead a message digest of the message is "signed" or encrypted with the private key. The encrypted "signature" is appended to the message. Once again, because it

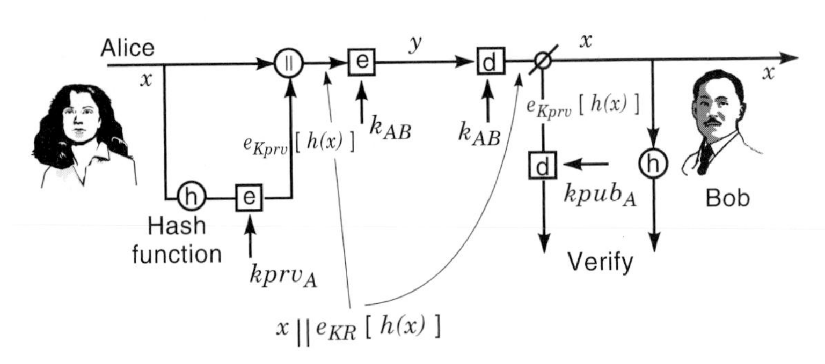

Figure 6B.1 Digital signatures

is computationally impossible to derive a message from the hash, the signature and the message are bound together. If the document needs to be confidential, the usual encryption procedures can be employed after the signature is applied. Figure 6B.1 shows a schematic of how digital signatures are applied. Here, k_{AB} is a session key that is used to keep the document confidential.

QUESTIONS

6.1 What three operational issues are important in wireless networks compared with wired networks? Why?

6.2 Name the two important issues in mobility management.

6.3 What is location management? What are the three components of location management? What are the tradeoffs between them?

6.4 Name three location update mechanisms.

6.5 Name three paging mechanisms. Explain what blanket paging is.

6.6 What are the two steps in handoff?

6.7 Explain three traditional handoff techniques.

6.8 Differentiate between mobile-controlled and mobile-assisted handoff.

6.9 Explain a general handoff procedure. Explain the entities involved with GSM as an example.

6.10 What is agent advertisement? Why is it important in Mobile IP?

6.11 What are smooth handoffs? What application(s) may benefit from them?

6.12 Why is power control important in wireless networks?

6.13 What are the differences in power control for voice-oriented and data-oriented networks?

6.14 Differentiate between open-loop and closed-loop power control.

6.15 Differentiate between centralized and distributed power control.

6.16 Name two types of power-saving mechanisms.

6.17 Differentiate between sleep modes in IS-136 and IEEE 802.11.

6.18 What intelligent protocol features are available in IEEE 802.11 and HIPERLAN to save battery power?

6.19 Describe some energy efficient software approaches.

6.20 What are the privacy and authentication requirements of wireless networks?

6.21 How are public-key and secret-key algorithms different?

6.22 Explain the importance of key sizes in the security of an encryption algorithm.

6.23 What is a challenge-response scheme? How does it work in IS-136?

PROBLEMS

6.1 A mobile terminal samples signals from four BSs as a function of time. The times and signal strengths (in dBm) from the samples are given in Table 6.3. Assume the mobile terminal is initially attached to BS 1 (BS_1). The mobile makes handoff decisions by considering the signals from the BSs after each sampling time. For example, if just RSS is used, just after $t = 12.5$s, the mobile terminal would be connected to BS_3. On a plot, show the handoff transitions between BSs for each of the following algorithms as a function of time. If a condition is met for more than one BS, assume the best one (strongest RSS) is selected.

a. Received signal strength (RSS)

b. RSS + threshold of –60 dBm

c. RSS + hysteresis of 10 dB

d. RSS + hysteresis of 5 dB + threshold of –55 dBm

Table 6.3 RSS from Four Base Stations

Time(s)	0	2.5	5	7.5	10	12.5	15	17.5	20
BS_1	−47	−57	−52	−55	−60	−62	−60	−65	−64
BS_2	−59	−56	−55	−54	−52	−51	−49	−60.5	−52
BS_3	−70	−72	−75	−70	−58	−50	−60.5	−62	−75
BS_4	−72	−71	−65	−60	−55	−53	−50	−49	−56

6.2 In Problem 6.1, which technique is the best in terms of reducing the number of unnecessary handoffs? What parameters will you change to reduce the number of unnecessary handoffs? If the minimum required RSS for good signal quality is −55 dBm, would your answers change?

6.3 A mobile node has a home address of 136.142.117.21 and a care-of address of 130.216.16.5. It listens to agent advertisements periodically.

a. The agent advertisement indicates that the care-of address is 130.216.45.3. What happens? Why?

b. The agent advertisement indicates that the care-of address is 136.142.117.1. What happens? Why?

6.4 Mobile terminals in seven co-channel cells (labeled A) of a cellular system are transmitting on the same frequency channel as shown in Figure 6.18. Due to a glitch in the handoff mechanism, the transmit power of the mobile in the center cell increases without control as it moves away from a BS, and it continues to be connected to it even after

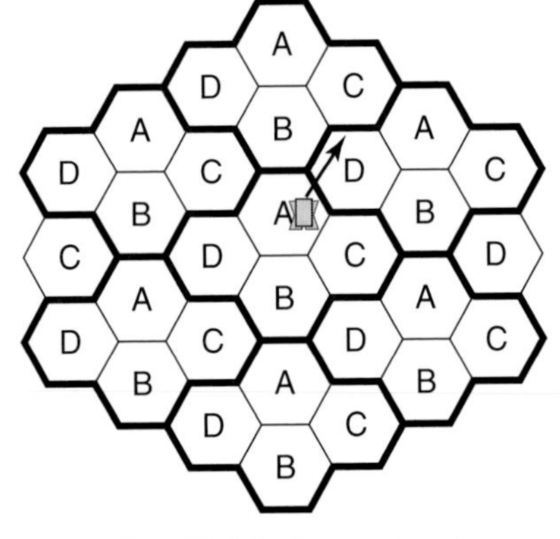

Figure 6.18 Mobile terminals of a cellular system.

it moves out of cell A into cell D as shown in the figure so that its transmit power is now three times as the rest. Assume that the frequency reuse factor is $N = 4$. Determine the interference suffered by the other six mobiles. You can approximate the distance between mobiles with the distance between the cells they are in. Comment on your results. What are the implications of this in termobile terminal of power control?

6.5 The energy per bit in a transmission system is 10 dBm. Using Eq. (6.2), Determine the energy efficiency of error correction codes that have a code rate of 1/5, 1/3, and 1/2 respectively if there are NO errors on the channel. Next consider a harsh radio channel where a stop and wait protocol is employed, i.e., packets are retransmitted if they are not received correctly one by one. About 30% of packets are damaged on this channel on average. The rate 1/5, 1/3 and 1/2 codes can respectively repair 80%, 40%, and 10% of the damaged packets. Determine the energy efficiency of the error correcting codes in this case. Assume that all events are independent.

6.6 A not-so-rich hacker uses an old computer and brute force to break into some wireless systems. It takes him 1 ms on average to test a key to see if it is the right one for an encryption independent of the algorithm employed. How long will it take him to break into an IEEE 802.11 system in the worst case? How long will it take him to break into an IS-136 system on average?

6.7 In Problem 6.6, the hacker realizes that the last six bits of the keys used in a private 802.11 LAN are always zeros. In what time can he break into the system in the worst case?

6.8 In Problem 6.6, the hacker manages to buy a second old computer that can test a key in 1.5 ms. With the two computers, in what time can he break into the system in the worst case? Then he upgrades his computer so that he can test a key in 1 microsecond. In what time can he break into the system in the worst case?

PART THREE Wireless WANs

The third part of the book consists of three chapters devoted to description of important voice- and data-oriented wireless WANs.

CHAPTER 7 GSM AND TDMA TECHNOLOGY

GSM is a complete standard that includes specifications of the air-interface as well as fixed wired infrastructure to support the services. Other TDMA digital cellular standards, such as IS-136 or JTC, are very similar to GSM. Chapter 7 is devoted to detailed description of the GSM standard. The study of this chapter introduces the reader with the complexity and diversity of the issues involved in development of a wireless standard, including elements of the network architecture; mobility support mechanisms such as registration, call establishment, handoff, and security; and details of formation and transmission of packets.

CHAPTER 8 CDMA TECHNOLOGY, IS-95, AND IMT-2000

W-CDMA air-interface is the favorite choice for the IMT-2000 3G standard. The only 2G CDMA standard is the IS-95 air-interface, which is the foundation of the W-CDMA technology. Chapter 8 is devoted to the IS-95 and W-CDMA technology used in IMT-2000. First, we provide a brief description of North American interim standards IS-41 and IS-634, used for MSC-BSC and MSC-MSC communications, respectively. Then we describe details of the IS-95 air interface standard followed by the 3G W-CDMA air interfaces used in IMT-2000.

CHAPTER 9 MOBILE DATA NETWORKS

During the evolution of 2G systems, mobile data or data-oriented wireless WANs emerged as independent connectionless networks serving mobile computers over a large geographical area. As we got closer to the 3G systems, mobile data services became integrated with the cellular voice services. Chapter 9 is devoted to various aspects of this fragmented technology. We first provide an overview of the major mobile data networks and classify them into logical groups. Then we provide details of CDPD and GPRS networks to demonstrate the operation of two popular methods for implementation of the mobile data services. The last sections in the chapter describe SMS and wireless data applications.

CHAPTER 7

GSM AND TDMA TECHNOLOGY

7.1 Introduction

7.2 What is GSM?

7.2.1 GSM Services
7.2.2 Reference Architecture

7.3 Mechanisms to Support a Mobile Environment

7.3.1 Registration
7.3.2 Call Establishment
7.3.3 Handover
7.3.4 Security

7.4 Communications in the Infrastructure

7.4.1 Layer I: Physical Layer
7.4.2 Layer II: Data Link Layer
7.4.3 Layer III: Networking Layer

Questions

Problems

7.1 INTRODUCTION

In this book we first provided the path of evolution of the wireless information networking industry and identified forces that have shaped this evolution. All the major standards that have emerged in this evolution were introduced, and they were logically categorized into different groups at different stages of evolution. In the second part of the book, we introduced principles of operation of these networks by dividing technical aspects into logical categories and explaining each aspect with examples from existing systems. In the rest of the book, we discuss the example systems to provide the reader with a deeper understanding of the details of how a variety of networks operate. This description is also divided into two groups, systems for WANs and those for LANs. Among WANs we first discuss voice-oriented networks that employ TDMA and CDMA for channel access. Our detailed example of TDMA systems is GSM, and for CDMA systems we discuss IS-95 and IMT-2000. Then we examine data-oriented WANs—those integrated with the voice-oriented networks by using the same air-interface and those that have their own air-interface. The example for the first group is GPRS, and the example for the second group is CDPD.

The rest of this chapter is devoted to TDMA systems with GSM as the example system. As we described in Chapter 1, a number of TDMA systems emerged in the 1980s during the evolution of the 2G systems. These systems included the Pan-European GSM, the North American IS-136, and the Japanese JDC digital cellular standards, and CT-2, DECT, PHS, and PACS standards for the then so-called PCS services. In the 1990s, with the advancements in battery technology for handheld terminals and the overwhelming popularity of cellular telephones, the differentiation between digital cellular and PCS systems disappeared. The increasing demand for capacity diverted the attention of the service providers to the availability of spectrum rather than technology. With the emergence of the new PCS bands at around 2 GHz, most service providers expanded their cellular services by upgrading the frequency of operation of their digital cellular systems from around 1 GHz to around 2 GHz, without using the so-called PCS standards. Today, GSM is by far the most popular TDMA standard in the world; it is used both in the cellular and PCS bands. The structure of the system is also very clear and useful for educational purposes. Therefore, we use GSM as our example for TDMA systems.

Wireless networks are complex multidisciplinary systems, and a description of their standards is often very long and tedious. Our objective in explaining standards is to provide the reader with an adequate understanding of the overall objectives of a system, a view of the architecture of the hardware and software elements of the network, and details of protocols and algorithms to understand how information transfer works. In this chapter we first define the objectives and architecture of the GSM. Then we discuss mechanisms that are designed to support mobility to the GSM terminals, and at last we describe protocols used for communication among the elements of the infrastructure. For further details and other presentations of the GSM standard, we refer the readers to [SIE95], [MEH97], [TIS98], [GOO97]. To minimize the difficulties in understanding the details of all standards, we have made a conscious effort to follow a similar pattern in describing other standards in the rest of the book. Following a similar format will help the reader grasp a better overall picture and be more comfortable in reading the details. However, this effort will not completely eliminate our difficulties because each standard uses its own reference model and a number of acronyms that are different from one standard to another. To further help the reader, we have also provided a number of examples that describe certain features of a standard in depth. This way we have preserved the flow of the depth of the text, while some important features are treated in more details.

7.2 WHAT IS GSM?

The Global System for Mobile (GSM) is an ETSI standard for 2G pan-European digital cellular with international roaming. In 1982, frequency bands of 890–915 MHz and 935–960 MHz were allocated for the Pan-European Public Land Mobile Network (PLMN), and the GSM was formed. The main charter of the group was to develop a 2G standard to resolve the roaming problem in the six existing different 1G analog cellular systems in Europe. After evaluating several options, the committee decided to go for a unified new digital standard as it would facilitate roaming and at the same time provide for large-volume production. In 1986, the task force was formed, and in 1987 a memorandum of understanding was signed. In 1989, ETSI included GSM in its domain, and the name of the group was changed to Special Mobile Group (SMG). The resulting standard was named the Global System for Mobile (GSM) communications. In 1991, the specification of the standard was completed, and in 1992, the first deployment started. By the year 1993, 32 operators in 22 countries adopted the GSM standard, and by 2001, close to 150 countries [GSMweb] had adopted GSM for cellular operation.

Although the original goals of the GSM could be met only by defining a new air-interface, the group went beyond just the air-interface and defined a system that complied with emerging ISDN-like services and other emerging fixed network features. To this end, the committee also defined a number of other interfaces between the hardware and software elements of the network, making GSM a complete digital cellular standard that is very suitable for pedagogical purposes. One of the interesting ironies of this evolution is that GSM, and later on all other 2G digital cellular systems, brought ISDN-like mobile digital services to all users while the original wired ISDN lost its popularity and never found a massive acceptance with users. This reflects the real multidisciplinary nature of the telecommunication industry in which the behavior of the market is not always as predictable as more focused industries such as component design.

7.2.1 GSM Services

The first step in understanding a multipurpose system is to identify the services that are provided by that network because the entire network is designed to support these services. Analog cellular systems were developed for a single application—voice—and in a manner similar to analog access to the PSTN, other data services such as fax and voice-band modems were defined as overlay services on top of the analog voice service. GSM is an integrated voice-data service that provides a number of services beyond cellular telephone. Table 7.1 [RED95] shows the GSM Phase 1 services and Table 7.2 [RED95] shows the GSM Phase 2 services. These services are divided into three categories: teleservices, bearer services, and supplementary services.

Teleservices provide communication between two end user applications according to a standard protocol. As shown in Table 7.1, Phase 1 GSM bearer services were telephony, emergency speech calls, Group 3 facsimile, teletex, short messages (unicast and multicast), and videotex. The upper-most layer of the protocol stack of the standard should be specified so that it could communicate with protocols used in these applications.

Bearer services provide capabilities to transmit information among user-network-interfaces or APs. Traditional bearer services include a variety of asynchronous and synchronous data access to PSTN/ISDN and packet switched public data networks as shown in Table 7.1. To implement bearer services, the lower layers and frame format of the standard should specify how these transmissions would be implemented over the air-interface.

Supplementary services are not stand-alone services but they are services that supplement a bearer- or teleservice. Supplementary services in Phase 1 GSM were call forwarding and call barring. They were applied to both bearer and teleservices. Other supplementary services include call waiting and calling number identification. These services are usually implemented at the wired infrastructure of the cellular network. Table 7.2 provides a wider range of services for GSM Phase 2 which demonstrate how services can evolve with different phases in time.

7.2.2 Reference Architecture

Description of a wireless network standard is a complex process that involves detailed specification of the terminal, fixed hardware backbone, and software databases that are needed to support the operation. To describe such a complex system, a reference model or overall architecture is needed to provide an overall understanding of the network elements and operation and divide the system into subsys-

Table 7.1 GSM Phase 1 Services

Service Category	Service	Comment
Teleservices	Telephony	Full rate at 13 kbps voice
	Emergency calls	"112" is GSM-wide emergency number
	Short messaging service	Point to point (between two users) and cell broadcast types
	Videotext access	
	Teletex, FAX, etc.	
Bearer Services	Asynchronous data	300–9,600 bps (transparent/nontransparent)
	Synchronous data	2400–9,600 bps transparent
	Synchronous packet data	
	Others	
Supplementary Services	Call forwarding	All calls, when subscriber is not available
	Call barring	Outgoing calls with specifications

Table 7.2 GSM Phase 2 Additional Services

Service Category	Service	Comment
Teleservices	Half-rate speech coder	Optional implementation
	Enhanced full rate	
Supplementary Services	Calling line identification	Presentation or restriction of displaying the caller's ID
	Connected line identification	Presentation or restriction of displaying the called ID
	Call waiting	Incoming call during current conversation
	Call hold	Put current call on hold to answer another
	Multiparty communications	Up to five ongoing calls can be included in one conversation
	Closed user group	
	Advice of charge	Online charge information
	Operator determined call barring	Restriction of certain features from individual subscribers by operator

tems. Our presentation of the GSM system is organized in three major segments shown in Figure 7.1. These segments are mobile station (MS), base station subsystem (BSS), and network and switching subsystem (NSS). Figure 7.2 provides a more physical representation of the architectural elements of GSM and the relation among these elements. This division of the architectural elements was adopted from [HAU94], and we follow that for the description of the system elements in the following section.

7.2.2.1 Mobile Station

The MS communicates the information with the user and modifies it to the transmission protocols of the air-interface to communicate with the BSS. The user information is communicated with the MS through a microphone and speaker for

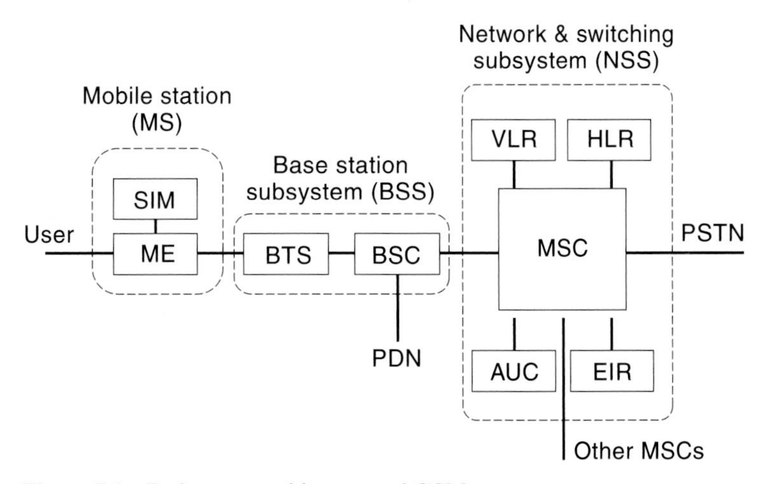

Figure 7.1 Reference architecture of GSM.

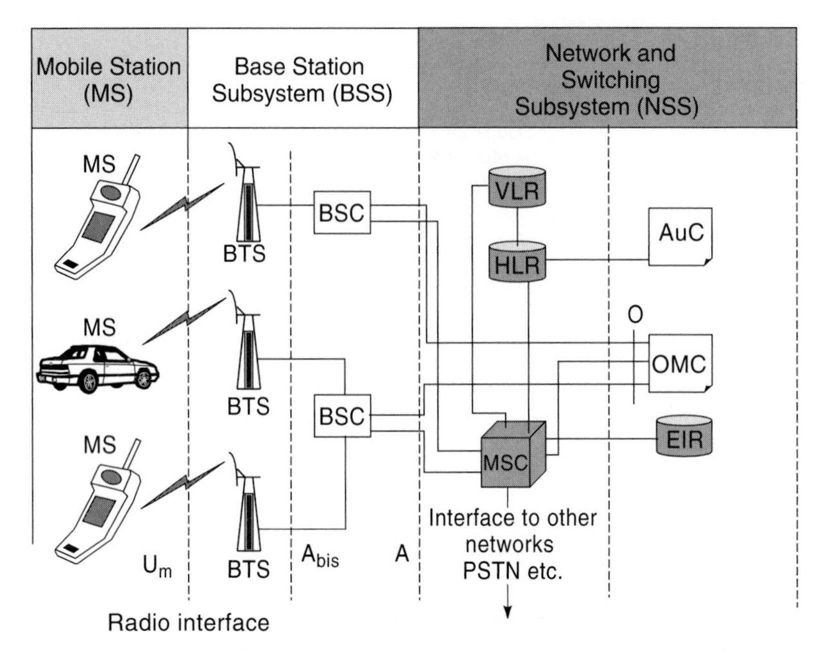

Figure 7.2 A different view of the reference architecture for GSM.

the speech, keypad, and display for short messaging, and the cable connection for other data terminals. The MS has two elements. The first element is *mobile equipment* (ME), which is a piece of hardware that the customer purchases from the equipment manufacturer or their dealers. This hardware piece contains all the components needed for the implementation of the protocols to interface with the user and the air-interface to the BSS. The components include speaker, microphone, keypad, and the radio modem. Therefore, the ME is an expensive piece of hardware. To encourage more users to subscribe to the wireless services, a number of service providers in the early days of the cellular industry, and even today, subsidize the price of the MEs.

The second element of the MS in the GSM is the *subscriber identity module* (SIM) that is a smart card issued at the subscription time identifying the specifications of a user such as address and type of service. The calls in the GSM are directed to the SIM rather than the terminal. Short messages are also stored in the SIM card. Using SIM cards was not a possibility with the analog cellular systems, and the existing North American digital cellular standards have not implemented this option. Although implementing a SIM is a fairly simple concept, it has a significant impact on the way that a user transacts with the service provider. A SIM card carries every user's personal information which enables a number of useful applications.

Example 7.1: Roaming and SIMs

People visiting different GSM-enabled countries who are not keen on making calls at their home number can always carry their own terminal and purchase a

SIM card in every country that they visit. This way they avoid roaming charges and the expense of having a different contact number.

Example 7.2: Sharing a Single Terminal and SIMs

Several users can share a terminal with different SIM cards. At the Telecommunication Laboratory, University of Oulu, Finland, they have a number of GSM MEs that they loan to visitors for use with their own SIMs. Therefore, visitors from the United States and Canada can obtain a cellular service for their personal use without investing in a terminal that may not be useful at home.

Because SIM cards carry the private information for a user, a security mechanism is implemented in the GSM that asks for a four-digit PIN number to make the information on the card available to the user.

7.2.2.2 Base Station Subsystem

The BSS communicates with the user through the wireless air-interface and with the wired infrastructure through the wired protocols. In other words, it translates between the air-interface and fixed wired infrastructure protocols. The needs for the wireless and wired media are different because the wireless medium is unreliable, bandwidth limited, and needs to support mobility. As a result, protocols used in the wireless and wired mediums are different. The BSS provides for the translation among these protocols.

Example 7.3: Speech Conversion

The user's speech signal is converted into 13 kbps-digitized voice with a speech coder and communicated over the air-interface to provide a bandwidth efficient air-interface. The backbone wired network uses a 64 kbps PCM digitized voice in the PSTN hierarchy. Conversion from analog to 13 kbps voice takes place at the MS, and the change from 13 to 64 kbps coding takes place at the BSS.

Example 7.4: Signaling in GSM

The signaling format to establish a connection in wired networks is a multitone frequency scheme used in POTS. GSM, on the other hand, establishes the call through the exchange of a number of packets. The translation of this communication into a dialing signal is made in the BSS.

As with speech coding and dialing, explained in these examples, data transmission protocols over the air-interface are different from that of the wired infrastructure. All these translations are performed at the BSS. As we will see in the description of GPRS in Chapter 9, to implement packet data services on the same air-interface as GSM, the BSS also separates packet switching data from the PSTN traffic and directs it to the packet switched data networks. There are two architectural elements in the BSS.

The *BTS* is the counterpart of the MS for physical communication over the air-interface. The BTS components include a transmitter, a receiver, and signaling equipment to operate over the air-interface, and it is physically located in the center of the cells where the BSS antenna is installed. One BSS may have from one up to several hundred BTSs under its control [RED95].

The second architectural element of the BSS is the *BSC,* that is a small switch inside the BSS in charge of frequency administration and handover among the BTSs inside a BSS. The hardware of the BSC in single BTS site is located at the antenna and in the multi-BTS systems in a switching center where other hardware elements of NSS are located.

7.2.2.3 Network and Switching Subsystem

The NSS is responsible for network operation. It provides for communications with other wired and wireless networks, as well as support for registration and maintenance of the connection with the MSs. The NSS could be interpreted as a wireless specific switch that communicates with other switches in the PSTN and at the same time supports functionalities that are needed for a cellular mobile environment. The NSS in the GSM interconnects to the PSTN through ISDN protocols. Indeed, in the development of GSM a conscious effort has been made to use ISDN compatible protocols. The NSS is the most elaborate element of the GSM network, and it has one hardware, MSC, and four software elements: visitor location register (VLR), home location register (HLR), equipment identification register (EIR), and authentication center (AUC).

A *MSC* is the hardware part of the wireless switch that can communicate with PSTN switches using the signaling system-7 (SS-7) protocol, as well as other MSCs in the coverage area of a service provider. Sometime the MSC that communicates with the PSTN is referred to as Gateway MSC (GMSC) [RED95]. The MSC also provides to the network the specific information on the status of the mobile terminals.

HLR is database software that handles the management of the mobile subscriber account. It stores the subscriber's address, service type, current location, forwarding address, authentication/ciphering keys, and billing information. In addition to the ISDN telephone number for the terminal, the SIM card is identified with an international mobile subscriber identity (IMSI) number that is totally different from the ISDN telephone number. The IMSI is used totally for internal networking applications.

Example 7.5: Numbering Schemes in GSM

The telephone number of a subscriber in Finland could be 358-40-770-5246. The first three digits are the country code; the next two are the digits for the specific MSC, and the rest are the telephone number. The IMSI of the same user can be 244-91 followed by a 10-digit number that is totally different from the ISDN telephone number. The first three digits of the IMSI identify the country, Finland, and the next two digits, the billing company (SONERA, formerly Finnish Telecom).

VLR is a temporary database software similar to the HLR identifying the subscribers visiting inside the coverage area of an MSC. The VLR assigns a temporary mobile subscriber identity (TMSI) that is used to avoid using IMSI on the air. Maintenance of two databases at home and at the visiting site allows a mechanism to support call routing and dialing in a roaming situation where the MS is visiting the coverage area of a different MSC. As discussed in Chapter 6 and as we will see in the later chapters, the mechanism of holding two databases to support mobility is used almost in all mobile networks.

The *AUC* holds different algorithms that are used for authentication and encryption of the subscribers. Different classes of SIM cards have their own algorithms, and the AUC collects all of these algorithms to allow the NSS to operate with different terminals from different geographic areas.

The *EIR* is another database managing the identification of the mobile equipment against faults and theft. This database keeps the *international mobile equipment identity* (IMEI) that reveals the manufacturer, country of production, and terminal type. Such information can be used to report stolen phones or check if the phone is operating according to the specification of its type. The implementation of the EIR is left optional to the service provider.

7.3 MECHANISMS TO SUPPORT A MOBILE ENVIRONMENT

Now that we have described all the hardware and software elements of the GSM network, we can describe how different functionalities of the network is implemented with these elements. Four mechanisms are embedded in all voice-oriented wireless networks that allow a mobile to establish and maintain a connection with the network. These mechanisms are registration, call establishment, handover (or hand-off), and security. Registration takes place as soon as one turns the mobile unit on, call establishment occurs when the user initiates or receives a call, handover helps the MS to change its connection point to the network, and security protects the user from fraud and eavesdropping. General description of these mechanisms are in Chapter 6; in this section we describe the details of their implementation over the GSM architecture that was described in the last section. To illustrate the complexity of wireless networks, when we discuss registration and call establishment in GSM we compare these mechanisms with their counterpart in POTS.

7.3.1 Registration

When we subscribe to a POTS service, the telephone company brings a pair of wires to our home that is connected to a port of a switch in a PSTN end office. Then our telephone number is registered in a database in the network, and our registration is fixed. Therefore, connection and registration process for a wired access to the network is a one-shot operation, and after that connection is active and registration is valid as long as subscription to service is valid. With wireless access to a cellular network, each time that we turn the MS on we need to establish a new connection and possibly establish a new registration with the network. We may actually connect to the network at different locations through a BS that may not be owned by our service provider. Therefore, a wireless network needs a registration process that is far more complex than the registration in wired networks.

Technically speaking, as we turn on an MS it passively synchronizes to the frequency, bit, and frame timings of the closest BS to get ready for information exchange with the BS.

After this preliminary setup, the MS reads the system and cell identity to determine its location in the network. If the current location is not the same as before, the MS initiates a *registration* procedure. During a registration procedure, network provides the MS with a channel for preliminary signaling. The MS provides its identity in exchange for the identity of the network, and finally the network authenticates the MS. The simplest connection takes place if the MS is turned on in the previous area, and the most complex registration process occurs when the mobile is turned on in a new MSC area which needs changes in the entries of the VLR and HLR. The following example illustrated the complexity of the registration process of the GSM when a mobile is turned on in a new MSC.

Example 7.6: The Registration Procedure

Figure 7.3 shows the 12-step registration process in the GSM that takes place when an MS is turned on in a new MSC area. In the first four steps, a radio channel is established between the MS and BSS to process the registration. In the next four steps, the NSS authenticates the MS. In the next three steps, a TMSI is assigned, and adjustments are made to the entries in the VLR and HLR. In the final step, the temporary radio channel for communication is released, and transmission starts over a traffic channel.

Steps	MS	BTS	BSC	MSC	VLR	HLR
1. Channel request	→	→				
2. Activation response		←				
3. Activation ACK		→				
4. Channel assigned	←	←				
5. Location update request	→	→	→			
6. Authentication request	←	←	←			
7. Authentication response	→	→	→			
8. Authentication check				↔		
9. Assigning TMSI	←	←	←			
10. ACK for TMSI	→	→	→			
11. Entry to VLR and HLR				←—	—→	
12. Channel release	←	←				

Figure 7.3 Registration procedure.

7.3.2 Call Establishment

Call establishment in POTS starts with a dialing process that transfers the number to the nearest PSTN switch where a routing algorithm finds the best connection through intermediate switches to the destination. After establishment of the link, the last switch (end office) at the destination sends a signal back to the source to announce whether the destination is available or busy that is signaled to the user at the source. When the destination POTS terminal is off-hook, another signal is sent to the source end-office to stop the waiting tone and establish the traffic line. In the mobile environment we have two separate call establishment procedures for mobile-to-fixed and fixed-to-mobile calls. Mobile-to-mobile calls are a combination of the two. The following two examples provide the detailed procedure in the GSM network for both types of call establishment.

Example 7.7: Mobile Originated Call

The five-step procedure in POTS for call setup changes to a 15-step mobile originated call establishment procedure in the GSM. As shown in Figure 7.4, the first five steps are similar to the registration process in GSM, except that these are done to prepare for call establishment. The next two steps start ciphering (encryption) to provide a protection against eavesdropping. The rest of the steps are similar to those in wired networks except that we have an additional traffic channel assignment procedure.

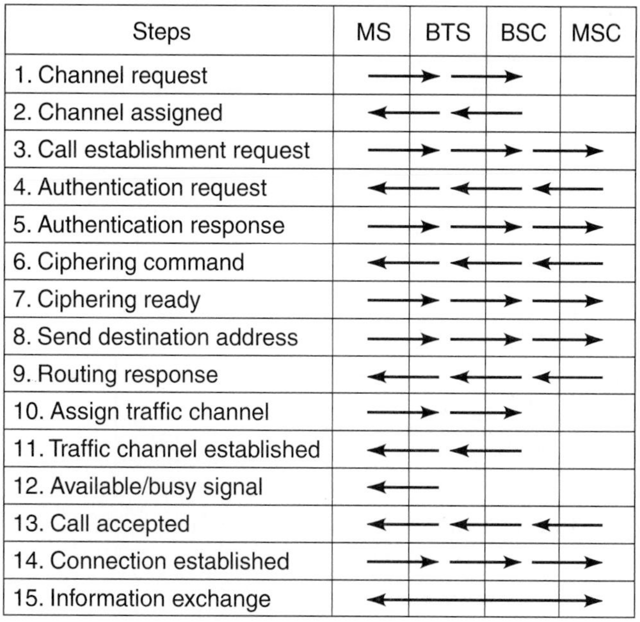

Steps	MS	BTS	BSC	MSC
1. Channel request	→	→		
2. Channel assigned	←	←		
3. Call establishment request	→	→	→	
4. Authentication request	←	←	←	
5. Authentication response	→	→	→	
6. Ciphering command	←	←	←	
7. Ciphering ready	→	→	→	
8. Send destination address	→	→	→	
9. Routing response	←	←	←	
10. Assign traffic channel	→	→		
11. Traffic channel established	←	←		
12. Available/busy signal	←			
13. Call accepted	←	←	←	
14. Connection established	→	→	→	
15. Information exchange	←——	————	——→	

Figure 7.4 Mobile originated call.

Example 7.8: Mobile Terminated Call

The most complicated call establishment is for the situation where a fixed telephone dials a mobile visiting another MSC. As shown in Figure 7.5, after dialing, the PSTN directs the call to the MSC identified by the destination address. The MSC requests routing information from the HLR. Because, in this case, the mobile is roaming in the area of a different MSC, the address of the new MSC is given to MSC, and it contacts the new MSC. At the destination MSC, the VLR initiates a paging procedure in all BSSs under the control of the MSC holding the registration. After a reply from the MS, the VLR sends the necessary parameters to the MSC to establish the link to the MS.

7.3.3 Handoff

Handoff in the United States is referred to as handover in Europe and hence in GSM. The procedures for handoff broadly follow the procedures described in Chapter 6 that dealt with mobility management in general. There are two types of handover—internal and external. Internal handover is between BTSs that belong to the same BSS, and external handovers are between two different BSSs belonging to the same MSC. Sometimes there are handoffs between BSSs that are controlled by two different MSCs. In such a case, the old MSC continues to handle call management. Roaming between two MSCs in two different countries is prohibited, and the call simply drops.

Handoff is initiated because of a variety of reasons. Signal strength deterioration is the most common cause for handoff at the edge of a cell. Other reasons include traffic balancing where the handoff is network oriented to ease traffic congestion by moving calls in a highly congested cell to a lightly loaded cell. The handoff could be synchronous where the two cells involved are synchronized or it may be asynchronous. Because the MS does not have to resynchronize itself in the

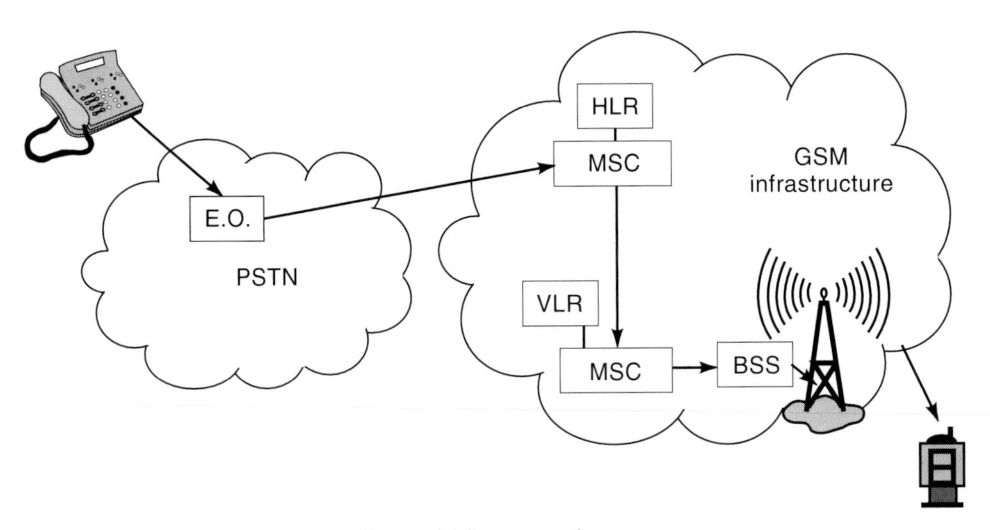

Figure 7.5 Mobile terminated call in a visiting network.

former scenario, the handoff delay is much smaller (100 ms against 200 ms in the asynchronous case).

Figure 7.6 shows the handoff procedure between two BSSs that are controlled by one MSC. The BTS provides the MS with a list of available channels in neighboring cells via the BCCH. The MS monitors the RSS from the BCCHs of these neighboring cells and reports these values to the MSC using the SACCH. This is called *mobile-assisted* handoff as discussed in Chapter 6. The BTS also monitors the RSS from the MS to make a handoff decision. Proprietary algorithms are used to decide when a handoff should be initiated. If a decision to make a handoff is made, the MSC negotiates a new channel with the new BSS and indicates to the MS that a handoff should be made using a handoff command. Upon completion of the handoff, the MS indicates this with a handoff complete message to the MSC.

7.3.4 Security

As discussed in Chapter 6, security in cellular systems is implemented to prevent fraud via authentication, avoid revealing the subscriber number over the air, and encrypt conversations where possible. All these are achieved using proprietary (secret) algorithms in GSM. The SIM cards discussed in Example 7.1 have a microprocessor chip that can perform the computations required for security purposes. A secret key K_i is stored on the SIM card, and it is unique to the card. This key is used in two algorithms—A3 and A8—that are used for authentication and confidentiality, respectively. For authentication purposes, the secret key K_i is used in a challenge response protocol using the A3 algorithm between the BSS and the MS. The secret key K_i is used to generate a privacy key K_c that is used to encrypt messages (voice or data) as the case may be using the A8 algorithm. The control channel signals are encrypted using a third encryption algorithm called A5. The size of the secret key K_i is 128 bits, and the response to the challenge is 32 bits long. Consequently, it is not very secure.

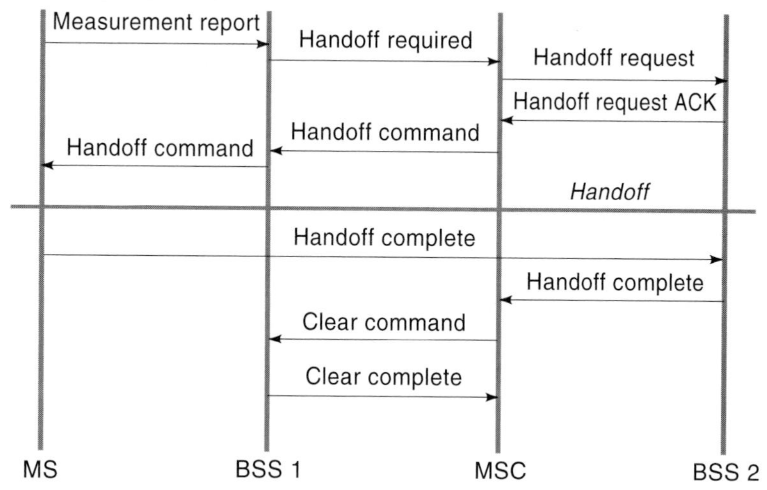

Figure 7.6 Handoff involving a single MSC and two BSSs.

Another aspect of security in GSM is that the secret key information is not shared between systems. Instead a triple consisting of the random number used in the challenge, the response to the challenge, and the data encryption key K_c is exchanged between the VLR and the HLR. The VLR verifies if the response generated by the MS is the same. The algorithms A8 and A3 are secret and not shared between different systems.

7.4 COMMUNICATIONS IN THE INFRASTRUCTURE

In the previous sections of this chapter, we introduced the GSM services and architectural elements, as well as an overview of the mechanisms that allows this architecture to support mobile operation. In this section we provide the description of how these elements and mechanisms are integrated with one another to implement the services. Elements of a network communicate with each other through a protocol stack that is specified by the standard committee. The GSM standard specifies the interfaces among all the elements of the architecture that was discussed earlier. Figure 7.7 shows the protocol architecture for communication between the main hardware elements and the associated interfaces.

The air-interface U_m, which specifies communication between the MS and BTS, is the most detailed and wireless related interface. The A-bis interface between the BTS and BSC and the A interface between BSC and MSC draw signifi-

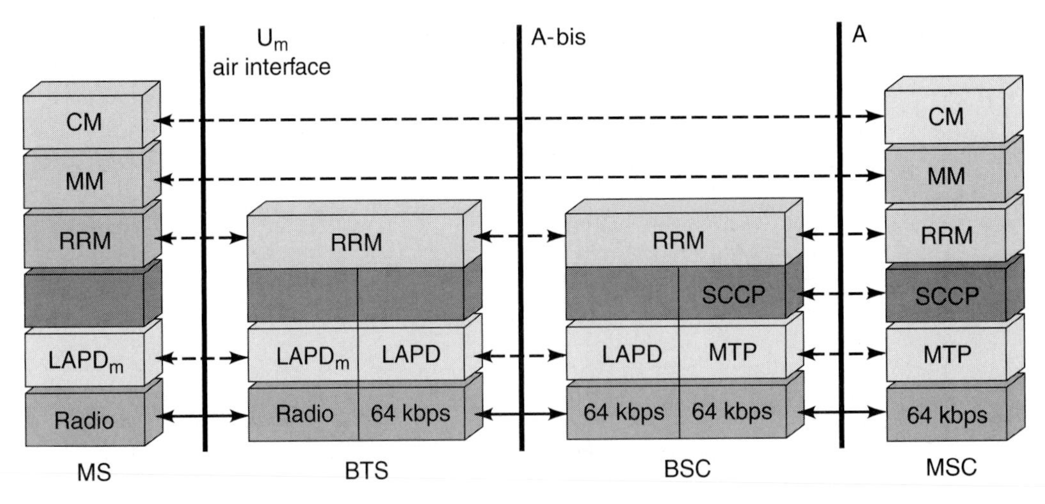

Figure 7.7 The GSM protocol architecture.

cantly on the existing ISDN protocols. The protocol stack is divided into three layers:

Layer 1: Physical Layer

Layer 2: Data Link Layer (DLL)

Layer 3: Networking or Messaging Layer

Messages between the BTS and BSC flow through the A-bis interface. The support on this interface is for voice traffic at 64 kbps and data/signaling traffic at 16 kbps. Both types of traffic are carried over LAPD (which is a data link protocol used in ISDN). The A-interface is used for message transfer between different BSCs to the MSC. The physical layer is a 2 Mbps CCITT connection and employs SS-7 protocols for communication. The Message Transport Protocol (MTP) and the Signaling Connection Control Part (SCCP) of SS-7 are used for error-free transport and logical connection, respectively. The applications that employ the SS-7 protocols deal with direct transfer of data and management (via the BSS application part—BSSAP) for radio resource handling and operation and maintenance information (via the BSS operation and maintenance application part—BSSOMAP) for the operation and maintenance communication messages.

In the following three sections, we cover more details of the three layers with specific examples to provide the readers with an understanding of how a GSM system operates to support different services.

7.4.1 Layer I: Physical Layer

The physical layer of the A and A-bis interfaces follow the ISDN standard with 64 kbps digital data per voice user. The new physical layer defined in the GSM specifications is for the U_m air-interface. This layer specifies how the information from different voice and data services are formatted into packets and sent through the radio channel. It specifies the radio modem details, structure of traffic and control packets in the air, and the packaging of a variety of services into the bits of a packet. This layer specifies modulation and coding techniques, power control methodology, and time synchronization approaches which enable establishment and maintenance of the channels.

7.4.1.1 Power and Power Control

As discussed in Chapter 6, power management is an important issue in wireless networks in general. Power management in cellular telephone networks helps the service provider to control the interference among the users and minimize the power consumption at the terminal. Therefore, power management has direct impact on QoS and the life of the batteries that are the extremely important to the users.

There are three major classes of mobile stations: vehicle mounted, portable, and handheld terminals. Mobile mounted uses the car battery, portables use larger rechargeable batteries, and handheld uses smaller rechargeable batteries. The antenna for the mobile is mounted outside the car, which is away from the user's body, whereas the antenna in the handheld terminals is next to the ear and brain of the user which raises health concerns for high-radiated powers. GSM cells have radii ranging from 300 m to 35 km. The size of the cells also plays a role in the required transmitted power for the BTS and the MS. To allow manufacturers and service providers to accommodate the diversified requirements for different MS and BSS subsystems, a number of radiated power classes are identified by the GSM standard. There are five power classes for the mobile terminal from 29 dBm (0.8 W) up to 44 dBm (20 W) with a 4 dB separation between consecutive mobile classes. There are eight classes for the BTS power ranging from 34 dBm (2.5 W) up to 55 dBm (320 W) in 3-dB steps.

Transmitted radio frequency power in the MS is always controlled to its minimum required value to minimize the cochannel interference among different cells and maximize the life of the battery. The MS is allowed to reduce its peak output power down to 20 mW in 2 dB steps. The BSS calculates the power level for individual MSs by monitoring the interference and received signal strength and sends this information through control signaling packets to the MS.

7.4.1.2 Physical Packet Bursts

GSM uses 890–915 MHz for the uplink (reverse) and 935–960 MHz for the downlink (forward) channels. As shown in Figure 7.8, the 25 MHz band for each direction is divided into 124 channels, each occupying 200 kHz with 100 kHz guard band at two edges of the spectrum. Each carrier supports eight time slots for the TDMA operation. The data rate of each carrier is 270.833 kbps that is provided with a GMSK modem with a normalized bandwidth expansion factor of 0.3. With this data rate, the duration of each bit is 3.69 μsec. The user transmission packet burst is fixed at 577 μsec, which accommodates information bits and a time gap between the packets for duration equivalent to 156.25 times the bit duration of 3.69 μsec.

GSM supports four types of bursts for traffic and control signaling. Figure 7.9 shows all four bursts types. The *normal burst (NB)*, shown in Figure 7.9a, consists of three tail bits (TBs) at the beginning and at the end of the packet, equivalent to 8.25 bits of gap period (GP), two sets of 58 bits encrypted bits (a total of 116 bits), and a 26-bits training sequence. The TBs are 3 zero bits providing a gap time for the digital radio circuitry to cover the uncertainty period to ramp on and off for the radiated power and to initiate the convolutional decoding of the data. The 26-bit

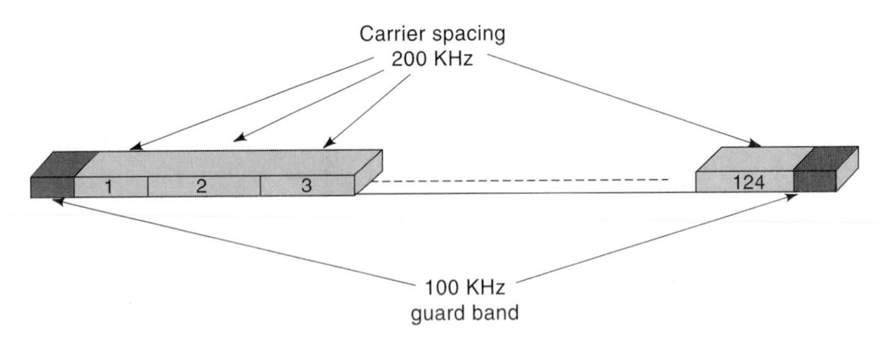

Figure 7.8 Frequency bands in GSM.

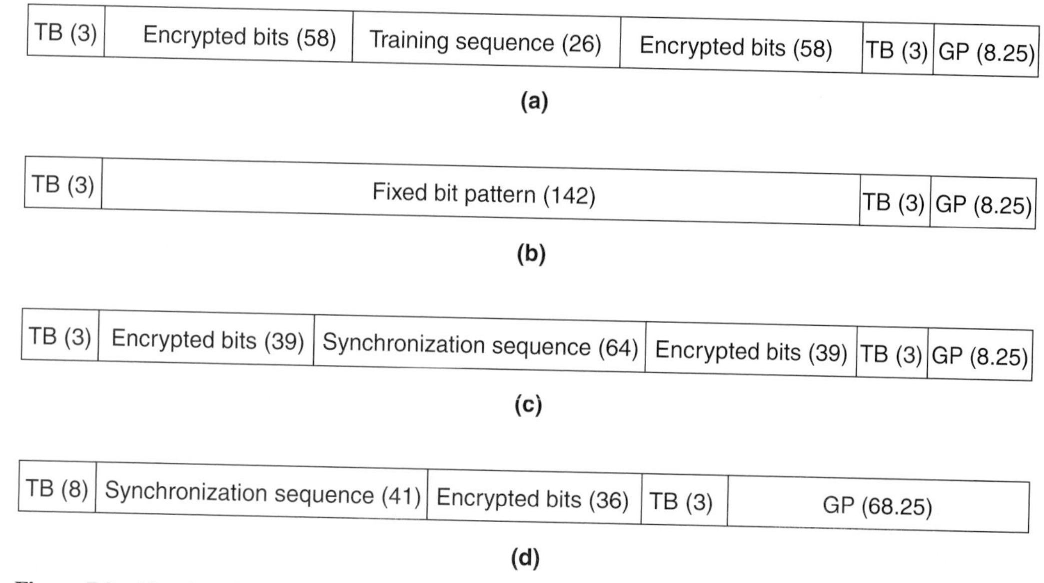

Figure 7.9 The four burst types in GSM: (a) normal burst, (b) frequency correction burst, (c) synchronization burst, and (d) random access burst.

training sequence is used to train the adaptive equalizer at the receiver. Because the channel behavior is constantly changing during the transmission of the packet, the most effective place for the training of the equalizer is in the middle of the burst. The 116 encrypted data bits include 114 bits of data and two flag bits at the end of each part of the data which indicates whether data is user traffic or signaling and control. The user traffic data arrives in frames of length 456 bits as shown in Figure 7.10. They are interleaved into the transmitted normal bursts in blocks of 57 bits plus one flag bit. The purpose of interleaving is to improve the performance for the users by distributing the effects of fade hits among several users. The 456 bits are produced every 20 ms. Therefore, the equivalent of 20 ms of arriving information is mapped into 456 bits. The standard specifies the method that maps the 20 ms of the traffic into the 456 bits.

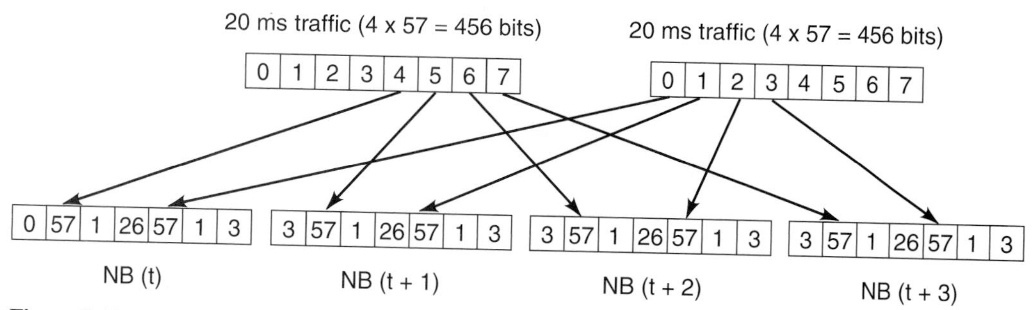

Figure 7.10 Interleaving traffic frames onto TDMA GSM frame in the air.

Example 7.9: Packets of Voice Traffic

Figure 7.11 shows how the 456-bit packets are formed from the speech signal. Each 20 ms of the coded speech at 13 kbps forms a 260-bit packet. The first 50 most significant bits receives a 3-bit CRC code protection first, and then they are added to the second group of 132 bits with lower importance and a 4-bit tail that is all zeros. The resulting 132 + 53 + 4 = 189 bits are then encoded with a ½ convolutional encoder that doubles number of bits to 378. The convolutional code provides for error correction capabilities. The 378 coded bits are added to the 78 least important speech-coded bits to form a 456 bits packet every 20 ms. The 456 bits packets are used to form normal transmission bursts shown in Figure 7.9. In this encoding scheme, we have three classes of speech coded bits. The first class of 50 bits receives both CRC error detection and the rate ½ convolutional error correcting coding protection. The second 132 bits receive only the convolutional encoding protection, and the last 78 bits receives no protection. Therefore, the speech coder can protect the more important bits representing larger values of voltages by assigning them into different categories.

Example 7.10: Packets of Data Traffic

Figure 7.12 shows the formation of the 456 bit packets for 9,600 bps data. The 192 bits of information are accompanied by 48 bits of signaling information and 4 tail bits to form a 244 bits packet that is then expanded to 456 bits using a ½ rate punctured convolutional encoder. Punctured coding can eliminate the need for doubling the number of transmitted bits by eliminating (puncturing) a certain number of bits [PRO01]. The resulting 456 bits are turned to NBs similar to the speech packets. The interesting point is that the 13 kbps speech coded signal and 9,600 bps data modem both occupy the same transmission resources on the air-interface. More channel coding bits are allocated to the data modem packets that are expected to provide better error rate performance.

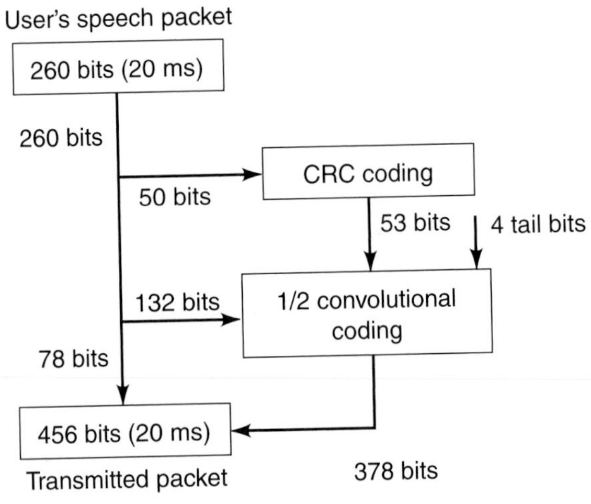

Figure 7.11 Coded speech packets in GSM.

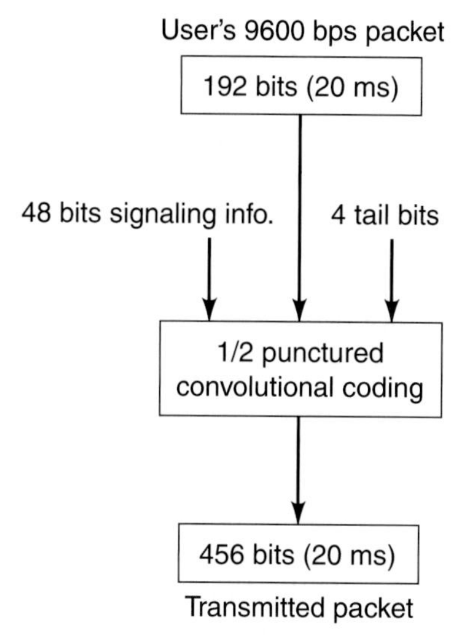

Figure 7.12 Coded data packets in GSM.

Example 7.11: Packets of Signaling Channel

In addition to the traffic channels, we need a number of signaling or control channels that are used to determine how the traffic packets should be routed in the network. Signaling channels using the NB as the channel over-the-air-interface (shown in Figure 7.13) use 184 signaling bits to convey the signaling message. These bits are first block coded with 40 additional parity check bits and 4 tail bits to form a 228-bits block. The 228-bits block is then coded with a ½ rate convolutional encoder to form a 456 bits packet occupying a 20 ms slot that is turned to a burst for transmission as shown in Figure 7.10.

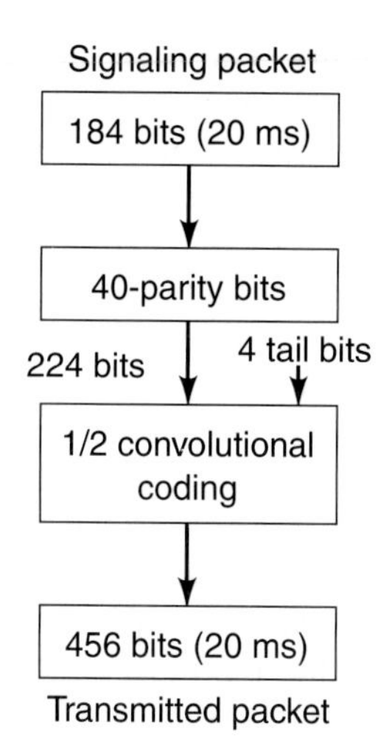

Figure 7.13 Coded signaling packets in GSM.

To describe logical channels in the GSM network, we first divide these channels into two principal categories: *traffic channels* (TCHs) and *control channels* (CCHs). Traffic channels are two-way channels carrying the voice and data traffic between the MS and BTS. TCH logical channels are implemented over the NB physical bursts shown in Figure 7.9(a). There are two types of TCH channels:

- The *full-rate traffic channel* (TCH/F) uses a 13 kbps speech-coding scheme and 9,600 bps, 4,800 bps, and 2,400 bps data. Figures 7.11 and 7.12 show the procedures to create the frames for 13 kbps speech and 9,600 bps data, respectively. As we saw earlier, when we include signaling overhead each channel has a gross bit rate of 22.8 kbps for the network.
- The *half-rate traffic channel* (TCH/H): GSM also supports half-rate speech coding traffic channels. The TCH/H channel uses 16 slots per frame that has a gross bit rate of 11.4 kbps. The half-rate TCH supports 4.8 kbps and 2.4 kbps data.

There are three classes of control channels: *broadcast channels* (BCH), *common control channels* (CCCH), and *dedicated control channels* (DCCH). The BCH channels are broadcast from the BTS to MSs in the coverage area of the BTS. There are three broadcast channels:

- The *frequency control channel* (FCCH) used by the BTS broadcasts carrier synchronization signals. An MS in the coverage area of a BTS uses the broadcast FCCH to synchronize its carrier frequency and bit timing. The physical FCB shown in Figure 7.9(b) is used to implement the logical FCCH.
- The *synchronization channel* (SCH) used by the BTS to broadcast frame synchronization signals to all MSs. Using SCH, MSs will synchronize their counters to specify the location of arriving packets in the TDMA hierarchy. The physical SBs shown in Figure 7.9(c) are used to implement SCH.
- The *broadcast control channel* (BCCH) is used by BTS to broadcast synchronization parameters, available services, and cell ID. Once the carrier, bit, and frame synchronization between the BTS and MS are established, the BCCH informs the MS about the environment parameters associated with the BTS covering that area. The BCCH is physically implemented over the NBs. The BCCH is also a continuously keyed channel, and it is used for signal strength measurements for handoff.

The CCCH channels are also one-way channels used for call establishment. There are three CCCH logical channels:

- The *paging channel* (PCH) is used by the BTS to page the MS for an incoming call is a broadcast channel implemented on a NB.
- The *random access channel* (RCH) is used by the MS to access the BTS for call establishment. The RCH is used for implementation of a slotted-ALOHA protocol, which is used by mobile stations to contend for one of the

available slots in the GSM traffic frames. The RCH is implemented on the short RABs shown in Figure 7.9(d).

- The *access grant channel* (AGCH) is used for implementation of the acknowledgement from the BTS to the MS after a successful attempt by MS using RCH. This channel is implemented on an NB and indicates the TCH for access to the GSM network.

The DCCH are two-way channels supporting signaling and control for individual users. There are three DCCH logical channels:

- The *stand-alone dedicated control channel* (SDCCH) is a two-way channel assigned to each terminal to transfer network control information for call establishment and mobility management. The physical channel for SDCCH occupies four slots in every 51 control-multiframes with an approximated gross data rate of 2 kbps per terminal.
- The *slow associated control channel* (SACCH) is a two-way channel assigned to each TCH and SDCCH channels. The SACCH is used to exchange the necessary parameters between the BTS and the MS to maintain the link. The gross data rate of the SACCH channel is half of that of the SDCCH.
- The *fast associated control channel* (FACCH) is a two-way channel used to support fast transitions in the channel when SACCH is not adequate. The FACCH is physically multiplexed with the TCH or SDCCH to provide additional support to the SACCH.

A more detailed description of the logical channels and GSM operation is available in [GOO97], [RED95]. At this stage we provide an example of using logical channels to implement an operation in a GSM network.

Example 7.13: Logical Channels Used for Call Establishment

Figure 7.15, which is similar to Figure 7.4, represents the 15 steps for mobile initiated call establishment procedure. In Figure 7.15 the logical channel used for each step is also identified. Call establishment is made through the common control RACH and AGCH. Calling number and security is through the dedicated SDCCH, signaling for connection status through dedicated FACCH and traffic exchange through TCH.

7.4.2 Layer II: Data Link Layer

Any connection-based network can be considered to be two networks: one used for traffic and the other for signaling and control. The signaling and control may be through the same physical channels or through separate physical channels. In traffic channels for GSM, as we saw in Figures 7.11 and 7.12, the information bits are encoded with strong error detection and correction codes to form packets of length 456 that are then sent with four normal bursts. Signaling and control data are conveyed through Layer II and Layer III messages. The overall purpose of DLL

Steps	MS	BTS	BSC	MSC
1. Channel request (RACH)	→	→		
2. Channel assigned (AGCH)	←	←		
3. Call establishment request (SDCCH)	→	→	→	
4. Authentication request (SDCCH)	←	←	←	
5. Authentication response (SDCCH)	→	→	→	
6. Ciphering command (SDCCH)	←	←	←	
7. Ciphering ready (SDCCH)	→	→	→	
8. Send destination address (SDCCH)	→	→	→	
9. Routing response (SDCCH)	←	←	←	
10. Assign traffic channel (SDCCH)	→	→		
11. Traffic channel established (FACCH)	←	←		
12. Available/busy signal (FACCH)	←			
13. Call accepted (FACCH)	←	←	←	
14. Connection established (FACCH)	→	→	→	
15. Information exchange (TCH)	↔	↔	↔	

Figure 7.15 Call establishment in GSM using logical channels.

(Layer II) is to check the flow of packets for Layer III and allow multiple service access points (SAP) with one physical layer. In GSM the DLL checks the address and sequence number for Layer III and manages acknowledgments for transmission of the packets. In addition, the DLL allows two SAPs for signaling and short messages (SMS). Unlike other GSM data services that are carried through traffic channels, the SMS traffic channel in the GSM is not communicated through traffic channels. In GSM, the SMS is transmitted through a fake signaling packet that carries user information over signaling channels. The DLL in GSM provides this mechanism for multiplexing the SMS data into signaling streams.

As we saw in Figure 7.13, signaling packets delivered to the physical layer are each 184 bits, this number conforming with the length of the DLL packets in the LAPD protocol used in the ISDN networks. In fact, as shown in Figure 7.7, the LAPD protocol is used for the A and A-bis interfaces connecting the BTS to BSS and BSS to MSC, respectively. The DLL for the U_m air-interface is LAPDm where m refers to the modified version of LAPD adapted to the mobile environment. The length of the LAPDm packets, shown in Figure 7.16, is the same as LAPD, but the format is slightly adjusted to fit the mobile environment. The synchronization bits

Fixed at 184 bits

Address field (8 bits)	Control field (8 bits)	Length indicator (8 bits)	Information bits (variable)	Fill-in bits (variable)

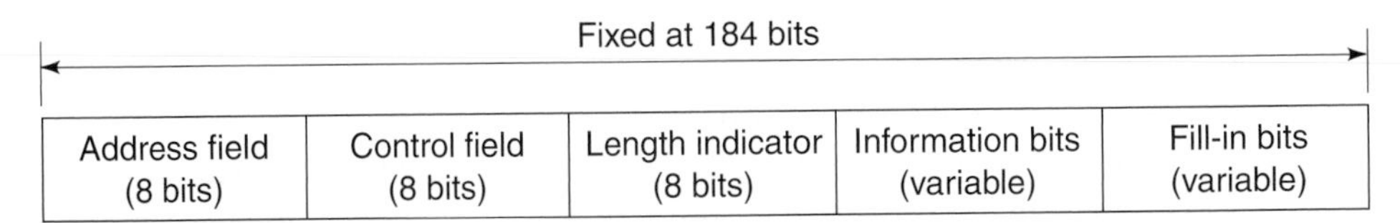

Figure 7.16 Frame format of the DLL in LAPDm.

and CRC codes in LAPD are eliminated in the LAPDm because GSM has the time synchronization and strong coding at the physical layer. The address field is optional, and it identifies the SAP, protocol revision type, and nature of the message. The control field is optional, and it holds the type of the frame (command or response) and the transmitted and received sequence numbers. The length indicator identifies the length of the information field. The information field carries the Layer III payload. Fill-in bits are all "1" bits to extend the length to the desired 184 bits. In peer-to-peer Layer II communications, such as DLL acknowledgments, there is no Layer III payload and fill-in bits cover this field.

The pure or peer-to-peer Layer II messages are set asynchronous balanced mode, disconnect, unnumbered acknowledgment, receiver ready, receiver not ready, and reject. These messages do not have Layer III information bits and are referred to as Layer II messages. The information bits in Layer II packets specify Layer III operations implemented on the logical signaling channels. These information bits are different for different operations.

Example 7.14: Information Field in Layer II Packets

The PCH, AGCH, and BCCH are each 176 bits. The DLL packets for these signaling channels only have an eight bit length of the field that makes a total length of 184 bits encoded into 456 bits transmitted over four physical NBs. The SDCCH and FACCH are each 160 bits with three 8-bits used for address, control, and length of the information fields. The SACCH has 144 bits that needs 16 fill-in bits in addition to the other three fields each carrying 8 bits.

7.4.3 Layer III: Networking Layer

As we discussed in Section 7.3, there are a number of mechanisms needed to establish, maintain, and terminate a mobile communication session. The networking or signaling layer implements the protocols needed to support these mechanisms. The networking layer in GSM is also responsible for control functions for supplementary and SMS services. The traffic channels, as we saw earlier, are mapped into the TCH and carried by normal bursts in different formats associated with different speech or data services. The signaling information uses other bursts and more complicated DLL packaging. A signaling procedure or mechanism or protocol, such as the registration process shown in Figure 7.2, is composed of a sequence of communication events or *messages* between hardware elements of the systems that are implemented on the logical channels encapsulated in the DLL frames illustrated in the last example of the last section. Layer III defines the details of implementation of messages on the logical channels encapsulated in DLL frames. Among all messages communicated between two elements of the network only a few, such as DLL acknowledgment, do not carry Layer III information.

Example 7.15: Format of Layer II and Layer III Messages

Figure 7.17 shows the typical format of Layer II and Layer III messages in a procedure between two elements of the network. They start with simple pure Layer II messages without Layer III information bits to initiate a procedure. Then a number

Layer II messages
Set asynchronous balanced mode
Unnumbered acknowledgement

Layer III RRM, MM, and CM messages started
....
....
....
(Layer III messages ended)

Layer II messages
Disconnect
Unnumbered acknowledgement

Figure 7.17 Typical format of the messages in a procedure used for implementation of a network operation mechanism.

of Layer II messages with Layer III information follow to complete the necessary operation for the procedure. At the end, a couple of pure Layer II messages disconnect the session between the two elements.

Information bits of the Layer II packets, shown in Figure 7.17, specify the operation of a Layer III message. As shown in Figure 7.18, these bits are further divided into several fields. The transaction identifier (TI) field is used to identify a procedure or protocol that consists of a sequence of messages. This field allows multiple procedures to operate in parallel. The protocol discriminator (PD) identifies the category of the operation (management, supplementary services, call control, and test procedure). The message type (MT) identifies the type of message for a given PD. Information elements (IE) is an optional field for the time that an instruction carries some information that is specified by an IE identifier (IEI).

The number of Layer III messages is much larger than the number of pure Layer II messages. To further simplify the description of the Layer III messages, GSM standard divides the messages into three subcategories or sublayers: Radio Resource Management (RRM), Mobility Management (MM), and Connection Management (CM) messages.

The RRM sublayer of Layer III manages the frequency of operation and the quality of the radio link. This sublayer does not have an equivalent in wired networks because there is no frequency assignment issue in the wired networks. The main responsibilities of the RRM are to assign the radio channel and hop to new channels in

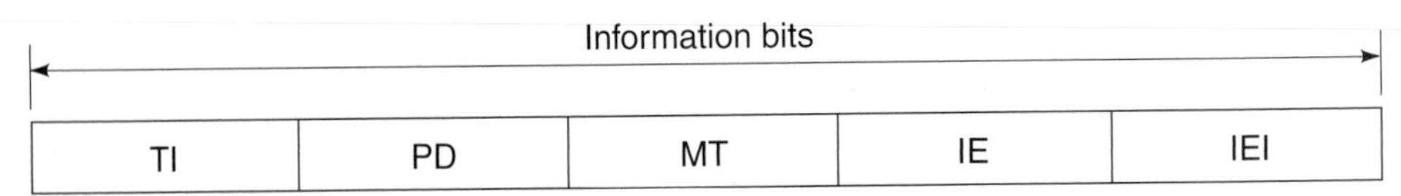

Figure 7.18 The typical Layer III message format.

implementation of the slow frequency hopping option, to manage handover procedure and measurement reports from MS for handover decision, to implement power control procedure, and to adapt to timing advance for synchronization.

The MM sublayer handles mobility issues that are not directly related to the radio. Major responsibilities of this sublayer are location update, authentication procedure, TMSI handling, and attachment and detachment procedures for the IMSI. The CM sublayer establishes, maintains, and releases the circuit-switched connection and helps in SMS. Specific procedures for the CM sublayer are mobile-originated and -terminated call establishment, change of transmission mode during the call, control of dialing using dual-tones, and call reestablishment after MM interruption

An explanation of the details of coding of each message and a complete list of the GSM messages are beyond the scope of this book. For the complete list of the messages used in Layer III of the GSM, the reader can refer to [GOO97], and for further detail of the operation to [RED95] or [GAR99]. We complete this section with an example of a procedure and division of the taskes among different sublayers.

Example 7.16: Call Establishment

Figure 7.19 shows the 15-step mobile initiated call establishment procedure that was discussed earlier in Figures 7.15 and 7.5. The first column identifies the message. The second column identifies the logical channel that is used to carry the message. The third column identifies the sublayer of the Layer III in which GSM standard describes the message. Note that Layer III does not handle the traffic message, and therefore we have no sublayer association for that part of the procedure.

Message name	Logical channel	Category
1. Channel request	RACH	RRM
2. Immediate assignment	AGCH	RRM
3. Call establishment request	SDCCH	CM
4. Authentication request	SDCCH	MM
5. Authentication response	SDCCH	MM
6. Ciphering command	SDCCH	RRM
7. Ciphering ready	SDCCH	RRM
8. Send destination address	SDCCH	CM
9. Routing response	SDCCH	CM
10. Assign traffic channel	SDCCH	RRM
11. Traffic channel established	FACCH	RRM
12. Available/busy signal	FACCH	CM
13. Call accepted	FACCH	CM
14. Connection established	FACCH	CM
15. Information exchange	TCH	

Figure 7.19 Layer III sublayer categories for mobile-assisted call establishment

QUESTIONS

7.1 What are the differences between a mobile digital telephone and POTS?

7.2 Name the three subsystems in the GSM architecture.

7.3 Name the three types of services provided by GSM.

7.4 What is the importance of the framing structure in GSM?

7.5 What are the data services provided by GSM?

7.6 What are the incentives for power control in a TDMA network? Name the elements of the GSM system that are involved in handling power control.

7.7 What are VLR and HLR, where they are physically located, and why we need them?

7.8 What is the difference between registration and call establishment?

7.9 What are the reasons to perform handoff?

7.10 What is the difference between network-decided and mobile-assisted handovers?

7.11 What is the difference between a logical and physical channel?

7.12 Name the five most important logical channels in GSM.

7.13 How does GSM convert 456 bits of the speech, data, or control signal into a normal burst of 156.25 bits?

PROBLEMS

7.1 **a.** Using the bit and time durations in Figure 7.11, show that the speech coding rate for GSM is 13 kbps and the effective transmission rate to support one 13 kbps coded voice channel is 22.8 kbps.

b. What is the required transmission bandwidth for eight slots of the GSM system?

c. Give the overall overhead rate of the system; that is, the difference between the required transmission rate for the traffic and the actual transmission rate of the GSM.

d. Determine the efficiency of the system that is the ratio of the overhead over raw transmission rate.

7.2 **a.** Consider the multiframe transmission in GSM depicted in Figure 7.14. Use the overall structure of the multiframe, frame, and slot to show that the transmission rate of the GSM is indeed 270.833 kbps.

b. In each GSM multiframe 24 frames are used for traffic and two for associated control signaling. Considering the detailed burst frame and multiframe infrastructure, show that the effective transmission rate for each GSM voice traffic is 22.8 kbps.

c. The slow association control channel uses 114 bits of one slot of each 26-slot traffic multiframe. What is the transmission rate for this channel in bits per second?

7.3 The stand-alone dedicated control channel (SDCCH) uses four time slots per each 51-control multiframe shown in Figure 7.14. Use the superframe timing to determine the effective data rate of this logical channel.

7.4 Considering Figure 7.12 give the net data rate (data plus signaling) and the effective transmission rate of a 9,600 bps GSM data service.

7.5 **a.** Considering the frequency allocation strategy of Fig 7.8 for the GSM systems, give the total number of traffic channels per 50 MHz of bandwidth used for two-way GSM communications.

b. Give the total number of GSM channel per MHz of bandwidth.

c. Give the number of channels per cell for frequency reuse factors of N = 4 and N = 3.

7.6 Repeat Problem 7.5 for the IS-136 assuming that this system replaces an AMPS system with 395 traffic channel and a frequency reuse factor of $N = 7$.

7.7 **a.** What is the allowable power ramping time for GSM receivers? (*Hint:* The time gaps of normal, frequency correction, and synchronization bursts, shown in Figure 7.9, are designed to allow power ramping.)

b. The time gap of the random access burst, shown in Fig. 7.9, is designed to assure this packet does not collide with the normal bursts. What is the maximum coverage, the distance between the BS and MS of a GSM base station? Assume that this gap is reserved for two-way travel and radio wave travel at 300,000 Km/sec.

c. The length of the synchronization sequence in synchronization burst is designed to allow time advance for two-way bit synchronization. Use this parameter to calculate the maximum coverage of GSM. Compare your results with that of part (b).

CHAPTER 8

CDMA TECHNOLOGY, IS-95, AND IMT-2000

8.1 Introduction

8.2 Reference Architecture for North American Systems

8.2.1 The IS-634 Standard for MSC-BS Interface
8.2.2 The IS-41 Standard for the MSC-MSC Interface

8.3 What is CDMA?

8.3.1 The IS-95 CDMA Forward Channel
8.3.2 The IS-95 CDMA Reverse Channel
8.3.3 Packet and Frame Formats in IS-95
8.3.4 Mobility and Radio Resource Management in IS-95

8.4 IMT-2000

8.4.1 Forward Channels in W-CDMA and cdma2000
8.4.2 Reverse Channels in W-CDMA and cdma2000
8.4.3 Handoff and Power Control in 3G Systems

Questions

Problems

8.1 INTRODUCTION

In the cellular telephone industry, CDMA is primarily an air-interface or radio transmission technology and access technique that is based on direct sequence spread spectrum techniques described in Chapter 3.[1] Error control coding, spreading of the spectrum, soft handoffs, and strict power control play a very important role in the design and operation of CDMA-based systems. Although the air interface is significantly different in the case of CDMA compared with TDMA techniques, the core fixed network infrastructure that supports the wireless interface is very similar to the structure of the GSM core network. In fact, the core network for North American CDMA and TDMA systems are more or less identical.

In the previous chapter, we discussed the GSM in some detail. The GSM standard came about in Europe as a result of initiatives by ETSI toward a unified digital cellular system. Although the original goals of GSM could be met only by defining a new air-interface, the group went beyond just the air-interface and defined a system that complied with emerging ISDN-like services and other emerging fixed network features. To this end, the committee also defined a number of other interfaces between the hardware and software elements of the network, making GSM a complete digital cellular standard.

Unlike GSM in the EU, in the United States, standards activities are based on developed technologies and have considerable input from the industry. The Telecommunications Industry Association (TIA), or the T1P1 Committee of the Alliance for Telecommunications Industry Solutions (ATIS), develops North American wireless standards. The so-called *Interim Standards* (IS) developed by the TIA form the basis for deploying cellular systems until they are formally specified as TIA or ITU standards. The AMPS was the predominant analog cellular service in the 1980s in the United States. Although AMPS specified the air-interface, very little standardization was available in the backbone infrastructure leading to proprietary implementations and lack of interoperability. Roaming across system boundaries was very complicated, requiring subscriber intervention. To solve these problems, the TIA worked on the IS-41 standard that specifies an open communications interface between two AMPS systems. Digital cellular services evolved in two different directions for the air-interface—TDMA and the IS-136 standard and CDMA and the IS-95 standard. Interoperability between these standards was impossible over the air-interface except via dual mode telephones that are actually implementing two separate radio systems and coordinating the mobile terminal with the available wireless service in an area. The backbone infrastructure specified by the IS-41 standard, however, has now evolved to support both the IS-136 and IS-95 standards.

IS-95 is the North American digital cellular standard that employs CDMA as the access method as well as the air-interface. This technology was developed by Qualcomm around 1990 and is also called cdmaOne today. In 1989, Qualcomm

---

[1]Bluetooth technology uses CDMA with frequency hopping spread spectrum, and UWB technology uses CDMA with pulse position modulation.

first demonstrated its technology and developed the common air-interface specifications in 1991. In 1993, the TIA published Qualcomm's common air-interface specifications as the interim standard IS-95. Since then, this standard has undergone several revisions, such as IS-95a and IS-95b.

Third-generation (3G) systems are being standardized all over the world currently by the ITU under the banner of *International Mobile Telecommunications beyond 2000* (IMT-2000), and there are once again two major paths that are emerging—the first being an evolution of GSM to IMT-2000 and the second an evolution of the North American IS-136 and IS-95 standards to IMT-2000. In both cases, CDMA is the air-interface. The former is being specified by the Third Generation Partnership Project (3GPP) and the latter by a second association called 3GPP2. The 3GPP specified standard is commonly referred to as WCDMA, and the 3GPP2 specification is called CDMA2000. CDMA2000 is meant to be backward compatible with IS-95 or cdmaOne standards.

In this chapter, we look at the North American reference architecture for cellular telephony and consider the air-interface aspects of CDMA through both the 2G IS-95 standard and the emerging 3G standards. The treatment of the reference model and the networking aspects is very brief because of its close similarity to that of the GSM described in the previous chapter. The main emphasis is on the description of the CDMA air-interface specifications used in cdmaOne and IMT-2000. There are several similarities between all these systems in terms of the CDMA air-interface. We present a unified treatment by pointing out the differences. Once again, the reader is referred to several references [e.g., GAR00] for additional details.

8.2 REFERENCE ARCHITECTURE FOR NORTH AMERICAN SYSTEMS

The reference architecture for North American Systems is based on interim standards developed by TIA. The reference architecture is very similar to the GSM reference architecture with a few differences. The Committee TR-45 develops performance, compatibility, interoperability, and service standards for mobile and PCSs in the 800 MHz and 1,800 MHz spectrum bands. These standards consider, among other things, service information, wireless terminal equipment, wireless base station equipment, wireless switching office equipment, and intersystem operations and interfaces. TR-45.3 deals with TDMA technology and TR 45.5 with CDMA technology. The TR45 committee is also closely working with the 3GPP2 group to specify the CDMA2000 standard. TR-46 is the adaptation of TR-45 for the PCS bands, which includes some minor changes and interchanging the names of some elements. Figure 8.1 shows the TR-45/46 reference model.

Example 8.1: The TR-45/46 Reference Model versus the GSM Reference Model

In Chapter 7, we discussed the GSM reference architecture. Figures 7.1 and 7.2 shows the GSM reference architecture; Figure 8.1 shows the TR-45/46 reference architecture. The two are very similar with the exception of a few interfaces and

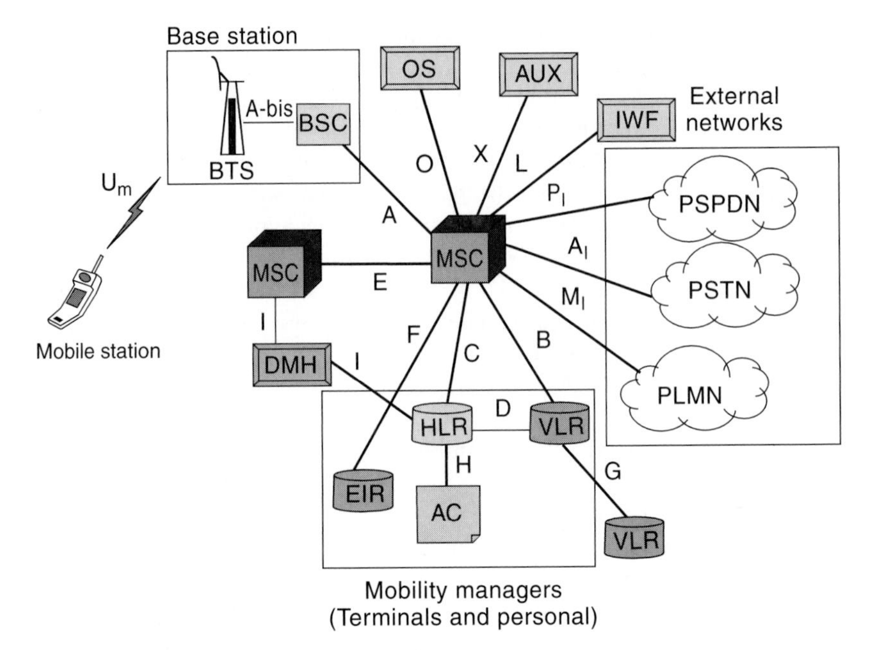

Figure 8.1 TR-45/46 reference model.

elements. In particular, the data message handler (DMH) that collects billing information, the interworking function (IWF) that allows an MSC to connect to other networks, and the auxiliary (AUX) equipment that can connect to an MS are new. The authentication center (AuC) in GSM is shown as the AC in Figure 8.1 and the operation and maintenance center (OMC) is shown as the operation system (OS).

As in the case of GSM, messaging in the infrastructure is carried out by protocols very similar to SS-7 or modifications thereof. In particular, two interim standards are important in the backbone handling of messages. The IS-634 is an open interface standard between the mobile switching center and the radio base station subsystem in a cellular network. The IS-41 protocol permits intersystem roaming and specifies basic and supplementary services.

8.2.1 The IS-634 Standard For MSC-BS Interface

The IS-634 is now formalized as TIA/EIA-634 MSC-BS interface for public wireless communications systems standard. It defines the functional capabilities, including services and features of the messages communicated along the MSC-BS interface, their sequencing and timers at the BS and MSC. In Figure 8.1, this interface is labeled the *A-interface*. The idea behind the standard is that a standard interface allows the BS and MSC equipment to evolve independently and to be provided by multiple vendors. It partitions the tasks between the BS and the MSC without dictating implementation, and it supports all services provided by North American standards. Previously this interface was not standardized, and, hence, both the BS and MSC elements had to be provided by a single vendor. Figure 8.2 shows the layered architecture specified by IS-634.

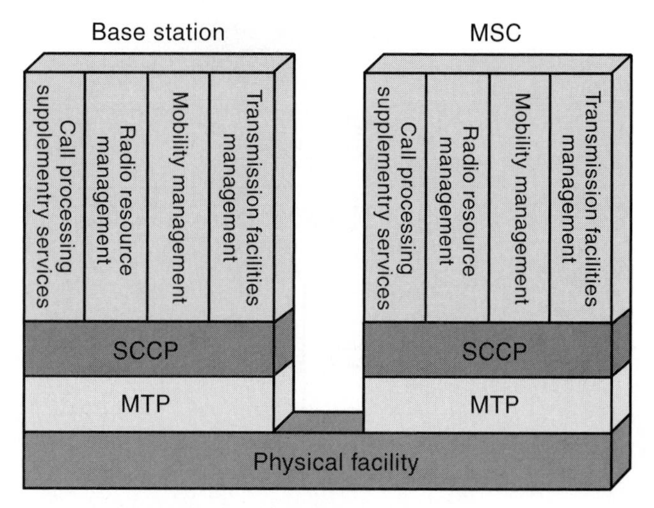

Figure 8.2 IS-634 layered architecture.

In IS-634, the underlying physical layer is based on ISDN over one or multiple T1 carriers at 1.544 Mbps, and in that sense it is similar to GSM. The physical link can be used to transport either signaling or user traffic. The message transport part (MTP) and the signaling connection control part (SCCP) are similar to those in SS-7 and are used only for error-free transport of signaling messages. The IS-634 applications include call processing, radio resources management, mobility management, and transmission facilities management.

Example 8.2: Services Supported over IS-634

Call processing and mobility management procedures occur between the MS and the MSC. Call processing and supplementary services supported over IS-634 include mobile originated and terminated calls, call release, call waiting, and so on. Several messages are defined in IS-634 to support these services. Mobility management supports the usual registration, deregistration, authentication, and voice privacy.

Radio resource management and transmission facilities management procedures occur between the BS and the MSC. RRM, as defined in Chapter 6, is responsible for maintaining a good radio link by supervision, management, and handoff initiation. In the case of CDMA, it has to additionally support soft handoff. The transmission facilities management handles the terrestrial circuits that carry voice, data, or signaling information (e.g., call blocking, overload, etc.).

In Figure 8.3, the functional architecture of IS-634 is shown. Here we see that several BTSs will connect to a BSC. As discussed later on, this is also a result of soft handoff when an MS connects to several BTSs simultaneously. However, from an IS-634 perspective, the BSC is the BS. The transcoder converts the speech from the air-interface QCELP format to the wireline PCM format. This has a bearing on the size of data because QCELP has voice encoded at 13 kbps and PCM at 64 kbps. Depending on where the transcoder is located, the quantity of data to be

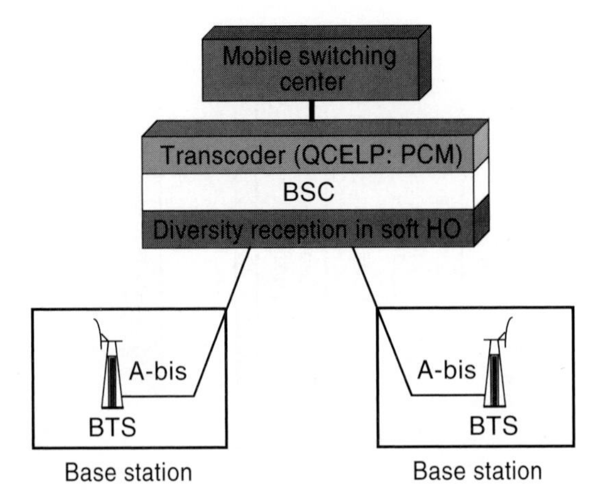

Figure 8.3 IS-634 functional architecture.

transported from the BS to the MSC may be different. IS-634 allows the transcoder to be placed either at the BSC or at the MSC or somewhere in between.

8.2.2 The IS-41 Standard for the MSC-MSC Interface

The TIA/EIA-41 specification, formalized from IS-41, is primarily used in the core network to provide services such as automatic roaming, authentication, intersystem handoff, short message service, and so on. All wireless network elements such as the MSC, HLR, VLR, EIR, and AUC, use this messaging protocol to communicate among themselves. The IS-41 standard provides an open architecture similar to GSM to allow two different North American systems to accommodate roaming.

In its preliminary form, IS-41 simply specified general handoff procedures when it was first introduced in 1988. Since then, Revision A in January 1991 specifies roaming; Revision B in December 1991 specifies dual mode AMPS/TDMA handoff operations; Revision C in 1997 the authentication, CDMA handoff, and short messaging services; Revision D, aspects of international roaming; and Revision E, QoS handling and multimedia services.

With reference to Figure 8.1, the IS-41 standard specifies communications between the core network elements such as the MSC, VLR, HLR, AC, and so on. The signaling protocol stack of IS-41 is very similar to SS7, an out-of-band common channel signaling scheme used in wired telephony, as well as to communicate between different entities in the network through multiple nodes. Only certain application services are different. In addition to SS7's transaction capabilities application part (TCAP), IS-41 also supports the *mobile applications part* or MAP. The MAP defines various messages transported between the core network elements.

If handoff involves two base station subsystems connected to the same MSC, the wireline network is not involved to the extent when the two BSSs are controlled by different MSCs. In the former case, IS-41 has very little role to play because the messaging is restricted to one MSC that also controls the HLR and VLR. In the latter case, IS-41 is involved in so-called *intersystem handoffs*. In this case, the current MSC will request a RSS measurement from the candidate MSC. Once RSS measurements indicate the candidate BSS as suitable for handoff, the two MSCs will complete the intersystem handoff. There are three types of handoffs—handoff forward (transfer from one MSC to another MSC of a new system), handoff backward (transfer from the new MSC back to the old MSC), and handoff third (transfer from a MSC in a second system to a MSC of a third system). In the last case, procedures may also be carried out to reduce the total number of trunks utilized because of the handoff (essentially re-creating the call circuit if necessary).

During handoff, IS-41 signaling messages will carry terminal information (the IMSI, the electronic serial number [ESN], system identifier, the MS capabilities, etc.), the call information (billing ID, trunks carrying the call between the two MSCs), and the air-interface information (serving and destination cells and channels). It also performs authentication procedures between two systems.

8.3 WHAT IS CDMA?

CDMA is both an access method and an air-interface. The rest of the network and system is very similar to any other TDMA system such as GSM that was described in Chapter 7. Radio resource management, mobility management, and security of the CDMA systems are all implemented in the same way as in TDMA systems. There are differences in terms of handling the power control and employing soft handoffs. In this chapter we only address these differences and concentrate on the implementation of the air-interface. For more details of the fixed infrastructure for CDMA and other North American systems, the reader is referred to [GAR00], [GOO97].

With CDMA, all user data, and in most implementations the control channel and signaling information, are transmitted at the same frequency at the same time. All the CDMA systems employ direct sequence spread spectrum and powerful error control codes. The primary significance of CDMA is that by employing a variety of physical layer schemes, it is possible to reuse frequencies in all cells unlike the traditional cellular telephony described in Chapter 5. These include spread spectrum with processing gain, RAKE diversity gain, powerful error correcting codes, variable rate voice coders that provide a gain from pauses in natural conversation, a fast power control mechanism to minimize interference, and soft handoff. This is made possible by reducing the required signal to noise ratio (E_b/N_t) for proper operation. In Chapter 3, we discussed the basics of spread spectrum and error control coding techniques. In this chapter we discuss the details of implementation in CDMA systems. Depending on the system, the actual implementation of CDMA may be different. There are three CDMA systems that need consideration—the IS-95 2G digital cellular standard that specifies the U_m interface in the 800–900 MHz bands for cellular radio in the United States, the JTC-Std-008 for the 1,900 MHz PCS bands, and the emerging WCDMA and CDMA2000 standards for IMT-2000.

CDMA, as it has been implemented in the IS-95 standard, has demonstrated an increase in system capacity compared with the analog and TDMA systems. CDMA improves quality of voice by using a better voice coder, has resistance to multipath and fading, implements soft handoffs, has less power consumption (6–7 mW on average), that is, about 10 percent of analog or TDMA phones because of implementation of power control, and does not require frequency planning because all cells employ the same frequency at the same time. This has resulted in CDMA becoming the popular choice for 3G systems. Although CDMA does provide an inherent flexibility for multimedia traffic, its disadvantage lies in the necessity for power control and implementation complexity. Analysis of the capacity of CDMA systems is presented in Chapter 4 with a comparison among CDMA, FDMA, and TDMA systems.

The air-interface in CDMA systems is by far the most complex of all systems, and it is not symmetrical on the forward and reverse channels unlike TDMA systems. The way spreading the spectrum and error control coding are employed on the forward and reverse channels are different. In the forward channel, transmissions originate at a single transmitter (the BS) and transmissions for all users are synchronized. It is thus possible to employ *orthogonal* spreading codes to minimize the interference between users. On the reverse channel, mobile terminals transmit whenever they have to. As the transmissions are not synchronized, the way spreading is employed is to use the same orthogonal codes for orthogonal modulation to reduce the error rate. In the following sections, we look at the specifics of the forward and reverse channels. We discuss the IS-95 as an example first and discuss the variations in the 3G standards such as WCDMA and CDMA2000 later.

8.3.1 The IS-95 CDMA Forward Channel

The CDMA forward channel is between the base station and the mobile station. The forward link in IS-95 occupies the same frequency spectrum as AMPS and IS-136 North American TDMA standards. Each carrier of the IS-95 occupies a 1.25 MHz of band, whereas carriers of AMPS and IS-136 each occupy 30 kHz of bandwidth. The IS-95 forward channel consists of four types of logical channels—pilot channel, synchronization channel, paging channel, and traffic channels. As shown in Figure 8.4, each carrier contains a pilot, a synchronization, up to seven paging, and a number of traffic channels. These channels are separated from one another using different spreading codes. The modulation scheme employed for transmission of spread signal in the forward channel is QPSK.

The fundamental format of the spreading procedure for all channels is shown in Figure 8.5. Any information contained in the form of *symbols* (after coding, interleaving, etc.) is modulated by *Walsh Codes* which are obtained from *Hadamard Matrices* discussed in Chapter 3. Each Walsh code identifies one of the 64 forward channels. After the channel symbols are spread using the orthogonal codes, they are further *scrambled* in the in-phase and quadrature phase lines by what are called the *short PN-spreading codes*. The PN spreading codes are not orthogonal, but possess excellent autocorrelation and cross-correlation properties to minimize interference among different channels. The PN-spreading codes are M-sequences generated by LFSRs of length 15 with a period of 32,768 chips. The orthogonal codes are used to isolate the transmissions between different channels *within* a cell, and the PN spreading codes are used to separate the transmissions between different cells. In effect the PN sequences are used to differentiate between several BSs

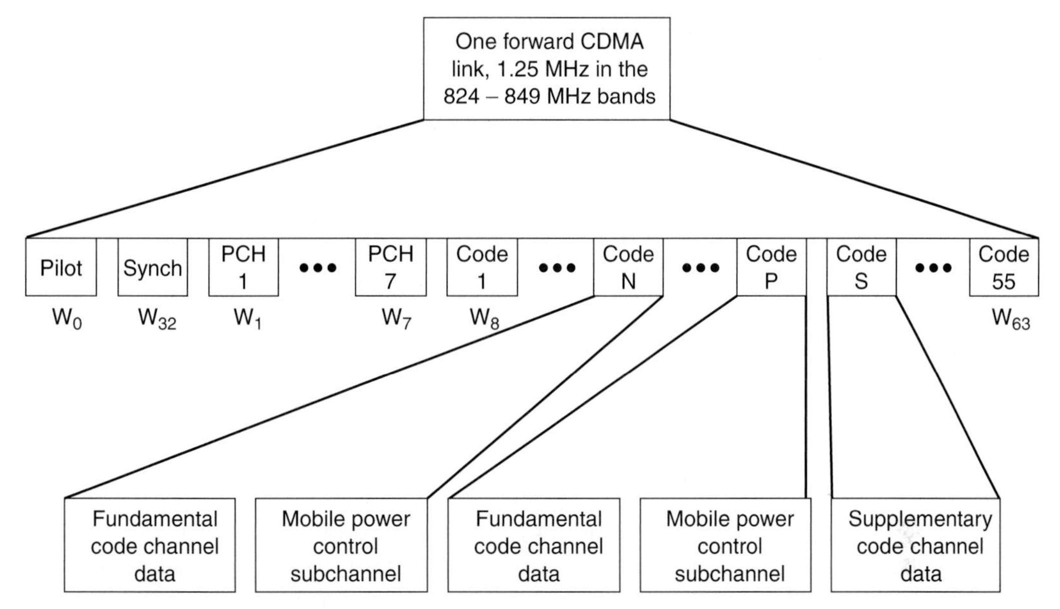

Figure 8.4 IS-95 forward channels.

in the areas that are all employing the same frequency. The same PN sequence is used in all BSs, but the PN sequence of each BS is *offset* from those of other BSs by some value. For this reason, BSs in IS-95 have to be synchronized on the downlink. Such synchronization is achieved using GPS.

The rows of the Hadamard matrix form Walsh codes such that each row in the matrix is orthogonal to every other row. It is possible to generate a Hadamard matrix recursively.

Example 8.3: Recursive Generation of Hadamard Matrices and Walsh Codes

The Hadamard matrix of order 2 is defined as $H_2 = \begin{bmatrix} 0 & 0 \\ 0 & 1 \end{bmatrix}$. All other higher order Hadamard matrices can be obtained via the recursion $H_{2N} = \begin{bmatrix} H_N & H_N \\ H_N & \overline{H_N} \end{bmatrix}$. Here the matrix $\overline{H_N}$ is the matrix H_N with all zeros and ones interchanged.

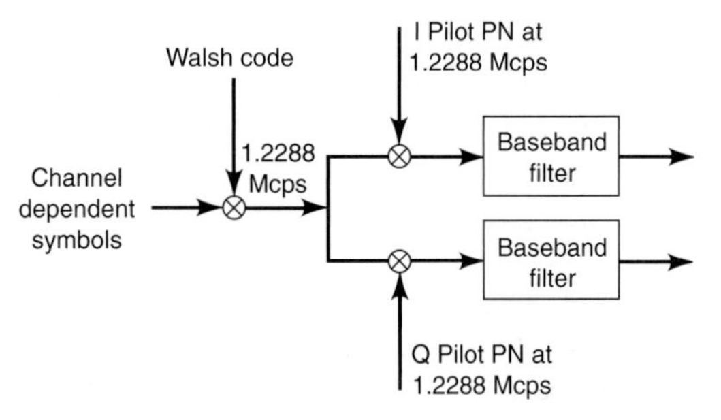

Figure 8.5 Basic spreading procedure on the forward channel in IS-95.

Proceeding in this fashion, it is easy to generate H_{64} that is employed in IS-95. Each row of the Hadamard matrix corresponds to a Walsh code. Consider

$$H_8 = \begin{bmatrix} 0 & 0 & 0 & 0 & 0 & 0 & 0 & 0 \\ 0 & 1 & 0 & 1 & 0 & 1 & 0 & 1 \\ 0 & 0 & 1 & 1 & 0 & 0 & 1 & 1 \\ 0 & 1 & 1 & 0 & 0 & 1 & 1 & 0 \\ 0 & 0 & 0 & 0 & 1 & 1 & 1 & 1 \\ 0 & 1 & 0 & 1 & 1 & 0 & 1 & 0 \\ 0 & 0 & 1 & 1 & 1 & 1 & 0 & 0 \\ 0 & 1 & 1 & 0 & 1 & 0 & 0 & 1 \end{bmatrix}$$

The first Walsh code from this matrix is $W_0 = [0\ 0\ 0\ 0\ 0\ 0\ 0\ 0]$, that is, the all zero code. The last Walsh code is $W_7 = [0\ 1\ 1\ 0\ 1\ 0\ 0\ 1]$. Note that all pairs of Walsh codes are orthogonal.

Various Walsh codes are used for spreading various logical channels in IS-95. The pilot channel employs the all zero Walsh code W_0. The synchronization channel is assigned the Walsh code W_{32} and so on. The assignment of Walsh codes is shown in Figure 8.4.

Example 8.4: The Pilot and Sync Channels

The way the pilot channel is created is shown in Figure 8.6(a). The pilot channel is intended to provide a reference signal for all MSs within a cell that provides the phase reference for coherent demodulation. It is about 4-6 dB stronger than all other channels. The pilot channel is used to lock onto all the other logical channels. It is also used for signal strength comparison. It uses the all zero Walsh code and contains no information except the RF carrier. It is also spread using the PN-spreading code to identify the BS. The way to identify the BS is to *offset* the PN sequence by some number of chips. In IS-95, the PN sequences are used with offsets of 64 chips that provide 512 possible spreading code offsets, providing for unique BS identification in dense microcellular areas as well.

The sync channel is used to acquire initial time synchronization and the way in which it is formed is shown in Figure 8.6(b). It uses the Walsh code W_{32} for spreading. Note that it uses the same PN spreading codes for scrambling as the pilot channel. The sync channel data operates at 1,200 bps. After a rate ½ convolutional encoding, the data rate is increased to 2,400 bps, repeated to 4,800 bps, and then block interleaving is employed. The synch message includes the system and network identification, the offset of the PN short code, the state of the PN-long code (see next section), and the paging channel data rate (4.8 or 9.6 kbps).

The paging channel, as in the case of GSM, is used to page the MS when there is an incoming call, and to carry the control messages for call setup. Figure 8.7 shows how a paging channel message is created. It employs Walsh codes 1 to 7 so that there may be up to seven paging channels. There is no power control for the pilot, synch, and paging channels. The paging channel is additionally scrambled by the PN long code as shown in Figure 8.7. The long code is generated using a paging

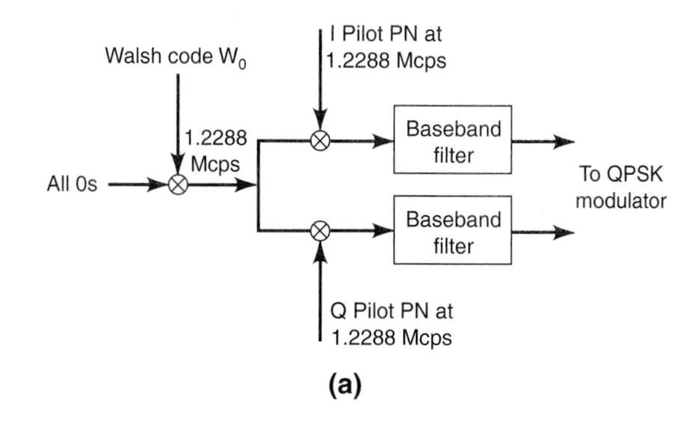

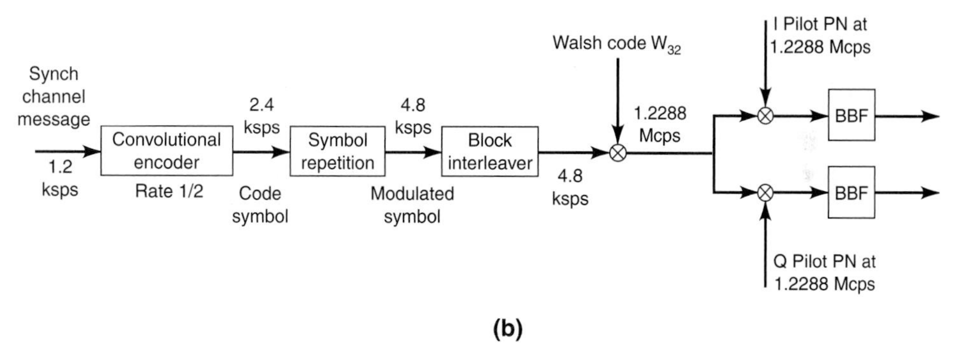

Figure 8.6 (a) Pilot and (b) sync channel processing in IS-95.

channel long code mask of length 42. This means that the PN long code is generated by an LFSR of length 42 and has a period of 2^{42} chips.

The traffic channels carry the actual user information (i.e., digitally encoded voice or data). The forward traffic channel has two possible *rate sets* called RS1 and RS2. RS1 supports data rates of 9.6, 4.8, 2.4, and 1.2 kbps. RS2 supports 14.4, 7.2, 3.6, and 1.8 kbps. RS1 has mandatory support for IS-95, and RS2 can be optionally supported. The way in which the symbols are processed for the two rate sets is shown in Figure 8.8 and 8.9, respectively. Walsh codes W_2 through W_{31} and W_{33}

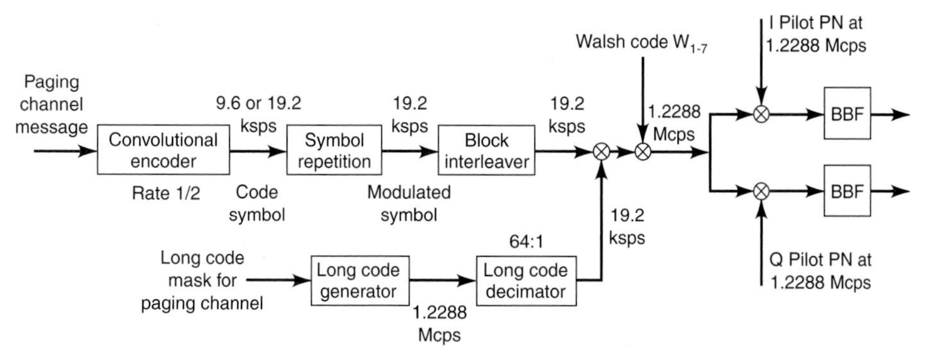

Figure 8.7 Paging channel processing in IS-95.

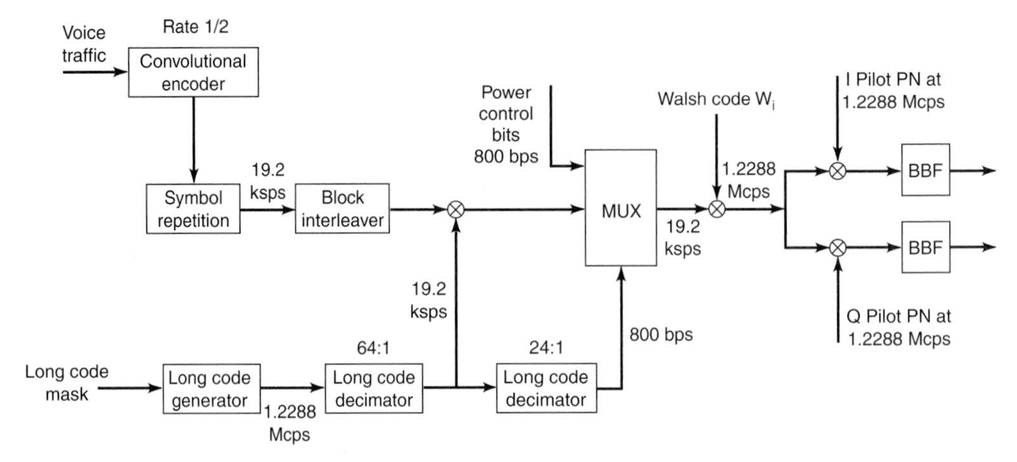

Figure 8.8 Forward traffic channel processing in IS-95 (rate set 1).

through W_{63} can be used to spread the traffic channels depending on how many paging channels are supported in the cell.

Example 8.5: Forward Traffic Channels

On the forward traffic channels, a rate ½ convolutional encoder is used that effectively doubles the data rate. In the case of RS1, symbol repetition is used to increase the data rate to 19.2 kbps. There is no repetition for 9.6 kbps-encoded voice and a repetition of four times for voice at 2.4 kbps. In the case of RS2, the rate at the output of the symbol repeater is 28.8 kbps that is *punctured* by selecting only four out of every six bits. This reduces the data rate to 19.2 kbps at the input of the block interleaver. The forward traffic channels are multiplexed with power control information for the reverse link as shown in Figures 8.8 and 8.9. Power control bits are multiplexed with the scrambled voice bits at 800 bps. Note

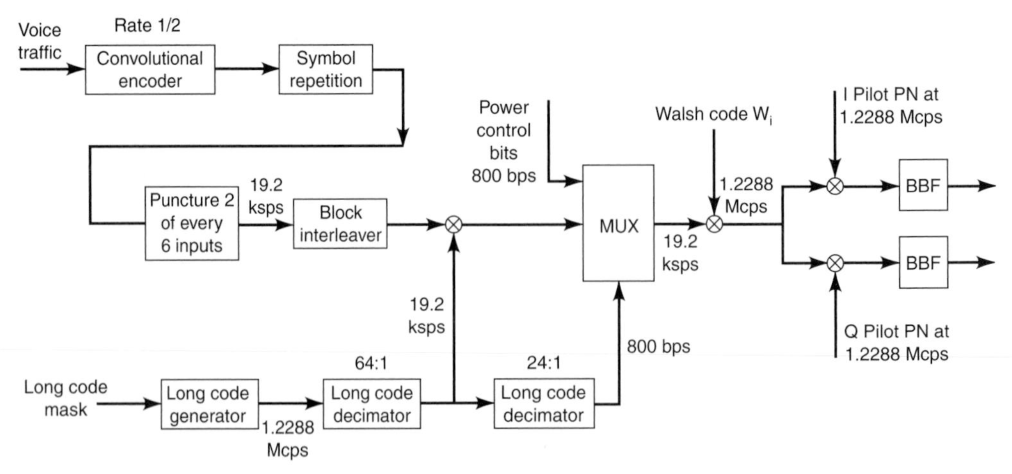

Figure 8.9 Forward traffic channel processing in IS-95 (rate set 2).

that the traffic channels are scrambled with both the PN long code and the PN short codes to further reduce interference among channels.

8.3.2 The IS-95 CDMA Reverse Channel

The CDMA reverse channel is fundamentally different from the forward channel. It employs OQPSK rather than QPSK used in the forward channel. The OQPSK is closer to a constant envelop modulation. As described in Chapter 5, constant envelop modulation techniques provide for a more power efficient implementation of the transmitter at the MS. The QPSK modulation is easier for demodulation again at the MS. The overall structure of the reverse channels in IS-95 is shown in Figure 8.10.

Compared with the forward channel, there is no spreading of the data symbols using orthogonal codes. Instead, the orthogonal codes are used for *waveform encoding*. This means that the reverse link employs an orthogonal modulation scheme that consumes bandwidth but reduces the error rate performance of the system.

Example 8.6: Waveform Encoding in IS-95

As a simple example of waveform encoding, consider the example of the Hadamard matrix H_8. There are eight orthogonal Walsh codes. We can perform a mapping between inputs of three bits to one of eight waveforms as shown in Figure 8.11. A different mapping scheme is employed in IS-95. Consider the Walsh codes of length 64. There are 64 such codes, and they are orthogonal to one another. If these codes are used as waveforms to represent a group of information bits, we can *encode* $\log_2 64 = 6$ bits using a Walsh code. For example, an input

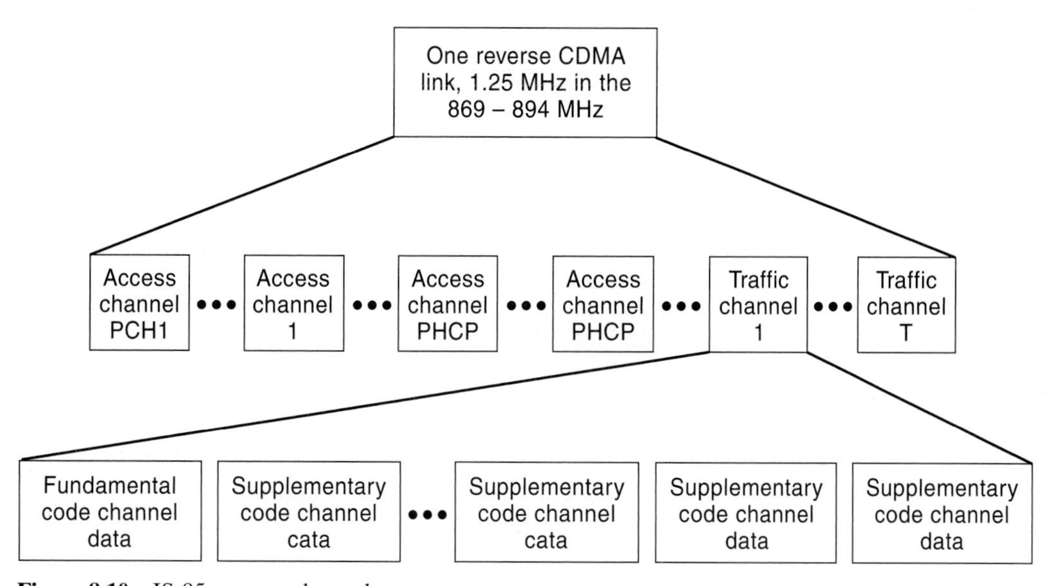

Figure 8.10 IS-95 reverse channels.

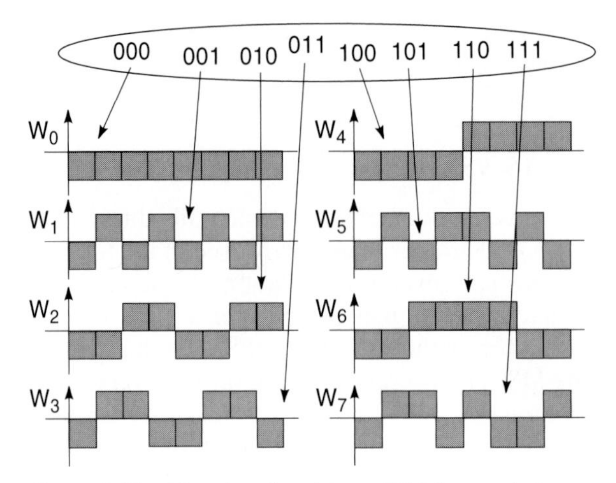

Figure 8.11 Mapping data bits to Walsh encoded symbols.

data stream 0 0 0 0 0 0 can be transmitted using the all zero Walsh code W_0. This is something like a 64-ary modulation scheme where there are 64 symbols or alphabets for transmission. Cross correlation at the receiver is employed to detect the alphabets. In IS-95, the Walsh code that is used for encoding is determined by the equation

$$i = c_0 + 2c_1 + 4c_2 + 8c_3 + 16c_4 + 32c_5$$

where c_5 is the most recent bit. For instance, if the input six bits are (1 1 1 0 1 0), the Walsh code selected is $i = 1 + 2 \times 1 + 4 \times 1 + 8 \times 0 + 16 \times 1 + 32 \times 0 = 23$, i.e., W_{23} is transmitted.

There are basically two types of reverse channels in IS-95—the access channels and the reverse traffic channels.

Example 8.7: The Access Channel in IS-95

The MS transmits control information such as call origination, response to a page, and so on to the BS via the access channels. The data rate over the access channels is fixed at 4,800 bps. It is sent through a rate 1/3 convolutional encoder that increases the data rate to 14.4 kbps. Symbol repetition is employed to increase the data rate to 28.8 kbps. Every six bits is now mapped into 64 bits using the 64-ary orthogonal modulator. The long PN code is used to distinguish between different access channels. It spreads each of the bits at the output of the 64-ary orthogonal modulator by a factor of four that yields a chip rate of 1.288 Mcps. Details are shown in Figure 8.12.

The reverse traffic channel, like the forward traffic channel, supports voice data at two rate sets—RS1 and RS2. In either case, as Figures 8.13 and 8.14 show the data burst after coding and interleaving, but just before the 64-ary orthogonal modulation is at a rate of 28.8 kbps. The output of the 64-ary orthogonal modulator is $28.8 \times 64 / 6 = 307.2$ kcps. After spreading by the long PN code by a factor of

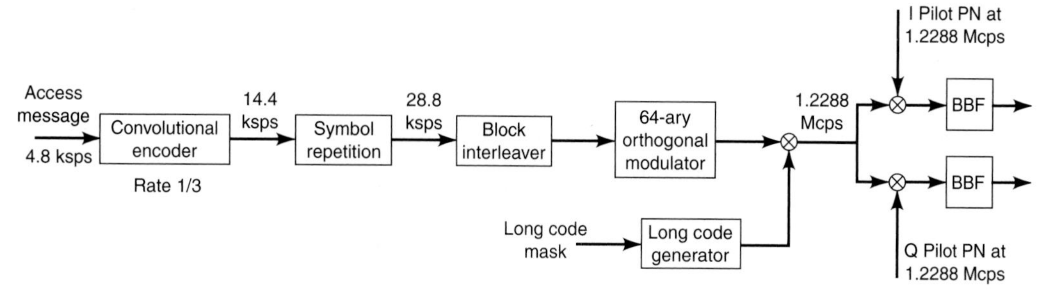

Figure 8.12 Access channel processing in IS-95.

four, the final chip rate is 307.2 × 4 = 1.2288 Mcps. A data randomizer is used in the fundamental code channel to mask out redundant data in case of symbol repetition. More about this is discussed in the section on power control. The reverse traffic channel sends information related to the signal strength of the pilot and frame error rate statistics to the BS. It is also used to transmit control information to the BS such as a handoff completion message and a parameter response message.

8.3.3 Packet and Frame Formats in IS-95

As discussed in the previous sections, the forward logical channels are of four types—the pilot, the synch, the paging and the traffic channels. The reverse channels are either access channels or traffic channels. The forward traffic channel carries user data (either data bits or encoded voice) at 9,600, 4,800, 2,400 or 1,200 bps in RS1 and 14,400, 7,200, 3,600, or 1,800 bps in RS2. The forward traffic channel frame is 20 ms long. Table 8.1 shows the number of information bits, frame error control check bits, and tail bits in each case.

The synch channel provides the MS information about the system identification (SID), the network ID (NID), PN short sequence offset, the PN long code state, and

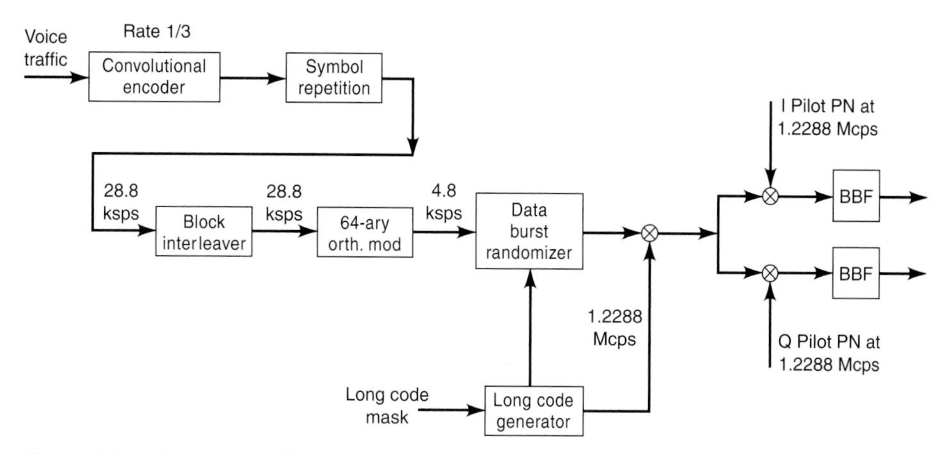

Figure 8.13 Reverse traffic channel processing for fundamental code channel in IS-95.

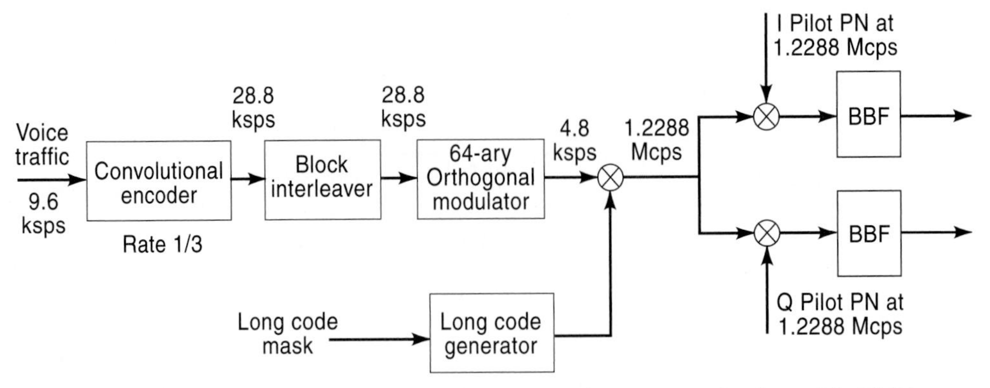

Figure 8.14 Reverse traffic channel processing for supplementary code channel in IS-95.

the system time among other things. Such *messages* can be long and are fragmented into *synch channel frames* of 32 bits shown in Figure 8.15(a). Three of the synch channel frames are combined into a synch channel superframe of 96 bits. The start of message (SOM) bit is 1 for the first synch channel frame and zero for subsequent ones that belong to the same message. The message itself (shown in the top part) consists of the message length, the data, an error checking code, and some padding. Padding with zeros is used to ensure that every new message starts in a new superframe.

The paging channel, shown in Figure 8.15(b) announces a number of parameters to the MS that includes the traffic channel information, the temporary mobile subscriber identity, response to access requests, and list of neighboring base stations and their parameters. Paging can be slotted or unslotted. In the former case, which enables the MS to save on battery power, the channel is divided into 80 ms slots. The paging channel message is similar in structure to the synch channel message (it has a message length, data, CRC, etc.). Because it is too long for transmission in one slot, it is fragmented into 47 or 95 bits (data rate of 4,800 or 9,600 bps) and transmitted over a paging channel *half-frame* (10 ms long). The half-frame has one bit called the synchronization capsule indicator (SCI) that has functionality similar to the SOM bit. In this case however, a message can start anywhere (not necessarily in a half-frame) and a zero value for the SCI could indicate that one paging message ends and another starts within the same half-frame. Eight paging half-frames are combined into one paging slot of 80 ms.

Table 8.1 Frame Contents for Forward Traffic Channels

Rate Set 1				Rate Set 2			
Data Rate	**Information Bits**	**CRC Bits**	**Tail Bits**	**Data Rate**	**Information Bits**	**CRC Bits**	**Tail & Reserved**
9,600 bps	172	12	8	14,400	267	12	9
4,800 bps	80	8	8	7,200	125	10	9
2,400 bps	40	0	8	3,600	55	8	9
1,200 bps	16	0	8	1,800	21	6	9

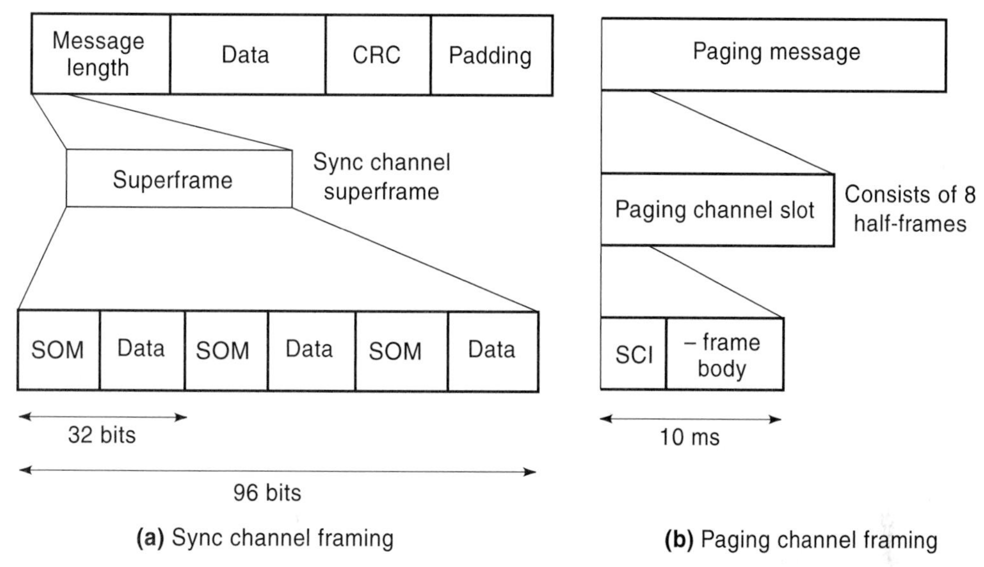

Figure 8.15 Framing in IS-95 forward channels.

Example 8.8: Number of Bits in the Paging Channel Half-Frame and Slot

The number of bits depends upon the data rate. If the data rate is 9,600 bps, a 10 ms half-frame will carry 96 bits (one bit is the SCI) and 48 bits if the data rate is 4,800 bps. Consequently, a paging slot, that has 8 half-frames together, will contain $96 \times 8 = 768$ bits at 9,600 bps and $48 \times 8 = 784$ bits at 4,800 bps.

The access channel data rate is 4,800 bps and each access channel message (very similar in structure to a synch message) is composed of several access channel frames lasting 20 ms. Thus an access channel frame is 96 bits long. An *access channel preamble* always precedes an access channel message, and it consists of several 96-bit frames with all bits in the frame equal to zero. The actual message itself is fragmented into 96-bit frames that have 88 bits of data and 8 tail bits set to zero.

The reverse traffic channel is once again broken into 20 ms traffic channel frames. The frame is further divided into 1.25 ms *power control groups* (PCGs). There are thus 16 PCGs in one frame. A data burst randomizer randomly masks out individual PCGs depending on the data rate that results in less interference on the reverse channel. For instance, at 4.8 kbps (half the data rate), eight PCGs are masked. In addition to voice traffic, the traffic channel can also be used to transfer signaling or secondary data. In the *blank and burst* case, the entire frame carries data. In the *dim and burst* case, part of the frame carries voice and part of it data. The frame structures for the reverse traffic channel are very similar to that of the forward traffic channel.

8.3.4 Mobility and Radio Resource Management in IS-95

Of all the 2G cellular systems, the IS-95 standard is the most complex because of the use of spread spectrum that brings with it a set of advantages not available to TDMA-based systems. These include a frequency reuse factor of one, robust performance in the presence of interference and multipath, and the ability to increase capacity. Operation with a RAKE receiver is an important characteristic of CDMA. It provides inherent diversity in the presence of multipath fading to improve voice quality. The fingers of a RAKE receiver can select either a multipath signal or a signal from another base station if it is within the range of the MS. This ability is employed in IS-95 to perform what are known as *soft handoffs,* which improve voice quality during handoff. Mobility management outside of soft-handoff is based on the general mobility management procedures discussed in Chapter 6. In the case of CDMA specific messages are additionally included.

Using spread spectrum has a disadvantage in that the near-far effect becomes predominant and in order to prevent the signal from one user overwhelming that of another user, strict power control has to be implemented. The advantage of implementing strict power control is that the MS can operate at the minimum *required* E_b/N_t for adequate performance. This increases battery life and reduces the size and weight of the mobile terminal.

8.3.4.1 Soft Handoff

Soft handoff refers to the process by which an MS is in communication with multiple candidate BSs before finally deciding to communicate its traffic through one of them. The reason for implementing soft handoff has its basis in the near-far problem and the associated power control mechanism. If an MS moves far away from a BS and continues to increase its transmit power to compensate for the near-far problem, it will very likely end up in an unstable situation. It will also cause a lot of interference to MSs in neighboring cells. To avoid this situation and ensure that an MS is connected to the BS with the largest RSS, a soft handoff strategy is implemented. An MS will continuously track all BSs nearby and communicate with multiple BSs for a short while if necessary before deciding which BS to select as its point of attachment.

In the IS-95 standard, three types of soft handoffs are defined that are depicted in Figure 8.16. In the *softer* handoff case shown in Figure 8.16(a), the handoff is between two sectors of the same cell. In the *soft* handoff case of Figure 8.16(b), the handoff is between two sectors of different cells. In the *soft-softer* handoff case, illustrated in Figure 8.16(c), the candidates for handoff include two sectors from the same cell and a third sector from a different cell. In all cases, the handoff decision mechanism is more or less the same. Whether the connection in the infrastructure needs to be torn down and set up again depends on the sectors involved in the final handoff.

The soft handoff procedure involves several base stations. A controlling primary BS coordinates the addition or deletion of other base stations to the call during soft handoff. The primary BS uses a handoff direction message (HDM) to indicate the pilot channels to be used or removed as part of the soft handoff process. At some point of time, the primary BS is also changed after handoff. The signals from multiple BSs are combined in the BSC or MSC and processed as a single call. This process is achieved using a *frame selector join* message. Figure 8.17 shows the setup and ending of handoff in a two-way soft handoff. The MS detects a pilot signal from a new BS and informs the primary BS. After a traffic channel is

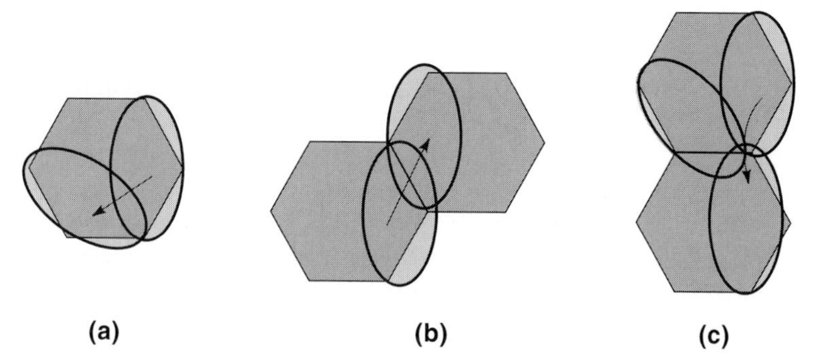

Figure 8.16 (a) Softer, (b) soft, and (c) soft-softer handoff.

set up with the new BS, the frame selector join message is used to select signal from both BSs at the BSC/MSC. After a while, the pilot signal from the old BS starts falling, and the MS will request its removal, which is achieved via a *frame selector remove* message.

The pilot channels of each cell are involved in the handoff mechanism. The reason behind this is that this is the only channel not subject to power control and provides a measure of the RSS. The MS maintains a list of pilot channels that it can hear and classifies them into the following four categories. The *active set* consists of pilots that are being continuously monitored or used by the MS. The MS has three RAKE fingers in IS-95 that allows it to monitor or use up to three pilots. The active set pilot

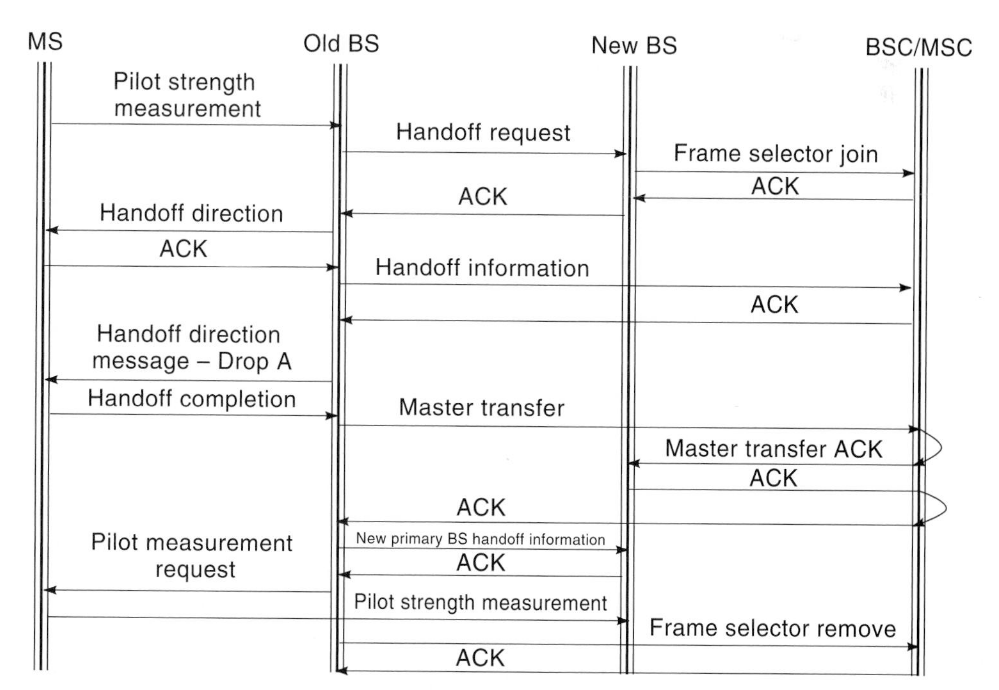

Figure 8.17 Setup and ending of soft handoff.

channels are indicated in the HDM on the downlink by the BS. The *candidate set* can have at most six pilots, and these refer to pilots that are not in the active set but that have sufficient RSS to be demodulated and used in demodulating the associated traffic channels. The *neighbor set* contains pilots that belong to neighboring cells and are intimated to the MS by a system parameters message on the paging channel. The *remaining set* contains all other possible pilots in the system. Because the receiver uses a RAKE to capture multipath components, it employs *search windows* to track each of the sets of pilot channels. The search windows are large enough to capture all the multipath components of the pilot from a BS but small enough to minimize searching time. The multipath delays are a function of the distance between the MS and the BS, and, consequently, the search windows are also affected.

Several thresholds are used in the soft handoff procedure. These are similar to the RSS thresholds discussed in Chapter 6. Details are available in [GAR00]. Whenever the strength in a pilot falls below a threshold, the MS starts a dwell timer. Unless the pilot strength goes back above the threshold before the timer expires, the MS will drop it from a given set. There is a trade-off in setting high or low values for these thresholds and timers.

Example 8.9: Pilot Detection Threshold in IS-95

The MS maintains a list of pilots that are being used in the active set. Initially the MS is connected to one BS and only its pilot and the multipath components of the pilot are in the active set (and indicated by the HDM). As the MS moves away, the pilot of the adjacent cell becomes stronger. If its strength is above the *pilot detection threshold* (T_ADD), this pilot must be added to the active set and the MS enters a soft handoff region. If the pilot detection threshold is too small, there may be false alarms caused by noise or interfering signals. If the pilot detection threshold is too large, useful pilots are not detected, and the call may be dropped. Thus a trade-off is required in the value of this threshold.

Example 8.10: Using Various Thresholds in Soft Handoffs

Figure 8.18 shows an example (from [GAR00]) of how the handoff thresholds work. As soon as the strength of the pilot exceeds T_ADD, it is transferred to the candidate set (1), and the MS sends the pilot strength measurement to the BS that is transmitting the pilot. The BS sends a handoff direction message to the MS (2) at which time the pilot is transferred to the active set. The MS acquires a traffic channel and sends a handoff completion message (3). After the pilot strength drops below T_DROP, the handoff drop timer is started. If the strength is still below T_DROP after the timer expires, the MS sends another pilot strength measurement to the BS associated with the pilot (5). When it receives the corresponding HDM without the pilot in it, the MS moves the pilot to the neighbor set (6) and sends a handoff completion message (7). At some point, the BS may send it a neighbor update list message that no longer contains the pilot and it is moved into the remaining set (8).

All signal strength measurements are based on the pilot channel. The handoff is also a mobile assisted handoff because the MS reports the signal strength mea-

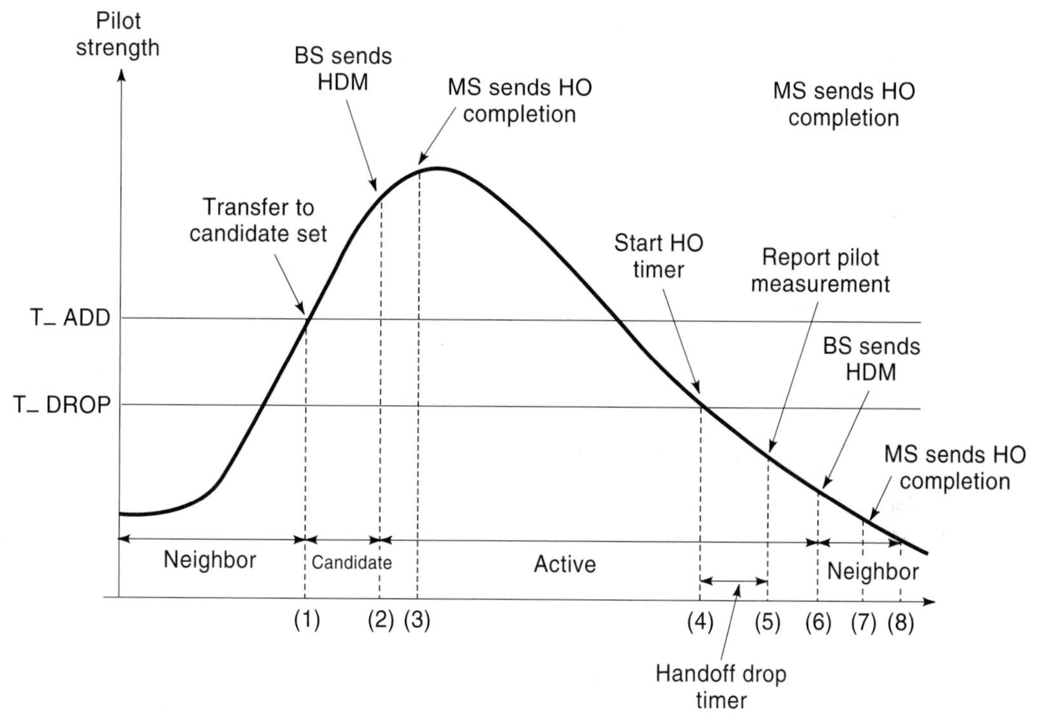

Figure 8.18 Handoff thresholds in IS-95.

surements to the network. The handoff thresholds may be adjusted dynamically to provide improvement in system performance.

8.3.4.2 Power Control

Like all cellular telephony systems, CDMA is also interference limited. However, co-channel and adjacent channel interference are not the major problems here. Instead the interference is from other users transmitting in the same frequency band at the same time. In order to avoid the near-far effect, it is important to implement good power control. Also, in order to maintain a good link quality, effects such as fading and shadowing need to be countered by increasing the transmit power. In the case of CDMA, an important factor is that the signal strength may be good, but frames are still received in error because of interference. Consequently, using the frame error rate (FER) for power control decisions is preferred over using the signal strength used in other systems. It is usually assumed that a FER of 1 percent with maximum error bursts of two frames is optimum and a range of 0.2 percent to 3 percent is allowed, with error bursts of up to four frames.

In IS-95, power control is very important especially on the reverse link where non-coherent detection is employed. Two types of power control are implemented—an open loop and a closed loop as discussed in Chapter 6. A slow mobile assisted power control is employed on the forward link.

Example 8.11: Open Loop Reverse Link Power Control in IS-95

Before a traffic channel is assigned, there is no closed loop power control in CDMA because the closed loop power control involves feedback from the BS that is delivered on the traffic channel 800 times per second. For this reason and in order to prevent the sudden fall of signal strength, an open loop power control scheme is implemented. The rule here is to use a transmit power that is inversely proportional to the received signal strength of pilots from all BSs. On the access channel, the MS sends a request using a weak signal if the pilot is strong. An acknowledgment may not be received because of collisions or because the transmit power was low. If no acknowledgement is received, a stronger access probe is transmitted. This is continued a few times, and then the attempt is stopped after a maximum power level is reached. Then the process is repeated after a back-off delay. Up to 15 attempts can be made to obtain a traffic channel. The disadvantages of the open loop power control are the assumption that the forward and reverse link characteristics are identical, slow response times (30 ms), and using the total power received from all BSs in calculating the required transmit power.

Example 8.12: Closed Loop Reverse Link Power Control in IS-95

On the downlink traffic channel, a power control bit is transmitted every 1.25 ms (800 times per second) as shown in Figure 8.19. A zero bit indicates that the MS should increase its transmit power and a 1 that the MS should decrease its transmit power. Every 1.25 ms, in the BS, the receiver determines the received E_b/I_t (the signal to interference ratio) by sampling it 16 times, and if it is above a preset target, the MS is instructed to reduce its power by 1 dB. If it is not above the target, the MS is instructed to increase its power by 1 dB. This is called the *inner-loop power control*

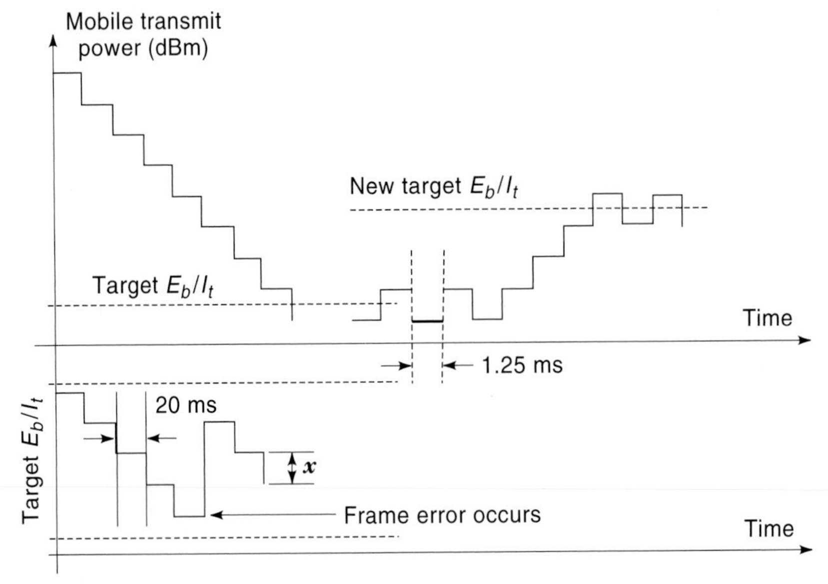

Figure 8.19 Inner- and outer-loop closed loop power control on the reverse link in IS-95.

as it enables changing the transmit power value in the MS. The target value in the base station controls the long-term frame error rate. The FER is not linearly dependent on the E_b/I_t, but is also a function of the velocity, fading, environment, and so on. The target E_b/I_t is also varied over time to reflect accurate values. It is reduced by a value of x dB every 20 ms if the FER is small enough. Typically, the value of $100x$ is 3 dB. The target value may be rapidly increased if the FER starts to increase. This mechanism to change the target E_b/I_t is called *the outer-loop power control.*

Example 8.13: Forward Link Power Control

Power control on the forward link is employed to reduce intercell interference. Within a cell, multiple users employ orthogonal sequences, and the primary source of interference is from users of other cells or from multipath. A *mobile assisted* power control is used. The MS periodically reports the FER on the forward link to the BS station, which will then adjust its transmit power accordingly. Maximum and minimum transmit power values are preset to prevent excessive interference and to avoid allowing voice quality to drop respectively.

8.4 IMT-2000

The ITU is on an accelerated pace to specify the 3G mobile communications standards. The primary standard for 3G systems is referred to as the International Mobile Telecommunications beyond the year 2000 (IMT-2000)—the goal of which is to support higher data rates that can support multimedia applications, provide a high spectral efficiency, make as many of the interfaces standard as possible, and provide compatibility to services within the IMT-2000 [ZEN00]. Although voice traffic will continue to be the main source of revenue, packet data for Internet access, advanced messaging services such as multimedia email, and real-time multimedia for applications such as telemedicine and remote security are envisaged in IMT-2000. The requirements for IMT-2000 include improved voice quality (wireline quality), data rates up to 384 kbps everywhere and 2 Mbps indoor, support for packet and circuit switched data services, seamless incorporation of existing 2G and satellite systems, seamless international roaming, and support for several simultaneous multimedia connections.

In the late 1980s, ITU formed a group to study wireless standards for very high-speed data and multimedia services. In 1992, the world administrative radio conference (WARC) assigned 230 MHz of spectrum globally for IMT-2000. In June 1998, 15 proposals were received by ITU-R (the radio communications sector of ITU) for candidate radio transmission technologies (RTTs). Most of the proposals were based on CDMA as CDMA provides a better voice quality and is more flexible for customized multimedia applications. In order to avoid multiple standards, efforts were made to harmonize a single converged global standard. Backward compatibility with legacy systems is also a major issue with support for the GSM-MAP and ANSI-41 (the core GSM and IS-41 backbone infrastructures) essential. Both of these core networks are supposed to evolve towards a common 3G-core network that will be very likely all IP.

As far as the RTTs are concerned, there were two major competing proposals—the W-CDMA based on the UMTS Terrestrial Radio Access (UTRA) FDD and TDD proposals and the CDMA2000 proposal that is backward compatible with IS-95. The main differences can be summarized as follows [ZEN00].

1. Although CDMA2000 proposes multiples of 1.2288 Mcps chip rates to allow greater compatibility with IS-95 (in particular, 3.6864 Mcps is suggested), W-CDMA employs 3.84 Mcps.
2. In IS-95 and CDMA2000, the BSs operate synchronously by obtaining timing from GPS. W-CDMA advocates asynchronous operation to enable deploying picocells within buildings where GPS is not available.
3. The frame length of W-CDMA is 10 ms to ensure small end-to-end delays, though it is 20 ms in CDMA2000.

The harmonization activities were initiated via a 3GPP that consisted of members from industry and standards bodies to work on the core network, the radio access network, service and system aspects, and the mobile terminal. To include non-GSM technologies, a 3GPP2 was initiated in parallel by ANSI to prepare technical specifications for a 3G mobile system based on CDMA2000 and the IS-41 based core network. Both 3GPP and 3GPP2 are expected to cooperate in harmonization and consolidation. Meanwhile, an operators harmonization group (OHG) set up at the end of 1998 agreed on a further harmonized Global 3G (G3G) standard (Figure 8.20) that has the following components:

1. Three air-interface standards—two frequency division duplex modes: a direct sequence (DS) mode based on W-CDMA at 3.84 Mcps chip rate, a multi-carrier (MC) mode based on CDMA2000 with a chip rate of 3.6864 Mcps, and one time division duplex mode operating at 3.84 Mcps

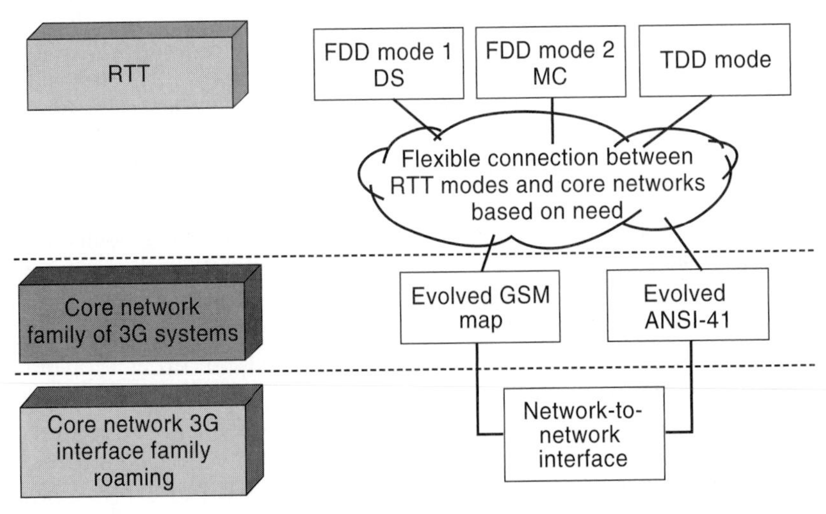

Figure 8.20 The G3G proposal.

2. Support for both GSM-MAP and ANSI-41 with all air-interface modes
3. Support for functionality based synchronous operation
4. Seamless handoff between DS and MC modes, as well as interoperability of sorts between the UMTS core network and ANSI-41

The idea is also to minimize the complexity of multimode terminals that include all of the standards. Table 8.2 describes the characteristics of the two air-interfaces [ZEN00], [HOL00].

8.4.1 Forward Channels in WCDMA and CDMA2000

The primary requirements of 3G systems are that they should be able to support a variety of application data rates (from 384 kbps circuit switched connections to 2 Mbps in indoor areas) and operation environments. This means that there must be support for quality of service and operation from megacells to picocells. The

Table 8.2 The Major 3G RTT Proposals

	3GPP	**3GPP2**
Parameters	**UTRA DS-FDD/TDD**	**CDMA2000**
Multiple access	DS-CDMA	Uplink: DS-CDMA Downlink: MC-CDMA/DS-CDMA
Chip rate	3.84 Mcps	$N \times 1.2288$ Mcps $N = 1,3,6,9,12$
Pilot structure	Dedicated pilots on the uplink and common or dedicated pilots on the downlink	Uplink: Code-divided continuous dedicated pilot Downlink: Code-divided continuous common pilot and dedicated or common auxiliary pilots
Frame length	10 ms with 15 slots	5, 10, 20, 40, 80 ms
Modulation	Uplink: Dual channel QPSK Downlink: QPSK	Uplink: BPSK; Downlink: QPSK
Spreading modulation	QPSK both directions	Uplink: HPSK Downlink: QPSK
Detection	Coherent pilot aided	Coherent pilot aided
Channelization codes	OVSF codes	Uplink: Walsh codes Downlink: Walsh or quasi-orthogonal codes
Scrambling codes	Uplink: Short code (256 chips from S[2]) or long code (38,400 chips Gold code) Downlink: Gold code	Long code ($2^{42}-1$ chips) Short code ($2^{15}-1$ chips)
Access schemes	Random access with power ramping with acquisition indication	RsMa—Flexible • Basic access • Power controlled access • Reserved access • Designated access (initiated by BS)
Inter-base station operation	Asynchronous Synchronous (optional)	Synchronous

forward channels are referred to as *transport channels* in the UTRA W-CDMA standard proposed by 3GPP. The forward channel modifications are as follows.

In W-CDMA, the BSs can operate in a asynchronous fashion that obviates the need of GPS availability to synchronize base stations. W-CDMA employs what is known as the orthogonal variable spreading factor (OVSF) codes. OVSF codes allow a variable spreading factor technique that maintains orthogonality between spreading codes of different lengths. The logical channels are called *transport channels* in W-CDMA.

cdma2000 employs multiple carriers to provide a higher data rate compared with W-CDMA. It employs N carriers ($N = 1, 3, 6, 9$) for an overall chip rate of $N \times 1.2288$ which is 3.6864 Mcps for $N = 3$. Alternatively, a single carrier can be employed to chip at the larger chipping rate. The former mode of operation is suitable for overlaying cdma2000 over existing IS-95 systems. Walsh codes from 128 chips to 4 chips are employed to provide variable spreading and processing gains. All N carriers use the same single code for scrambling. The BSs still need to be synchronized and use the PN-code offsets for differentiation as before. Pilot channels are used for fast acquisition and handoff as before. QPSK modulation is employed before spreading with the Walsh codes to increase the number of usable Walsh codes. In addition to the pilot, synch, and paging channels, auxiliary pilot channels can be used to supply beam-forming information if smart antennas are employed. A dedicated MAC channel (DMCH) may be available for sending MAC layer messages to specific mobiles. To support QoS at different rates, a fundamental channel (FCH) for signaling and a supplemental channel (SCH) for traffic can be made available. Turbo codes are employed on the forward supplemental channels for high data rates.

8.4.2 Reverse Channels in W-CDMA and cdma2000

Support for variable data rates and operation in a variety of environments once again governs the implementation of the reverse link for 3G systems. In W-CDMA, Gold codes and S(2) codes are used for scrambling on the uplink. The periodicity of the Gold code is 38,400 chips for using a RAKE receiver in the BS and that of the S(2) codes is 256 chips for employing multiuser detection.

In cdma2000, the reverse link is made more symmetrical with the forward link in many aspects. For instance, a *reverse pilot channel* is employed between each mobile and the BS for initial acquisition, time tracking, and power control measurement. More powerful codes are used such as a rate ¼ convolutional code with a constraint length of 9. Turbocodes are employed on the reverse supplementary channels. Variable rate spreading is supported to enable better error correction capability and a variety of data rates.

8.4.3 Handoff and Power Control in 3G Systems

cdma2000 is very similar to IS-95 in terms of power control and handoff procedures. In W-CDMA, a fast power control scheme is used at 1,500 bps as compared with 800 bps with IS-95 and cdma2000. In W-CDMA, the handoff procedure is somewhat different. Once again, different sets of pilots are maintained, and the active set corresponds to the pilot channels being used for completing the call. Relative threshold values are employed instead of absolute values as in IS-95 (i.e., the pilot strengths are compared with each other instead of T_ADD and T_DROP values that need tuning depending on the environment in IS-95). The working of the algorithm is illustrated by the following example from [HOL00].

Example 8.13: Soft Handoff in W-CDMA

> Figure 8.21 shows an example of soft handoff in W-CDMA. The events indicated along the abscissa correspond to adding a pilot to the active set if the active set is not full or removing a pilot from the active set. At event 1A, a pilot is added to the active set because its strength is greater than the strength of the best pilot minus a reporting range plus a hysteresis margin for more than a time ΔT. A pilot is removed from the active set (event 1B) if its strength is below that of the best pilot by the sum of a reporting range and a hysteresis margin for a time greater than ΔT. Event 1C corresponds to a combined addition and deletion of pilots. This happens when the active set is full and the worst pilot in the set is smaller than the best pilot minus a hysteresis margin for a time ΔT. In this case, the worst pilot is deleted, and the best candidate pilot is added to the active set. The reporting range is a threshold for soft handoff.

It should be noted that when the signal strength comparisons are made, averaged values are used and not instantaneous samples.

In WCDMA, closed loop power control is implemented in a manner similar to IS-95 with the power control bits transmitted 1,500 times a second. This allows a very fast control of power and provides significant capacity gains in W-CDMA, especially at pedestrian speeds. Both inner and outer loop power control mechanisms

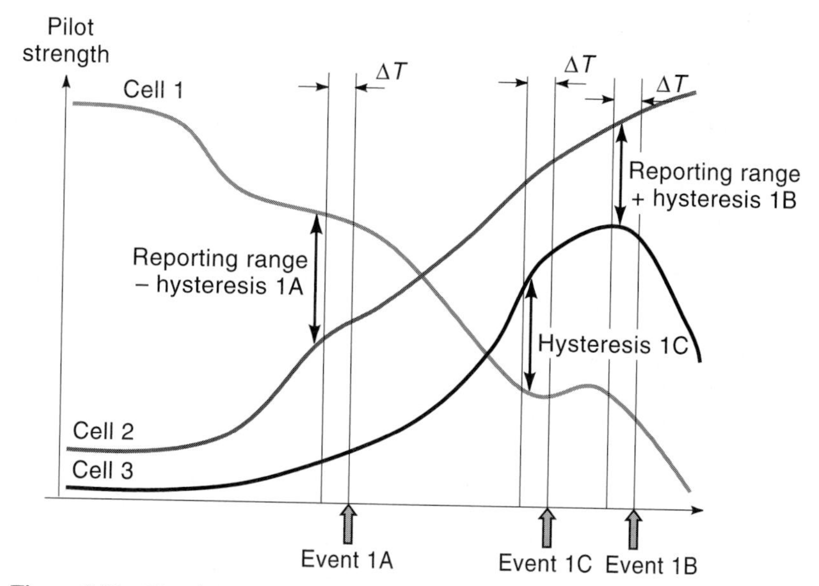

Figure 8.21 Handoff thresholds in W-CDMA.

are employed. The interested reader is referred to [HUB00], [DAHL98], [SWE00], [3GPPweb], [3GPP2web] and [GAR02] for details of 3G systems.

QUESTIONS

8.1 What is IS-634? What are its functionalities?

8.2 What is IS-41? Why is it important in North American cellular systems?

8.3 What is IS-95?

8.4 What is TR-45?

8.5 What are the bandwidths and chip rates used in WCDMA and how they compare with cdmaOne?

8.6 How many physical channels are available in each IS-95 carrier? What type of coding separates these channels from one another?

8.7 Name the forward and reverse channels used in IS-95.

8.8 How are Walsh codes employed in the cdmaOne forward and reverse channels? Explain the difference.

8.9 Why does WCDMA use Walsh codes in forward and reverse channels for separating users, while cdmaOne uses them only in the forward channel?

8.10 What are the bit rates of the data services supported by IS-95?

8.11 What is soft handoff and how does it compare with hard and seamless handoffs? Give one example system for each of these three handoff methods.

8.12 Why is power control important in CDMA?

8.13 What forward channels are involved in IS-95 for power control?

8.14 Handoff decisions in wireless networks are performed using received signal strength measurements. Name the forward channel in IS-95 that is used for this purpose.

8.15 Why are several pilot channels monitored in IS-95? When does a pilot channel from a base station move from an active set to a candidate set?

PROBLEMS

8.1 **a.** Sketch all four of the 4-bit Walsh codes.

b. Sketch the autocorrelation function of all four codes.

c. Sketch the cross-correlation function of the first and the second code.

8.2 Repeat Problem 8.1 for 16-bit Walsh codes.

8.3 The M-sequence is a class of spread spectrum sequences used in IS-95 CDMA systems. Compute the autocorrelation of an M-sequence, which is given by the following vector. Assume that the chip duration is T_c and the duration of the M-sequence "pulse" is T. The M-sequence is: [1 −1 −1 1 1 1 −1 1 1 1 −1 −1 −1 −1 1]. You can use the `xcorr` function in Matlab to verify your results.

8.4 **a.** Using Table 3A.1, calculate required γ_b for the BER of 10^{-3} for the QPSK modulation used in IS-95. (*Hint:* you can use the Matlab function `erfc` for calculation of the complementary error function.)

b. Use Eq. (4.4) with γ_b (the same as S_r) determined in part (a) to calculate the number of simultaneous users, M, in a cell operating in one sector of a 3-sector antenna with one IS-95 carrier. Assume a data transmission rate of R = 9,600 bps and a performance improvement factor of K = 4 (6 dB).

c. Repeat parts (a) and (b) for different values of BER between 10^{-2} and 10^{-12} to produce a computer plot of BER in logarithmic scale vs. number of users, M, in linear scale. Using this curve, explain the effects of error rate requirement on the capacity of a CDMA system.

d. Repeat the plot in part (c) for a normalized number of users per MHz of band. If we change the system to an IMT-2000 system, does this plot or the plot in part (c) change? Explain why.

8.5 **a.** Assume we want to support a 19.2 kbps data service with minimum required error rate of 10^{-3} over a W-CDMA system. What is the minimum chip rate and bandwidth needed to support 100 simultaneous users with one carrier of this W-CDMA system? Assume a performance improvement factor of K = 4 (6 dB) using Table 3A.1 and Eq. (4.4).

b. What would be the bandwidth requirement in (a) if number of users was increased to 200?

c. What would be the bandwidth requirement in (a) if the data rate requirement were increased to 192 kbps?

d. What would be the bandwidth requirement in (a) if the error rate requirement were increased to 10^{-4}?

CHAPTER 9

MOBILE DATA NETWORKS

9.1 Introduction

9.1.1 What Is Mobile Data?
9.1.2 Independent Mobile Data
9.1.3 Shared Mobile Data
9.1.4 Overlay Mobile Data

9.2 The Data-Oriented CDPD Network

9.2.1 What Is CDPD?
9.2.2 Reference Architecture in CDPD
9.2.3 Mobility Support in CDPD
9.2.4 Protocol Layers in CDPD

9.3 GPRS and Higher Data Rates

9.3.1 What is GPRS?
9.3.2 Reference Architecture in GPRS
9.3.3 Mobility Support in GPRS
9.3.4 Protocol Layers in GPRS

9.4 Short Messaging Service in GSM

9.4.1 What is SMS?
9.4.2 Overview of SMS Operation

9.5 Mobile Application Protocols

9.5.1 Wireless Application Protocol (WAP)
9.5.2 i-Mode

Questions

Problems

9.1 INTRODUCTION

In Part III of the book we provide detailed examples of important wide area wireless systems. In the previous two chapters, we provided examples of TDMA and CDMA voice-oriented networks. This chapter provides an overview of the lower speed wide area wireless data services that we refer to as *mobile data networks* or *wireless wide area networks.* In the remainder of the book, our emphasis is on the data-oriented networks and Part IV, following this chapter, is devoted to the details of broadband and ad hoc wireless local networks. Wireless data networks are becoming increasingly important in the light of wireless access to the Internet, development of wireless-oriented consumer products, and wireless home networking. By far, voice services have been the major revenue generators for a long time, both in the wireless and wired networks. Long distance and local telephony have provided the biggest revenue for telecommunications networks over the last century. In the 1990s, the emergence of the PC and the Internet have shifted the focus toward data services over the same media as voice services and alternative media like coaxial cable and wireless as well. The driving applications and revenue generators have included content-based services that draw revenue from advertisements, electronic commerce (E-commerce), and, more recently, mobile commerce (M-commerce). Although they were started in 1983, until recently mobile data services were considered expensive, unreliable, and slow. In the past few years, despite all market predictions of failures since the original inception of this industry, the same services that have drawn attention in the wired data world have drawn a new surge of extensive attention in wireless data services. In the past year or so the unpredicted success of SMS in Finland and Japan has catalyzed this new growth of the data oriented industry. Cellular service providers, who once were only focused on voice applications as the main source of income, today envision the future as being comprised of more and more data applications. The mobile data industry that started with ARDIS (in 1983) as a private network for IBM [PAH94] is now being assimilated as a public service into the next generation of the cellular networks and plays an important role in shaping the future of this industry.

More recently there are new (emerging) applications based on the WAP and i-Mode that are driving the usage of mobile data networks that we need to address briefly. In this chapter we first provide an overview of the mobile data services, then we provide detailed description of a couple of mobile data services to familiarize the reader with implementation of these systems, and finally we finish the chapter with a short description of WAP and i-Mode.

9.1.1 What Is Mobile Data?

By mobile data networks we refer to those services, technologies, and standards that are related to data services over wide area coverage areas spanning more than the local area or campus. Examples include metropolitan area wireless data services such as Metricom's Ricochet service and those that operate over the same coverage areas as cellular networks such as CDPD or GPRS. The history of evolution of mobile data services was provided in Table 1.2 and a comparative description of major alternatives in Table 1.7.

In addition to traditional mobile data services, discussed in Chapter 1 and summarized in Table 1.7, SMS services can also be considered a part of these systems. Short messaging services are embedded in digital cellular systems such as GSM. These services use the 10-digit keypad of the mobile terminal to type and display a message and use the digital cellular network facilities to deliver the message. In traditional mobile data services, the subscriber uses computer keyboards to enter the message. Considering this larger picture for mobile data services or wireless WANs, as shown in Figure 9.1 we can classify mobile data networks into three categories: independent, shared, and overlay networks based on the way they relate to the cellular infrastructure.

9.1.2 Independent Mobile Data

Independent networks have their own spectrum that is not coordinated with any other service and their own infrastructure that is not shared with any other service. These networks are divided into two groups according to the status of their operating frequency band. The first group uses independent spectrum in licensed bands. Examples of such networks are ARDIS and Mobitex, and historically they were the first mobile data services that were introduced. Such networks were not economically successful because the revenues generated, mostly from the vertical applications, could not justify the cost of the implementation of the infrastructure. For these networks to survive, either a sizable vertical market is needed, or the cost of implementation of the infrastructure should be reduced, or a horizontal killer application is needed. The TETRA network (see Table 1.7) is designed for public safety application that is a prosperous vertical market. It was defined by the ETSI

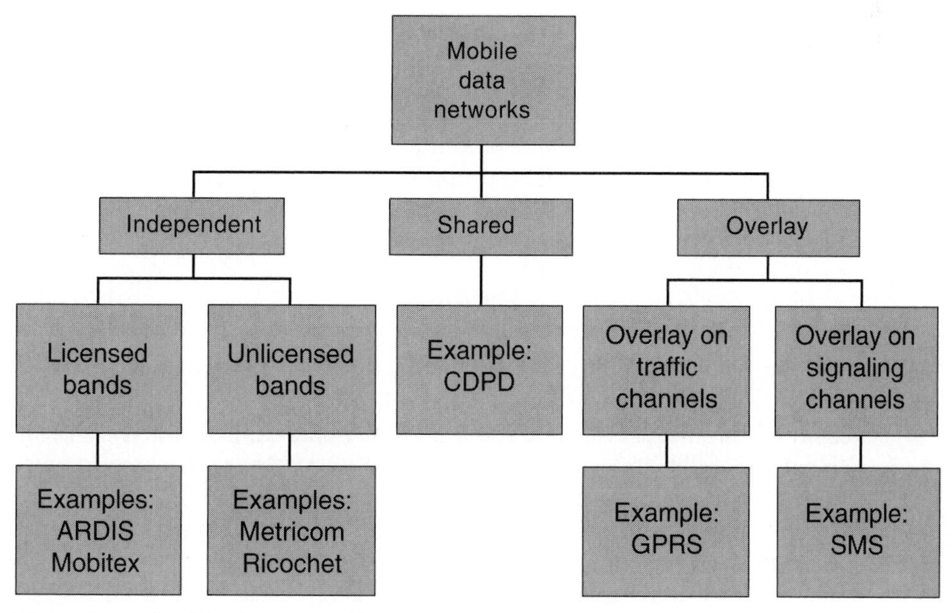

Figure 9.1 Classification of mobile data networks.

to meet the needs of professional mobile radio users. As a result, in spite of the fates of the previously mentioned mobile data projects, at the time of this writing, a number of companies in EU are engaged in manufacturing TETRA equipment. These companies have already reached agreements with the public safety organizations in Europe to use manufactured equipment in their future public land mobile data networks. Reduction of the cost of the infrastructure and killer applications are envisioned in the shared and overlay systems that are discussed in the following two sections.

The second group of independent mobile data networks makes use of unlicensed spectrum that is shared among a variety of applications and users. Metricom's Ricochet service, which used the 915 MHz unlicensed ISM band spectrum, was an example of this service. This service was deployed in airports and some metropolitan areas for wireless Internet access.

9.1.3 Shared Mobile Data

These networks share the spectrum and part of the infrastructure with an already existing voice-oriented analog service. The services operate in the same radio channels used for analog voice, but they have their own air-interface and MAC protocols. In addition to dedicated channels for data, these mobile data services can also use the available unused voice channels. These systems share an existing system infrastructure, therefore the initial investment is not huge, and it could be made as gradually as possible. Initial deployment could be made in areas where there is subscriber demand and subsequent penetration into other areas is considered as the customer base enlarges. The CDPD service (see Table 1.7 for overview), which shares spectrum and part of the infrastructure with AMPS, is an example of such networks. It does have an independent air-interface and MAC layer, along with additional infrastructure required for operation of data services. The CDPD standard was completed in the early 1990s, and the expectation was that by the year 2000, nearly 13 million subscribers would be using it. However, this expectation was never met, and generating income remained as the main obstacle for this service. Recently CDPD services have picked up with the availability of modems for palm computing devices. In the next section, we provide the details of implementation of CDPD to familiarize the reader with the implementation aspects of a data-oriented service.

9.1.4 Overlay Mobile Data

The last group of mobile data networks is an *overlay* on existing networks and services. This means that the data service will not only make use of the spectrum allocated for another service but also the MAC frames and air-interface of an existing voice-oriented digital cellular system. GPRS and GSM's SMS are examples of such overlays. They make use of free time slots available within the traffic channels and signaling channels in GSM. This way, the amount of new infrastructure required is reduced to a bare minimum. Most of the extra components required are implemented in software, making it inexpensive and easy to deploy. GPRS type of services uses computer keyboards to communicate longer messages, and SMS use the cellular phone dialing keypad to communicate short messages. Sections 9.3 and 9.4 of this chapter provide further details on GPRS and SMS.

9.2 THE DATA-ORIENTED CDPD NETWORK

CDPD [TAY97], [CDPD95], [SAL99a,b] has been one of the longest surviving wide area mobile data technologies worldwide. It is a shared mobile data network that shares part of the infrastructure and the entire spectrum with AMPS in the United States. It is, however, an open standard, making its implementation easier and more widespread. CDPD initially had mixed success because the coverage was not universal, the data rates low, and prices prohibitive. CDPD has been more popular with vertical applications such as inventory for vending machines, fleet management, and so on. With the emergence of new handheld computers and palm-based devices, CDPD is making resurgence as a service for low data rate text-based Web, email, and short messaging horizontal applications. For example, modems are available for popular PalmOS and Windows CE devices that provide unlimited CDPD access.

In 1991, McCaw and IBM came up with the idea of developing a packet data overlay on AMPS. By the end of 1991, a telephone network–oriented prototype architecture was in place. This preliminary version was discussed in 1992 at a CDPD technical conference in Santa Clara by which point of time, CDPD also had the support of several regional telephone companies such as Ameritech, Bell Atlantic, GTE, McCaw, Nynex, PacTel, Southwest Bell, and US West. The first field trials were held in the San Francisco Bay area in 1992. At the same time, there was a recognition of the complexity of the telephony oriented architecture, and so open standards based on data networking standards and the OSI model were investigated around this time. Ultimately the latter approach based on data networking standards was selected, and the first official specifications were released in July 1993. The specifications were based on an open architecture, and the mobility management technique closely follows and is a precursor of the Mobile-IP standard. This specification was accepted by several of the major cellular telephony service providers. The CDPD forum was created in 1994, which attracted about 100 members. In 1995, release 1.1 of the CDPD standard came out. CDPD has been deployed by many of the former regional Bell operating companies as well as AT&T wireless.

9.2.1 What Is CDPD?

The design of CDPD was based on several design objectives that are often repeated in designing overlay networks or new networks. Some of these design goals are discussed. A lot of emphasis was laid on open architectures and reusing as much of the existing RF infrastructure as possible. The design goals of CDPD included location independence and independence from service provider, so that coverage could be maximized; application transparency and multiprotocol support, interoperability between products from multiple vendors, minimal invention, and use of COTS technology as far as possible; and optimal usage of RF where air-interface efficiency is given priority over other resources. CDPD used primitive RF technology for cost reasons, and for this purpose the well-known GMSK modulation scheme was chosen. The raw signaling rate is 19.2 kbps, and with Reed-Solomon (RS) coding the effective data rate is 14.4 kbps full duplex before control overhead. The design was intended to be evolutionary and based on the OSI model with support for native IP, so that if new transport layer or application layer protocols were implemented there was no need for changes. The prominent features of CDPD have been its openness and freedom from all proprietary technology, support for multivendor interoperability, and simplicity in design. There is, however, constraint on the design of CDPD because it is a shared mobile data network and AMPS has priority over the usage of the spectrum. CDPD employs a technique called RF sniffing to detect whether an AMPS call is trying to access a frequency channel, and hopping to move from such a band to another to give the voice call priority. About 20 percent of AMPS frequency channels that can be used for CDPD are idle at a given time.

Example 9.1: Reuse of AMPS Infrastructure in CDPD

The AMPS infrastructure has a BS that transmits and receives RF signals. The demodulated voice signals are digitized and multiplexed on T1 connections that terminate in an MSC. The CDPD system aims at minimizing changes to existing infrastructure. In addition to the radio spectrum, the physical plant and communication links are precious resources. The MDBS (mobile data base station) (see following paragraph) handles RF communications and reuses existing antenna feeds and towers. The MDBS is a small compact box that can fit in existing cell sites and recent designs can be deployed in microcellular environments as well. The data is then relayed using *idle* channels in the T1 link to the MD-IS (mobile data intermediate system) that is usually colocated with the MSC.

Example 9.2: Channel Hopping in CDPD

When the telephone system selects a new channel for a voice call, if CDPD is using that channel it should exit within 40 ms. The MDBS should find an alternative channel and allocate it to the mobile end-system. The MDBS informs the mobile that the downlink is being changed to another channel and hops to this channel. A timed hop is a mandatory channel hop that is planned to happen after a fixed time. A forced hop is due to a voice call request. If no channels are available, CDPD enters "blackout."

9.2.1.1 CDPD Services

In a manner similar to our discussion of GSM, we start with the services that CDPD offers. *Network services* are the basic form of services offered by CDPD. This is simply support for transfer of data from one location to another via popular or standard network layer protocols. In particular, CDPD supports connectionless layer 3 protocols (IP or connectionless network protocol—CLNP) and in that sense acts simply as a wireless extension to the Internet. *Network support services* are services necessary to maintain the operation of the mobile data network such as management, accounting, security, and so on. These services include mobility and radio resource management and are usually transparent to the user. Such services add "intelligence" to the network. The last category of services includes *network application services*. These are value added services such as limited size messaging on top of the network services and need explicit subscription.

9.2.2 Reference Architecture in CDPD

Figure 9.2 shows the reference architecture for CDPD. There are three key CDPD interfaces that form logical boundaries for a CDPD service provider's network. They are essential for the proper operation of CDPD. There are some interfaces internal to the "cloud." Such interfaces are only recommended, and a service provider can implement them differently. Each interface specifies a protocol stack corresponding to the OSI model and primitives are defined at each "layer" that can request and obtain services from the layer below.

9.2.2.1 Interface Details

The *A-Interface* is the air link interface, and parts 400–409 of the CDPD specifications specify it. The *E-Interface* is the external interface, and it is the means by which CDPD operates with the rest of data network. Over this interface IP and CLNP are supported, and IPv6 will be supported as it becomes deployed. Other protocols are supported by encapsulation because they are outside the CDPD specs. Mobility is transparent to the network beyond the E-interface. The *I-Interface* is the interservice provider interface. The North American market is partitioned into a multiplicity of service providers and the I-interface enables seamless nationwide service. It supports all of the E-interface protocols *plus* two CDPD specific protocols—the *mobile network location protocol* (MNLP) which is the

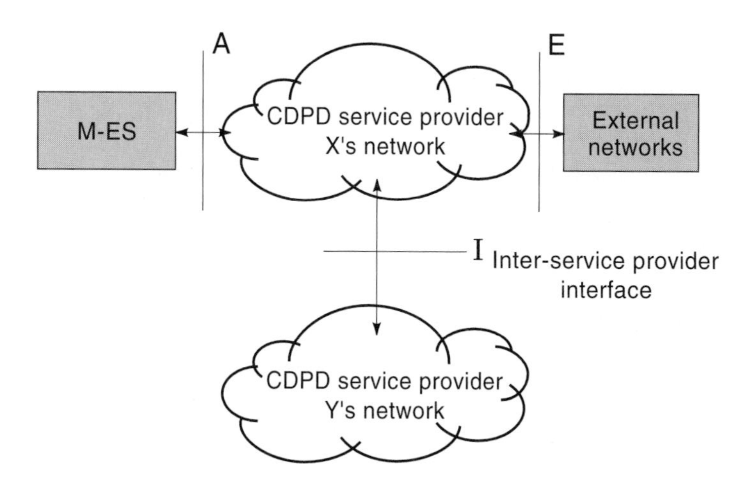

Figure 9.2 Reference architecture for CDPD.

protocol by which mobile users from one system are supported by another system and is a key piece of CDPD mobility management scheme. Network management and accounting protocols are also defined at the I-interface. All the protocols are based on CLNP (except reverse channel IP packets).

9.2.2.2 Physical Architecture

The physical elements of the CDPD architecture and their relationship with the three interfaces are shown in Figure 9.3. These elements are the mobile end system, the mobile data BS, and the mobile data intermediate system, along with some servers for accounting and network management and databases called the mobile home and mobile serving functions.

The mobile-end system (M-ES) is the ultimate source and destination of protocol data units (PDUs). It is equipped with a CDPD radio and software, and example M-ESs are telemetry devices, laptops, vending machines, and so on. In the M-ES, protocols are specified up to layer 3. An M-ES can be full duplex or half duplex and supports all standard APIs. Communication is implemented via conventional means like sockets, NDIS, and so on. An M-ES has three functional units: subscriber unit (SU), subscriber identity module (SIM), and mobile application subsystem (MAS). The subscriber unit establishes and maintains data communication, executes CDPD air-interface protocols, and includes administrative and management layers. The SIM is a repository of identity and authentication credentials, and the SIM card is very similar to GSM. In fact it is based on GSM standards. The MAS deals with the higher layer protocols and can perform remote database access, email, vending machine inventory, and so on. The MASs span a wide range of

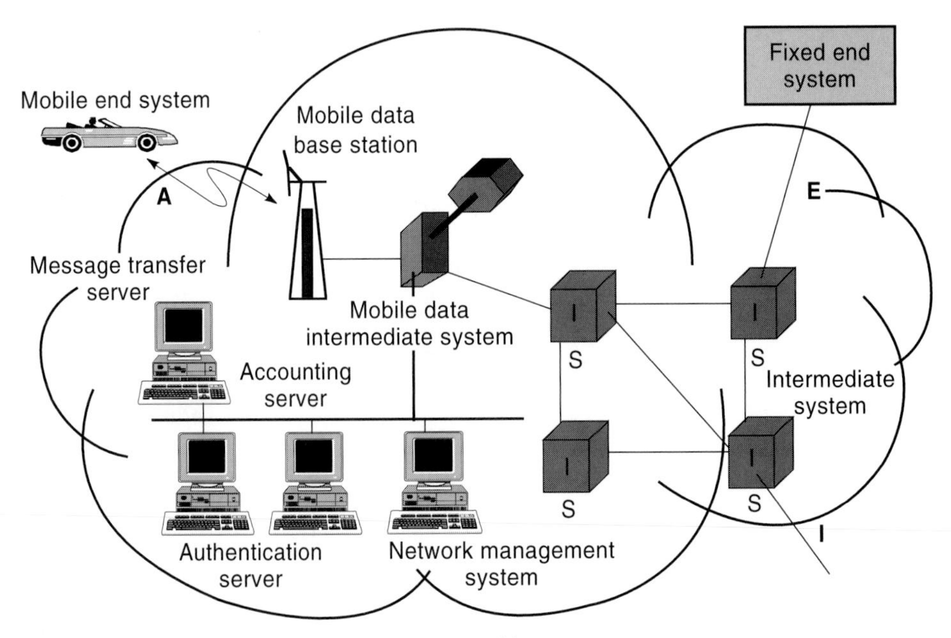

Figure 9.3 Physical elements in the CDPD architecture.

features. Some support both CDPD and an AMPS modem. Others can support voice communications as well. The M-ES employs a variety of power sources such as internal batteries, vehicular power source, or a laptop-based PC card.

The MDBS is the system end (network side) of the MAC sublayer over the air interface. It is equivalent to the BSS in GSM and is pretty much the BS electronics for CDPD. It is also like an Ethernet hub and acts as a link layer relay or bridge. It communicates with the M-ES through the A-interface and performs modulation of data bits and demodulation of the RF signal. It actively participates in the medium access scheme called digital sense multiple access (DSMA) (see Chapter 4). It is Layer 3 addressable for network management.

The MD-IS is the focal point of CDPD mobility management and packet forwarding. It has a mobile serving function (MSF) that serves as the foundation for registration of a mobile and a mobile home function (MHF) that serves as an anchor point for locating a mobile. It tracks the local point of access of the mobile devices and is responsible for presenting the external interface on behalf of the M-ES and for routing all traffic to and from the M-ES. It also performs accounting services. The MDBS to MD-IS interface is not specified though there are recommendations. Because it is internal to the CDPD cloud, proprietary implementations are possible although most service providers follow the recommendations. The intermediate system (IS) is a fancy name for a router, and it handles packet forwarding. Border ISs are further required to provide security filtering and access control functions that are *not* part of the CDPD specifications.

The fixed-end system (F-ES) is a conventional network node that includes most PCs, workstations, and so on that are transport layer peers of the M-ES. An Internal F-ES operates within the boundaries of the CDPD network and is under the control of the service provider. It usually operates functionally in the role of administrative servers and value-added servers. This is an example of the flexibility of CDPD.

Example 9.3: Internal F-ESs in CDPD

Internal F-ESs include an accounting server (AS), an authentication server, a directory server, a network management system, and a message transfer system. For example, the AS is in charge of collection and distribution of accounting data such as packet count, packet size, source and destination address, geographic information, time of transmission, and so on. CDPD employs pay by the byte (airlink usage) charging rather than time of connection. The network management system is based on general network management schemes of the OSI model using the common management interface protocol (CMIP), or optionally, simple network management protocol (SNMPv2).

9.2.3 Mobility Support in CDPD

As in the case of most mobile networks, mechanisms are in place in CDPD to support the mobile environment. We consider radio resources, mobility management, and security in this section.

9.2.3.1 Mobility Management

Handoff in CDPD occurs when an M-ES moves from one cell to another or if the CDPD channel quality deteriorates, the current CDPD channel is requested by an AMPS voice call (forced hop), or the load on CDPD channels in the current cell is much more than the load on the channels in an overlapping cell.

The physical layer of CDPD provides the ability to tune to a specific RF channel, the ability to measure the received signal strength indication (RSSI) of the received signal, the ability to set the power of the M-ES transmitted signal to a specified level, and the ability to suspend and resume monitoring of RF channels in the M-ES. Both uplink and downlink channels are slotted. There is no contention on the downlink, and the MDBS will transmit link layer frames sequentially. On the uplink, a DSMA/CD (digital sense multiple-access with collision detection) protocol is employed. Collision detection is at the BS and informed to the MHs on the downlink. On the downlink, multiple *cell configuration messages* are broadcast, including for the given cell and its neighbors, the cell identifier, a reference channel for the cell, a value that provides the difference in power between the reference channel, and the actual CDPD data channel, a RSS bias to compare the RSS of the reference channels of the given cell and adjacent cells, and a list of channels allocated to CDPD within the given cell. RSS measurements are always done on the reference channel because the CDPD channel list may keep changing [CDPD95].

Upon powering on, the MH scans the air and locks on to the strongest "acceptable" CDPD channel stream it can find and *registers* with the mobile-data intermediate system (MD-IS) that serves the base station. This is done via the mobile network registration protocol (MNRP) whereby the MH announces its presence and also authenticates itself. Registration protects against fraud and enables CDPD network to know the mobile location and update its mobility databases. The MH continues to listen to the CDPD channel unless it (or the CDPD network) initiates a handoff.

CDPD mobility management is based on principles similar to mobile-IP. The details are shown in Figure 9.4. The MD-IS is the central element in the process. An MD-IS is logically separated into a home MD-IS and a serving MD-IS. A home MD-IS contains a subscription database for its geographic area. Each subscriber is registered in his home MD-IS associated with his home area. The IP address of a subscriber points to his home MD-IS. At the home MD-IS, an MHF maintains information about the current location of MHs associated with (homed at) that home MD-IS. The MHF also encapsulates any packet that is addressed to an M-ES homed with it directing it to an MSF associated with the serving MD-IS, whose serving area the M-ES is currently visiting. A serving MD-IS manages one serving area. Mobile data BSs that provide coverage in this area are connected to the serving MD-IS, whose MSF contains information about all subscribers currently visiting the area and registered with it. The MSF employs the mobile network location protocol (MNLP) to notify the MHF about the presence of the M-ES in its service area. The channel stream in which a subscriber is active is also indicated. The MSF decapsulates forwarded packets and routes them to the correct channel stream in the cell.

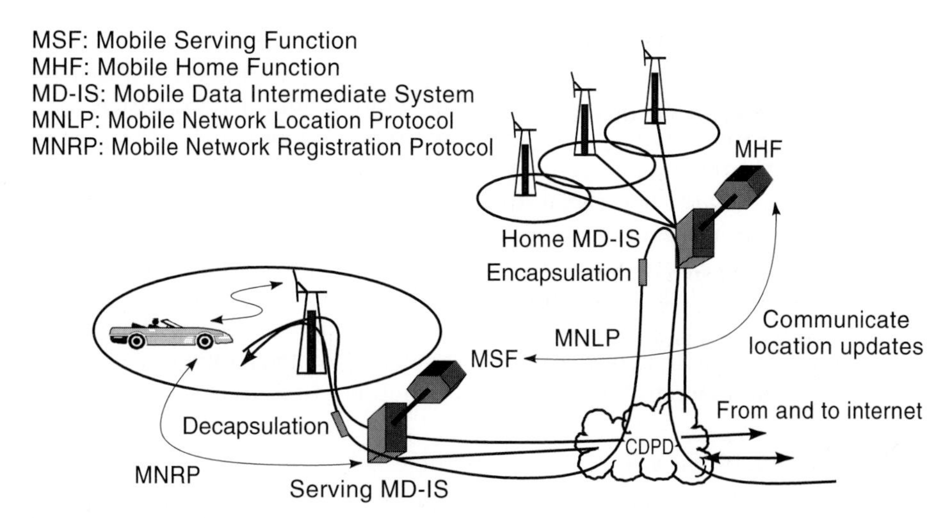

Figure 9.4 Mobility management in CDPD.

The handoff procedure in CDPD is shown in Figure 9.5. The handoff initiation and decision in CDPD are as follows. The handoff is mobile controlled. The M-ES always measures the signal strength of the reference channel [SAL99b]. An M-ES scans for alternative channels when its signal deteriorates. Because certain cells may have large shadowing effects within them, the operator can set a RSSI scan value to determine when a M-ES should start scanning for alternative channels. An M-ES will ignore a drop in signal level if the RSSI scan value is large enough or start scanning

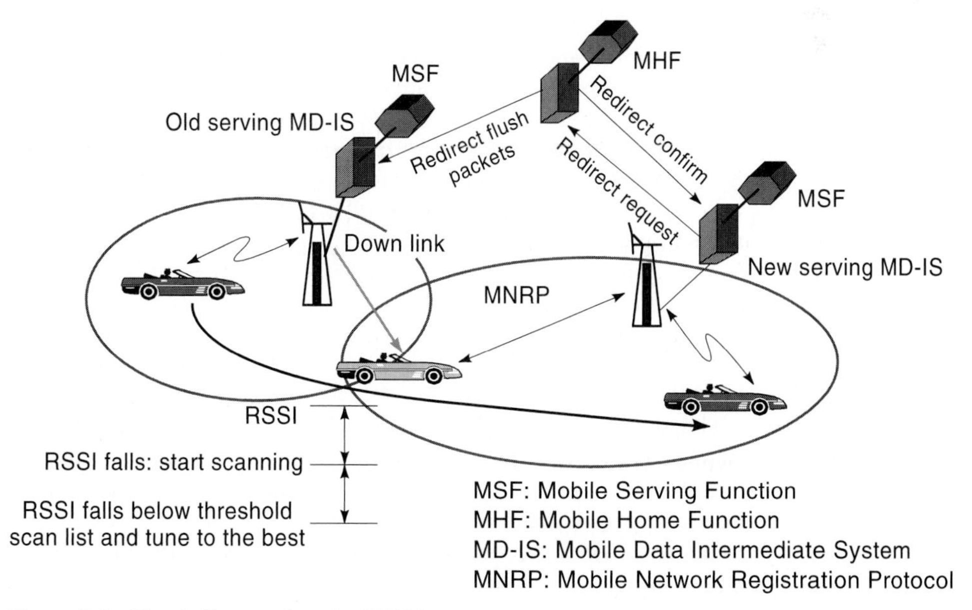

Figure 9.5 Handoff procedure in CDPD.

for alternative channels if it is small. This value is also useful (and should be made small) when the signal strength does not drop even when the M-ES has moved well into a neighboring cell. When additional thresholds for RSSI hysteresis, block error rate (BLER) and/or symbol error rate (SER) are reached, the M-ES will go through a list of channels of adjacent cells that the current BS is broadcasting and tune in to the one with the best signal strength. The M-ES informs the new BS that it has entered its cell. The mobile serving function of the new MD-IS uses a redirect-request and redirect-confirm procedure with the mobile home function of M-ES. The mobile home function also informs the old serving MD-IS about the handoff and directs it through its MSF to redirect packets it may have received for the M-ES to the new serving MD-IS or flush them. Depending on the nature of handoff (interoperator or intraoperator), the delay of registration and traffic redirection will vary.

9.2.3.2 Security

The security functions in CDPD are limited to data link confidentiality and M-ES authentication. There are some mechanisms for key management, the ability to upgrade, access control based on location, a network entity identifier (NEI), and screening lists.

CDPD authentication is performed by the *mobile network registration protocol management entity* (MME) that exists in both the MD-IS and M-ES. It uses the NEI along with an authentication sequence number (ASN) and an authentication random number (ARN) for authentication. The M-ES and network both maintain two sets of ARN/ASN-tuples (in case a fresh ARN is lost due to poor radio coverage). The shared historical record is the basis for authentication. It is updated every 24 hours. Authentication may be initiated at any time by the network.

CDPD confidentiality is based on encrypting all data using a secret key that is different in each session. The usual concept of using public key for exchanging keys and secret keys for block data encryption is employed. The session key generation is based on "Diffie-Hellman" key exchange with 256 bit values. It is, however, susceptible to the man-in-the-middle attack. RC-4 is used for block data encryption, and it is not a very secure secret key algorithm. Consequently, security limitations exist in CDPD. The CDPD network is not authenticated to the mobile as masquerading as a CDPD network is assumed to be impossible. Data confidentiality is not end-to-end, and it is assumed to be a higher layer issue. There are no mechanisms for data integrity, nonrepudiation, or traffic flow confidentiality.

9.2.3.3 Radio Resource Management

RRM is handled by a management layer in CDPD [SAL99b] and contains the procedures to handle the dynamically changing RF environment. In particular it takes care of (1) acquiring and releasing channels due to competition between CDPD and AMPS, and (2) handoff from one cell to another or from one channel to another. Its function is to continuously provide the best possible RF channel between the M-ES and the fixed network. The procedures are distributed between the M-ES and the network. The elements, algorithms, and procedures reside in the MDBS on the network side. The RRM also ensures that transmission power levels are set dynamically to minimize cochannel interference and optimize communication quality on the reverse channel. In the MDBS, the RRM handles distribution of network configuration data, and power control data help the M-ES to track channel hops, perform handoffs, and satisfy power control requirements. On the M-ES side, the RRM has the algorithms and procedures to acquire and track CDPD forward channel transmissions, maintain the best possible CDPD channel in the area where the M-ES is located, and keep transmission power at the required level. All these may be assisted by data transmitted by the network. Unlike voice networks, the M-ES is supposed to handle RRM because the nature of transmission is extremely bursty. As such, the RRM functionality can be provided *with* or *without* the assistance of data provided by the network.

Example 9.4: Messaging for Power Control in CDPD

The channel stream identification (CSI) message (see Figure 9.6) on the downlink provides information about the current channel and also contains the following parameters: a cell identifier, a channel stream identifier, the service provider identity (SPI), a wide area service identifier (WASI), and finally a power product and maximum power level. The transmission power of the M-ES is calculated via the formula [SAL00]:

$$\text{Transmission power (dBW)} = \text{Power product (dB)} - 143\ \text{dBW} - \text{RSSI (dBW)}$$

RSSI is the received signal strength indication that is calculated from the received signal.

9.2.4 Protocol Layers in CDPD

The CDPD standard specifies a protocol stack for CDPD as shown in Figure 9.7. There are four layers: the physical layer, the MAC layer, the data link layer, and the subnetwork dependent convergence protocol (SNDCP) layer.

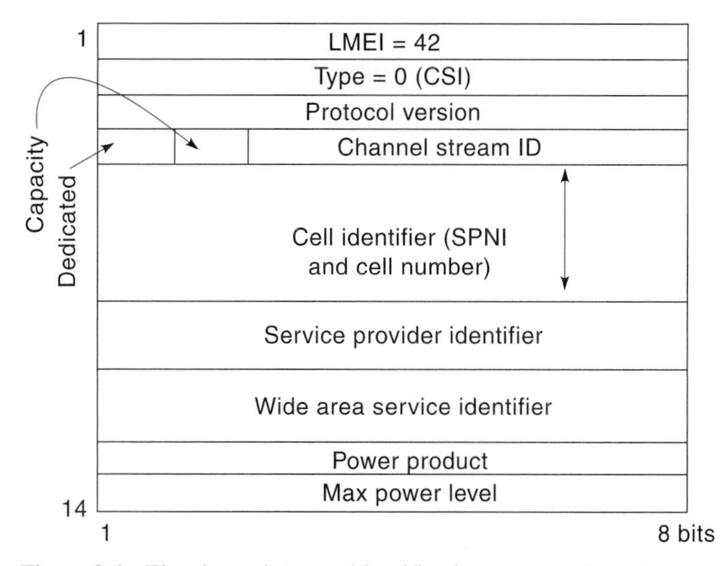

Figure 9.6 The channel stream identification message format.

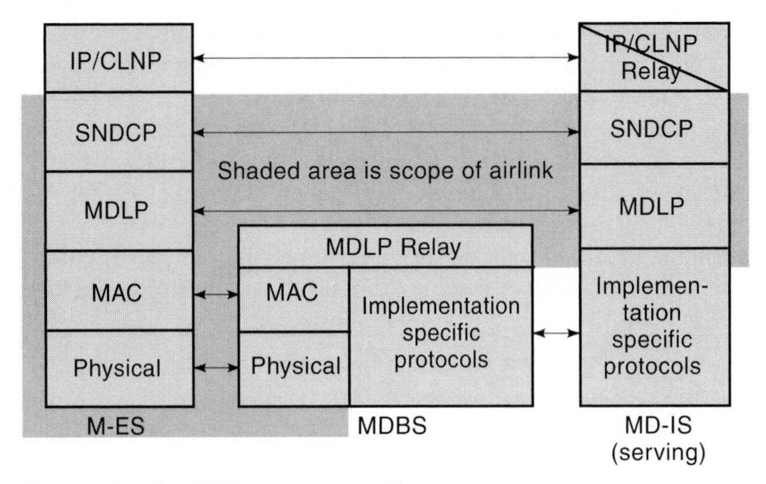

Figure 9.7 The CDPD protocol architecture.

The *physical layer* specifies two distinct one-way RF channels. The *forward channel* is from the MDBS to the M-ES and the *reverse channel* is from the M-ES to the MDBS. They are shown in Figures 9.8 and 9.9, respectively. Each channel is 30 kHz wide and corresponds to the same frequencies as AMPS. The transmission is digital GMSK at 19.2 kbps.

Example 9.5: Error Control Coding in CDPD

The physical layer is robust by employing a (63,47) RS coding. On the forward channel, one RS-block is transmitted every 21.875 ms, so that the raw bit rate is 420 bits/21.875 ms = 19.2 kbps. Up to eight symbol errors can be corrected with this code, but the CDPD specifications suggest correcting only up to seven symbol errors. The undetected symbol error rate is 2.75×10^{-8}.

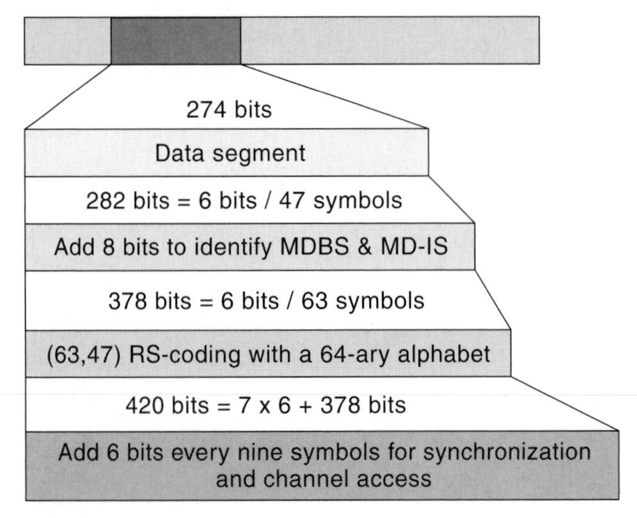

Figure 9.8 The forward channel in CDPD.

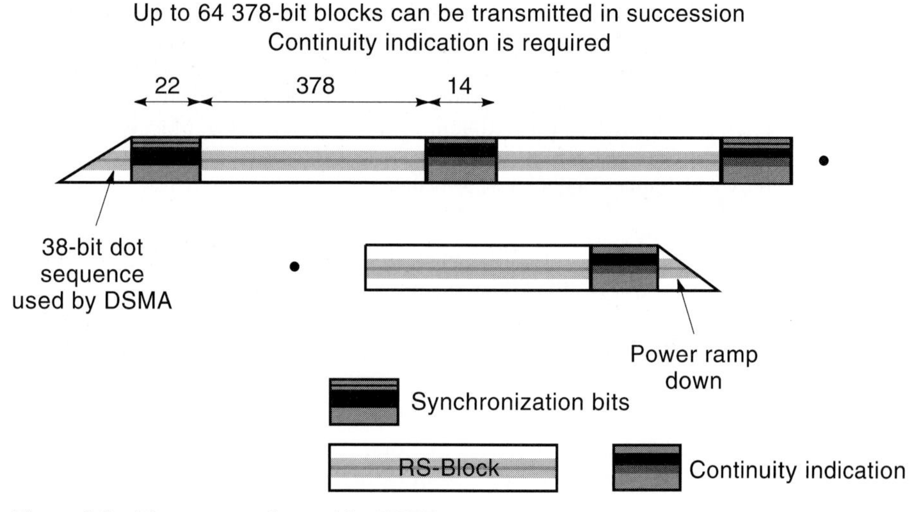

Figure 9.9 The reverse channel in CDPD.

The MAC layer follows the familiar CSMA/CD scheme typified by Ethernet. The forward channel has only one transmitter, the MDBS, and there is no contention for this channel. The reverse channel has multiple M-ESs competing for access. An M-ES may transmit on a channel whenever it has data to send and the channel is *not* already occupied by another transmission. The following steps are followed: (1) An M-ES must assess the state of the channel to see if it is available; (2) If the channel is occupied, there is a random back-off period (nonpersistent); (3) The M-ES transmits if the channel is free; and (4) If two M-ESs find the channel free and transmit simultaneously, there is a collision that must be detected to ensure retransmission. On a wireless link, there are problems with collision detection (see Chapter 4). Carrier sensing is easier at baseband than at RF frequencies. The receive and transmit frequencies are different. Also, transmissions from ground level can be detected at a tower but not at the ground level. Circuitry cost and power consumption become prohibitive for collision detection by a M-ES, so the channel-busy status and collision detection is enabled by the MDBS. A digital indicator is provided on the forward channel in order to indicate reverse channel status. This indicator is called the BUSY/IDLE indicator and it is set to BUSY whenever the MDBS senses reverse channel transmissions. The channels are slotted so that transmissions occur only in time slots. Another flag on the forward channel called *decode status* indicates whether the transmission was successfully decoded. Five bits are used to indicate busy/idle status and M-ESs use a simple majority to decide whether the channel is busy or idle.

Example 9.6: Flexibility in the MAC Protocol

The MAC protocol specifies a Min_idle_Time that is the minimum amount of time a mobile must wait after one burst to transmit again to ensure fairness. Two counts Min_count and Max_count specify limits of back-off delay periods in the case where there are too many or too few users. For example, if Min_count = 4, on the first observation that the channel is busy or the first collision, the mobile backs off for a random number uniformly distributed between zero and $2^4 = 16$ slots. On a consecutive observation of busy channel, the distribution changes to twice the number of slots, for instance, random in (0,32) slots. The value Max_count = 8 implies, the maximum distribution interval is $2^8 = 256$ slots. Max_blocks is the maximum number of RS-blocks that can be transmitted in a burst. It's default value is 64, which implies that at most a 2 kbyte packet can be sent.

The *data link layer* uses the mobile data link protocol (MDLP) to connect the MD-IS and the M-ES. This is at the LLC (logical link control) layer equivalent of the 802.3 protocol. The "logical link" is identified by a "temporary equipment identifier" (TEI). The TEI reduces load on the air-interface and ensures privacy of the user. MDLP is similar to the LAPD of ISDN, and it uses a sliding window protocol. A selective reject mechanism is used. The sender has to retransmit only that frame that is acknowledged by an SREJ message. There is, however, no CRC check as in LAPD and the error control is shifted to the MAC layer.

Example 9.7: Other Data Link Features

Zap frames are used by the MD-IS to disable transmissions from a mobile for a given period of time. This is in case a M-ES does not follow the CDPD protocol in backing off or is not correctly working. A sleep mode is also available to power down if not transmitting to save battery life. The link layer is maintained in suspended mode, and all timers are saved. If a mobile does not transmit for a time period called T203 it is assumed to have gone to sleep. T203 is used to determine the "wake-up" time. Every T204 seconds the network broadcasts a list of TEIs that have outstanding data packets to receive. Before sleeping the mobile tracks the last T204 broadcast and wakes up at the next T204 to listen to the broadcast. After N203 TEI broadcasts, if a mobile is not up, the network discards the packets. Packet forwarding is also possible by storing it in the network for future transmissions.

The SNDCP maps MDLP services to those expected by IP or CLNP. It manages the difference between maximum data link frame size of 130 bytes, network packet size of up to 2048 bytes, and multiple network connections using the same MDLP. The functions of SNDCP include segmentation and reassembly, multiplexing, header compression, TCP/IP header compression using the Van Jacobson method, and the CLNP header compression using similar process and data encryption. Details of CDPD protocols, message formats, and so on can be found in [TAY97].

9.3 GPRS AND HIGHER DATA RATES

GPRS is an overlay on top of the GSM physical layer and network entities. It extends data capabilities of GSM and provides connection to external packet data networks through the GSM infrastructure with short access time to the network for independent short packets (500–1,000 bytes). There are no hardware changes to the BTS/BSC (compared with CDPD), easy to scale, support for voice/data and data only terminals, high throughput (up to 21.4 kbps), and user-friendly billing.

9.3.1 What Is GPRS?

GPRS [KAL00], [CAI97], [BRA97] is an enhancement of the GSM. It uses exactly the same physical radio channels as GSM, and only new logical GPRS radio channels are defined. Allocation of these channels is flexible: from one to eight radio interface timeslots can be allocated per TDMA frame. The active users share timeslots, and uplink and downlink are allocated separately. Physical channels are taken from the common pool of available channels in the cell. Allocation to circuit switched services and GPRS is done dynamically according to a "capacity on demand" principle. This means that the capacity allocation for GPRS is based on the actual need for packet transfers. GPRS does not require permanently allocated physical channels. GPRS offers permanent connections to the Internet with volume based charging that enables a user to obtain a less expensive connection to the Internet. The GPRS MSs (terminals) are of three types. Class A terminals operate GPRS and other GSM services simultaneously. Class B terminals can monitor all services, but operate either GPRS or another service, such as GSM, one at a time. Class C terminals operate only GPRS service. This way there are options to have high-end or low-end terminals.

GPRS has some limitations in that there is only a limited cell capacity for all users and speeds much lower in reality. There is no store and forward service in case the MS is not available. The more popular short messaging service provides this feature, as we shall see in the next section.

The adaptation of GPRS to the IS-136 TDMA cellular standard is called GPRS-136. It is very similar to GPRS except that it uses 30 kHz physical channels instead of 200 kHz physical channels. Also there is no separate BSC. It can use coherent 8-PSK in addition to $\pi/4$-DQPSK to increase throughput over a limited area. This concept is similar to the 2.5G data service called *enhanced data rates for global evolution* (EDGE). Hooks in the standard allow the possibility of 16-QAM, 16-PSK, or 16-DQPSK in the future [SAR00].

9.3.1.1 GPRS Network Services

GPRS provides the following network services—point-to-multipoint (PTM-M) that is a multicast service to all subscribers in a given area, point-to-multipoint (PTM-G) that is a multicast service to predetermined group that may be dispersed over a geographic area, and point-to-point (PTP) service which is packet data transfer. This is of two types: connectionless based on IP and CLNP called PTP-CLNS and connection-oriented based on X.25 (PTP-CONS). GPRS also provides a bearer service for GSM's SMS discussed later in this chapter. There is also an anonymous access for MS at no charge. This is for example similar to an 800 number service where an agency that charges toll could allow an MS to access its credit card verification service for free.

GPRS has parameters that specify a QoS based on service precedence, a priority of a service in relation to another service (high, normal, and low), reliability, and transmission characteristics required. Three reliability cases are defined and four delay classes. Here delay is defined as the end-to-end delay between two MSs or between an MS and the interface to the network external to GPRS. The reliability and delay classes are outlined in Tables 9.1 and 9.2. Transmission characteristics are specified by the maximum and mean bit rates. The maximum bit rate value can be between 8 kbps and 11 Mbps. The mean bit rate value is 0.22 bps to 111 kbps.

9.3.2 Reference Architecture in GPRS

As already mentioned, GPRS reuses the GSM architecture to a very large extent. There are a few new network entities called GPRS support nodes (GSN) that are responsible for delivery and routing of data packets between the mobile station and external packet network. There are two types of GSNs, the *serving GPRS support node* (SGSN) and the *Gateway GPRS support node* (GGSN). These are comparable to the MD-IS in CDPD. There is also a new database called the GPRS register (GR) that is colocated with the HLR. It stores routing information and maps the IMSI to a PDN address (IP address for example). Figure 9.10 shows this reference architecture.

The U_m interface is the air-interface and connects the MS to the BSS. The interface between the BSS and the SGSN is called the G_b interface and that between the SGSN and the GGSN is called the G_n interface. The SGSN is a router that is similar to the foreign agent in Mobile-IP. It controls access to MSs that may be attached to a group of BSCs. This is called a *routing area* (RA) or *service area* of the SGSN. The SGSN is responsible for delivery of packets to the MS in its service area and from the MS to the Internet. It also performs the logical link management, authentication, and charging functions. The GGSN acts as a logical interface to the Internet. It maintains routing information related to a MS, so that it can route packets to the SGSN servicing the MS. It analyses the PDN address of the MS and converts it to the corresponding IMSI and is equivalent to the HA in Mobile-IP.

9.3.3 Mobility Support in GPRS

In a manner similar to GSM and CDPD, there are mechanisms in GPRS to support mobility. We will discuss these issues in the following sections.

Table 9.1 Reliability Classes

	Probability for			
Class	***Lost Packet***	***Dupilcated Packet***	***Out-of-Sequence Packet***	***Corrupted Packet***
1	10^{-9}	10^{-9}	10^{-9}	10^{-9}
2	10^{-4}	10^{-5}	10^{-5}	10^{-6}
3	10^{-2}	10^{-5}	10^{-5}	10^{-2}

Table 9.2 Delay Classes

	128 Byte Packet		1,024 Byte Packet	
Class	***Mean Delay***	***95% Delay***	***Mean Delay***	***95% Delay***
1	< 0.5s	< 1.5s	< 2s	< 7s
2	< 5s	< 25s	< 15s	< 75s
3	< 50s	< 250s	< 75s	< 375s
4	Best Effort	Best Effort	Best Effort	Best Effort

9.3.3.1 Attachment Procedure

Before accessing GPRS services, the MS must register with the GPRS network and become "known" to the PDN. The MS performs an attachment procedure with an SGSN that includes authentication (by checking with the GR). The MS is allocated a temporary logical link identity (TLLI) by the SGSN and a PDP (packet data protocol) context is created for the MS. The PDP context is a set of parameters

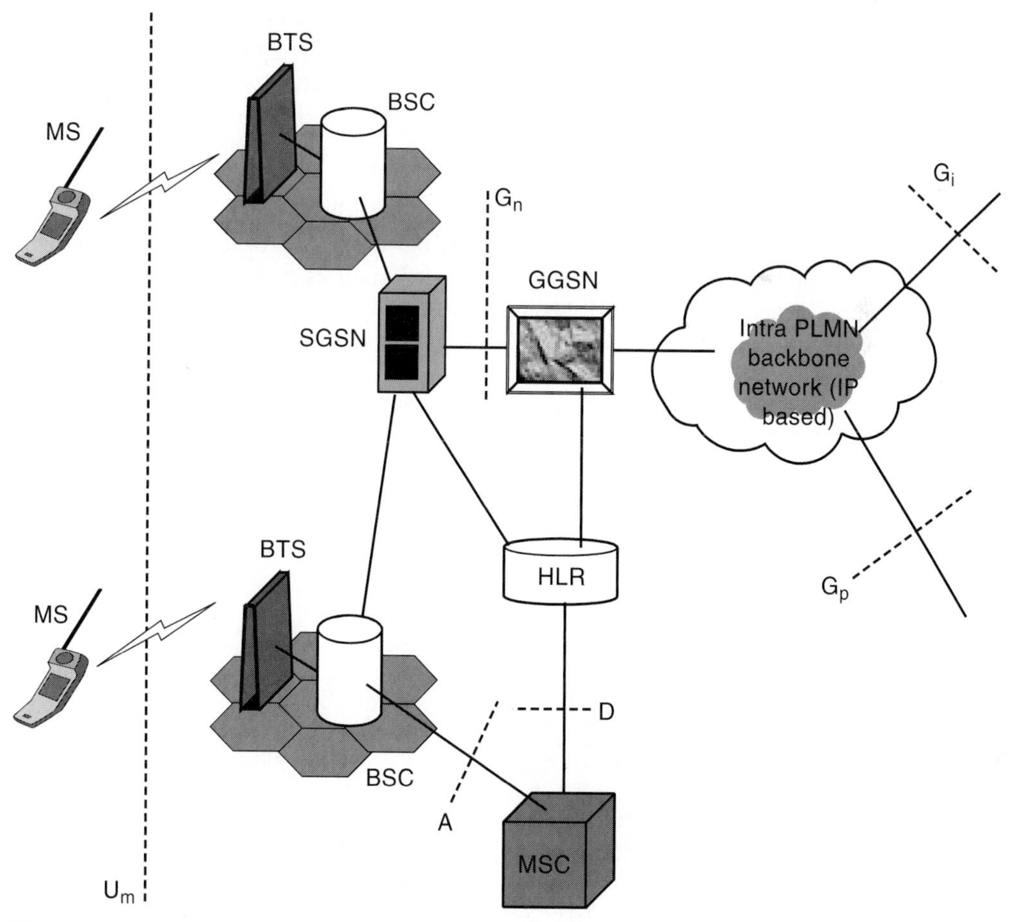

Figure 9.10 GPRS system architecture.

created for each session and contains the PDP type, such as IPv4, the PDP address assigned to the MS, the requested QoS parameters, and the GGSN address that serves the point of access to the PDN. The PDP context is stored in the MS, the SGSN, and the GGSN. A user may have several PDP contexts enabled at a time. The PDP address may be statically or dynamically assigned (static address is the common situation). The PDP context is used to route packets accordingly.

9.3.3.2 Location and Handoff Management

The location and mobility management procedures in GPRS are based on keeping track of the MSs location and having the ability to route packets to it accordingly. The SGSN and the GGSN play the role of foreign and HAs (visiting and home databases) in GPRS.

Location management depends on three states in which the MS can be (Figure 9.11). In the IDLE state the MS is not reachable, and all PDP contexts are deleted. In the STANDBY state, movement across routing areas is updated to the SGSN but not across cells. In the READY state, every movement of the MS is indicated to the SGSN. The reason for the three states is based on discussions similar to those in Chapter 6. If the MS updates its location too often, it consumes battery power and wastes the air-interface resources. If it updates too infrequently, a systemwide paging is needed; this is also a waste of resources. A standby state focuses the area to the service area of the SGSN. In the standby state, there is a medium chance of packets addressed to the MS. The ready state pinpoints the area when the chances of packets reaching are high.

Routing area updates that are part of the standby state are of two types. In the intra-SGSN RA update, the SGSN already has the user profile and PDP context. A new temporary mobile subscriber identity is issued as part of routing area update "accept." The HLR need not be updated. In an inter-SGSN RA update, the new RA is serviced by a new SGSN. The new SGSN requests the old SGSN to send the PDP contexts of the MS. The new SGSN informs the home GGSN, the GR, and other GGSNs about the user's new routing context.

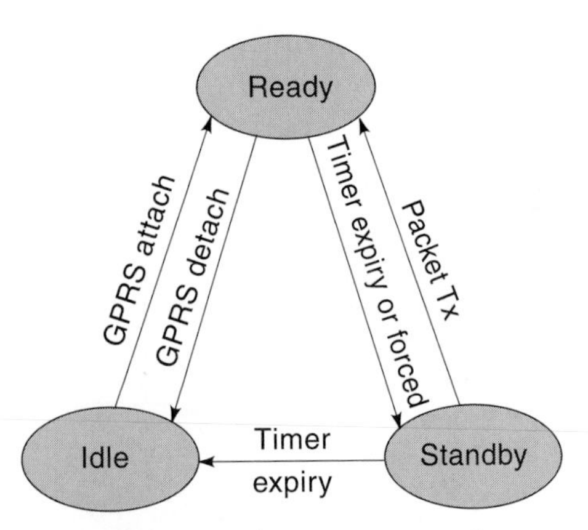

Figure 9.11 Location management in GPRS.

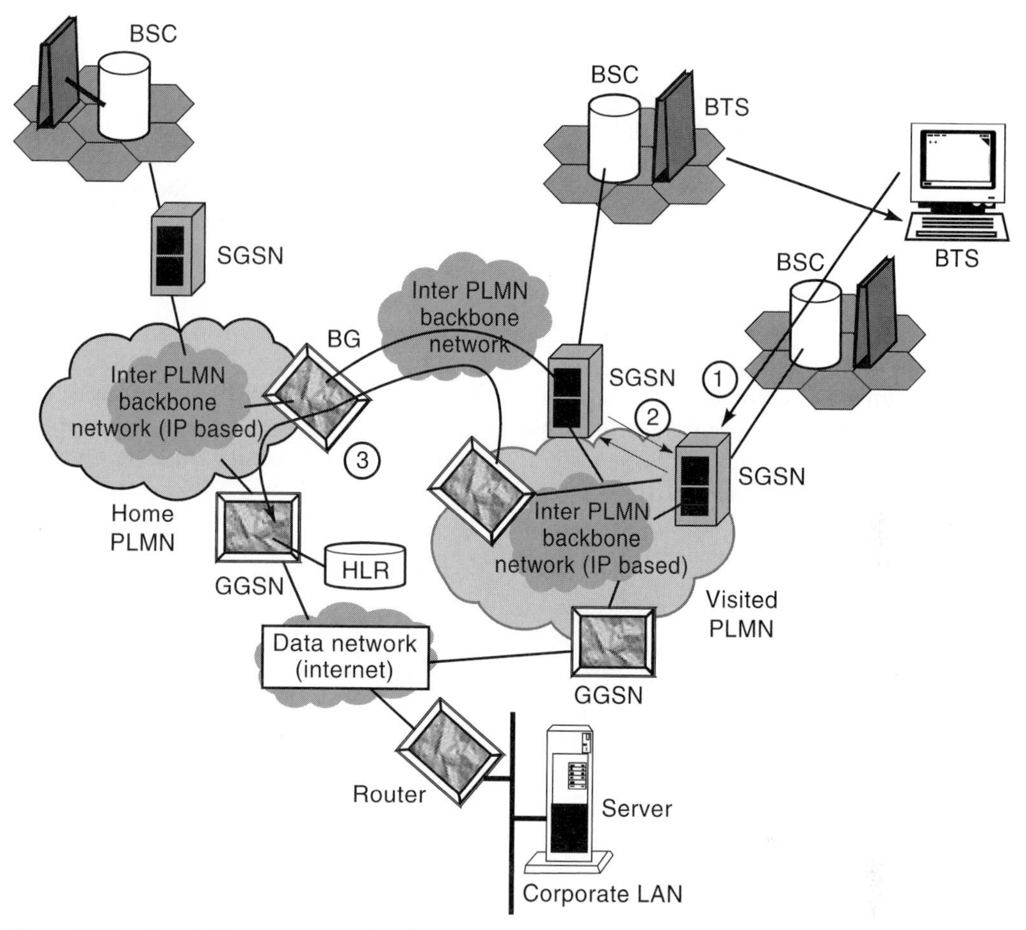

Figure 9.12 Handoff management in GPRS.

Mobility management in GPRS starts at handoff initiation. The MS listens to the BCCH and decides which cell it has to select. Proprietary algorithms are employed that use RSS, cell ranking, path loss, power budget, and so on. The MS is responsible for cell reselection independently, and this is done in the same way as GSM. The MS measures the RSS of the current BCCH and compares it with the RSS of the BCCH of adjacent cells and decides on which cell to attach it to. There is, however, an option available to operators to make the BSS ask reports from the MH (as in GSM), and then the handoff is done as in GSM (MAHO). Plain GPRS specific information can be sent in a *packet BCCH* (PBCCH), but the RSS is always measured from the BCCH. There are also other principles, which may be considered in handoff decision such as path loss, cell-ranking, and so on. The handoff procedure is very similar to mobile IP.

The location is updated with a routing update procedure, as shown in Figure 9.12. When a MS changes a routing area (RA), it sends an RA update request containing the cell identity and the identity of previous routing area, to the new SGSN (1). Note that an intra-SGSN routing area update (as discussed above) is also possible when the same SGSN serves the new RA. The new SGSN asks the old SGSN to provide the routing context (GGSN address and tunneling information) of the MS (2). The new SGSN then updates the GGSN of the home network with the new SGSN address and new tunneling information (3). The new SGSN also updates the HLR. The HLR cancels the MS information context in the old SGSN and loads the subscriber data to the new SGSN. The new SGSN acknowledges the MS. The previous SGSN is requested to transmit undelivered data to the new SGSN.

9.4.3.3 Power Control and Security

Power control and security mechanisms are very similar to the way in which they are implemented in GSM (see Chapter 7). The ciphering algorithm is used to provide confidentiality and integrity protection of GPRS user data used for PTP mobile-originated and mobile-terminated data transmission and point-to-multipoint group (PTM-G) mobile terminated data transmission. The algorithm is restricted to the MS-SGSN encryption.

9.3.4 Protocol Layers in GPRS

In order to transport different network layer packets, GPRS specifies a protocol stack like CDPD and GSM (see Figure 9.13). This is the transport plane (where user data is transferred over the GPRS/GSM infrastructure). There is also a GPRS signaling plane to enable signaling between various elements in the architecture (like messaging between the SGSN and BSS etc.). GPRS employs out-of-band signaling in support of actual data transmission. Signaling between SGSN, HLR, VLR, and EIR is similar to GSM and extends only the GPRS related functionality. So it is based on SS-7. Between the MS and SGSN, a GPRS mobility management and session management (GMM/SM) protocol is used for signaling purposes.

The GPRS transport plane has different layers in different elements. The physical layers between the MS-BSS, BSS-SGSN, and SGSN-GGSN are indicated

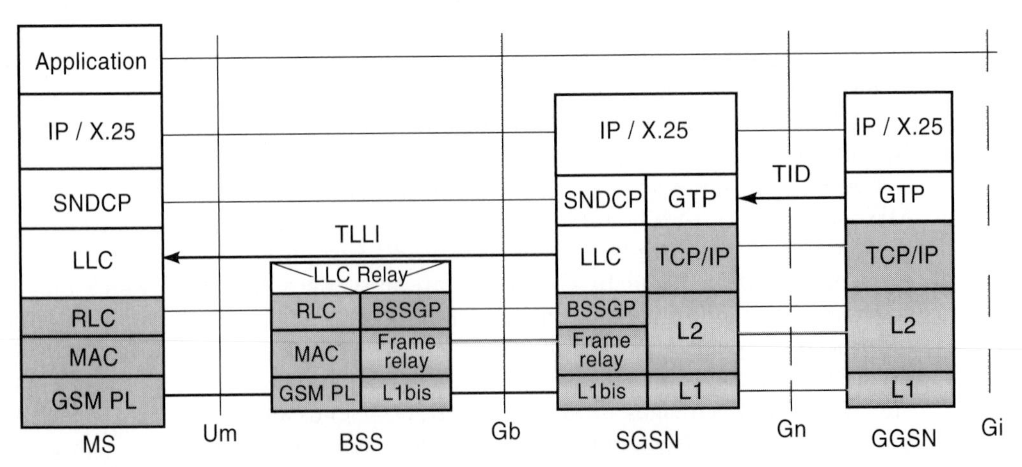

Figure 9.13 GPRS transport plane.

in Figure 9.13. Over the air, the physical layer is the same as GSM (i.e., it uses GMSK). Its functionalities include forward error correction and indication of uncorrectable code words, interleaving of radio "blocks," synchronization, monitoring of radio link signal quality, and so on. All other functions are similar to GSM. GPRS allows a MS to transmit on multiple time slots of the same TDMA frame unlike GSM. A very flexible channel allocation is possible since 1–8 time slots can be allocated per TDMA frame to a single MS. Uplink and downlink slots can be allocated differently to support asymmetric data traffic. Some channels may be allocated solely for GPRS. These are called packet data channels (PDCH).

Allocation of radio resources is also slightly different compared with GSM. A cell may or may not support GPRS and if it does support GPRS, radio resources are dynamically allocated between GSM and GPRS services. Any GPRS information is broadcast on the CCHs. PDCHs may be dynamically allocated or deallocated by the network (usually the BSC). If an MS is unaware that the PDCH has been deallocated, it may cause interference to a voice call. In such a case, fast release of PDCHs is achieved by a broadcast of a deallocation message on a PACCH.

The uplink and downlink transmissions are independent. The medium access protocol is called "Master-Slave Dynamic Rate Access" or MSDRA. Here, the organization of time-slot assignment is done centrally by the BSS. A "master" PDCH includes common control channels that carry the signaling information required to initiate packet transfer. The "slave" PDCH includes user data and dedicated signaling information for an MS. Several logical traffic and control GPRS channels are defined analogous to GSM. For example, PDTCH is the packet data traffic channel and PBCCH is the packet broadcast control channel. For random access to obtain a traffic channel, the packet random access, access grant, and paging channels are called PRACH, PAGCH, and PPCH, respectively. Additionally there is a packet notification channel (PNCH) that notifies arrival of a packet for the MS and a packet associated control channel (PACCH) used to send ACKs for received packets. A packet timing-advance control channel (PTCCH) is used for adaptive frame synchronization.

The packet transfer on the uplink and downlink are shown in Figures 9.14 and 9.15, respectively. They are quite similar to the process in GSM. Some of the differences are as follows. If a MS does not get an ACK, it will back off for a random time and try again. On the uplink, the Master-Slave mechanism utilizes a 3-bit uplink status flag (USF) on the downlink to indicate what PDCHs are available. A list of PDCHs and their USFs are specified in this USF. The packet resource or immediate assignment message indicates what USF state is reserved for the mobile on a PDCH. Channel assignment can also be done so that a MS can send packets uninterrupted for a predetermined amount of time. On the downlink, data transmission to a mobile can be interrupted if a high-priority message needs to be sent. Instead of paging, a resource assignment message may be sent to the MS if it is already in a "ready" state.

GPRS supports IP and X.25 packets at the network layer to be used by end-to-end applications. The SNDCP supports a variety of network protocols (IP, X.25, CLNP, etc.). All network layer packets share the same SNDCP. It multiplexes and demultiplexes the network layer payload and forms the interface between the link layer (LLC) and the network layer. Also the SNDCP handles packets based on QoS.

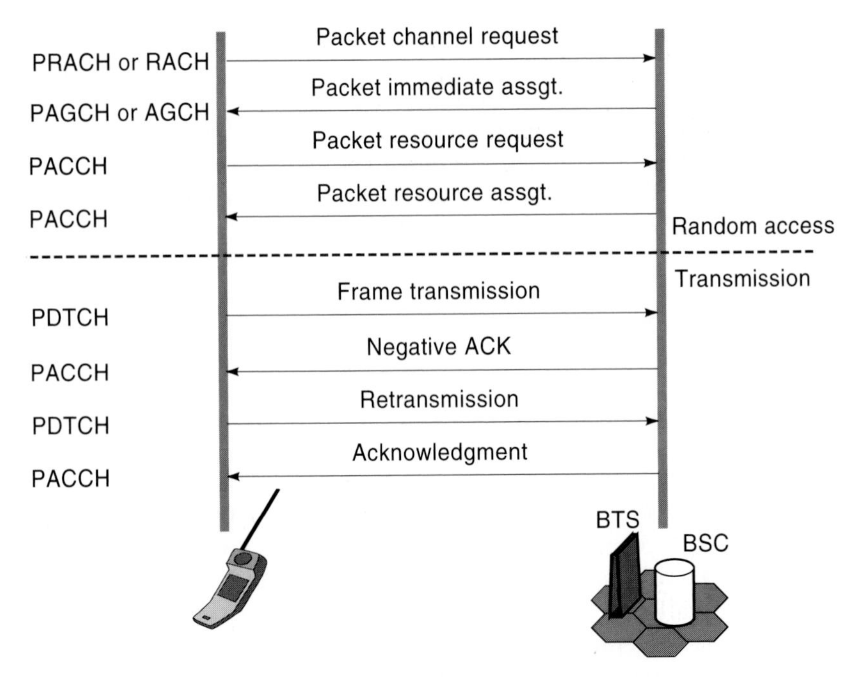

Figure 9.14 Uplink data transfer.

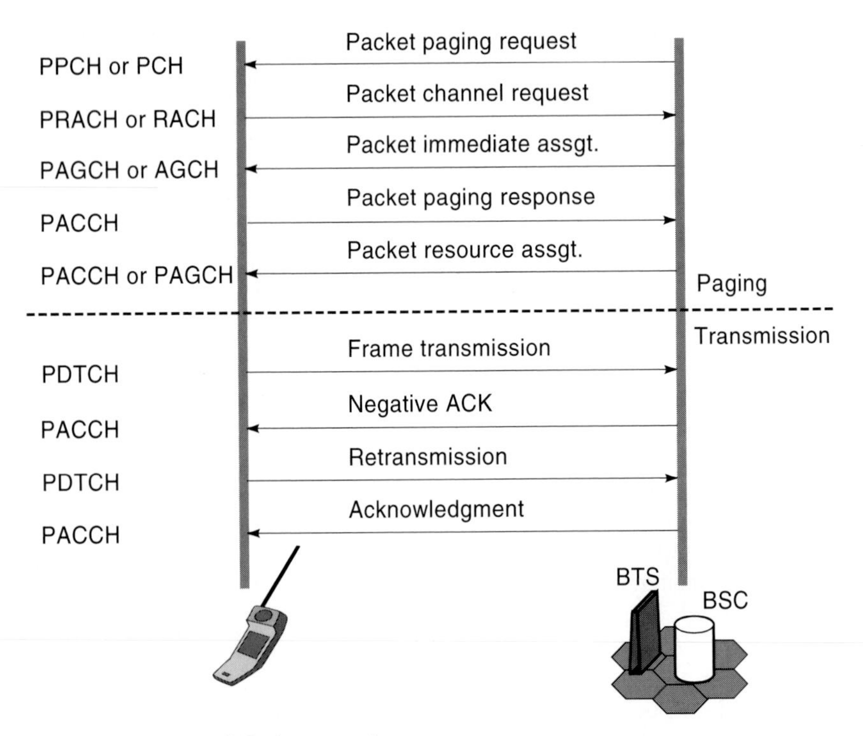

Figure 9.15 Downlink data transfer.

The LLC layer forms a logical link between the MS and the SGSN (similar to CDPD's MDLP). Each MS has a temporary logical link identity (TLLI) to identify itself in the LLC header. The LLC performs sequence control, error recovery, flow control, and encryption. It has an acknowledged mode (with retransmission for network layer payloads) and an unacknowledged mode (for signaling and SMS). The LLC also supports various QoS classes. Figure 9.16 shows how packets flow from higher layers, applications, and signaling levels to the SNDCP and the LLC. The packet transformation data flow is shown in Figure 9.17. The end result is blocks of 114 bits that are transmitted in bursts similar to GSM.

There are two levels of connections (tunneling mechanisms) implemented in the GPRS infrastructure as shown in Figures 9.13 and 9.18, one between the MS and the SGSN and the second between the SGSN and the GGSN. The two-level tunneling mechanism corresponds to a two-level mobility management: LLC "tunnels" (or virtual circuits) correspond to small area mobility, while GPRS tunneling protocol (GTP) tunnels correspond to wide area mobility. A new logical link is created each time the MS makes a handoff in the ready state between itself and the SGSN. If the SGSN does not change, the tunneling of the packet beyond the SGSN remains the same with the same GTP.

The BSS Gateway protocol (BSSGP) operates between the BSS and the SGSN relaying the LLC packets from the MS to the SGSN. Many MS LLCs can be multiplexed over one BSSGP. Its primary function is to relay radio related, QoS, and routing information between the BSS and SGSN and paging requests from SGSN to the BSS. It supports flushing of old messages from BSS. The data transfer is unconfirmed between BSS and SGSN.

The GTP allows multiprotocol packets to be tunneled through the GPRS backbone. A tunnel ID (TID) is created using the signaling plane that tracks the PDP context of each MS session. GTP can multiplex various payloads. The GTP

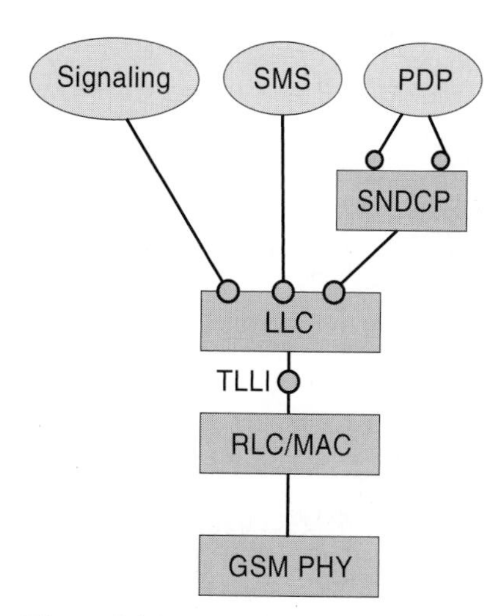

Figure 9.16 SNDCP and LLC in GPRS.

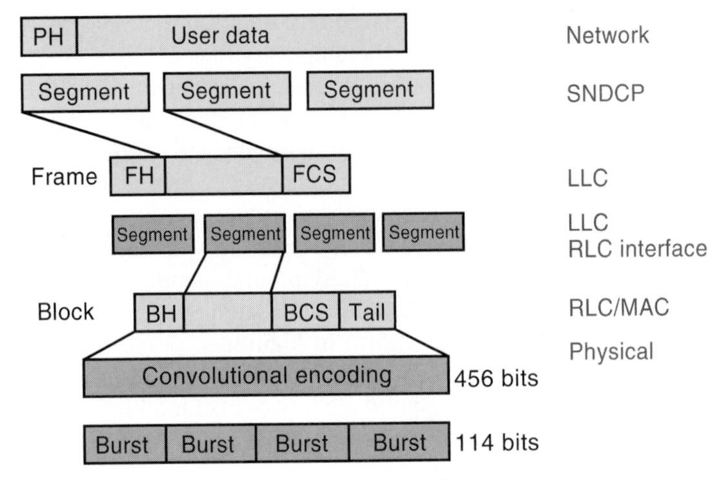

Figure 9.17 Packet transformation data flow.

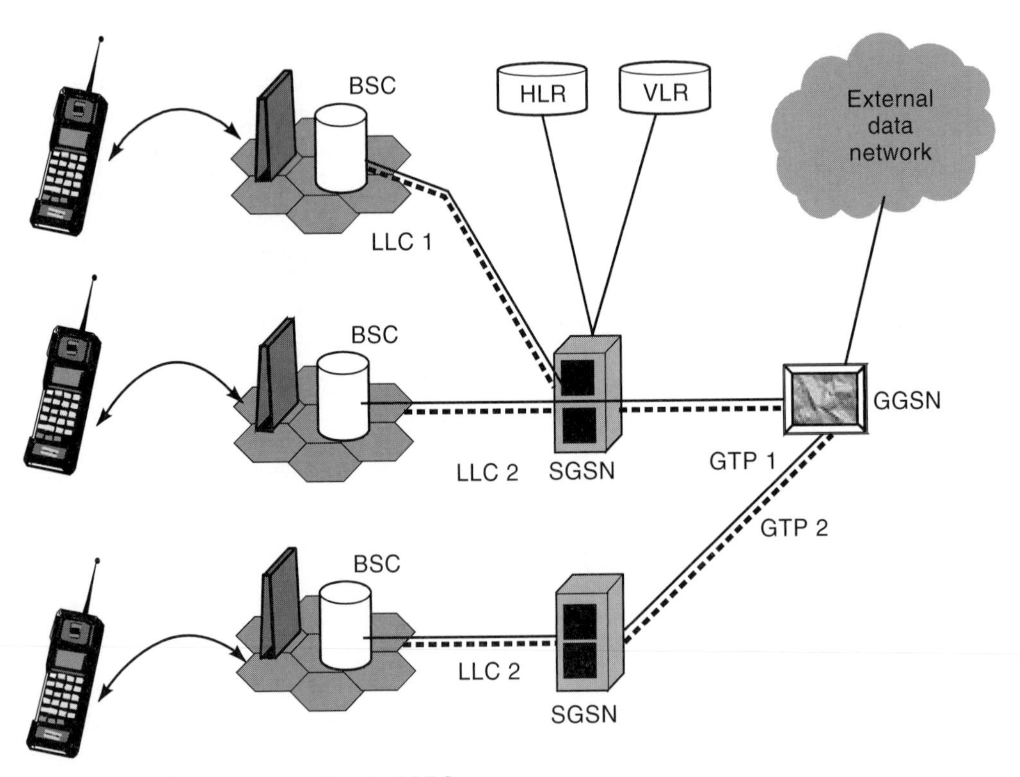

Figure 9.18 Two-level tunneling in GPRS.

packet is carried by either UDP/IP or TCP/IP, depending on whether the payload is IP or X.25, respectively.

9.4 SHORT MESSAGING SERVICES IN GPRS

The proliferation of GSM enabled the introduction of the *short messaging service* (SMS), which has become extremely popular in Europe. It is similar to the peer-to-peer instant messaging services on the Internet. Users of SMS [PEE00a], [PEE00b] can exchange alphanumeric messages of up to 160 characters (mapped into 140 bytes) within seconds of submission of the message. The service is available wherever GSM exists and makes it a very attractive wide area data service.

9.4.1 What Is SMS?

SMS was developed as part of GSM Phase 2 specifications (see Chapter 7). It operates over all GSM networks making use of the GSM infrastructure completely. It uses the same network entities (with the addition of a SMS center—SMSC), the same physical layer, and intelligently reuses the logical channels of the GSM system to transmit the very short alphanumeric messages.

9.4.1.1 Service Description

SMS has both an almost instant delivery service if the destination MS is active or a store and forward service if the MS is inactive. Two types of services are specified: In the *cell broadcast* service, the message is transmitted to all MSs that are active in a cell and that are subscribed to the service. This is an unconfirmed, one-way service used to send weather forecasts, stock quotes, and so on. In the *PTP* service, an MS may send a message to another MS using a handset keypad, a PDA, or a laptop connected to the handset, or by calling a paging center. Recently, SMS messages can be transmitted via dial-up to the service center and the Internet as well [PEE00a].

A short message (SM) can have a specified priority level, future delivery time, expiration time, or it might be one of several short predefined messages. A sender may request acknowledgment of message receipt. A recipient can manually acknowledge a message or may have predefined messages for acknowledgement.

An SM will be delivered and acknowledged whether a call is in progress because of the way logical channels in GSM are used for SMS. We discuss this in a later section.

9.4.2 Overview of SMS Operation

The SMS makes use of the GSM infrastructure, protocols, and the physical layer to manage the delivery of messages. Note that the service has a store-and-forward nature. As a result, each message is treated individually. Each message is maintained and transmitted by the SMSC. The SMSC sorts and routes the messages appropriately. The short messages are transmitted through the GSM infrastructure using SS-7. Details of the packet formats and messaging can be found in [PEE00a].

Figure 9.19 shows the reference architecture and the layered protocol architecture for SMS. There are two cases of short messages: a mobile originated SM and a mobile terminated short message. A SM originating from an MS has to be first delivered to a service center. Before that, it reaches an MSC for processing. A dedicated function in the MSC called the *SMS-interworking MSC* (SMS-IWMSC) allows the forwarding of the SM to the SMSC using a global SMSC ID. An SM that terminates at the MS is forwarded by the SMSC to the *SMS-gateway MSC* (SMS-GMSC) function in an MSC. As in the case of GSM, it either queries the HLR or sends it to the SMS-GMSC function at the home MSC of the recipient. Subsequently, the SM is forwarded to the appropriate MSC that has the responsibility of finally delivering the message to the MS. This delivery is performed by querying the VLR for details about the location of the MS, the BSC controlling the BTS providing coverage to the MS, and so on.

There are four layers in SMS—the application layer (AL), the transfer layer (TL), the relay layer (RL), and the link layer (LL). The AL can generate and display the alphanumeric message. The SMS-TL services the SMS-AL to exchange SMs and receive confirmation of receipt of SMs. It can obtain a delivery report or status of the SM sent in either direction. The RL relays the SMS PDUs through the LL. There are six PDUs types in SMS that convey the short message—from the SMSC to the MS and vice versa, convey a failure cause, and convey status reports and commands.

Over the air, the SMs are transmitted in time slots that are freed up in the control channels. If the MS is in an idle state (see Chapter 10), the short messages are sent over the SDCCH at 184 bits within approximately 240 ms. If the MS is in the active state (i.e., it is handling a call), the SDCCH is used for call set-up and maintenance. In that case, the SACCH has to be used for delivering the SM. This occurs at around 168 bits every 480 ms and is much slower. Failures can occur if there is a state change when the SM is in transit. The short message will have to be transmitted later.

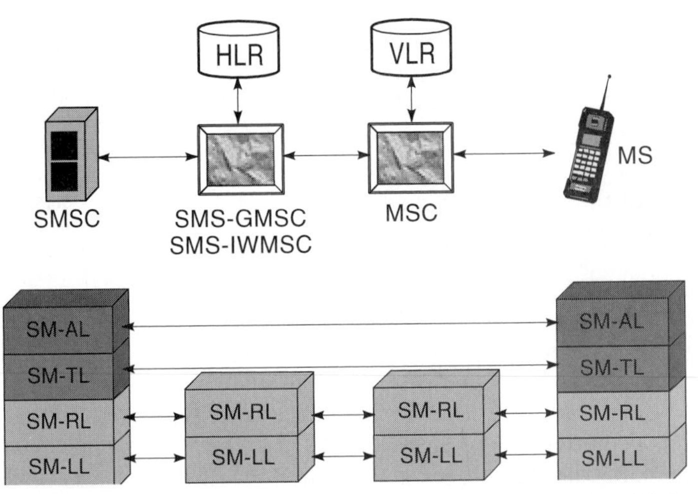

Figure 9.19 Reference and layered protocol architecture for SMS.

In the case of cell broadcast, a cell broadcast entity and a cell broadcast center are used to send the weather forecast or other broadcast SMs to multiple BSCs for delivery. The broadcasts contain the data and identities of MSs that are to receive the message. The cell broadcast is on the cell broadcast channel (CBCH).

9.5 MOBILE APPLICATION PROTOCOLS

Mobile applications are becoming very important in the age of the Internet. Data networks and dial-up services were mostly restricted to research laboratories and educational institutions in the early 1980s and became a household service by the late 1990s because of the emergence of applications such as email, e-commerce, and the World Wide Web. Recently, efforts provide similar applications on cellular networks. A major problem with providing such services on cellular networks has been the lack of resources such as bandwidth, processing power, memory, display sizes, interfaces like keypads, and so on that makes the service expensive. The constraints are also larger such as more latency, less connection stability, and less predictive availability. The wireless application protocol (WAP) and the i-mode service offered by NTT-DoCoMo of Japan are examples of how Internet based applications are being adapted to the cellular systems.

9.5.1 Wireless Application Protocol (WAP)

Initiated in 1997 by Nokia, Ericsson, Motorola and Phone.com, the WAP is an industry standard developed by the WAP Forum [WAPweb] for integrating cellular telephony and the Internet by providing Web content and advanced services to mobile telephone users. The WAP protocol is expected to help implementation of a variety of applications that include Internet access, m-commerce, multimedia email, tele-medicine, and mobile geo-positioning. The WAP application framework can run over several transport frameworks that include SMS, GPRS, CDPD, IS-136, and circuit switched wireless data services. Using WAP the wireless network technology will allow development of variety of applications. WAP orients the display toward text and material suitable for very small screens, thereby reducing both the load on the network and on the mobile terminal. In a nutshell, WAP attempts to optimize the Web and existing tools for a wireless environment.

WAP provides an extensible and scaleable platform for application development for mobile telephones. However, support for WAP on color PDA terminals is not very good. WAP also does not support seamless roaming between different link level bearer services such as CDPD and GSM, for example. Also, multimedia communications are not supported very well on WAP [FAS99]. For all these limitations, modifications to WAP will be necessary.

9.5.1.1 WAP Programming Architecture

How WAP does the extensions and modifications to the existing worldwide Web architecture is as follows. WAP introduces a *gateway* in between the wireless client and the rest of the Internet which manages the delivery of content to the mobile terminal. Such an architecture is shown in Figure 9.20.

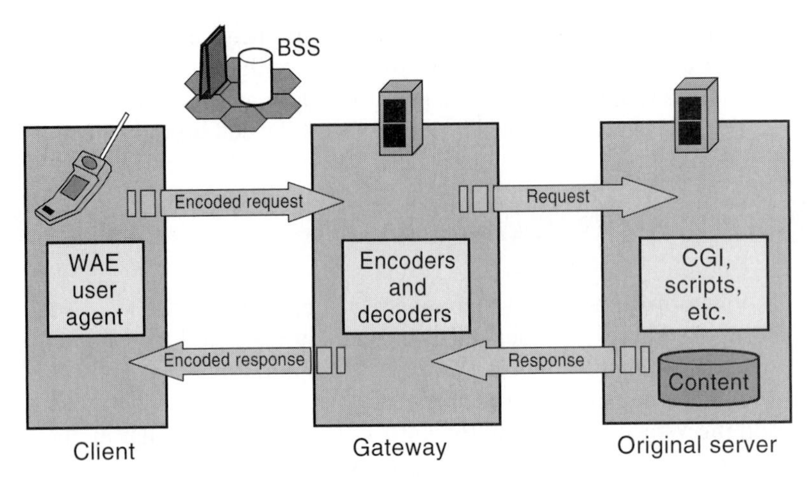

Figure 9.20 WAP programming architecture.

Most current Web content resides in Web servers and consists of a variety of material such as HTML pages, JavaScript interactive Web pages, images, and multimedia. These contents are developed for full-screen computer displays. A mobile terminal cannot access all this material because the screen size is smaller, the data rate for access is lower, and the cost of the access is higher than the wired access. To overcome this problem, the request for content is first made to the WAP Gateway, shown in Figure 9.20. The WAP request is made using a binary format of a *wireless markup language* (WML) that has some kind of a correspondence with HTML. That is, HTML pages can be converted into WML content. WML was derived from the extensible markup language (XML) and describes menu trees or decks through which a user can navigate with the microbrowser. The binary request conserves bandwidth by compressing the data. The request is decoded by the gateway and transmitted to the original server as an HTTP request. The server responds with the content in HTML. The content is filtered into WML, encoded into binary, and transmitted to the handset. A microbrowser in the handset coordinates the user interface and is similar to the usual Web browser. In essence, the gateway acts as a proxy device within the network. An example of a WAP network is shown in Figure 9.21. The WTA stands for wireless telephony application that provides an interface to a wireless application environment for network related activities such as call control, access to local address books, network events, and so on.

9.5.1.2 Protocol Layers in WAP

The layered protocol architecture specified for WAP is shown in Figure 9.22. The *wireless application environment* (WAE) is an application development platform that combines aspects of the Web and mobile telephony. It includes a microbrowser, WML, a scripting language similar to JavaScript called WMLscript,

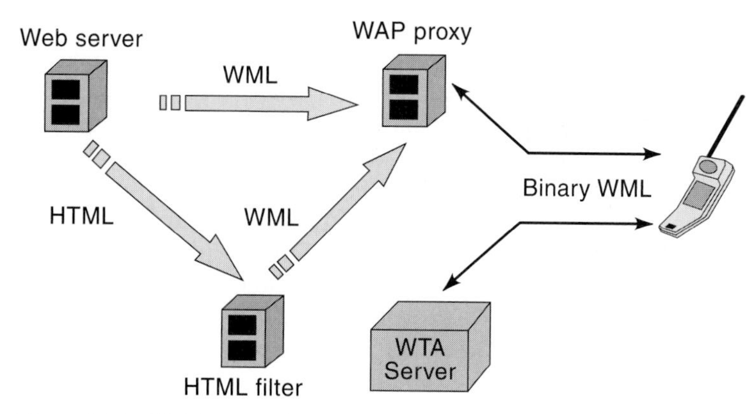

Figure 9.21 WAP example network.

several data formats such as images, calendars, and records, and telephony applications. The *wireless session protocol* (WSP) provides both a connection oriented and a connectionless service to the WAE. The connection-oriented service operates on top of the *wireless transport protocol* (WTP) and the connectionless service on top of the *wireless datagram protocol* (WDP). These are similar to Internet's TCP and UDP, respectively. WSP can suspend or migrate sessions unlike protocols on the wired link, and it is optimized for low bandwidth links.

The *wireless transport layer security* (WTLS) is based on the IETF standard transport layer security and the secure socket layer. It provides data integrity, privacy, authentication and techniques to reject replay attacks.

9.5.2 i-Mode

i-Mode is a service that tries to eliminate the use of a gateway and provide direct access to the Internet to the extent possible. With this goal in mind, Japan's NTT-DoCoMo has introduced this extremely popular service in Japan in 1999. By

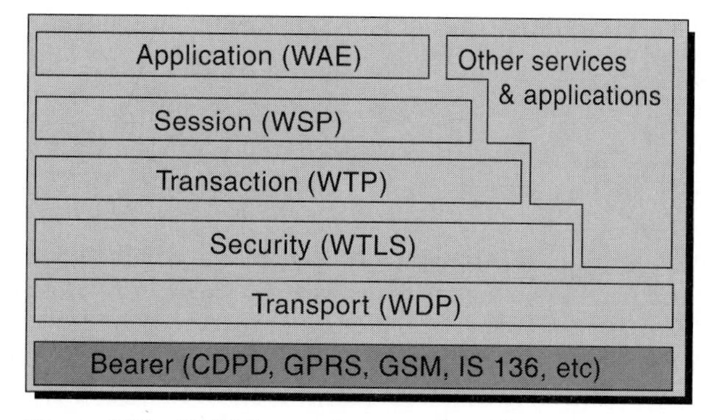

Figure 9.22 WAP layered protocol architecture.

November 2000 i-Mode had 14.9 million subscribers in Japan and Hong Kong. These terminals transmit data at 9,600 bps that allows graphics and small text messaging on a larger screen than the WAP. This display allows six to 10 lines of text at 16 to 20 characters per line that can be color or monochrome. i-Mode telephones can access HTML files across the web using C-HTML without a protocol like WAP. i-Mode is more similar to HTML which allows different computers to exchange information. i-Mode extends this to communicate with PDAs and i-Mode enabled cellular phones. One of the major features of i-Mode is that its charging mechanism is based on packet transmission rather than connection time which makes it less expensive for most users applications. The next version of WAP is expected to include CHTML as an alternative mechanism so as to include i-Mode as an option within itself.

QUESTIONS

9.1 What are the new elements added to the GSM infrastructure to support GPRS?

9.2 What are the new elements added to the AMPS infrastructure to support CDPD?

9.3 What are the duties of SGSN and GGSN in GPRS?

9.4 Why is the channel spacing in ARDIS and Tetra 25 kHz and in CDPD 30 kHz?

9.5 How does GPRS provides a variety of data rates?

9.6 How many classes of QoS are supported by GPRS and what are the differences between them?

9.7 Name the connectionless- and connection-oriented services provided by the GPRS.

9.8 What is the transmission band of the GPRS? Does it change when the data rate is changed?

9.9 What is the number of bits in each burst of GPRS and how does it differ from a GSM burst?

9.10 What are the expected latencies for circuit-switched and packet-switched data networks?

9.11 What is the difference between charging techniques for a packet data and a circuit-switched network?

9.12 With dedicated packet data networks such as ARDIS or MOBITEX, why would one choose CDPD?

9.13 What are the differences between GGSN and SGSN in GPRS, and MD-IS and MDBS in the CDPD?

9.14 What are the differences between the MAC layers of GPRS and CDPD?

9.15 What are the differences between the air interfaces of GPRS and CDPD?

9.16 What are the differences between the Mobile IP protocol and CDPD's method to support mobility?

9.17 What is the difference between full-duplex and half-duplex CDPD operations?

9.18 What is the maximum network layer throughput in CDPD for reverse and forward channels and why they are not the same?

9.19 What are the equivalents of CDPD's MHF and MSF in the Mobile IP protocol?

9.20 What are the MTP and PTP, and SMS?

9.21 Differentiate between the two types of independent mobile data networks.

9.22 Of the design goals of CDPD, which three do you consider important? Why?

9.23 Give three reasons why it is difficult to detect collisions at the transmitter in wireless networks.

9.24 What is the importance of color codes in CDPD?

9.25 Assume that you have a single mobile end system in a CDPD cell. What value of Min_Idle_Time would you suggest? Why?

9.26 Draw the protocol stack of CDPD at the M-ES, at the MDBS, and at the MD-IS. Show the communication between different peer layers.

9.27 What is GPRS-136? How does it differ from GPRS?

PROBLEMS

9.1 The fade rate of a channel is the number of times on average that the signal crosses an acceptable threshold. If the fade rate is larger than the packet rate, a packet cannot usually be received correctly over a wireless link. Packet sizes and data rates determine the critical fade rate that can be tolerated by a given system. Assume that there is no error correction or interleaving and the packet size in CDPD is 128 bytes at 19.2 kbps. If packets are continuously transmitted, what fade rate will destroy all packets? Suppose the data rate is increased to 56 kbps. How does the critical fade rate value change?

9.2 In CDPD, the M-ES has three base stations in its CCIB: BS1, BS2, and BS3. The quantities associated with these base stations are indicated by the subscripts 1, 2, and 3 in the following. The M-ES measures the RSS from a reference channel. To correct the measured value to reflect the actual RSS on the CDPD channel, it uses an ERP_Delta value, i.e., ERP_Delta = RSSI (reference channel) − RSSI (CDPD channel). An RSSI_bias value biases the handoff in favor of either the current cell or the neighboring cell. The RRME evaluates neighbor cells if

RSSI < PRSSI – RSSI_Scan_Delta for a time larger than RSSI average time *or*
RSSI > PRSSI + RSSI_Scan_Delta for a time larger than RSSI average time

This means the M-ES will start scanning for adjacent cells if the RSSI changes by RSSI_Scan_Delta for RSSI_Average_Time. This is called a non-critical condition. A critical condition implies for example that the BLER is higher than the threshold for a time larger than the BLER_average_time. The handoff procedure is as follows:

- Step 1: Close current channel. Compute RSSI_eff for each cell in the CCIB using:
  - RSSIeff = RSSI – ERP_delta + RSSI_Bias
  - Select best neighbor cells (with highest RSSI_eff)
- Step 2: If non-critical condition, compare current cell with best neighboring cells.
- If RSSIcurrent + RSSI_Hysteresis > RSSI_eff, stick with current cell else make a handoff.
- Skip Step 2 if a critical condition occurs.

Assume that the values recorded are as follows for a time larger than

RSSI_Average_Time. RSSI_Hysteresis = 3 dB

for the current base station (BS1).

RSSI 1 = −45 dBm RSSI2 = −42 dBm RSSI_bias2 = −6 dB
PRSSI1 = −50 dBm RSSI3 = −57 dBm RSSI_bias3 = 3 dB
BLER_Threshold = 10^{-3} ERP_delta1 = 0 dB ERP_delta2 = 0 dB
ERP_delta3=0 dB

a. Explain what happens if the RSSI_Scan_delta is 6 dB and the measured BLER for a time greater than BLER_average_time is 10^{-4}? Why?

b. Explain what happens if the RSSI_Scan_delta is 4 dB and the measured BLER for a time greater than BLER_average_time is 10^{-4}? Why?

c. Explain what happens if the RSSI_Scan_delta is 6 dB and the measured BLER for a time greater than BLER_average_time is 1.5×10^{-3}? Why?

In each case explain whether a handoff is made, possibly to which base station, and why.

9.3 We know that the raw data rate at the physical layer in CDPD is 19.2 kbps. Calculate the actual data rate if you assume that the data is

a. before adding the control bits.

b. before performing the Reed-Solomon coding.

c. before frame delimiting.

What are your conclusions?

9.4 In CDPD, a (63,47) Reed-Solomon (RS) code is employed. The Reed-Solomon code encodes blocks of 47 symbols each carrying six bits into codewords of 63 symbols (also each of six bits). A codeword can correct up to seven symbol errors. A block of 63 symbols is transmitted roughly every 21 ms. What is the maximum fade duration that can be corrected by the RS code? How many codewords should be interleaved to combat fade durations of 105 ms? Explain how you arrive at your answers.

9.5 In 2.5G wireless data networks, sometimes called Enhanced Data rates for Global Evolution (EDGE), data throughput speeds of up to 384 kbps are proposed using existing GSM infrastructure. The idea with this approach is to use higher level modulation schemes like 8-PSK instead of GMSK or QPSK. This approach has moderate implementation complexity since the carrier spacing and TDMA frame structure remain the same. Only the channel coding and interleaving are different. Depending on the code rate and modulation scheme, the data rates provided are changed. Several combinations have been proposed for this purpose.

a. Determine the data rates that can be provided for the following cases:

i. CS-1 where GMSK is used with a code rate of 0.49

ii. CS-3 where GMSK is used with a code rate of 0.73

iii. PCS-3 where 8-PSK is used with a code rate of 0.6

iv. PCS-6 where 8-PSK is used with a code rate of 1

b. Discuss the implications of using the above combinations.

9.6 One of the problems with EDGE (see Problem 9.5) is the increased SNR requirement if higher level modulation schemes are employed. Assume that 8-PSK and QPSK are used without any coding in one implementation of EDGE. Using the Okumura-Hata model, determine what will be the coverage of a base station if it uses 8-PSK signal and if it uses QPSK. Assume that a bit error rate of 10^{-4} is required in either case. (*Hint:* Use the bit error rate formulas from Chapter 3. Assume that the transmit power is 30 dBm, the antenna heights are 75 m and 1 m, and the frequency of operation is 900 MHz.)

PART FOUR Local Broadband and Ad Hoc Networks

The last part of the book is devoted to the short-range local wireless networks. This part consists of five short chapters, three of them devoted to WLAN, one to WPAN, and one to wireless geolocation principles.

CHAPTER 10 INTRODUCTION TO WLANs

Emergence and growth of the WLAN industry relied heavily on the application domain for this technology, and these applications have changed with time. On the other hand, a number of transmission techniques, access methods, and frequency of operations have been examined for implementation of WLANs. A good understanding of the WLAN industry requires an overview of the applications and an explanation of the relation between the applications and the variety of technologies. Chapter 10 analyzes the evolutionary path of the WLAN industry and explains how it emerged as a technology for office and manufacturing environments and how it is currently heading towards home and personal area networking.

CHAPTER 11 IEEE 802.11 WLANs

Chapter 11 provides details of the IEEE 802.11 standard to demonstrate the technical aspects of a data-oriented wireless standard operating in unlicensed bands. The medium access technology for the IEEE 802.11 is CSMA/CD that sets this standard as a connectionless data-oriented standard. This feature eases the Internet access either by direct connection or connection through an existing wired LAN. Chapter 11 describes the objective of the IEEE 802.11 standard, explains specifications of the PHY and MAC layer alternatives supported by this standard, and provides the details of mobility support mechanisms such as registration, handoff, power management, and security.

CHAPTER 12 WATM AND HIPERLAN

Chapter 12 is devoted to WATM activities and HIPERLAN standards. HIPERLAN-1 is similar to IEEE 802.11 and is considered a data-oriented WLAN standard. ATM and HIPERLAN-2 are voice-oriented WLANs. The medium access for these standards pays special attention to maintenance of the QoS, which makes them better suited for integration into the existing voice-oriented PSTN and cellular telephone networks. We start the chapter with an overview of the technical aspects of WATM and a short description of HIPERLAN-1. The rest of the chapter is devoted to a detailed description of the HIPERLAN-2 standard.

CHAPTER 13 AD HOC NETWORKING AND WPANs

WPANs are ad hoc networks designed to connect personal equipment to one another. Chapter 13 describes WPANs with particular emphasis on details of Bluetooth technology. The chapter starts with describing the IEEE 802.15 WPAN committee and Bluetooth and homeRF standardization activities. This discussion is followed by a detailed description of key aspects of Bluetooth technology and a detailed analysis of the interference between the IEEE 802.11 and Bluetooth technologies.

CHAPTER 14 WIRELESS GEOLOCATION SYSTEMS

Our final chapter is devoted to indoor geolocation and cellular positioning as emerging technologies to complement the WLAN and wireless WAN services. This chapter provides a generic architecture for wireless geolocation services, describes alternative technologies for implementation of these systems, and gives examples of evolving location-based services. Location-based services provide a fertile environment for the emergence of e- and m-commerce applications.

CHAPTER 10

INTRODUCTION TO WIRELESS LANs

10.1 Introduction

10.2 Historical Overview of the LAN industry

10.3 Evolution of the WLAN industry

10.3.1 Early Experiences
10.3.2 Emergence of Unlicensed Bands
10.3.3 Products, Bands, and Standards
10.3.4 Shift in Marketing Strategy

10.4 New Interest from Military and Service Providers

10.5 A New Explosion of Market and Technology

10.6 Wireless Home Networking

10.6.1 What Is a HAN?
10.6.2 Why Do We Need a HAN?
10.6.3 HAN Technologies
10.6.4 Home-Access Networks

Questions

Problems

10.1 INTRODUCTION

In the first part of the book, we provided an overview of wireless networks and the fundamental principles involved in their deployment and operation. In the third part of the book, we provided examples of the leading voice-oriented TDMA and CDMA cellular networks, as well as an overview of wide area data networks mostly operating in the licensed bands and designed for wide area coverage. In the fourth part of the book, we look into the wideband local wireless networks operating in unlicensed bands. WLAN were the first technology that was examined for wideband local access. More recently wireless personal area networks (WPAN) have attracted serious attention for local ad hoc wireless networking. Current standardization bodies differentiate a WLAN and WPAN based on their coverage area, data rate, and battery consumption. In practice, however, often WLANs and WPANs are considered as competing options for the implementation of various wireless indoor applications. Wireless LANs evolved in the data-oriented LAN industry and support data rates of up to several tens of Mbps in a coverage area of around 100 meters. The low-powered WPANs cover an order of magnitude smaller area with a maximum data rate of about order of magnitude lower than WLANs. The MAC layer of the popular standards for WLANs and WPANs are either voice- or data-oriented, which also plays a role in selection of a technology for a particular application.

Today the major standards for WLANs are IEEE 802.11 and HIPERLAN. The WPAN activitiy is under IEEE 802.16, which has chosen Bluetooth as the base standard, whereas HomeRF is another option for WPANs. In this chapter we discuss the overview of the WLANs and home networking industries. The next chapter is devoted to the IEEE 802.11 standard followed by a chapter on wireless ATM and HIPERLAN. The WLAN industry emerged as an extension to the wired LAN industry. To understand forces that shaped this industry, we first provide a short overview of the evolution of the wired LAN industry.

10.2 HISTORICAL OVERVIEW OF THE LAN INDUSTRY

The cost of infrastructure in WANs is very high, and the coverage is very wide. As a result, WANs are offered as a charged *service* to the user. The service provider invests a large capital for the installation of the infrastructure and generates revenue through monthly service charges. Local networks are sold as end products to the user, and there is no service payment for local communications.

Example 10.1: PBXs within Organizations

PBX networks are owned by the companies for their local communications. The only time that a company owning a PBX pays the PSTN service provider is when a call goes out of the local area using the service provider's infrastructure. Operation of LANs is very similar to a PBX in that the user owns them and pays monthly charges to the wide area Internet service providers for wide area communications.

The LAN industry emerged during the 1970s to enable sharing of expensive resources such as printers and to manage the wiring problem caused by increasing number of terminals in offices. By the early 1980s three standards were developed: Ethernet (IEEE 802.3), Token Bus (IEEE 802.4), and Token Ring (IEEE 802.5); they specified three distinct PHY and MAC layers and different topologies for networking over thick cable medium but shared the same management and bridging (IEEE 802.1) and LLC (IEEE 802.2). With the growing popularity of LANs in the mid-1980s, high-installation costs of thick cable in the office buildings moved the LAN industry toward using thin cables that are also referred to as "cheapernet." Cheapernet covered shorter distances of up to 185 m compared with the 500 m coverage of thick cables. In the early 1990s the star topology (often referred to as Hub and Spoke LANs), using easy-to-wire TP wiring with coverage of 100 m, was introduced. Figure 10.1 depicts the evolution of the wired LAN from thick to thin and finally TP networks. The interesting observation is that this industry has made a compromise on the coverage to obtain a more structured solution that is also easier to install. Twisted pair wiring, also used by PSTN service providers for telephone wiring distribution in homes and offices for over a hundred years, is much easier to install. The star network topology opened an avenue for structured hierarchical

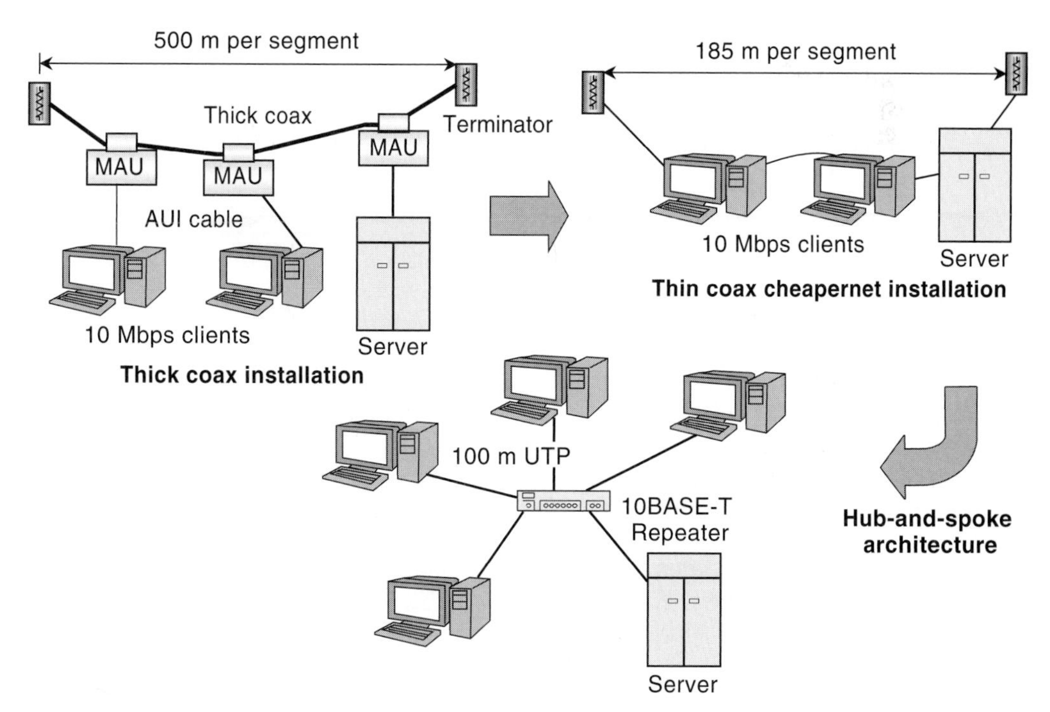

Figure 10.1 Evolution of the LANs from thick to thin cable and then to star topology using TP.

wiring, also similar to the telephone network topology. Today, IEEE 802.3 (Ethernet) using TP wiring is the dominant wired LAN technology.

The data rates of legacy Ethernet—thick, thin, and TP—were all 10 Mbps. The need for higher data rates emerged from two directions; (1) there was a need to interconnect LANs that are located in different buildings of a campus to share high-speed servers, and (2) computer terminals became faster and more capable of running high-speed, multimedia applications. To address these needs, several standards for higher data rate operations were introduced. The first fast LAN operating at 100 Mbps was the fiber distributed data interface (FDDI) that emerged in the mid 1980s as a backbone medium for interconnecting LANs. The ANSI published this standard directly. In the mid-1990s, 100 Mbps fast Ethernet was developed under IEEE 802.3 and 100VG-AnyLAN under IEEE 802.12. In the late 1990s, IEEE 802.3 approved the Gigabit Ethernet. All these high-speed LANs use multiple TP wiring to support faster transmission.

Example 10.2: Hierarchical TP Wiring in LANs

> Figure 10.2 shows an example of a hierarchical wiring of a LAN. A variety of 10 and 100 Mbps terminals are connected with two levels of switches and repeaters to a router that connects the LAN to the rest of the world.

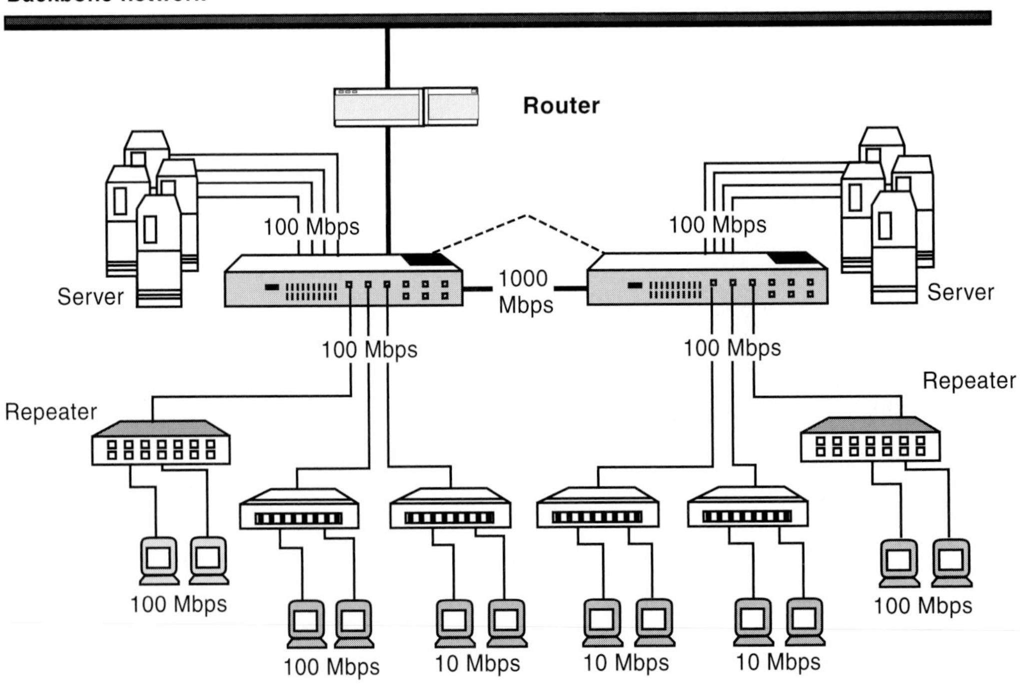

Figure 10.2 Hierarchical LANs [STA00].

In the mid 1990s when most people in the telecommunication industry believed that ATM would take over the entire emerging multimedia communications industry, ATM LAN emulation (LANE) was initiated. The purpose of ATM LANE was to adapt the existing legacy LAN infrastructures and applications to the then perceived end-to-end ATM network. The main technical challenge in implementing LANE was the adaptation of a connectionless legacy LAN to a connection-oriented ATM network. Details of the variety of LANs are available in [STA00]. More recently, a new wave of LANs is emerging for home applications that will be treated separately later on in this chapter.

In summary, the LAN industry has developed a number of standards, mostly under the IEEE 802 community. Figure 10.3 shows an overview of the important IEEE 802 community standards. The 802.1 and 801.2 parts are common for all the standards, 802.3, 802.4 and 802.5 are wired LANs, and 802.9 is the so called ISO-Ethernet that supports voice and data over the traditional Ethernet mediums. IEEE 802.6 corresponds to metropolitan area networking and the IEEE 802.11, 15, and 16 are related to wireless local networks. The IEEE 802.14 is devoted to cable modem–based networks providing Internet access through cable TV distribution networks operating over coaxial cable wiring and fiber originally installed for TV distribution. The IEEE 802.10 is concerned with security issues and operates at higher layers of the protocols. The existing LANs can be logically divided into four generations:[1] first generation legacy LANs, 802.3 and 802.5, that provided terminal-to-host connectivity and client-server architectures at moderate data rates of around 10 Mbps in offices; second generation LANs such as FDDI that responded to the need for backbone LANs and support of high-performance workstations;

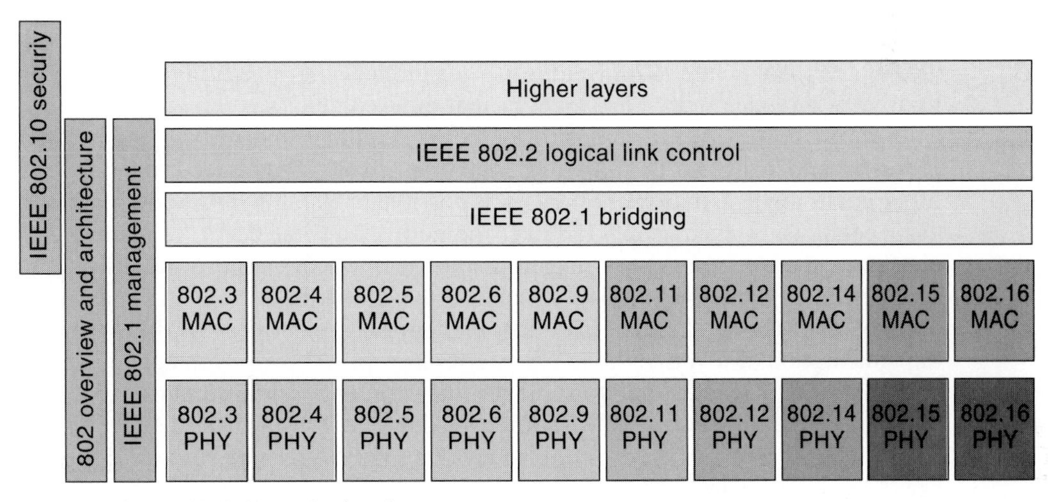

Figure 10.3 IEEE 802 standard series.

[1]The reader should not confuse these LAN generations with the generations in cellular systems.

third generation LANs such as ATM-LANE, Fast Ethernet, Gigabit Ethernet, and 100VG-AnyLAN that were designed for high throughput with delay control for multimedia applications; and fourth generation LANs such as IEEE 802.14 and the home phone-networking alliance (HPNA) that support networking in residences and home access and distribution. The two major drives of this industry have always been the ease of installation and increase of data rate.

10.3 EVOLUTION OF THE WLAN INDUSTRY

During the past two decades, as the vision of the WLAN industry evolved, WLANs were implemented based on a variety of innovative technologies and raised a lot of hopes for development of a sizable market several times. Today, the major differentiation of WLANs from wide area cellular services is the method of delivery of data to users, data rate limitations, and frequency band regulation. Cellular data services are delivered by operating companies as services, whereas the WLAN users belong to the organization that owns the network. At the time when the 3G cellular industries are striving for 2 Mbps packet data services, WLAN standards are working on 54 Mbps services. Another differentiation with other radio networks is that today almost all WLANs operate in the unlicensed bands where frequency regulations are loose and there is no charge or waiting time to obtain the band. To obtain a deeper understanding of all these issues, it is very educational to go over the history of the WLAN industry to see how all of these unique issues evolved.

10.3.1 Early Experiences

Gfeller at the IBM Rüschlikon Laboratories in Switzerland first introduced the idea of a WLAN in late 1970s [GFE80]. The number of terminals in manufacturing floors was growing and wiring them in that environment was difficult. In offices, wires are normally snaked under the suspended ceilings and through the interior partitions and walls, but these options are not available in manufacturing floors. In office environments, in extreme cases, it is possible to install wiring under the floor, using conduits, or even simply left over the floor with some cover. In manufacturing floors, the environment is rugged; under floor wiring is more expensive; and simply leaving wires on the floor is can be dangerous because heavy machinery may roll over them. The diffused IR technology was selected at IBM labs for the implementation of a WLAN to avoid interference with the electromagnetic signals radiating from machinery and to avoid dealing with long-lasting administrative procedures with frequency administration agencies. The principal researcher of this project abandoned the project because the goal of 1 Mbps with reasonable coverage did not materialize.

Ferrert at HP's Palo Alto Research Laboratories in California performed the second project on WLANs around the same time [FER80]. In this project a 100 kbps DSSS WLAN operating around 900 MHz was developed for office areas that used CSMA as the method of access. This project was conducted under an experimental license agreement from the FCC. The principal of this project failed to obtain the necessary frequency bands from the FCC, and discouraged with the administrative complexity, he also abandoned his project. A couple of years later, Codex, Motorola attempted to implement a WLAN at 1.73 GHz, and that project was also abandoned after negotiations with the FCC.

Although all the pioneering WLAN projects were abandoned, WLANs continued to attract attention, and negotiations continued with the FCC to secure frequency bands for this purpose [PAH85]. These projects revealed several important challenges facing the WLAN industry that remain to this day:

1. *Complexity and cost:* The alternatives for implementing WLANs such as IR, spread spectrum, or traditional radios are far more complex and diversified than the wired LANs.
2. *Bandwidth:* Data rate limitations of the wireless medium are more serious than those of wired media.
3. *Coverage:* The coverage of a WLAN operating within a building is less than that of a single cable (bus or ring) or even TP-based LANs.
4. *Interference:* WLANs are subject to interference from other overlaid WLANs or other users operating in the same frequency bands.
5. *Frequency administration:* Radio-based WLANs are subject to expensive and untimely frequency regulations.

10.3.2 Emergence of Unlicensed Bands

Wireless LANs need a bandwidth of at least several tens of MHz though they have not yet shown a market compatible in strength with the cellular voice industry that originally started with two pieces of 25 MHz bands that produced a huge market. Comparable sizes of bands for PCS applications were auctioned in the United States for tens of billions of dollars though the market for WLANs has not yet passed a billion dollars per year. The dilemma for the frequency administration agencies was to justify a frequency allocation for a product with a weak market.

In the mid-1980s, the FCC found two solutions for this problem. The first and the simplest solution was to avoid the 1–2 GHz bands used for the cellular telephone and PCS applications and approve higher frequencies at several tens of GHz where plenty of unused bands were available. This solution was first negotiated between Motorola and the FCC and resulted in Motorola's Altair, the first wireless LAN product operating in licensed 18–19 GHz bands. Motorola had actually established headquarters to facilitate user negotiation with the FCC for the usage of WLANs in different areas. A user who changed the location of operation of his/her WLAN substantially (from a town to another) contacted Motorola, and they would manage the necessary frequency administration issues with the FCC.

The second and more innovative approach was resorting to unlicensed frequency bands as the solution. In response to the applications for bands for WLAN projects mentioned in the previous section and motivated by studies for various implementations of wireless LANs [PAH85], Mike Marcus of the FCC initiated the release of the unlicensed ISM bands in May 1985 [MAR85]. The ISM bands were the first unlicensed bands for consumer product development and played a major role in the development of the WLAN industry. In simple words licensed and unlicensed bands can be compared with private backyards and public gardens. If one can afford it, he or she can own a private backyard (licensed band) and arrange a barbeque dinner (a wireless product). If one cannot afford to buy a house with a backyard, he or she simply moves the barbeque party to the public park (unlicensed band) where he/she should observe certain rules or *etiquette* that allows others to share the public resource as well. The rules enforced on ISM bands restricted the transmit power to 1 W and enforced the modems radiating more than 1 mW to employ spread spectrum technology. It was believed that spread spectrum communications would restrict interference and allow coexistence of several wireless applications in the same band. Table 10.1 provides a summary of the important features of the ISM bands.

10.3.3 Products, Bands, and Standards

Encouraged by the FCC ruling [MAR85] and some visionary publications in wireless office information networks summarizing previous works and addressing the future directions in this field [PAH85, PAH88, KAV87], a number of WLAN product development projects mushroomed almost exclusively over the North American continent. By the late 1980s the first WLAN products using three different technologies, licensed bands at 18–19 GHz, spread spectrum in the ISM bands around 900 MHz, and IR, appeared in the market. At around the same time, a standardization activity for WLANs under IEEE 802.4L was initiated that was soon converted into the independent IEEE 802.11 that was finalized in 1997! The early products consisted of shoe box sized APs and receiver boxes or PC installed cards that could connect workstations to LANs wherever wiring difficulties for the LANs justified using a more expensive WLAN connection. Today, we call this application LAN-extension [PAH94], [PAH95]. Market predictions at that time were estimating a shift of around 15 percent of the LAN market to WLANs that would generate a few billion dollars of sale per year by the first few years of the 1990s. In May 1991, to create a scientific forum for the exchange of knowledge on WLANs, the first IEEE-sponsored WLAN workshop was organized concurrent to the 802.11 meeting, in Worcester, Massachusetts [WOR91].

In 1992, as a follow up to the initial momentum for WLAN developments, lead by Apple, an industrial alliance called WINForum was formed aiming at obtaining more unlicensed bands from FCC for the so called Data-PCS activities. WINForum finally succeeded in securing 20 MHz of bandwidth in the PCS bands that was divided into two 10 MHz bands—one for isochronous (voice like) and one for asynchronous (data type) applications. The original aim of WINForum was to secure 40 MHz for asynchronous applications. WINForum also defined a set of rules or *etiquettes* for these bands that would allow the coexistence. Figure 10.4

Table 10.1 Properties of the ISM Bands

- Frequencies of operation: 902–928 MHz; 2.4–2.4835 GHz; 5.725–5.875 GHz
- Transmit power limitation of 1 W for DSSS and FHSS
- Low power with any modulation

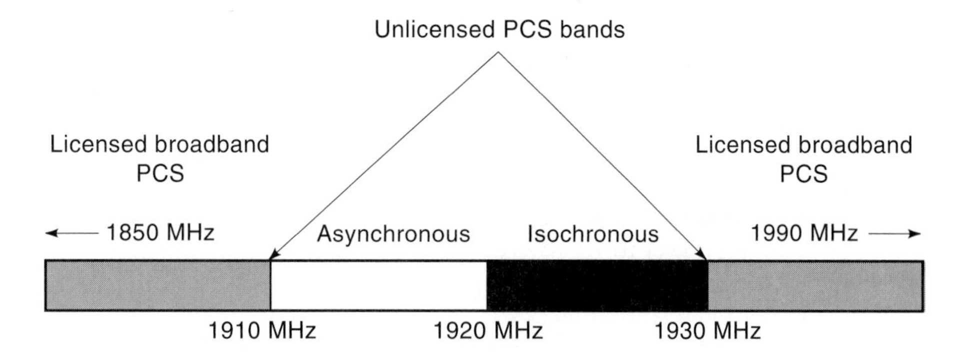

Three basic rules

1. Listen before talk (or transmit) *LBT protocol*
2. Low transmitter power
3. Restricted duration of transmission

Figure 10.4 Unlicensed PCS bands and their spectrum etiquette.

shows the unlicensed PCS bands and the spectrum etiquette associated with them. The WINForum etiquette is based on CSMA rather than CDMA and spread spectrum communications used in ISM bands. This was a better choice because implementation of CDMA needed power control and larger bandwidth that was not feasible in uncoordinated, multiuser, multivendor WLANs and spread spectrum without CDMA offers a less bandwidth efficient solution.

Another standardization activity started in 1992 was the HIPERLAN. This ETSI based standard aimed at high performance LANs with data rates of up to 23 Mbps that was an order of magnitude higher than the original 802.11 data rates of 2 Mbps. To support these data rates, the HIPERLAN community was able to secure two 200 MHz bands: 5.15–5.35 GHz and 17.1–17.3 GHz for WLAN operation. This encouraged the FCC to release the so-called U-NII bands in 1997 when the original HIPERLAN standard (now called HIPERLAN-1) was completed. Table 10.2

Table 10.2 Properties of the U-NII bands

Band of Operation	Maximum Tx Power	Max. Power with Antenna Gain of 6 dBi	Maximum PSD	Applications: Suggested and/or Mandated	Other Remarks
5.15–5.25 GHz	50 mW	200 mW	2.5 mW/MHz	Restricted to indoor applications	Antenna must be an integral part of the device
5.25–5.35 GHz	250 mW	1,000 mW	12.5 mW/MHz	Campus LANs	Compatible with HIPERLAN
5.725–5.825 GHz	1,000 mW	4,000 mW	50 mW/MHz	Community networks	Longer range in low-interference (rural) environs

summarizes the U-NII bands and their restrictions. The WINForum etiquette was evaluated for the U-NII bands, but it was found not be suitable because research activities around that time favored wireless ATM that could not operate on a listen-before-talk etiquette. Today, U-NII bands are used by IEEE 802.11a and HIPERLAN-2 projects for the implementation of 54 Mbps OFDM-based WLANs. Table 1.8 in Chapter 1 provides a summary of the WLAN standards under IEEE 802.11 and HIPERLAN activities.

10.3.4 Shift in Marketing Strategy

In the first half of the 1990s, WLAN products were expecting a sizable market of around a few billions of dollars per year for shoebox-size products used for LAN-extension application in indoor areas. This did not materialize. Under this situation, two new directions for product development emerged. The first and the most simple approach was to take the existing shoebox-type WLANs, boost up their transmitted power to the maximum allowed under regulations, and equip them with directional antennas for outdoor interbuilding LAN interconnects. These technically simple solutions would allow coverage of up to a few tens of kilometers with suitable rooftop antennas. The new inter-LAN wireless bridges could connect corporate LANs that were within range. The cost of the inter-LAN wireless solution was much cheaper than the wired alternative, T1-carrier lines, leased from the PSTN service providers. The second alternative was to reduce the size of the design to a PCMCIA WLAN card to be used with laptops that were enjoying a sizable growth and demanded mobility for LAN connectivity. This approach was not available for all existing products, and it was more suitable for the spread spectrum products operating in lower frequencies. Figure 10.5 illustrates these three applications for the WLANs. Recently, there are new low-cost products for LAN extension that can convert a serial port or Ethernet connector to a WLAN interface for desktop PCs and workstations that operate at 11 Mbps.

The original marketing strategy for LAN-extension application was indeed a horizontal one aiming at selling individual WLAN components directly to the customers. Another major shift in the marketing strategy of a few successful companies in the mid 1990s was the move toward vertical markets where a wireless network was sold as a complete solution to an application. The major vertical markets approached by the WLAN industry were *"barcode" industries* providing a wireless inventory check and tracking in warehouses and manufacturing floors, *financial services* providing for wireless financial updates in large stock exchanges, *health care* networks providing wireless mobile services inside the hospitals, and *wireless campus area networks* (WCANs) providing for wireless classrooms and offices. All these efforts boosted the market for the WLAN to above half a billion dollars per year over the last few years of the 1990s.

Example 10.3: A WCAN in WPI

Figure 10.6 illustrates the schematic of an experimental NSF-sponsored WCAN that was design as a testbed for performance monitoring of WLAN products at CWINS, WPI in 1996. The testbed connects five buildings with inter-LAN bridges using different technologies. Inside each building APs provide coverage to the laptops that are carried by the students. The professor broadcasts his image

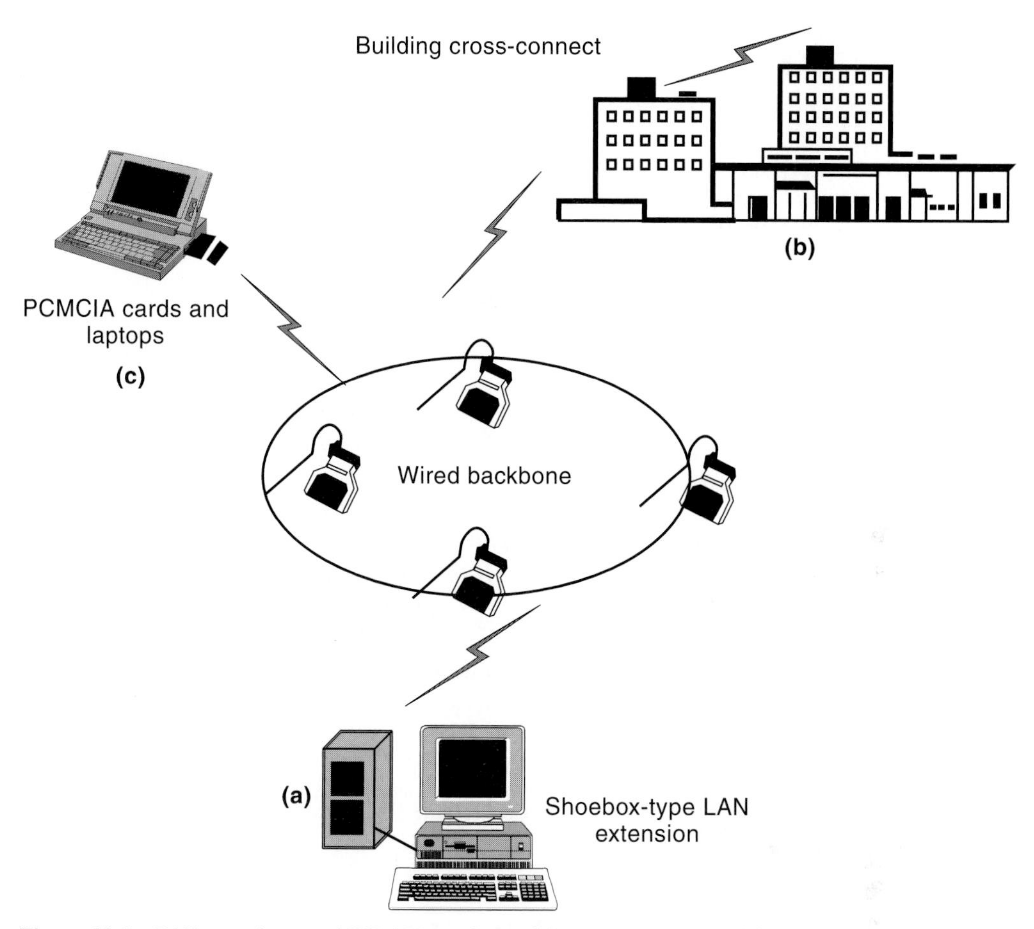

Figure 10.5 Different forms of WLAN products: (a) LAN-extension, (b) Inter-LAN bridge, and (c) PCMCIA cards for laptops.

and writing on the electronic board to allow students to participate in the wireless classroom from different buildings in the campus. The entire wireless network is connected to the backbone through a router to isolate the traffic for traffic monitoring experimentations.

Today the horizontal market for the WLAN industry is mainly focused on WLANs as an alternative to wiring additional LAN segments wherever the cost of the WLAN is justifiable. One example of this situation is installation with frequent relocations where the additional cost of the WLAN solution is justified by the relocation costs of the wired solution. Temporary networking situations such as registration sites in conferences or fairs (jobs, food, etc.) is another example where a wireless solution is preferred to the expensive but more reliable wired alternative. Buildings with difficult or impossible-to-wire situations, such as marble buildings or historical monuments where drilling for wiring is not favored, provide another example of situations where WLANs are justified. The most popular incentive for WLANs is the general use in the laptops in the home and in offices.

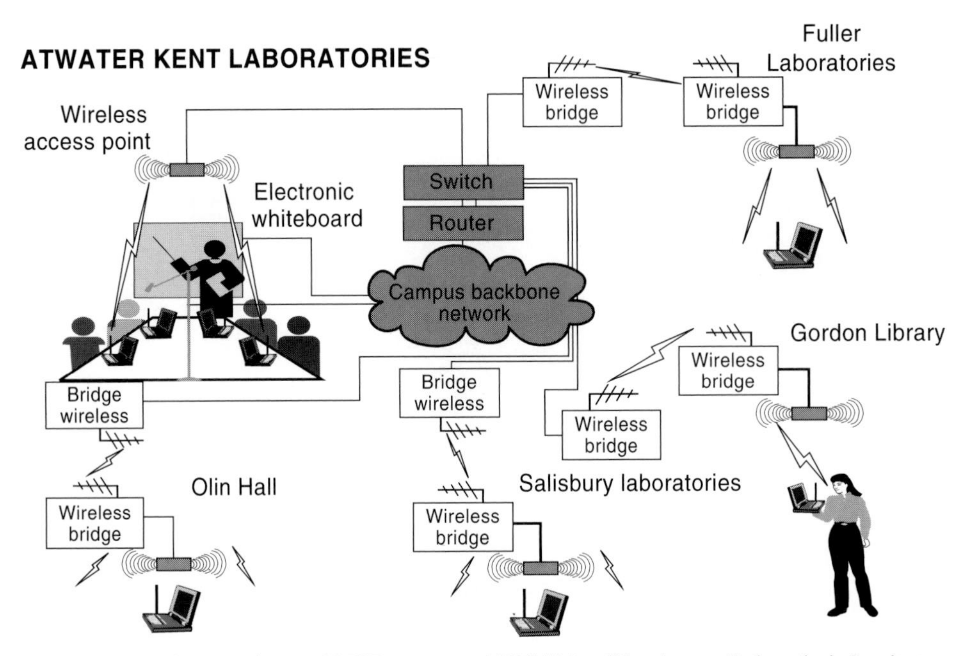

Figure 10.6 The experimental NSF-sponsored WCAN at Worchester Polytechnic Institute.

10.4 NEW INTEREST FROM MILITARY AND SERVICE PROVIDERS

In the mid-1990s, when the WLAN industry was struggling to find a market, a new wave of interest for WLAN was initiated in the United States for military applications and in the European Union (EU) for commercial applications. These projects poured a considerable amount of research investments in this field that further brightened the future of this industry [PAH97]. The incentive for the military was to discover new horizons for implementation of global mobile military networks that support integration with computing and positioning systems. Some examples of these projects will further clarify the situation.

Example 10.4: The InfoPAD Project

The InfoPAD project in the University of California, Berkeley was one of the early WLAN DARPA projects. Figure 10.7 represents the general application concept in this project. The environment is like a battleship equipped with a number of computing facilities. Soldiers in the environment are carrying InfoPADs that are small asymmetric communication devices carrying user instructions to the computing backbone to initiate computational operations whose results are downloaded to the PAD. The challenge in this project was the implementation of a PAD with a reasonable size and integration of multimedia applications in such a device [BRO98].

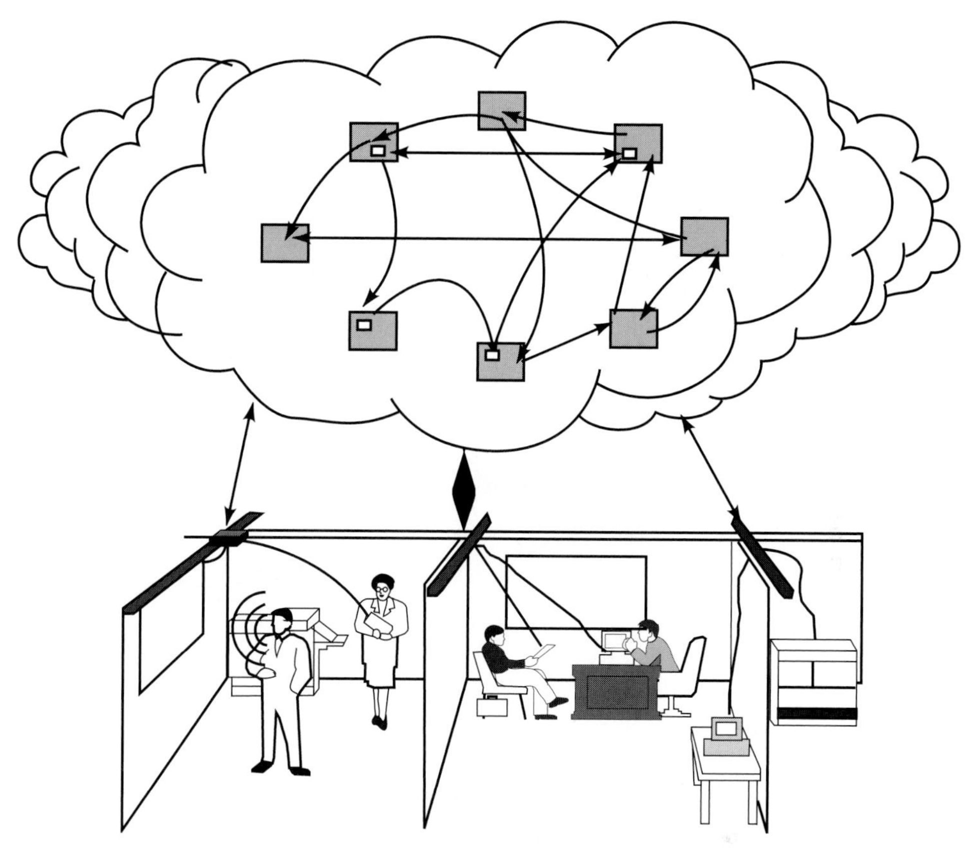

Figure 10.7 Fusion of computers and communications in the InfoPAD project at the University of California, Berkeley.

Example 10.5: The BodyLAN Project

The BodyLAN [BAR96] project sponsored by DARPA was initiated at BBN in Cambridge, Massachusetts. Figure 10.8 illustrates the operation of a BodyLAN. This project intended to design a low-power network that can monitor the vital information about the body conditions (heartbeat, temperature, etc.) and communicate it with nearby soldiers. As we will see later, the concept of this project motivated some of the work in the IEEE 802.15 WPAN activities.

Example 10.6: The SUO/SAS Project

A more recent DARPA project was the Small Unit Operations Situation Awareness Systems (SUO/SAS) [WIL01] that was aiming at an integrated telecommunication and geolocation network for modern fighting scenarios. Among the technical challenges in this project was the accurate indoor positioning [PAH98] that allows situation awareness information to be communicated with the war

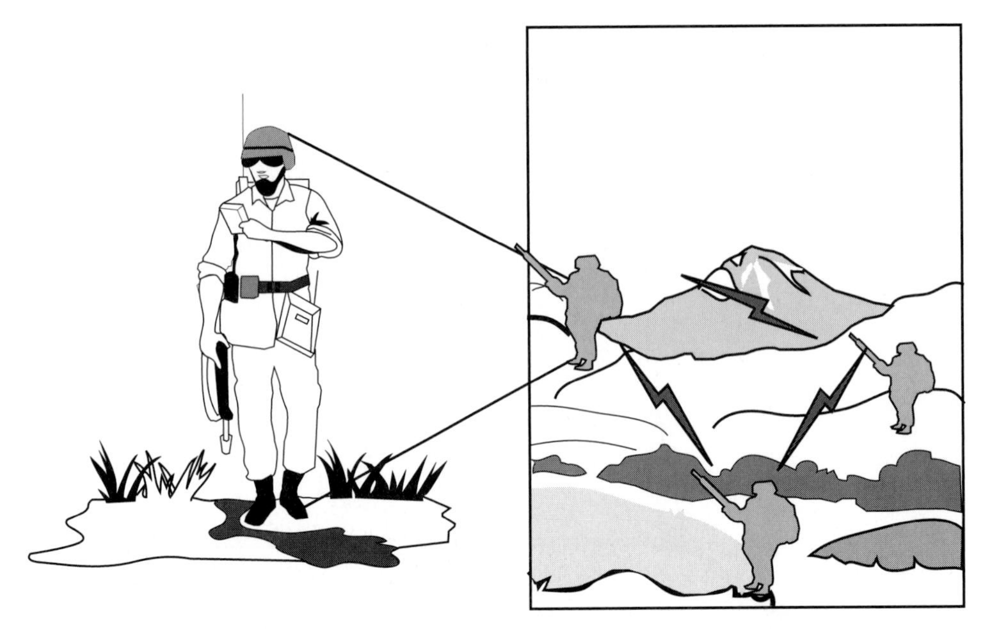

Figure 10.8 BodyLAN or wearable LAN.

fighters. Figure 10.9 provides a scenario for operation of such systems in urban area fighting. This system expects to provide a full communication and positioning link to a soldier operating inside a building.

Equipment manufacturers for service providers initiated the commercial interest of the European Union in WLANs. They were keen on incorporation of higher data rate services into the evolving rich cellular industry. In the mid-1990s, both commercial service providers and military network designers believed that the future of the backbone networks would be an end-to-end ATM based network. The technical impact of this perception of the wideband local networking industry was the start of the wireless ATM movement that we discuss in detail in the following chapters. From the application point of view, service providers intend to integrate WLAN products into their existing services. One popular scenario used by the HIPERLAN-2 project to represent the service providers' point of view is shown in Figure 10.10. In this scenario it is assumed that a WLAN user carries his/her laptop in the office, home, and public places (airports, train stations, etc.). In the home and the office, a laptop is connected to the free network whose infrastructure is owned by the user or his company. In public buildings or other places, either the service provider has a WLAN AP that provides high-speed access or there is a backbone wireless WAN that supports the connection with a lower data rate. In all public places, the service provider who owns the infrastructure will charge the user. One of the technical challenges for implementation of this scenario is the vertical roaming among different networks [PAH00a]. Another challenge is

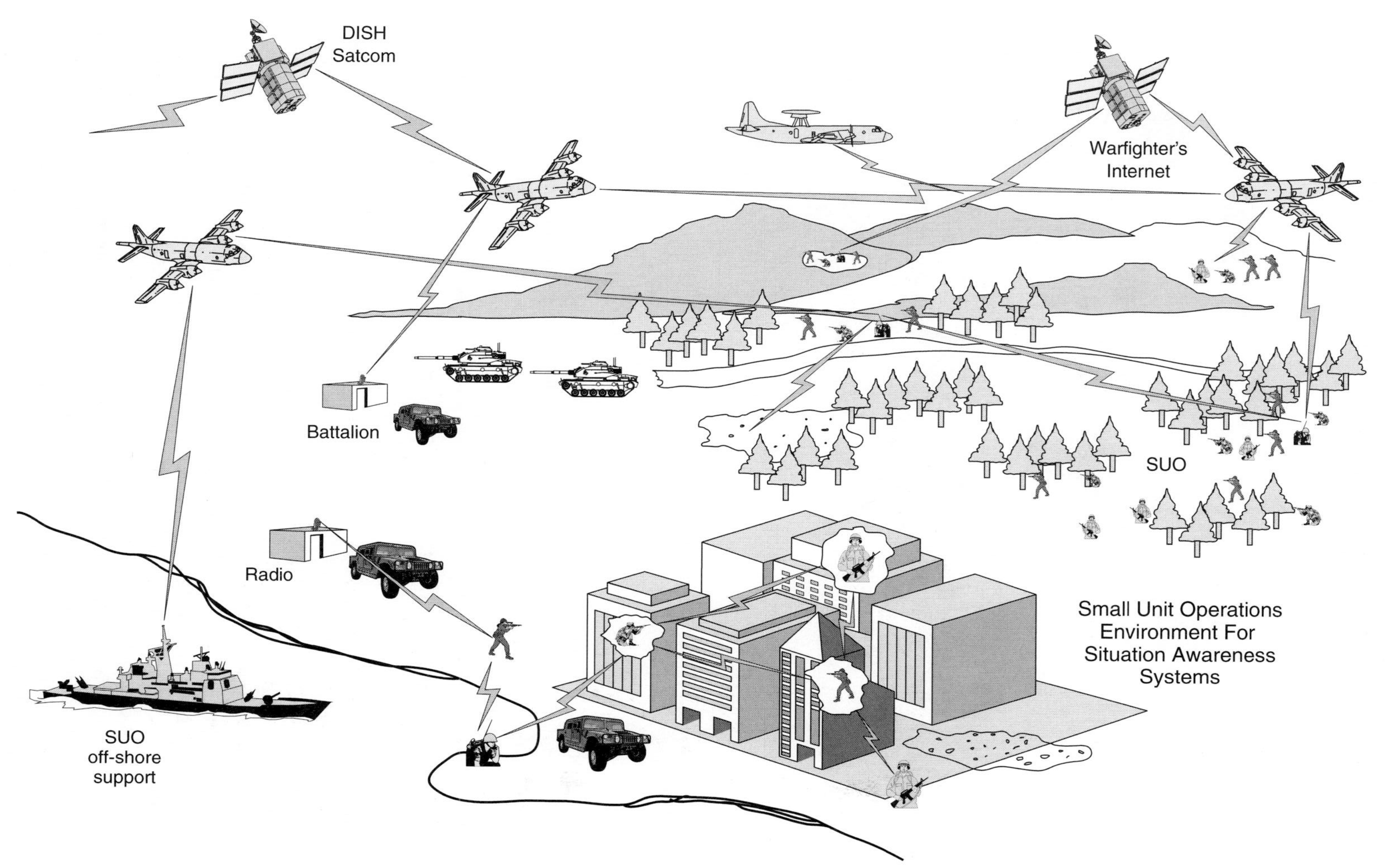

Figure 10.9 The urban/outskirts combat scenario for the SUO-SAS project.

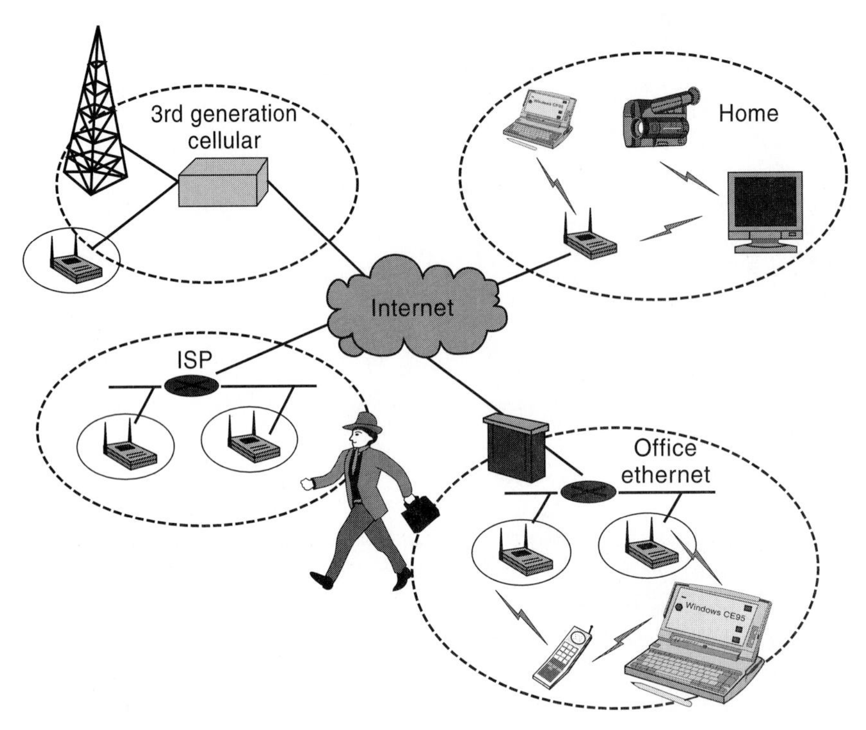

Figure 10.10 Service providers' view of the LANs.

to incorporate roaming and tariff mechanisms for WLANs. These issues are currently under investigation. It is interesting that commercial service with WLANs is now available in big airports in North America such as Dallas–Fort Worth.

10.5 A NEW EXPLOSION OF MARKET AND TECHNOLOGY

In the last couple of years of the 1990s, a new explosive growth of interest stimulated the WLAN industry. The WLAN industry that owned an almost exclusively North American market with an income only equal to a fraction of the cellular industry has suddenly attracted widespread attention in Japan and the European Union and renewed interest in the United States hoping for a sizable market comparable with that of the cellular industry. In Japan, small office spaces promoted usage of laptops replacing desktop PCs. A natural networking solution for laptops is nothing but WLANs. In the European Union, the rich cellular industry started considering WLANs as a part of their next generation of high-speed packet data services. The interest is twofold, WLANs provide a practical higher speed solution, and they operate in the unlicensed bands that are free of charge though the cost of the licensed bands are constantly increasing. In the North American continent the successful growth of broadband Internet access to homes has opened a new window for

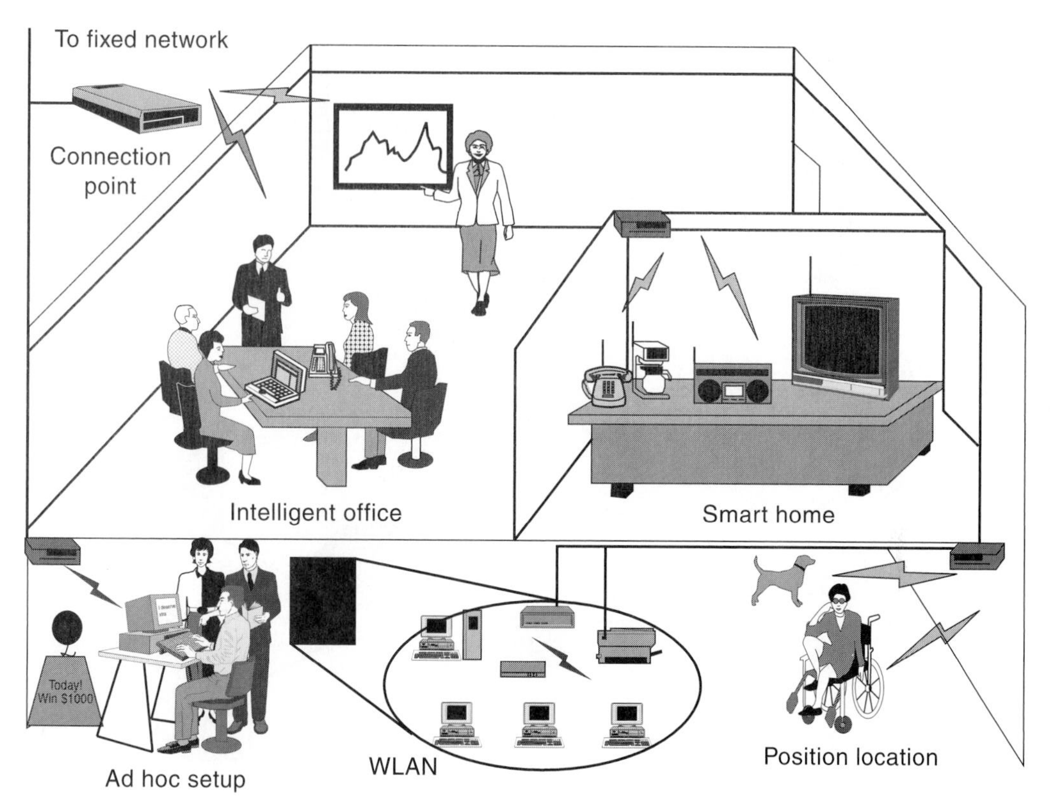

Figure 10.11 Wireless networks: 2000 and beyond.

a sizable market in home networking. These trends have been further catalyzed by the emergence of new low power personal area ad hoc wireless networking technologies such as Bluetooth and UWB for local distribution, LMDS for home access, and indoor positioning for a variety of applications. Availability of low-power, low-cost wireless chip sets started a new revolution in consumer product development raising hopes of sales exceeding hundreds of millions of these chip sets per year. All together these hopes initiated a *Gold Rush* in chip manufacturing for WLAN and WPAN applications that still continues. As far as technical directions in this industry are concerned, they continue to be toward providing higher data rates, comprehensive coverage, less interference, and lower cost. Figure 10.11 provides a visionary figure of the evolving applications for WLAN and WPAN networks.

10.6 WIRELESS HOME NETWORKING

Figure 10.12 illustrates the typical networking connections in most residences. The residence is connected to the PSTN for telephone services, the Internet for Web access, and a cable network for multichannel TV services. Within the home,

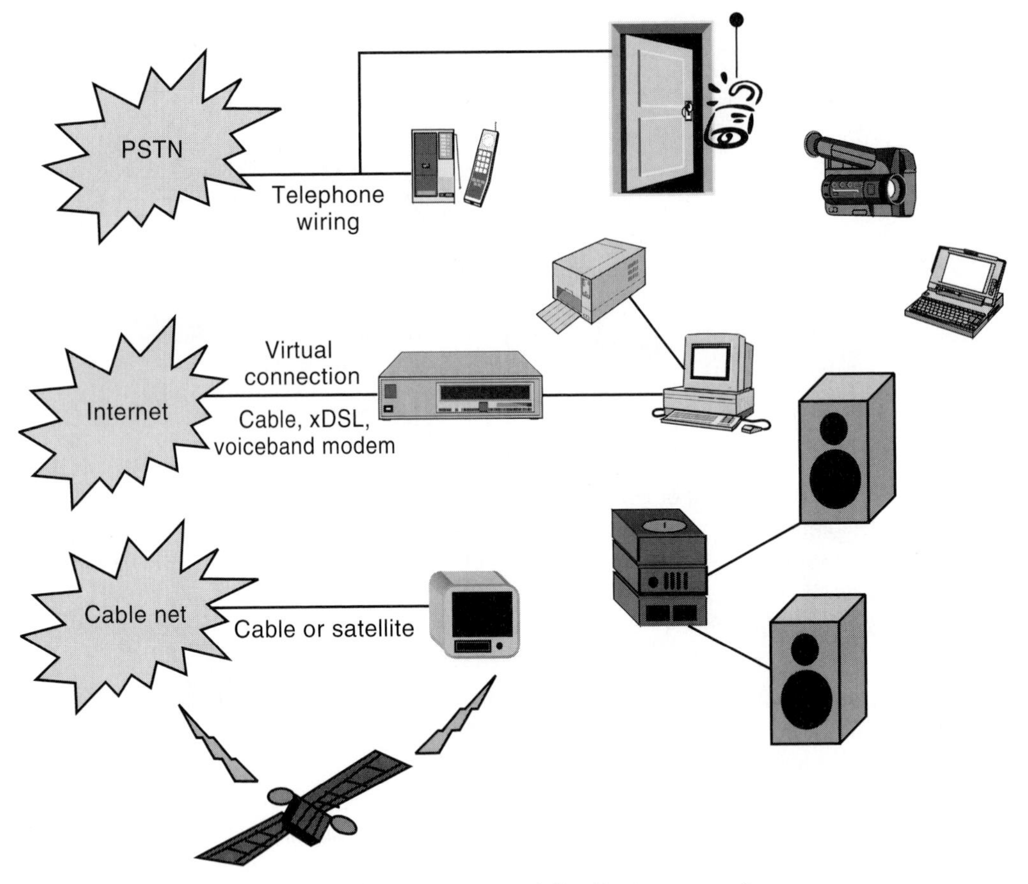

Figure 10.12 Today's fragmented home access and distribution networks.

computers and printers are connected to the Internet through voiceband modems, xDSL services, or cable modems. The telephone services and security systems are connected through PSTN wiring. The TV is connected to multichannel services through HFC cables or satellite dishes. The audio and video entertainment equipment such as video cameras and stereo systems and other computing systems such as laptops are either isolated or have proprietary wired connections. This fragmented networking environment has prompted a number of recent initiatives to create a unified home network. The home networking industry started in the last years of the 1990s by the design of the so-called home or residential gateways to connect the increasing number of information appliances at home through a single Internet connection.

Figure 10.13 shows the growth prediction for the home networks. The number of home networks in the United States is expected to almost double each year. As shown in Figure 10.14, this industry has two distinct segments, home access and home distribution. The home access technology employs different wireless and

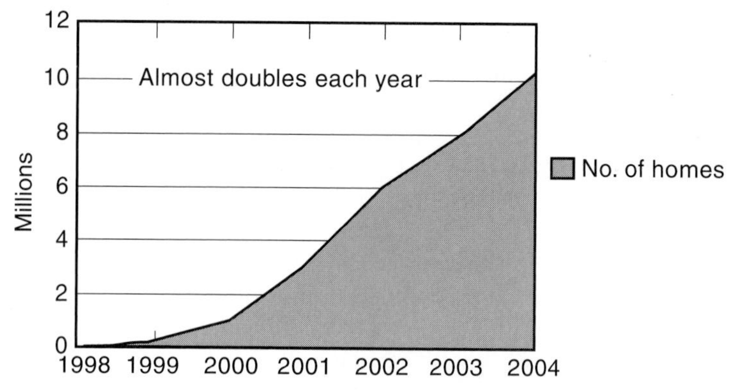

Figure 10.13 Growth of the home networking industry.

wired alternatives to secure a broadband Internet access to the home gateway to be distributed to the user's information appliances. The home distribution or home area network (HAN) interconnects all home appliances and connects them to the Internet through the home gateway. As far as access is concerned, it is expected that 80 percent of U.S. households will have a broadband data access by the year 2002. It is expected that the number of sold "information appliances" will exceed the sold number of PCs by the year 2002. It is also expected that to interconnect PCs and information appliances to the broadband services, 10 million home networks will be installed by the year 2004.

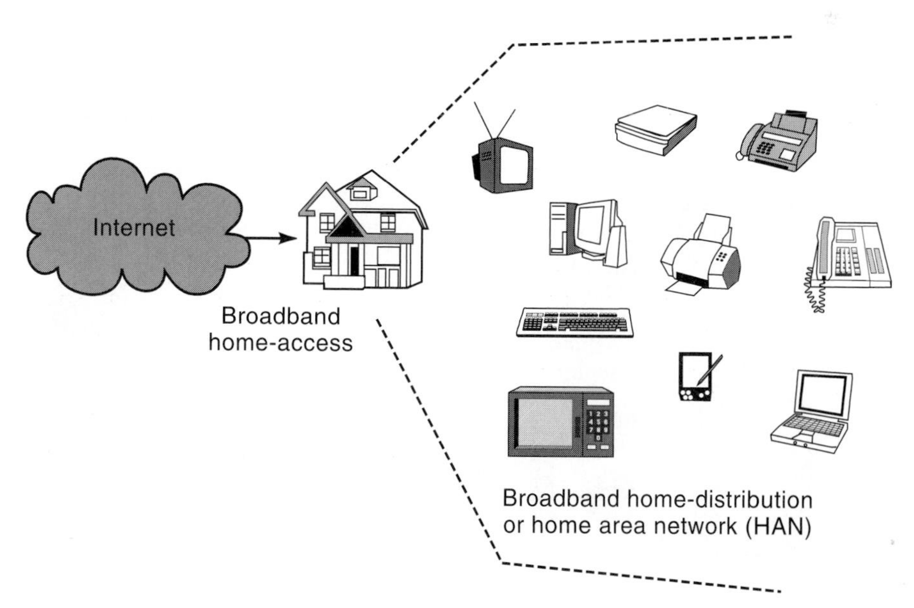

Figure 10.14 Two basic technologies needed for home networking.

10.6.1 What Is a HAN?

The HAN provides an infrastructure to interconnect a variety of home appliances and enable them to access the Internet through a central home gateway. A number of home appliances are emerging in the market that are in need of a HAN. Figure 10.15 provides an overview of these appliances classified into logical groups.

Home computing equipment, used for computing and Internet transaction interface access, includes PCs, laptops, printers, scanners, and Web cameras. If there is no home distribution network, all the equipment is connected together either through the PC or laptop ports. A home computing network allows multiple computers as well as multiple devices to connect with a network protocol. A wireless network allows flexibility in installation and relocation of these devices in different rooms of a home.

Figure 10.15 Classification of home equipment demanding networked operation.

Phone appliances, used for two-way conversations, include cordless telephones, intercom devices, and standard wired telephone sets. All telephone services have an interface to communicate with the PSTN. Currently they are connected through the home telephone wiring to the PSTN. With a HAN, these devices can share the home access medium (cable or TP), allowing one service provider to support both data and voice services.

Entertainment and audiovisual appliances include TV sets, stereo systems, CD players, VCRs, DVD players, tape recorders, camcorders, speakers, and headphones. These devices communicate through their own protocols such as IEEE 1394 or Firewire [FIRweb].

Security systems include motion detectors, door pins, system control panels, cameras, and alarms that are networked separately using protocols such as EIA-600 CEBus or EIA-709 Lon Works [LONweb]. Currently, the access to these systems is through the telephone lines that initiate emergency request alarms to the police stations.

Appliance manufacturers are working on *smart appliances* capable of communication. This intelligence allows remote checking test for maintenance (e.g., receiving alarm that a refrigerator's heat-pump needs to be replaced) and remote control of operation (e.g., dimming lights remotely). These devices use protocols such as Lon Works or WRAP. Another wave of interest in home network stems from utility companies for distance *utility metering.* The electricity, gas, water, and fuel companies would like to remotely read the utility meters at home for billing or other needs (e.g., refilling the gas tank). One of the solutions is to communicate this information through the HAN and its access to the Internet. More recently, a number of start-up companies are designing *indoor locating systems* that can be used for distance monitoring of children, elderly people, and pets or for navigating the blind. These systems are expected to be integrated with the home networks to provide access through the Internet.

10.6.2 Why Do We Need a HAN?

The existing LANs designed for office environment do not provide a good solution for home networking. The application diversity, network requirement, building infrastructure, and market size of the HANs are distinctly different from the LANs. In homes, the number of users of the network is much smaller than in offices but the diversity of device types and their bandwidth requirements is much larger than that in offices. The diversity of the bandwidth requirement of the large appliance list introduced in the previous section includes multichannel video up to monthly meter reading. Offices mostly employ computing devices. The home environment includes new applications such as positioning/navigation and audio/video broadcasting that were not applications for traditional office LANs. Offices are larger than residential homes, and they are made of more concrete material than homes. Therefore physical wiring and wireless coverage in homes is easier than in offices. Homeowners are more reluctant to allow service workers to enter their home, and they cannot afford a network manager to operate their network. The number of homes is orders of magnitude larger than the offices so the market for home networking is expected to be much larger than the LAN market.

These specific requirement on home applications imposes certain constrains on the design of HANs. A HAN needs to be *user-friendly* because it is used and managed by nonprofessionals with limited technical skills and a small budget size. A HAN must be low cost, easy to install and relocate, and easy to upgrade. In terms of *performance* a HAN should enable multimedia applications and be capable of accommodating legacy voice and data services. A HAN also needs to be *flexible* and *scalable* to allow location independence and easy reconfigurability of networks without significant performance degradation. To avoid eavesdropping or session hijacking, a HAN also needs *security* and *privacy* provisions.

10.6.3 HAN Technologies

In an office, a company sees the need for networking, decides on the installation of a network, opens a budget for expensive wiring, and installs the LAN infrastructure. In the case of home networking, users gradually build their networks at their leisure with an investment that is spread over a relatively long period. An average consumer does not spend a sizable budget on wiring to develop an infrastructure. Therefore, the trend for today's HANs is to either use existing wires or try a wireless solution. The existing wiring at homes consists of TP telephone wires, powerlines, and cable from cable TV. The *phone line* wiring has a relatively good distribution and in most modern homes at least you can locate one telephone outlet in every room. The wiring for phone lines is voice-grade TP that is suitable for Ethernet connections. However, the same line is also used for telephony and xDSL transmissions which interfere with the Ethernet signal. *Power line* wiring is even better distributed because every room has several power outlets. However, the quality of the line is poor, and the level of the noise in power lines much more than TP phone line wiring. Figure 10.16 shows a typical transfer function and noise level in a power line. The power line is a frequency selective channel that suffers from impulsive noise in the frequency domain. These characteristics impose limitations on the data rate that can be overcome only by using more complex transmission techniques. Existing *cable TV* wiring has a very restricted distribution, and only a few outlets are available in each household. This wiring is used for multichannel TV distribution that interferes with the baseband data signal. The expensive broadband cable TV modems can be used to overcome this problem. Because of its limited distribution and expensive modem requirement, cable TV wiring is not considered seriously for home distribution. A *wireless* solution appears ideal for home networking. The ease of installation and relocation provides an excellent solution. Challenges for wireless are reliability, bandwidth, coverage, security, and interference. Comparing wired and wireless solutions, wired HANs can be implemented over less expensive cards, and they can support higher data rates. Wireless HANs provide an ideal ad hoc solution that supports portability.

10.6.3.1 HPNA

All networked home computers have Ethernet cards, and PCMCIA laptop cards are almost exclusively made for Ethernet. However, as we mentioned before, TP phone line wiring at home is used for analog voice and distribution access and DSL access. The HPNA [HPNAweb] is an Ethernet-compatible LAN over

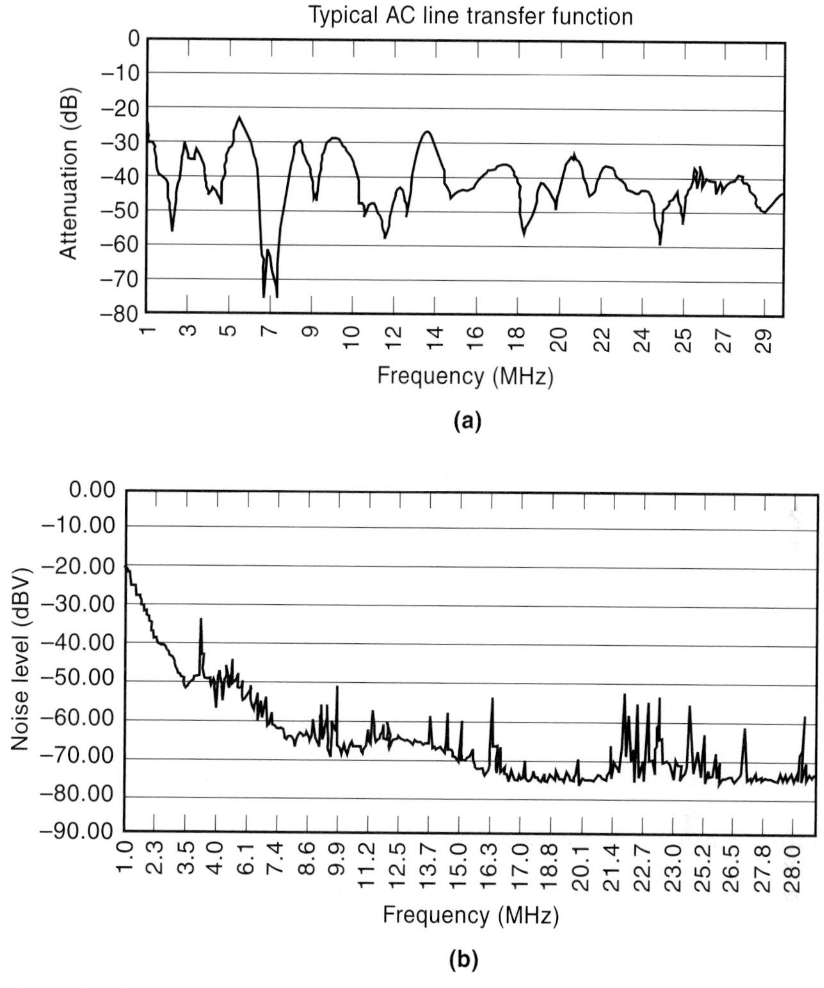

Figure 10.16 (a) Typical power line transfer function and (b) typical noise level in the power lines.

random-tree home phone lines. It uses a stand-alone adapter to connect any device having an Ethernet 10Base-T interface card, operating at 10 Mbps over TP lines, directly to the in-home telephone jacks. Figure 10.17 shows an example of HPNA applications where a number of data terminals and analog voice terminals share the home TP telephone wiring. The data equipment also receives Internet access through a home gateway.

Figure 10.18 illustrates how HPNA shares the medium through FDM with POTS and DSL. In the same home wiring, POTS uses the 20 Hz–3.4 kHz band for analog voice transmission, xDSL uses the 25 kHz–1.1 MHz band to provide high-speed Internet access to the home, and HPNA uses the 2 MHz–30 MHz band for home distribution networking. HPNA is based on a patented physical layer that up converts the differential Manchester-coded Ethernet physical layer to the HPNA band. This physical layer is more immune to high-noise conditions in the home

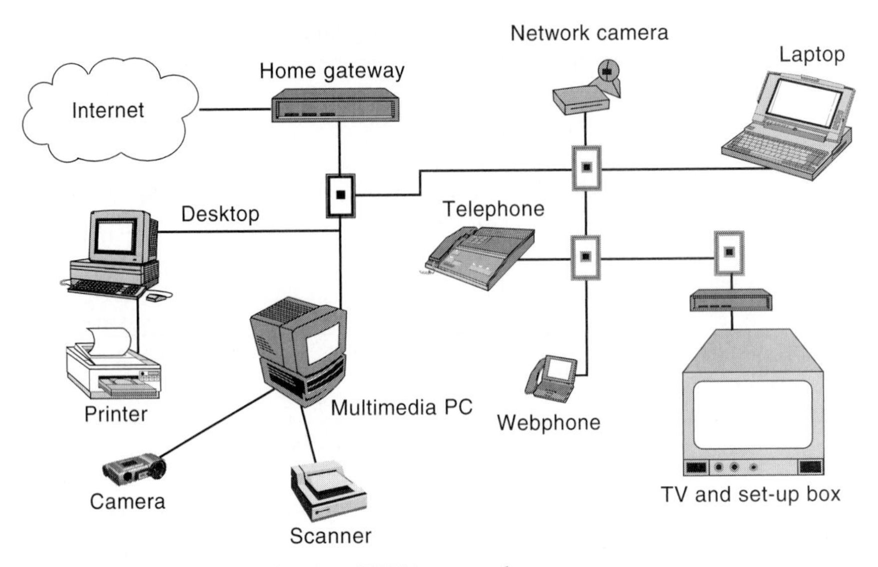

Figure 10.17 An example of an HPNA network.

telephone wiring. The MAC layer for the HPNA is the same as the MAC layer of the IEEE 802.3 Ethernet. From the user's point of view, the HPNA network accommodates the existing legacy Ethernet software and hardware. Only an adaptor is placed between the Ethernet connection and the phone plugs. The next step for the HPNA is to boost the data rate to accommodate video applications.

10.6.3.2 Power Line Modems

Compared with telephone wiring, power lines have the best wiring distribution in homes because they reach every corner of a building. In outdoor areas, the PSTN network has two parts: the TP analog access wiring and the backbone digital

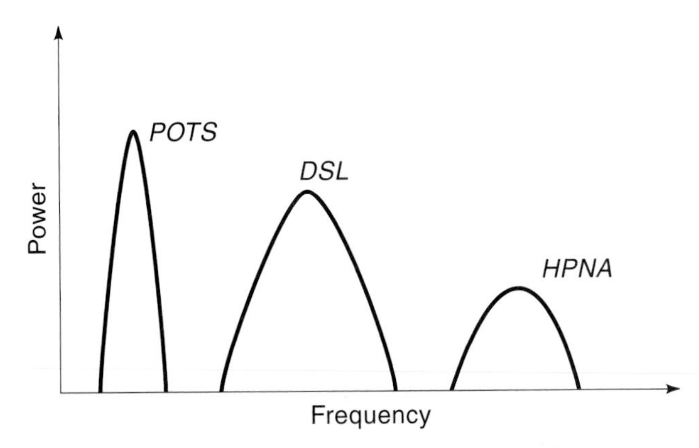

Figure 10.18 Phone line wirings shared among three technologies using FDM: (a) POTS uses 20 Hz–3.4 kHz, (b) DSL uses 25 kHz–1.1 MHz, and (c) HPNA uses 2–30 MHz.

wiring connecting switches together. The digital segment of the PSTN fully utilizes the wiring. The traditional access wiring uses around 4 kHz of the band for analog POTS and the rest was unexplored until DSL and HPNA were introduced. Comparing external PSTN wiring with the power line wiring, we can say that the entire power line wiring is used for transmissions of 50–60 Hz waveforms, and the rest is available for other applications. For many years power lines were used for low data rate (< 100 kbps) control and security networks operating below 500 kHz. These systems were using X-10, CEBus/CAL, and Lonworks standards [LONweb]. More recently, power lines are being considered for high data rates (> 1 Mbps) that operate above a frequency of 1 MHz to provide adequate speeds for computer networking. This area is still in the preliminary stages of development with no clear standard initiative. The only constraint by regulation on power lines is that they should not interfere with AM radio systems. AM radios in the United States operate between 9 and 490 kHz and in the EU between 3 and 148.5 KHz. Figure 10.19 shows the typical band separation in the power line applications. Low frequencies of up to a few kHz are used for low data rate applications such as security. The frequencies in the range of the AM radios are not used to avoid interference with AM radios. The frequency band of 1–30 MHz is used for high-speed data communications.

Recently smart appliances are emerging in the market that have some built-in intelligence, can sense other appliances on the power lines, and can be accessed through the Internet. Electric companies are investigating using the outside AC lines to deliver various services such as meter reading, energy management, and even Internet access to homes. The main current research thrust is to enable the access to control and security systems through the Internet. Figure 10.20 illustrates a number of potential applications that are emerging for the wire line systems. These applications include traditional low-speed security systems, remote control through the Internet, smart appliances, and meter reading operating in lower bands of the power lines and high-speed computing networks operating in higher bands. Power lines suffer from tremendous interference from electrical appliances, high attenuation, reflection caused from varying input impedance, and multipath phenomena that makes communication over this medium as challenging as communication

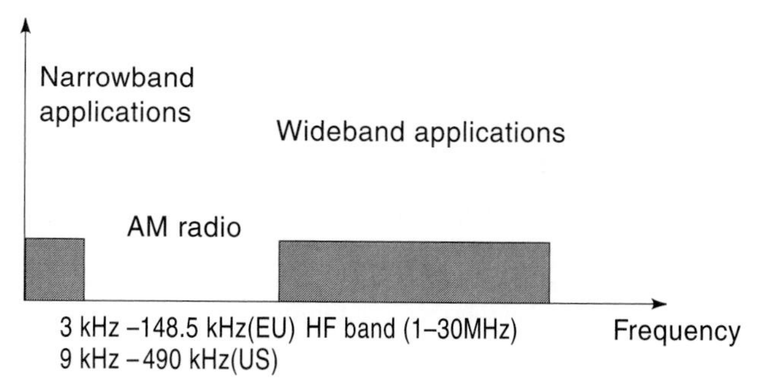

Figure 10.19 Frequency bands for low- and high-speed data communications over power lines.

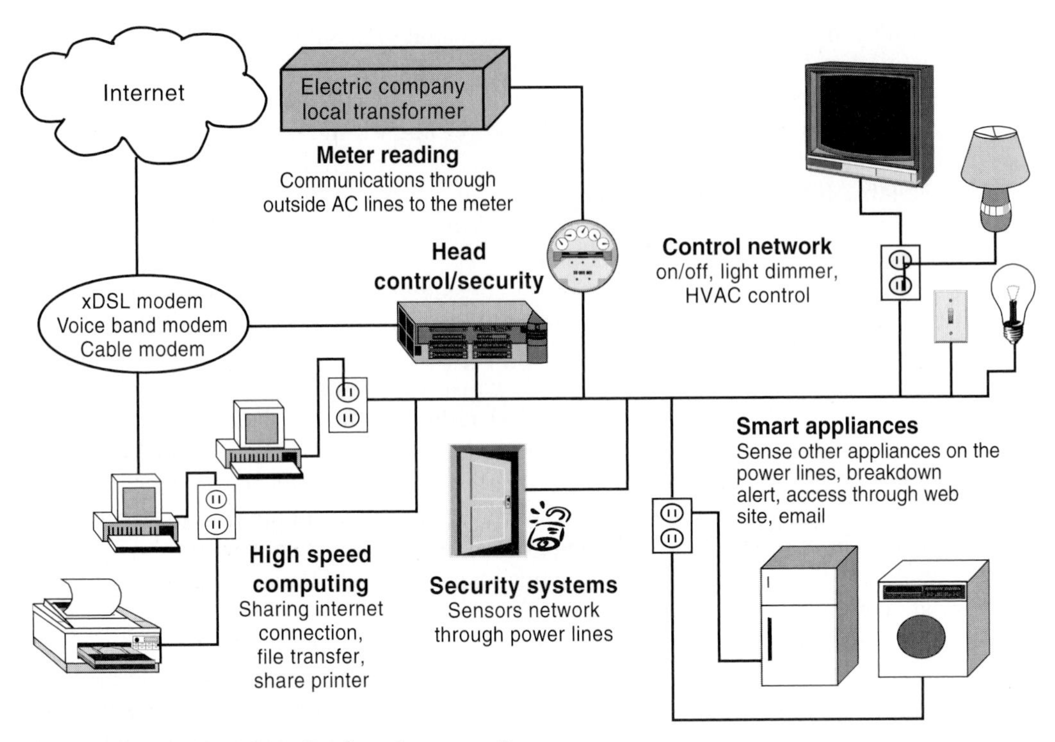

Figure 10.20 Potential applications for power lines.

over radio channels. As a result, a variety of complex transmission techniques and medium access protocols has been examined for different power line applications. Traditional FSK and QPSK are used in lower frequency bands and more complex spread spectrum and OFDM modems are used in the higher bands. DSMA and CSMA access methods, discussed in wireless data networks, are commonly used in the power line systems. The difficulty of the medium has kept the cost at higher rates which is a draw back in the growth of the market in this medium.

10.6.3.3 Wireless Solution Alternatives

A common advantage of wireless solutions over a wired network is that the wireless solution provides mobility and it covers the home as well as the yard.

Example 10.7: Coverage of a Wireless Cordless Telephone

As we explained in Chapter 1, cordless telephones were very successful from the early days of their appearance in the market. Because cordless telephones provide mobility and cover a larger area at home, users have paid higher prices to purchase cordless telephones in addition to a wired phone.

Other advantages of the wireless solution are that wireless is easy to install, relocate, scale, and maintain.

Example 10.8: Wireless Home Security Solutions

Introduction of wireless solutions to home security systems resulted in a sizable growth of that industry. From the user's point of view, wireless security systems are installed very quickly without additional holes in the walls or wires distributed in the home. Besides, selecting the location for placing a wireless product is more flexible which allows a better blending with the decoration of the home. Other advantages are that the wireless solutions are easier for expansion, moving to new homes, maintenance, and upgrading.

These examples are selected applications where a wireless solution is certainly necessary. As we discussed earlier, we have numerous home applications, and wireless may not be as important for all of them.

Example 10.9: Applications Where a Wireless Solution May Not Be Important

Smart appliances need a very low data rate for communication. These days manufacturers prefer to use power lines for networking so that by plugging the appliance to the wall, the user connects to the power and the network at the same time.

Certainly it is not difficult to find examples for cases where phone lines are the preferred home networking solution. The important conclusion from this discussion is that in home networking, we want to avoid new wiring, and therefore we need to use all the existing media that provide a nonhomogeneous environment for networking. Figure 10.21 provides a near future visionary portrayal of the evolving home networking solutions. Appliances are connected through a power line network, fixed computing equipment through HPNA, laptops and cordless phones through evolving wireless home networking, and security systems use their own proprietary network. A home server box connects all these networks together and to the Internet. Figure 10.22 represents a broader view of the home networking from the point of view of the now defunct Home RF group at the IEEE 802.15 standards group that will be discussed later on in Chapter 13. This is a more comprehensive view which also includes integration of entertainment systems using the IEEE 1394 standard. In this vision, wireless is only used for computing and cordless phones. New approaches for wireless networking was the scope of the EU workshop on wireless home networking [WIR00].

Wireless solutions for home networking include WLANs and WPANs which are discussed in detail in the following chapters. There are a number of home specific challenges for wireless networking. Using multiple devices allows operation of several different wireless solutions operating in the same unlicensed band. Handling interference in such a case is an important issue in wireless home networks. As we move to higher frequencies of operation at 5 GHz to support higher data rates, the coverage of the wireless solution may become a challenge. Wireless home networking needs designing inexpensive reconfigurable devices and Internet working between different media using different protocols. Incorporating cable TV into the network demands high transmission rates and a change in the access method to deliver video to the TV set. Today the cable that delivers the signal to the cable

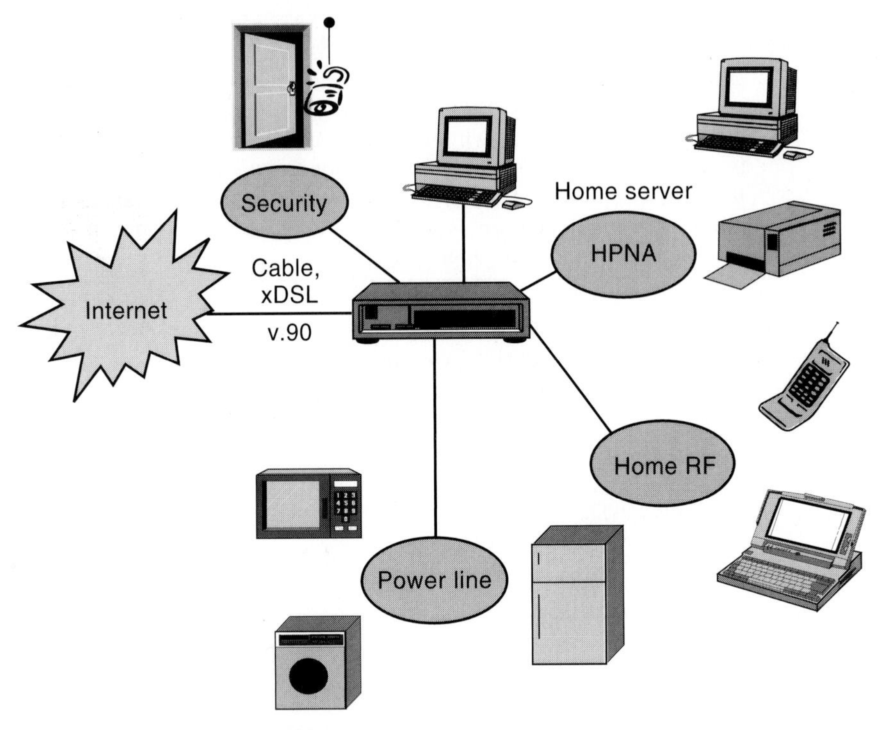

Figure 10.21 Evolving HANs.

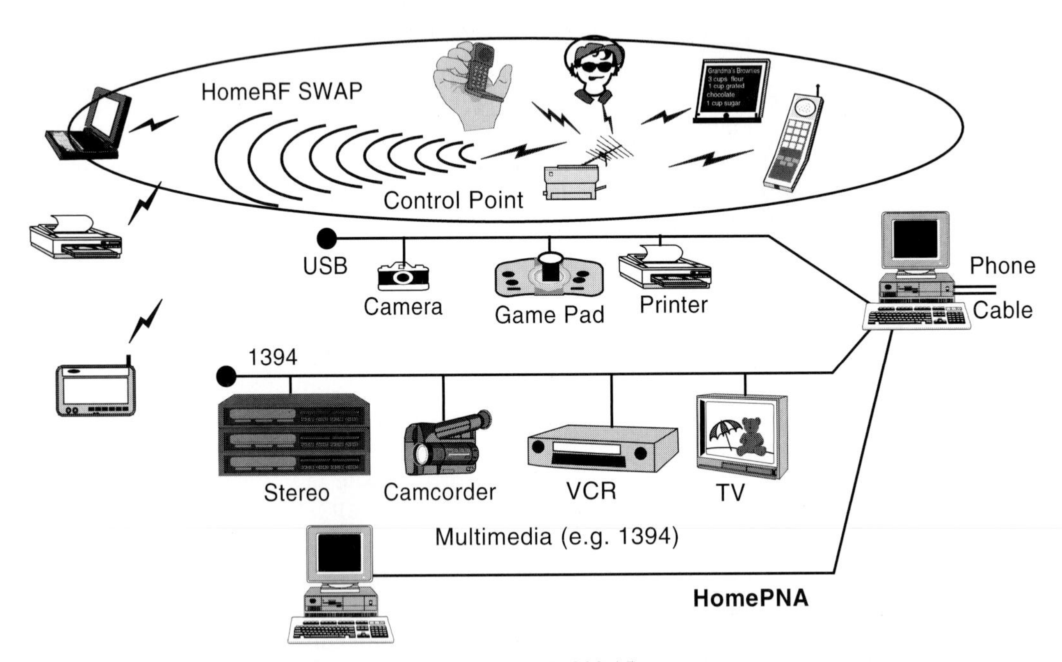

Figure 10.22 Vision of the home RF group at IEEE 802.15.

box, carries around a hundred analog video channels. Such bandwidths are not feasible over the wireless medium. Therefore, the entire system needs to be redesigned.

10.6.4 Home-Access Networks

The early home access technology was based on voice-band modems. Today broadband home access (with data rates on the order of 10 Mbps) is provided through cable modems and DSL services over the telephone lines. Cable modems operate on the cable TV wiring. The bandpass channel allocated to the TV channels is used by the cable modem which uses QAM modulation to provide higher data rate for transmission. The cable distribution in the residential areas has a bus topology that is optimally designed for one-way TV signal distribution. The bus carries all the stations in the neighborhood, and the cable TV box selects the station for the TV. If it is a scrambled paid channel, it also descrambles the signal. To control the set-top boxes, a reverse channel is available in all modern cable wiring. Broadband cable services use one of the video channels and the reverse channel to establish a two-way communication and access to Internet. The xDSL services use the 25 kHz–1.1 MHz bands on the telephone wirings and multisymbol QAM modulation to support high data rates to the users. The topology of the telephone line is a

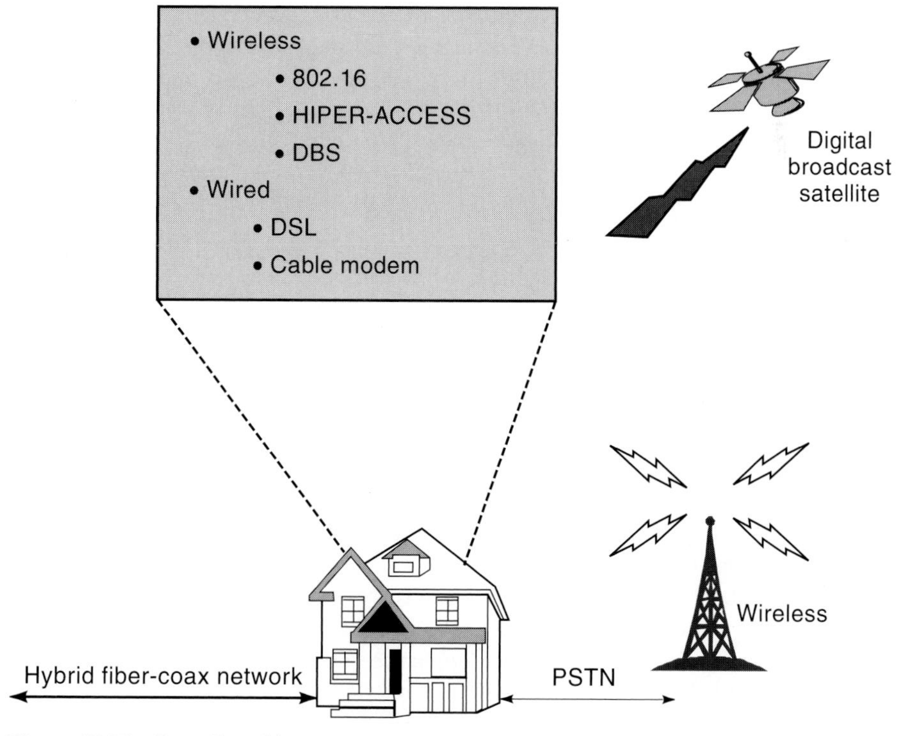

Figure 10.23 Broadband home access alternatives.

star topology that connects every user directly to the end-office where the DSL data is directed to Internet through a router.

Higher speed wireless home access uses LMDS or even existing WLAN inter-LAN bridges to provide the service. The advantage of using fixed-wireless solutions is that it does not involve wiring the streets. If there is no available wiring in the neighborhood, a wireless solution is certainly the main solution. The IEEE 802.16 in the United States and HIPER-ACCESS program in the EU are studying the next generation of networks in this area. Other wireless alternatives are direct satellite TV broadcasting and 3G wireless networks. Direct broadcast usually suffers from lack of a reverse channel and high delay that challenges the implementation of broadband services on this medium. The high-speed 3G wireless packet data services are expected to provide up to 2 Mbps which is very suitable for Internet access. The data rates on these systems are lower, and they are using licensed bands that ultimately may be expensive. Figure 10.23 summarizes the existing solutions for the home access technologies.

QUESTIONS

10.1 What is the difference between nomadic access and ad hoc networking?

10.2 Name three categories of unlicensed bands used in the United States and compare them in terms of size of the available band and coverage.

10.3 Explain the differences between WLAN and WPAN.

10.4 What are the IEEE 802.3, 802.4, and 802.5 standards and what are the differences between them?

10.5 Name the three IEEE 802 standard groups that are working on wireless networks and briefly explain the objective of each of them.

10.6 How does the current state of the art data rate of the wired and wireless LANs compare with one another?

10.7 Name the five major challenges for implementation of wireless LANs that existed from the beginning of this industry.

10.8 Why are unlicensed bands essential for the WLAN and WPAN industries?

10.9 Explain the difference between wireless Inter-LAN bridges and WLANs.

10.10 What are the differences between IEEE 802.11 and HIPERLAN standards?

10.11 Explain the differences among WCAN and WHAN technologies.

10.12 Why does the military show more interest in the wireless ATM approach?

10.13 Name three military projects related to broadband wireless local access.

10.14 Why did service providers become interested in WLAN and WPAN?

10.15 Name different alternatives for Internet access to the home and different mediums for home distribution.

10.16 What are the differences between LANs and HANs?

10.17 Explain the specifics challenges for the design of a HAN.

10.18 Name the classes of home appliances that are emerging in the home networking market.

10.19 What are the differences between HPNA and Ethernet?

10.20 How do HPNA, DSL, and POTS share the telephone wirings?

10.21 Compare telephone wiring, cable TV wiring, and power lines as media for home networking.

10.22 Compare wireless and wired solutions for home access and in-home distribution.

PROBLEMS

10.1 The IEEE 802.11 operates at 2.4 GHz, transmits 100 mW, and the minimum acceptable received signal strength for its proper operation is −80 dBm.

a. Using the JTC model, determine its approximated coverage in a three-floor office building.

b. If it is used as an Inter-LAN bridge in an open area in a large city to connect two 30 m tall buildings that are 1 km apart, what is the path loss that the transmitted signal will suffer? Can this Inter-LAN bridge operate properly? (Assume that antennas are installed on top of the roof and use the Okumura-Hata macro-cell model for prediction).

10.2 We want to install a LAN in a 5-story office building with identical floor plans. Each floor of the building is a 80m × 80m square with a height of 4 m. There are 15 terminals on each floor of the building and the external wiring comes to the first floor.

a. What is the total cost of wiring, equipment, and installation of the entire network if an IEEE 802.3 star network with one 240-port switch in the first floor connects all terminals to each other and the external connection? Assume a charge of $150 per run of wiring between two locations, a $6,000 cost for the switch, and a $35 per terminal cost for the network interface card.

b. To avoid wiring costs, assume that we use an IEEE 802.11 WLAN access point with a 100 mW (20dBm) transmitter power and a −80dBm receiver sensitivity in the center of the third floor. What is the total cost of the WLAN if the access point is $800 and each network interface card is $120?

c. Use the JTC model to calculate the coverage of the access point. Can we cover the entire building with one access point?

d. Compare the advantages and disadvantages of the two solutions.

10.3 You are designing a WLAN for an office building. You are not able to perform measurements or site surveys and have to rely on statistical models and certain other information. There are also certain constraints on where you can actually place the access point(s). You have the following information available to you:

Maximum number of walls between an access point and a mobile terminal = 4
Maximum number of floors between an access point and the mobile terminal = 2
Transmit power possibilities = 250 mW and 100 mW
Sensitivity of receiver is −80 dBm
Maximum distance from access point to building edge = 30m
Building has office walls, brick walls, and metallic doors
Shadow fading margin = 8 dB

What would be a conservative estimate of the number of access points required for the WLAN set-up? State your assumptions, models, and provide reasons for all your assumptions and calculations. Explain why. (*Hint:* Use path loss models from chapter 2 that are applicable to indoor areas.)

CHAPTER 11

IEEE 802.11 WLANs

11.1 Introduction

11.2 What Is IEEE 802.11?

11.2.1 Overview of IEEE 802.11
11.2.2 Reference Architecture
11.2.3 Layered Protocol Architecture

11.3 The PHY Layer

11.3.1 FHSS
11.3.2 DSSS
11.3.3 DFIR
11.3.4 IEEE 802.11a, b
11.3.5 Carrier Sensing

11.4 MAC Sublayer

11.4.1 General MAC Frame Format
11.4.2 Control Field in MAC Frames

11.5 MAC Management Sublayer

11.5.1 Registration
11.5.2 Handoff
11.5.3 Power Management
11.5.4 Security

Questions

Problems

11.1 INTRODUCTION

In this part of the book, we consider broadband local access using WLANs and WPANs. These are successful activities that were evolving outside of the voice-oriented 3G systems as broadband distribution systems for wireless local access. The last chapter provided an overview of the evolution of broadband wireless local access from the LAN industry and then the recent expansion to the home networking market. The following three chapters provide the detailed technical aspects of the important standards in broadband local access.

In the last chapter, we classified broadband local access systems into WLANs and WPANs. There are two camps in the WLAN standardization activities based on the orientation of their MAC layers. The IEEE 802.11 camp is a *connectionless* WLAN camp that evolved from data-oriented computer communications. Its counterpart is the HIPERLAN-2 camp that is more focused on *connection-based* WLANs addressing the needs of voice-oriented cellular telephony. The IEEE 802.11 has settled upon the MAC layer and is working on different physical layers to support higher data rates and today it holds the entire market. There are some activities addressing MAC enhancements in 802.11. The HIPERLAN-2 standard uses the same PHY layer as 802.11a with a MAC that supports the needs of the cellular telephone industry in supporting mechanisms for tariff, integration with existing cellular systems, and providing QoS. We present the IEEE 802.11 standard in this chapter and wireless ATM and HIPERLAN activities in the following chapter. The reader can refer to [CRO97], [STA01], [VAN99], [VAL98] for details on IEEE 802.11.

11.2 WHAT IS IEEE 802.11?

The IEEE 802.11 is the first WLAN standard and so far the only one that has secured a market. The IEEE 802.11 standardization activity originally started in 1987 as a part of the IEEE 802.4 token bus standard under the group number IEEE 802.4L. The IEEE 802.4 is a counterpart of the IEEE 802.3 and 802.5, which pays special attention in supporting factory environments. As we saw in the history of the WLANs in the last chapter, one of the early motives for using WLANs was in factories for control of and communication between equipment. For this reason, car manufacturers such as GM were actively participating in the IEEE 802.4L activities in the early days of this industry. In 1990 the 802.4L WLAN group was renamed as IEEE 802.11, an independent 802 standard, to define the PHY and MAC layers for WLANs. The first IEEE 802.11 standard for 1 and 2 Mbps, completed in 1997, supported DSSS, FHSS, and diffused infrared (DFIR) physical layers. Since completion of this first standard, new PHY layers supporting 11 Mbps using CCK (called IEEE 802.11b) and 54 Mbps using OFDM (called IEEE 802.11a) have been defined. All three versions share the same MAC layer that uses CSMA/CA for contention data, a request-to-send/clear-to-send (RTS/CTS) mechanism to accommodate the hidden terminal problem, and an optional mechanism called point coordination function (PCF) to support time-bounded applications. The IEEE 802.11

standard supports both infrastructure WLANs connecting through an AP and ad hoc operation allowing peer-to-peer communication between terminals.

The IEEE 802.11 standard was the first WLAN standard facing the challenge of organizing a systematic approach for defining a standard for wireless wideband local access. Compared with wired LANs, WLANs operate in a difficult medium for communication, and they need to support mobility and security. The wireless medium has serious bandwidth limitations and frequency regulations. It suffers from time and location dependent multipath fading. It is subject to interference from other WLANs as well as other radio and nonradio devices operating in the vicinity of a WLAN. Wireless standards need to have provisions to *support mobility* that is not shared in the other LAN standards. The IEEE 802.11 body had to examine connection management, link reliability management, and power management—none of which were concerns for other 802 standards. In addition, WLANs have no physical boundaries, and they overlap with each other, and therefore the standardization organization needed to define provisions for the *security* of the links. For all these reasons and because of several competing proposals, it took nearly 10 years for the development of IEEE 802.11 which was far longer than other 802 standards designed for wired mediums. Once the overall picture and the approach became clear, it only took a reasonable time to develop the IEEE 802.11b and IEEE 802.11a enhancements.

11.2.1 Overview of IEEE 802.11

To start a voice-oriented connection-based standard, such as GSM, the first step is to specify the services. Then the reference system architecture and its interfaces are defined, and at last the detailed layered interfaces are specified to accommodate all the services. The situation in connectionless data-oriented networks, such as IEEE 802.11, is quite different. The IEEE 802.11 standard provides for a general PHY and MAC layer specification that can accommodate any connectionless applications whose transport and network layers accommodate the IEEE 802.11 MAC layer. Today, TCP/IP is the dominant transport/network layer protocol hosting all popular connectionless applications such as Web access, e-mail, FTP, or telnet, and it works over all MAC layers of the LANs, including IEEE 802.11. Therefore, the IEEE 802.11 standard does not need to specify the services. However, IEEE 802.11 provides for local and privately owned WLANs with a number of competing solutions. In this situation, the first step in the standardization is to group all the solutions into one set of requirements with a reasonable number of options. The next step, in a manner similar to connection-based standards, is to define a reference system model and its associated detailed interface specifications.

11.2.1.1 Requirements

The number of participants in the IEEE 802.11 standards soon exceeded a hundred suggesting a number of alternative solutions. The finalized set of requirements, which did not came about easily, indicated that the standard should provide:

- Single MAC to support multiple PHY layers
- Mechanisms to allow multiple overlapping networks in the same area

- Provisions to handle the interference from other ISM band radios and microwave ovens
- Mechanism to handle "hidden terminals"
- Options to support time-bounded services
- Provisions to handle privacy and access control

In addition it was decided that the standard would not be concerned with licensed band operations. These requirements set the overall direction of the standard in adopting different alternatives. However, as it often happens in these types of standards, the actual adoptions were based on successful products that were already available in the market.

Example 11.1: Origins of PHY Layer Solutions

The DSSS solution for IEEE 802.11 is based on WaveLAN that was designed at NCR, Netherlands [TUCH91], [WOR91]. The FHSS solution was highly affected by RangeLAN designed by Proxim, CA, and products from Photonics, CA, and Spectrix, IL [SPECweb] affected the DFIR standard.

11.2.2 Reference Architecture

The reference model in connection-based WANs, such as GSM described in Chapter 7, contains a relatively sizable wired infrastructure with a number of hardware and software elements and many interfaces among all these elements. The connectionless IEEE 802.11 local network defines two topologies and several terminologies to start the standardization process. Figure 11.1 illustrates the infrastructure and ad hoc topologies that are the two configurations that the IEEE 802.11 standard considers. In the infrastructure configuration, wireless terminals are connected to a backbone network through APs. In the ad hoc configuration, terminals communicate in a peer-to-peer basis.

In IEEE 802.11 terminology the AP provides access to distribution services through the wireless medium. The *basic service area* (BSA) is the coverage area of one access point. The *basic service set* (BSS) is a set of stations controlled by one access point. The *distribution system* (DS) is the fixed (wired) infrastructure used to connect a set of BSS to create an *extended service set* (ESS). IEEE 802.11 also defines *portal(s)* as the logical point(s) at which non-802.11 packets enter an ESS.

In a typical application a number of laptops are connected through a WLAN to a backbone wired LAN. Each laptop carries a card and the connection point to the backbone is an AP with another card that forms a practical AP consisting of AP and a portal. Figure 11.2 illustrates this practical situation. The cards in the laptop and the AP support the MAC and PHY layers of the IEEE 802.11, the rest of the AP device acts as a bridge to convert the 802.11 protocol to MAC and the PHY layer of the backbone DS, that is typically an IEEE 802.3 Ethernet LAN. Laptops connect to the LAN through the AP to communicate with other devices, such as

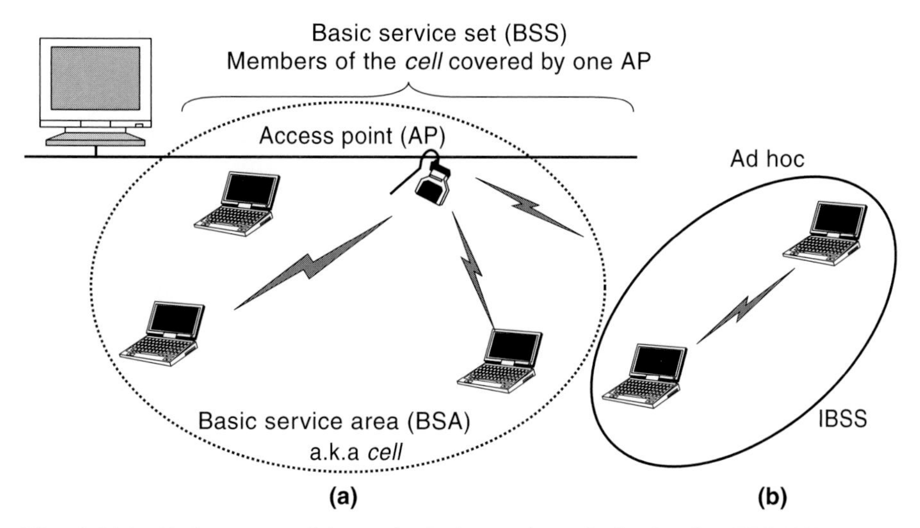

Figure 11.1 Reference model, terminologies, and topologies for the IEEE 802.11: (a) Infrastructure network and (b) ad hoc network.

the server shown in Figure 11.2. Typical ESS is formed by installing APs at different locations to connect to the backbone DS and cover the area.

11.2.3 Layered Protocol Architecture

Figure 11.3 shows the entities in the protocol stack of the IEEE 802.11 standard. The traditional simple MAC and PHY layers definitions in the IEEE 802 substandards are broken into other sublayers to make the specification process easier. The MAC layer is divided into MAC sublayer and MAC management sublayer entities.

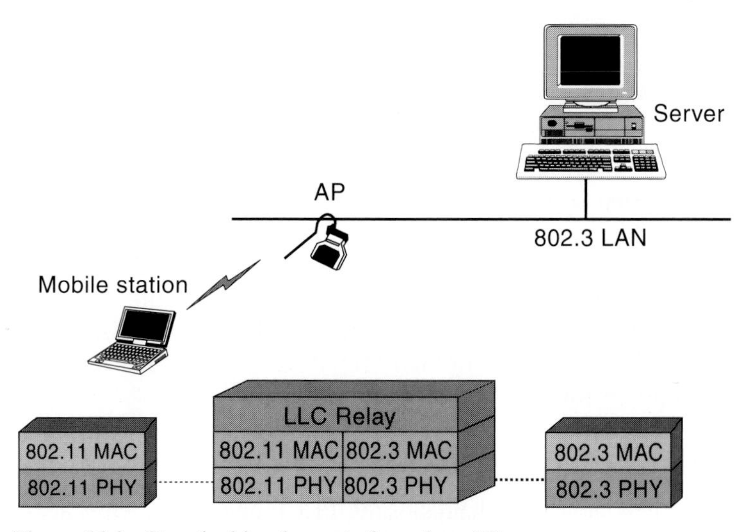

Figure 11.2 Practical implementation of an AP.

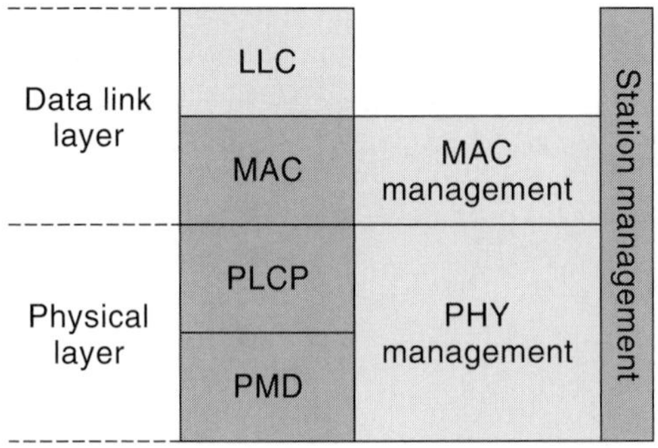

Data link layer	LLC		Station management
	MAC	MAC management	
Physical layer	PLCP	PHY management	
	PMD		

PLCP: Physical layer convergence protocol
PMD: Physical medium dependent

Figure 11.3 Protocol entities for the IEEE 802.11.

The *MAC* sublayer is responsible for access mechanism and fragmentation and reassembly of the packets. The *MAC layer management* sublayer is responsible for roaming in ESS, power management, and association, dissociation, and reassociation processes for registration connection management. The PHY layer is divided into three sublayers: PHY layer convergence protocol (PLCP), PHY medium dependent (PMD) protocol, and the PHY layer management sublayer. The *PLCP* is responsible for carrier sensing assessment and forming packets for different PHY layers. The *PMD* sublayer specifies the modulation and coding technique for signaling with the medium, and *PHY layer management* decides on channel tuning to different options for each PHY layer. In addition IEEE 802.11 specifies a *station management* sublayer that is responsible for coordination of the interactions between MAC and PHY layers.

11.3 THE PHY LAYER

When the MAC protocol data units (MPDU) arrive to the PLCP layer, a header is attached that is designed specifically for the PMD of the choice for transmission. The PLCP packet is then transmitted by PMD according to the specification of the signaling techniques. In the original IEEE 802.11, there are three choices of FHSS, DSSS, and DFIR for PMD transmission and therefore IEEE 802.11 defines three PLCP packet formats to prepare the MPDU for transmission. The rest of this section provides the details of the PMD, PLCP, and PHY layer management sublayers of all options for the IEEE 802.11 and IEEE 802.11b.

11.3.1 FHSS

Figure 11.4 shows the details of the PLCP header, which is added to the whitened MAC PDU to prepare it for transmission using FHSS physical layer specifications of the IEEE 802.11. There are two data rates for transmission of the information at

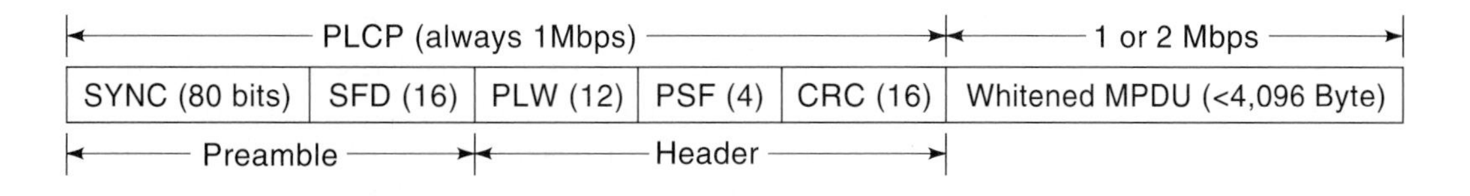

SYNC: Alternating 0,1
SFD: 0000110010111101
PLW: Packet length width
PSF: Data rate in 500 kbps steps
CRC: PLCP header coding

Figure 11.4 PLCP frame for the FHSS of the IEEE 802.11.

1 and 2 Mbps using two- and four-level GFSK modulation, respectively. The lower data rate always provides a simpler environment for synchronization between various adaptive parts of the receiver. Besides, with all packets starting with the same format, it is easier for the receiver to start a dialog with the transmitter. As a result, the PLCP header is always transmitted with the lower data rate of 1 Mbps using the simpler two-symbol GFSK modulation. In layman terms this allows a slower warm-up and initial negotiations for the receiver. The MPDU, however, is transmitted either by existing 1 or 2 Mbps transmission rate, or it can be at any other rate that may evolve in the future.

The PLCP additional bits consist of a preamble and a header. The *preamble* is a sequence of alternating 0 and 1 symbol for 80 bits that is used to extract the received clock for carrier and bit synchronization. The start of the frame delimiter (*SFD*) is a specific pattern of 16 bits, shown in the figure, indicating start of the frame. The next part of the PLCP is the header that has three fields. The 12-bit packet length width (PLW) field identifies the length of the packet that could be up to 4 kbytes. The 4 bits of the packet-signaling field (PSF) identifies the data rate in 0.5 Mbps steps starting with 1 Mbps.

Example 11.2: Specification of Data Rate on the Physical Layer

The existing 1 Mbps is represented by 0000 as the first step. The 2 Mbps by 0010 is 2×0.5 Mbps + 1 Mbps = 2 Mbps. The maximum 3-bit number represented by this system is 0111 that is associated with $7 \times 0.5+1 = 4.5$ Mbps. If all four bits are used, we have $15 \times 0.5+1 = 8.5$ Mbps. These limitations impose a need for changes in this field to accommodate next generation systems using data rates higher than 10 Mbps.

The rest of the rates are reserved for the future. The 16-bit CRC code is added to protect the PLCP bits. It can recover from errors of up to 2 bits, and otherwise identify whether the PLCP bits are corrupted. The total overhead of the PCLP is 16 bytes (128 bits) that is less than 0.4 percent of the maximum MPDU load, justifying the low impact of running the PCLP at lower data rates.

The FHSS PMD hops over 78 channels of 1 MHz each in the center of the 2.44 GHz ISM bands. The modulation technique is the GFSK that is described in Chapter 3. For 1 Mbps two levels and for 2 Mbps 4 levels of GFSK are employed.

Each BSS can select one of the three patterns of 26 hops given by (0,3,6,9,...,75), or (1,4,7,10,...,76) or (2,5,8,11,...,77) shown in Figure 11.5. The selection process is the responsibility of the PHY layer management sublayer. The IEEE 802.11 actually specifies specific random hopping patterns for each of these frequency groups that facilitates multivendor interpretability. Multiple BSSs can coexist in the same area by up to three APs using different frequency groups because these sets are nonoverlapping. In practice if three APs are installed next to one another in the same geographic area, we can provide a throughput of up to 6 Mbps. If APs use different frequency hopping patterns using CDMA in the same frequency group, the throughput can be increased substantially. The minimum hop rate of the IEEE 802.11 FHSS system is 2.5 hops per second, which is rather slow. The maximum recommended transmitted power is 100 mW.

The received MPDU is passed through a scrambler to be randomized. Randomization of the transmitted bits, which is also called whitening because the spectrum of a random signal is flat, eliminated the dc bias of the received signal. A scrambler is a simple shift register finite state machine with special feedback that is used both for scrambling and descrambling of the transmitted bits.

11.3.2 DSSS

The PMD of the DSSS version of the IEEE 802.11 uses a Barker code of length 11 that was described in Example 3.20. The reader should remember that the FHSS receiver is a narrowband receiver (1 MHz bandwidth) whose center frequency hops over 76 MHz, but DSSS communicates using nonoverlapping pulses at the chip rate of 11 Mcps which occupies around 26 MHz of bandwidth. The modulation techniques used for the 1 and 2 Mbps are DBPSK and DQPSK, respectively, which send one or two bits per transmitted symbol. The ISM band at 2.4 GHz is divided into 11 overlapping channels spaced by 5 MHz, shown in Figure 11.6, to provide several choices for coexisting networks in the same area. A PHY layer management sublayer of the AP covering a BSS can select one of the five choices according to the level of interference in different bands. The maximum transmitted power is the same as the FHSS and is recommended at 100mW. The FHSS is easier for implementation because the sampling rate is at the orders of the symbol rate of 1 Msps. The DSSS implementation requires sampling rates at the orders of 11 Mcps. Because of the wider bandwidth, DSSS provides a better coverage and a more stable signal.

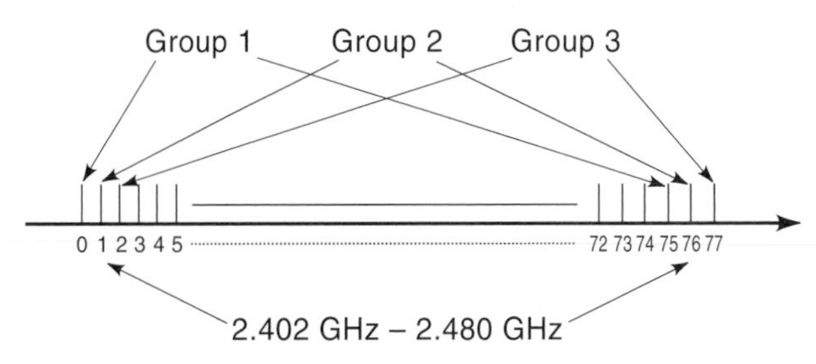

Figure 11.5 Three frequency groups for the FHSS in the IEEE 802.11.

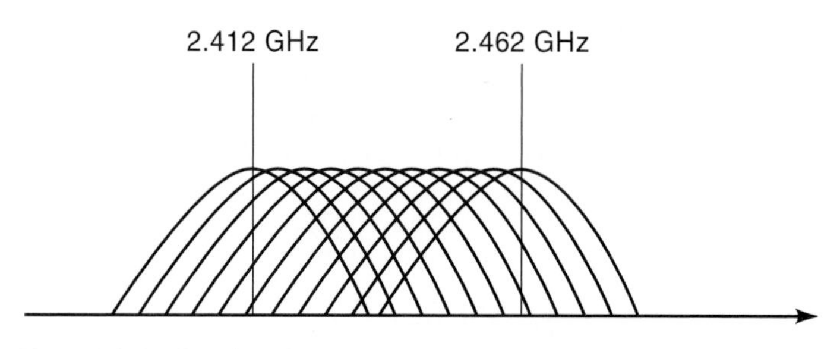

Figure 11.6 Overlapping frequency bands for the DSSS in the IEEE 802.11.

Figure 11.7 shows the details of the PLCP frame for the DSSS version of the IEEE 802.11. The overall format is similar to the FHSS, but the length of the fields is different because transmission techniques are different and different manufacturers designed the model product for development of the FHSS and DSSS standards. The MPDU from the MAC layer is transmitted either at 1 or 2 Mbps, however, analogous to the FHSS version of the standard, the PLCP of the DSSS version also uses the simpler BPSK modulation at 1 Mbps all the time. The MPDU for the DSSS does not need to be scrambled for whitening because each bit is transmitted as a set of random chips that is a whitened transmitted signal. The length of the SYNC in the DSSS is 128 bits that is longer than FHSS because DSSS needs a longer time to synchronize. The format of the SFD of the DSSS is identical to that of the FHSS but the value of the code, shown in Figure 11.7, is different. The PSF field of the FHSS is called *signal* field, and it uses 8 bits to identify data rates in steps of 100 kbps (five times more precision than FHSS).

Example 11.3: Frame Formats for Various Data Rates in IEEE 802.11

Using the above encoding, we represent 1 Mbps for DSSS by 00001010 (10 × 100 kbps), the 2 Mbps by 00010100 (20 × 100 kbps), and the 5.5 Mbps 11 and Mbps (used in IEEE 802.11b) by 001101110 (55 × 100 kbps) and 01101110 (110 × 100 kbps). The maximum number in this system is 11111111, which represents 255 × 100 kbps = 25.5 Mbps.

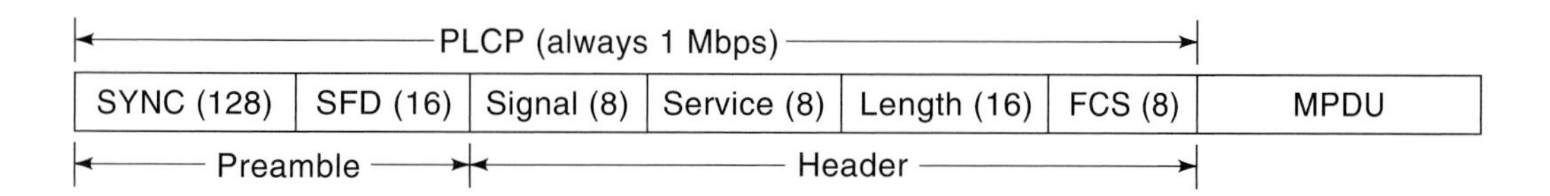

SYNC: Alternating 0,1
SFD: 1111001110100000
Signal: Data rate in 100 KHz steps
Service: Reserved for future use
Length: Length of MPDU in microsecond
FCS: PLCP header coding

Figure 11.7 PLCP frame for the DSSS of the IEEE 802.11.

The *service* field in the DSSS is reserved for future use, and it does not exist in the FHSS version. The *length* field of the DSSS is analogous to the PLW in the FHSS, however, length field of DSSS specifies the length of the MPDU in microseconds. The frame correction sequence (*FCS*) field of the DSSS is identical to the CRC field of the FHSS.

11.3.3 DFIR

The PMD of DFIR operates based on transmission of 250 ns pulses that are generated by switching the transmitter LEDs on and off for the duration of the pulse. Figure 11.8 illustrates the 16-PPM and 4-PPM modulation techniques recommended by the IEEE 802.11 for 1 and 2 Mbps, respectively. In the 16-PPM, blocks of 4 bits of the information are coded to occupy one of the 16 slots of a 16-bit length sequence according to their value. In this format each 16×250 ns = 4,000 ns carries 4 bits of information that supports a 4 bits/4,000 ns = 1 Mbps transmission rate. For the 2 Mbps version, every 2 bits are PPM-modulated into 4 slots of duration 4×250 ns = 1,000 ns that generates data at 2 bits/1,000 ns = 2 Mbps. The peak transmitted optical power is specified at 2W with an average of 125 or 250 mW. The wavelength of the light is specified at 850 nm to 950 nm.

The PLCP packet format for the DFIR is shown in Figure 11.9. The PLCP signals are shown in the unit of slots of 250 ns for one basic pulse. The synch and SDF fields are shorter than FHSS and DSSS because noncoherent detection using photosensitive diode detectors do not need carrier recovery or elaborate random code synchronizations. The three-slot data rate indication field starts by 000 for 1 Mbps and 001 for 2 Mbps. The length and FCS are identical to the DSSS. The only new field is the DC level adjustment (DCLA) that sends a sequence of 32 slots, al-

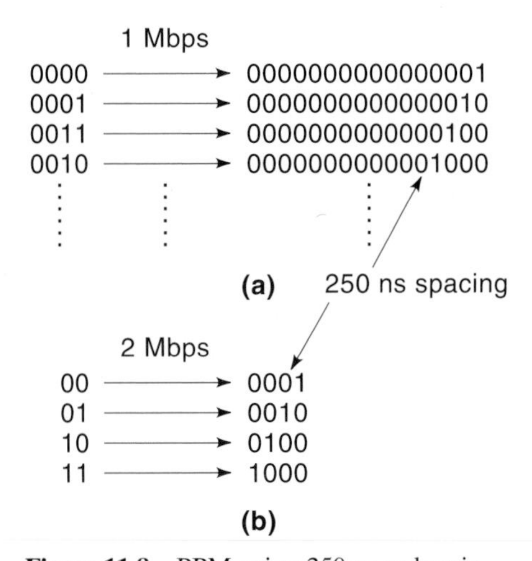

Figure 11.8 PPM using 250 ns pulses in the DFIR version of the IEEE 802.11: (a) 16 PPM for 1 Mbps and (b) 4 PAM for 2 Mbps.

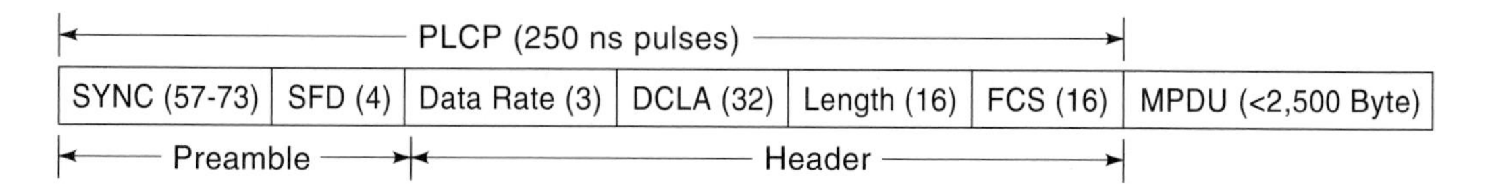

SYNC: Alternating 0,1 pulses
SFD: 1001
Data rate: 000 and 001 for
DCLA: DC level adjustments sequences
Length: Length of MPDU in microsecond
FCS: PLCP header coding

Figure 11.9 PLCP frame for the DFIR of the IEEE 802.11.

lowing the receiver to set its level of the received signal to set threshold for deciding between received zeros and 1s. The MPDU length is restricted to 2,500 bytes.

11.3.4 IEEE 802.11a, b

The PHY layer of the IEEE 802.11a is based on an OFDM transmission that operates in the 5 GHz U-NII bands. The principle of operation of OFDM was discussed in Chapter 3. The specifics of the IEEE 802.11a OFDM system are shared with the HIPERLAN-2 standard, and we provide it in the next chapter. The IEEE 802.11a and b MAC layer remains the same as that of the other IEEE 802.11 standards.

At 2.4 GHz, the IEEE 802.11b standard specifies a new PHY layer, called CCK, to support data rates of 5.5 and 11 Mbps. The IEEE 802.11b uses the same PLCP as the IEEE 802.11 DSSS standard. The principles of operation of the CCK were provided in Chapter 3. There are several unique features for the IEEE 802.11b that are worth mentioning. The IEEE 802.11b uses Walsh codes with complementary codes for M-ary orthogonal data transmission. This scheme interoperates with existing 1 and 2 Mbps networks using the same preamble and header. Therefore, it can be easily integrated in the existing IEEE 802.11 DSSS modems.

In the CCK method described in Example 3.17, we explained that the CCK maps blocks of eight bits of the incoming data into blocks of eight-QPSK complex symbols to derive an 11 Mbps data transmission system with the same chip transmission rate of the 11 Mcps. This mapping was explained to be performed by a CCK encoder that uses the 8-bit input block as the address of one of the 256 orthogonal eight-QPSK symbol sequence to be transmitted. The transformation that makes this mapping possible was given by

$$c = \{e^{j(\varphi_1+\varphi_2+\varphi_3+\varphi_4)}, e^{j(\varphi_1+\varphi_3+\varphi_4)}, e^{j(\varphi_1+\varphi_2+\varphi_4)}, -e^{j(\varphi_1+\varphi_4)}, e^{j(\varphi_1+\varphi_2+\varphi_3)}, -e^{j(\varphi_1+\varphi_3)}, -e^{j(\varphi_1+\varphi_2)}, e^{j(\varphi_1)} \qquad (11.1)$$

where $(\varphi_1, \varphi_2, \varphi_3, \varphi_4)$ are four phases associated with the eight-bit arriving block with each two bits combined in a four-phase complex number. After finding four complex phases associated to an eight-bit arriving block, we replace them in the earlier equation to obtain one of the 256 complex orthogonal CCK codes. All the terms of the earlier equation share the first phase; if we factor that out we have

$$c = \{e^{j(\varphi_2+\varphi_3+\varphi_4)}, e^{j(\varphi_3+\varphi_4)}, e^{j(\varphi_2+\varphi_4)}, -e^{j(\varphi_4)}, e^{j(\varphi_2+\varphi_3)}, e^{j(\varphi_3)}, -e^{j(\varphi_2)}, 1\}e^{j(\varphi_1)} \tag{11.2}$$

This change suggests that our 256 transformation matrix can be decomposed into two transformations, one a unity transformation that maps two bits (one complex phase) directly and the other one that maps six bits (three phases) into an 8-element complex vector with 64 possibilities determined by the inner function of the above equation. The above decomposition leads to a simplified implementation of the CCK system that is shown in the Figure 11.10. At the transmitter the serial data in multiplied into 8-bit addresses. Six of the eight bits are used to select one of the 64 orthogonal codes produced as one of the 8-complex code, and two bits are directly modulated over all elements of the code that are transmitted sequentially. The receiver is actually comprised of two parts: one the standard IEEE 802.11 DSSS decoder using Barker codes and one a decoder with 64 correlators for the orthogonal codes and an ordinary demodulator for IEEE 802.11b. By checking the PLCP data rate, the field receiver knows which decoder should be employed for the received packets. This scheme provides an environment for implementation of a WLAN that accommodates both 802.11 and 802.11b devices.

The IEEE 802.11b also supports 5.5 Mbps as a backup for the 11Mbps operation. Figure 11.11 shows the format of the bits for both data rates. In the 5.5 Mbps mode, blocks of four bits, rather than eight bits, are used for multiplexing, and two of the four bits are used to select one of the four possible complex orthogonal vectors.

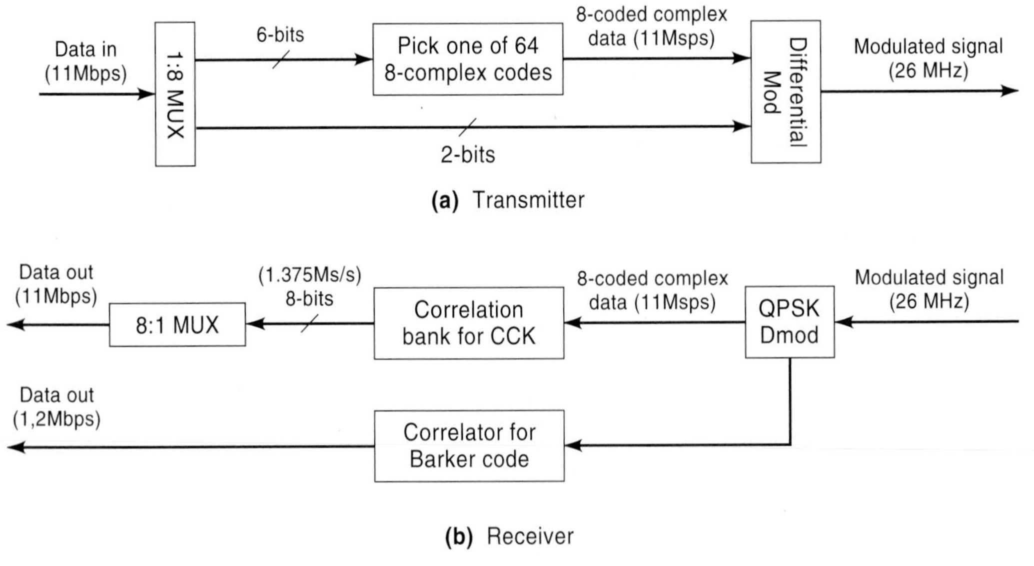

Figure 11.10 Simplified implementation of the CCK for IEEE 802.11b.

$$c = \{e^{i(\varphi_1+\varphi_2+\varphi_3+\varphi_4)}\ e^{i(\varphi_1+\varphi_3+\varphi_4)},\ e^{i(\varphi_1+\varphi_2+\varphi_4)},$$
$$-e^{i(\varphi_1+\varphi_4)},\ e^{i(\varphi_1+\varphi_2+\varphi_3)}\ e^{i(\varphi_1\varphi_3)},\ -e^{i(\varphi_1+\varphi_2)},\ e^{i\varphi_1}\}$$

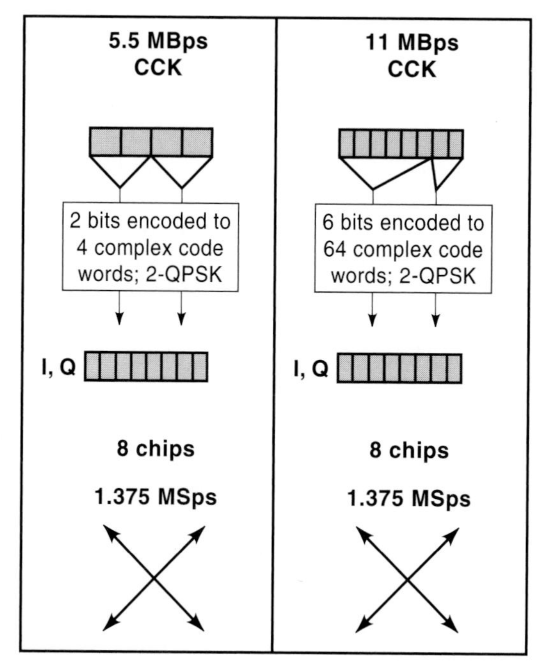

Figure 11.11 Multiplexing of the input data for 5.5 and 11 Mbps IEEE 802.11b systems.

11.3.5 Carrier Sensing

In IEEE 802.3 sensing the channel is very simple. The receiver reads the peak voltage on the wire of cable and compares it against a threshold. In the IEEE 802.11, the sensing mechanism is more complicated, and it is performed either physically or virtually. The PHY sensing is through the clear channel assessment (CCA) signal produced by PLCP in the PHY layer of the IEEE 802.11. The CCA is generated based on "real" sensing of the air-interface either by sensing the detected bits in the air or by checking the RSS of the carrier against a threshold. Decisions based on the detected bits are made slightly slower, but they are more reliable. Decisions based on the RSS may create a false alarm caused by measuring the level of the interference. The best designs take advantage of both carrier sensing and detected data sensing. In addition to PHY sensing, the IEEE 802.11 also provides for the virtual carrier sensing. Virtual carrier sensing is based on a network allocation vector (NAV) signal supported by the RTS/CTS and PCF mechanisms in the MAC layer. In either of these MAC operations, the system allows generation of a NAV signal that acts like a sensed channel to prevent transmission of contention data for a preset duration of time. A "length" field in the MAC layer is used to specify the amount of time that must elapse before the medium can be freed.

11.4 MAC SUBLAYER

The overall MAC layer responsibilities are divided between MAC sublayer and MAC layer management sublayer. The major responsibilities of the MAC sublayer are to define the access mechanisms and packet formats. The MAC management sublayer defines roaming support in the ESS, power management, and security.

The IEEE 802.11 specifies three access mechanisms that support both contention and contention-free access. The contention mechanism is supported by CSMA/CA protocol that was explained in the Example 4.18. There are two cases for contention-free transmission. The first case is the RTS/CTS, discussed in Section 4.3, which is used to solve the hidden terminal problem. The second case is the implementation of the PCF for time-bounded information.

To allow coordination of a number of options for the MAC operations, IEEE 802.11 recommends three inter-frame spacings (IFSs) between the transmissions of the packets. These IFS periods provide a mechanism for assigning priority that is then used for implementation of QoS support for time-bounded or other applications. After completion of each transmission, all terminals having information packets wait for one of the three IFS periods according to the level of priority of their information packet. These interframing intervals are DCF-IFS (DIFS) used for contention data spacing that has the lowest priority and longest duration. Short IFS (SIFS), used for highest priority packets such as ACK and CTS (clear to send), has the lowest duration of time. The PCF IFS (PIFS), designed for PCF operation, has the second priority rate with a duration between DIFS and SIFS.

In CSMA/CA, as soon as the MAC has a packet to transmit, it senses the channel to see if the channel is available both physically and virtually. If the channel is virtually busy because a NAV signal is turned on, the operation is delayed until the NAV signal has disappeared. When the channel is virtually available, the MAC layer senses the PHY condition of the channel. If the channel is idle, as shown in Figure 11.12, the terminal waits for a DIFS period and transmits the data. If the channel is sensed busy, MAC runs a random number generator to set a *backoff* clock. During the transmission of the packet and its associated DIFS, contention is differed but sensing continues. When the channel becomes available, as shown in Figure 11.12, a contention window start in which all terminals having packets for transmission run down their *backoff* clocks. The first terminal that expires its clock starts transmission. Other terminals sense the new transmission and freeze their clock to be restarted after the completion of the current transmission in the next contention period. This mechanism reduces the collisions, but it cannot

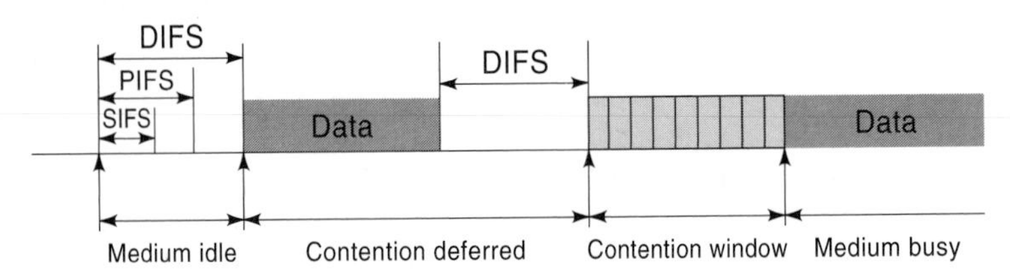

Figure 11.12 Primary operation of the CSMA/CA in the IEEE 802.11.

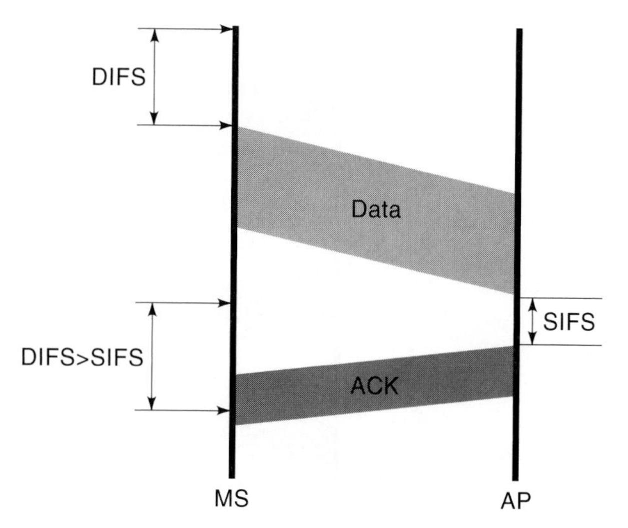

Figure 11.13 Implementation of the CSMA/CA with ACK in an infrastructure network.

eliminate it. To reduce the probability of repeated collisions, in a manner similar to IEEE 802.3 described in Example 4.17, the length of backoff time is exponentially increased as the terminal goes through successive retransmissions.

The IEEE 802.11 recommends both CSMA/CA based on the clear channel assignment signal from the PHY layer and CSMA/CA with ACK for MAC recovery. Note that Ethernet does not provide for MAC recovery, and this feature is unique to the IEEE 802.11. Figure 11.13 illustrated a sample operation of the CSMA/CA with ACK in a communication between a terminal and an AP. When the AP receives a packet of data, it waits for a SIFS and sends the ACK. Because SIFS is smaller than DIFS, all the other terminals must wait until transmission of the ACK to the MS is completed. The CSMA with ACK is not implemented on most of the IEEE 802.11 products, and ACK is left for other layers of the protocol that are implemented on software.

In the RTS/CTS mechanism in the IEEE 802.11 MAC sublayer, shown in Figure 11.14, a terminal ready for transmission sends a short RTS packet (20 bytes) identifying the source address, destination address, and the length of the data to be transmitted. The destination station will respond with a CTS packet (16 bytes) after a SIFS period. The source terminal receives the CTS and sends the data after another SIFS. The destination terminal sends an ACK after another SIFS period. Other terminals hearing RTS/CTS that is not addressed to them will go to the virtual carrier-sensing mode for the entire period identified in the RTS/CTS communication, by setting their NAV signal on. Therefore, the source terminal sends its packet with no contention. After completion of the transmission, the destination terminal sends an acknowledgment packet, and the NAV signal is terminated, opening the contention for other users. This method provides a unique access right to a terminal to transmit without any contention. It helps in the situations when a terminal with low received power at the AP is shadowed with terminals with much higher received power.

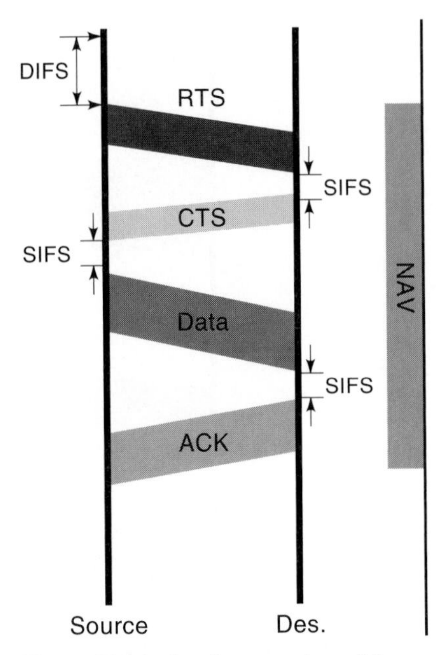

Figure 11.14 Implementation of the RTS/CTS mechanisms in the IEEE 802.11.

The PCF mechanism in the IEEE 802.11 MAC sublayer, shown in Figure 11.15, is built on top of the DCF using CSMA to support contention-free time bounded and asynchronous transmission operations. This is another optional MAC service that has complicated the MAC layer, and all manufacturers have not chosen to implement it in their products. In the PCF operation, only available for

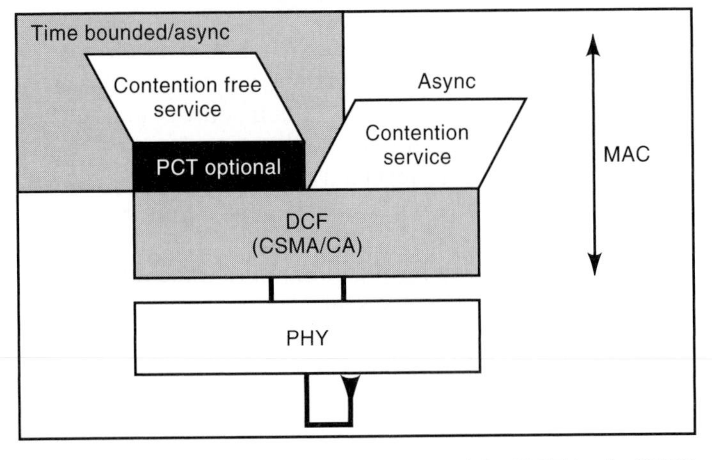

Figure 11.15 Implementation PCF on top of the DCF in the IEEE 802.11.

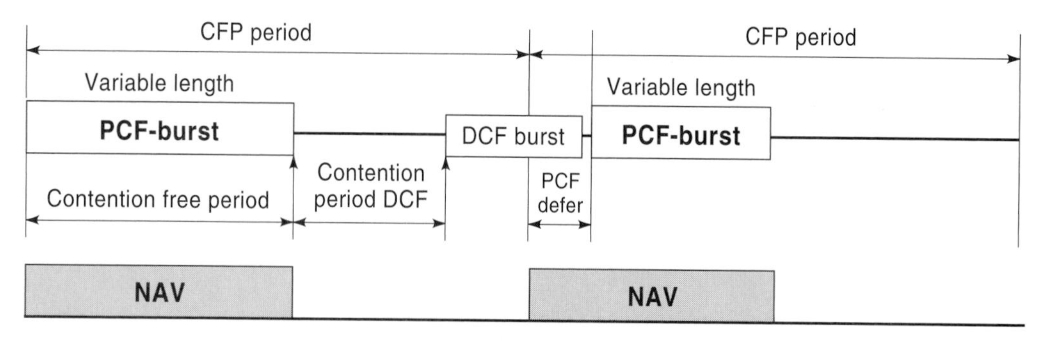

Figure 11.16 Alternation of contention-free and contention periods under PCF control from AP.

infrastructure networks, the AP takes charge of the operation to provide the service to all terminals involved. The AP, playing the point coordinator, stops all other terminals and polls other stations in a semiperiodic manner. Figure 11.16 illustrates the operation of the PCF. The AP organizes a periodical contention-free period (CFP) for the time-bounded information. It coordinates time-bounded data to be transmitted at the beginning of each CFP, and during those periods it arranges an NAV signal for all other terminals. The length of the PCF period is variable, and it only occupies a portion of the CFP. The rest of the CFP is released for contention and DCF packets. If a DCF packet occupies the channel and does not complete before the start of the next CFP, the starting time of the CFP will defer (see Fig. 11.16). However, the NAV signal for all other terminals goes to operation at the beginning of the CFP.

11.4.1 General MAC Frame Format

To discuss the packet format, it is instructional to start with the IEEE 802.3 Ethernet packet format. The original Ethernet standard defines only one frame format shown in Figure 11.17. The control and management in the network are so simple that one frame format accommodates the entire operation of the network. Each packet starts with a preamble, alternating 1 and 0 values that are used for synchronization. An SFD sequence of eight bits having the bit configuration 10101011 indicates the start of the frame. The destination address (DA) and source address (SA) for the MAC are either two bytes or six bytes but are practically always implemented in six bytes. A length-field indicates the number of bytes in the subsequent MAC client data field. The MAC client data field and a pad field contain the data transferred from the source station to bring the frame size up to the minimum length for carrier sensing. The last field is the frame check sequence, which contains a 4-byte CRC for error checking. As we saw earlier in this section, in the IEEE 802.11 the preamble, SFD, and the length of the packet fields were defined

Preamble (7)	Start delimiter (1)	DA (2 or 6)	SA (2 or 6)	Length of data (2)	Data (0–1500)	Pad (0–46)	Checksum (4)

Figure 11.17 Frame format of the IEEE 802.3 Ethernet.

DATA

Field	Bytes
Frame control	2
Duration/ID	2
Address 1	6
Address 2	6
Address 3	6
Sequence control	2
Address 4	6
Frame body	0–2312
CRC	4

Figure 11.18 General MAC frame format of the IEEE 802.11.

in the PLCP of the PHY layer packet format. Therefore the MAC frame format of the 802.11 is equivalent to the remaining five fields of the 802.3 handling addressing, MAC client data, and MAC error correcting bits. The 802.11 frame format is much more complicated than 802.3 because it should also accommodate a number of management and control packets.

Figure 11.18 shows the general MAC frame format of the IEEE 802.11. It starts with the *frame control* field. This field carries instructions on the nature of the packet. It distinguishes data from control and management frame and specifies the type of control or management signaling that the packet is meant to do. The details of frame control and its format are shown in Figure 11.19. The *duration/ID* field is used to identify the length of the fragmented packets to follow. The four *address fields* in the 802.11 frame format, rather than two in the 802.3, identify the source, destination, and APs that they are connected to which are identified by a six-byte (48-bit) Ethernet-like address. The *sequence control* is used for fragmenta-

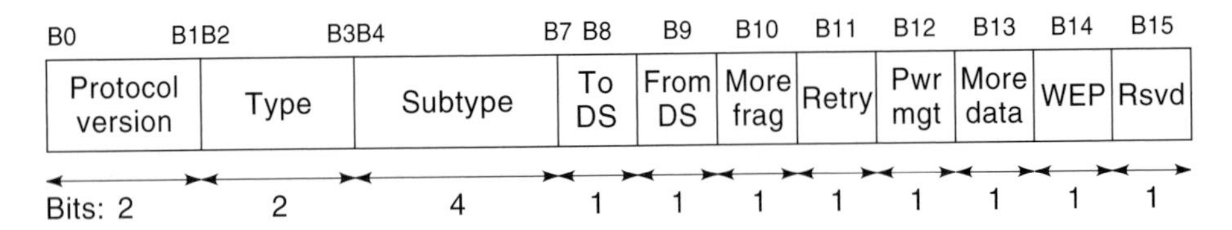

Protocol version: currently 00, other options reserved for future
To DS/from DS; "1" for communication between two APs
More fragmentation: "1" if another section of a fragment follows
Retry: "1" if packet is retransmitted
Power management: "1" if station is in sleep mode
Wave data: "1" more packets to the terminal in power-save mode
Wired equivalent privacy: "1" data bits are encrypted

Figure 11.19 Details of the frame control field in the MAC header of the IEEE 802.11.

tion numbering to control the sequencing. The sequence control and duration/ID are only available in the 802.11 to support fragmentation and the reassembly feature of the MAC protocol. The *frame body* of the 802.11 contains 0–2,312 bytes as compared with 802.3 that have 46–1,500 bytes. Similar to the IEEE 802.3, the IEEE 802.11 uses a four-byte CRC for protection of the MAC client information. Note that we had shorter CRC codes in PLCP that were used to protect the PLCL header.

Example 11.4: Packet Formats

Figure 11.20 shows three examples of the short packets RTS, CTS, and ACK. Not all the fields are included in all packets. But the general format follows the same pattern.

11.4.2 Control Field in MAC Frames

Compared with Ethernet, IEEE 802.11 is a wireless network that needs to have control and management signaling to handle registration process, mobility management, power management, and security. To implement these features, the frame format of the 802.11 should accommodate a number of instructing packets, similar to those we described in WANs. The capability of implementing these instructions is embedded in the control field of the MAC frames. Figure 11.20 shows the overall format of the control field in the 802.11 MAC frame with description of all fields except type and subtype. These two fields are very important because they specify

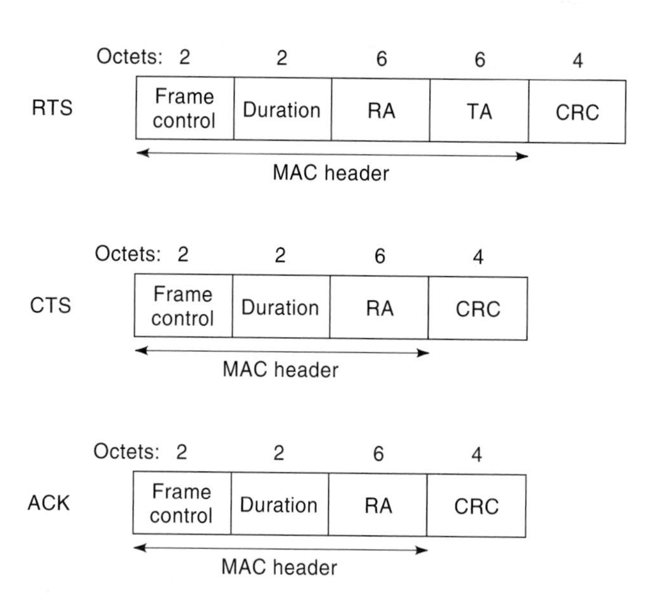

Figure 11.20 Details of the frame control field in the MAC header of the IEEE 802.11.

Table 11.1. Type and Subtype Fields and Their Associated Instructions

- Management Type (00)
  - Association Request/Response (0000/0001)
  - Reassociation Request/Response (0010/0011)
  - Probe-Request/Response (0100/0101)
  - Beacon (1000)
  - ATIM: Announcement Traffic Indication Map (1001)
  - Dissociation (1010)
  - Authentication/Deauthentication (1011/1100)
- Control Type (01)
  - Power Save Poll (1010)
  - RTS/CTS (1011/1100)
  - ACK (1101)
  - CF End/CF End with ACK (1110/1111)
- Data Type (10)
  - Data/Data with CF ACK/No Data (0000/0001)
  - Data Poll with CF/Data Poll with CF and ACK (0010/0011)
  - No Data/CF ACK (0100/0101)
  - CF Poll/CF Poll ACK (0101/0110)

various instructions for using the packet. The 2-bit *type* field specifies four options for the frame type:

- Management Frame (00)
- Control Frame (01)
- Data Frame (10)
- Unspecified (11)

The four-bit *subtype* provides an opportunity to define up to 16 instructions for each type of frame. Table 11.1 shows all used six bits for the type and subtypes in the frame control field. Combinations that are not used provide an opportunity to incorporate new features in the future.

11.5 MAC MANAGEMENT SUBLAYER

The MAC management sublayer handles establishment of communications between stations and APs. This layer handles the mechanisms required for a mobile environment. The same issues that exist in other wireless systems exist here to a large extent. In general, a MAC management frame has the format shown in Figure 11.21. A variety of MAC management frames are used for different purposes.

11.5.1 Registration

The *beacon* is a management frame that is transmitted quasi-periodically by the AP to establish the timing synchronization function (TSF). It contains information such as the BSS-ID, timestamp (for synchronization), traffic indication map (for sleep mode), power management, and roaming. RSS measurements are made on the beacon message. The beacon is used to identify the AP, the network, and so on.

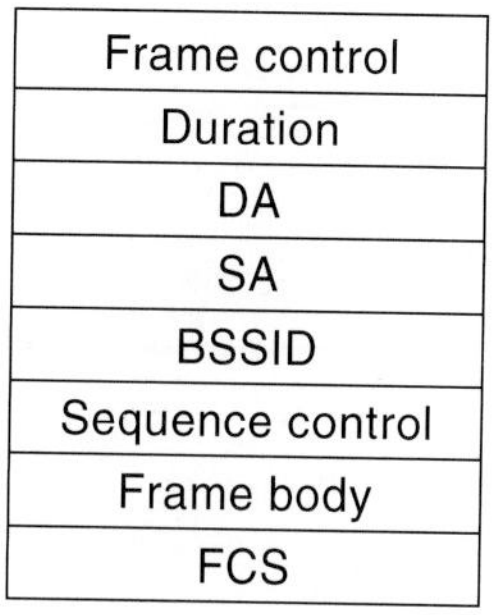

Frame control
Duration
DA
SA
BSSID
Sequence control
Frame body
FCS

Figure 11.21 MAC management frame format.

In order to deliver a frame to an MS, the distribution system must know which AP is serving the MS. *Association* is a procedure by which an MS "registers" with an AP. Only after association can an MS send or receive packets through an AP. How the association information is maintained in the distribution system is *not* specified by the standard. An MS sends the *association request* frame to the AP if it wants to associate with the AP. The AP grants permission to the MS via an *association response* frame. MAC management frames similar to these two frames are made use of in handoff.

11.5.2 Handoff

There are three mobility types in IEEE 802.11. The "no transition" type implies that the MS is static or moving within a BSA. A "BSS transition" indicates that the MS moves from one BSS to another within the same ESS. The most general form of mobility is "ESS transition" when the MS moves from one BSS to another BSS that is part of a new ESS. In this case upper layer connections may break (it will need a mobile IP for continuous connection).

The *reassociation* service is used when an MS moves from one BSS to another within the same ESS. The MS always initiates it, and it enables the distribution system to recognize the fact that the MS has moved its association from one AP to another. The *dissociation* service is used to terminate an association. It may be invoked by either party of an association (the AP or the MS), and it is a notification and not a request. It cannot be refused. MSs leaving a BSS will send a dissociation message to the AP that need not be always received.

The handoff procedures in a WLAN are as shown in Figure 11.22. The AP broadcasts a beacon signal periodically (typically the period is around 100 ms). An MS that powers on scans the beacon signal and *associates* itself with the AP with the strongest beacon. The beacon contains information corresponding to the AP such as a timestamp, beacon interval, capabilities, ESS ID, and traffic indication map (TIM). The MS uses the information in the beacon to distinguish between different APs.

The MS keeps track of the RSS of the beacon of the AP with which it is associated, and when the RSS becomes weak, it starts to scan for stronger beacons from neighboring APs. The scanning process can be either active or passive. In passive scanning, the MS simply listens to available beacons. In active scanning, the MS

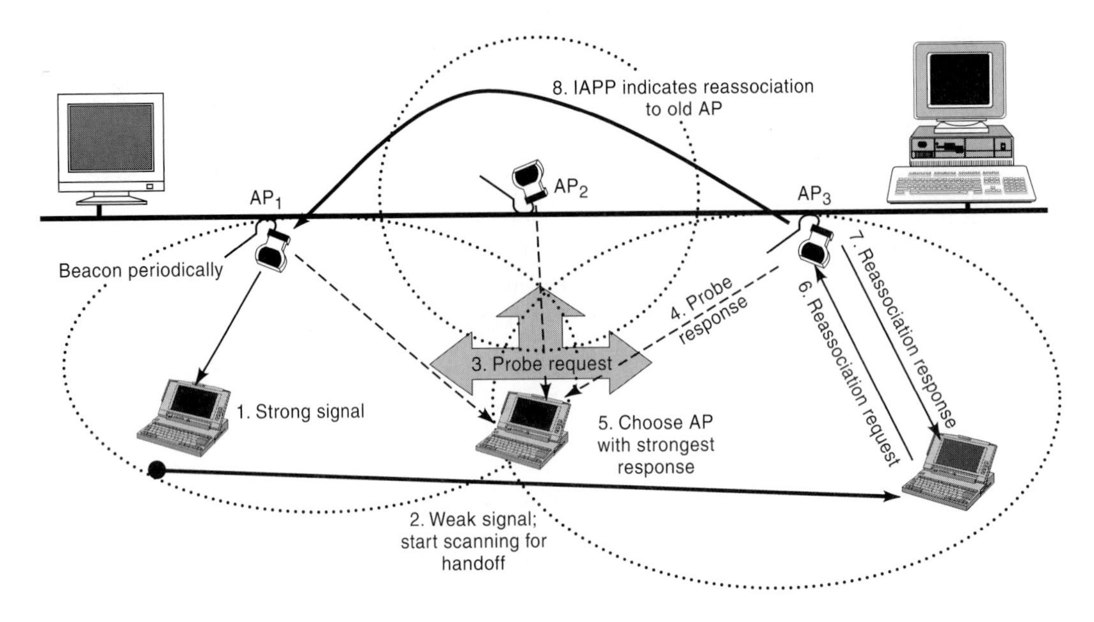

Figure 11.22 Handoff procedure in IEEE 802.11.

sends a probe request to a targeted set of APs that are capable of receiving its probe. Each AP that receives the probe responds with a probe response that contains the same information that is available in a regular beacon with the exception of the TIM. The probe response thus serves as a "solicited beacon." The mobile chooses the AP with the strongest beacon or probe response and sends a *reassociation request* to the new AP. The reassociation requests contains information about the MS as well as the old AP. In response, the new AP sends a *reassociation response* that has information about the supported bit rates, station ID, and so on needed to resume communication. The old AP is not informed by the MS about the change of location. So far, each WLAN vendor had some form of proprietary implementation of the emerging IAPP (inter-access point protocol) standard for completing the last stage of the handoff procedure (intimating the old AP about the mobile host's change of location). The IAPP protocol employs two PDUs to indicate that a handoff has taken place. These PDUs are transferred over the wired network from the new AP to the old AP using UDP-IP (Figure 11.23). If the AP does not have an IP address, an 802.11 subnetwork access protocol (SNAP) is employed for transferring the PDUs. IAPP is also used to announce the existence of APs and to create a database of APs within each AP.

11.5.3 Power Management

The power conservation problem in WLANs is that stations receive data in bursts but remain in an idle receive state constantly, which dominates the LAN adaptor power consumption. The challenge is how we can power off during the idle periods and maintain the session. The IEEE 802.11 solution is to put the MS in sleeping

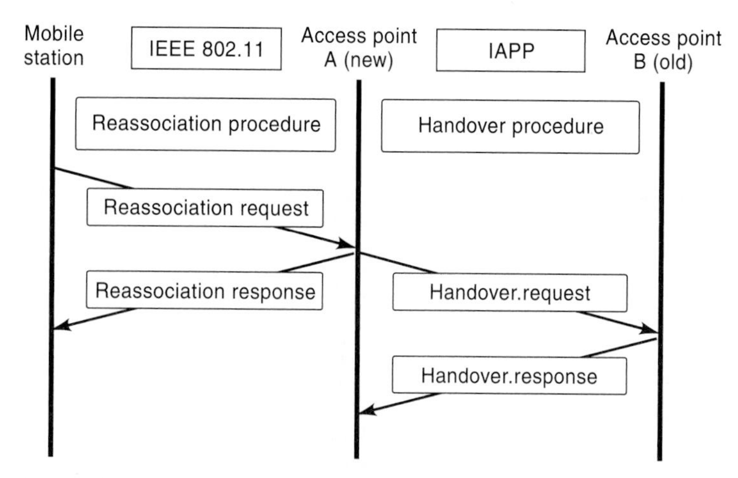

Figure 11.23 Role of IAPP in handoff.

mode, buffer the data at AP, and send the data when the MS is awakened. Compared with the continuous power control in cellular telephones, this is a solution tailored for bursty data applications.

Since using TSF, all MSs are synchronized, and they will wake up at the same time to listen to the beacon (see Fig. 11.24). MS uses the power-management bit in the frame control field to announce its sleep/awake mode. With every beacon a TIM is sent that has the list of stations having buffered data. The MS learns that it has a buffered data by checking beacon and TIM. The MS with buffered data sends a power-save poll to AP. The AP sends the buffered data when the station is in active mode.

11.5.4 Security

There are provisions for authentication and privacy in IEEE 802.11. There are two types of *authentication* schemes in IEEE 802.11. The open system authentication is the default. Here the request frame sends the authentication algorithm ID for "open system." The response frame sends the results of request. The shared key authentication provides a greater amount of security. The request frame sends the authentication frame ID for the "shared key" using a 40-bit secret code that is shared between itself and the AP. The second station sends a challenge text (128 bytes). The first station sends the encrypted challenged text as the response. The

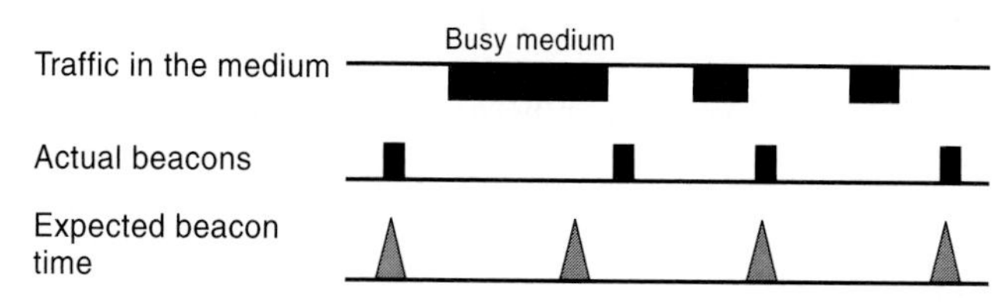

Figure 11.24 Listening to the beacon for power management.

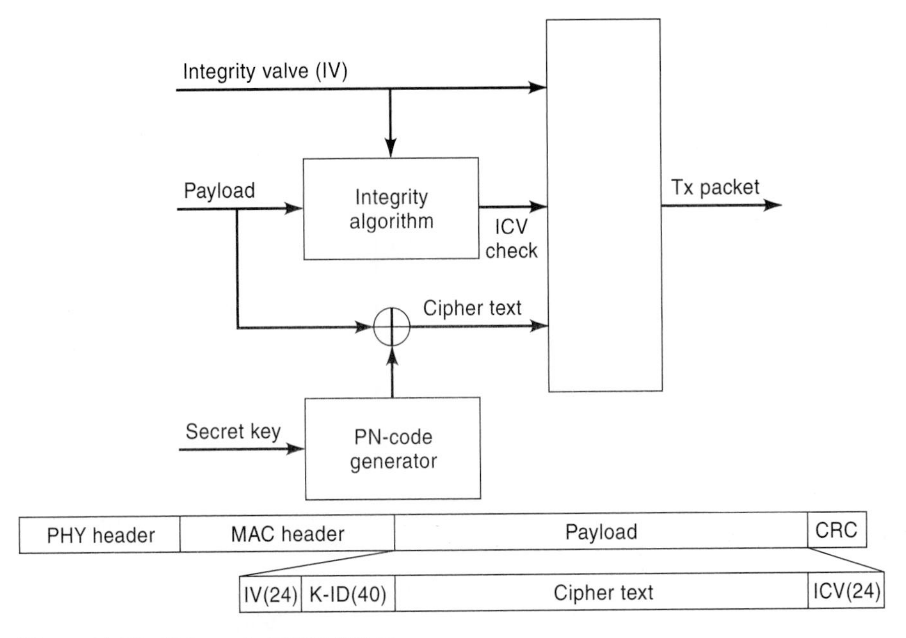

Figure 11.25 Privacy in IEEE 802.11.

second station sends the authentication results. This is exactly like the challenge response identification protocols discussed in Chapter 6. Note that the 40-bit key provides very little security. The secret key algorithm usually employed in all systems is RC-4, although some products employ DES. *Privacy* can also be maintained in IEEE 802.11 via the wired-equivalent privacy (WEP) specification. A pseudorandom generator is used (see Fig. 11.25) along with the 40-bit secret key to create a key sequence that is simply XOR-ed with the plaintext message. This offers very little security and is quite susceptible to planned attacks.

QUESTIONS

11.1 Name four major transmission techniques considered for WLAN standards and give the standard activity associated with each of them.

11.2 Compare OFDM and spread spectrum technology for the WLAN application.

11.3 Give the physical specification summary of the DSSS and FHSS used by the IEEE 802.11.

11.4 What are the MAC services of IEEE 802.11 that are not provided in the traditional LANs such as 802.3?

11.5 Why does the MAC layer of the 802.11 have four address fields, compared with 802.3, which has two?

11.6 What is the PCF in 802.11, what services does it provide, and how is it implemented?

11.7 What is the RTS/CTS in 802.11, what services does it provide, and how is it implemented?

11.8 What is the difference between an ESS and a BSS in the IEEE 802.11?

11.9 Explain why an AP in the 802.11 also acts as a bridge.

11.10 What are the differences between carrier sensing in the 802.11 and 802.3?

11.11 What are the responsibilities of the MAC management sublayer in 802.11?

11.12 What is the difference between the backoff algorithms in the 802.11 and 802.3?

11.13 What is the purpose of PIF, DIF, and SIF time intervals and how are they used in the IEEE 802.11?

11.14 Why do we have four addressing slots in the 802.11 MAC and only two in the 802.3?

11.15 What is the 802.11 MAC frame that has the type/subtype frame control code of 001000? What are its responsibilities?

11.16 What is the difference between a probe and a beacon signal in 802.11?

11.17 Explain how the timing of the beacon signal in 802.11 operates.

11.18 What is the difference between power control in 802.11 and power control in cellular systems?

11.19 What was the mission of the IAPP group?

PROBLEMS

11.1 The original WaveLAN, the basis for the IEEE 802.11, uses an 11-bit Barker code of $[1,-1,1,1,-1,1,1,1,-1,-1,-1]$ for DSSS.

a. Sketch the ACF of the code.

b. If we use the system using random codes with the same chip length in a CDMA environment, how many simultaneous data users can we support with an omnidirectional antenna and one access point?

11.2 **a.** If in the PPM-IR PHY layer used for the IEEE 802.11 instead of PPM we were using baseband Manchester coding, what would be the transmission data rate? Your reasoning must be given.

b. What is the symbol transmission rate in the IEEE 802.11b? How many complex QPSK symbols are used in one coded symbol? How many bits are mapped into one transmitted symbol? What is the redundancy of the coded symbols (the ratio of the coded symbols to total number of choices)?

11.3 **a.** Use the equation for generation of CCK to generate the complex transmitted codes associated with the data sequence $\{0,1,0,0,1,0,1,1\}$.

b. Repeat (a) for the sequence $\{1,1,0,0,1,1,0,0\}$.

c. Show that the two generated codes are orthogonal.

11.4 Redraw the timing diagram of Example 4.18, shown in Figure 4.15, assuming that Terminal B uses the RTS/CTS mechanism to send its packet.

11.5 A voice-over IP application layer software generates a 64 kbps coded voice packet every 50 ms. This software is installed in two laptops with WLAN PCMCIA cards communicating with an AP connected to a Fast Ethernet (100 Mbps).

a. Use Figure 4.37 to give the buffer length at the receivers.

b. What is the length of the voice packets in ms, if the PCMCIA cards were DSSS IEEE 802.11?

c. If the two terminals start to send voice packets almost at the same time, give the timing diagram to show how the first packets are delivered though the wireless medium to the AP using the CSMA/CA mechanism.

d. Repeat (b) and (c) if 802.11b was used instead of DSSS 802.11.

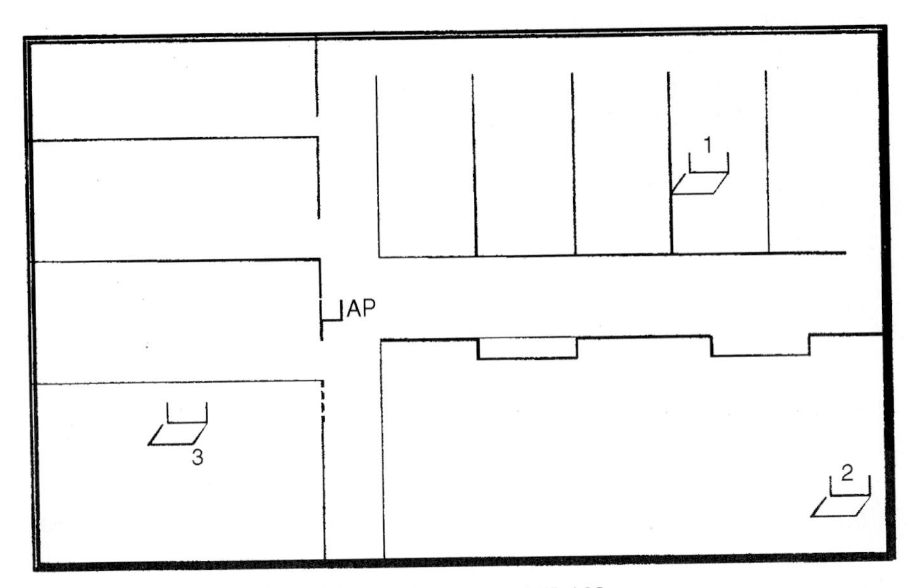

Figure 11.26 Layout of an office building with a WLAN.

11.6 Figure 11.26 shows the layout of an office building. If the distance between the AP and the MSs 1, 2, and 3 are 50, 65, and 25 meters, respectively, determine the path loss for between the AP and MSs:

a. Using a loss of 3dB per wall.

b. Using the JTC model.

c. Using the 802.11 transmitted and received power specifications to determine whether a single AP can cover the entire building.

d. If the minimum SNR requirement for proper operation of the IEEE 802.11 modems is 10 dB, find out which of the three MSs in the previous problem may be hidden with another terminal (use the JTC model for propagation).

CHAPTER 12

WIRELESS ATM AND HIPERLAN

12.1 Introduction

12.2 What Is Wireless ATM?

12.2.1 Reference Model
12.2.2 Protocol Entities
12.2.3 PHY and MAC Layer Alternatives
12.2.4 Mobility Support

12.3 What Is HIPERLAN?

12.3.1 HIPERLAN-1 Requirements and Architecture
12.3.2 HIPERLAN-1 PHY and MAC Layers

12.4 HIPERLAN-2

12.4.1 Architecture and Reference Model
12.4.2 PHY Layer
12.4.3 DLC Layer
12.4.4 Convergence Layer
12.4.5 Security
12.4.6 Overall Comparison with 802.11

Questions

Problems

12.1 INTRODUCTION

In Chapter 10 we provided an overview of the wideband local access technologies and divided them into WLAN and WPAN divisions. We further divided WLAN standards into connectionless, represented by the IEEE 802.11, and connection-based represented by HIPERLAN-2. Connection-based networks are voice-oriented networks that in early 1990s were perceived to turn into the so-called end-to-end ATM solutions. This perception initiated a couple of activities related to the LANs. The first activity was the LAN emulation (LANE) using ATM that intended to integrate legacy LAN applications in an ATM setting [STA00]. The second activity was wireless ATM (WATM) that intended to design a wideband local access that integrated with the ATM backbone to provide an end-to-end ATM solution. Both these activities lost the heat of their publicity by the turn of the last century, after the success of the Internet and IP-based networks to provide a cheaper solution for the backbone connections. ATM never made it to the desktop in any case.

The WATM research activities addressed the important issue of how to implement QoS in a mobile environment. The QoS support has two features, the first providing multirate services on demand with guaranteed bandwidth, delay, and so on, and the second providing implementation of a variable user-charging mechanism according to their usage of the resources. Currently data-oriented Internet services are provided at a flat rate to the fixed users, but the voice-oriented services are charged according to the usage of the network. As was shown in Figure 10.10, the vision of the service providers for the future of the wideband local access is to provide wireless Internet access services that carry different charging mechanisms where, for example, no charge will be associated for home or office connections, but there is a tariff for public usage of the AP owned by the service provider. These public APs could be broadband WLAN APs or a GPRS BS. Therefore, the argument is that we need different charging mechanisms because we are providing different QoS to the user. To incorporate this charging mechanism into the wideband local access, service providers are keen on supporting multi-QoS operation under wideband local access. As a result, the successful EU cellular industry has supported HIPERLAN-2 to provide the transmission rate of the WLANs but incorporate a connection-based charging mechanism that is useful for variable QoS support and in particular the tariff mechanism for that support.

In this chapter first we provide an overview of the WATM activities and then we explain ETSI BRAN's initiatives in wideband local access. Symbolically one can think of the IEEE activities as the representative of successful Internet industry in the United States and HIPERLAN-2 as a representative of the European Union's successful wireless cellular telephone industry. We can then view the previous chapter as being concerned with the standards activities in the United States, and this chapter considers the standardization in the European Union's.

12.2 WHAT IS WIRELESS ATM?

Wireless ATM was first introduced in 1992 [RAY92], and it meant to provide for an integrated broadband application programming interface (API) to the ATM network for a variety of mobile terminals. In 1996, a WATM working group was formed under the ATM forum that drew around a hundred participants in their first meeting in Helsinki. Figure 12.1 illustrates the vision of the end-to-end ATM network that was used by the ATM forum. In the mid-1990s, a number of experimental projects at NEC Laboratories, Nokia, AT&T, Olivetti (now AT&T), and other research labs developed prototypes for implementation of this concept, but in the late 1990s, the heat of the WATM cooled down significantly [PAH97], [RAY99]. In the year 2000, the ATM forum regrouped to pursue this matter by cooperating with other WLAN standards activities, in particular HIPERLAN-2. However, research efforts for implementation of these testbeds discovered a number of interesting issues related to broadband wireless access that has had an impact on the development of new standards in this field. A comprehensive overview of WATM activities is available in [RAY99].

The first fundamental challenge for the implementation of a WATM system is that the ATM was designed for fast switches connecting extremely wideband and reliable fiber transmission channels. The wireless medium, however, is very unreliable and has serious limitations on wideband operations. This imposes problems in the basic transmission mechanism.

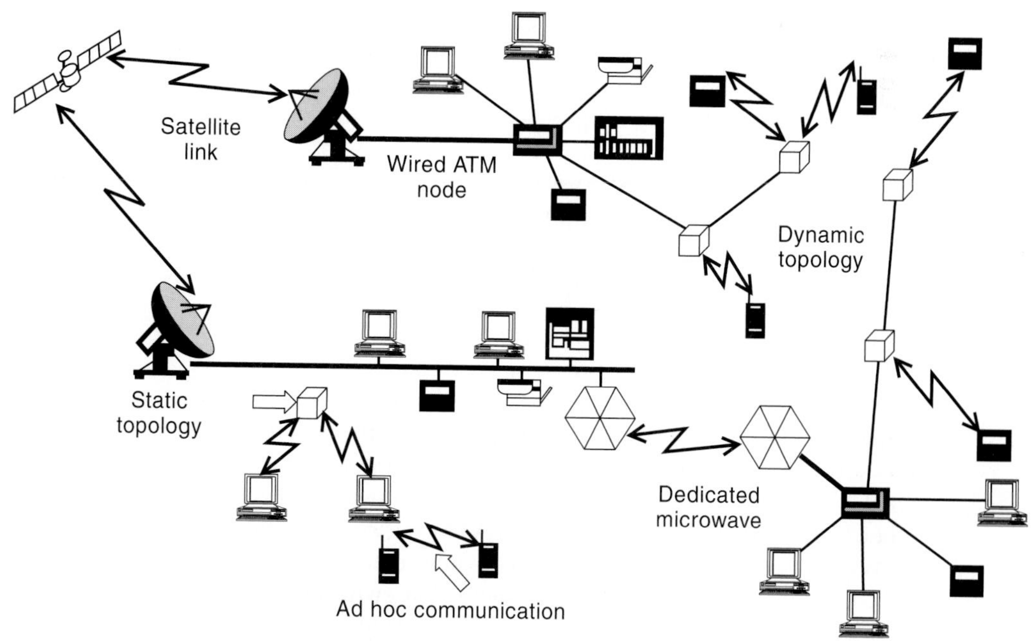

Figure 12.1 Vision of the ATM forum's Wireless ATM working group for an end-to-end ATM network [DEL96].

Wireless header	ATM header (5 bytes)	ATM payload (48 bytes)	Wireless trailer

Figure 12.2 Typical packet frame format for the WATM.

Example 12.1: Format of WATM Packets

Consider the packet format for the WATM system shown in Figure 12.2. ATM packets (cells) have a fixed size of 53 bytes with a 48-byte payload. The benefit of the fixed packet size is that it facilitates fast switching in a multimedia environment. The ATM cells were considered for operation on reliable optical channels that do not need acknowledgment. When we use the same packet format in a wireless environment, for example, FHSS 802.11, we will have another additional 16 bytes for the PLCP header and a few more for a wireless MAC layer that makes the overhead so large that a 48-byte length of payload makes the transmission inefficient. On the other hand, with the unreliable fading environment in wireless channels, we need to add acknowledgment to ensure safe transmission of the packets. If we change the packet format and add acknowledgments, then it is difficult to call this protocol wireless ATM. This would be a wireless method to interact with ATM switches. So the name WATM is not really appropriate. After all we don't call 802.11 a wireless 802.3.

The second fundamental challenge is that the ATM switches are designed to support QoS based on a basic negotiation with the user terminal that is maintained throughout the session. In a wireless environment, because of the fluctuations in channel conditions, a continual support of a negotiated QoS is impossible. Besides, when a terminal roams from one AP to another, it needs to renegotiate its contract, and the new AP may not be able to honor the old contract. Therefore, the basic promise of the ATM that is honoring a negotiated service would need a new definition.

Other challenges facing WATM are to find methods to provide faster and more reliable air-interfaces, to find a method to distribute the additional complexities of the network among the network elements, and to find an efficient way to cope with IP applications developed for connectionless environments.

12.2.1 Reference Model

The basic elements of the traditional ATM networks are ATM switches and ATM terminals. The ATM forum defines, among other things, two protocols, the user–network interface (UNI) protocol connecting user terminals to the switches, and the network–network interface (NNI) protocol connecting two switches together [STA98b]. The elements of a WATM network are shown in Figure 12.3. In the WATM environment, we have two new elements, the WATM-MS and the ATM-BS, and we need to upgrade the ATM switches to support mobility. The WATM environment needs to define a new UNI air-interface "W" for wireless operation and a new UNI-NNI interface protocol, "M" that connects the ATM-BS to the ATM switches. The *WATM-MS* is physically implemented on a radio NIC (net-

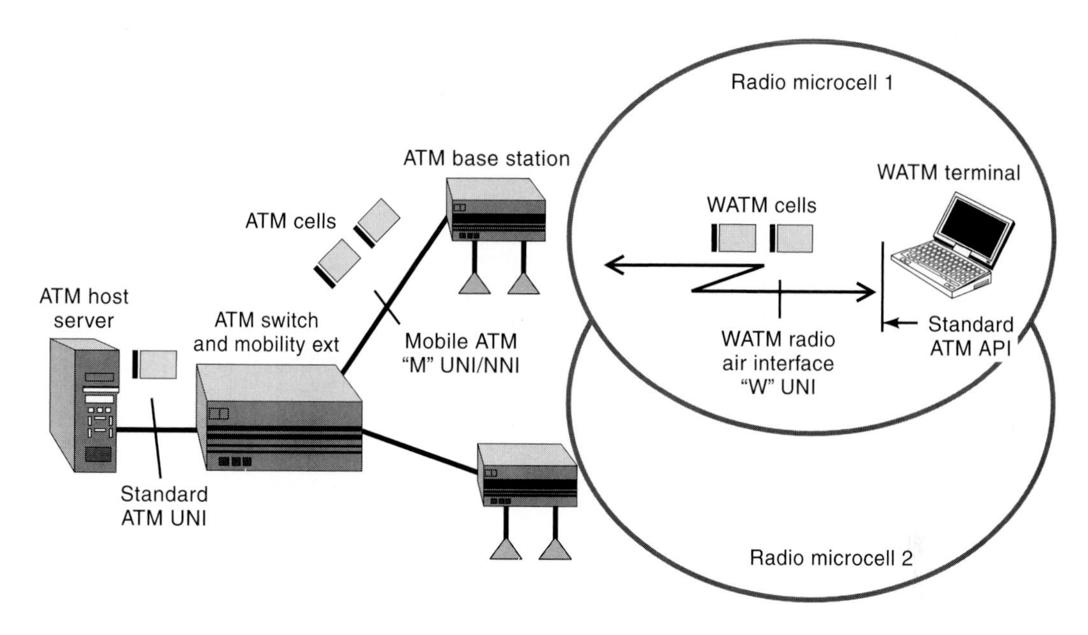

Figure 12.3 Basic architecture of a WATM network [RAY99].

work interface card) hardware and mobility or radio enhanced UNI software for handling call initiation and traffic handling. The *ATM-BS* is a new small ATM switch with ports for wireless and wired connections. This switch also needs enhanced UNI-NNI software which supports mobility in the wireless medium. The *ATM switch* with mobility software is a normal switch with upgraded UNI-NNI software that handles mobility.

12.2.2 Protocol Entities

Figure 12.4 represents a general description of the protocol entities in the ATM that includes the WATM entities. The PHY layer specifies the transmission medium and the signal encoding technique. In the wired part, the PHY layer standard specifies high-speed connection to SONET, TAXI, and UTP-3 [STA00] media for optical and wired copper media. In the wireless part, the standard has not specified anything yet, but the experimental systems use variety of technologies commonly used in WLAN standards and products. The *ATM* network layer defines transmission in fixed-size cells and the use of connections for wired parts. Additional WATM data link control (DLC) and MAC layers are needed to adapt the system to the wireless environment. The ATM adaptation layer (AAL) is a service-dependent layer that maps higher-layer protocol packets (such as AppleTalk, IP, NetWare) into ATM cells. The ATM forum defines five different AAL specifications, AAL-5 being the most popular in LAN applications that is also suitable for WATM operation. The *user plane* provides control (flow, error, etc.) over information transfer. The *control plane* provides for call establishment and control functions. The specified Q.2931 signaling for the ATM needs further modifications

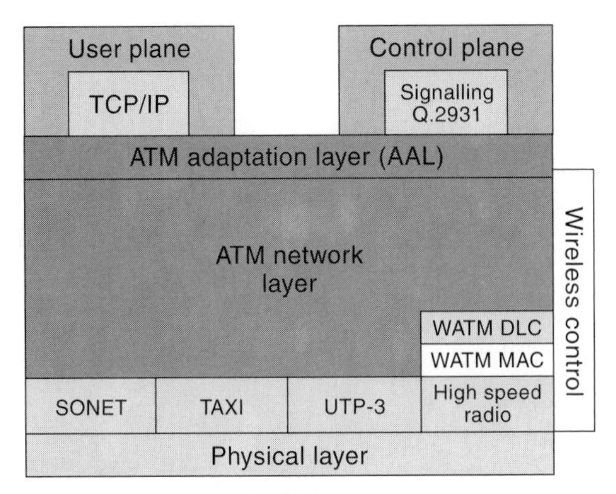

Figure 12.4 Protocol entities in an ATM environment that supports wireless mobility.

to support wireless mobile environments. In addition, a wireless control layer is needed to coordinate all the additional functionalities added to support wireless operation.

Figure 12.5 represents three possibilities for communication between a mobile terminal (MT) and a fixed terminal (FT). Figure 12.5(a) is a normal WLAN-LAN communication. Two applications in the two terminals communicate through the TCP/IP protocol. The IP packets in the wireless terminal use the MAC and PHY layer of IEEE 802.11, and the IP packets at the wired terminal use the MAC and PHY layer of the IEEE 802.3. Protocol conversion takes place at the wireless AP. This situation is the same as normal WLAN operation described earlier in Figure 11.2. Figure 12.5(b) represents a case for communication between a WLAN and an ATM environment. The wireless side is the same as before, but the wired terminal applications run on top of the ATM protocol stack. The application on the ATM terminal could be a native ATM application or IP application. The AP in this case needs LANE software to interface the AAL-5 packets to the 802.11 MAC. Figure 12.5(c) represents the third case when a WATM terminal communicates with an ATM terminal. The PHY and MAC layers of WATM are not yet specified, so the system is an experimental system using a proprietary design.

Figure 12.5 clarifies the importance of two issues. Regardless of the technology for the air-interface, a local network works to run legacy applications over available backbones. If ATM switches move inside offices or homes and native ATM applications become popular, then there is a need for a full WATM service. If LANs, already installed in all offices and penetrating all homes, become the predominant local access mechanisms, then WATM will not find any application. Today, as we discussed earlier, hopes for a WATM type of operation is in the public APs. At home and in offices, the existing legacy LANs appear to be adequate. One solution in this type of environment is to use the operation as in Figure 12.5(a) for the home and office and Figure 12.5(b) for public access.

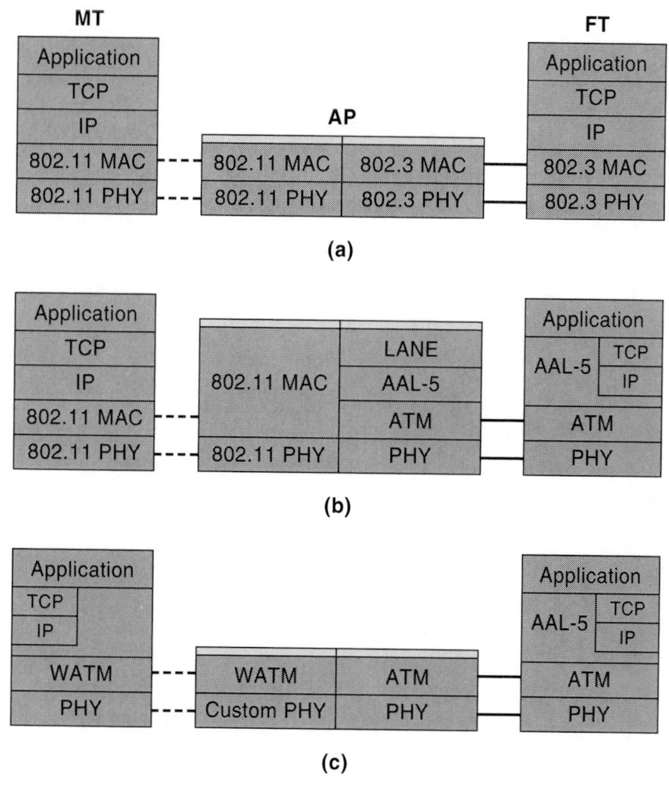

Figure 12.5 Different methods to run applications: (a) WLAN to LAN, (b) WLAN to ATM, and (c) WATM to ATM.

12.2.3 PHY and MAC Layer Alternatives

The ATM forum's WATM working group has not made any specification for a standard. To illustrate the PHY and MAC layer options for WATM, we provide a comparative overview of a few test beds used for implementation of the WATM concept. Table 12.1 summarizes various features of five major projects in this field [PAH97], [AGR96], [AYA96], [ENG95], [RAY97], [WAN96]. Of these, the SWAN and MII/BAHAMA were developed at AT&T/Lucent Research Labs, Olivetti in the United Kingdom, NEC was the leading project (in their laboratory in New Jersey), and the Magic WAND was supported by the ACTS research program in EC that involved a number of participants led by Nokia. These prototypes operated in the ISM and U-NII bands. The data rates ranged from less than 1 Mbps to 24 Mbps. Transmission techniques included FH-SS, traditional QPSK with DFE, and OFDM. The access methods were token passing, reservation-slotted ALOHA, or TDMA/TDD, all providing a controlled environment for the support of the QoS. Today, to support higher data rates of up to 54 Mbps, HIPERLAN-2 and IEEE 802.11a standards use OFDM modulation at 5 GHz. Prior to these standards, the Magic WAND and MII/BAHAMA had adopted the same solution to provide data

Table 12.1 Overview of Several WATM Experimental Projects

WATM System	SWAN	MI/BAHAMA	Olivetti	NEC	WAND
Frequency bands	2.4 GHz ISM bands	900 MHz (Proposed 5 GHz :U-NII)	2.4 GHz ISM bands	2.4 GHz ISM bands	5.2 GHz
Data rate	625 kbps	2–20 Mbps	10 Mbps	8 Mbps	24 Mbps
Modulation scheme	Frequency hopping	(suggested OFDM or GMSK with LMS/RLS)	QPSK	$\pi/4$-QPSK with DFE	16-channel OFDM
Medium access	Each mobile has a fixed channel; token passing	Distributed queue reservation updated multiple access (DQRUMA)	Reservation with Slotted Aloha and piggy-backing on data cells	TDMA/TDD with Slotted Aloha	Reservation with Slotted Aloha
Handoffs	Mobile initiated	Mobile initiated	Mobile initiated (with Mobile Manager) Infrastructure initiated (with Mobile Representative)	Mobile initiated	Mobile initiated
Techniques for reliability	FEC with (8, 4) linear codes	FEC (proposed Reed-Solomon codes for real-time traffic and FEC/retransmissions for data	16 bit CRC and ARQ	Data link control for error recovery	FEC
QoS	MAC supports QoS	Supported	Priority for certain traffic	Fixed slots available for QoS support	Worst case QoS estimate to be used

rates around 20 Mbps. The access method of HIPERLAN-2 is similar to TDMA/TDD which was experimented in NEC's WATM prototype earlier. All these testbeds have tried different approaches to ensure certain levels of QoS. These efforts have helped the understanding of the complexity of QoS in a wireless mobile environment, and have discovered partial solutions for this problem. These studies have laid a ground for the HIPERLAN-2 standard to work on implementation of QoS in a WLAN standard.

12.2.4 Mobility Support

As we saw in previous examples of connection-oriented and connectionless networks, a wireless mobile operation requires a number of functionalities. The MS needs to support location management to identify where the MS is. It needs a registration process and a handoff procedure to register the terminal to an AP each time it is turned on and manage the switching of connections to other APs. It needs authentication and ciphering to provide security and power management to save in the life of the battery. We have provided detailed examples of these functionalities in connection-oriented (GSM and CDMA systems) as well as connectionless networks (WLAN and mobile data services). The overall structure of these functionalities for WATM is very similar to the others. However, handling the connection in an environment where there is a contract for QoS is challenging. In the case of ATM, ATM cells must be received in sequence, and they all follow the same route (virtual circuit—VC). When an MS moves from one AP to another, the existing VC is broken. The VC should either be extended or reconstructed to satisfy the negotiated QoS. This problem is quite challenging.

12.3 WHAT IS HIPERLAN?

The HIPERLAN stands for HIgh PErformance Radio LAN and was initiated by the RES-10 group of the ETSI as a pan-European standard for high-speed wireless local networks. The so-called HIPERLAN-1, the first defined technology by this standard group, started in 1992 and completed in 1997. Unlike IEEE 802.11, which was based on products, HIPERLAN-1 was based on certain functional requirements specified by ETSI. In 1993, CEPT released spectrum at 5 and 17 GHz for the implementation of the HIPERLAN. The HIPERLAN 5.15–5.35 GHz band for unlicensed operation was the first band that was used by a WLAN standard at 5 GHz. These bands being assigned for HIPERLAN in the European Union was one of the motives for the FCC to release the U-NII bands in 1996, which stimulated a new wave of developments in the WLAN industry. During the standardization process, a couple of HIPERLAN-1 prototypes were developed; however, no manufacturer adopted this standard for product development. For that reason, those involved in the EU standardization process consider this effort an unsuccessful attempt. Later on HIPERLAN standardization moved under the ETSI BRAN project with a new and more structured organization. Figure 12.6 [WIL96] shows the overall format of the HIPERLAN activities after completion of the HIPERLAN-1. In addition to

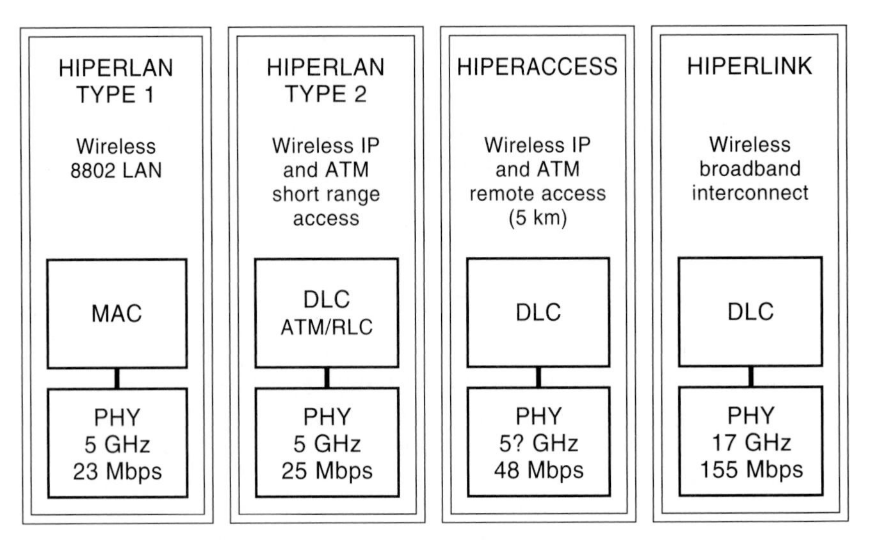

Figure 12.6 Divisions of the HIPERLAN activities.

HIPERLAN-1, we have HIPERLAN-2, which aims at higher data rates and intends to accommodate ATM as well as IP type access. This standardization process is under development. They have coordinated with the IEEE 802.11a in the PHY layer specification and current work on the MAC to support QoS is under progress. Other versions of HIPERLAN are HIPER-ACCESS for remote access and HIPER-LINK to interconnect switches in the backbone. In the United States, these activities are under IEEE 802.16 for LMDS. Only HIPERLAN-1 and -2 are considered WLANs and will be discussed in this chapter. Most of the emphasis is on HIPERLAN-2 which has attracted significant support from cellular manufacturers such as Nokia and Ericsson.

12.3.1 HIPERLAN-1 Requirements and Architecture

The original "functional requirements" for the HIPERLAN-1 were defined by ETSI. These requirements were

- Data rates of 23.529 Mbps
- Coverage of up to 100 m
- Multi-hop ad hoc networking capability
- Support of time-bounded services
- Support of power saving

The frequency of operation was 5.2 GHz unlicensed bands that were released by CEPT in 1993, several years before release of the U-NII bands. The difference between this standard and the IEEE 802.11 was perceived to be the data rate, which was an order of magnitude higher than the original 802.11 and emphasis on ad hoc networking and time-bounded services.

Figure 12.7 shows the overall architecture of an ad hoc network. In HIPERLAN-1's ad hoc network architecture, a multihub topology is considered that also allows overlay of two WLANs. As shown in this figure, the multihop routing extends the HIPERLAN communication beyond the radio range of a single node. Each HIPERLAN node is either a forwarder, designated by "F," or a nonforwarder. A nonforwarder node simply accepts the packet that is intended for it. A forwarder node retransmits the received packet, if the packet does not have its own node address, to other terminals in its neighborhood. Each nonforwarder node should select at least one of its neighbors as a forwarder. Inter-HIPERLAN forwarding needs bilateral cooperation and agreement between two HIPERLANs. To support routing and maintain the operation of a HIPERLAN, the forwarder and nonforwarder nodes need to periodically update several databases. In Figure 12.7, solid lines represent peer-to-peer communications between two terminals and dashed lines represent the connections for forwarding. Three of the terminals, 1, 4, and 6, are designated by letter "F" indicating that they have forwarding connections. There are two overlapping HIPERLANs, A and B, and terminal 4 is a member of both WLANs which can also act as a bridge between the two. This architecture does not have an infrastructure, and it has a large coverage through the multihop operation.

As we mentioned earlier, HIPERLAN-1 did not generate any product development, but it had some pioneering impact on other standards. The use of 5 GHz

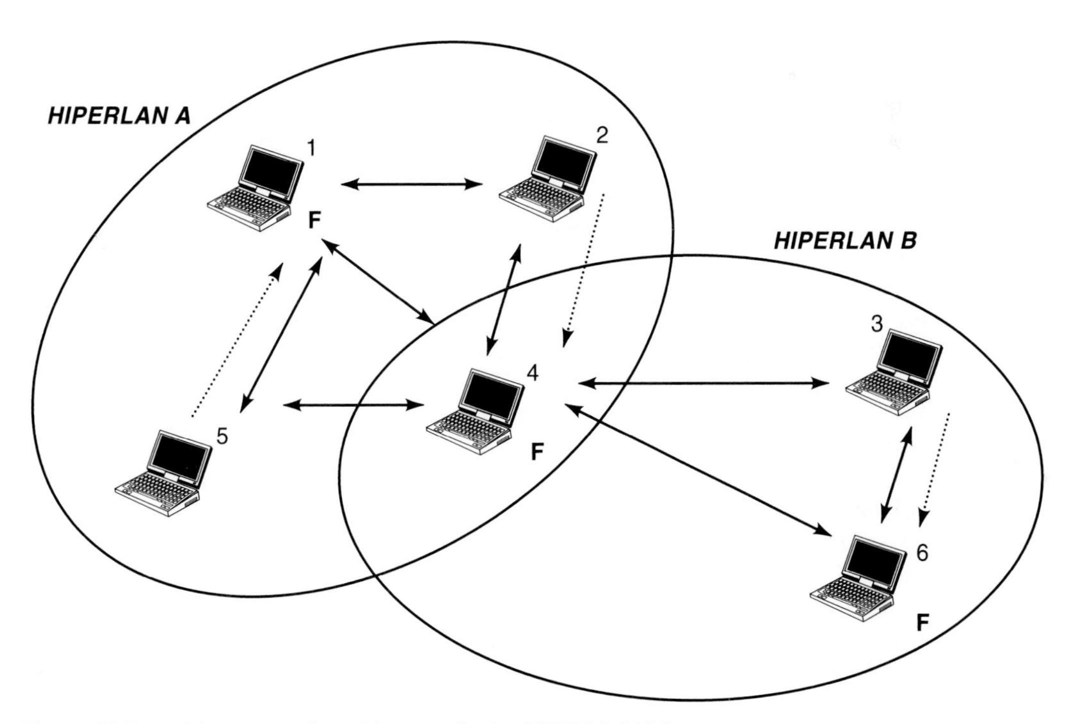

Figure 12.7 Ad hoc network architecture in the HIPERLAN-1.

unlicensed bands, first considered in HIPERLAN-1, is used by IEEE 802.11a and HIPERLAN-2. The multihop feature of the HIPERLAN-1 is considered in the HIPERLAN-2 to be used in an environment with a connection to wired infrastructure.

12.3.2 HIPERLAN-1 PHY and MAC Layers

The PHY layer of the HIPERLAN-1 uses 200 MHz at 5.15–5.35 GHz, which is divided in 5 channels (40 MHz spacing) in the European Union and 6 channels (33 MHz spacing) in the United States. In the United States, there are 3 more channels at 5.725–5.825 GHz bands. The transmission power can go as high as 1 W (30 dBm), and modulation is the single carrier GMSK that can support up to 23 Mbps. To support such high data rates receivers would include a DFE. As we discussed in Chapter 3, DFE consumes considerable electronic power. Using GMSK with the DFE is also challenging for the implementation of fallback data rates. The multi-symbol QAM modulation techniques embedded in the OFDM systems allow simple implementation of fallback data rates. In QAM systems fallback is implemented by simple reduction of the number of transmitted symbols per symbol interval while the symbol interval is kept constant. The PHY layer of the HIPERLAN-1 codes each 416 bits into 496 coded bits with a maximum of 47 codewords per packet and 450 bits per packet for training the equalizer.

The nonpreemptive multiple access (NPMA) protocol used in HIPERLAN is a listen before talk protocol, similar to CSMA/CA used in 802.11, which supports both asynchronous and isochronous (voice-oriented) transmissions. Carrier sensing in HIPERLAN-1 is active, rather than passive as in 802.11, and contention resolution and ACKing is mandatory. The HIPERLAN MAC defines a priority scheme and a lifetime for each packet, which facilitates the control of QoS. In addition to the routing, the MAC layer also handles the encryption and power conservation. The MAC address of the HIPERLAN-1 uses six bytes to support IEEE 802.2 LLC and to be compatible with other 802 standards. Each packet has six address fields that identify source, destination, and immediate neighbor (for multihop implementation) transmitters and receivers. IEEE 802.11 had four address fields because it does not support the multihop operation.

Figure 12.8 shows the basic principles of the HIPERLAN-1 MAC protocol. If a terminal senses the medium to be free for at least 1,700 bit durations, it immediately transmits. If the channel is busy, the terminal access has three phases when the channel becomes available. These phases, shown in Figure 12.8, are prioritization phase, contention phase, and transmission phase. During the *prioritization* phase, competing terminals with the highest priority, among the five available priority levels, will survive, and the rest will wait for the next time that the channel is available. The combing algorithm that was described in Chapter 4 is used in five slots, each 256 bits long, to implement this phase. At the end of the prioritization period, all the terminals listen to the asserted highest priority to make sure that all terminals have understood the asserted priority level. This way MSs with the highest priority survive and contend for the next phase, and others are eliminated from the contention. This prioritization mechanism is a counterpart of the three priority level mechanism that was implemented in the 802.11 using SIFS, PIFS, and DIFS interframing intervals. The combing algorithm is more structured and active which

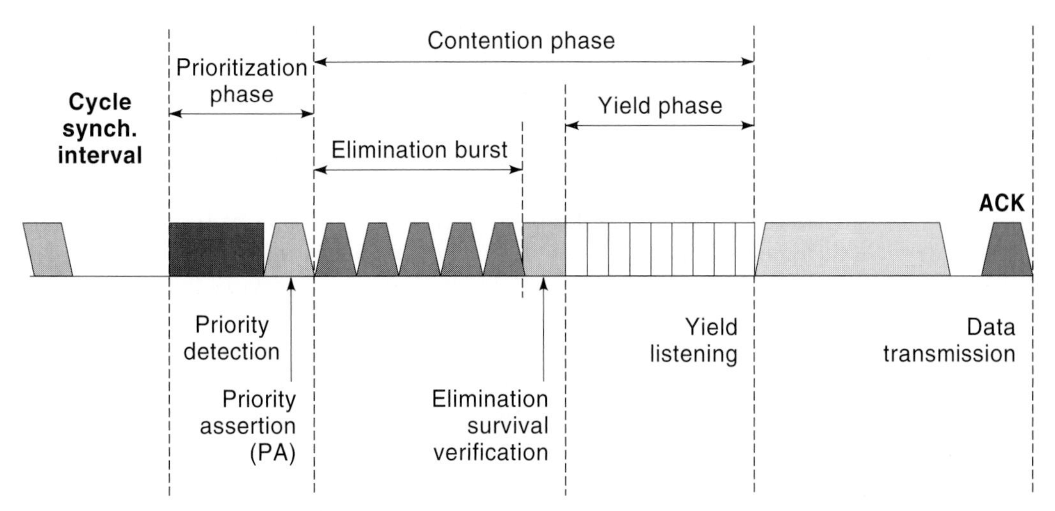

Figure 12.8 Channel access cycle in the HIPERLAN-1.

will provide for a more robust prioritization process. However, the reader should note that prioritization has not shown to be an important issue in the implementation of the WLAN products. All the existing 802.11 products don't implement PCF or any sort of prioritization.

The *contention* phase of the HIPERLAN-1 has two periods, elimination and yield. During the *elimination* period each terminal runs a random number generator to select one of the 12 available slots in which it sends a continuous burst of 256 bits. After sending a burst, an MS listens to the channel for 256-bit durations. If it does not hear any other burst after its transmission, it will send another burst after the twelfth slots in the elimination survival verification interval to ensure everyone that there are survivors. The terminals that hear a burst in this period eliminate themselves. The remainder of the terminals go to the so-called yield part of the contention interval. In the *yield* period, the remaining MSs have a random yield period that is similar to the 802.11 waiting counters. Each MS will "listen" to the channel for the duration of its yield period which is determined from an exponentially distributed random variable, rather than a uniformly distributed random variable used in 802.11. The exponential distribution reduces the average waiting time for running the counter. If an MS senses the channel to be idle for the entire yield period, it has survived, and it will start transmitting data that automatically eliminates other MSs that are listening to the channel. Here the contention process is more complicated and has active as well as passive parts while contention in the 802.11 was entirely passive.

12.4 HIPERLAN-2

Today HIPERLAN-1 is not considered a successful standard by the European Union, but the HIPERLAN-2 project is very popular in and out of the European Union. The HIPERLAN-2 standardization process coordinated with IEEE 802.11 in defining the transmission technique and is working on new higher layers that facilitate the integration of WLANs into the next-generation cellular systems. Integration of WLAN into the cellular systems requires two features: (1) support for vertical roaming between local area and wide areas as well as between corporate and public environments, and (2) support for QoS control for integration into multiservice voice-oriented backbone PSTN which includes ATM switches and other facilities.

While it is still under development, HIPERLAN-2 aims at IP and ATM type services at high data rates for indoor and possibly outdoor applications. It expects to support both connectionless and connection-oriented services which will make its MAC layer far more complicated than 802.11 and HIPERLAN-1 that supports only connectionless services. Connection-based services facilitate integration into the voice-oriented networks. The HIPERLAN-2 that started as a WATM type activity now aims at connecting to IP-based as well as UMTS and ATM networks. In HIPERLAN-2 the ad hoc architecture of HIPERLAN-1 is expanded to support centralized access by using APs in a manner similar to IEEE 802.11. The TDMA/TDD-based MAC layer is similar to the PCS voice-oriented access methods that were previously used in DECT, and this provides a comfortable environment for traditional methods for support of QoS. This feature is carried from the WATM activities that we discussed earlier in this chapter. The OFDM modem operating at 5 GHz is the same as 802.11a. Support of data rates of up to 54 Mbps with this PHY layer opens an environment for innovative wireless video applications that is very crucial for development of integrated home networks. In the next few sections, we provide the details of the HIPERLAN-2 standard that are finalized by the time of this writing. The HIPERLAN-2 standard activities group has four subgroups in interoperability, regulatory, application, and marketing. For more detailed and up-to-date information, the reader can refer to [HIPweb] or [JON99].

12.4.1 Architecture and Reference Model

The overall architecture of the HIPERLAN-2 is shown in Figure 12.9. Like IEEE 802.11, HIPERLAN-2 supports centralized and ad hoc topologies. In the centralized topology of HIPERLAN, shown in Figure 12.9(a), connection between the MS and the AP is similar to that in 802.11, but communication between the APs are different. The IEEE 802.11 with IAPP protocol allows two AP connected to an IP-based subnet to communicate with one another. HIPERLAN-2 allows both handover in a subnet and IP-based handover in a nonhomogeneous network. This generic architecture allows seamless interoperation with Ethernet, point-to-point protocol connection (e.g., over dial-up modem connections), UMTS cellular networks, IEEE 1394 (e.g., Firewire, i.LINK) for entertainment systems, and ATM-based networks. These features allow manufacturers to support vertical roaming capability over a number of networks. The ad hoc networking in the HIPERLAN-2 is expected to support multihop topology that provides for a better coverage.

Features considered for HIPERLAN-2 are far more complex and detailed than the features of the data-oriented IEEE 802.11. As a result, HIPERLAN-2 uses a new protocol stack architecture that is similar to the voice-oriented cellular networks. Figure 12.10 illustrates the simplified protocol stack of the HIPERLAN-2.

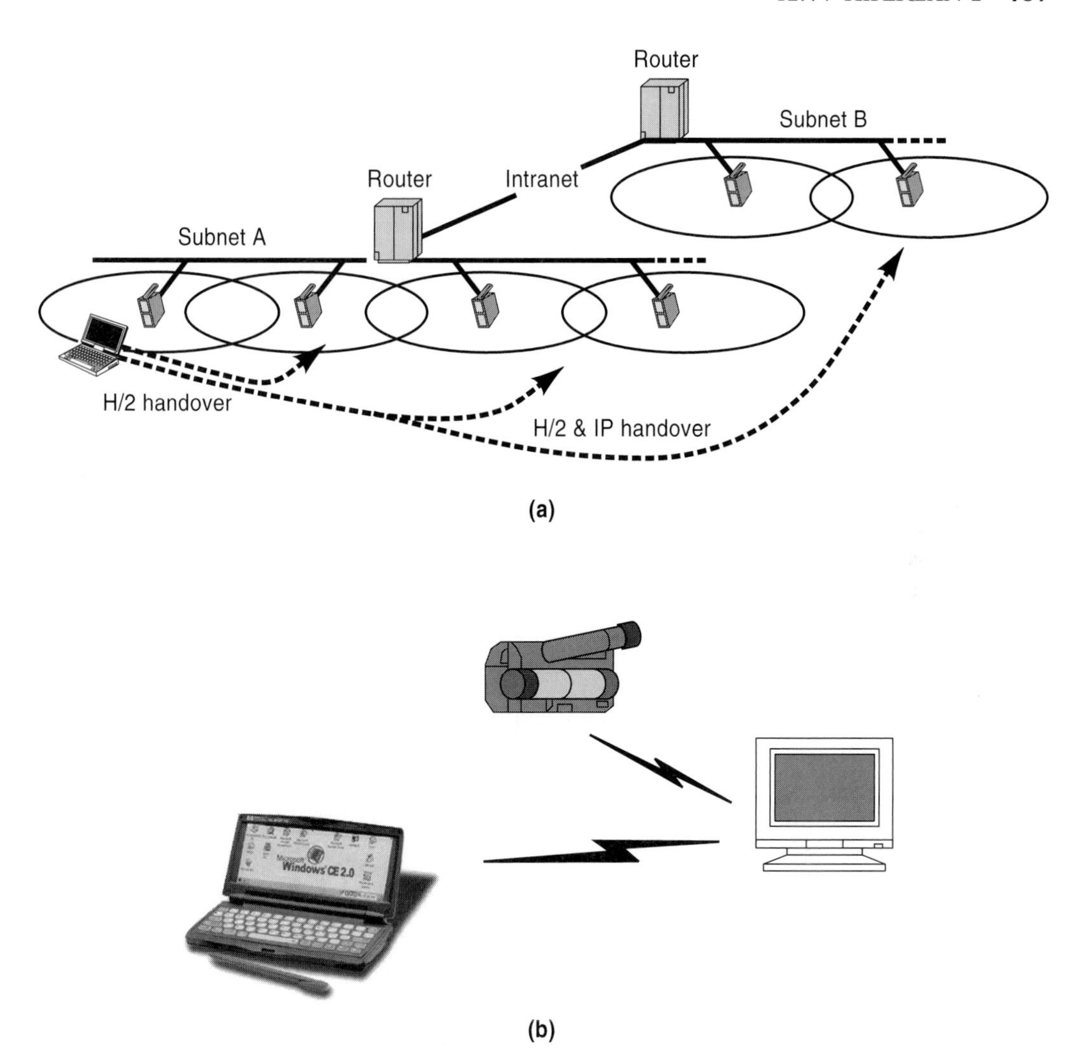

Figure 12.9 Topologies in the HIPERLAN-2: (a) infrastructure and (b) ad hoc.

Basically there are three layers: PHY, DLC, and convergence. Multiple convergence layers, operating one at a time, map a number of higher layer protocol (PPP/IP, ATM, UMTS, Firewire, Ethernet) packets to DLC. The DLC layer provides for the logical link between an AP and the MTs and includes functions for both medium access and communication management for connection handling. The DLC provides for a logical structure to map the convergence layer packets carrying a number of different application protocols onto a single PHY layer.

12.4.2 PHY Layer

The PHY layer of the HIPERLAN-2 uses OFDM modulation that was described in Chapter 3. The specific details of the 802.11a/HIPERLAN-2 transmission system were illustrated in Example 3.12. The PHY layer of the HIPERLAN-2 standard

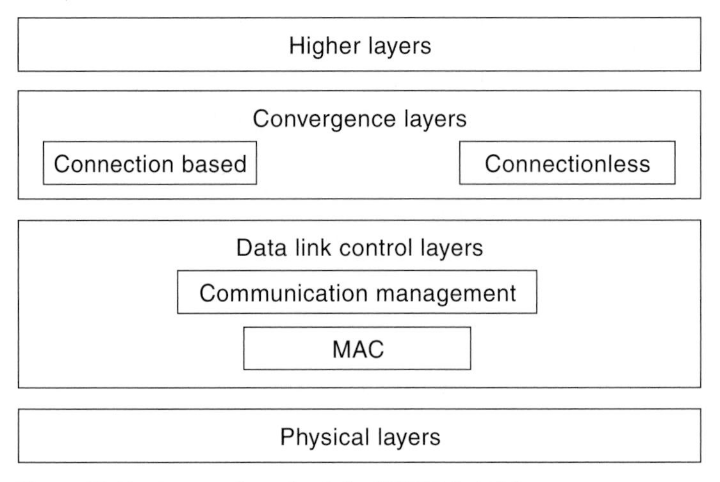

Figure 12.10 Protocol stack of the HIPERLAN-2.

adds the preamble of the DSSS IEEE 802.11, described in the previous chapter, to its own DLC packets. Then by defining a number of logical channels, similar to GSM or TDMA systems, transmits the packet as OFDM modulated bursts.

Figure 12.11 shows the detailed block diagram of the modem. Like IEEE 802.11 FH-SS standard, the received data in HIPERLAN-2 is first scrambled for the whitening process. It is important to remind the reader that IEEE 802.11 DSSS does not go through the whitening process because the DS-SS process whitens the transmitted symbols when it turns the bits to chips. Like IEEE 802.11a, the scrambled data in the HIPERLAN-2 is then passed through a convolutional coder that uses one of the rates—$1/2$, $9/16$, or $3/4$ that are used for different modulation techniques. The coded data is then interleaved to improve the reliability over temporal fading. The interleaved data is then modulated using BPSK, QPSK, 16-QAM, or 64-QAM,

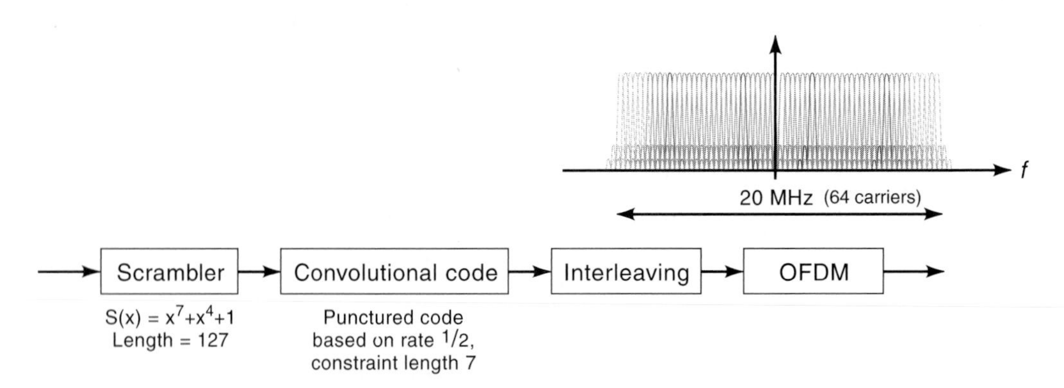

Figure 12.11 Block diagram of the OFDM modem.

to support a variety of data rates. Figure 12.12 shows all the data rates and corresponding modulation and coding schemes that are adopted by the IEEE 802.11a and HIPERLAN-2 standards. As we explained in Example 3.8, there are 64 subcarriers in the OFDM modem of the IEEE 802.11/HIPERLAN-2 of which 48 are used for user data. To support multiple user data rates, modulation and coding in the subcarriers are changed, but the symbol transmission rate is kept at 250 kSps. By keeping the same symbol transmission rate for all data rates, the sampling rate of the signal and other signal processing filters at the receiver remain the same for all rates, but the coding of the bits and number of bits per symbol are changed digitally.

Example 12.2: Data Rates in HIPERLAN-2

For the 6 Mbps user data rate, each carrier carries 6 Mbps/48 = 125 kbps of data using rate ½ convolutional encoder. The rate ½ convolutional encoder requires a 250 kbps transmission rate to support 125 kbps user data. The 250 kSps user data is modulated over a BPSK modem that transmits one symbol per each coded bit. Therefore, the symbol or pulse transmission rate of the system is 250 kSps.

Example 12.3: QAM and Rate ¾ Convolutional Coding

When we use 64-QAM modulation (six bits per symbol) with a rate ¾ convolutional coder the effective data rate will be 250 kbps/carrier × 4/3 × 6 bits/symbols × 48 carriers = 54 Mbps.

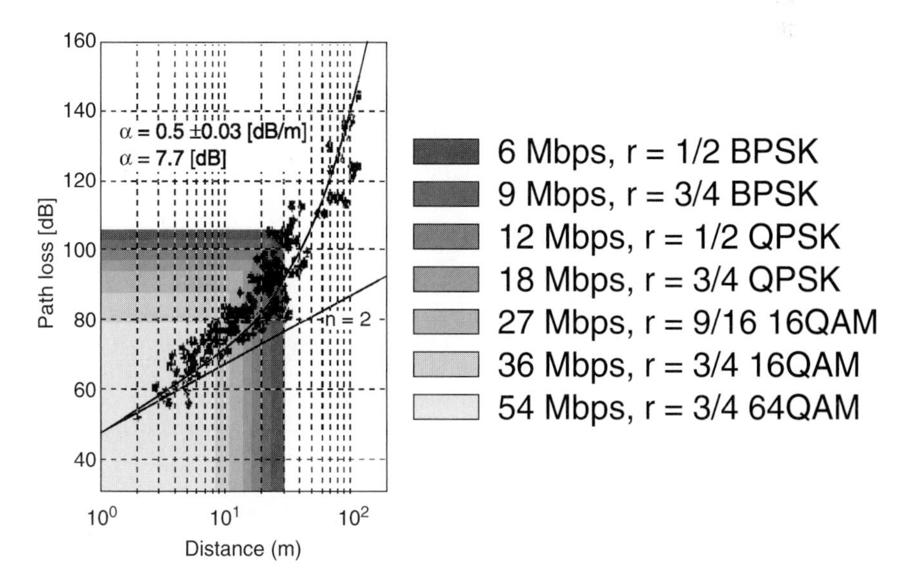

Figure 12.12 Data rates and corresponding tone modulation.

The symbol transmission rate of 250 kSps has a symbol transmission duration of 1/250 kSps = 4,000 ns. Therefore, each symbol is sent like a pulse of duration 4,000 ns with 800 ns time guard between two symbols. This time-gating process improves the resistance to multipath delay spread by preventing the intersymbol interference. The standard also allows an optional 400 ns guard time for shorter distances where delay spread is smaller. Providing for a multirate transmission is a key feature of the 802.11a/HIPERLAN-2 transmissions. Multirate transmission provides for adaptation to the radio link quality and support of different DLC requests for transportation rates.

12.4.3 DLC Layer

The DLC layer provides for a logical link between an AP and the MTs over the OFDM PHY layer. Figure 12.13 illustrates the details of the DLC layer in the HIPERLAN-2. The MAC protocol and frame format for logical and transport channels are the major elements of the DLC layer. Using MAC protocol multiple users share the medium for information transmission and control signaling using transport channels. Using logical channels, similar to those used in voice-oriented networks, HIPERLAN-2 implements four protocols for proper operation of the network. These protocols are radio link control (RLC) protocol, DLC connection control (DCC), radio resource control (RRC), and association control function (ACF). DLC also supports the error control (EC) mechanism over logical channels to improve the reliability of the link.

The MAC layer protocol is dynamic TDMA/TDD, which was described in Chapter 4 under the voice-oriented fixed assignment access method, and it is similar to the access method used in DECT. This protocol supports AP to MT unicast and multicast communication, as well as MT-MT peer-to-peer transmissions. The

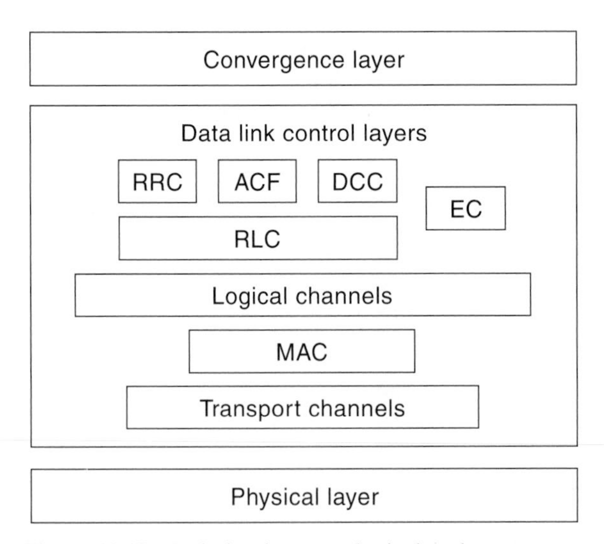

Figure 12.13 Relation between logical and transport channels in HIPERLAN-2.

centralized AP scheduling is expected to provide for dynamic resource distribution, QoS support, and collision-free transmission. Compared with IEEE 802.11, the first two are available only in HIPERLAN-2, and the third one is similar to the PCF in 802.11. Random access for reservation has a specific channel, similar to GSM, and uses slotted ALOHA with exponential backoff and ACKs.

Figure 12.14 represents the overall format of the MAC protocol. Communication between the AP and the MT is based on 2-ms MAC frames. Each frame is divided into broadcast control (BCH), frame control (FCH), access control (ACH), down link data, uplink data, and random access (RCH) time slots. The uplink and downlink are also divided into short and long channels (SCH/LCH) that are used for data transportation of lengths 9 and 54 bytes. The BCH contains broadcast control information for all the MTs. It provides for general information such as the network and AP identifiers, transmission power levels, and FCH and RCH length and wake-up indicator. The FCH contains details of distribution of resources among the fields of each packet. The ACH conveys information on previous access attempts made in the RCH. The RCH is commonly shared among all MTs for random access and contention. If collisions occur the results from RCH access are reported back to the MTs in ACH. Except for the RCH, all other slots are dedicated to specific users. Except for BCH, the duration of the other slots is dynamically adapted to the current traffic situation. BCH, FCH, and ACH are down link channels, RCH is an uplink channel, and SCH/LCH are used in both directions. The HIPERLAN-2 standard refers to all channels shown in Figure 12.14 as *transport channels*.

Like other voice-oriented networks, HIPERLAN-2 defines a set of logical channels for signaling, control, and information transfer. Logical channels in the

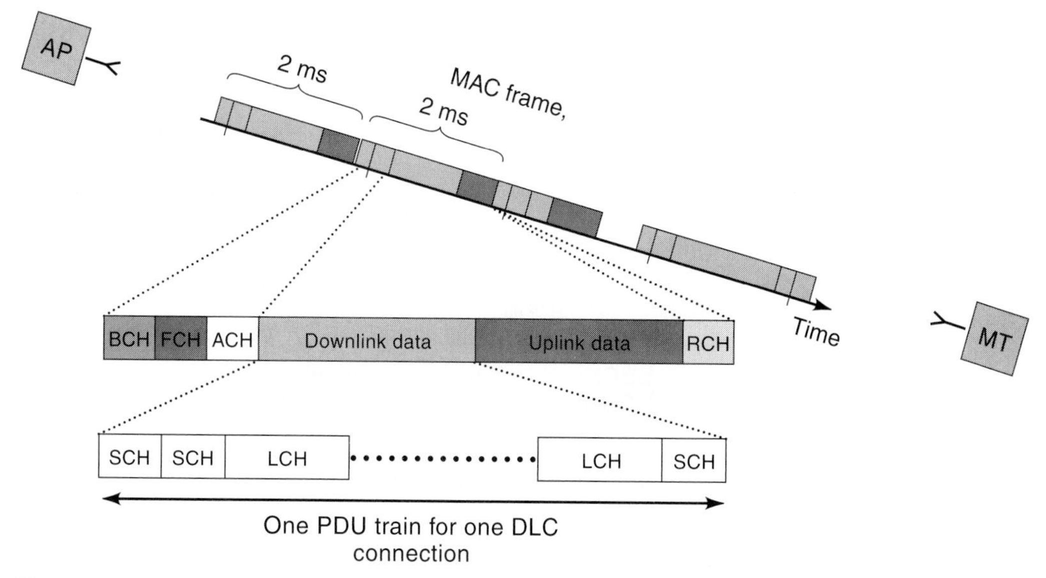

Figure 12.14 The TDMA/TDD MAC structure of the HIPERLAN-2.

HIPERLAN-2 are mapped on to the SCH, LCH, and RCH transport channels. Figure 12.15 illustrates the relation between the logical and transport channels in HIPERLAN-2 standard. The SBCH is used only in downlink to broadcast control information related to the cell, whenever needed. It assists in handover, security, association, and radio link control functions. The DCCH conveys RLC sublayer signals between an MT and the AP. The UDCH carries DLC PDU for convergence layer data. The LCCH is used for error control functions for a specific UDCH. The ASCH is used for association request and reassociation request messages.

Using the logical channels, HIPERLAN-2 implements the protocols for the proper operation of the network, shown in Figure 12.13. The RLC protocol gives a transport service for the signaling entities of the three other algorithms. These four entities provide for the DLC control plane to implement signaling messages. The ACF protocol handles association and dissociation to the network. To *associate* with the network, the MT listens to the BCH from different APs and selects the AP with the best radio link quality. The MT then continues with listening to the broadcast of a globally unique network operator in the SBCH as to avoid association to a network, which is not able or allowed to offer services to the user of the MT. If the MT decides to continue the association, the MT will request and be given a MAC-ID from the AP. This is followed by an exchange of link capabilities using the ASCH and establishing the PHY and convergence layer connection, as well as authentication and encryption procedures. After association, the MT can request for

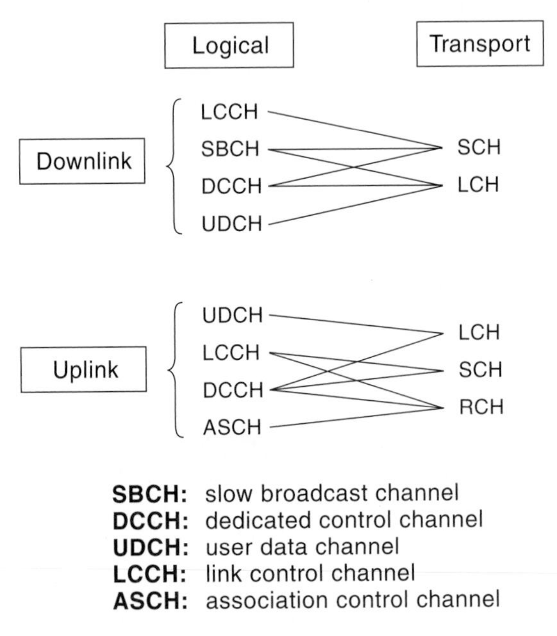

Figure 12.15 Relation between logical and transport channels in HIPERLAN-2.

a DCCH that it uses to set up a DLC user connection with a unique support for QoS. For *disassociation* from the network, either MT notifies the AP that it no longer wants to communicate or AP realizes that the MT is no more active and it is out of the network. In either case, the AP will release all resources allocated for that MT. To implement the DCC algorithm, terminals request a DLC user connection by transmitting signaling messages over the DCCH. The resource for the connection gets allocated, and after an ACK signal, the DLC connection is ready for traffic. The DCCH controls the resources for specific MAC entities. The algorithm also supports procedure for ending the connection or defining a new connection.

The RRC protocol handles handover, dynamic frequency selection, and sleeping mode and power saving operation. Like 802.11, the *handover* in HIPERLAN-2 starts with passive scanning that can be followed by an active request for handover. The difference between the 802.11 and HIPERLAN-2 is that HIPERLAN-2 provides two alternatives for passing the information for handover to the new AP. In the first approach, similar to 802.11, the new AP retrieves connection status and association information from the MT. In the second approach, MT provides the old AP address to the new AP, and the information is exchanged over the wire between the old and new APs. The second approach is faster, because the backbones always have higher bandwidth and capacity, and it does not add to the air traffic that is always desirable. The RRC in HIPERLAN-2 supports mechanism to measure the power and communicate with neighboring APs that allows *dynamic frequency selection* (DFS). Similar to 802.11, the RRC of the HIPERLAN-2 supports mechanisms for the AP to allow the MTs to go to sleeping mode to save in power consumption. The DLC layer of the HIPERLAN-2 also supports the error control mechanism to detect the errors in the arriving packets and arrange the retransmission through ACK/NACK signaling.

To support QoS HIPERLAN-2 recommends changing the periodicity of the transmitted messages that are illustrated in Figure 12.16. There are three periodic operations shown in this figure, the longest belonging to the broadcast period, the medium to Terminal A, and the shortest to Terminal B. Apparently, the delay associated with the packets from Terminal B is the shortest and packets from Terminal A have medium delay as they are compared with the normal broadcast messages. This mechanism allows a delay-controlled environment that is fertile for the implementation of the QoS control.

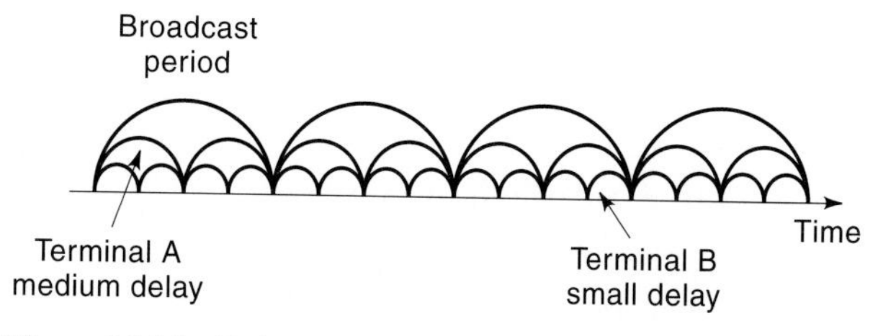

Figure 12.16 Delay control mechanism in HIPERLAN-2 for QoS support [HIPweb].

12.4.4 Convergence Layer

The main two responsibilities of the CL are adapting the service request from a higher layer to the DLC capabilities and to perform fragmentation and reassembly of different size packets from a variety of application protocols to HIPERLAN-2 packet format. Multiple convergence layers operate one at a time to map connection-based and connectionless higher layers such as PPP/IP, ATM, UMTS, Firewire, and Ethernet packets to HIPERLAN-2 DLC packets. To implement all these features, the CL of the HIPERLAN-2 provides a number of services. These services include segmentation and reassembly, priority mapping from 802.1p, address mapping from 802, multicast/broadcast handling, and flexible QoS classes [KHA00].

12.4.5 Security

Comprehensive security mechanisms are seen for the first time in the HIPERLAN-2 system compared with other wireless standards. When contacted by an MT, the AP will respond with a subset of supported PHY modes, a selected convergence layer (only one), and a selected authentication and encryption procedure. As always, there is an option not to use any authentication or encryption. If encryption is agreed upon, the MT will initiate the Diffie-Hellman key exchange to negotiate the secret session key for all unicast traffic between the MT and the AP. The Diffie-Hellman key exchange is discussed in Appendix 6A. In all other wireless systems, key management is a big issue. It is, however, not clear what the computational burden of the Diffie-Hellman key exchange is on wireless devices. Encryption is based on stream ciphers generated using a mechanism similar to the output feedback mode of DES [STI95].

HIPERLAN-2 supports both the use of DES and triple-DES (that is the de facto standard, while AES is in the standardization process) algorithms for strong encryption. Broadcast and multicast traffic can also be protected by encryption through the use of common keys distributed in an encrypted manner through the use of a unicast encryption key. All encryption keys must be periodically refreshed to avoid flaws in the security as discussed in Chapter 6.

Secret and public key algorithms can be employed for authentication. Authentication is possible using message authentication codes based on MD5, HMAC, and digital signatures based on RSA. Mutual authentication is supported for authentication of both the AP and the MT. HIPERLAN-2 supports a variety of identifiers for identification of the MS, via the network access identifier, IEEE address, and X.509 certificates. Challenge response mechanisms are also employed for identification.

12.4.6 Overall Comparison with 802.11

There are several hundreds MHz bands that are available for the 802.11a/HIPERLAN-2 networks which can provide a comfortable multichannel operation for these standards. Availability of these bands and licensing conditions, however, is different in the United States, Europe, and Japan. Figure 12.17 shows the available spectrum for the operation of the 802.11a/HIPERLAN-2 networks in typical

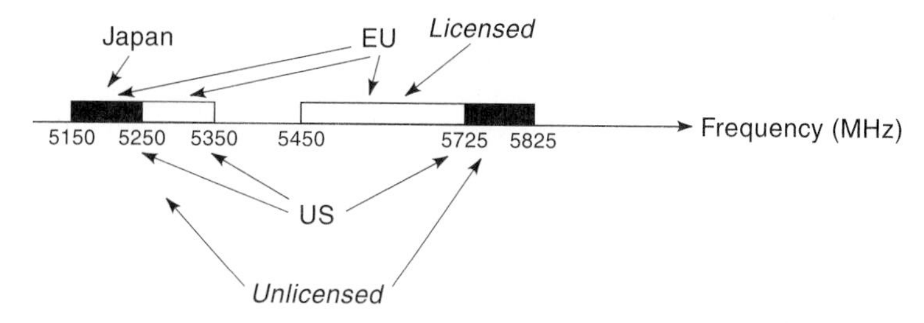

Figure 12.17 Frequency bands for HIPERLAN-2/802.11a.

countries around the world. There are 100 MHz unlicensed bands at 5,150–5,250 MHz that are available in the United States, Europe, and Japan. Another 100 MHz of unlicensed bands at 5,250–5,350 MHz are also available in Europe and the United States. In the United States only, there is another 100 MHz of unlicensed bands at 5.725–5.825 MHz. Finally, there is 255 MHz of licensed bands at 5,470–5,725 MHz in Europe that are assigned for outdoor operation. As we discussed in Chapter 11, the ISM bands in 2.4 GHz are only 84 MHz wide. For this reason, recently 5 GHz developments have dominated the attention of the wideband wireless local access industry. However, the reader must note that the penetration and consequently coverage at 2.4 GHz are better than at 5 GHz.

Table 12.2 provides an overall comparison between all aspects of the 802.11 and HIPERLAN-2 standards [JOH99]. The physical characteristics of HIPERLAN-2 and 802.11a are the same. The access method in HIPERLAN-2 is a voice-oriented access method that allows for better integration into voice-oriented backbones such as UMTS and ATM networks. Connection-orientation, compulsory authentication, link adaptation, dynamic frequency selection, and support of QoS make

Table 12.2 Detailed Comparison of 802.11 and HIPERLAN-2

	802.11	**802.11b**	**802.11a**	**HIPERLAN-2**
Frequency	2.4 GHz	2.4 GHz	5 GHz	5 GHz
Max trans. rate	2 Mbps	11 Mpbs	54 Mbps	54 Mbps
Max throughput	1.2 Mbps	5 Mbps	32 Mbps	32 Mbps
Freq. management	None			Dynamic selection
Medium access	Through sensing			Centralized scheduling
Authentication	None			NAI/IEEE Add/X.509
Encryption	40-bit RC4			DES, 3DES
QoS Support	PCF			ATM/802.1p/RSVP
Wired backbone	Ethernet			Ethernet/ATM/UMTS/ FireWire/PPP/IP
Connectivity	Connectionless			Connection-oriented
Link quality control	None			Link adaptation

HIPERLAN-2 look like a next-generation cellular network that supports high data rates and provides IP services. The main distinction with a cellular system would be the use of unlicensed bands for which a service provider cannot predict the interference. The IEEE 802.11a is an IP-based network that draws from LAN backbone.

QUESTIONS

12.1 What are the differences between the 802.11a and HIPERLAN-2 ?

12.2 Why can't cellular service providers incorporate existing IEEE 802.11 LANs into their networks?

12.3 Explain the general differences between the packet format of the ATM and WATM.

12.4 Explain the general difference between the packet format of the WATM and IEEE 802.11.

12.5 What are the major challenges in implementing WATM that did not exist for data-oriented Ethernet like IEEE 802.11?

12.6 Compare the WTAM reference model of Figure 12.3 with the IEEE 802.11 reference model in Figure 11.1 in terms of functionality of elements, connection to the backbone, and the changes needed in the infrastructure to support mobility.

12.7 Explain the differences between the protocol stacks of the WATM given in Figure 12.4 and that of IEEE 802.11 given in Figure 11.3.

12.8 What were the aspects of WATM trials that impacted the formation of HIPERLAN-2 standard?

12.9 Which WLAN standard first adopted the 5 GHz band operation?

12.10 Explain the similarities between the HIPERLAN-1 and IEEE 802.11.

12.11 Explain the similarities between the HIPERLAN-1 and HIPERLAN-2.

12.12 Explain the differences between NPMA and CSMA/CA medium access control mechanisms used in HIPERLAN-1 and IEEE 802.11 respectively.

12.13 How is priority implemented in HIPERLAN-1 and what is its difference from the priority schemes in the IEEE 802.11 and HIPERLAN-2?

12.14 Explain the architectural differences between HIPERLAN-2 and IEEE 802.11.

12.15 Explain the differences between the protocol stacks of HIPERLAN-2 given in Figure 12.10 and that of IEEE 802.11 given in Figure 11.3.

12.16 What are the purposes of scrambler and interleaver in the HIPERLAN-2 modem?

12.17 What is the basic difference between the medium access control of the HIPERLAN-2 and IEEE 802.11?

12.18 How many transport channels and logical channels are implemented in the HIPERLAN-2 DLC layer?

12.19 What is the length of the frame in the HIPERLAN-2 MAC and how is it divided into traffic and signaling messages?

12.20 What is the difference between a logical and a transport channel in HIPERLAN-2?

12.21 What are BCH and FCH channels in HIPERLAN-2 and what are their functionalities?

12.22 Explain similarities between the medium access control of the HIPERLAN-2 and DECT.

12.23 What is the symbol duration and guard time of the IEEE 802.11a/HIPRLAN-2 OFDM modems? What is the purpose of the guard time?

PROBLEMS

12.1 Compare the overhead of an IEEE 802.11 frame with that of an ATM packet. Do your calculations for the maximum and minimum frame lengths in the case of 802.11. Comment on your results.

12.2 Consider the HIPERLAN-2 standard that uses BPSK and $r = 3/4$ codes for 9 Mbps information transmission and 16-QAM with the same coding for the actual payload data transmission rate of 36 Mbps.

a. Calculate the coded symbol transmission rate per subcarrier for each of the two modes. What is the bit transmission rate per subcarrier for each of the two modes?

b. If one switches from 36 Mbps mode to 9 Mbps mode, how much more (in dB) of the path-loss can it afford?

c. If the system was covering up to 50 meters with 36 Mbps, what would be the coverage with 9 Mbps mode? (*Hint:* use the distance power gradient of the JTC model for an office to calculate this distance.)

12.3 Experimental measurements indicate that the coverage of a 5 GHz WLAN is 47 m at 11 Mbps and 25 m at 54 Mbps. How do these numbers compare with the results of Problem 12.2? If there are any discrepancies, what might be the reason?

12.4 The following parameters are available for HIPERLAN-1 mobile stations trying to access the wireless medium after a busy-period. Explain clearly what happens during each part of the channel access cycle and which MS survives which phase. Which of the mobile stations is ultimately able to transmit data? Under what different circumstances would a collision occur?

Station	Priority	Elimination burst (slots)	Yield time (slots)
MS 1	1	7	4
MS 2	3	3	3
MS 3	2	12	5
MS 4	1	5	1
MS 5	2	12	6
MS 6	1	7	3

12.5 IEEE 802.11a/HIPERLAN-2 use BPSK and $r = 1/2$ convolutional coding for 9 Mbps information transmission and 64-QAM with $r = 3/4$ convolutional coding for 54 Mbps.

a. What is the difference in maximum acceptable path loss (in dB) between the 9 Mbps and 54 Mbps modes of operation? Assume that the stronger $r = 1/2$ codes provide about 1 dB advantage over the weaker $r = 3/4$ codes.

b. If the modem operating at 54 Mbps covers up to 30 meters in a home, what would be the coverage in 9 Mbps mode? Assume JTC model for residential areas is valid for power calculation of this scenario.

CHAPTER 13

AD HOC NETWORKING AND WPAN

13.1 Introduction

13.2 What is IEEE 802.15 WPAN?

13.3 What is HomeRF?

13.4 What is Bluetooth?

13.4.1 Overall Architecture
13.4.2 Protocol Stack
13.4.3 Physical Connection
13.4.4 MAC Mechanism
13.4.5 Frame Formats
13.4.6 Connection Management
13.4.7 Security

13.5 Interference between Bluetooth and 802.11

13.5.1 Interference Range
13.5.2 Probability of Interference
13.5.3 Empirical Results

Questions

Problems

13.1 INTRODUCTION

In the last three chapters, we provided an overview of the wideband wireless local access techniques. We divided these activities into WLANs and WPANs and provided the details of WLAN standards. In this chapter we provide an overview of the WPAN activities. At the time of this writing, WPANs are differentiated from the WLANs with their smaller area of coverage, ad hoc only topology, plug and play architecture, support of voice and data devices, and low-power consumption. WPANs started as BodyLANs which connect sensors and information devices attached to the body to the neighbors for the military application and as personal networks to connect personal equipment such as the laptops, notepads, and cell phones of a person in commercial applications.

The very first personal area network to be announced was the BodyLAN which emerged from a DARPA project in the mid-1990s. This was a low-power, small-size, inexpensive WPAN with modest bandwidth that could connect personal devices within a range of around five feet [DEN96]. Motivated by the BodyLAN project, a WPAN group originally started in June 1997 as a part of the IEEE 802.11 standardization activity. In January 1998, the WPAN group published the original functionality requirement. In May 1998, the study group invited participation from the WATM, Bluetooth, HomeRF, BRAN (HIPERLAN), IrDA (IR short-range access), IETF (Internet standardization), and WLANA (a marketing alliance for WLAN companies in the United States). Only the HomeRF and Bluetooth groups responded to the invitations. In March 1998, the Home RF group was formed. In May 1998, the Bluetooth development was announced, and a Bluetooth special group was formed within the WPAN group [SIE00]. In March 1999, the IEEE 802.15 was approved as a separate group in the 802 community to handle WPAN standardization. At the time of this writing, IEEE 802.15 WPAN has four subcommittees on Bluetooth, coexistence, high data rate, and low data rate. Bluetooth has been selected as the base specification for IEEE 802.15. In the rest of this chapter, we provide an overview of the WPAN, HomeRF, and Bluetooth activities.

13.2 WHAT IS IEEE 802.15 WPAN?

The 802.15 WPAN group is focused on development of standards for short distance wireless networks used for networking of portable and mobile computing devices such as PCs, PDAs, cell phones, printers, speakers, microphones, and other consumer electronics. The WPAN group intends to publish standards that allow these devices to coexist and interoperate with one another and other wireless and wired networks in an internationally acceptable frequency of operation.

The original functional requirement published in January 22, 1998, was based on the BodyLAN project and specified devices with [HEI98]:

- Power management: low current consumption
- Range: 0–10 meters

- Speed: 19.2–100 kbps
- Small size: .5 cubic inches without antenna
- Low cost relative to target device
- Should allow overlap of multiple networks in the same area
- Networking support for a minimum of 16 devices

As we will see later on in this chapter, these specifications fit the Bluetooth specification that was announced after this premier announcement. The initial activities in the WPAN group included HomeRF and Bluetooth group. Today HomeRF maintains its own Web site at [HomeRFweb] and IEEE 802.15 WPAN has four task groups. Task group one is based on Bluetooth and defines PHY and MAC specifications for wireless connectivity with fixed, portable, and moving devices within or entering a personal operating space (POS). A POS is the space about a person or object that typically extends up to 10 meters in all directions and envelops the person whether stationary or in motion. The proposed project will address QoS to support a variety of traffic classes [IEE00].

Task group two is focused on coexistence of WPAN and 802.11 WLANs. This group is developing a coexistence model to quantify the mutual interference and a coexistence mechanism to facilitate coexistence of an IEEE 802.11 WLAN and an IEEE 802.15 WPAN device. A goal of the WPAN Group will be to achieve a level of interoperability that could allow the transfer of data between a WPAN device and an 802.11 device.

Task group three of the IEEE P802.15 works on PHY and MAC layers for high-rate WPANs that operate at data rates higher than 20 Mbps. This standard will provide for low-power, low-cost solutions addressing the needs of portable consumer digital imaging and multimedia applications. This standard aims at providing compatibility with the Bluetooth specification of the task group one and expects to be completed by early 2002.

Task group four is chartered to investigate an ultralow complexity, ultralow power consuming, ultralow cost PHY and MAC layer for data rates of up to 200 kbps. Potential applications are sensors, interactive toys, smart badges, remote controls, and home automation. The project may also address the location tracking capabilities required to support uses of smart tags and badges.

13.3 WHAT IS HomeRF?

According to [HOM00] the mission of the HomeRF working group is to provide the foundation for a broad range of interoperable consumer devices by establishing an open industry specification for wireless digital communication between PCs and consumer electronic devices anywhere in and around the home. Figure 13.1 represents the overall vision of the HomeRF working group. Like the general architecture of home networks described in Chapter 10, the HomeRF working group architecture supports both ad hoc and infrastructure networks. In the more popular infrastructure network, the home Internet and PSTN access arrives at a control HomeRF distribu-

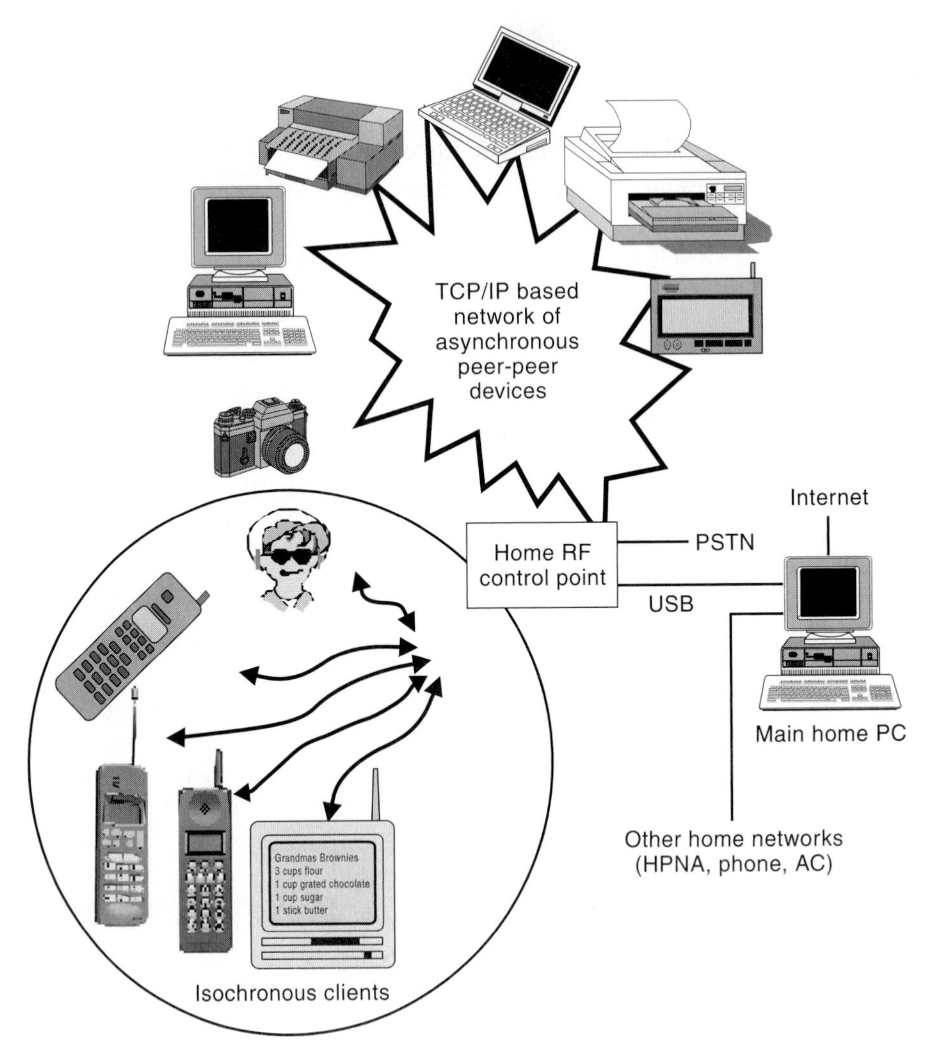

Figure 13.1 The overall architecture of the HomeRF System [HomeRFweb].

tion box that supports wireless as well as HPNA networks. The wireless part supports an isochronous network interconnecting up to six cordless telephone devices and an asynchronous network interconnecting a number of data devices. The two major competitors of this technology are HIPERLAN-2 and Bluetooth. Compared with HIPERLAN-2, the HomeRF solution provides a narrower bandwidth (up to 2 Mbps against 54 Mbps in HIPERLAN-2) that cannot support video for TV and VCR applications. HomeRF has a higher data rate than Bluetooth, but the latter was introduced as an inexpensive chip set that soon attracted a large alliance.

The HomeRF working group has developed a specification for wireless communications in the home called shared wireless access protocol (SWAP). The SWAP specification defines a new common interface that supports wireless voice and data networking in the home. The SWAP specification is an extension of

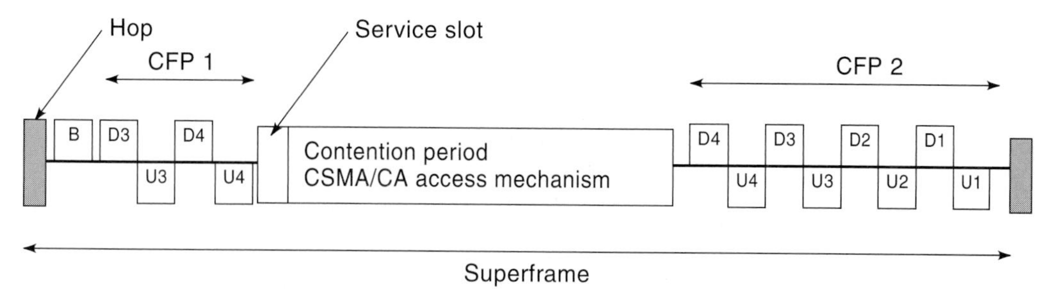

Figure 13.2 SWAP frame specification.

DECT (using TDMA) for voice and a relaxed 802.11 (CSMA/CA) for high-speed data applications. Figure 13.2 shows the MAC frame structure of the SWAP. Each superframe packet has a length of 20 ms and is transmitted in one hop (50 hops per second) of the FHSS system that supports 1 and 2 Mbps using two- and four-level FSK like IEEE 802.11. The superframe has a beacon period for control functions, two contention-free periods (CFPs) for voice traffic, and a contention period (CP) for data traffic. The second CFP is shared by four 2-way (TDD) voice users. The first CFP is used for two 2-way voice channels that can also be used for retransmission of the lost voice packets in the first two channels. The TDMA/TDD access mechanism used in this part is the same as that of the DECT. After completion of the first CFP, the channel will be available through CSMA/CA protocol for data access. Then it is left for the CFP voice transmission. The reader interested in more details on HomeRF can refer to [NEG00], [HOM00].

13.4 WHAT IS BLUETOOTH?

Bluetooth is an open specification for short range wireless voice and data communications that was originally developed for cable replacement in personal area networking to operate all over the world. In 1994 the initial study for development of this technology started at Ericsson, Sweden. In 1998, Ericsson, Nokia, IBM, Toshiba, and Intel formed a special interest group (SIG) to expand the concept and develop a standard under IEEE 802.15 WPAN. In 1999, the first specification was released and then accepted as the IEEE 802.15 WPAN standard for 1 Mbps networks. At the time of this writing, over 1,000 companies have participated as members in the Bluetooth SIG, and a number of companies all over the world are developing Bluetooth chip sets. Marketing forecasts indicate penetration of Bluetooth in more than 100 million cellular phones and several millions of other consumer devices. The IEEE 802.15 standard is also studying coexistence among and interference between Bluetooth and IEEE 802.11 products operating at 2.4 GHz.

The story of the origin of the name Bluetooth is interesting and worth mentioning. "Bluetooth" was the nickname of Harald Blaatand (A.D. 940–981), king of Denmark and Norway. When the Bluetooth specification was introduced to the public, a stone carving, shown in Figure 13.3, erected from Harald Blaatand's capital

Figure 13.3 Picture of Bluetooth on the stone [BLUweb].

city Jelling, was also presented [BLU00]. This strange carving was interpreted as Bluetooth connecting a cellular phone and a wireless notepad in his hands. The picture was used to symbolize the vision in using Bluetooth to connect personal computing and communication devices. Bluetooth, the king, was also known as a peacemaker and a person who brought Christianity to Scandinavians to harmonize their beliefs with the rest of Europe. That fact is used to symbolize the need for "religious" harmony among manufacturers of WPANs around the world to support the growth of WPAN industry.

Bluetooth is the first popular technology for short-range, ad hoc networking that is designed for an integrated voice and data applications. Unlike WLANs, Bluetooth has a lower data rate, but it has an embedded mechanism to support voice applications. Unlike 3G cellular systems, Bluetooth is an inexpensive personal area ad hoc network operating in unlicensed bands and owned by the user.

The Bluetooth SIG considers three application basic scenarios that are shown in Figure 13.4 [BLU00]. The first scenario, shown in Figure 13.4(a), is the wire replacement to connect a PC or laptop to its keyboard, mouse, microphone, and notepad. As the name of the scenario indicates, it avoids the multiple short-range wiring surrounding today's personal computing devices. The second scenario is ad hoc networking of several different users at very short range in an area such as a conference room. As we saw in the last three chapters, WLAN standards and products also commonly consider this scenario. The third scenario is to use Bluetooth as an AP to the wide area voice and data services provided by the cellular networks, wired connection, or satellite links. The 802.11 community also considers this over-

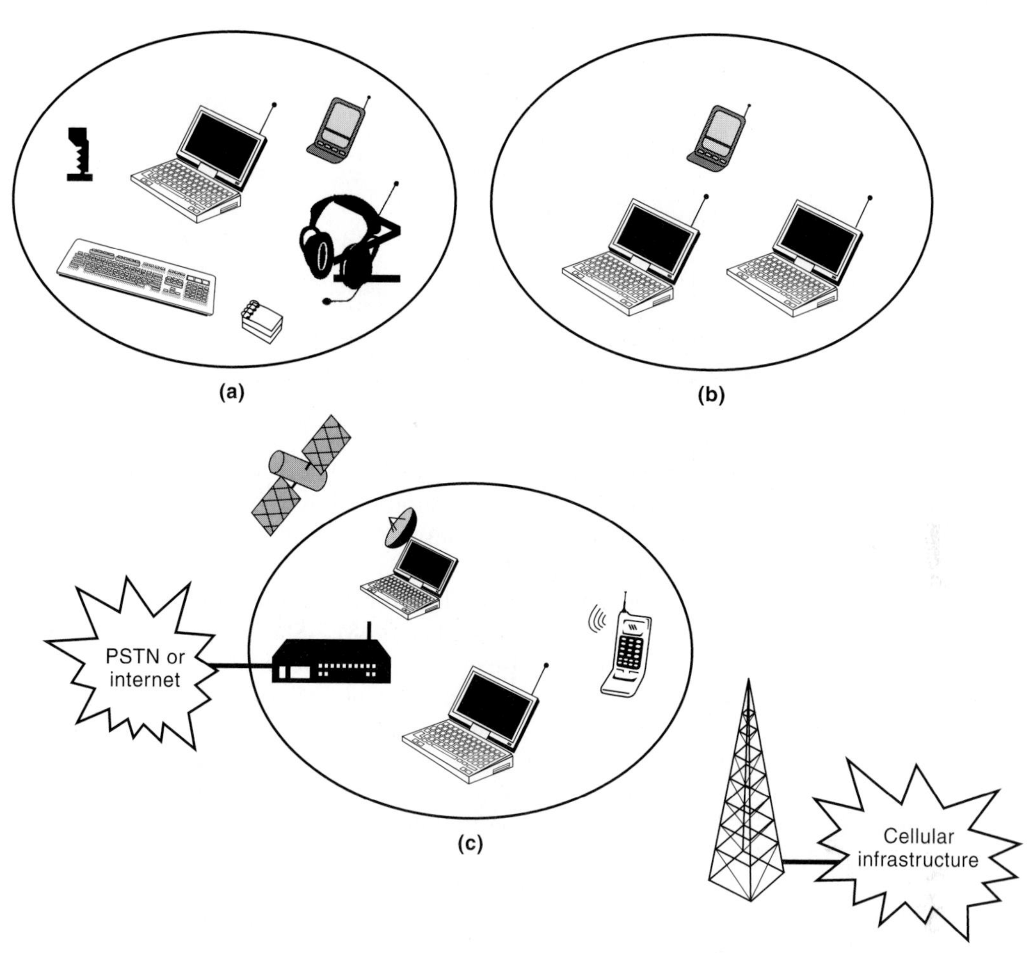

Figure 13.4 Bluetooth application scenarios: (a) cable replacement, (b) ad hoc personal network, and (c) integrated AP.

all concept of the AP. However, the Bluetooth AP is used in an integrated manner to connect to both voice and data backbone infrastructures. The HIPERLAN-2 standard is expected to provide a more comprehensive version of similar connections that supports a larger number of users and wider bandwidths.

13.4.1 Overall Architecture

The topology of the Bluetooth is referred to as *scattered ad hoc topology* that is illustrated in Figure 13.5. In a scattered ad hoc environment, a number of small networks support a few terminals to coexist or possibly interoperate with one another. To implement such a network, we need a plug-and-play environment. The network should be self-configurable, providing an easy mechanism to form a new small network and a procedure for participation in an existing one. To implement that environment, the system should be capable of providing different states for connecting

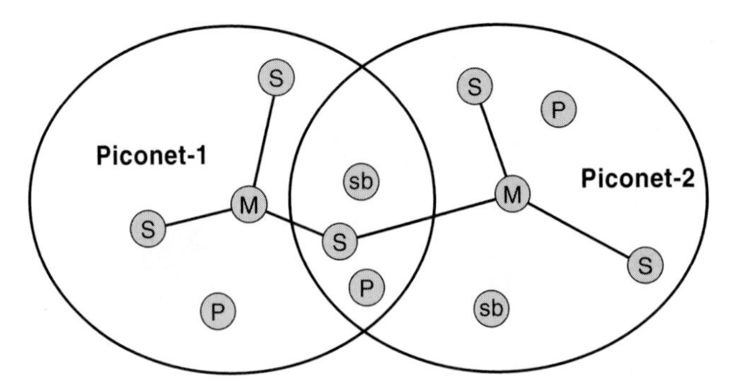

Figure 13.5 Bluetooth's scattered ad hoc topology.

to the network. The terminals should have options to associate with multiple networks at the same time. The access method should allow formation of small, independent ad hoc cells, as well as the possibility of interacting with large voice and data networks considered by Bluetooth.

To accommodate these features, the Bluetooth specification defines a small cell as a *piconet* and identifies four states, Master "M," Slave "S," Stand By "SB," and Parked/Hold "P" for a terminal. Like other ad hoc topologies, such as the one supported by IEEE 802.11, each terminal can be an "M" or an "S." As shown in Figure 13.5, the Bluetooth topology, however, allows "S" terminals to participate in more than one piconet. An "M" terminal in the Bluetooth can handle seven simultaneous and up to 200 active slaves in a piconet. If access is not available, a terminal can enter the "SB" mode waiting to join the piconet later. A radio can also be in a parked/hold, "P," in a low power connection. In the parked mode, the terminal releases its MAC address, while in the "SB" state it keeps its MAC address. Up to 10 piconets can operate in one area [BLU00]. Bluetooth specifications have selected the unlicensed ISM bands at 2.4 GHz for operation. The advantage is the worldwide availability of the bands and the disadvantage is the existence of other users, in particular IEEE 802.11 and 802.11b products in the same band. At the time of this writing, a subcommittee of the IEEE 802.15 is working on the interference issues related to the Bluetooth and IEEE 802.11.

13.4.2 Protocol Stack

One of the distinct features of the Bluetooth is that it provides a complete protocol stack that allows different applications to communicate over a variety of devices. Other wireless local networks, such as IEEE 802.11, specify the three lower layers for communications. The protocol stack for voice, data, and control signaling in Bluetooth is shown in Figure 13.6 [HAA00]. The *RF layer* specifies the radio modem used for transmission and reception of the information. The *baseband layer* specifies the link control at bit and packet level. It specifies coding and encryption for packet assembly and frequency hopping operation. The *link management protocol* (LMP) configures the links to other devices by providing for authentication and encryption, state of units in the piconet, power modes, traffic scheduling, and packet format. The *logical link control and adaptation protocol* (L2CAP) provides

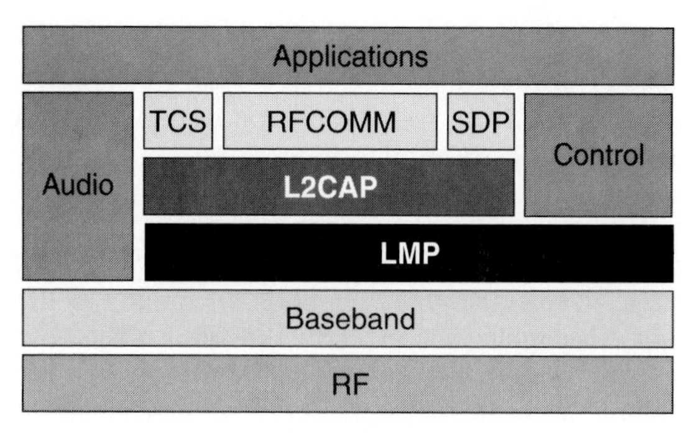

Figure 13.6 Protocol stack for Bluetooth.

connection-oriented and connectionless data services to the upper layer protocols. These services include protocol multiplexing, segmentation and reassembly, and group abstractions for data packets up to 64 kB in length. The audio signal is directly transferred from the application to the Baseband. Also LMP and the application exchange control messages interact to prepare the physical transport to the application.

Different applications may use different protocol stacks but nevertheless all of them share the same physical and data link control mechanisms. There are three other protocols above the L2CAP. The *service discovery protocol* (SDP) finds the characteristics of the services and connects two or more Bluetooth devices to support a service such as faxing, printing, teleconferencing, or e-commerce facilities. The *telephony control protocol* (TCP) defines the call control signaling and mobility management for the establishment of speech for cordless telephone application. Using these protocols legacy telecommunication applications can be developed.

Example 13.1: Telephony Control Protocol in Bluetooth

Figure 13.7 shows the protocol stack for implementation of the cordless telephone application. The audio signal is directly transferred to the Baseband layer while SDC and TCP protocols operating over L2CAP and LMP handle signaling and connection management.

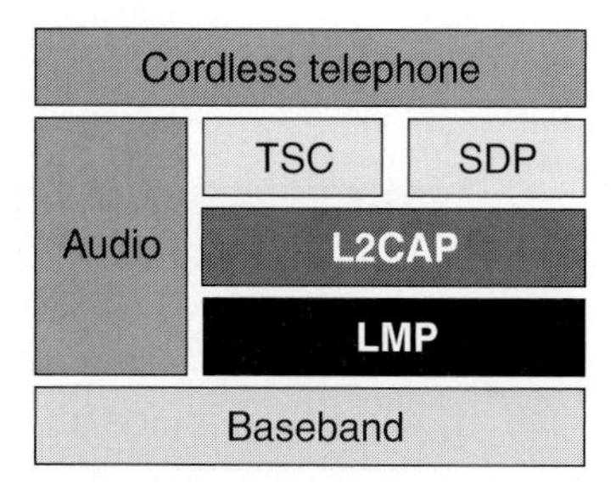

Figure 13.7 Protocol stack for implementation of cordless telephone over Bluetooth.

The *RFCOMM* is a "cable replacement" protocol that emulates the standard RS-232 control and data signals over Bluetooth baseband. Using RFCOMM a number of non-Bluetooth specific protocols can be implemented on the Bluetooth devices to support legacy applications.

Example 13.2: Lightweight Applications in Bluetooth

Figure 13.8 shows the implementation of a vCard application for credit card verification. This application protocol runs over object exchange protocol (OBEX) that is accommodated by the RFCOMM protocol in the Bluetooth protocol stack. Therefore, the sequence of protocols for implementation of credit card verification over Bluetooth is vCard–OBEX–RFCOMM–L2CAP–Baseband–RF. This protocol stack implementation contains both internal object representation convention of vCard and over-the-air transport protocols of the Bluetooth.

Example 13.3: WAP over Bluetooth

Figure 13.9 shows the implementation of a wireless application environment (WAE) protocol that defines applications over the wireless access protocol (WAP). The WAP packets use the TCP/UDP protocols for Internet access on top of the point-to-point protocol (PPP) that runs over the RFCOMM.

The overall Bluetooth protocols can be divided into three classes. The Bluetooth SIG developed the core exclusively Bluetooth-specific protocols for Baseband, LMP, L2CAP, and SDP. The protocols that are also developed by the Bluetooth SIG but based on existing protocols include RFCOMM and TCP. The third group consists of existing protocols that are adopted by Bluetooth SIG. At the time of this writing, these protocols include PPP, UDP/TCP/IP, OBEX, WAP, vCard, vCal, IrMC-1, and WAE. Bluetooth specification is open, and other legacy protocols such as HTTP and FTP can be accommodated on top of the existing protocol stack.

Example 13.4: FTP over Bluetooth

Figure 13.10 provides a protocol stack for implementation of the FTP application. OBEX and RFCOMM manage the data transfer, whereas SDP provides for the establishment of the link.

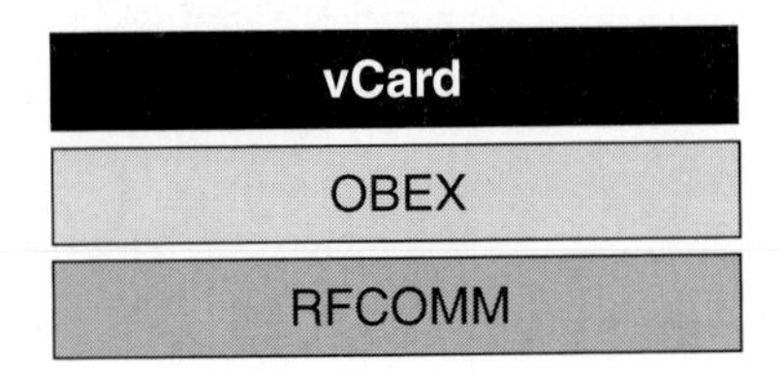

Figure 13.8 Protocol stack for implementation of vCard over Bluetooth.

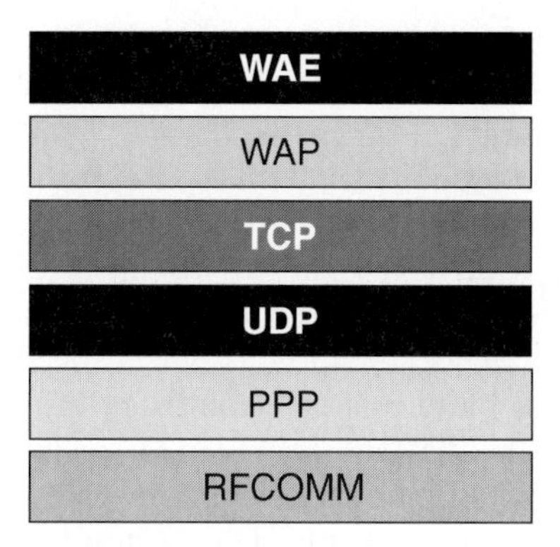

Figure 13.9 Protocol stack for implementation of WAE over Bluetooth.

The overall structure of the protocol stack in the Bluetooth does not clearly follow the OSI model and its acronyms. Therefore, the division of the following section may appear somehow different from other wireless local networks described in the last two chapters. However, we make our every effort to make them as close as what we had in the previous chapters to provide a fluency that comforts the reader in understanding the details and relating them to other details for similar systems.

13.4.3 Physical Connection

The OSI equivalent PHY layer of the Bluetooth is embedded in the RF and Baseband layers of the Bluetooth protocol stack. The physical connection of Bluetooth uses a FHSS modem with a nominal antenna power of 0 dBm (10 m coverage) that has an option to operate at 20 dBm (100 meter coverage). Like the 1 Mbps option of the IEEE 802.11 FHSS standard, the Bluetooth specification uses a two-level GFSK modem with a transmission rate of 1 Mbps that hops over 79 channels in the ISM bands starting at 2.402 GHz and stopping at 2.480 GHz. The hopping rate and pattern and number of hops used in Bluetooth, however, are different from IEEE 802.11. The Bluetooth hopping rate is 1,600 hops per second (625 μs dwell time) as

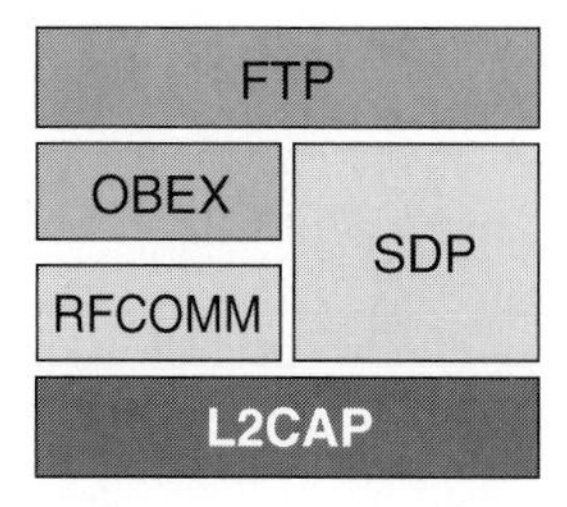

Figure 13.10 Protocol stack for implementation of FTP over Bluetooth.

compared with the 2.5 hops per second (400 ms dwell time) system adopted by the 802.11. The two-level GFSK modem allows simple noncoherent detection implementation using simple FM demodulators. The 0 dBm modem with the Bluetooth hopping pattern complies with the FCC rules in the United States, and due to local regulations, the bandwidth is reduced in Japan, France, and Spain. An internal software switch (which allows an environment for implementation of a system that works universally) handles this transition.

The Bluetooth specification assigns a specific frequency-hopping pattern for each piconet. This pattern is determined by the piconet identity and master clock phase residing in the master terminal in the piconet. Figure 13.11 illustrates the element of frequency hopping strategy in the Bluetooth. The overall hopping pattern is divided into 32 hop segments. The 32-hop pseudorandom hopping pattern segment is generated based on the master identity and clock phase. The 79 frequency hops at the ISM bands are arranged in odd and even classes. Each 32-hop sequence starts at a point in the spectrum and hops over the pattern that covers 64 MHz because it hops either on odd or even frequencies. After completion of each segment, the sequence is altered, and the segment is shifted 16 frequencies to the forward direction. The 32 hops are concatenated, and the random selection of the index is changed for each new segment. This way segments slide through the carrier list to maintain the average time each frequency is used at an equal probability. Change of the clock or identity of the piconets will change the sequence and segment mapping, allowing different piconets to operate with different set of random codes. These codes are not orthogonal to one another, but they are randomized against each other. With 79 hops it is difficult to find a large number of orthogonal codes anyway [HAA00].

To protect the integrity of the transmitted data Bluetooth uses two error-correction schemes in the baseband controllers. An FEC code is always applied to the header information, and if needed it is extended to the payload data for the voice-oriented synchronous packets. The FEC code generally reduces the number of retransmissions. It is always applied to the header because header information is short and important. The flexibility of using FEC for payload provides an option to avoid overhead in favor of increased throughput when the channel is good and error free. An unnumbered ARQ scheme is also applied by the baseband layer for

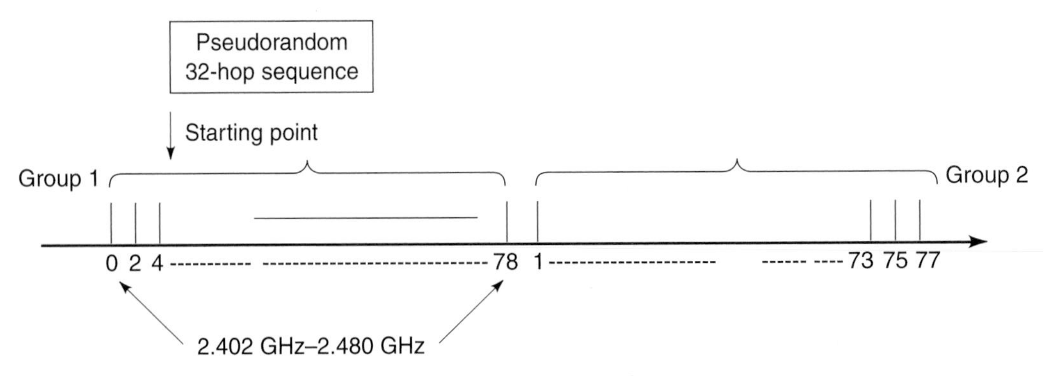

Figure 13.11 The hopping sequence mechanism in Bluetooth.

the asynchronous data-oriented information in which the recipient acknowledges data transmitted. For data transmission to be acknowledged, both the header error check and the payload check, if applied, must indicate no error condition. These functionalities implemented in the Baseband layer of the Bluetooth protocol stack are often implemented in the data link layer of the OSI reference model for networks complying with that model.

13.4.4 MAC Mechanism

Although the modulation technique and frequency of operation of the Bluetooth radio system closely follows that of the FHSS 802.11, the MAC mechanism in the Bluetooth is widely different from the 802.11. The Bluetooth access mechanism is a voice-oriented innovative system that is neither identical to the data-oriented CSMA/CA type nor voice-oriented CDMA or TDMA access methods and yet has elements that are somehow related to these access methods. The medium access mechanism of Bluetooth is a fast FH-CDMA/TDD system that employs polling to establish the link. The fast hopping of 1,600 hops per second allows short time slots of 625 μs (625 bits at 1 Mbps) for one packet transmission that allows a better performance in the presence of interference. Bluetooth is a CDMA system that is implemented using FHSS. In the Bluetooth CDMA, each piconet has its own spreading sequence, whereas in the DSSS/CDMA system used for digital cellular systems each user link is identified with a spreading code. The DSSS/CDMA is not selected for Bluetooth because DSSS/CDMA needs central power control that is not possible in the scattered ad hoc topology envisioned for Bluetooth applications. Without any need to centralized power control for CDMA operation, the FH/CDMA in Bluetooth allows tens of piconets to overlap in the same area providing an effective throughput that is much larger than 1 Mbps. As we discussed in Chapter 11, the FHSS 802.11 operates in the same 79 hops as Bluetooth with only three sets of hopping patterns. The throughput of the Bluetooth FH/CDMA system, however, is less than 79 Mbps which could be achieved in a coordinated FDM or OFDM system employed as in 802.11a and HIPERLAN-2 operating in 5.2 GHz U-NII bands. In Bluetooth, the FH/CDMA is selected over simple FDM or OFDM because ISM bands at 2.4 GHz only allow spread spectrum technology. The access method in each piconet of Bluetooth is TDMA/TDD. The TDMA format allows multiple voice and data terminals to participate in a piconet. The TDD eliminates cross talk between the transmitter and the receiver, allowing a single chip implementation in which a radio alternates between transmitter and receiver modes. To share the medium among a larger number of terminals, at each slot "M" decides and *polls* a "S." Polling is used rather than contention access methods because contention provides too much overhead for the short packets (625 bits) that were selected for the implementation of a fast FHSS system.

13.4.5 Frame Formats

The Bluetooth packet format is based on one packet per hop and a basic 1-slot packet of 625 μsec that can be extended to three slots (1,875 μsec) and five slots (3,125 μsec). This frame format and the FH/TDMA/TDD access mechanism allow an "M" terminal to poll multiple "S" terminals at different data rates for voice and data applications to form a piconet.

Example 13.5: Operation of Piconets

Figure 13.12 illustrates several examples of Bluetooth operation in a piconet. In Figure 13.12(a), an "M" terminal is communicating with three "S" terminals. The TDMA/TDD format allows simultaneous operation of the three terminals assigning 625 μs (equivalent to 625 bits at 1 Mbps) for transmission and a time gap between the two packets in each direction. Terminals may run different applications (voice or data at different rates), but applications should run on one of the one-slot detailed packet formats that are specified by the Bluetooth SIG. The time gap is specified at 200 μs to allow a terminal to switch from transmitter to receiver mode for the TDD operation [HAA00]. Figure 13.12(b) shows an asymmetric communication in which the "M" uses a higher speed three-slot link, whereas the "S" operates at a lower rate with one-slot packets. Figure 13.12(c) represents a symmetric higher speed three-slot communication, and Figure 13.12(d) an asymmetric very high-speed five-slot with a return low speed one-slot link.

The overall packet structure of the Bluetooth is shown in Figure 13.13. There are 74 bits for the access code field, 54 bits for the header field and up to 2,744 bits for different payloads that can be as long as five slots. In IEEE 802.11 FHSS packets, the preamble and header of the PHY layer, shown in Figure 11.4, were 96 and 32 bits, respectively, whereas the payload could be as long as $4096 \times 8 = 32{,}768$ bits. The size of the overhead is more or less in the same range, but the maximum payload of 802.11 is at least an order of magnitude larger. Apparently Bluetooth uses more flexible shorter packets for ease of integration and better performance

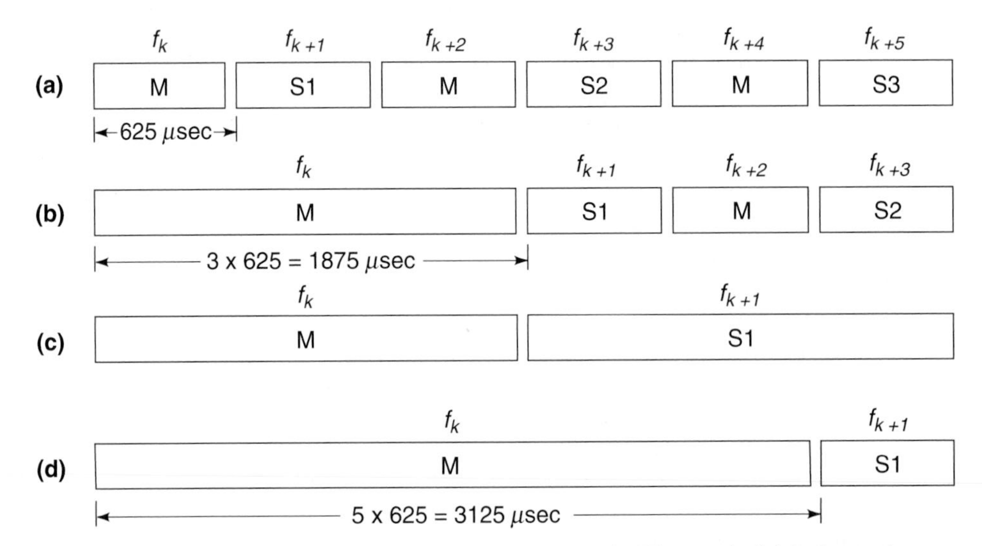

Figure 13.12 FH/TDMA/TDD multislot packet formats in Bluetooth: (a) 1-slot packets, (b) asymmetric 3-slots, (c) symmetric 3-slots (1,875 μsec), and (d) asymmetric 5-slots (3,125 μsec).

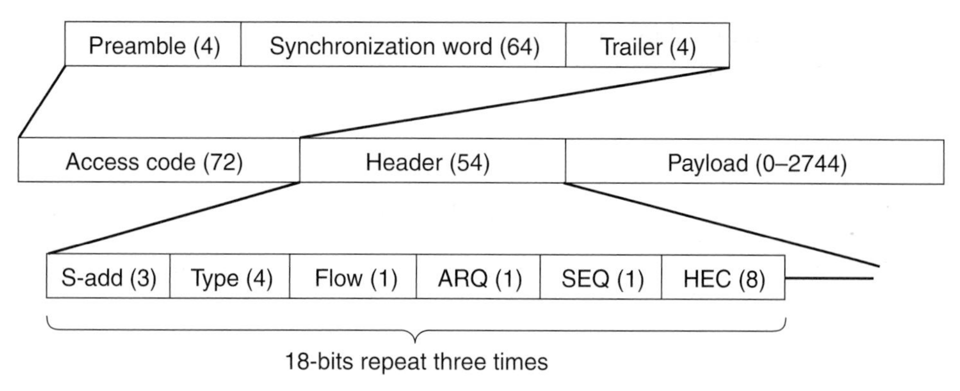

Figure 13.13 Overall frame format of the Bluetooth packets.

in fading, but these gains are at the expense of a higher percentage of overhead that reduces the throughput.

As shown in Figure 13.13, the access code field consists of a four-bit preamble and a four-bit trailer plus a 64-bit synchronization PN-sequence with a large number of codes with good autocorrelation and cross-correlation properties. The 48-bit IEEE MAC address unique to every Bluetooth device is used as the seed to derive the PN-sequence for hopping frequencies of the device. There are four different types of access codes. The first type identifies a "M" terminal and its piconet address. The second type of access code specifies an "S" identity that is used to page a specific "S." The third type is a fixed access code reserved for the inquiry process that is explained later. The fourth type is the dedicated access code that is reserved to identify specific set of devices such as fax machines, printers, or cellular phones.

As shown in Figure 13.13, the header field has 18 bits that are repeated three times with a 1/3 FEC code. The 18-bit starts with a 3-bit "S" address identifying, 4-bit packet type, 3-bit status reports, and an 8-bit error check parity for the header. The 3-bit S-ADD allows addressing the seven possible active "M"s in a piconet. The 4-bit packet type allows 16 choices for different grade voice services, data services at different rates, and four control packets. The 3-bit status reports are used to flag overflow of the terminal with information, acknowledgment of successful transmission of a packet, and sequencing to differentiate the sent and resent packets.

The Bluetooth SIG specifies different payloads and associated packet type codes that allow implementation of a number of voice and data services. Different master-slave pairs in a piconet can use different packet types, and the packet type may change arbitrarily during a communication session. The four-bit packet type identifies 16 different packet formats for the payloads of the Bluetooth packets. Six of these payload formats are asynchronous connectionless (ACL), primarily used for packet data communications. Three of the payload formats are synchronous connection oriented (SCO), primarily used for voice communications. One is an integrated voice (SCO) and data (ACL) packet, and four are control packets common for both SCO and ACL links.

The three SCO packets, shown in Figure 13.14, are high-quality voice (HV) packets numbered as HV1, 2, and 3 to designate the level of quality of the service.

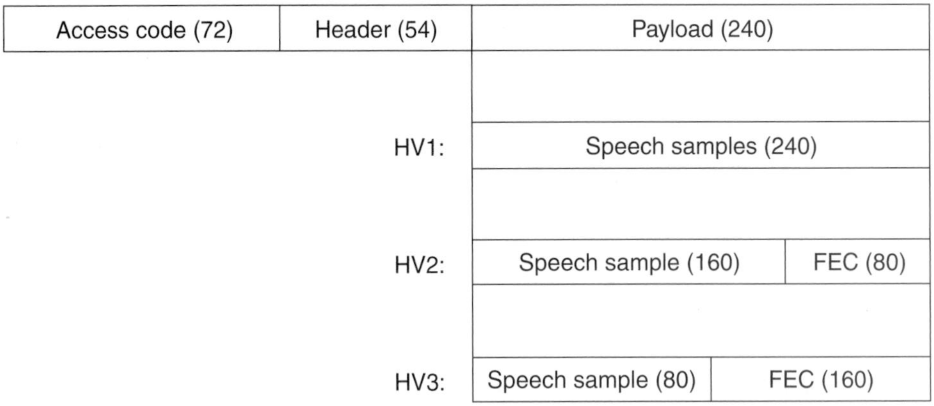

Access code (72)	Header (54)	Payload (240)

	Payload
HV1:	Speech samples (240)
HV2:	Speech sample (160) | FEC (80)
HV3:	Speech sample (80) | FEC (160)

Figure 13.14 SCO 1-slot packet frame formats.

The SCO packets are all single slot packets, the length of the payload being fixed at 240 bits, and they do not use the status report bits, but they are transmitted over reserved periodic duplex intervals to support 64 kbps per voice user. HV1 uses all 240 bits for the user voice samples, HV2 uses 160 bits for user voice samples and 80 bits of parity for a $1/3$ FEC code, and HV3 uses 80 bits of user voice samples and 160 bits of parity for a $2/3$ FEC code. To keep the data rate for voice samples at 64 kbps, the HV1, HV2, and HV3 packets in each direction are sent every six, four, and two slots, respectively.

Example 13.6: Data Rate of High Quality Voice Packets

The HV1 packets are 240 bits long, and so they are sent every six slots. The packets are 1-slot packets sent at the rate of 1,600 slots/sec. Therefore, we have

$$\frac{1{,}600\,(slots/\text{sec})}{6\,(slots)} \times 240\,(bits) = 64\text{ kbps}$$

The overall format of the payload for the six ACL packets is shown in Figure 13.15. The payload has its own 8- or 16-bit header, payload, and 16-bits CRC code. The header has information on the length and identity of the packet. If we want to compare the headers with those of 802.11, we may compare the overhead with the MAC overhead of the 802.11 shown in Figure 11.19. This time the overhead of Bluetooth is significantly lower than the 34 bytes (272 bits) overhead of the 802.11 MAC frames. Most of the saving in the overhead of Bluetooth occurs because 802.11 employs four addresses—source, destination of the device, and the intermediate APs. Bluetooth uses one 48-bit IEEE MAC address to identify a device that is embedded in the access code and is not needed in the payload.

The six ACL packets are data medium (DM) and data high (DH) rate packets numbered as DM or DH1, 3, or 5 according to the length of the slot they take. Figure 13.15 shows the overall frame format of all DM and DH data-oriented packets. DM packets use a rate of $2/3$ FEC that improves the quality of the service. DH packets do not employ coding to achieve higher data rates. Using a different number of slots for a packet data payload size, exercising the coding option and changing

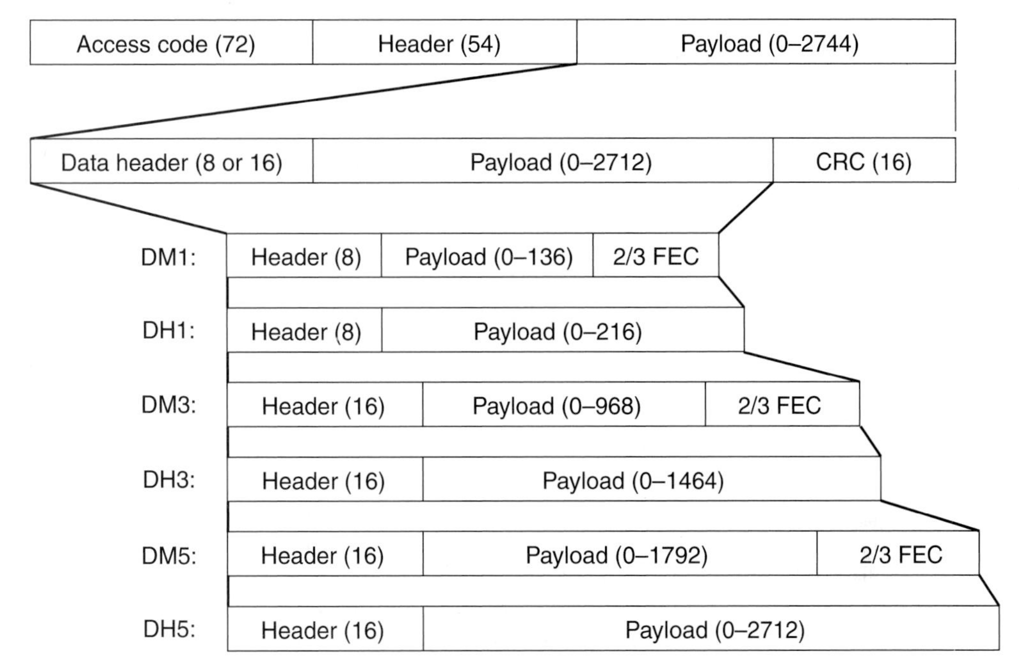

Figure 13.15 ACLs 1-, 3-, and 5-slot packet frame formats.

the symmetric nature of the transmitted packets in each direction, a number of packet data links can be implemented in the Bluetooth specification.

Example 13.7: High Data Rate in Bluetooth

A symmetric 1-slot DH1 link between an "M" and an "S" terminal carries 216 bits per slot at a rate of 800 slots per second (every other slot) in each direction. The associated data rate is 216 $(bits/slot) \times 800\ (slots/\text{sec}) = 172.8$ Kbps.

Example 13.8: Medium Data Rate in Bluetooth

The asymmetric DM5 link, shown in Figure 13.15(d) uses five-slot packets carrying 1,792 bits per packet by the "M" and 1-slot packet carrying 136 bits per packet by the "S" terminal. The number of packets per second in each direction is 1,600/6 packets per second. Therefore, the data rate from "M" is given by:

$$1{,}792\ (bits/packet) \times \frac{1{,}600}{6}\ (packets/\text{sec}) = 477.8 \text{ Kbps}$$

The data rate of the "S" terminal in this asymmetric connection is:

$$136\ (bits/packet) \times \frac{1{,}600}{6}\ (packets/\text{sec}) = 36.3 \text{ Kbps}$$

Table 13.1 shows all 12 symmetric and asymmetric data links that are supported with the frame format of the Bluetooth specification. The maximum data rate of 723.2 kbps is available in an asymmetric channel for a single user, with a

Table 13.1 ACL Packet Types and Associated Data Rates in Symmetric and Asymmetric Modes

Type	Symmetric	Asymmetric	
DM1	108.8	108.8	108.8
DH1	172.8	172.8	172.8
DM3	256.0	384.0	54.4
DH3	384.0	576.0	86.4
DM5	286.7	477.8	36.3
DH5	432.6	721.0	57.6

reverse channel carrying 57.6 kbps. The reader should remember that data applications operate in bursts, and therefore, even if an "M" node communicates with the maximum seven "S" data terminals still most of the time only one of the "S" terminals will communicate with the "M." When more than one "S" terminal simultaneously attempts to communicate with an "M" terminal, the QoS provided to the "S" terminals has to be compromised either by sharing the throughput or by providing additional delays. The decision-making process to reach a compromise in the voice-oriented access methods, such as the one used in Bluetooth, needs a complex algorithm to handle the QoS as negotiated at the start of a session. Comparing this situation with CSMA/CA used in 802.11, there is no negotiation at the starting point. When more than one terminal attempts to communicate with a single AP, the medium is shared, and the compromise is made automatically through the CSMA/CA access method described in Chapter 4. Apparently, for the data-only applications, CSMA/CA is more appropriate, and that is why it was developed by the data-oriented networking industry. However, when voice applications become dominant, the TDMA/TDD type access methods can guarantee QoS for the voice though CSMA/CA cannot do it easily.

The only remaining traffic packet in Bluetooth is a data voice (DV) packet that is a mixed SCO and ACL packet with the same access code and overall header that must be transmitted in regular intervals. The voice part carries 80 bits of voice payload without any coding, and the data part is a short packet of 0–72 bits with a 16-bit $2/3$ CRC coding and an eight-bit data payload header. This packet also uses the three status report bits.

The Bluetooth specification also defines four control packets: ID, NULL, POLL, and FHS. The ID packet occupies only half of a slot, and it carries the access code with no data or even a packet type code. This packet is used before connection establishment to only pass an address. The NULL and POLL packet have the access code and the header, and so they have packet type codes and status report bits. The NULL packet is used for ACK signaling, and there is no ACK packet for it. The POLL packet is similar to the to the NULL packet, but it has an ACK. "M" terminals use the POLL packet to find the "S" terminals in their coverage area. The frequency hop synchronization (FHS) packet carries all the information necessary to synchronize two devices in terms of access code and hopping timing. This packet is used in the inquiry and paging process that is explained later.

13.4.6 Connection Management

The link manager (LM) layer and L2CAP layer of the Bluetooth perform the link setup, authentication, and link configuration. An important issue in a truly ad hoc network is how to establish and maintain all the connections in a network whose elements appear and disappear in an ad hoc manner, and there is no central unit transmitting signals to coordinate these terminals. In both digital cellular systems and WLANs, there is a common control or a beacon signal that allows a new terminal to lock to the network and exchange its identity with the networks identity. The Bluetooth specification achieves initiation of the network through a unique inquiry and page algorithm.

The overall state diagram of the Bluetooth is shown in Figure 13.16. In the beginning of the formation of a piconet, all devices are in SB mode, then one of the devices starts with an inquiry and becomes the "M" terminal. During the inquiry process, the "M" terminal registers all the SB terminals that then become "S" terminals. After the inquiry process, identification, and timing of all "S" terminals is

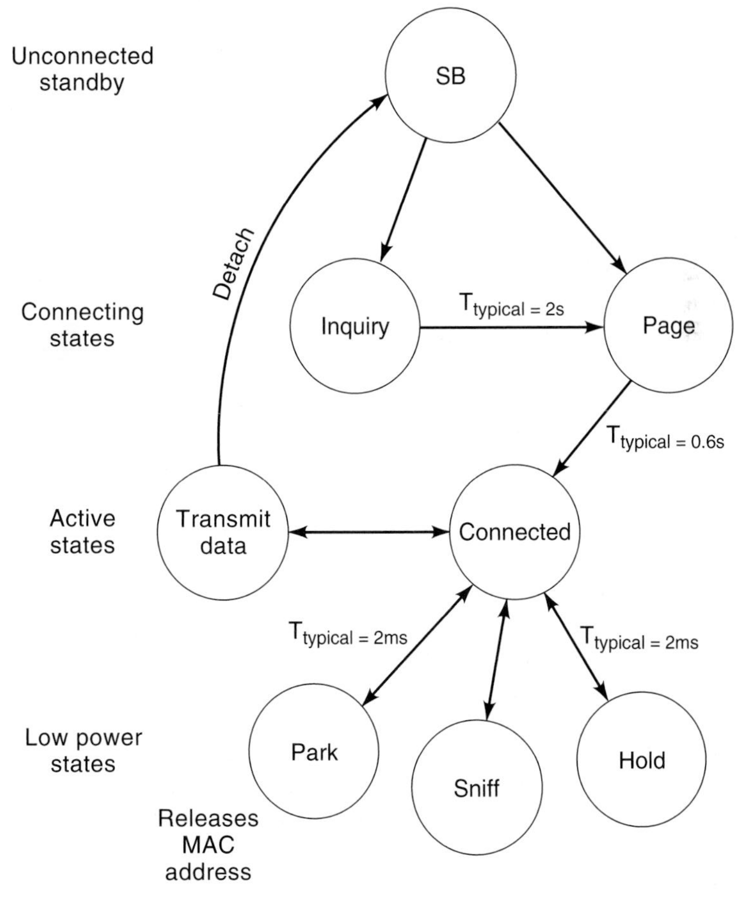

Figure 13.16 Functional overview of the Bluetooth specification.

sent to the "M" terminal using the FHS packets. A connection starts with a PAGE message with which the "M" terminal sends its timing and identification to the "S" terminal. When the connection is established, the communication session takes place, and at the end, the terminal can be sent back to the SB, Hold, Park, or Sniff states. Hold, Park, and Sniff are power-saving options. The Hold mode is used when connecting several piconets together or managing a low-power device. In the Hold mode, data transfer restarts as soon as the unit is out of this mode. In the Sniff mode, a slave device listens to the piconet at reduced and programmable intervals according to the application needs. In the Park mode, a device gives up its MAC address but remains synchronized to the piconet. A Parked device does not participate in the traffic but occasionally listens to the traffic of the "M" terminal to resynchronize and check on broadcast messages.

The main innovative part of the inquiry and paging algorithms in Bluetooth is a searching mechanism for two terminals that are not synchronized, but they both know a common address. The following example explains this algorithm.

Example 13.9: Search Algorithm for Synchronization

The two Bluetooth devices knowing a common 48-bit IEEE 802 address of an "M" terminal first use the common address to generate a common FH pattern of 32 hops and a common PN-sequence for the access code of all their packets. Then they start their operation as depicted in Figure 13.17. In the initial state, Terminal 1 sends two ID packets carrying the common access code every half slot on a dif-

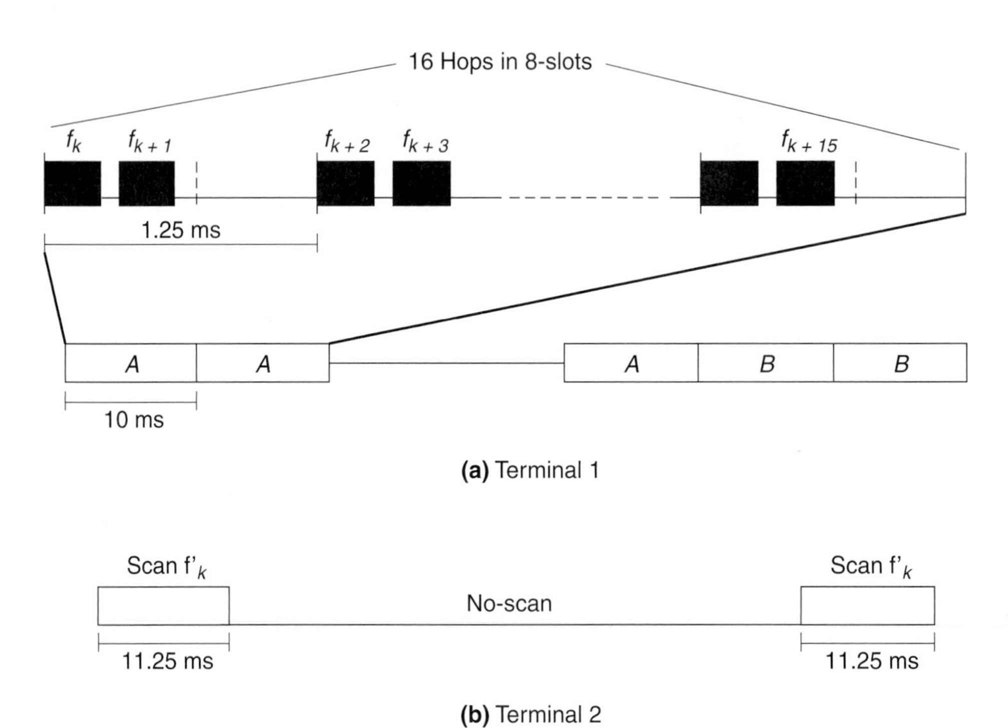

Figure 13.17 Basic search for paging algorithm in the Bluetooth.

ferent hop frequency associated with the common frequency hopping pattern and listens to the response of the slave in the next slot. If there was no response, it continues broadcasting the ID packets on the two new frequencies in the common hop pattern and repeats this procedure eight times for a period of 10 ms (eight 2-slot times). During these 10 ms, the common ID is broadcast at 16 of the total 32 different hop frequencies. If there is no response, Terminal 1 assumes that Terminal 2 is in sleep mode and repeats the same broadcast again and again until the period of transmission becomes longer than the expected sleeping time of Terminal 2. At this time Terminal 1 assumes that Terminal 2 has scanned, but its scan frequency was not among the 16 hops, designated by A in Figure 13.17 and continues its broadcast with the second half of the 32 hop frequencies, designated by B in the figure. Terminal 2 is in sleeping mode; it wakes up periodically for a period of 11.25 ms to scan the channel at a given frequency for its desirable access code and sleeps again. In each scan period of 11.25 ms, the sliding correlator in Terminal 2 hears the desired address at 16 different frequencies. If one of these frequencies is the same as the scanning frequency, the correlator peaks and synchronization is signaled. Depending on the operation, Terminal 2 can scan the second time at the same frequency or at a new frequency for verification. In either case the objective is to maximize the probability of hitting the same frequency as the broadcast frequency.

The basic principle explained in Example 13.10 is used during the inquiry and paging processes. The following two examples explain these applications for the above mechanism.

Example 13.10: Paging

As in the previous example, the "M" terminal broadcasts repeating ID page trains carrying the access code of the paged terminal two per slot, waits for the response in the next slot, repeats at new hopping frequencies of the paged terminal to cover 16 frequencies every 10 ms, and repeats this for the estimated length of the sleeping time. The "S" terminal scans for 11.25 ms with one of the 32 frequencies of its hopping pattern, then sleeps and scans at the next hopping frequency. When frequencies are the same, a peak appears at the correlator output of the "S" terminal, and the slave responds by sending its own ID packet as an acknowledgment for detection of frequency hoping timing. The "M" terminal then stops broadcasting ID packets and sends an FHS packet containing its own ID and timing information. The "S" terminal responds with another ID packet at the timing of the "M" terminal and then connection is established, and the slave joins the piconet for information exchange. Usually, the "M" terminal knows the approximate timing of the hopping pattern, and 16 most probable hops are adequate to establish the connection. In case this estimate is not correct, like the previous example, the "M" terminal resorts to the second half of the 16 hops when there is no response after the estimated sleeping time.

Example 13.11: Inquiry

The Inquiry message is typically used for finding Bluetooth devices, including public printers, fax machines, and similar devices with an unknown address. The general format of the inquiry process is very similar to the paging mechanism. A unique access code and FH pattern are reserved for inquiry. In other words, the inquiry process is universally identified with all attributes of a device. Like paging, inquiry starts with an "inquirer" broadcasting an ID packet every half slot at a different hop frequency, covering 16 frequencies every 10 ms, and repeats the same process until it receives responses. The "inquiree" scans with the sliding correlator for 11.25 ms. When the frequencies are the same, the sliding correlator peaks in all devices that are scanning. To avoid collision a device detecting the Inquiry ID runs a random number generator and waits for the length of the outcome before it scans the channel again. When the peak appears the second time after random waiting time, the inquiree terminal sends an FHS packet, allowing the inquirer to learn its ID and timing information. After process is completed, the inquirer's radio has device IDs and clocks of all radios in its range of coverage. After completion of the first inquiry, the inquired device changes its scan frequency and continues scanning for the next inquiry and follow-up FHS signaling.

13.4.7 Security

Bluetooth specifications provide usage protection and information confidentiality. Bluetooth has three modes of operation—nonsecure, service-level, and link-level security. Devices also can be classified into trusted and distrusted. It makes use of two secret keys (128 bits for authentication and 8 to 128 bits long for encryption), a 128-bit long random number, and the 48-bit MAC address of devices. Any pair of Bluetooth devices that wish to communicate will create a session key (called the link key) using an initialization key, the device MAC address, and a PIN number. This protocol has been shown to have several vulnerabilities [WET01] by which a malicious entity could obtain the PIN numbers and keys depending on how the session initialization of the communication protocol is performed [BRA01].

13.5 INTERFERENCE BETWEEN BLUETOOTH AND 802.11

Obviously, when two wireless network overlap in their coverage and operate at the same frequency at the same time without any access coordination, they will interfere with one another. The literature on military communication systems offers many detailed analyses of the performance of communication systems in the presence of various intentional interferers or jammers [SIM85]. These jammers are designed to disrupt the operation of a system, and they can employ relatively sophisticated techniques, such as multitone jamming and pulsed jamming. In civilian applications, the interference is neither intentional nor sophisticated. Most often, the interferer is simply another system designed to operate in a portion of or the entire band of operation of our system, and the users are generally willing to cooperate so as to minimize the mutual interference. Depending on the level of coordination of the overlapping wireless network since the early days of the IEEE 802.11 [HAY91], [WOR91], the WLAN industry has specified three levels of overlapping: interference, coexistence, and interoperation.

Multiple wireless networks are said to *interfere* with one another if colocation causes significant performance degradation of any of the devices. Multiple wireless networks are said to *coexist* if they can be colocated without significant impact on the performance of any of the devices. Coexistence provides for the ability of one system to perform a task in a shared frequency band with other systems that may or may not be using the same set of rules for operation. *Interoperability* provides for an environment for multiple overlapping wireless systems to perform a given task using a single set of rules. In an interoperable environment multiple wireless networks exchange and use the information among each other. Interoperability is an important issue for wired as well as wireless networks. Coexistence and interference are issues mainly consuming the attention of wireless network designers, and it becomes more important for the case of ad hoc networks. This terminology for unlicensed bands was first discussed in the IEEE 802.11 community [HAY91]. Later on when WINForum approached the FCC to obtain unlicensed PCS bands they came up with *etiquettes* or rules of coexistence in unlicensed PCS bands [PAH97] that were introduced in Chapter 10. More recently, the IEEE 802.15 WPAN group is engaged in interference analysis in its task group number two. They have performed introductory interference analysis between Bluetooth and IEEE 802.11 devices operating in 2.4 GHz ISM bands and at the time of this writing are working on practical coexistence and interoperability methods [IEE01, ENN98].

Bluetooth is a fast frequency hopping (1,600 hops per second at 1Mbps) wireless system operating in the 84 MHz of bandwidth that is available in the 2.4 GHz ISM bands that are also used for DSSS IEEE 802.11 (1 and 2 Mbps) and CCK IEEE 802.11b (5.5 and 11 Mbps), as well as slower FHSS (2.5 hops per second at 1 and 2Mbps) IEEE 802.11 systems. Therefore, the interaction between a Bluetooth system and a colocated 802.11 WLAN system needs an analysis of the interference between the FHSS and DSSS, as well as fast FHSS and slow FHSS systems.

13.5.1 Interference Range

The first issue in interference is the *interference range*, which is the distance between two terminals in order to interfere, in case they operate at the same frequency and at the same time. The range of interference is related to propagation characteristics of the environment, processing gain of the receivers, and the transmitted power from different devices. Figure 13.18 illustrates an interference scenario between a Bluetooth (BT-1) device and a receiving FHSS IEEE 802.11 MS colocated in an area. The IEEE 802.11 AP is usually located on the wall to provide better coverage; as a result usually they are less likely to be interfered by the BT devices. The interference takes place both when MS is receiving information from the AP, and BT-1 is transmitting information to BT-2; or when the MS is transmitting and BT-1 is receiving. For our analysis we assume that interference from the AP to the BT devices and interference of the BT-2 device to 802.11 devices are negligible. Following the same analysis for interference presented in Chapter 5, when the MS is receiving and BT-1 is transmitting the signal to interference level at the MS is given by:

$$S_r = \frac{KP_{AP}d^{-\alpha}}{KP_{BT}r^{-\alpha}} = \frac{P_{AP}}{P_{BT}}\left(\frac{r}{d}\right)^{\alpha} \qquad (13.1)$$

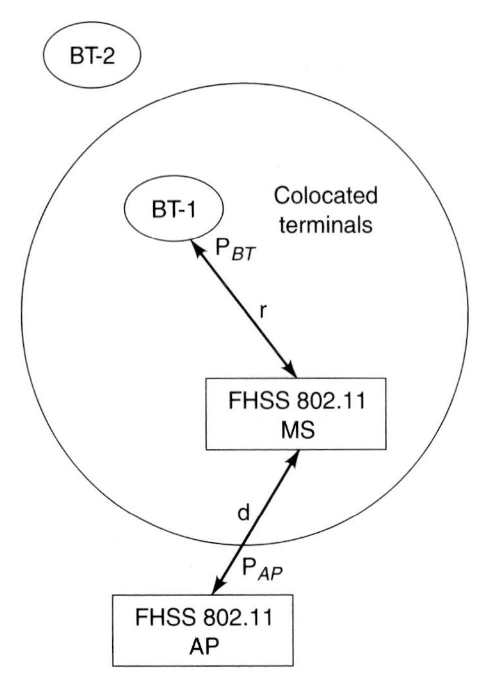

Figure 13.18 The basic interference scenario between Bluetooth and IEEE 802.11 FHSS.

where d and r are the distances between the MS and AP and Bluetooth device, respectively. Also, P_{AP} and P_{BT} represent the transmitted power by the AP and the Bluetooth device, respectively, and α is the distance power gradient of the propagation environment. Therefore the *range of interference* between the Bluetooth and the MS is given by:

$$r_{int} = d\sqrt[\alpha]{S_{min}P_{BT}/P_{AP}} \tag{13.2}$$

where r_{int} is the maximum distance at which the two terminals interfere, and S_{min} is the minimum acceptable received signal to noise ratio needed for proper operation of the MS. In other words, the range of interference of the BT-1 terminal to the MS is directly related to the distance to the AP, required signal-to-noise ratio for proper operation of the MS, and transmit power of BT-1, and it is inversely related to the transmit power of the AP. In general, as we discussed in Chapter 5, the value of α may change from less than two in hallways and open areas up to around six in building with metal partitioning. Depending on the location of the Bluetooth device, the path loss gradients may be different as well. In open areas with no walls, which include a number of scenarios involved with short-range devices, the environment is close to free space propagation and α is often close to 2 [PAH95]. Although the coverage of the 802.11 devices is estimated to be 100 meters (at 20 dBm transmit power), in practice 802.11 APs are installed every 20–40 meters allowing maximum distances of d = 10–20 meters between an AP and a MS. The low power

(0 dBm transmit power) Bluetooth devices are used for WPAN applications where, r, the distance between the devices, is only a few meters. Bluetooth also allows 20 dBm operation that can cover up to 100 m.

Example 13.12: Interference of Bluetooth with FHSS 802.11

a) Assuming an open area with $\alpha = 2$, $S_{min} = 10$ (10 dB), $P_{AP} = 100$ mW (20 dBm), $P_{BT} = 1$ mW (0 dBm), and $d = 20$ m, we have $r_{int} = 6.4$ m. That means if the frequencies of the BT-1 and MS are the same and BT-1 transmits at the same time that the MS receives, a BT-1 device that is closer than 6.4 m to the MS will interfere and destroy the received packets.

b) In a partitioned environment with $\alpha = 4$ we will have $r_{int} = 10.2$ m.

c) If the Bluetooth device is at its maximum transmit power of 100 mw (20dBm) in the same partitioned area, then $r_{int} = 17.7$ m, which is an order of magnitude larger than the value in the 0 dBm mode.

Figure 13.19 illustrates the simple scenario for the interference of FHSS 802.11 to Bluetooth terminals. In this case the MS is transmitting to the AP, and BT-1 is receiving from BT-2. If we assume that the transmitting MS terminal is at a distance r from BT-1 and the two Bluetooth devices are a distance d apart, we have:

$$r_{int} = d\sqrt[\alpha]{S_{min}P_{MT}/P_{BT}} \tag{13.3}$$

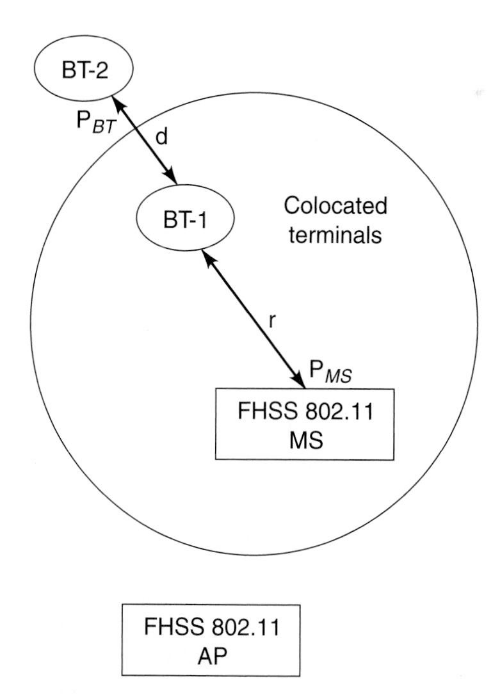

Figure 13.19 The basic interference scenario between FHSS IEEE 802.11 and Bluetooth.

Again the range of interference is directly proportional to the distance between desired terminals, the minimum acceptable signal-to-noise ratio of the receiving terminal, and the power of the interfering terminal, and inversely proportional to the power of the desired transmitter.

Example 13.13: Interference of FHSS 802.11 with Bluetooth

a) With a typical values of 2 m for the distance between the two Bluetooth devices, P_{BT} = 1mW (0 dBm), S_{min} = 10 (10 dB), and P_{MT} = 100 mW (20 dBm), we will have r_{int} = 63.2 m. This is because the 802.11 device is radiating 100 times more power.

b) If the Bluetooth device operates at 20 dBm, with the same power as the 802.11 then r_{int} = 6.32 m.

If instead of FHSS we use DSSS in the scenario of Figure 13.18, as we discussed in Chapter 4, the minimum required received signal to interference ratio at the MS is reduced by a factor equivalent to the value of the processing gain of the DSSS, N. Then, the interference range (BT-1 interfering with MS) becomes:

$$r_{int} = d\sqrt[\alpha]{S_{min}P_{BT}/P_{AP}N} \tag{13.4}$$

For the case of MS interfering with the Bluetooth device, the spectral height of the DSSS is reduced by the value of the processing gain that results in a similar effect and a range of:

$$r_{int} = d\sqrt[\alpha]{S_{min}P_{MT}/P_{BT}\,N} \tag{13.5}$$

Example 13.14: Interference of Bluetooth with DSSS 802.11

a) Assume an open area with $\alpha = 2$, S_{min} = 10 (10 dB), P_{AP} = 100 mW (20 dBm), P_{BT} = 1 mW (0 dBm), and d = 20m. For a processing gain of N = 11, used in IEEE 802.11, the interference range will reduce to around r_{int} = 1.9 m (from 6.4 m) (BT-1 interfering with the MS).

b) For P_{BT} = 100 mW (20 dBm), we have an interference range of r_{int} = 19 m.

c) With a typical values of 2 m distance between the two Bluetooth devices, P_{BT} = 1 mW (0dBm), S_{min} = 10 (10 dB), P_{MT} = 100 mW (20 dBm), and N = 11, we will have r_{int} = 19 m (the MS interfering with BT-1).

d) If the Bluetooth device operates at 20 dBm option, then r_{int} = 1.9 m.

The conclusion from these simple examples is that considering the 10 m range of operation of a Bluetooth piconet and a 100 m range of operation of the 802.11 devices, if a Bluetooth hop coincides with frequency of a FHSS or DSSS IEEE 802.11 WLAN, the interference is serious. The DSSS reduces the interference of the narrowband systems and interference to the narrowband system by the value of its processing gain. This results in a $\sqrt[\alpha]{1/N}$ reduction in the range of interference compared with an FHSS system. However, the spectrum of DSSS is much wider and the probability of frequency coincidence of the DSSS and Bluetooth is much higher than the probability of a hit of a FHSS system and Bluetooth. In the next section, we quantify this statement further.

13.5.2 Probability of Collision

In the last section, we showed that the range of interference of Bluetooth and IEEE 802.11 DSSS is smaller than the range of interference of Bluetooth and FHSS 802.11 systems. However, FHSS is a narrowband signal that changes its frequency of operation randomly while a DSSS is a true wideband system. A narrowband Bluetooth transmitter will interfere with the reception of a wideband DSSS signal with a greater probability than it will with the reception of a FHSS signal on a different narrowband channel. Therefore, the probability of interference between Bluetooth and 802.11 DSSS or 802.11b CCK devices is much higher than the probability of interference between a Bluetooth device and a FHSS 802.11 system.

To further analyze the interference we first pay attention to interference between Bluetooth and FHSS 802.11 devices. Both Bluetooth and FHSS 802.11 are frequency-hopping systems using the 78 carrier frequencies in the 2.4 GHz ISM bands shown in the vertical axis of Figure 13.20. Bluetooth packets are normally shorter than 802.11 packets and hop at a much slower rate of 2.5 hops per second. When a terminal is in the interference range of the other terminal and the hopping

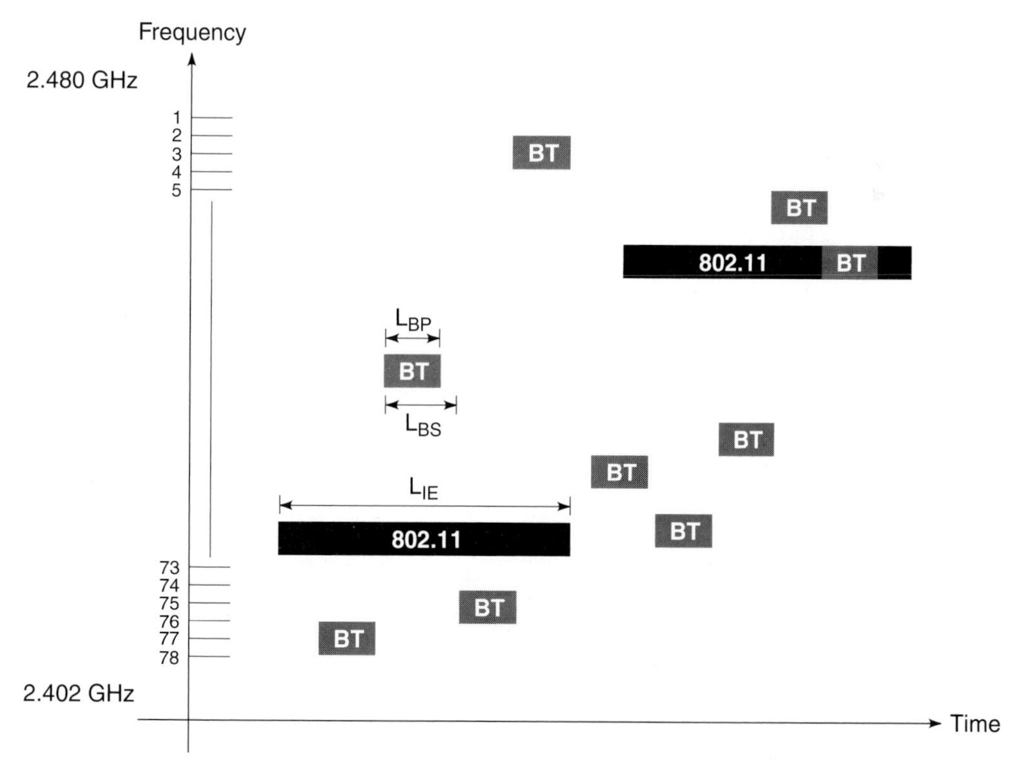

Figure 13.20 Time-frequency characteristics of the FHSS IEEE 802.11 and Bluetooth.

frequencies are the same, packets collide and get destroyed. To analyze this situation, we need to find the probability of collision in time and in frequency.

Because Bluetooth packets are shorter than 802.11 packets, during transmission of one 802.11 packet, the colocated Bluetooth device hops and sends one packet per hop several times. Assuming L_{IE} is the length of the IEEE 802.11 packet and L_{BS} the length of a Bluetooth slot, the minimum number of Bluetooth hops occurring during transmission of one 802.11 packet is $n = [\, L_{IE} / L_{BS} \,]$ where [x] represents the smallest integer greater than or equal to x. The maximum number of Bluetooth hops occurring in duration of an 802.11 packet is $[\, L_{IE} / L_{BS} \,] + 1$. It can be easily shown [ENN98] that the probability of an 802.11 packet overlap with $n = [\, L_{IE} / L_{BS} \,]$ Bluetooth dwell periods of duration L_{BS} is

$$P_n = L_{IE}/L_{BS} - [\, L_{IE}/L_{BS} \,]$$

The probability that it overlaps with $n + 1 = [\, L_{IE} / L_{BS} \,] + 1$ dwell periods is

$$P_{n+1} = 1 - L_{IE}/L_{BS} + [\, L_{IE}/L_{BS} \,]$$

Example 13.15: Overlap between Bluetooth and FHSS 802.11

If $L_{IE} / L_{BS} = 4.3$, the probability of overlap of 802.11 packet with $n = 4$ Bluetooth dwell periods is 30 percent and the probability of overlap with $n+1 = 5$ dwell periods is 70 percent.

Considering these expressions, the probability of an 802.11 packet surviving BT interference, $P_{survive}$, is approximated by:

$$P_{survive} = (1 - P_{hit})^n \, P_n + (1 - P_{hit})^{n+1} \, P_{n+1}$$

where P_{hit} is the probability of having the same frequency for both 802.11 and Bluetooth. The probability of collision is given by $P_{collision} = 1 - P_{survive}$.

Example 13.16: Collision between FH-SS 802.11 and Bluetooth

The probability of a Bluetooth hop to occur at the operating frequency of the FHSS system is $P_{hit} = 1/79 = .013$. For a 1,000 byte 802.11 packet at 2 Mbps,

$$L_{IE} = \frac{1{,}000(bytes) \times 8(bits/byte)}{2(Mbits/s)} = 4ms$$

If Bluetooth is sending 1-slot packets $L_{BS} = 625$ μsec. Therefore,

$$n = \left\lceil \frac{4m \text{ sec}}{625 \ \mu\text{sec}} \right\rceil = 6$$

and $P_n = 0.4$ that result in $P_{n+1} = 0.6$. Therefore,

$$P_{survive} = (1 - 0.013)^6 \times 0.4 + (1 - 0.013)^7 \times 0.6 = 0.92$$

and the collision probability is 0.08 or 8 percent.

Example 13.17: Collision between DSSS 802.11 and Bluetooth

Figure 13.21 shows the mechanism with which the frequency-hopping pattern of Bluetooth and the spectrum of the DSSS 802.11 or CCK 802.11b hit one another. The probability of a Bluetooth hop to occur at the operating frequency of the DSSS system is $P_{hit} = 26/78 = 0.33$. For a 1,000 byte 802.11 packet at 2 Mbps, all the other parameters remain the same as the last example, and we have:

$$P_{survive} = (1 - 0.33)^6 \times 0.4 + (1 - 0.33)^7 \times 0.6 = 0.072$$

The probability of collision is 0.928 or 92.8 percent as compared with 8 percent for the FHSS 802.11 example.

Example 13.18: Bluetooth Interference with 802.11b

The IEEE 802.11b uses the same band as 802.11 DSSS to transmit at 11 Mbps. Therefore, again we have $P_{hit} = 26/79 = 0.33$. However, for a 1,000 byte 802:11 packet at 11 Mbps, we have

$$L_{IE} = \frac{1000(bytes) \times 8(bits/byte)}{11(Mbits/s)} = 727\ \mu s$$

With Bluetooth 1-slot packets we have $n = \left\lceil \frac{727m\sec}{625\ \mu\text{sec}} \right\rceil = 1$ and $P_n = 0.16$, which results in $P_{n+1} = 0.84$. Therefore,

$$P_{survive} = (1 - 0.33)^1 \times 0.16 + (1 - 0.33)^2 \times 0.84 = 0.49$$

The collision probability is 0.51 or 51 percent which is substantially better than 802.11 DSSS and much worse than the 802.11 FHSS.

13.5.3 Empirical Results

The analysis in the last section is at the PHY layer, but a more thorough analysis including the effects of all layers should be done experimentally. A group of undergraduate students at Worcester Polytechnic Institute developed a testbed for the experimental analysis of the interference between the IEEE 802.11b and Bluetooth voice and data channels for their senior undergraduate project [CHA00b]. In this project they considered a number of scenarios and measured the overall packet

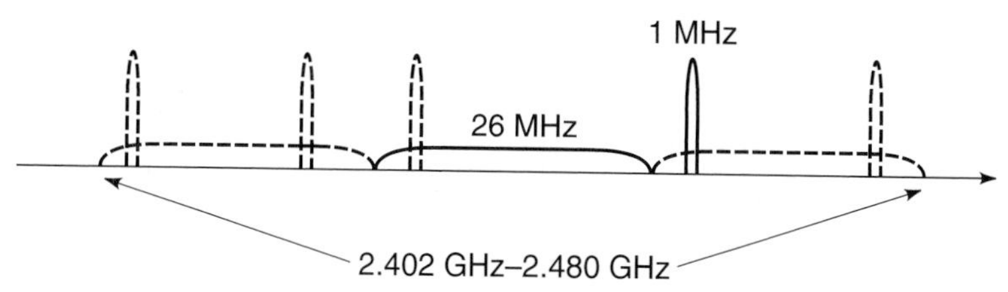

Figure 13.21 Overlapping DSSS IEEE 802.11 and FHSS Bluetooth spectrum.

loss, throughput, and delay characteristics of the interfering Bluetooth and 802.11 devices, as well as cordless telephones. In this section we provide some of their results and conclusions that are related to the scenarios described in Figures 13.18 and 13.19 which relate the performance of interfering 802.11b and BT terminals to the distance between the devices.

Example 13.19: Packet Loss Rate (PLR) in Bluetooth with Interfering 802.11b

Figure 13.22 shows the floor plan and one of the measurement scenarios in which two 20 dBm Bluetooth equipped laptops (triangles) are separated by 10 meters and an 802.11b laptop (circle) is moved from a distance of 1 to 10 meter from the Bluetooth laptop. The 802.11b station is communicating with another laptop that is far away and does not interfere significantly with the Bluetooth device. Figure 13.23 shows the PLR of the Bluetooth device. As the distance of the interfering 802.11 device increases the packet loss reduces. When the Bluetooth and 802.11b interferers are next to one another, the PLR is 70 percent, as the distance increases to five meters, there is no interference effect. The lengths of the 802.11 packets are approximately 1,000 bytes and Bluetooth data packets are 366 bits long. Figure 13.24 shows the delay characteristics evaluated using ping messages that measure the round trip delay [CHA00].

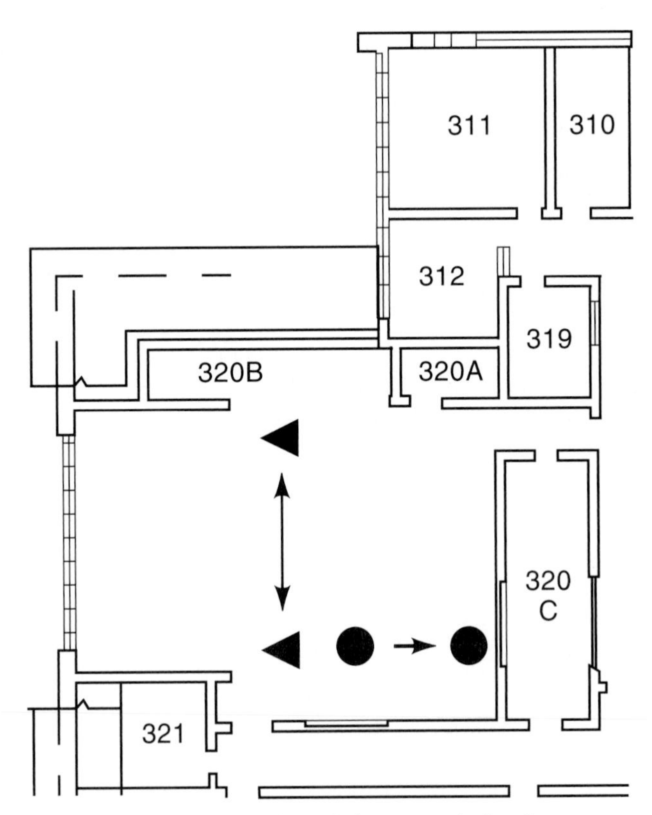

Figure 13.22 Bluetooth interfering scenario for the experiemental interference analysis.

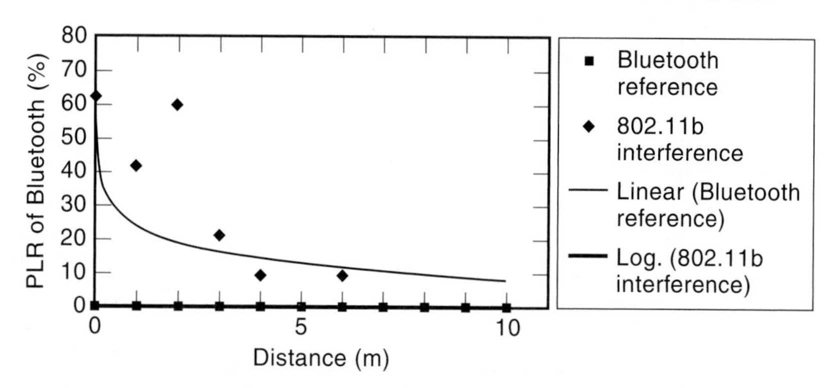

Figure 13.23 Packet loss rate (PLR) of Bluetooth with and without 802.11b interfering terminal.

Example 13.20: PLR in 802.11b with Interfering Bluetooth

Figure 13.25 shows the PLR of 802.11b in a scenario that is the opposite of the last example shown in Figure 13.21. In this example two 802.11b devices are located at a distance of 10 m, and an interfering Bluetooth terminal is moved from one to 10 meters from them. At close distances, the PLR is close to 45 percent, and as the distance of the interfering Bluetooth device increases beyond 3 m, the effects of interference are negligible.

The general conclusion of these studies is that the interference between FHSS 802.11 and BT devices is negligible, however, DSSS 802.11 devices will

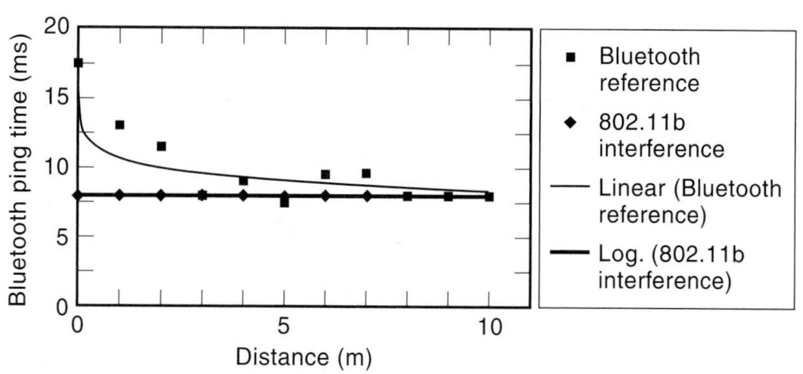

Figure 13.24 Bluetooth delay characteristics with and without 802.11b interference.

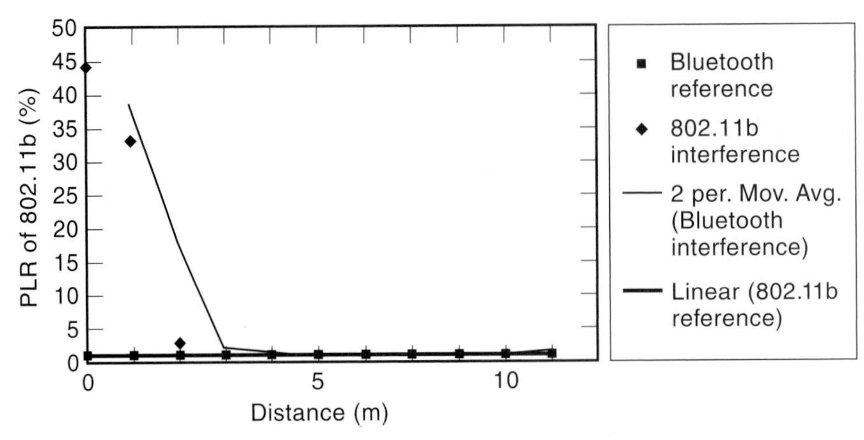

Figure 13.25 802.11b PLR with and without interfering Bluetooth device.

interfere significantly with the BT devices. The IEEE 802.15 is currently working on this issue to find remedies for the coexistence of these systems [IEE00].

QUESTIONS

13.1 What is IEEE 802.15 and what is its relation to the Bluetooth and HomeRF?

13.2 What are the differences between IEEE 802.15 device specification and the device specification of the IEEE 802.11 devices?

13.3 Divide home networking applications discussed in Figure 10.15 into those which can and those which cannot be supported by IEEE 802.15 HomeRF technology.

13.4 Name the four states that a Bluetooth terminal can take and explain the difference among these states.

13.5 Name the three classes of applications that are considered for Bluetooth technology and identify those which can also be 802.11 and HIPERLAN WLAN technologies.

13.6 What are the similarities and differences between the FHSS used in the IEEE 802.11 and Bluetooth in terms of data rate, modulation technique, available frequencies for hopping, speed of the hop, and the number and pattern of the hops?

13.7 What are the differences between ad hoc solutions offered by 802.11 and 802.15?

13.8 What is the difference between the MAC protocol of the Bluetooth and the IEEE 802.11?

13.9 What are the two standard MAC protocols that are combined in the HomeRF SWAP protocol?

13.10 Which IEEE 802.11 standards interfere with Bluetooth and which of these standards has more serious interference condition with it?

13.11 How many different voice services does Bluetooth support and how they are differentiated from one another?

13.12 How many different symmetric and asymmetric data services does Bluetooth support?

13.13 What is the maximum supported asymmetric packet data rate by Bluetooth? How many slots per hop does it use? What is its associated data rate in the reverse channel?

13.14 Compare the header and access code of the Bluetooth with the PLCP header of the FHSS IEEE 802.11.

13.15 What is the maximum data rate of an overlay Bluetooth network? How does it compare with the maximum data rate of the overlay FHSS IEEE 802.11?

13.16 What are the differences between the implementation of paging and inquiry algorithms in Bluetooth?

13.17 Using Figures 13.23 and 13.25 to explain the nature of interference between Bluetooth and IEEE 802.11b.

PROBLEMS

13.1 Give the complete stack protocol for the implementation of an email application over Bluetooth.

13.2 Considering that the encoded voice in Bluetooth is at 64 Kbps in each direction:

a. Use packet format for the HV1 channels to show that these packets are sent every six slots.

b. Use packet format for the HV2 channels to find how often these packets are sent.

c. Repeat (b) for HV3 packets.

13.3 **a.** What is the hopping rate of Bluetooth and how many bits are transmitted in each one slot packet transmission?

b. If each frame of the HV3 voice packets in Bluetooth carries 80 bits of the samples speech, what is the efficiency of the packet transmission (ratio of the overhead to overall packet length)?

c. Determine how often HV3 packets have to be sent to support 64 kbps in each direction.

d. The DH5 packets carry 2,712 bits per each five-slot packet. Determine its effective data rate in each direction.

13.4 Repeat Examples 13.7 and 13.8 for all other data rates supported by Bluetooth shown in Table 13.1.

13.5 Consider the Bluetooth and FHSS IEEE 802.11 interference scenario of Figure 13.18:

a. Assuming that the acceptable error rate for the MT is 10^{-5}, determine the S_{min} that supports this error rate (use the FSK formulas in Chapter 3 for approximate calculation of S_{min}).

b. Using S_{min} of (a) and Eq. (13.2), calculate r_{in} for $d = 10$ m, $\alpha = 2$, $P_{BT} = 20$ dBm and $P_{AP} = 20$ dBm.

c. Produce a computer plot to illustrate the relation between r_{in} and acceptable error rates between 10^{-2} and 10^{-7}(in logarithmic form). Using the computer plot, discuss the impact of error rate requirement on the range of interference between Bluetooth and FHSS IEEE 802.11. Assume the rest of parameters are the same as (b).

d. Produce a computer plot to illustrate the relation between r_{in} and distance-power gradient of the medium for values of α between 1.5 and 6. Using the computer plot, discuss the impact of medium on the range of interference between Bluetooth and FHSS IEEE 802.11. Assume the rest of parameters are the same as (b).

e. Repeat (c) and (d) for $P_{BT} = 10$ dBm. Compare the results with associated results in the previous parts and discuss the effects of power level in the interference.

13.6 Consider the FHSS IEEE 802.11 and Bluetooth interference scenario of Figure 13.19.

a. Assuming that the acceptable error rate for the Bluetooth is 10^{-5}, determine the S_{min} that supports this error rate. Use equations in Table 3A.1 for calculation of S_{min}.

b. Using S_{min} of (a) and Eq. (13.3), calculate r_{in} for $d = 10$m, $\alpha = 2$, $P_{BT} = 0$ dBm and $P_{AP} = 20$ dBm.

c. Produce a computer plot to illustrate the relation between r_{in} and acceptable error rates between 10^{-2} and 10^{-7}(in logarithmic form). Using the computer plot discuss the impact of error rate requirement on the range of interference between FHSS IEEE 802.11 and Bluetooth. Assume the rest of parameters are the same as (b).

d. Produce a computer plot to illustrate the relation between r_{in} and distance-power gradient of the medium for values of α between 1.5 and 6. Using the computer plot discuss the impact of medium on the range of interference between FHSS IEEE 802.11 and Bluetooth. Assume the rest of parameters are the same as (b).

13.7 Repeat Problem 13.5 if the FHSS IEEE 802.11 device is replaced by a DSSS IEEE 802.11 device.

13.8 An FHSS IEEE 802.11 and a Bluetooth device are operating in close vicinity to each other. Generate a computer plot illustrating the probability of collision of their packets versus the size of the FHSS packet. Using the results of computer plots, explain the impact of packet length on the probability of collision between FHSS IEEE 802.11 and Bluetooth. Note that the maximum length of the 802.11 packets is specified by the standard.

13.9 Repeat Problem 13.7 for interference analysis between the DSSS IEEE 802.11 and Bluetooth.

13.10 Repeat Problem 13.7 for interference analysis between the CCK IEEE 802.11b and Bluetooth.

CHAPTER 14

WIRELESS GEOLOCATION SYSTEMS

14.1 Introduction

14.2 What is Wireless Geolocation?

14.3 Wireless Geolocation System Architecture

14.4 Technologies for Wireless Geolocation

14.4.1 Direction-Based Techniques
14.4.2 Distance-Based Techniques
14.4.3 Fingerprinting-Based Techniques

14.5 Geolocation Standards for E-911 Services

14.6 Performance Measures for Geolocation Systems

Questions

Problems

14.1 INTRODUCTION

Geolocation, position location, and radiolocation are terms that are widely used today to indicate the ability to determine the location of an MS. Location usually implies the coordinates of the MS that may be in two or three dimensions, and usually includes information such as the latitude and longitude where the MS is located. In indoor areas and within buildings, alternative coordinates and visualization techniques may be employed. Geolocation technologies are gaining prominence in the wireless market for several reasons, primarily the FCC mandate requiring all wireless cellular carriers to be able to provide the location of emergency 911 callers to a *public safety answering point.* However, geolocation technology has proved to be significant for both military and commercial applications in general beyond emergency location. Commercial applications include the need for hospitals to locate patients and equipment in a timely fashion, in homes to locate children and pets, and in the evolution toward 4G networks, the need to provide *location aware* services. In the military and public sector, enabling soldiers, policemen, and firefighters with knowledge of their location and the location of other personnel, victims, exits, dangers, and so on proves to be invaluable. The GPS has been the most successful positioning technique in outdoor areas, and we now see the GPS receiver as an inexpensive commonplace gadget. Although GPS has been hugely successful, it also has several drawbacks for use in the applications that we have discussed so far, especially in indoor areas. In this chapter, we discuss position location issues in today's wireless networks, alternative technologies that are being investigated and standardized for position location in outdoor and indoor areas, and trends in this area.

14.2 WHAT IS WIRELESS GEOLOCATION?

The term "location-based service" is used to denote services provided to mobile users based on their geographic location, position, or known presence. These are primarily based on a geolocation infrastructure and system put in place to obtain location information of users. Positioning systems have found a variety of applications both in the civilian and military environments. There are numerous such applications that are already available today such as mapping services (that provide driving directions), information services (that provide local news, weather, traffic, etc.), and concierge services (for making dinner reservations, movie tickets, directory services, etc.). Commercially, content, advertising, and personalization services that are location dependent are being deployed today. We discuss example indoor and outdoor applications that are becoming increasingly important.

Indoor geolocation applications traditionally have been directed toward locating people and assets within buildings. Finding mentally impaired patients in hospitals and portable equipment such as projectors, wheelchairs, and so on that are often moved and never returned to a tractable location are two common examples. The so-called *personal locator services* (PLS) [KOS00], which could also operate outdoors, employ a *locator device* that resides with a person whose location is to be determined. There are two possible scenarios—in the first case, someone requests a service to provide them with the location of the individual and appropriate steps are taken to determine the person's location. In the second case, the person is lost or in some other dire predicament and can employ a panic button to request help. Here the locator service will determine the location and provide the requested assistance. For locating equipment, only the former scenario applies.

Existing communications and computing environments, both in residential areas and offices, have been statically configured, making the task of reconfiguration extremely complex and cumbersome and requiring manual intervention. To overcome this inconvenience, smart spaces and smart office environments are being considered for deployment that can automatically change their functionality depending on the context [WAR97]. Such *context-aware* networks are based on awareness of who or what is present around them. With location awareness, computing devices ranging from small PDAs to desktops and Internet appliances could personalize and adapt themselves to their current set of users, each requiring their own services from the smart environment. For this purpose, not only should the smart space be aware of who is present, but it should also be aware of where the user is located and whether there are other mobile devices in the vicinity. For instance, a handheld computer should be able to automatically determine the closest printer to print a document in an office environment. Such nontraditional applications also demand geolocation services.

There are several outdoor geolocation applications, the most common of which is simply the application of locating one's own self using GPS while traveling on the road. Information technology has increased the number of applications far beyond this simple self-location application.

The term *telematics* is used to imply the convergence of telecommunications and information processing, and it has since then evolved to refer to automobile systems that combine the GPS location mechanism with wireless communications for services such as automatic roadside assistance, remote diagnostics, and content delivery (information and entertainment) to the automobile. A good example of such a system is General Motors's OnStar system [OnSweb].

Intelligent transportation systems (ITS) refer to the ability to autonomously navigate vehicles while making use of the latest traffic information, road conditions, travel duration, and so on. This includes fleet management as well as the automatic steering of vehicles. In order to obtain relevant information from service providers or servers across a network or the Internet, the vehicles should be able to provide their location and destination information. Alternatively, the service provider should be able to determine the vehicle's location.

Wireless enhanced-911 or E-911 services, by far, have provided to be the biggest catalyst for investment and development of geolocation technology suitable for cellular communications. A caller on a wired telephone to an E-911 service is immediately located because the location of the fixed telephone is known with an accuracy of within a couple of rooms in a building. If the same caller is on a mobile telephone, there does not currently exist any technology that can obtain the location except that the caller was connected to a particular BS. In order to improve emergency response, the FCC had mandated that all cellular telephones, PCS handheld communicators, and specialized mobile radios should provide geolocation services by October 1, 2001. (This date has since been postponed.) The mandate requires that a public safety answering point (PSAP) be able to locate the mobile device to within 50 m, for 67 percent of E-911 calls and 150 m for 95 percent of the calls if a *handset-based* geolocation technology is used and to within 100 m (300 m) for 67 percent (95 percent) of calls if *network-based* geolocation technology is employed. We discuss handset-based and network-based technologies in subsequent sections.

14.3 WIRELESS GEOLOCATION SYSTEM ARCHITECTURE

A functional architecture of a geolocation system is shown in Figure 14.1 [CHA99]. The two essential functional ingredients for position location are the location estimation of the MS U and this information with appropriate attributes shared with the network $\boldsymbol{N}$. Geolocation systems measure parameters of radio signals that travel from a mobile and a fixed set of receivers or from a fixed set of transmitters to a mobile receiver. There are thus two ways in which the actual estimate of the location of the MS can be obtained. In a *self-positioning system,* the MS locates its own position using measurements of its distance or direction from known locations of transmitters (for example GPS receivers). In some cases, *dead reckoning*, a predictive method of estimating the position of the mobile by applying the course and distance traveled since to a previously determined position, could be employed. Self-positioning systems are often referred to as mobile-based or terminal-centric [MEY96] positioning systems. In *remote positioning systems,* receivers at known locations on a network together compute the location of a mobile transmitter using the measurements of the distance or direction of this mobile from each of the re-

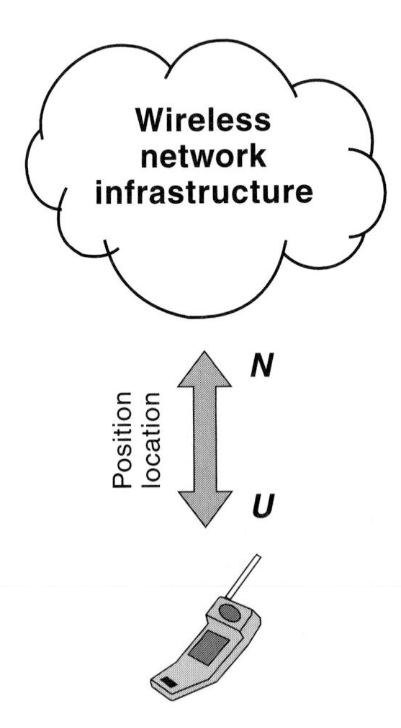

Figure 14.1 Functional architecture of a geolocation system.

ceivers [DRA98]. Remote positioning systems are also called network-based or network-centric [MEY96] positioning systems. Network-based positioning systems have the advantage that the MS can be implemented as a simple transceiver with small size and low-power consumption for easy carrying or attachment to valuable equipment as a tag. In addition, it is possible to have *indirect* remote or self-positioning systems where the mobile may transmit information about its location to a location control center, or the location control center transmits the location of each mobile to itself through an appropriate communications channel.

An example of a geolocation system architecture [KOS00] is shown in Figure 14.2. A geolocation service provider provides location information and location aware services to subscribers. Upon a request from a subscriber for location

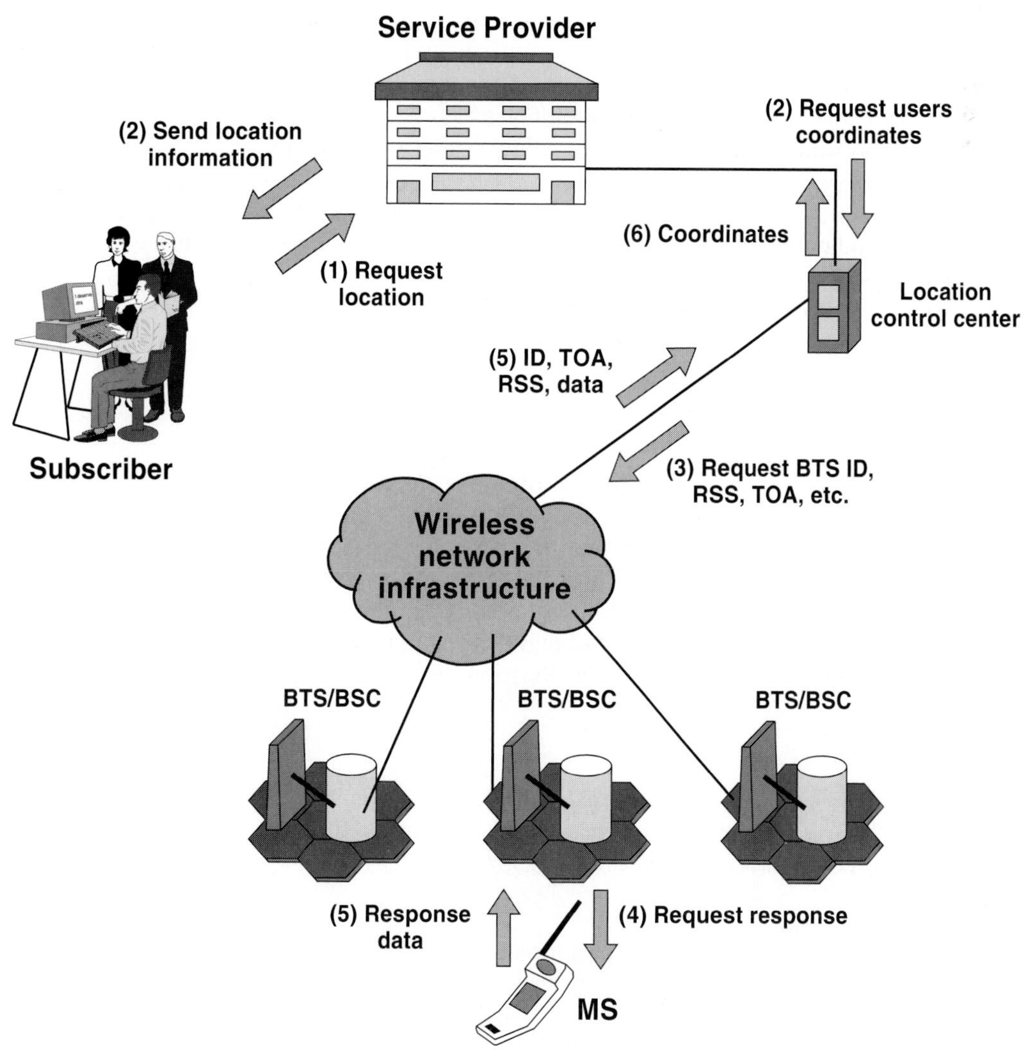

Figure 14.2 General architecture of a geolocation system.

information about an MS, the service provider will contact a location control center querying it for the coordinates of the MS. This subscriber could be a commercial subscriber desiring to track a mobile device or a PSAP trying to answer an E-911 call. The location control center will gather information required to compute the MS's location. This information could be parameters such as received signal strength, BTS ID, TOA of signals, and so on that we discuss later. Depending on past information about the MS, a set of BSs could be used to page the MS, and directly or indirectly obtain the location parameters. These are sometimes called *geolocation base* stations (GBSs). Once this information is collected, the location control center can determine the location of the mobile with certain accuracy and convey this information to the service provider. The service provider will then use this information to visually display the MS's location to the subscriber. Sometimes the subscriber could be the MS itself, in which case the messaging and architecture will be simplified, especially if the application involves self-positioning.

Example 14.1: Indirect Remote Positioning

In indirect remote positioning, an E-911 PSAP requires the location information of a caller. If a mobile-based positioning system is used, the MS determines its own position either using GPS or signals from multiple BSs. This information has to be transmitted to the location control center by the mobile terminal through one of the BSs.

14.4 TECHNOLOGIES FOR WIRELESS GEOLOCATION

The location of an MS can be determined as follows. Let us consider, for example, a remote-positioning system where the GBSs are together determining the MS's position. A similar approach is applicable for self-positioning systems. It is possible to exploit characteristics of radio signals transmitted by an MS to fixed receivers of known location to determine the location of the MS. The GBSs measure certain signal characteristics and make an estimate of the location of the MS based on the knowledge of their own location. The general problem can be stated as follows:

> The locations of N receivers (GBSs) are known via their coordinates (x_i, y_i) for i = 1, 2, 3, . . ., N. We need to determine the location of the MS (x_m, y_m) using characteristics of the signals received by these transmitters.

Clearly, in order to determine (x_m, y_m), the distance or direction (or both) of the MS must be estimated by several of the GBSs from their received signals. Distances can be determined using properties of the received signal such as the signal strength, the signal phase, or the time of arrival. The directions of the MS can be determined from the angle of arrival of the received signal.

14.4.1 Direction-Based Techniques

The angle of arrival (AOA) geolocation technique uses the direction of arrival of the received signal to determine the location of the MS as shown in Figure 14.3. The receiver measures the direction of received signals (i.e., the AOA with respect to a fixed direction [east in Figure 14.3]) from the target transmitter using directional antennas or antenna arrays. If the accuracy of the direction measurement (roughly the beam width of the antenna array) is $\pm\theta_s$, the AOA measurement at the receiver will restrict the transmitter position around the LOS signal path with an angular spread of $2\theta_s$. Two such AOA measurements will provide a position fix as illustrated in Figure 14.3.

The accuracy of the position estimation depends on where the transmitter is located with respect to the receivers. If the transmitter lies in between the two receivers along a straight line, AOA measurements will not be able to provide a position fix. As a result, more than two receivers are usually needed to improve the location accuracy. For macrocellular environments, where the primary scatterers

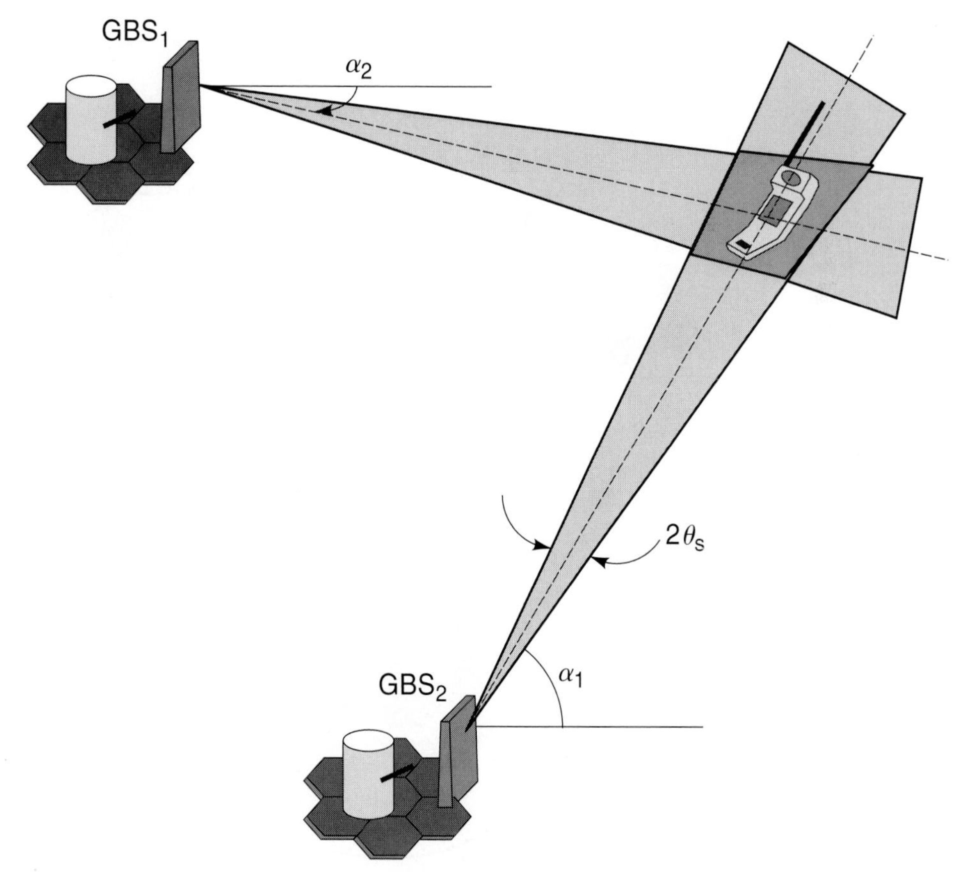

Figure 14.3 AOA technique for geolocation.

are located around the transmitter and far away from the receivers (the GBSs), the AOA method can provide acceptable location accuracy [CAF98]. But dramatically large location errors occur if the LOS signal path is blocked and the AOA of a reflected or a scattered signal component is used for estimating the direction. In indoor environments, surrounding objects or walls mostly block the LOS signal path. Thus the AOA technique is not suitable for indoor geolocation systems. In addition, this requires placing expensive array antennas at the receivers to track the direction of arrival of the signal. Although this is a feasible option in the next generation cellular systems where smart antennas are expected to be deployed to increase capacity, it is in general not a good solution for low-cost indoor applications.

14.4.2 Distance-Based Techniques

With the received signal strength, the TOA or the time difference of arrival techniques, the distances between the MS and receiver are estimated. In such a case, three measurements are required to estimate the position of the mobile in two dimensions, and four measurements are required for estimating the position in three dimensions. In Figure 14.4, the need for three measurements for estimating the po-

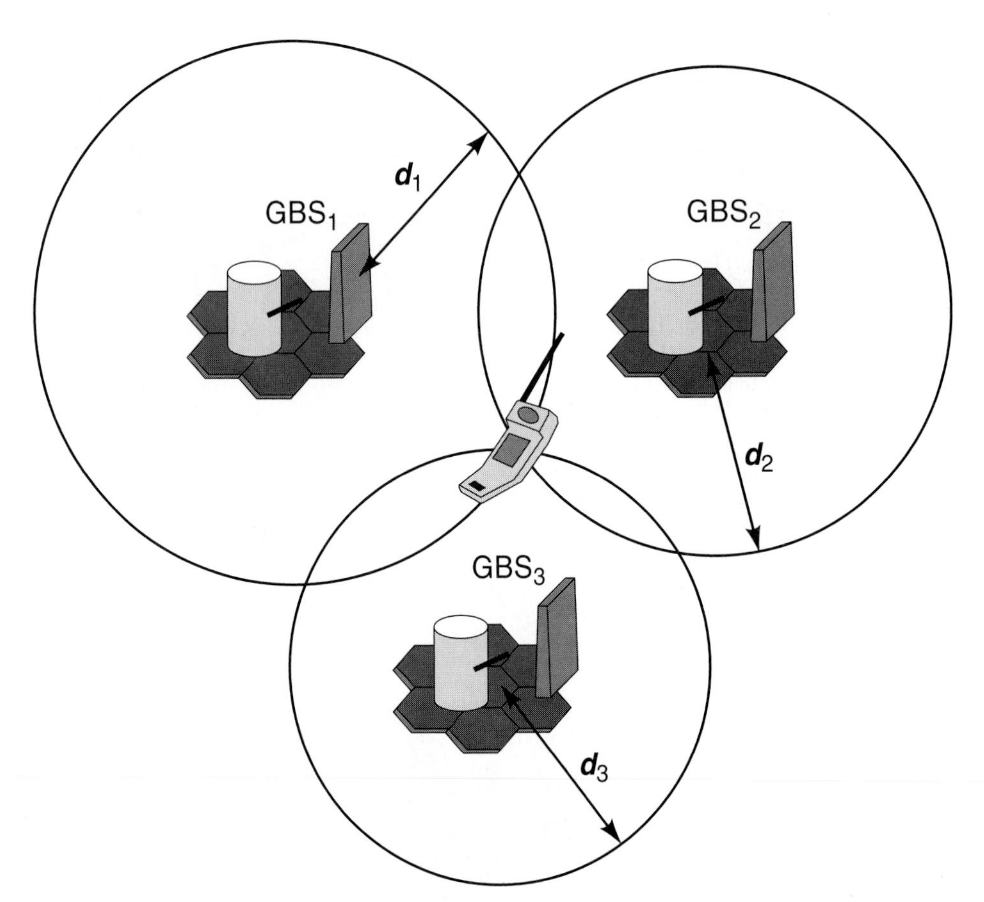

Figure 14.4 Direction-based geolocation technique.

sition in two dimensions is illustrated. If the distance between the receiver and the mobile is estimated to be d, it is obvious that the mobile could be located on a circle of radius d centered on the receiver. A second measurement reduces the position ambiguity to the endpoints of the chord that is common to the two circles. The third measurement provides a fix on the location of the mobile.

14.4.2.1 Arrival Time Methods

A transmitted signal travels 3×10^8 m in one second in air or free space, and this property can be exploited to determine the distance between the transmitter and receiver. This is the TOA technique that is employed with some modifications in current GPS receivers [KAP96], as well as certain E-911 location systems. When a GBS detects a signal, its absolute TOA is determined. If the time at which the MS transmitted the signal is known, the difference in the two times will give an estimate of the time taken by the signal to arrive at the GBS from the MS. Three distinct measurements (in two dimensions) and four distinct measurements (in three dimensions) can be employed to determine the location of the mobile. The TOA technique provides circles centered on the mobile or fixed transceiver as described previously (see Figure 14.4).

Example 14.2: Commercial Indoor Location Systems Based on TOA

Recently, a few commercial products for indoor geolocation appeared in the market [WER98]. The overall system architecture of a system is shown in Figure 14.5. These systems use simple-structured *tags* that can be attached to valuable assets or personnel badges. Indoor areas are divided into cells with each cell being served by a *cell controller*. The cell controller is connected to a number of antennas (16 in [WER98]) located at known positions. To locate the tag position, a cell controller transmits a 2.4 GHz spread spectrum signal in the unlicensed ISM bands through different antennas in time division multiplexed mode. Upon receiving signals from the cell controller antenna network, tags simply change the frequency of the received signal to another portion of the available unlicensed bands, either in 2.4 GHz or 5.8 GHz, and transmit the signal back to the cell controller with tag ID information phase-modulated onto the signal. The distance between tag and antenna is determined by measuring round trip time of flight. With the measured distances from tag to antennas, the tag position can be obtained using the TOA method. A host computer is connected to each cell controller, through a TCP/IP network or other means, to manage the location information of the tags. Because the cell controller generates the signal and measures round trip time of flight, there is no need to synchronize the clocks of tags and antennas.

The multipath effect is one of the limiting factors for indoor geolocation (see Chapter 2 for a discussion). Without multipath signal components, the TOA can be easily determined from the autocorrelation function of the spread spectrum signal. The autocorrelation is two chips wide, and the time to rise from the noise floor to the peak is one chip. If the chipping rate were 1 MHz, it would take 1,000 ns to rise from the noise floor to the peak, providing a "ruler" with a thousand 30-cm increments. In this manner, a 40 MHz chipping rate, chosen for the PinPoint system, provides a ruler of 25 ns that provides real-world increments of about 3.8 m. Because of regulatory restrictions in the 2.4 and 5.8 GHz unlicensed bands, faster chipping rates are not easy to achieve, and signal-processing tech-

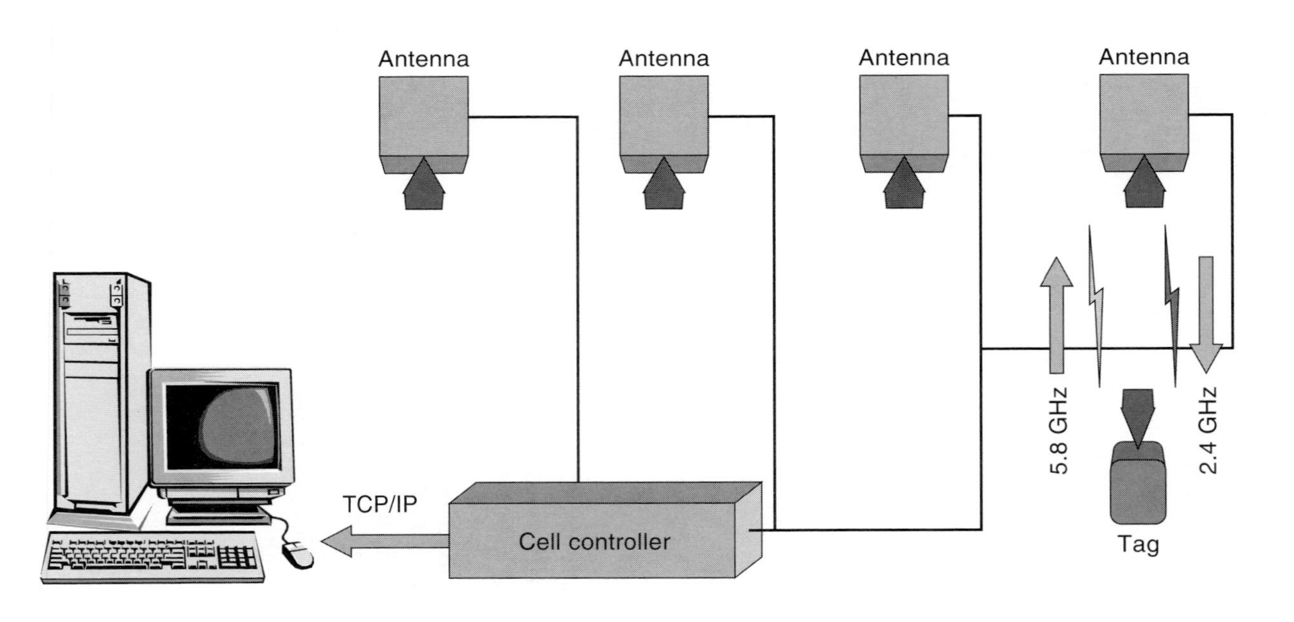

Figure 14.5 The PinPoint local positioning system.

niques must be used to further improve the accuracy. If different frequency bands are used for uplink and downlink communications, the interference between the channels can be further isolated.

In GPS, the *time difference of arrival* (TDOA) technique is used where the differences in the TOAs are used to locate the mobile. The TDOA technique defines hyperbolas (rather than circles) on which the transmitter must be located with foci at the receivers. Three or more TDOA measurements provide a position fix at the intersection of hyperbolas. Even though geometric interpretation can be used to calculate the intersection of circles or hyperbolas, when there are errors, estimates have to be used. Exact solutions and Taylor series approximations are available [COM98] for solving these equations. Compared with the TOA method, the main advantage of the TDOA method is that it does not require the knowledge of the transmit time from the transmitter. As a result, strict time synchronization between the MS and the GBSs is not required. However, the TDOA method requires time synchronization among all the receivers used for geolocation.

A recursive least squares estimate is used when there are errors in the distance measurements. Let the distance d_i from the ith GBS be determined from the absolute TOA of the signal it receives as $d_i = c \times \tau$ where c is the velocity of light and τ is the time taken by the signal to reach the GBS. If the location of the ith GBS is (x_i, y_i) and the location of the mobile is (x,y), we have N equations of the form:

$$f_i(x, y) = (x - x_i)^2 + (y - y_i)^2 - d_i^2 = 0 \tag{14.1}$$

for $i = 1, 2, \ldots, N$. As a geolocation problem, extensive research has been done to improve upon the accuracy of algorithms that are used to estimate the position of a mobile. Especially when N is more than three or four (thus providing redundancy in the measurements), information in the redundant measurements can be used to reduce errors that are introduced by noise, environment, multipath, and so on, [KAP96]. Figure 14.6 shows the example of using a recursive least squares technique to arrive at the location of the MS. Signals from seven receivers are combined to arrive at this location.

Wireless systems that employ the TOA (or TDOA) technique employ pulse transmission, phase information, or spread spectrum techniques to form time estimates. For instance, the time difference between two signals received for either self-positioning or remote positioning, can be estimated from their cross-correlation. As already mentioned, TOA techniques are generally superior compared with AOA techniques. In [CAF98], it has been reported that TOA techniques outperform the AOA technique by approximately 100 m in absolute position error in cellular systems with three BSs, and the incremental returns are worse if more BSs are used.

14.4.2.2 Signal Strength Method

If the transmitted power at the MS is known, measuring the RSS at the GBS can provide an estimate of the distance between the transmitter and the receiver using known mathematical models for radio signal path loss that depend on distances (see Chapter 2). As with the TOA method, the measured distance will determine a circle, centered on the receiver, on which the mobile transmitter must lie.

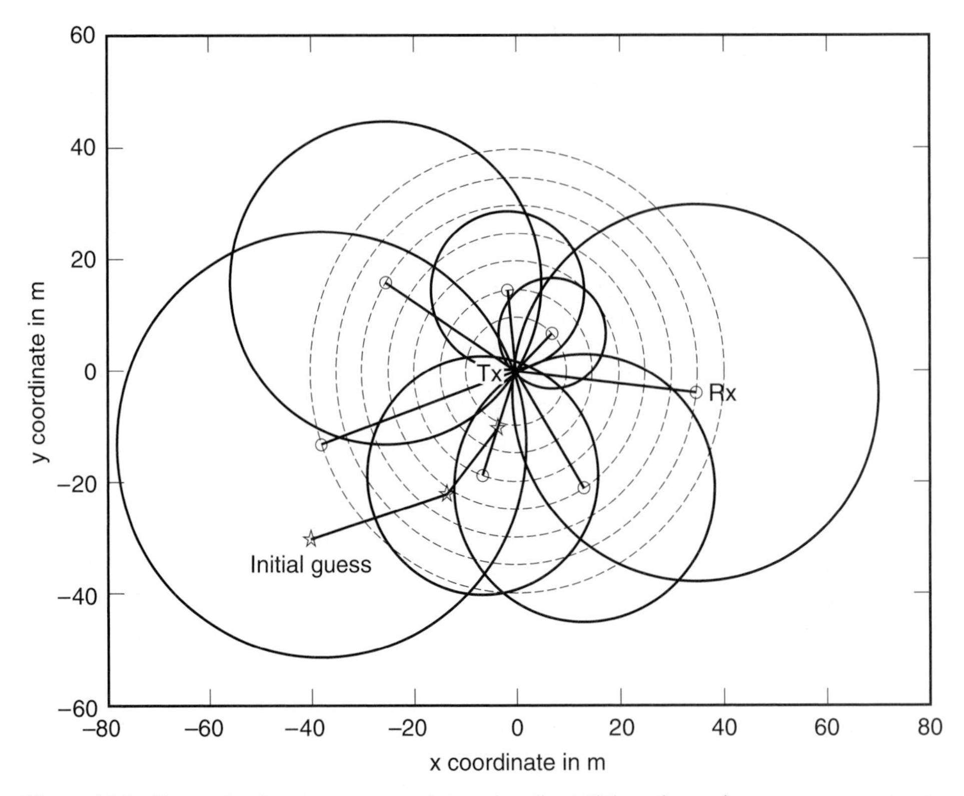

Figure 14.6 Recursive least squares to determine the MS location using measurements at seven GBSs.

This technique results in a low complexity receiver for a self-positioning system. This method is, however, very unreliable because of the wide variety in path loss models and the large standard deviations in the errors associated with these models due to shadow fading effects. Receivers do not distinguish between signal strength in the LOS path and in reflected paths [MEY96]. Especially indoors, the power distance gradients can vary anywhere between 15–20 dB/decade to as high as 70 dB/decade. Also, these gradients and other parameters employed in path loss models are site specific. As a result, this technique cannot be employed in situations where the required accuracy is a few meters. The accuracy of this method can be improved by utilizing premeasured RSS contours centered at the receiver [FIG69] and multiple measurements at several BSs [KOS00]. A fuzzy logic algorithm was shown in [SON94] to be able to significantly improve the location accuracy.

Example 14.3: A Commercial Geolocation System Based on RSS

The infrastructure of Paltrack indoor geolocation system [PALweb], developed by Sovereign Technologies Corp., consists of tags, antennas, cell controllers, and an administrative software server system. The PalTrack system utilizes a network structure that resides on an RS-485 node platform. A network of transceivers is located at known positions within the serving area while the transmitter tags are attached to assets. The tag transmitters transmit a unique identification code at 418 MHz frequency band to a network of transceivers when on motion or at predefined time intervals. Transceivers estimate the tag location by measuring RSS and utilize a robust RSS-based algorithm patented by Sovereign Technologies Corp. The master transceiver collects measured information from the transceivers and relays it to a PC-based server system. The accuracy for PalTrack is 0.6 to 2.4 m. The key component of the PalTrack system is the RSS-based geolocation algorithm.

14.3.2.3 Received Signal Phase Method

The received signal phase is another possible geolocation metric. It is well known that with the aid of reference receivers to measure the carrier phase, differential GPS (DGPS) can improve the location accuracy from about 20 m to within 1 m compared with standard GPS, which only uses range measurements [KAP96]. One problem associated with the phase measurements lies in the ambiguity resulting from the periodic property (with period 2π) of the signal phase while the standard range measurements are unambiguous. Consequently, in DGPS, the ambiguous carrier phase measurement is used for fine-tuning the range measurement. A complementary Kalman filter is used to combine the low-noise ambiguous carrier phase measurements and the unambiguous but noisier range measurements [KAP96]. For an indoor geolocation system, it is possible to use the signal phase method together with the TOA/TDOA or RSS method to fine-tune the location estimate. However unlike DGPS, where the LOS signal path is always observed, the serious multipath condition of the indoor geolocation environment causes more errors in the phase measurements.

14.4.3 Fingerprinting-Based Techniques

Recently, signal fingerprinting has been adopted as yet another technique for position location [KOS00]. The received signal is extremely site-specific because of its dependence on the terrain and intervening obstacles. So the multipath structure of the channel is unique to every location and can be considered as a fingerprint or signature of the location if the same RF signal is transmitted from that location. This property has been exploited in proprietary systems to develop a "signature database" of a location grid in specific service areas. The received signal is measured as a vehicle moves along this grid and recorded in the signature database. When another vehicle moves in the same area, the signal received from it is compared with the entry in the database, and thus its location is determined. Such a scheme may also be useful for indoor applications where the multipath structure in an area can be exploited.

Example 14.4: A Commercial Geolocation System Based on Fingerprinting

A company called U.S. Wireless Corporation has designed and implemented a system based on this scheme called RadioCamera in downtown Oakland in California [RADweb]. The mobile transmits RF signals, which scatter because of multipath conditions. RadioCamera™ takes measurements of the RF signals and collects all the multipath rays. A *location pattern signature* is developed using the multipath rays. The location signature is compared with a learned database, and a location is determined. Continuous measurements of the location pattern signature provide tracking.

14.5 GEOLOCATION STANDARDS FOR E-911 SERVICES

Commercial geolocation products have already appeared in the market outside of traditional GPS systems that are extremely popular. The enhanced 911 services are still the primary driving force for geolocation services. The options for E-911 services when they were first mandated included traditional GPS or a network centric approach based on TDOA techniques. GPS, especially after the elimination of selective availability of the signal by the U.S. government, provides sufficient accuracy for E-911 systems. The one disadvantage of GPS is that the time to first fix (TTFF) can be very long depending on what satellite constellation a MS may be able to see. There is also the problem of using GPS in urban canyons. However, compared with stand-alone GPS, network-centric approaches can provide a faster TTFF but are unreliable and inaccurate.

To solve this problem, a new technique called *assisted GPS* (AGPS) has been proposed whereby an entity in the cellular network is enabled with a GPS receiver that can see the same satellites as the MS. By predicting what signal an MS may see and sending that information to the MS, the network entity can enable a faster TTFF, shortening it from minutes to a second or less [DJU01]. Assisted GPS also enables the network entity to detect signals with a weaker signal strength than an MS and send a *sensitivity assistance message* to the MS. It can also reduce the satellite search space for an MS by informing the MS of what satellites may be visible and what will be their expected code phase for synchronization.

Assisted GPS is now the technology of choice and is being standardized for AMPS and IS-95 based cellular telephones. It is being incorporated in the TIA 45.1 and 45.5 standards and will likely be approved for the TDMA (TIA TR45.3) standard as well. For GSM telephones, ETSI and T1P1.5 are looking at AGPS, network-based TOA, and enhanced-observed time difference (E-OTD) technologies [ZHA00].

Example 14.5: TDOA in GSM

In GSM, the MS already observes bursts from different BSs and determines the time interval between the receptions of bursts from two different BTSs. This difference is called the observed time difference (OTD). The real time difference (RTD) between BSs is also known. These two values can be used in a TDOA technique similar to the one discussed earlier to determine the location of the MS.

Example 14.6: TDOA in CDMA

In CDMA, the BSs are already synchronized and time difference measurement is easier than in GSM. By measuring the phase delay between CDMA pilot signals, the MS can obtain range information, and this method is called *advanced forward link trilateration*. The calculations can either be done in the MS or the phase differences could be reported to the network, which will then calculate the location.

Architectural changes will be required to the backbone of the cellular telephone networks to provide geolocation services. Figure 14.7 shows one such architecture [MEY96]. The MSC communicates the location parameters to the E-911 selective router that routes the information to the appropriate PSAP. An automatic location information (ALI) database is queried by the PSAP with the location parameters and receives the location back along with information related to the mobile station. A geographic information service (GIS) can be added to the PSAP to provide a visual display pinpointing the roads, streets, addresses, jurisdiction, and so on. Eventually a wireless intelligent network or data backbone could be incorporated to interface the geolocation technology system with the wireless network backbone. Several possibilities are being considered to convey data from the geolocation BSs to the PSAP. One approach is to use SS-7 as proposed in [MEY96]. Another approach is to use a data backbone such as GPRS [CHA99].

14.6 PERFORMANCE MEASURES FOR GEOLOCATION SYSTEMS

In this section, we consider performance benchmarking of geolocation systems [TEK98] and compare their relative advantages and disadvantages. Wireless systems have traditionally focused on telecommunications performance issues such as QoS, grade of service, BERs, capacity, reliability, and coverage. For geolocation systems, some of these performance issues are still valid although new performance benchmarks need to be introduced.

Table 14.1 compares performance measures for telecommunications and geolocation systems based on [TEK98].

One of the most important performance measures of a geolocation system is the accuracy with which the location is determined. This is similar to the BER or packet error rate requirements in telecommunications systems. As in the case of BER, the actual benchmark values may be different depending on the application in question. For example, voice packets can tolerate a BER of 1 percent, but data packets need a BER of at least 10^{-6}. In the same way, outdoor position location applications demand a lower accuracy compared with indoor applications.

Location system accuracy is often defined as the area of uncertainty around the exact location where a percentage of repeated location measurements are reported. For example, 67 percent of the measurements of the location of an MS lie within 50 m of the actual location or 95 percent of the measurements lie within 1 m

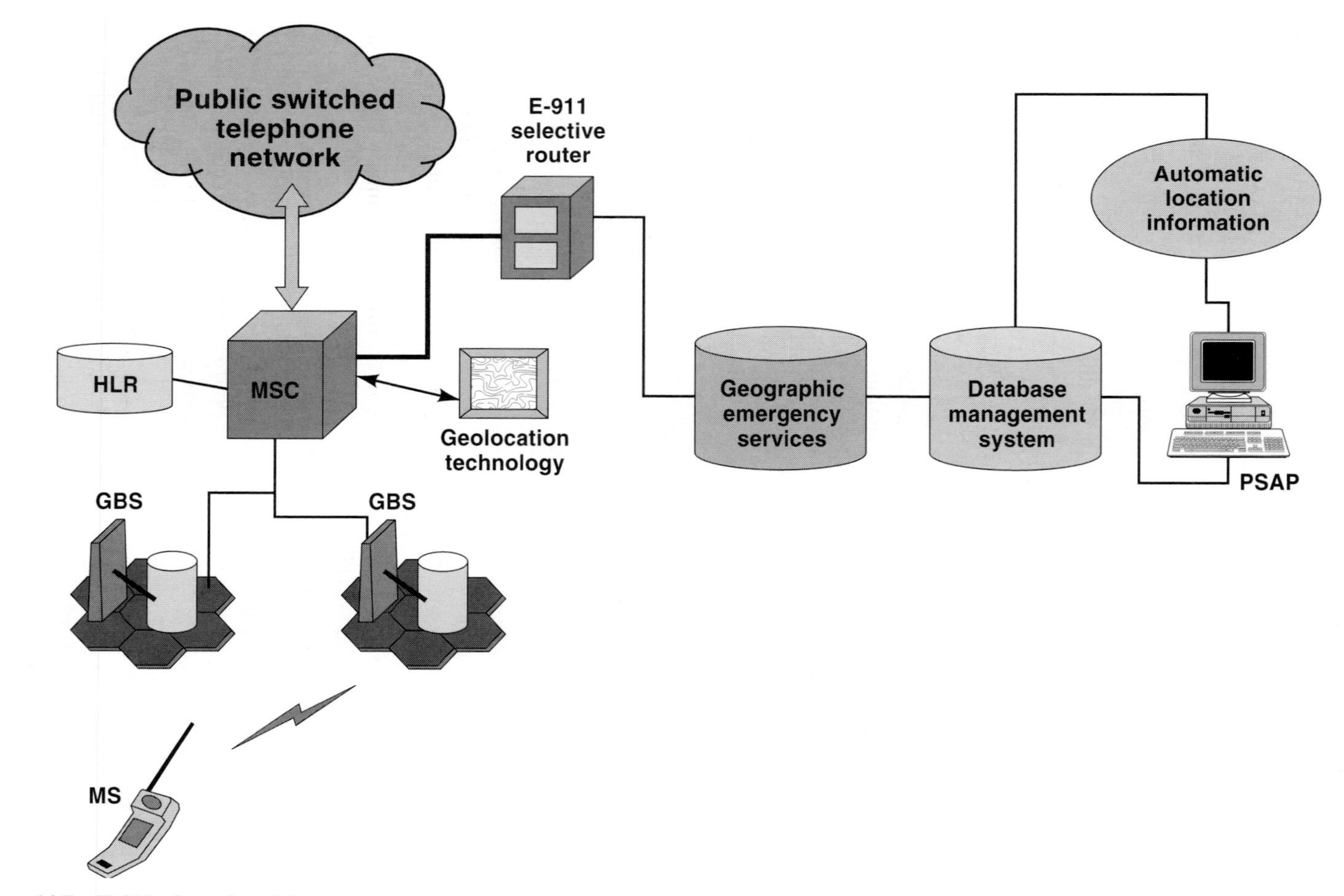

Figure 14.7 E-911 phase 2 architecture.

Table 14.1 Comparison of Performance Measures for Telecommunications and Geolocation Systems

Telecommunications Systems	Geolocation Systems
Quality of service	*Accuracy of service*
• Signal to interference ratio	• Percentage of calls located to within an accuracy of δ meters
• Packet error rate	• Distribution of distance error at a geolocation receiver
• Bit error rate	
Grade of service	*Location availability*
• Call blocking probability	• Percent of location requests not fulfilled
• Availability of resources	• Unacceptable uncertainty in location
• Unacceptable quality	
Coverage	*Coverage*
Capacity	*Capacity*
• Subscriber density that can be handled	• Location requests/frequency that can be handled
Miscellaneous	*Miscellaneous*
• Delay in call setup	• Delay in location computation
• Reliability	• Reliability
• Database look-up time	• Database look-up time
• Management and complexity	• Management and complexity

of the actual location. This accuracy heavily depends on the radio propagation environment, receiver design, noise and interference characteristics, number of redundant measurements available for the same location, and the complexity of signal processing performed. In [PAH98], [KRI98], the distribution of the distance error that ultimately affects the area of uncertainty is discussed in detail based on measurements and simulations.

The grade of service for telecommunication systems is usually the call-blocking rate during the peak hour. In a similar manner, the probability that a location request will not be fulfilled is a measure of the grade of service for a geolocation system. The location request will not be fulfilled if TOA, AOA, or other measurements are not available in sufficient number or the measurements lead to unacceptable location accuracy.

Coverage in telecommunications systems is related to the service area where, at a bare minimum, access to the wireless network is possible. For geolocation systems, coverage corresponds to the availability of a sufficient number of TOA, AOA, RSS, or fingerprint measurements to perform a location computation.

Finally several other issues [TEK98] are also important in geolocation systems in a manner similar to telecommunications systems—delay in triggering a location measurement, location algorithm calculation time, network transmission delay, database look-up time, end-to-end delay between the time a location request is made and the location information is received, and so on. Reliability—the mean time between failures and the mean time to repair—management, and complexity are also important.

Table 14.2 Comparison of Geolocation Techniques

Geolocation Technique	Coverage	Accuracy	Delay	Complexity/Cost/Others
Mobile Centric based on GPS	No indoor coverage Fails in radio shadows	Good in rural and open areas Poor in urban and indoor areas	Low backhaul requirements—forward only location estimate Long time to first fix	No changes to network, but changes to all handsets System upgrades limited by deployed handset base Privacy is user controlled Robust under failure of some network components
Network Centric AOA	Good in outdoor areas	Poor	Low: only angular information needs to be transmitted	High-BS cost due to complex antenna array systems BSs need to be changed No changes to handsets Privacy is network controlled
Network Centric TOA/TDOA	Good	Medium, sometimes unreliable	Low to high depending on how many samples need to be transmitted	Additional investment in infrastructure No changes to handsets Privacy is network controlled
Assisted GPS	Good; indoor coverage is suspect	Superior	Short time to first fix Lots of signaling between network and mobile	System evolves with network upgrades, but both the network and handsets need changes Privacy is partly user controlled Interoperability issues between network and mobile terminals Computation is offloaded to the network so that there is minimum impact on handset battery life

Table 14.2 compares different geolocation approaches based on some of the performance measures discussed earlier [DJU01], [ZAG98]. The table is self-explanatory.

QUESTIONS

14.1 Explain the differences between GPS, wireless cellular assisted GPS, and indoor geolocation systems.

14.2 Differentiate between remote and self-positioning systems.

14.3 Why is RSS not a very good measure of the distance between a transmitter and a receiver?

14.4 Compare mobile-centric and network-centric geolocation techniques in terms of complexity and accuracy.

14.5 Why are cellular service providers interested in location-dependent services? Give some examples of location-dependent services.

14.6 What are E-911 services and who has mandated these services?

14.7 What are the basic elements of a wireless geolocation system?

14.8 Name the three major metrics used for location finding and explain how they are implemented in a system.

14.9 Why are AOA techniques not popular in indoor geolocation applications?

PROBLEMS

14.1 Two base stations located at (500, 150) and (200, 200) are measuring the angle of arrival of the signal from a mobile terminal with respect to the x-axis. The first base station measures this angle as 45 degrees and the second as 75 degrees. What are the coordinates of the mobile terminal?

14.2 In Problem 14.1, what happens if the first base station incorrectly measures the AOA from the mobile terminal as 50 degrees? 30 degrees?

14.3 Base stations A, B, and C located at (50, 50), (300, 0), and (0, 134) are found to be at distances 90, 200, and 100 m from a mobile terminal. Draw circles corresponding to these values and try to determine the location of the mobile terminal.

14.4 In Problem 14.2, what happens if the mobile incorrectly measures the distance from base station B as 100 m? 300 m?

ACRONYMS AND ABBREVIATIONS

π/4-DQPSK π/4 differential quadrature phase shift keying
1G First generation cellular
2G Second generation cellular
3G Third generation cellular
AAL ATM adaptation layer
ACF Association control function
ACH Access feedback channel
ACI Adjacent channel interference
ACTS Advanced communications technologies and services
ADPCM Adaptive differential pulse code modulation
AGCH Access grant channel
ALI Automatic location information
AM Amplitude modulation
AMPS Advanced mobile phone system
ANSI American National Standards Institute
AOA Angle of arrival
AP Access point
ARIB Association of Radio Industries and Businesses
ARN Authentication random number
ARP Address resolution protocol
ARQ Automatic repeat request
ASCH Association control channel
ASK Amplitude shift keying
ASN Authentication sequence number
ATM Asynchronous transfer mode
AuC Authentication center
AWGN Additive white Gaussian noise
BCCH Broadcast control channel
BCH Broadcast channel
BER Bit error rate
BG Border gateway
BLER Block error rate

BRAN Broadband radio access networks
BS Base station
BSA Basic service area
BSC Base station controller
BSIC Base station identity code
BSS Basic service set
BSS Base station subsystem
BSSGP BSS gateway protocol
BTMA Busy tone multiple access
BTS Base transceiver subsystem
CCA Clear channel assignment
CCH Control channel
CCI Co-channel interference
CCITT International Telegraph and Telephone Consultative Committee
CCK Complementary code keying
CDMA Code division multiple access
CDPD Cellular digital packet data
CELP Code excited linear prediction
CEPT Committee of European Post and Telecommunications
CFP Contention free period
CLI Calling line identification
CLNP Connectionless network protocol
CN Correspondent node
COA Care of address
COFDM Coded orthogonal frequency division multiplexing
CONS Connection oriented
COST Co-operative for scientific and technical research
CPC Centralized power control
CSMA/CA Carrier sense multiple access with collision avoidance
CSMA/CD Carrier sense multiple access with collision detection
CT Cordless telephony
CW Contention window
DAB Digital audio broadcast
DBPSK Differential binary phase shift keying
DCA Dynamic channel allocation
DCC DLC connection control
DCF Distributed coordination function
DDCA Distributed dynamic channel assignment
DDP Dominant direct path
DECT Digital enhanced cordless telephone
DES Data encryption standard
DFE Decision feedback equalizer
DFIR Diffused IR

DFS Dynamic frequency selection
DGPS Differential GPS
DIFS DCF inter-frame spacing
DL Discrete logarithm
DLC Data link control
DLL Data link layer
DLOS Direct line of sight
DPC Distributed power control
DQPSK Differential quadrature phase shift keying
DSL Digital subscriber line
DSMA Digital sense multiple access
DSS Digital signature standard
DSSS Direct sequence spread spectrum
DVCC Digital verification color code
EDGE Enhanced data rates for global evolution
EFR Enhanced full rate
EIR Equipment identity register
EOTD Enhanced observed time difference
ESS Extended service set
ETACS Enhanced total access communications system
ETSI European Telecommunications Standards Institute
EU European Union
FA Foreign agent
FACCH Fast associated control channel
FCA Fixed channel allocation
FCC Federal communications commission
FCH Frequency correction channel
FCS Frame correction sequence
FDD Frequency division duplexing
FDMA Frequency division multiple access
FER Frame error rate
F-ES Fixed end system
FFT Fast Fourier transform
FHSS Frequency hopping spread spectrum
FM Frequency modulation
FSK Frequency shift keying
FT Fixed terminal
FTP File transfer protocol
GBS Geolocation base station
GFSK Gaussian frequency shift keying
GGSN Gateway GPRS support node
GIS Geographic information system
GMSC Gateway MSC

GMSK Gaussian minimum shift keying
GPRS General packet radio service
GPS Global positioning system
GR GPRS register
GSM Global system of mobile communications
GTP GPRS tunnel protocol
GUI Graphical user interface
HA Home agent
HAN Home area network
HCA Hybrid channel assignment
HDM Handoff direction message
HDR High data rate
HFC Hybrid fiber coax
HIPERLAN High performance radio LAN
HLR Home location register
HMAC Hashed message authentication code
HPN Home phone networking
HPNA Home phone network alliance
HSCSD High speed circuit switched data
IAPP Inter-access point protocol
ICMP Internet control message protocol
IEEE Institute of Electrical and Electronics Engineers
IETF Internet Engineering Task Force
IFS Inter frame spacing
IMEI International mobile equipment identity
IMT-2000 International mobile telecommunications beyond 2000
IP Internet protocol
IR Infrared
IrDA Infrared data association
ISDN Integrated services digital network
ISI Inter symbol interference
ISM Industrial, scientific, and medical
ISP Internet service provider
ITS Intelligent transportation system
ITU International Telecommunication Union
ITU-R Radio sector of ITU
ITU-T Telecom sector of ITU
IV Initial vector
JDC Japanese digital cellular
JTACS Japanese total access communications system
JTC Joint technical committee
L1, L2, etc. Layer-1, Layer-2, etc.
LA Location area

LAN Local area network
LANE LAN emulation
LAPD Link access protocol-D
LBT Listen before talk
LCCH Link control channel
LCH Long transport channel
LEO Low earth orbiting (for satellites)
LFSR Linear feedback shift register
LLC Logical link control
LMDS Local multipoint distribution service
LOS Line of sight
LPC Linear predictive coding
LR Location register
LTE Linear transversal equalizer
MAC Medium access control
MAHO Mobile assisted handoff
MAP Mobile application part
MCHO Mobile controlled handoff
MCM Multi carrier modulation
MDBS Mobile data base station
MD-IS Mobile data intermediate system
MDLP Mobile data link protocol
ME Mobile equipment
M-ES Mobile end system
MLSE Maximum likelihood sequence estimator
MME MNRP management entity
MN Mobile node
MNLP Mobile network location protocol
MNRP Mobile network registration protocol
MOS Mean opinion score
MPDU MAC PDU
MSC Mobile switching center
MSK Minimum shift keying
NCHO Network controlled handoff
NDDP Non-dominant direct path
NIC Network interface card
NID Network ID
NIST National Institutes of Standards and Technology
NLOS Non-line of sight
NMT Nordic Mobile Telephony
NNI Network network interface
NPMA Non-preemptive multiple access
NSS Network subsystem

NTT Nippon Telephone and Telegraph
OFDM Orthogonal frequency division multiplexing
OLOS Obstructed line of sight
OSI Open system interconnection
PACCH Packet associated control channel
PACS Personal access communications system
PAGCH Packet AGCH
PAM Pulse amplitude modulation
PBCCH Packet BCCH
PBX Private branch exchange
PCF Point coordination function
PCH Paging channel
PCS Personal communications services
PDCH Packet data channel
PDF Probability density function
PDM Pulse duration modulation
PDN Public data network
PDP Packet data protocol
PDU Protocol data unit
PHP Personal handy phone
PHS Personal handyphone system
PHY Physical layer
PIFS PCF inter-frame spacing
PIN Personal identification number
PLCP Physical layer convergence protocol
PLS Personal locator services
PLW Packet length width
PMD Physical medium dependent
PNCH Packet notification channel
POTS Plain old telephone service
PPM Pulse position modulation
PPP Point-to-point protocol
PPRCH Packet paging response channel
PRACH Packet RACH
PRMA Packet reservation multiple access
PSAP Public safety answering point
PSF Packet signaling field
PSK Phase shift keying
PSPDN Packet switched public data networks
PSTN Public switched telephone network
PTCH Packet traffic channel
PTM-G Point-to-multipoint group
PTM-M Point-to-multipoint multicast

PTP Point-to-point
PWM Pulse width modulation
QAM Quadrature amplitude modulation
QoS Quality of service
QPSK Quadrature phase shift keying
RA Routing area
RACE Research in advanced communications in Europe
RACH Random access channel
RAI RA identity
R-ALOHA Reservation ALOHA
RFC Request for comments
RFCH Radio frequency channel
RLC Radio link control
RPE Regular pulse excitation
RRC Radio resource control
RRM Radio resource management
RS1 Rate set 1
RSS Received signal strength
RSVP Reservation protocol
RTS/CTS Request to send/clear to send
RTT Radio transmission technology
RTT Round trip time
SACCH Slow associated control channel
SAS Sectored antenna system
SAS Situation awareness system
SCH Synchronization channel
SCH Short transport channel
SDCCH Stand alone dedicated control channel
SDMA Space division multiple access
SDP Service discovery protocol
SER Symbol error rate
SFD Start of frame delimiter
SGSN Serving GPRS support node
SHA Secure hash algorithm
SID System ID
SIFS Short inter-frame spacing
SIM Subscriber identity module
SIR Signal to interference ratio
SMS Short messaging service
SMSC SMS center
SNDCP Subnetwork dependent convergence protocol
SNR Signal to noise ratio
SS-7 Signaling system-7

STP Shielded twisted pair
SUO Small unit operations
TACS Total access communications system
TASI Time assignment speech interpolation
TCH Traffic channel
TCP/IP Transmission control protocol/Internet protocol
TDD Time division duplex
TDL Tapped delay line
TDMA Time division multiple access
TDOA Time difference of arrival
TEI Temporary equipment identifier
TETRA Terrestrial trunked radio
TIA Telecommunications Industry Association
TIM Traffic indication map
TLLI Temporary logical link identity
TMSI Temporary mobile subscriber identity
TOA Time of arrival
TP Twisted pair
TTFF Time to first fix
TX Transmitter/transmission
UDP User datagram protocol
UDP Undetected direct path
UMTS Universal mobile telecommunications system
UNI User network interface
U-NII Unlicensed national information infrastructure
USF Uplink status flag
UTP Unshielded twisted pair
VAD Voice activity detection
VC Virtual circuit
VLR Visitor location register
VoIP Voice over IP
VSELP Vector sum excitation linear prediction
WAN Wide area network
WATM Wireless ATM
WCAN Wireless campus area network
WCDMA Wideband CDMA
WLAN Wireless local area network
WLL Wireless local loop
WML Wireless markup language
WPAN Wireless personal area network
WTA Wireless telephony application
WTLS Wireless transport layer security
XML Extensible markup language

References

[3GPPweb] Third generation partnership project website: http://www.3gpp.org

[3GPP2web] Third generation partnership project-2 website: http://www.3gpp2.org

[AES00] NIST website on AES: http://csrc.nist.gov/encryption/aes/aes_home.htm

[AGR96] P. Agrawal et al., "SWAN: A Mobile Multimedia Wireless Network," *IEEE Pers. Commun.,* Apr. 1996, pp. 18–33.

[AGR98] P. Agrawal, "Energy Efficient Protocols for Wireless Systems," *Proceedings of PIMRC'98*, pp. 564–569, September 1998.

[AKY98] I.F. Akyildiz et al., "Mobility management in current and future communications networks," *IEEE Network Magazine*, pp. 39–50, July/August 1998.

[AYA96] E. Ayanoglu et al., "Mobile information infrastructure," *Bell Labs Tech. J.*, pp. 143–63, Autumn 1996.

[BEL69] P.A. Bello, "A troposcatter channel model," *IEEE Trans. Comm. Tech.*, Vol. 17, pp. 130–137, 1969.

[BEL84] P.A. Bello and K. Pahlavan, "Adaptive equalization for SQPSK and SQPR over frequency selective microwave LOS channels," *IEEE Trans. Comm.*, Vol. 32, pp. 609–615, 1984.

[BEL88] P.A. Bello, "Performance of some RAKE modems over the non-disturbed wideband HF channel," *IEEE MILCOM*, pp. 89–95, 1988.

[BER00] H.L. Bertoni, *Radio Propagation for Modern Wireless Systems*, Prentice Hall, New Jersey, 2000.

[BER87] D. Bertsekas and R. Gallagher, *Data Networks*, Prentice Hall, 1987.

[BER93] W. Honcharenko, H.L. Bertoni, and J. Dailing, "Mechanisms governing radio propagation between different floors in buildings," *IEEE Trans. Ant. Prop.*, Vol. 41, No. 6, pp. 787–790, June 1993.

[BER94] H. L. Bertoni, W. Honcharenko, L.R. Maciel, H.H. Xia, "UHF propagation prediction for wireless personal communications," *Proceedings of the IEEE*, Vol. 82, No. 9, pp. 1333–1359, September 1994.

[BER99] D. Har, H.H. Xia, and H.L. Bertoni, "Path-loss prediction model for microcells," *IEEE Transactions on Vehicular Technology*, Vol. 48, No. 5, pp. 1453–1462, September 1999.

[BIN90] J.A.C. Bingham, "Multicarrier modulation for data transmission: An idea whose time has come," *IEEE Communications Magazine*, Vol. 28, No. 5, pp. 5–14, May 1990.

[BLU00] Bluetooth Special Interest Group, "Specifications of the Bluetooth System, vol. 1 v. 1.1, 'Core' and vol. 2 v. 1.0 B 'Profiles'," 2000.

[BLUweb] Bluetooth website: http://www.bluetooth.com

[BRA01] J. Bray and C.F. Sturman, *Bluetooth: Connect without cables*, Prentice Hall, New Jersey, 2001.

[BRA97] B. Walke and G. Brasche, "Concepts, services, and protocols of the new GSM phase 2+ general packet radio service," *IEEE Communications Magazine*, Vol. 35, No. 8, pp. 94–104, August 1997

[BRO98] R. Broderson, "The InfoPad project: Review and lessons learned," *Proc. 11th International conference on VLSI Design,* pp. 2–3, 1998.

[BUD97] K.C. Budka, H.J. Jiang, and S.E. Sommars, "Cellular Packet Data Networks," *Bell Lab Technical Journal*, Summer 1997.

[BUR00] E. Buracchini, "The software radio concept," *IEEE Communications Magazine,* pp. 138–143, September 2000.

[CAC95] R. Cacares and L. Iftode, "Improving the performance of reliable transport protocols in mobile computing environments," *IEEE JSAC*, pp. 850–857, June 1995.

[CAF98] J. Caffery, Jr. and G.L. Stuber, "Subscriber location in CDMA cellular networks," *IEEE Trans. Veh. Technol.*, Vol. 47, No. 2, pp. 406–416, May 1998.

[CAI97] J. Cai and D.J. Goodman, "General packet radio service in GSM," *IEEE Communications Magazine*, Vol. 35, No. 10, pp. 122–131, October 1997.

[CAR98] C. Carroll, Y. Frankel, and Y. Tsiounis, "Efficient key distribution for slow computing devices," *IEEE Symposium on Security and Privacy*, pp. 66–76, May 1998.

[CDPD95] CDPD Specifications, 1995.

[CHA00] A. Chandra, V. Gummalla, and J. O. Limb, "Wireless medium access control protocols," *IEEE Communications Surveys,* Vol. 3, No. 2, Second Quarter 2000.

[CHA00b] M.V.S. Chandrashekhar, P. Choi, K. Maver, R. Sieber, and K. Pahlavan, "Evaluation of interference between IEEE 802.11b and Bluetooth in a typical office environment," *Proc. PIMRC 01*, San Diego, 2001.

[CHA75] D. Chase and P.A. Bello, "A combined coding and modulation approach for high speed data transmission over troposcatter channel," *Proc. NTC*, pp. 28–32, December 1975.

[CHA99] S. Chakrabarti and A. Mishra, "A network architecture for global wireless position location services," *Proc. ICC'99*, pp. 1779–1783, 1999.

[COM98] IEEE Communications Magazine on Geolocation Applications

[CON78] W.J. Conner, "ANTRC-170, A new digital troposcatter communication system," *Proc. IEEE ICC*, pp. 40–43, 1978.

[COU01] Leon W. Couch II, *Digital and analog communication systems*, Prentice Hall, New Jersey, 2001.

[COX96] Donald Cox, "Pattern recognition handoff algorithms for cellular communication systems," *Project Description submitted to Stanford Center for Telecommunications*, September 1996.

[COX99] A. Lozano and D.C. Cox, "Integrated dynamic channel assignment and power control in TDMA mobile wireless communication systems," *IEEE JSAC*, Vol. 17, No. 11, pp. 2031–2040, November 1999.

[CRO97] B. P. Crow, I. Widjaja, L.G. Kim, and P.T. Sakai, "IEEE 802.11 Wireless Local Area Networks," *IEEE Communications Magazine*, Vol. 35, No. 9, pp. 116–126, September 1997.

[DAH98] E. Dahlman, B. Gudmundson, M. Nilsson, and J. Skold, "UMTS/IMT-2000 based on Wideband CDMA," *IEEE Communications Magazine*, pp. 70–80, September 1998.

[DAR97] DARPA tactical technology office: Situation awareness system open review #1 and #2.

[DEH00] S. Dehgan, D. Lister, R. Owen, and P. Jones, "W-CDMA capacity and planning issues," *Electronics and Communications Engineering Journal*, June 2000.

[DEL96] L. Dellaverson, "High speed wireless ATM," Congress ATM '96.

[DEN96] L. R. Dennison, "BodyLAN: A wearable personal network," *Second IEEE Workshop on WLANs*, Worcester, MA, 1996.

[DIN98] E.H. Dinan and B. Jabbari, "Spreading codes for direct sequence CDMA and wideband CDMA cellular networks," *IEEE Communications Magazine*, September 1998.

[DJU01] G.M. Djuknic and R.E. Richton, "Geolocation and assisted GPS," *IEEE Computer*, February 2001.

[DRA98] C. Drane, M. Macnaughtan, and C. Scott, "Positioning GSM telephones," *IEEE Communications Magazine*, Vol. 36, No. 4, pp. 46–54, April 1998.

[EMC01] EMC World Cellular Database; June 2001 forecast. Available at: http://www.emc-database.com/

[ENG94] P. K. Enge, "Global positioning systems: Signals, measurements and performance," *Int'l. J. Wireless Info. Networks*, Vol. 1, No. 2, April 1994.

[ENG95] K. Y. Eng et al., "BAHAMA: A broadband Ad-Hoc wireless ATM local area network," *Proc. IEEE ICC '95*, 1995, pp. 1216–23.

[ENN98] G. Ennis, Doc: IEEE P802.11-98/319, *Impact of Bluetooth on 802.11 Direct Sequence*, September 15, 1998.

[ERT98] R.B. Ertel, P. Cardieri, K.W. Sowerby, T.S. Rappaport and J.H. Reed, "Overview of spatial channel models for antenna array communication systems," *IEEE Personal Communications*, February 1998.

[FAL76a] D.D. Falconer, "Analysis of a gradient algorithm for simultaneous passband equalization and carrier phase recovery," *Bell Systems Technical Journal*, Vol. 55, pp. 409–428, 1976.

[FAL76b] D.D. Falconer and F. R. Magee, "Evaluation of decision feedback equalization and Viterbi algorithm detection for voiceband data transmission: Part I," *IEEE Trans. Comm.*, Vol. 24, pp. 1130–1139, 1976.

[FAL76c] D.D. Falconer and F. R. Magee, "Evaluation of decision feedback equalization and Viterbi algorithm detection for voiceband data transmission: Part II," *IEEE Trans. Comm.*, Vol. 24, pp. 1238–1245, 1976.

[FAL85] D.D. Falconer et al., "Comparison of DFE and MLSE receiver performance on HF channels," *IEEE Trans. Comm.*, Vol. 33, pp. 484–486, 1985.

[FAL96] A. Falsafi, K. Pahlavan, and G. Yang, "Transmission techniques for radio LANs—A comparative performance evaluation using ray tracing," *IEEE Journal on Selected Areas in Communications*, Vol. 14, No. 3, pp. 477–491, April 1996.

[FAS99] A. Fasbender et al., "Any network, any terminal, anywhere," *IEEE Personal Communications,* pp. 22–30, April 1999.

[FEH91] K. Feher, "Modems for emerging digital cellular mobile radio systems," *IEEE Transactions on Vehicular Technology*, Vol. 40, pp. 355–365, 1991.

[FEI00] J. Feigin, K. Pahlavan, and M. Ylianttila, "Hardware-fitted modeling and simulation of VoIP over a wireless LAN," *52nd IEEE VTS Fall VTC, Vehicular Technology Conference*, Vol. 3, pp. 1431–1438, 2000.

[FEI99] J. Feigin and K. Pahlavan, "Measurement of characteristics of voice over IP in a wireless LAN environment," *IEEE International Workshop on Mobile Multimedia Communications (MoMuC '99)*, pp. 236–240, 1999.

[FER80] P. Ferert, "Application of spread spectrum radio to wireless terminal applications," *Proc. NTC'80*, pp. 244–248, Houston, TX, December 1980.

[FIG69] W. Figel, N. Shepherd and W. Trammell, "Vehicle location by a signal attenuation method," *IEEE Trans. Vehicular Technology*, vol. VT-18, pp. 105–110, Nov. 1969.

[FIRweb] IEEE Firewire website: http://www.1394ta.org

[FOR72] G.D. Forney, Jr., "Maximum likelihood sequence estimation of digital sequences in the presence of intersymbol interference," *IEEE Trans. Inf. Theory*, Vol. 18, pp. 363–378, 1972.

[GAN91] R. Ganesh and K. Pahlavan, "Modeling of the indoor radio channel," *IEE Proc. I: Comm., Speech and Vision*, No. 138, pp. 153–161, 1991.

[GAR00] V.K. Garg, *IS-95 and CDMA2000*, Prentice Hall, Upper Saddle River, 2000.

[GAR02] V. K. Garg, *Wireless Network Evolution: 2G to 3G*, Prentice Hall, New Jersey, 2002.

[GAR99] V.K. Garg and J.E. Wilkes, *Principles and Applications of GSM*, Prentice Hall, Upper Saddle River, NJ, 1999.

[GET93] I. A. Getting, "The Global Positioning System," *IEEE Spectrum*, pp. 36–47, December 1995.

[GFE80] F.R. Gfeller, "Infranet: Infrared microbroadcasting network for in house data communication," *IBM research report*, RZ 1068 (#38619), April 27, 1981.

[GOO89] D.J. Goodman, R.A. Valenzuela, K.T. Gayliard, and B. Ramamurthi, "Packet reservation multiple access for local wireless communications," *IEEE Transactions on Communications*, Vol. 37, No. 8, pp. 885–890, August 1989.

[GOO91] D.J. Goodman and S.X. Wei, "Efficiency of packet reservation multiple access," *IEEE Transactions on Vehicular Technology*, Vol. 40, No. 1 Part: 2, pp. 170–176, February 1991.

[GOO93] D.J. Goodman, J. Grandhi, and R. Vijayan, "Distributed dynamic channel assignment schemes," *Proc. of the 43rd IEEE Veh. Tech. Conf.*, pp. 532–535, 1993.

[GOO97] D.J. Goodman, *Wireless Personal Communications Systems*, Addison-Wesley, 1997.

[GRA93] S.A. Grandhi, R. Vijayan, D.J. Goodman, and J. Zander, "Centralized power control in cellular radio systems," *IEEE Transactions on Vehicular Technology*, Vol. 42, No. 4, pp. 466–468, November 1993.

[GRA94] S.A. Grandhi, R. Vijayan, and D.J. Goodman, "Distributed power control in cellular radio systems," *IEEE Transactions on Communications,* Vol. 42, pp. 226–228, February 1994.

[GRZ75] C.J. Grzenda, D.R. Kern and P. Monsen, "Megabit digital troposcatter subsystem," *Proc. NTC'75*, pp. 28-15 – 28-18, New Orleans, December 1975.

[GSMweb] GSM World Web Site: http://www.gsmworld.com

[GUE97] S. Guerin, Y.J. Guo, and S.K. Barton, "Indoor propagation measurements at 5 GHz for HIPERLAN," *10th Int. Conf. Ant. Prop.* , pp. 306–310, April 1997.

[HAA00] J.C. Haartsen and S. Mattisson, "Bluetooth—a new low-power radio interface providing short-range connectivity," *Proceedings of the IEEE*, Vol. 88, No. 10, pp. 1651–1661, October 2000.

[HAI92] J.L. Haine, P.M. Martin, and R.L.A. Goodings, "A European standard for packet mode mobile data," *Proc. PIMRC'92*, Boston, MA, October 1992.

[HAL83] S.W. Halpern, "Reuse partitioning in cellular systems," *Proc. of the IEEE Vehicular Technology Conference*, pp. 322–327, 1983.

[HAL96] C.J. Hall, and W.A. Foose, "Practical planning for CDMA networks: A design process overview," *Proc. Southcon'96*, pp. 66–71, 1996.

[HAM86] J.L. Hammond and P.J.P. O'Reilly, *Performance analysis of local computer networks*, Addison-Wesley, Reading, MA, 1986.

[HAT80] Masaharu Hata, "Empirical formula for propagation loss in land mobile radio services," *IEEE Transactions on Vehicular Technology*, Vol. VT-29, No. 3, pp. 317–324, August 1980.

[HAU94] T. Haug, "Overview of GSM: Philosophy and results," *International Journal of Wireless Information Networks*, January 1994.

[HAY00] Simon Haykin, *Communication Systems, 4th Edition*, John Wiley & Sons, NY, 2000.

[HAY91] V. Hayes, "Standardization efforts for wireless LANS," *IEEE Network*, Vol. 5, No. 6, pp. 19–20, 1991.

[HEI98] Robert Heille, *WPAN functional requirement*, Doc. IEEE 802.11/98/58, January 22, 1998.

[HIPweb] HIPERLAN2 website: http://www.hiperlan2.com

[HOL00] H. Holma and A. Toskala (Eds.), *WCDMA for UMTS: Radio access for third generation mobile communications*, John Wiley & Sons, NY, 2000.

[HOM00] HomeRF Special Interest Group (SIG) presentations.

[HomeRFweb] HomeRF website: http://www.homerf.org

[HOW90] S. J. Howard and K. Pahlavan, "Measurement and analysis of the indoor radio channel in the frequency domain," *IEEE Trans. Instr. Meas.*, No. 39, pp. 751–755, 1990.

[HOW91] S.J. Howard, "Frequency domain characteristic and autoregressive modeling of the indoor radio channel," Ph.D. thesis, Worcester Polytechnic Institute, Worcester, MA, May 1991.

[HOW92] S.J. Howard and K. Pahlavan, "Autoregressive modeling of wideband indoor radio propagation," *IEEE Trans. Communication*, Vol. 40, pp. 1540–1552, 1992.

[HPNAweb] Home phone networking alliance website: http://www.hpna.com

[HUB00] J.F. Huber, D. Weiler, and H. Brand, "UMTS, the mobile multimedia vision for IMT-2000: A focus on standardization," *IEEE Communications Magazine*, pp. 129–136, September 2000.

[IEE00] IEEE 802.15 Working Group: http://grouper.ieee.org/groups/802/15/

[IETweb] Internet Engineering Task Force website: http://www.ietf.org

[INFweb] Infopad project at University of California: http://infopad.eecs.berkeley.edu

[JAK94] W. Jakes and D. Cox (Eds.), *Microwave mobile communications*, IEEE Press, April 1994.

[JAY84] N.S. Jayant and P. Noll, *Digital coding of waveforms*, Prentice-Hall, Englewood Cliffs, NJ, 1984.

[JON99] M. Jonsson, "HiperLAN2 – The Broadband Radio Transmission Technology Operating in the 5 GHz Frequency Band," HIPERLAN-2 Global Forum White Paper.

[JTC94] JTC Technical Report on RF Channel Characterization and Deployment Modeling, Air Interface Standards, September 1994.

[KAL00] R. Kalden, I. Meirick, and M. Meyer, "Wireless Internet access based on GPRS," *IEEE Personal Communications*, Vol. 7, No. 2, pp. 8–18, April 2000.

[KAM81] M.A. Kamil, "Simulation of digital communication through urban/suburban multipath," Ph.D. Dissertation, EECS Department, University of California, Berkeley, June 1981.

[KAP96] E.D. Kaplan, *Understanding GPS: Principles and applications*, Artech House Publishers, 1996.

[KAT96] I. Katzela and M. Naghshineh, "Channel assignment schemes for cellular mobile telecommunication systems: A comprehensive survey," *IEEE Personal Communications*, pp. 10–31, June 1996.

[KAV87] M. Kavehrad and P.J. McLane, "Spread spectrum for indoor digital radio," *IEEE Communications Magazine*, Vol. 25, No. 6, pp. 32–40, 1987.

[KEI89] G.E. Keiser, *Local area networks*, McGraw-Hill, New York, 1989.

[KER00] J.P. Kermoal, L. Schumacher, P.E. Mogensen, and K.I. Pedersen, "Experimental investigation of correlation properties of MIMO radio channels for indoor picocell scenarios," *52nd IEEE Vehicular Technology Conference*, 2000, pp. 14–21, 2000.

[KHA00] J. Khan-Josh, "HIPERLAN-2 and home networking," *Wireless Home Networking Workshop*, Brussels, May 24, 2000.

[KLE75b] L. Kleinrock and S.S. Lam, "Packet switching in a multi-access broadcast channel: Performance evaluation," *IEEE Trans. Comm.*, Vol. 23, pp. 410–423, 1975.

[KOS00] H. Koshima and J. Hoshen, "Personal locator services emerge," *IEEE Spectrum*, pp. 41–47, February 2000.

[KRI98] P. Krishnamurthy, K. Pahlavan, and J. Beneat, "Radio propagation modeling for indoor geolocation applications," *Proceedings of IEEE PIMRC'98*, September 1998.

[KRI99a] P. Krishnamurthy and K. Pahlavan, "Analysis of the probability of detecting the DLOS path for geolocation applications in indoor areas," 49th IEEE Vehicular Technology Conference, Vol. 2, pp. 1161–1165, 1999.

[KRI99b] P. Krishnamurthy and K. Pahlavan, "Distribution of range error and radio channel modeling for indoor geolocation applications," *Proc. PIMRC'99*, Osaka, Japan, 1999.

[KRI99c] P. Krishnamurthy, "Analysis and modeling of the indoor radio channel for geolocation applications," Ph.D. Thesis, Worcester Polytechnic Institute, August 1999.

[KUE92] S.S. Kuek and W.C. Wong, "Ordered dynamic channel assignment scheme with reassignment in highway microcells," *IEEE Transactions on Vehicular Technology*, Vol. 41, No. 3, pp. 271–276, August 1992.

[KUM74] K. Kummerle, "Multiplexer performance for integrated line-and packet switched traffic," ICCC, Stockholm, 1974.

[LEE91] W.C.Y. Lee, "Smaller cells for greater performance," *IEEE Communications Magazine*, pp. 19–23, November 1991.

[LEE98] W. C.Y. Lee, *Mobile communications engineering : Theory and applications (McGraw-Hill Series on Telecommunications)*, McGrawHill Professional Publishing, 1999.

[LEH99] P.H. Lehne and M. Pettersen, "An overview of smart antenna technology for mobile communications systems," *IEEE Communications Surveys*, pp. 2–13, Vol. 2, Fourth Quarter, 1999.

[LIN83] S. Lin and D.J. Costello, *Error control coding,* Prentice Hall, 1983.

[LIO94] G. Liodakis and P. Stravroulakis, "A novel approach in handover initiation for microcellular systems," *Proc. Vehicular Tech. Conf. '94*, Stockholm, Sweden, 1994.

[LIU90] C.L. Liu and K. Feher, "Non-coherent detection of $\pi/4$-shift QPSK systems in CCI-AWGN combined interference environment," *Proc. 40th IEEE Vehicular Technology Conference*, Orlando, pp. 482–486, May 1990.

[LONweb] Lonworks website: http://www.echelon.com

[LOR98] J.R. Lorch and A.J. Smith, "Software strategies for portable computer energy management," *IEEE Personal Communications Magazine*, pp. 60–73, June 1998.

[MAC79] V.H. MacDonald, "The cellular concept," *The Bell System Technical Journal*, Vol. 58, No. 1, pp. 15–41, January 1979.

[MAR85] M.J. Marcus, "Recent US regulatory decisions on civil use of spread spectrum," *Proc. IEEE Globecom*, 16.6.1–16.6.3, New Orleans, December 1985.

[McD98] J.T. Edward McDonnell, "5 GHz indoor channel characterization: measurements and models," *IEE Coll. on Ant. and Prop. for future mobile communications*, 1998.

[MEH97] A. Mehrotra, *GSM system engineering*, Artech House Mobile Communications Series, Artech House, 1997.

[MES87] D. G. Messerschmitt, "Advanced digital communications: systems and signal processing techniques," Chapter on *Echo Cancellation*, Prentice Hall, Englewood Cliffs, NJ, 1987.

[MEY96] M.J. Meyer, T. Jacobson, M. Palamara, E. Kidwell, R. Richton, and G. Vanucci, "Wireless enhanced 9-1-1 service — Making it a reality," *Bell Labs Technical Journal*, Autumn 1996.

[MIC85] A.M. Michelson and A.H. Levesque, *Error-control techniques for digital communications,* Wiley-Interscience, 1985.

[MIT00] Joseph Mitola III, *Software radio architecture: Object-oriented approaches to wireless systems engineering*, John Wiley & Sons, 2000.

[MIT01] T. Mitchell, "Broad is the way," *IEE Review*, pp. 35–39, January 2001.

[MON77] P. Monsen, "Theoretical and measured performance of a DFE modem on a fading multipath channel," *IEEE Trans. Comm.*, Vol. 25, pp. 1144–1153, 1977.

[MON84] P. Monsen, "MMSE equalization of interference on fading diversity channels," *IEEE Trans. Comm.*, Vol. 32, pp. 5–12, 1984.

[MOS96] S. Moshavi, "Multi-user detection for DS-CDMA communications," *IEEE Communications Magazine*, Vol. 34, No. 10, pp. 124–136, Oct. 1996.

[MOT88a] A.J. Motley and J.M.P. Keenan, "Personal communication radio coverage in buildings at 900 and 1700 MHz," *IEE Electronic Letters*, pp. 763–64, 1988.

[MOT88b] A.J. Motley, "Radio coverage in buildings," *Proc. Nat. Comm. Forum*, Chicago, pp. 1722–1730, October 1998.

[NAK90] N. Nakajima et al., "A system design for TDMA mobile radios," *Proc. 40th IEEE Veh. Tech. Conf.*, pp. 295–298, May 1990.

[NEG00] K. Negus, A Stephens, and J. Lansfield, "HomeRF: Wireless networking for the connected home," *IEEE Personal Communcations,* Feb. 2000.

[NJW97] "Report on the New Jersey wireless enhanced 911 system trial," State of New Jersey, Dept. of Law and Public Safety, June 16, 1997.

[ONSweb] General Motors Onstar website: http://www.onstar.com

[PAH00a] K. Pahlavan, P. Krishnamurthy, et. al., "Handoff in hybrid mobile data networks," *IEEE Personal Communications Magazine*, April 2000.

[PAH00b] K. Pahlavan, X. Li, M. Ylianttila, R. Chana, and M. Latva-aho, "An overview of wireless indoor geolocation techniques and systems," MWCN'2000, Paris, May 2000.

[PAH80] K. Pahlavan and J.W. Matthews, "Performance of channel measurement techniques over fading channels," *Proc. IEEE NTC*, 58.5.1–58.5.5, 1980.

[PAH85] K. Pahlavan, "Wireless Communications for Office Information Networks," *IEEE Commun. Mag.,* Sept. 1985.

[PAH87] K. Pahlavan, "Spread spectrum for wireless local networks," *Proc. IEEE PCCC*, pp. 215–219, Phoenix, 1987.

[PAH88] Pahlavan K. and J. L. Holsinger, "Voice-band data communication modems a historical review: 1919–1988," *IEEE Communications Magazine*, Vol. 26, pp. 16–27, January 1988.

[PAH90a] K. Pahlavan and M. Chase, "Spread spectrum multiple access performance of orthogonal codes for indoor radio communications," *IEEE Trans. Comm.*, Vol. 38, pp. 574–577, 1990.

[PAH94] K. Pahlavan and A. H. Levesque, "Wireless data communications," *Proceedings of the IEEE*, Vol. 82, No. 9, pp. 1398–1430, Sept. 1994.

[PAH95] K. Pahlavan and A. Levesque, *Wireless information networks*, John Wiley & Sons, New York, 1995.

[PAH97] K. Pahlavan, A. Zahedi, and P. Krishnamurthy, "Wideband local access: Wireless LAN and wireless ATM," *IEEE Communications Magazine*, Vol. 35, No. 11, pp. 34–40, Nov. 1997.

[PAH98] K. Pahlavan, P. Krishnamurthy, and J. Beneat, "Wideband radio propagation modeling for indoor geolocation applications," *IEEE Comm. Magazine*, pp. 60–65, April 1998.

[PALweb] PalTrack Tracking Systems: http://www.sovtechcorp.com/

[PED00] K.I. Pedersen, J.B. Andersen, J.P. Kermoal, and P. Morgensen, "A stochastic multiple-input-multiple-output radio channel model for evaluation of space-time coding algorithms," *52nd IEEE Vehicular Technology Conference, 2000,* pp. 893–897, 2000.

[PEE00] G. Peersman and S. Cvetkovic, "The global system for mobile communications short messaging service," *IEEE Personal Communications*, pp. 15–23, June 2000.

[PEE00a] G. Peersman et al., "A tutorial overview of the short message service within GSM," *Computing and Control Engineering Journal*, April 2000.

[PER97] C. E. Perkins, *Mobile IP: Design principles and practices*, Addison Wesley Communications Series, 1997.

[PIC91] R.L. Pickholtz, L.B. Milstein, and D. Schilling, "Spread spectrum for mobile communications, *IEEE Transactions on Vehicular Technology*, Vol. 40, No. 2, pp. 313–322, May 1991.

[POL96] G.P. Pollini, "Trends in handover design," *IEEE Communications Magazine*, March 1996.

[PRA92] G.J.M. Janssen and R. Prasad, "Propagation measurements in an indoor radio environment at 2.4 GHz, 4.75 GHz, and 11.5 GHz," *Proc. of the 42nd IEEE VTC*, pp. 617–620, 1992.

[PRI58] R. Price and P.E. Green, "A communication technique for multipath channels," *Proc. IRE*, Vol. 46, pp. 555–570, 1958.

[PRO00] John G. Proakis, *Digital communications*, 4th ed., McGraw-Hill, 2001.

[QUR85] S.H. Qureshi, "Adaptive equalization," *Proc. IEEE,* Vol. 73, pp. 1349–1387, 1985.

[RADweb] http://www.uswcorp.com/USWCMainPages/Applications/E-911.htm

[RAP95] T.S. Rappaport, *Wireless communications*: *Principles and practice*, Prentice Hall, New Jersey, 1995.

[RAP96] S.Y. Seidel and T.S. Rappaport, "914 MHz path loss prediction models for indoor wireless communications in multifloored buildings," *IEEE Transactions on Antennas and Propagation*, Vol. 40, No. 2, pp. 207–217, February 1992.

[RAY92] D. Raychaudhuri and N. Wilson, "Multimedia personal communication

networks: System design issues," *Proc. 3rd WINLAB Workshop on 3rd Generation Wireless Information Networks*, pp. 259–288, April 1992.

[RAY97] D. Raychaudhuri et al., "WATMNet: A Prototype wireless ATM system for multimedia personal communication," *IEEE JSAC*, Vol. 15, No. 1, 1997, pp. 83–95 January.

[RAY99] D. Raychaudhuri, "Wireless ATM networks: Technology status and future directions," *Proceedings of the IEEE*, Vol. 87, No. 10, pp. 1790–1806, October 1999.

[RED95] S. Redl, M.K. Weber, M. Oliphant, and W. Mohr, *An Introduction to GSM,* The Artech House Mobile Communications Series, Artech House, 1995.

[REZ95] R. Rezaiifar, A.M. Makowski, and S. Kumar, "Optimal control of handoffs in wireless networks," *Proc. IEEE VTC 95*, pp. 887–891, 1995.

[SAL73] J. Salz, "Optimum mean-square decision feedback equalization," *Bell Systems Technical Journal*, Vol. 52, pp. 1341–1373, 1973.

[SAL99a] A.K. Salkintzis, "Packet data over cellular networks: The CDPD approach," *IEEE Communications Magazine*, Vol. 37, No. 6, pp. 152–159, June 1999.

[SAL99b] A.K. Salkintzis, "Radio resource management in cellular digital packet data networks," *IEEE Personal Communications*, Vol. 6, No. 6, pp. 28–36, Dec. 1999.

[SAR00] B. Sarikaya, "Packet mode in wireless networks: overview of transition to third generation," *IEEE Communications Magazine*, Vol. 38, No. 9, pp. 164–172, Sept. 2000.

[SCH00] M.Z. Win and R.A. Scholtz, "Ultra-wide bandwidth time-hopping spread-spectrum impulse radio for wireless multiple-access communications," *IEEE Transactions on Communications*, Volume: 48, Issue: 4, pp. 679–689, April 2000.

[SCH94] B. Schneier, *Applied Cryptography*, John Wiley & Sons, 1996.

[SDRweb] Software defined radio forum website: http://www.sdrforum.com

[SEX89a] T. Sexton and K. Pahlavan, "Channel modeling and adaptive equalization of indoor radio channels," *IEEE JSAC*, Vol. 7, pp. 114–121, 1989.

[SIE00] T. Siep, I. Gifford, R. Braley, and R. Heile, "Paving the way for personal area network standards: An overview of the IEEE P802.15 working group for wireless personal area networks," *IEEE Personal Communications*, Feb. 2000.

[SIL99] R.D. Silverman, "A cost-based security analysis of symmetric and asymmetric key lengths," *RSA Bulletin Number 13,* April 2000.

[SIM85] M.K. Simon et al., *Spread spectrum communication*, Computer Science Press, 1985.

[SKL01] B. Sklar, *Digital communications: Fundamentals and applications*, 2nd ed., Prentice Hall, 2001.

[SON94] H.-L. Song, "Automatic vehicle location in cellular communications systems," *IEEE Trans. Vehicular Technology*, avol. 43, No. 4, pp. 902–908, Nov. 1994.

[SPECweb] Spectrixcorp website: http://www.spectrixcorp.com

[STA00] W. Stallings, *Local and metropolitan area networks*, Prentice Hall, New Jersey, 2000.

[STA01] W. Stallings, "IEEE 802.11: Moving closer to practical wireless LANs," *IT Professional*, Vol. 3, No. 3, pp. 17–23, May–June 2001.

[STA98] W. Stallings, *Cryptography and network security*, Prentice Hall, 1998.

[STA98b] W. Stallings, *High-speed networks TCP/IP and ATM design principles*, Prentice Hall, 1998.

[STI95] D. Stinson, *Cryptography: Theory and practice*, CRC Press, 1995.

[STW00] S. Williams, "IrDA: Past, Present, and Future," *IEEE CommSoc Mag, Feb. 2000.*

[SWE00] W. Sweet, "Cell phones answer Internet's call," *IEEE Spectrum*, pp. 42–46, August 2000.

[TAK85] H. Takagi and L. Kleinrock, "Throughput analysis for persistent CSMA systems," *IEEE Trans. Comm.*, Vol. 33, pp. 627–638, 1985.

[TAN97] A. S. Tanenbaum, *Computer networks*, Prentice Hall, 1996.

[TAY97] M.S. Taylor, W. Waung, and M. Banan, *Internetwork mobility: The CDPD approach*, Prentice Hall, 1997.

[TDCweb] Time domain corporation website: http://www.timedomain.com

[TEK91] S. Tekinay and B. Jabbari, "Handover policies and channel assignment strategies in mobile cellular networks," *IEEE Communications Magazine*, Vol. 29, No. 11, 1991.

[TEK98] S. Tekinay, E. Chao, and R. Richton, "Performance benchmarking for wireless location systems," *IEEE Communications Magazine*, April 1998.

[TIN01] R.D. Tingley and K. Pahlavan, "Space-time measurement of indoor radio propagation," *IEEE Transactions on Instrumentation and Measurement*, Vol. 50, No. 1, pp. 22–31, Feb. 2001.

[TIS98] J. Tisal, *GSM cellular radio telephony,* John Wiley & Sons, 1998.

[TOB75] F.A. Tobagi and L. Kleinrock, "Packet switching in radio channels Part II: The hidden terminal problem in CSMA and the busy tone solution," *IEEE Trans. Comm.*, Vol. 33, pp. 1417–1433, 1975.

[TOB80] F.A. Tobagi, "Multi-access protocols in packet communication systems," *IEEE Trans. Comm.*, Vol. 28, pp. 468–488, 1980.

[TRI97] Nishith Tripathi, "Generic adaptive handoff algorithms using fuzzy logic and neural networks," *Ph.D Thesis*, Virginia Polytechnic Institute and State University, August 1997.

[TRI98] N.D. Tripathi, J.H. Reed, and H. F. VanLandingham, "Handoff in cellular systems," *IEEE Personal Communications Magazine*, December 1998.

[TUC91] B. Tuch, "An ISM band spread spectrum local area network: WaveLAN," *Proc. 1st IEEE Workshop on WLANs*, pp. 103–111, Worcester, MA 1991.

[VAL98] R.T. Valadas, A.R. Tavares, A.M.deO. Duarte, A.C. Moreira, and C.T. Lomba, "The infrared physical layer of the IEEE 802.11 standard for wireless local area networks," *IEEE Communications Magazine*, Vol. 36, No. 12, pp. 107–112, Dec. 1998.

[VAN99] R. van Nee et al., "New high-rate wireless LAN standards," *IEEE Communications Magazine*, Vol. 37, No. 12, pp. 82–88, Dec. 1999.

[VIT67] A. J. Viterbi, "Error bounds for convolutional codes and an asymptotically optimum decoding algorithm," *IEEE Trans. Inf. Theory*, Vol. 13, pp. 260–269, 1967.

[WAN96] WAND Annual Project Rev. Rep. AC085/NMP/MR/I/035/1.0, Feb. 1996.

[WAPweb] http://www.wapforum.org; The WAP Forum

[WAR97] A. Ward, A. Jones, and A. Hopper, "A new location technique for the active office," *IEEE Personal Communications*, October 1997.

[WER98] Jay Werb and Colin Lanzl, "Designing a positioning system for finding things and people indoors," *IEEE Spectrum*, Vol. 35, No. 9, September 1998.

[WET01] M. Jakobsson and S. Wetzel, "Security weaknesses in Bluetooth," *RSA Conference'01*, April 8-12, 2001.

[WIL01] L.J. Williams, "Technology advances from small unit operations situation awareness system development," *IEEE Personal Communications*, Vol. 8, No. 1, pp. 30–33, Feb. 2001.

[WIL95a] J.E. Wilkes, "Privacy and authentication needs of PCS," *IEEE Personal Communications*, August 1995.

[WIL95b] T. A. Wilkinson, T. Phipps, and S. K. Barton, "A report on HIPERLAN standardization," *International Journal on Wireless Information Networks,* Vol. 2, pp. 99–120, March 1995.

[WIL96] T. Wilkinson, "HIPERLAN," *2nd IEEE Wksp. Wireless LANs*, Worcester Polytech. Inst., Oct. 24–25, 1996.

[WIN85] J.H. Winters and Y.S. Yeh, "On the performance of wideband digital radio transmission within buildings using diversity," *Proc. IEEE Globecom*, 32.5.1–32.5.6, New Orleans, 1985.

[WIR00] *Wireless home networking workshop*, Brussels, May 24, 2000.

[WOE98] H. Woesner, J-P. Ebert, M. Schlager, and A. Wolisz, "Power saving mechanisms in emerging standards for wireless LANs: A MAC level perspective," *IEEE Personal Communications Magazine*, pp. 40–48, June 1998.

[WOL82] J. K. Wolf, A.M. Michelson, and A.H. Levesque, "On the probability of undetected error for linear block codes," *IEEE Trans. Comm.*, Vol. 30, pp. 317–324, 1982.

[WON00] V.W.S. Wong and V.C.M. Leung, "Location management for next generation personal communication networks," *IEEE Network Magazine*, pp.18–24, September/October 2000.

[WOR91] First IEEE Workshop on WLANs, Worcester, MA, 1991.

[YAN94] G. Yang and K. Pahlavan, "Sectored antenna and DFE modem for high speed indoor radio communications," *IEEE Trans. Vehic. Tech.*, November 1994.

[ZAG98] J.M. Zagami, S.A. Parl, J.J. Bussgang, and K.D. Melillo, "Providing universal location services using a wireless E-911 location network," *IEEE Communications Magazine,"* April 1998.

[ZAH00] A. Zahedi and K. Pahlavan, "Capacity of a wireless LAN with voice and data services," *IEEE Trans on Comm*, July 2000.

[ZAH97] A. Zahedi and K. Pahlavan, "Terminal distribution and the impacts of natural hidden terminal," *Electronics Letters*, Vol. 33, No. 9, pp. 750–751, April 1997.

[ZEN00] M. Zeng, A. Annamalai, V.K. Bhargava, "Harmonization of global third-generation mobile systems," *IEEE Communications Magazine*, pp. 94–104, December 2000.

[ZHA00] Y. Zhao, "Mobile phone location determination and its impact on intelligent transportation systems," *IEEE Transactions on ITS,* Vol. 1, No. 1, March 2000.

[ZHA89] M. Zhang and T-S.P. Yum, "Comparisons of channel assignment strategies in cellular mobile telephone systems," *IEEE Transactions on Vehicular Technology*, Vol. 38, No. 4, pp. 211–215, November 1989.

[ZHA90] K. Zhang and K. Pahlavan, "An integrated voice/data system for mobile indoor radio networks," *IEEE Trans. Vehicular Technology*, Vol. 39, pp. 75–82, 1990.

[ZHA91] M. Zhang and T-S.P. Yum, "The non-uniform compact pattern allocation algorithm for cellular mobile systems," *IEEE Transactions on Vehicular Technology*, Vol. 40, No. 2, pp. 387–391, May 1991.

[ZHA92] K. Zhang and K. Pahlavan, "Relation between transmission and throughput of slotted ALOHA local packet radio networks," *IEEE Trans. Comm.*, Vol. 40, pp. 577–583, 1992.

[ZOR97] M. Zorzi and R. Rao, "Energy-constrained error control for wireless channels," *IEEE Personal Communications Magazine,* pp. 27-33, December 1997.

[ZOR99] M. Zorzi and R. Rao, "Is TCP energy efficient?," *Proc. of MoMuC 99,* pp. 198–201, 1999.

INDEX

π/4-QPSK, 100
  Constellation, 106
  Use in various systems, 107
1G Wireless Systems, 13
2G Wireless Systems, 15
3GPP, 351
3GPP2, 351

A

Access Point (AP), 225
Active set, 367
Ad hoc topology, 226
  Comparison with infrastructure topology, 228
  Multi hop, 227
  Single hop, 226
Additive White Gaussian Noise (AWGN), 147
Adjacent Channel Interference (ACI), 89, 230
Advanced Encryption Standard (AES), 304
Advanced Mobile Phone System (AMPS), 9, 13, 14, 78, 350
  CDPD Overlay, 202
  FDMA with FDD, 163
  Power control, 285
  Migration to IS-95, 262
  Signal to Interference ratio, 239
AGCH (See GSM-Logical Channels)
Agent advertisement, 281
ALOHA, 180–184
  Dynamic slotted, 183
  Reservation, 182
  Throughput, 181, 193
Amplitude Shift Keying (ASK), 87
Analog to digital migration, 259
  AMPS to IS-95, 262
Anchor, 271, 273
Angle of Arrival (AOA), 64, 539–540
ARDIS, 12, 18, 380
Asynchronous Connectionless (ACL), 513
Authentication, 221, 299
  Message, 300, 308
Autocorrelation, 158
  Barker sequence, 117
  M-sequence, 141
Automatic Repeat Request (ARQ), 138, 156

B

Band splitting, 14, 248–249
Base Station (BS), 225
Base Station Subsystems (BSS), 325
Baseband transmission, 86
BCCH (See GSM-Logical Channels)
BCH Codes, 153
Bearer services, 322
Blanket paging, 270
Block codes, 152–155
Block Error Rate (BLER), 390
Block interleaving, 157, 294
Bluetooth, 11, 413
  And frequency hopping for diversity, 129
  Architecture, 505–507
  Connection Management, 517
  FHSS, 509–511
  Frame formats, 512–516
  History, 503–505
  Inquiry, 519–520
  Interference with 802.11, 520–530

Bluetooth (*cont.*)
Medium access, 511
Paging, 519
Security, 520
BodyLAN, 427
BSSGP, 403
Busy Tone Multiple Access (BTMA), 186

C

Cable TV Infrastructure, 33–34
Call blocking, 176
Call blocking-CDPD overlay on AMPS, 204–206
Candidate set, 368
Capture phenomena, 197–199
Care-of address, 280
Carrier modulation, 96
CCCH (See GSM-Logical Channels)
cdma2000, 351
Comparison with W-CDMA, 373
Cell, 230
Cell breathing, 261
Cell sectoring, 243–246
Cell splitting, 242
Cellular concept, 230–234
Capacity expansion, 240–262
Overlays, 242, 247–250
Cellular Digital Packet Data (CDPD), 12, 18, 318
Reed Solomon Coding, 153
History, 383
Mobility management, 388–390
Overlay on AMPS, 202, 384
Power control, 391
Protocol layers, 391–394
Reference architecture, 385
Security, 390
Services, 384
Cellular hierarchy, 233
Challenge-response protocols, 311
Channel allocation techniques, 221, 251–259
Dynamic Channel Allocation (DCA), 256–258
Fixed Channel Allocation (FCA), 252–256
Channel borrowing techniques, 254
Channel sounder, 69
Ciphering (see also encryption), 221
Cluster, 230
Cochannel interference, 230
Code Division Multiple Access (CDMA) (see also IS-95), 9, 168–172
Capacity, 169–172
Comparison with FDMA and TDMA, 172–175
Network planning, 260–261
Code rate, 153
Coherence bandwidth, 61
Complementary Code Keying (CCK), 118, 457–458
Constant envelope modulation, 90
Continuous phase modulation, 98
Convolutional codes, 155–157
Correlation, 158
COST-231 model, 53
Cross correlation, 158
CSMA, 185–191
Persistent and Non-persistent, 186
CSMA/CA, 160, 189–191
Combing in HIPERLAN-1, 190
In IEEE 802.11, 189–190, 460–463
CSMA/CD, 160, 188
Binary exponential back-off, 188

D

Data Encryption Standard (DES), 299, 303
DCF, 460
Decibel, 79
Differential PSK, 101
Diffie-Hellman key exchange, 307, 313–314
Diffraction, 45
Digital Audio Broadcast (DAB), 108
Digital European Cordless Telephone (DECT), 10, 17
TDMA with TDD format, 166–167

Digital Sense Multiple Access (DSMA), 187
Digital signatures, 314
Direct Line of Sight (DLOS), 64
Direct Sequence Spread Spectrum (DSSS), 111, 116–117
- And CDMA, 117
- And RAKE diversity, 121
- Barker code, 117
- CCK Modulation, 118
- Orthogonal codes, 118–119, 141
- PN Sequences, 139–141
- Processing gain, 116
- Pulse Position Modulation, 118

Directional antennas, 243
Discontinuous transmission, 292
Diversity techniques, 120–133
- Time diversity, 121–127
- Frequency diversity, 127–131

Doppler spectrum, 59
Doppler spread, 43
Dynamic Channel Allocation (DCA), 256–258
- Centralized DCA, 257
- Comparison of algorithms, 258
- Comparison with FCA, 259
- Distributed DCA, 257

E

E-911, 64, 535, 546–547
Encryption, 301
Energy efficient protocols, 285, 292–295
- And TCP, 294
- Go-back-N scheme, 294
- Link layer, 292
- Software approaches, 295–297

Equal gain diversity combining, 120
Equalization, 124–127
- DFE, 126–127
- LTE, 126
- MLSE, 125

Erlang B formula, 177
Erlang C formula, 177–178
Error control coding, 137–138, 153–158
- Block codes, 153

F

FACCH (See GSM-Logical Channels)
Fade margin, 51
FCCH (See GSM-Logical Channels)
Femtocell, 233
Fixed assignment channel access methods, 161–179
Fixed Channel Allocation (FCA), 252–256
- Comparison with DCA, 259

Foreign agent, 280
Free space propagation, 46
Frequency diversity, 127–131
- And frequency hopping, 129–130
- And multicarrier modulation, 130

Frequency Division Duplex (FDD), 17, 162
Frequency Division Multiple Access (FDMA), 163–165
- Comparison with TDMA and CDMA, 172–175
- Near-far problem, 164

Frequency Hopping Spread Spectrum (FHSS), 111–116
- And CDMA, 115
- Concept, 113
- In GSM, 115
- In IEEE 802.11, 114–115

Frequency reuse, 229, 232
Frequency reuse factor, 232, 236
Frequency Shift Keying (FSK), 87, 98

G

G3G proposal, 372
Gaussian Minimum Shift Keying (GMSK), 91, 96–100
- Spectrum, 99
- Time bandwidth product, 98
- Use in various systems, 99

General Packet Radio Service (GPRS), 12, 318
- Attachment procedure, 397
- Data transfer procedure, 400–403
- Mobility Management, 398–400
- Overlay on GSM, 395

Reference Architecture, 396
Reservation ALOHA and, 184
Services, 395
Geolocation, 414, 533–551
AOA method, 539–540
Architecture, 536, 537
Fingerprinting based method, 545
Performance Measures, 547–550
Signal phase method, 545
Signal strength method, 543
Techniques, 538–546
TOA method, 540–543
Geolocation Base Station (GBS), 538, 540
Global Positioning System (GPS), 64, 534–536
Global System for Mobile Communications (GSM), 9, 317
Blanket paging, 270
Discontinuous transmission, 290
Location updates, 269
Numbering scheme, 326
Power control, 288
Registration, 327
TDMA with FDD, 166
Power Control, 333
Protocol Architecture, 332
Reference architecture, 322–324
Call establishment, 329
Forming of packets, 336
Frame hierarchy, 338
Handoff, 330
History, 321
Logical Channels, 339
Message formats, 343–345
Security, 331
Services, 321–322
Frequency hopping for diversity, 129
Convolutional coding, 155
Gold codes, 374
GPRS Tunneling Protocol (GTP), 403

H

Hadamard matrix, 141, 356–357
Handoff, 175, 272–280
Decision time algorithms, 276–278
Hard (see Hard handoff)
Seamless (see Seamless handoff)
Soft (see Soft handoff)
GSM, 330
IEEE 802.11, 467–469
Handoff management, 267, 272–280
Generic process, 278
Hard handoff, 275
Hidden terminal problem, 186, 196–201
High Data Rate (HDR), 207
HIPERLAN-1, 11, 19, 413
Energy efficient transmission, 293
History, 481
Non-Preemptive Multiple Access (NPMA), 484–485
Protocol layers, 484
Requirements, 482
HIPERLAN-2, 19, 413
Background, 485–486
Comparison with IEEE 802.11, 494
Logical channels, 491–493
Protocol layers, 486–491
QoS, 493
Security, 494
Home agent, 280
Home Area Networks (HAN), 25, 434–443
Home database, 271, 279
Home Location Register (HLR), 271, 326
Home network, 271, 279
Home-access networks, 443–444
HomeRF, 25, 413, 441–443, 501–503
HPNA, 436–438, 502
Hybrid Fiber Coax (HFC), 5, 432
Hysteresis margin, 276

I

IAPP, 468
Identification, 299, 309
IEEE 802.11, 11, 12, 94, 413
Message authentication, 308
Sleep mode, 291
Stream ciphers, 304
And frequency hopping for diversity, 129
Time diversity, 123

Comparison with HIPERLAN-2, 494
CSMA/CA, 460–463
DFIR, 456–457
DSSS, 454–456
FHSS, 452–454
Handoff, 467–468
History, 448–449
Interference with Bluetooth, 520–530
IR standard, 94
MAC Frame formats, 463–467
Operation of collision avoidance, 189
PHY Layer, 452–459
Power management, 468
Protocol layers, 452–467
Reference Architecture, 450
Requirements, 449–450
Security, 469–470
IEEE 802.3, 160, 193, 215
IEEE 802.11a, 109, 424, 457
IEEE 802.11b, 119, 457
IEEE 802.15, 25, 441, 500–501
IEEE 802.16, 443–444
i-Mode, 380, 409–410
Impulse Radio (see Ultra Wide Band)
IMT-2000, 10, 16, 352, 371–376
Handoff and Power control, 374
In-building signal penetration, 41
Industrial, Medical and Scientific (ISM) Bands, 11, 41
InfoPAD, 426
Infrared-based WLANs, 92
Infrastructure topology, 225
Comparison with ad hoc topology, 228
Integrity, 300
Intelligent Transportation Systems (ITS), 535
Interim Standards, 350
Internet, 5, 31–33
Inter-symbol Interference (ISI), 61, 125
Due to multipath, 151
IP telephony, 210
IS-41, 354
IS-136, 15
Sleep mode, 291
Challenge response scheme, 310
TDMA with FDD format, 168
IS-95, 15, 317
Discontinuous transmission, 291
Power control, 288
Access channel, 362
Convolutional coding, 155
Data, 202–203
Forward channel, 356
Frame formats, 363–365
Inner loop power control, 370
Mobility management, 365–369
Pilot channel, 358
Power control, 369–371
Reverse channel, 361
Time diversity, 123
Waveform encoding, 361–362
IS-634, 352
Layered architecture, 353
Supported services, 353

J

Joint Technical Committee (JTC) model, 53, 79
Indoor areas, 56

K

Key sizes in wireless systems, 303

L

Lee's microcell method, 246
Line coding, 92
Manchester, 92
Linear feedback shift register, 140
Line-of-Sight (LOS), 42
Local Broadband and Ad hoc Wireless Networks, 20
Location area, 268, 271
Location information dissemination, 268
Location management, 266, 267–272
Issues, 272
Location update, 267
Algorithms, 268
In GSM, 269
State based, profile based, 269
Log normal distribution, 51

M

Macrocell, 233
Maximal ratio diversity combining, 120
Megacell, 233
Microcell, 233
Minimum Shift Keying (MSK), 98
Mobile assisted handoff, 175, 275
Mobile controlled handoff, 275, 279
Mobile Data Intermediate System (MD-IS), 387
Mobile Data Networks, 380–412
Mobile Data Services, 18
Mobile IP, 280–284
Mobile Network Location Protocol (MNLP), 385, 388
Mobile Network Registration Protocol (MNRP), 388
Mobility management, 221, 266–280
  CDPD, 388–390
  IS-95, 365–369
Mobitex, 12
  Reservation ALOHA in, 182
Modulation schemes, Comparison, 133–137
Moore's Law, 303
Multicarrier modulation, 108
Multipath delay spread, 42, 61
Multirate transmission, 111
Multisymbol modulation, 110

N

Neighbor set, 368
Network and Switching Subsystem (NSS), 326
Network controlled handoff, 275, 279
Non Line-of-Sight (NLOS), 42, 76
Normal burst, 334

O

Offset Quadrature Phase Shift Keying (OQPSK), 103, 106
Okumura-Hata model, 52, 77
Orthogonal codes, 118–119, 141
Orthogonal Frequency Division Multiplexing (OFDM), 87, 108–110
  Coded (COFDM), 109
  In WLANs, 109
  Use in DSL, 87
Orthogonal Variable Spreading Factor (OVSF) codes, 374

P

Packet Reservation Multiple Access (PRMA), 183–184
Paging, 267, 270–271
Path-loss gradient, 43, 46, 48
  Measurement, 49
Path-loss model, 43
  Femtocellular areas, 57
  Indoor areas, 55
  Macrocellular areas, 52
  Microcellular areas, 53
  Multifloor, 55
  Partition dependent, 56
  Picocellular areas, 55
PCF, 462
PCH (See GSM-Logical Channels)
Personal locator services, 534–535
Phase Shift Keying (PSK), 87, 100
Picocell, 233
Piconet, 506, 512
Ping-pong effect, 269, 272
Plain Old Telephone Service (POTS), 2, 3, 9, 176
PN Sequences, 139–141, 356–358
  Barker, 140
Power control, 175, 285–289
  Centralized, 288
  Closed loop, 288
  Distributed, 289
  In AMPS, 286
  Open loop, 287
  CDPD, 392
  GSM, 333
Power line modems, 438–440
Power saving mechanisms, 285, 289–291
Privacy, 298, 300, 301–303
Probability of error
  In AWGN, 148, 151

In frequency selective fading, 152
In Rayleigh fading, 150–151
Public key algorithms, 305
Public Safety Answering Point (PSAP), 535, 538
Public Switched Telephone Network (PSTN), 4, 28–30, 431
Extension to cellular services, 4
Pulse modulation techniques, 92
Pulse Position Modulation (PPM), 93
Pulse shaping, 103–105
Raised cosine pulse, 104
Raised cosine pulse roll-off factor, 104

Q

QCELP, 138–139, 353
Quadrature Amplitude Modulation (QAM), 87, 110
Quadrature Phase Shift Keying (QPSK), 91, 101
Block diagram, 102
Improvements to, 106
Pulse shaping, 103–105
Signal constellation, 102
Spectrum, 103
Quality of Service (QoS), 5, 208
In GPRS, 396
In HIPERLAN-2, 493

R

Radio channel models (see also path loss models), 64, 82–83
Geolocation model, 64
Hardware emulation, 74
SIMO and MIMO, 67, 68, 76
Wideband multipath model, 63
Radio propagation, 42
Diffraction, 45
Free space, 46
Measurement and modeling, 68
Mechanisms, 44
Mitigation methods, 63
Reflection and transmission, 44
Scattering, 45
Shadow fading, 48, 50
Simulation of channel, 71
Summary characteristics, 63
Two ray model, 47
Radio Resource Management (RRM), 221, 284–291
RAKE receivers, 121–124
Random channel access methods, 161, 179–201
Performance analysis, 192–196
Ray tracing, 71–73, 75
Rayleigh distribution, 58
Registration, 274, 275, 281
GSM, 327
Remaining set, 368
Remote positioning, 536
Reuse partitioning, 249
Reverse tunneling, 284
Ricean distribution, 58
RMS-Multipath delay spread, 61
RS Codes, 153
RSA, 307
RTS/CTS, 191, 461

S

SACCH (See GSM-Logical Channels)
Scattered ad hoc topology, 505–506
SCH (See GSM-Logical Channels)
SDCCH (See GSM-Logical Channels)
Seamless handoff, 275
Secret key algorithms, 302
Sectored antenna, 132–133
Sectorization gain, 171
Security, 297–311
Requirements in wireless networks, 298
Bluetooth, 520
CDPD, 390
GSM, 331
HIPERLAN-2, 494
IEEE 802.11, 469–470
Selection diversity, 120
Self-positioning, 536
Sequential paging, 271
Shadow fading, 48, 50, 51

Shared Wireless Access Protocol (SWAP), 502
Signal constellation, 100, 102
Signal-to-Interference ratio, 237–239
Signal-to-noise ratio, 146
Sleep mode, 291
Small Unit Operations (SUO), 64, 427
Small-scale fading, 58
Smart antennas, 251
Smooth handoff, 283
SMS, 405–407
  Services, 405
SNDCP, 391, 394, 403
Soft handoff, 175, 275, 366–369
  W-CDMA, 375
Soft softer handoff, 366
Softer handoff, 366
Software radio, 142
Space diversity, 131–133
  Using sectored antenna, 132
Space Division Multiple Access (SDMA), 251
Speech coding, 138–139
Spread Spectrum, 111
  Advantages, 112
  Direct Sequence, 111
  Frequency Hopping, 111
Staggered Quadrature Phase Shift Keying (SQPSK), 106
Standards Organizations, 34, 35
Subscriber Identity Module (SIM), 324, 386
Supplementary services, 322
Synchronous Connection-Oriented (SCO), 513

T

TASI, 206
T-carrier system, 165
TCH (See GSM-Logical Channels)
Teleservices, 322
Terrestrial European Trunked Radio (TETRA), 18
Time Difference of Arrival (TDOA), 543
Time diversity, 121–127
  Multipath reception, 123
  RAKE receiver, 123
Time Division Duplex (TDD), 17, 162
Time Division Multiple Access (TDMA), 162, 165–168
  Comparison with FDMA and CDMA, 172–175
Time of Arrival (TOA), 65–66, 540–543
TR-45/46 Reference Model, 351–352
Triangle routing, 282

U

Ultra Wide Band (UWB), 21, 91, 94
  Design of transmission waveform, 95
  Pulse shape, 95
Unlicensed bands, 41, 421
Unlicensed National Information Infrastructure (U-NII) Bands, 11, 41, 421–423

V

Visiting database, 271, 279
Visiting network, 271, 279
Visitor Location Register (VLR), 271, 326
Voice activity detection, 290
Voice activity factor, 171, 203, 290
Voice and data integration, 201–214
Voice over IP (VoIP), 201, 208
  Jitter in a WLAN, 212–214
Voice-band modem, 86

W

Walsh sequence, 141, 356–358
W-CDMA, 10
  Comparison with cdma2000, 373
Wireless Application Protocol (WAP), 380, 407–409
Wireless ATM, 414, 428, 479–481
  Background, 475
  Mobility support, 481
  Projects, 480
  Protocol architecture, 477–479
  Reference model, 476

Wireless Campus Area Networks, 424
Wireless Local Area Networks (WLAN), 7, 18
  Capacity with voice and data, 209
  Evolution, 420–426
  History, 416–420
  Standards, 19
Wireless Markup Language (WML), 408
Wireless modems, 87–90
  Bandwidth efficiency, 87
  Design considerations, 87
  Out of band radiation, 89
  Power efficiency, 88
  Resistance to multipath, 90
Wireless Personal Area Networks (WPAN) (see also Bluetooth), 7, 500–501

ABOUT THE AUTHORS

Kaveh Pahlavan is a Professor of Electrical and Computer Engineering and Computer Science and Director of the Center for Wireless Information Network Studies at Worcester Polytechnic Institute, Worcester, Massachusetts. He is also a visiting Professor of Telecommunication Laboratory and CWC at the University of Oulu, Finland. His area of research is broadband wireless indoor networks. He has contributed to numerous seminal technical publications in this field. He is the principal author (with Allen Levesque) of *Wireless Information Networks* (John Wiley and Sons, 1995), the first textbook in this field. He has been a consultant to a number of companies including CNR Inc., GTE Laboratories, Steinbrecher Corp., Simplex, Mercury Computers, WINDATA, SieraComm, 3COM, and Codex/Motorola in Massachusetts; JPL, Savi Technologies, and RadioLAN in California; Airnoet in Ohio; United Technology Research Center in Connecticut; Honeywell in Arizona; Nokia, LK-Products, Elektrobit, TEKES, and Finnish Academy in Finland; and NTT in Japan. He has also serves as a member of the board of several start-up wireless companies. Before joining WPI, he was the Director of Advanced Development at Infinite Inc., in Andover, Massachusetts, working on data communications. He started his career as an Assistant Professor at Northeastern University in Boston, Massachusetts. He is the Editor-in-Chief of the International Journal on Wireless Information Networks. He was the founder, program chairman, and organizer of the IEEE Workshop on Wireless LANs in 1991, 1996, and 2001, and the organizer and technical program chairman of the IEEE International Symposium on Personal, Indoor, and Mobile Radio Communications in Boston, Massachusetts in 1992 and 1998. He was selected as a member of the Committee on Evolution of Untethered Communication, US National Research Council, 1997, and has led the US review team for the Finnish research and development programs in Electronics and Telecommunications in 1999. For his contributions to the area of wireless networks he was the Westin Hadden Professor of Electrical and Computer Engineering at WPI from 1993 to 1996, was elected as a fellow of the IEEE in 1996, and became a fellow of Nokia in 1999. From May to December 2000 he was the first Fulbright-Nokia scholar at the University of Oulu, Finland. Because of his inspiring visionary publications and his international conference activities for the growth of the wireless LAN industry, he is referred to as one of the founding fathers of the industry. Details of his contributions to this field are available at *www.cwins.wpi.edu*.

Prashant Krishnamurthy is an Assistant Professor in the Department of Information Science and Telecommunications at the University of Pittsburgh, where he has been leading the development of the wireless information systems track for the Master of Science in Telecommunications curriculum in the Telecommunications program. His research interests are in the areas of wireless data networks, wireless network security, and radio propagation modeling. He has received several research and teaching grants related to wireless information systems from the National Science Foundation, The Commonwealth of Pennsylvania, AT & T Foundation, and the University of Pittsburgh. Prashant is also the principal investigator for a research project titled "A survivable and secure wireless information infrastructure" that is funded by the National Institute of Standards and Technology. He has been a member of the Technical Program Committees of several distinguished conferences. He chairs the Pittsburgh chapter of the IEEE Communications Society.